LECTURES ON K3 SURFACES

K3 surfaces are central objects in modern algebraic geometry. This book examines this important class of Calabi–Yau manifolds from various perspectives in eighteen self-contained chapters. It starts with the basics and guides the reader to recent breakthroughs, such as the proof of the Tate conjecture for K3 surfaces and structural results on Chow groups. Powerful general techniques are introduced to study the many facets of K3 surfaces, including arithmetic, homological, and differential geometric aspects. In this context, the book covers Hodge structures, moduli spaces, periods, derived categories, birational techniques, Chow rings, and deformation theory. Famous open conjectures, for example the conjectures of Calabi, Weil, and Artin–Tate, are discussed in general and for K3 surfaces in particular, and each chapter ends with questions and open problems. Based on lectures at the advanced graduate level, this book is suitable for courses and as a reference for researchers.

Daniel Huybrechts is a professor at the Mathematical Institute of the University of Bonn. He previously held positions at the Université Denis Diderot Paris 7 and the University of Cologne. He is interested in algebraic geometry, particularly special geometries with rich algebraic, analytic, and arithmetic structures. His current work focuses on K3 surfaces and higher dimensional analogues. He has published four books.

Lectures on K3 Surfaces

Daniel Huybrechts
University of Bonn

CAMBRIDGE
UNIVERSITY PRESS

CAMBRIDGE
UNIVERSITY PRESS

University Printing House, Cambridge CB2 8BS, United Kingdom

One Liberty Plaza, 20th Floor, New York, NY 10006, USA

477 Williamstown Road, Port Melbourne, VIC 3207, Australia

314-321, 3rd Floor, Plot 3, Splendor Forum, Jasola District Centre, New Delhi - 110025, India

79 Anson Road, #06-04/06, Singapore 079906

Cambridge University Press is part of the University of Cambridge.

It furthers the University's mission by disseminating knowledge in the pursuit of education, learning and research at the highest international levels of excellence.

www.cambridge.org
Information on this title: www.cambridge.org/9781107153042

© Daniel Huybrechts 2016

First published 2016

A catalogue record for this publication is available from the British Library

Library of Congress Cataloging in Publication data
Names: Huybrechts, Daniel.
Title: Lectures on K3 surfaces / Daniel Huybrechts, University of Bonn.
Description: Cambridge : Cambridge University Press, [2016] | Includes bibliographical references and index.
Identifiers: LCCN 2016019204 | ISBN 9781107153042 (hardback : alk. paper)
Subjects: LCSH: Surfaces, Algebraic. | Geometry, Algebraic.
Classification: LCC QA571 .H89 2016 | DDC 516.3/52–dc23 LC
record available at https://lccn.loc.gov/2016019204

ISBN 978-1-107-15304-2 Hardback

Contents

Preface

This book originates from a graduate course in algebraic geometry held in the summer of 2010. I introduced many fundamental techniques in algebraic geometry and explained in detail how they are applied to K3 surfaces. The diversity of the theory of K3 surfaces, touching upon so many topics in both algebraic geometry and other areas, including arithmetic, complex and differential geometry, homological algebra, and even mathematical physics, is fascinating. I hoped to convey some of this fascination and, at the same time, to demonstrate how the various aspects – ranging from Hodge theory to moduli spaces to derived categories – come together in a meaningful way when studied for K3 surfaces.

Over time, the original lecture notes have grown. They now cover large parts, but by no means all, of the theory of K3 surfaces. As the notes made available online appeared to be useful, it seemed worthwhile to turn them into this book. I hope it will serve as an introduction to the subject as well as a guide to the vast literature. The balance between these two goals turned out to be difficult to achieve. Some chapters are more or less self-contained, while others are certainly not and rather are meant as an invitation to consult the original sources.

Each chapter is devoted to a different topic and presents the relevant theory in a condensed form, accompanied by extensive references to the original articles and to the relevant textbooks. Sometimes the text can be read as a survey, while other times technical aspects particular to K3 surfaces are discussed in detail. Often, I try to give ad hoc arguments that work only for K3 surfaces, though a more powerful general theory may be available. The aim is to allow the chapters to be read independently of each other, encouraging a non-linear reading. Although this goal was not fully achieved, I hope I at least made it easy to navigate between the chapters. Also, I have not hesitated to revisit some aspects to emphasize different angles or give more details.

The choice of topics is a personal one and I am aware of many omissions. Also, I have deliberately tried to avoid overlap with the existing accounts of larger parts of the theory as in the book [33] by Barth et al. or in the seminar notes [54] by Beauville et al., both of which are excellent introductions.

Important topics that are not covered here include the role of K3 surfaces in Claire Voisin's approach to Green's syzygy conjecture [49], the dynamical aspects of K3 surfaces as studied by Curt McMullen [403], Viatcheslav Kharlamov's results on real K3 surfaces [293], the work of Davesh Maulik, Rahul Pandharipande, and Richard Thomas on curve and sheaf counting on K3 surfaces in connection with Gromov–Witten and Donaldson–Thomas theory [399], questions related to rational points of K3 surfaces defined over number fields and the Brauer–Manin obstruction, K3 surfaces and conformal field theory [25], and many others. What is actually covered here is revealed by a quick look at the table of contents.

Prerequisites: As an advanced course, familiarity with the basic notions of algebraic geometry (schemes, varieties, cohomology of coherent sheaves, curves, and surfaces) and complex geometry (Kähler manifolds, Hodge theory) will be helpful.

Cross-references and proofs: Cross-references of the type 'Theorem 1.2.3' refer to Theorem 2.3 in another chapter, here in Chapter 1, whereas 'Proposition 2.3' refers to Proposition 2.3 within the same chapter.

If a proposition or a theorem concludes with a □, then either the arguments or the main ideas of the proof have been given earlier. If there is neither a proof nor a □, then the result needs to be looked up in the literature. Sometimes a proof is more like a sketch of the main ideas that should, however, allow one to reconstruct most of the details.

Acknowledgements: The two classics [33, 54] taught me much about the fundamental concepts in the theory of K3 surfaces, and discussions with François Charles and Davesh Maulik over recent years have further shaped my way of thinking about K3 surfaces. I gratefully acknowledge their influence.

Many people have been very helpful with their comments and suggestions on preliminary versions of these notes: Jean-Louis Colliot-Thélène, Umesh Dubey, Hélène Esnault, Jürgen Hausen, Yuya Matsumoto, Davesh Maulik, Ben Moonen, D. S. Nagaraj, John Ottem, Emanuel Reinecke, Stefan Schreieder, Nick Shepherd-Barron, Ichiro Shimada, Andrey Soldatenkov, Lenny Taelman, Richard Thomas, Jonathan Wahl, and Olivier Wittenberg. I wish to thank them all.

I am truly grateful for the generous help of François Charles, Christian Liedtke, Giovanni Mongardi, Matthias Schütt, Pawel Sosna, and Chenyang Xu. They have come up with long lists of comments on various chapters and provided assistance with technical problems. I also wish to thank Alex Perry, who not only worked through large portions of the manuscript and pointed out mathematical inaccuracies, but also helped with linguistic aspects. Contacted for historical context, Sir Michael Atiyah readily shared some memories for which I am very grateful.

K3 at sunset, photographed by Vittorio Sella in 1909.[1]

The name 'K3 surface' was introduced by André Weil. In the comments to his 'Final report on contract AF 18(603)-57', written in 1958 and published in his collected papers [634], he writes:

> *il s'agit des variétés kählériennes dites K3, ainsi nommées en l'honneur de Kummer, Kähler, Kodaira et de la belle montagne K2 au Cachemire.*

In fact, a mountain K3 exists as well. It is also known as Broad Peak and at 8051 m it is the twelfth highest mountain in the world and the fourth highest in Pakistan. Incidentally its first ascent took place in June 1957, which was around the time when Weil thought about K3 surfaces, and the first winter ascent occurred only in March 2013.

The first time the name 'K3 surface' appeared in a published article seems to be in Kodaira's [308]. Weil himself acknowledges discussions about K3 surfaces with Kodaira and Spencer during his visits to Princeton in 1958 and gives credit to Nirenberg, Andreotti, and Atiyah (whom he met during his stay in Cambridge in 1953–1954). Andreotti has never published his results on the local Torelli theorem for (Kummer) K3 surfaces, or any other result on K3 surfaces, but Grauert refers to them in [215]. In his paper [26] from 1958, Atiyah shows that Kummer surfaces and quartics are deformation equivalent, but general K3 surfaces are not discussed nor is the name mentioned.

---

[1] From Il Principe Luigi Amadeo di Savoia, Duca degli Abruzzi, *La Spedizione nel Karakoram e nell'Imalaia occidentale 1909,* relazione del Dott. Filippo de Filippi, illustr. da Vittorio Sella (Bologna: Zanichelli, 1912).

1

Basic Definitions

Algebraic K3 surfaces can be defined over arbitrary fields. Over the field of complex numbers a more general notion exists that includes non-algebraic K3 surfaces. In Section 1, the algebraic variant is introduced and some of the most important explicit examples are discussed. Classical numerical invariants are computed in Section 2. In Section 3, complex K3 surfaces are defined and Section 4 contains more examples which are used for illustration in later chapters.

1 Algebraic K3 Surfaces

Let k be an arbitrary field. A *variety over k* (usually) means a separated, geometrically integral scheme of finite type over k.

Definition 1.1 A *K3 surface* over k is a complete non-singular variety X of dimension two such that[1]

$$\Omega^2_{X/k} \simeq \mathcal{O}_X \text{ and } H^1(X, \mathcal{O}_X) = 0.$$

Once the base field is fixed, we often simply write Ω_X instead of $\Omega_{X/k}$. The canonical bundle of a non-singular variety X, i.e. the determinant of Ω_X, shall be denoted K_X or ω_X, depending on whether we regard it as a divisor or as an invertible sheaf.

By definition, the cotangent sheaf Ω_X of a K3 surface X is locally free of rank two and $\omega_X \simeq \mathcal{O}_X$. Moreover, the natural alternating pairing

$$\Omega_X \times \Omega_X \longrightarrow \omega_X \simeq \mathcal{O}_X,$$

---

[1] By definition, a variety over a field k is complete if the given morphism $X \longrightarrow \text{Spec}(k)$ is proper and X over k is non-singular if the cotangent sheaf $\Omega_{X/k}$ is locally free of rank $\dim(X)$, which is equivalent to $X_{\bar{k}} := X \times_k \bar{k}$ being regular; see e.g. [375, Prop. 6.2.2].

of which we think as an algebraic symplectic structure, induces a non-canonical isomorphism

$$\mathcal{T}_X := \Omega_X^* := \mathcal{H}om(\Omega_X, \mathcal{O}_X) \simeq \Omega_X.$$

Remark 1.2 Any smooth complete surface is projective. So, with the above definition, K3 surfaces are always projective.

There are various proofs for the general fact. For example, Goodman (see [235]) shows that the complement of any non-empty open affine subset is the support of an ample divisor. The proof in [33], written for smooth compact complex surfaces, uses fibrations of the surface associated with some rational functions. See [375, Ch. 9.3] for a proof over an arbitrary field.

Example 1.3 (i) A smooth *quartic* $X \subset \mathbb{P}^3$ is a K3 surface. Indeed, from the short exact sequence

$$0 \longrightarrow \mathcal{O}(-4) \longrightarrow \mathcal{O} \longrightarrow \mathcal{O}_X \longrightarrow 0$$

on $\mathbb{P}^3$ and the vanishings $H^1(\mathbb{P}^3, \mathcal{O}) = H^2(\mathbb{P}^3, \mathcal{O}(-4)) = 0$ one deduces $H^1(X, \mathcal{O}_X) = 0$. Taking determinants of the conormal bundle sequence (see [236, II.Prop. 8.12])

$$0 \longrightarrow \mathcal{O}(-4)|_X \longrightarrow \Omega_{\mathbb{P}^3}|_X \longrightarrow \Omega_X \longrightarrow 0$$

yields the adjunction formula $\omega_X \simeq \omega_{\mathbb{P}^3} \otimes \mathcal{O}(4)|_X \simeq \mathcal{O}_X$. In local homogeneous coordinates with X given as the zero set of a quartic polynomial f, a trivializing section of ω_X can be written explicitly as the residue

$$\mathrm{Res}\left(\frac{\sum(-1)^i x_i dx_0 \wedge \cdots \wedge \widehat{dx_i} \wedge \cdots \wedge dx_3}{f} \right), \tag{1.1}$$

which, for example, on the affine chart $x_0 = 1$ with affine coordinates y_1, y_2, y_3 is

$$\mathrm{Res}\left(\frac{dy_1 \wedge dy_2 \wedge dy_3}{f(1, y_1, y_2, y_3)} \right). \tag{1.2}$$

A particularly interesting special case is provided by the *Fermat quartic* $X \subset \mathbb{P}^3$ defined by the equation

$$x_0^4 + x_1^4 + x_2^4 + x_3^4 = 0.$$

In order for it to be smooth one has to assume char$(k) \neq 2$.

(ii) Similarly, a smooth complete intersection of type $(d_1, \ldots, d_n)$ in $\mathbb{P}^{n+2}$ is a K3 surface if and only if $\sum d_i = n + 3$. Note that under the natural assumption that all $d_i > 1$ there are in fact only three cases (up to permutation): $n = 1, d_1 = 4$ (as in (i)); $n = 2, d_1 = 2, d_2 = 3$; and $n = 3, d_1 = d_2 = d_3 = 2$. This yields examples of K3 surfaces of degree four, six, and eight.

(iii) Let k be a field of char$(k) \neq 2$ and let A be an abelian surface over k.[2] The natural involution $\iota: A \longmapsto A$, $x \longmapsto -x$, has the 16 two-torsion points as fixed points. (They are geometric points and not necessarily k-rational.) The minimal resolution $X \longrightarrow A/\iota$ of the quotient, which has only rational double point singularities (cf. Section **14**.0.3), defines a K3 surface. K3 surfaces of this type are called *Kummer surfaces*. For details in the case of $k = \mathbb{C}$; see [43, Prop. VIII.11] and for a completely algebraic discussion [28, Thm. 10.6].[3]

An alternative way of describing X starts with blowing up the fixed points $\widetilde{A} \longrightarrow A$. Since the fixed points are ι-invariant, the involution ι lifts to an involution $\widetilde{\iota}$ of $\widetilde{A}$. The quotient $\widetilde{A} \longrightarrow X$ by ι is a ramified double covering of degree two. A local calculation shows that smoothness of X and $\widetilde{A}$ are equivalent (in characteristic $\neq 2$).

$$
\begin{array}{ccc}
\widetilde{\iota} \subset \widetilde{A} & \longrightarrow & A \\
\downarrow & & \downarrow \\
X & \longrightarrow & A/\iota.
\end{array}
$$

Moreover, the canonical bundle formulae for the blow-up $\widetilde{A} \longrightarrow A$ (cf. [236, V. Prop. 3.3]) and for the branched covering $\pi: \widetilde{A} \longrightarrow X$ (cf. [33, I.16] or [442, Ch. 6]) yield

$$
\omega_{\widetilde{A}} \simeq \mathcal{O}(\Sigma E_i) \text{ and } \omega_{\widetilde{A}} \simeq \pi^* \omega_X \otimes \mathcal{O}(\Sigma E_i).
$$

This shows $\pi^* \omega_X \simeq \mathcal{O}_{\widetilde{A}}$. Here, the E_i are the exceptional divisors of $\widetilde{A} \longrightarrow A$. Their images $\overline{E}_i$ in X satisfy $\pi^* \mathcal{O}(\overline{E}_i) \simeq \mathcal{O}(2E_i)$. Note that $\pi_* \mathcal{O}_{\widetilde{A}} \simeq \mathcal{O}_X \oplus L^*$, where the line bundle L is a square root of $\mathcal{O}(\sum \overline{E}_i)$, and hence $\pi^* \omega_X \simeq \mathcal{O}_{\widetilde{A}}$ implies $\omega_X \simeq \mathcal{O}_X$. Finally note that the image of the injection $H^1(X, \mathcal{O}_X) \hookrightarrow H^1(\widetilde{A}, \mathcal{O}_{\widetilde{A}}) = H^1(A, \mathcal{O}_A)$ is contained in the invariant part of the action induced by ι. Hence, $H^1(X, \mathcal{O}_X) = 0$. See Remark **14**.3.16 for a converse describing which K3 surfaces are Kummer surfaces.

The Fermat surface in (i) is in fact a Kummer surface, but this is not obvious to see; cf. Example **14**.3.18.

(iv) Consider a double covering

$$
\pi: X \longrightarrow \mathbb{P}^2
$$

branched along a curve $C \subset \mathbb{P}^2$ of degree six. Then $\pi_* \mathcal{O}_X \simeq \mathcal{O}_{\mathbb{P}^2} \oplus \mathcal{O}(-3)$ which in particular shows $H^1(X, \mathcal{O}_X) = 0$. Note that for char$(k) \neq 2$ the surface X is non-singular if C is. The canonical bundle formula for branched coverings shows

---

[2] The standard reference for abelian varieties is Mumford's [443], but the short introduction [407] by Milne is also highly recommended.

[3] The same construction works in characteristic 2 under additional assumptions on A; see [283, 561]. There are fewer fixed points (4, 2, or 1), but the singularities of the quotient A/ι are worse and the minimal resolution defines a K3 surface if and only if A is not supersingular. Recently the case of char$(k) = 2$ has been revisited by Schröer, Shimada, and Zhang in [532, 558].

$\omega_X \simeq \pi^*(\omega_{\mathbb{P}^2} \otimes \mathcal{O}(3)) \simeq \mathcal{O}_X$, and, therefore, for C non-singular X is a K3 surface (of degree two), called a *double plane*.

If C is the union of six generic lines in $\mathbb{P}^2$, a local calculation reveals that the double cover X has 15 rational double points. The 15 points correspond to the pairwise intersections of the six lines. Blowing up these 15 singular points produces a K3 surface X'. The canonical bundle does not change under the blow-up; see [33, III. Prop. 3.5].

2 Classical Invariants

We start by recalling basic facts on the intersection pairing of divisors on general smooth surfaces before specializing to the case of K3 surfaces.

2.1 Let X be an arbitrary non-singular complete surface over k. For line bundles $L_1, L_2 \in \mathrm{Pic}(X)$ the intersection form $(L_1.L_2)$ can be defined as the coefficient of $n_1 \cdot n_2$ in the polynomial $\chi(X, L_1^{n_1} \otimes L_2^{n_2})$ (Kleiman's definition; see [235, I. Sec. 5]) or, more directly, as (see [437, Lect. 12])

$$(L_1.L_2) := \chi(X, \mathcal{O}_X) - \chi(X, L_1^*) - \chi(X, L_2^*) + \chi(X, L_1^* \otimes L_2^*). \qquad (2.1)$$

Of course, both definitions define the same symmetric bilinear form with the following properties:

(i) If $L_1 = \mathcal{O}(C)$ for some (e.g. for simplicity integral) curve $C \subset X$, then $(L_1.L_2) = \deg(L_2|_C)$.

(ii) If $L_i = \mathcal{O}(C_i)$ for two curves $C_i \subset X$, $i = 1, 2$, intersecting in only finitely many points $x_1, \ldots, x_n$, then

$$(L_1.L_2) = \sum_{i=1}^{n} \dim_k(\mathcal{O}_{X,x_i}/(f_{1,x_i}, f_{2,x_i})).$$

Here, f_{1,x_i}, f_{2,x_i} are the local equations for C_1 and C_2, respectively, in x_i.

(iii) If L_1 is ample and $L_2 = \mathcal{O}(C)$ for a curve $C \subset X$, then

$$(L_1.L_2) = (L_1.C) = \deg(L_1|_C) > 0. \qquad (2.2)$$

(iv) The Riemann–Roch theorem for line bundles on surfaces asserts:[4]

$$\chi(X, L) = \frac{(L.L \otimes \omega_X^*)}{2} + \chi(X, \mathcal{O}_X). \qquad (2.3)$$

---

[4] Of course, this is a special case of the much more general Hirzebruch–Riemann–Roch theorem (or of the even more general Grothendieck–Riemann–Roch theorem), but a direct, much easier proof exists in the present situation; see [437, Lect. 12] or [236, V.1].

We often write $(L.C)$ and $(C_1.C_2)$ instead of $(L.\mathcal{O}(C))$ and $(\mathcal{O}(C_1).\mathcal{O}(C_2))$ for curves or divisors C, C_i on X. Instead of $(L.L)$, we often use $(L)^2$ and similarly $(C)^2$ instead of $(C.C)$.

The *Néron–Severi group* of an algebraic surface X is the quotient

$$\mathrm{NS}(X) := \mathrm{Pic}(X)/\mathrm{Pic}^0(X)$$

by the connected component of the Picard variety $\mathrm{Pic}(X)$, i.e. by the subgroup of line bundles that are algebraically equivalent to zero.

A line bundle L is *numerically trivial* if $(L.L') = 0$ for all line bundles L'. For example, any $L \in \mathrm{Pic}^0(X)$ is numerically trivial. The subgroup of all numerically trivial line bundles is denoted $\mathrm{Pic}(X)^\tau \subset \mathrm{Pic}(X)$ and yields a quotient of $\mathrm{NS}(X)$

$$\mathrm{Num}(X) := \mathrm{Pic}(X)/\mathrm{Pic}^\tau(X).$$

Clearly, $\mathrm{Num}(X)$ is a free abelian group endowed with a non-degenerate, symmetric pairing:

$$(\,.\,): \mathrm{Num}(X) \times \mathrm{Num}(X) \longrightarrow \mathbb{Z}.$$

Proposition 2.1 *The Néron–Severi group* $\mathrm{NS}(X)$ *and its quotient* $\mathrm{Num}(X)$ *are finitely generated. The rank of* $\mathrm{NS}(X)$ *is called the* Picard number $\rho(X) = \mathrm{rk}\,\mathrm{NS}(X)$.[5]

2.2 The signature of the intersection form on $\mathrm{Num}(X)$ is $(1, \rho(X) - 1)$. This is called the *Hodge index theorem*; cf. e.g. [236, V.Thm. 1.9]. Thus, $(\,.\,)$ on

$$\mathrm{NS}(X)_{\mathbb{R}} := \mathrm{NS}(X) \otimes_{\mathbb{Z}} \mathbb{R}$$

can be diagonalized with entries $(1, -1, \ldots, -1)$.

Remark 2.2 The Hodge index theorem has the following immediate consequences.

(i) The cone of all classes $L \in \mathrm{NS}(X)_{\mathbb{R}}$ with $(L)^2 > 0$ has two connected components. The *positive cone* $\mathcal{C}_X \subset \mathrm{NS}(X)_{\mathbb{R}}$ is defined as the connected component that is distinguished by the property that it contains an ample line bundle. See Chapter 8 for more on the positive cone of K3 surfaces.

(ii) If L_1 and L_2 are line bundles such that $(L_1)^2 \geq 0$, then

$$(L_1)^2(L_2)^2 \leq (L_1.L_2)^2. \qquad (2.4)$$

---

[5] In [348] Lang and Néron gave a simplified proof of Néron's original result. To prove that $\mathrm{Num}(X)$ is finitely generated, one can use an appropriate cohomology theory. See Section 3.2 for an argument in the complex setting. Numerically trivial line bundles form a bounded family, and, therefore, $\mathrm{NS}(X) \longrightarrow \mathrm{Num}(X)$ has finite kernel and, in particular, $\mathrm{NS}(X)$ is finitely generated as well. Also, $\rho(X) = \mathrm{rk}\,\mathrm{NS}(X) = \mathrm{rk}\,\mathrm{Num}(X)$.

Just apply the Hodge index theorem to the linear combination $(L_1)^2 L_2 - (L_1.L_2)L_1$ (written additively) which is orthogonal to L_1. Note that (2.4) is simply expressing the fact that the determinant of the intersection matrix

$$\begin{pmatrix} (L_1)^2 & (L_1.L_2) \\ (L_1.L_2) & (L_2)^2 \end{pmatrix}$$

is non-positive.

2.3 For a K3 surface X one has by definition $h^0(X, \mathcal{O}_X) = 1$ and $h^1(X, \mathcal{O}_X) = 0$. Moreover, by Serre duality $H^2(X, \mathcal{O}_X) \simeq H^0(X, \omega_X)^*$ and hence $h^2(X, \mathcal{O}_X) = 1$.[6] Therefore,

$$\chi(X, \mathcal{O}_X) = 2.$$

Remark 2.3 This can be used to prove that the (algebraic) fundamental group $\pi_1(X)$ of a K3 surface X over a separably closed field k is trivial. Indeed, if $\widetilde{X} \longrightarrow X$ is an irreducible étale cover of finite degree d, then $\widetilde{X}$ is a smooth complete surface over k with trivial canonical bundle such that

$$\chi(\widetilde{X}, \mathcal{O}_{\widetilde{X}}) = d\,\chi(X, \mathcal{O}_X) = 2d$$

and $h^0(\widetilde{X}, \mathcal{O}_{\widetilde{X}}) = h^2(\widetilde{X}, \mathcal{O}_{\widetilde{X}}) = 1$ (use Serre duality). Combined this yields $2 - h^1(\widetilde{X}, \mathcal{O}_{\widetilde{X}}) = 2d$ and hence $d = 1$.

The Riemann–Roch formula (2.3) for a line bundle L on a K3 surface X reads

$$\chi(X, L) = \frac{(L)^2}{2} + 2. \tag{2.5}$$

Recall that a line bundle L is trivial if and only if $H^0(X, L)$ and $H^0(X, L^*)$ are both non-trivial. Thus, as Serre duality for a line bundle L shows $H^2(X, L) \simeq H^0(X, L^*)^*$, the Riemann–Roch formula for non-trivial line bundles L expresses $h^0(X, L) - h^1(X, L)$ or $h^0(X, L^*) - h^1(X, L)$.

Also note that for an ample line bundle L the first cohomology $H^1(X, L)$ vanishes (we comment on this in Theorem **2.1.8** and Remark **2.1.9**) and hence (2.5) computes directly the number of global sections of an ample line bundle L:

$$h^0(X, L) = \frac{(L)^2}{2} + 2.$$

---

[6] In [236] Serre duality is proved over algebraically closed fields, but it holds true more generally. The pairing is compatible with base change, so one can pass to algebraically closed fields once the trace map is shown to exist over k. In fact, the trace map exists in much broader generality; see Hartshorne [234]. For our purposes working with Serre duality over an algebraically closed field is enough: by flat base change $H^2(X, \mathcal{O}_X) \otimes \bar{k} = H^2(X_{\bar{k}}, \mathcal{O}_{X_{\bar{k}}})$ and $X_{\bar{k}}$ is again a K3 surface.

Proposition 2.4 *For a K3 surface X the natural surjections are isomorphisms*[7]

$$\text{Pic}(X) \xrightarrow{\sim} \text{NS}(X) \xrightarrow{\sim} \text{Num}(X).$$

Moreover, the intersection pairing (.) *on* $\text{Pic}(X)$ *is even, non-degenerate, and of signature* $(1, \rho(X) - 1)$.

Proof Suppose L is non-trivial, but $(L.L') = 0$ for an ample line bundle L'. Then $H^0(X, L) = 0$ and $H^2(X, L) \simeq H^0(X, L^*)^* = 0$ by (2.2). Therefore, (2.5) yields $0 \geq \chi(X, L) = (1/2)(L)^2 + 2$ and thus $(L)^2 < 0$. In particular, L cannot be numerically trivial and, hence, $\text{Pic}(X) \xrightarrow{\sim} \text{NS}(X) \xrightarrow{\sim} \text{Num}(X)$. Moreover, the intersection form is negative definite on the orthogonal complement of any ample line bundle, which proves the claim on the signature. Finally, the Riemann–Roch formula $(L)^2 = 2\chi(X, L) - 4 \equiv 0\,(2)$ shows that the pairing is even. □

For a K3 surface X the lattice $(\text{NS}(X), (\,.\,))$ is thus even and non-degenerate, but rarely unimodular. For more information about lattices that can be realized as Néron–Severi lattices of K3 surfaces, see Section **14**.3.1 and Chapter **17**.

Remark 2.5 Even without using the existence of an ample line bundle, one can show that there are no non-trivial torsion line bundles on K3 surfaces. Indeed, if L is torsion, then by the Riemann–Roch formula $\chi(L) = 2$ and hence L (or its dual) is effective. However, if $0 \neq s \in H^0(X, L)$, then $0 \neq s^k \in H^0(X, L^k)$ for all $k > 0$, and, moreover, the zero sets of both sections coincide. Thus, if L^k is trivial, L is also trivial. The argument also applies to (non-projective) complex K3 surfaces.

The non-existence of torsion line bundle can also be related to the triviality of the (algebraic) fundamental group $\pi_1(X)$; see Remark 2.3. Indeed, the usual unbranched covering construction (see e.g. [33, I.17]) would define for any line bundle L of order d (not divisible by $\text{char}(k)$) a non-singular étale covering $\widetilde{X} \longrightarrow X$.

2.4 We shall next explain how to use the general Hirzebruch–Riemann–Roch formula to determine the Chern number $c_2(X)$ and the Hodge numbers

$$h^{p,q}(X) := \dim H^q(X, \Omega_X^p)$$

of a K3 surface X.

For a locally free sheaf (or an arbitrary coherent) sheaf F on a K3 surface X the Hirzebruch–Riemann–Roch formula reads

$$\chi(X, F) = \int \text{ch}(F)\,\text{td}(X) = \text{ch}_2(F) + 2\,\text{rk}(F). \tag{2.6}$$

---

[7] *Warning*: The second isomorphism does not hold for general complex K3 surfaces; see Section 3.2 and Example 3.3.2.

The general version of this formula can be found e.g. in [236, App. A]. For $F = \mathcal{O}_X$ the first equality is the *Noether formula*

$$\chi(X, \mathcal{O}_X) = \frac{c_1^2(X) + c_2(X)}{12} = \frac{c_2(X)}{12}$$

which yields $c_2(X) = 24$.

Next, by definition one knows $h^{p,q}(X) = 1$ for $(p,q) = (0,0), (0,2), (2,0), (2,2)$ and $h^{0,1}(X) = 0$ for any K3 surface. For the remaining Hodge numbers, (2.6) implies

$$2\,h^0(X, \Omega_X) - h^1(X, \Omega_X) - \mathrm{ch}_2(\Omega_X) \mid 4 = 4 \quad c_2(\Omega_X) = -20.$$

It is also known that $h^0(X, \Omega_X) = 0$ and hence $h^1(X, \Omega_X) = 20$. Using $\mathcal{T}_X \simeq \Omega_X$, this vanishing can be rephrased, maybe more geometrically, as $H^0(X, \mathcal{T}_X) = 0$, i.e. a K3 surface has no global vector fields. In positive characteristic this is a difficult theorem on which we comment later; see Sections 9.4.1 and 9.5.1.[8] For $\mathrm{char}(k) = 0$ it follows from the complex case to be discussed below and the Lefschetz principle. In any event, the *Hodge diamond* of any K3 surface looks like this:

$$
\begin{array}{ccccccc}
 & & & h^{0,0} & & & \\
 & & h^{1,0} & & h^{0,1} & & \\
 & h^{2,0} & & h^{1,1} & & h^{0,2} & \\
 & & h^{2,1} & & h^{1,2} & & \\
 & & & h^{2,2} & & &
\end{array}
\qquad
\begin{array}{ccccccc}
 & & & 1 & & & \\
 & & 0 & & 0 & & \\
 & 1 & & 20 & & 1 & \\
 & & 0 & & 0 & & \\
 & & & 1 & & &
\end{array}
\qquad (2.7)
$$

This holds for K3 surfaces over arbitrary fields and also for non-projective complex ones; see below.

3 Complex K3 Surfaces

Even if interested solely in algebraic K3 surfaces (and maybe even only in those defined over fields of positive characteristic), one needs to study non-projective complex K3 surfaces as well. For example, the twistor space construction, used in the proof of the global Torelli theorem (see Chapter 4), which is one of the fundamental results in K3 surface theory, always involves non-projective K3 surfaces. For this reason, we try to deal simultaneously with the algebraic and the non-algebraic theory throughout this book.

---

[8] Note, however, that it can often easily be checked in concrete situations. For example, it is easy to see that $H^0(X, \mathcal{T}_X) = 0$ for smooth quartics $X \subset \mathbb{P}^3$, complete intersection K3 surfaces, and Kummer surfaces (for the latter, see [28, Rem. 10.7]). Thanks to Christian Liedtke for pointing this out.

3.1 The parallel theory in the realm of complex manifolds starts with the following definition.

Definition 3.1 A *complex K3 surface* is a compact connected complex manifold X of dimension two such that $\Omega_X^2 \simeq \mathcal{O}_X$ and $H^1(X, \mathcal{O}_X) = 0$.

Serre's GAGA principle (see [546, 446]) allows one to associate with any scheme of finite type over $\mathbb{C}$ a complex space X^{an} whose underlying set of points is just the set of all closed points of X. Moreover, with any coherent sheaf F on X there is naturally associated a coherent sheaf F^{an} on X^{an}. These constructions are well behaved in the sense that, for example, $\mathcal{O}_X^{\mathrm{an}} \simeq \mathcal{O}_{X^{\mathrm{an}}}$ and $\Omega_{X/\mathbb{C}}^{\mathrm{an}} \simeq \Omega_{X^{\mathrm{an}}}$. Also, there exists a natural morphism of ringed spaces $X^{\mathrm{an}} \longrightarrow X$.

For X projective (proper is enough) the construction leads to an equivalence of abelian categories

$$\mathrm{Coh}(X) \xrightarrow{\sim} \mathrm{Coh}(X^{\mathrm{an}}).$$

In particular, $H^*(X, F) \simeq H^*(X^{\mathrm{an}}, F^{\mathrm{an}})$ for all coherent sheaves F on X and smoothness of X implies that X^{an} is a manifold.

These general facts immediately yield the following proposition:

Proposition 3.2 *If X is an algebraic K3 surface over $k = \mathbb{C}$, then the associated complex space X^{an} is a complex K3 surface.*

It is important to note that all complex K3 surfaces obtained in this way are projective, but that there are (many) complex K3 surfaces that are not. In this sense we obtain a proper full embedding

$$\{ \text{algebraic K3 surfaces over } \mathbb{C} \} \hookrightarrow \{ \text{complex K3 surfaces} \}.$$

The image consists of all complex K3 surfaces that are projective, i.e. that can be embedded into a projective space. This is again a consequence of GAGA, because the ideal sheaf of $X \subset \mathbb{P}^n$ is a coherent analytic sheaf and hence associated with an algebraic ideal sheaf defining an algebraic K3 surface. A natural question at this point is whether complex K3 surfaces are at least always Kähler. This is in fact true and of great importance, but not easy to prove. See Section 7.3.2.

Example 3.3 The constructions described in the algebraic setting in Example 1.3 work as well here. They define different incarnations of the same geometric objects. Only for Kummer surfaces do we gain some flexibility by working with complex manifolds. Indeed, abelian surfaces A (over $\mathbb{C}$) can be replaced by arbitrary complex tori of dimension two, i.e. complex manifolds of the form $A = \mathbb{C}^2/\Gamma$ with $\Gamma \subset \mathbb{C}^2$ a lattice of rank four. The surface X, obtained as the minimal resolution of A/ι or, equivalently, as the quotient of the blow-up of all two-torsion points $\widetilde{A} \longrightarrow A$ by the lift

$\tilde{\iota}$ of the canonical involution, is a complex K3 surface. Indeed all algebraic arguments explained in Example 1.3, (iii), work in the complex setting.

One can show that X is projective if and only if the torus A is projective, i.e. the complex manifold associated with an abelian surface. It is known that many (in some sense most) complex tori $\mathbb{C}^2/\Gamma$ are not projective; cf. [64, 139]. Thus, we obtain many K3 surfaces this way that really are not projective.

Describing other examples of non-projective K3 surfaces is very difficult, which reflects a general construction problem in complex geometry.

3.2 Many but not all of the remarks and computations in Section 2 are valid for arbitrary complex K3 surfaces. For complex K3 surfaces, however, we have in addition at our disposal singular cohomology which sheds a new light on some of the results.

First, the long cohomology sequence of the exponential sequence

$$0 \longrightarrow \mathbb{Z} \longrightarrow \mathcal{O} \longrightarrow \mathcal{O}^* \longrightarrow 0$$

yields the exact sequence

$$0 \longrightarrow H^1(X,\mathbb{Z}) \longrightarrow H^1(X,\mathcal{O}) \longrightarrow H^1(X,\mathcal{O}^*) \longrightarrow H^2(X,\mathbb{Z}) \longrightarrow$$
$$\longrightarrow H^2(X,\mathcal{O}) \longrightarrow H^2(X,\mathcal{O}^*) \longrightarrow H^3(X,\mathbb{Z}) \longrightarrow 0$$

which for a complex K3 surface X (where $H^1(X,\mathcal{O}) = 0$) shows

$$H^1(X,\mathbb{Z}) = 0$$

and by Poincaré duality also $H^3(X,\mathbb{Z}) = 0$ up to torsion. So, in addition to $H^0(X,\mathbb{Z}) \simeq H^4(X,\mathbb{Z}) \simeq \mathbb{Z}$, the only other non-trivial integral singular cohomology group of X is $H^2(X,\mathbb{Z})$. We come back to the computation of its rank presently.

From the above sequence and the usual isomorphism $\mathrm{Pic}(X) \simeq H^1(X,\mathcal{O}^*)$, one also obtains the exact sequence

$$0 \longrightarrow \mathrm{Pic}(X) \longrightarrow H^2(X,\mathbb{Z}) \longrightarrow H^2(X,\mathcal{O}). \tag{3.1}$$

In other words, $\mathrm{Pic}(X)$ can be realized as the kernel of $H^2(X,\mathbb{Z}) \longrightarrow H^2(X,\mathcal{O})$. As $\mathbb{C} \simeq H^2(X,\mathcal{O})$ and by Remark 2.5 also $\mathrm{Pic}(X)$ are both torsion free, one finds that also $H^2(X,\mathbb{Z})$ is torsion free. A standard fact in topology says that the torsion of $H^i(X,\mathbb{Z})$ can be identified with the torsion of $H^{\dim_\mathbb{R} X - i + 1}(X,\mathbb{Z})$, which in our case shows that $H^3(X,\mathbb{Z})$ is indeed trivial (and not only up to torsion).

The intersection form (.) on $\mathrm{Pic}(X)$ is defined as in the algebraic case. In the complex setting it corresponds, under the above embedding $\mathrm{Pic}(X) \hookrightarrow H^2(X,\mathbb{Z})$, to the topological intersection form on $H^2(X,\mathbb{Z})$. The inclusion also shows that

$$\mathrm{Pic}(X) \xrightarrow{\sim} \mathrm{NS}(X)$$

holds for complex K3 surfaces as well; cf. Proposition 2.4.

Remark 3.4 However, it can happen (but only for non-projective complex K3 surfaces) that the subgroup of numerically trivial line bundles $\text{Pic}(X)^\tau$ is not trivial and hence $\text{Pic}(X) \neq \text{Num}(X)$. Indeed, $\text{Pic}(X)$ could be generated by a class of square zero and hence $\text{Num}(X) = 0$, but $\text{Pic}(X) \simeq \mathbb{Z}$; see Example **3**.3.2.

Hence, the Hodge index theorem does not necessarily hold any longer on $\text{Pic}(X) \simeq \text{NS}(X)$. In fact, this can happen even when $\text{NS}(X) \simeq \text{Num}(X)$, which could be generated by a class of negative square. The Hodge index theorem for $H^{1,1}(X)$ of a complex surface X with $b_1 = 0$, however, still ensures that the intersection form on $\text{Pic}(X)$ has at most one positive eigenvalue. Of course, the Hodge index theorem holds whenever X is projective, because then it underlies an algebraic K3 surface and the two intersection pairings coincide.[9]

3.3 For an arbitrary compact complex surface the Hodge–Frölicher spectral sequence degenerates (see [33, IV]) and hence

$$H^1(X, \mathbb{C}) \simeq H^1(X, \mathcal{O}_X) \oplus H^0(X, \Omega_X).$$

For a K3 surface we have seen already that $H^1(X, \mathbb{Z}) = 0$ and hence $H^1(X, \mathbb{C}) = 0$. Thus, one gets $H^0(X, \Omega_X) = 0$ for free. In other words, a complex K3 surface has no non-trivial global vector fields; cf. comments in Section 2.4.

These arguments conclude the computation of all the Hodge numbers of a complex K3 surface, confirming (2.7), and in particular show $h^{1,1}(X) = 20$. This last Hodge number tells us also something about the Picard number. Indeed, by the Lefschetz theorem on $(1, 1)$-classes, which follows from (3.1),

$$\text{Pic}(X) \simeq H^2(X, \mathbb{Z}) \cap H^{1,1}(X). \tag{3.2}$$

Thus, $\text{Pic}(X) \subset H^2(X, \mathbb{Z})$ is (contained in) the intersection $H^2(X, \mathbb{Z}) \cap H^1(X, \Omega_X)$, the complexification of which is a subspace of the 20-dimensional $H^1(X, \Omega_X)$. Hence

$$\rho(X) = \text{rk}(\text{Pic}(X)) \leq 20. \tag{3.3}$$

In fact every Picard number between 0 and 20 is realized by some complex K3 surface.[10]

The Riemann–Roch computations in Section 2.4 remain valid for complex K3 surfaces. So, we still have $c_2(X) = 24$ which for a complex surface can be read as an equality for the topological Euler number

---

[9] Maybe this is a good point to recall the following general result on the algebraicity of complex surfaces (see [33, IV.Thm. 6.2]): a smooth compact complex surface X is projective if and only if there exists a line bundle L with $(L)^2 > 0$. See Remark **8**.1.3.

[10] In the case of algebraic K3 surfaces over an arbitrary base field k (in which case clearly $1 \leq \rho(X)$) one can replace singular cohomology by étale cohomology and find the upper bound $\rho(X) \leq 22$; see Remark 3.7. For more on the Picard group of K3 surfaces, see Chapters 17 and 18, where it is explained why $\rho = 21$ is impossible for algebraically closed fields (see [16, p. 544] or Remark **18**.3.12) and that $\rho(X)$ is always even for K3 surfaces over $\overline{\mathbb{F}}_p$; see Corollary **17**.2.9.

$$e(X) = c_2(X) = 24,$$

i.e. $\sum(-1)^i b_i(X) = 24$. Since $b_1(X) = b_3(X) = 0$ and $b_0(X) = b_4(X) = 1$, this shows

$$b_2(X) = 22,$$

which can also be deduced from the Hodge–Frölicher spectral sequence and the computation of the Hodge numbers above.

Thus, $H^2(X, \mathbb{Z})$ is a free abelian group of rank 22. It is also generally known that the intersection form $(\, . \,)$ on $H^2(X, \mathbb{Z})$ of a compact oriented real four-dimensional manifold (modulo torsion, which is irrelevant for a K3 surface) defines a unimodular lattice. For general facts on lattices and the relevant notation; see Chapter 14.

Proposition 3.5 *The integral cohomology $H^2(X, \mathbb{Z})$ of a complex K3 surface X endowed with the intersection form $(\, . \,)$ is abstractly isomorphic to the lattice*

$$H^2(X, \mathbb{Z}) \simeq E_8(-1) \oplus E_8(-1) \oplus U \oplus U \oplus U. \tag{3.4}$$

Proof Here, U is the hyperbolic plane, i.e. the lattice of rank two that admits a basis of isotropic vectors e, f with $(e.f) = 1$, and $E_8(-1)$ is the standard E_8-lattice with the quadratic form changed by a sign; see Section **14**.0.3. Due to the general classification of unimodular lattices (see e.g. [547] or Corollary **14**.1.3) it is enough to prove that $H^2(X, \mathbb{Z})$ is even of signature $(3, 19)$.

According to Wu's formula (see [277, Ch. IX.9] or [412]), the intersection product of a compact differentiable fourfold M is even if and only if its second Stiefel–Whitney class $w_2(M)$ is trivial. Moreover, $w_2(M) \equiv c_1(X) \, (2)$ for any almost complex structure X on M. Hence, its intersection form is even.[11]

The signature of the intersection pairing can be computed by the Thom–Hirzebruch index theorem which in dimension two says that the index is

$$\frac{\mathrm{p}_1(X)}{3} = \frac{c_1^2(X) - 2c_2(X)}{3} = -16.$$

Since $b_2(X) = 22$, the signature is therefore $(3, 19)$.   □

It would be interesting to exhibit a particular K3 surface for which the identification of $H^2(X, \mathbb{Z})$ as (3.4) can be seen easily, i.e. by writing down appropriate cycles and without using any abstract lattice theory. A good candidate is a Kummer surface; see Example 3.3 and Section **14**.3.3.

Remark 3.6 (i) Theorem 7.1.1 shows that all complex K3 surfaces are diffeomorphic to a quartic $X \subset \mathbb{P}^3$, e.g. the Fermat quartic, and hence in particular simply connected.

---

[11] Note that this confirms our earlier observation that the Riemann–Roch formula implies $(L)^2 \equiv 0 \, (2)$ line bundles L on X; see Proposition 2.4. Unfortunately, holomorphic line bundles do not span $H^2(X, \mathbb{Z})$ (at least not if the complex structure stays fixed), so that this is not quite enough to conclude. However, as was pointed out to me by Nick Addington and Robert Lipshitz, a similar argument works for complex line bundles which do account for all classes in cohomology, so that Wu's formula can be avoided at this point.

The unbranched covering trick, mentioned in Remark 2.3, shows only that the profinite completion of the topological fundamental group $\pi_1(X)$ is trivial. If one is willing to use the existence of a Kähler–Einstein metric on a K3 surface (Calabi conjecture; see Theorem **9.4.11**), then $\pi_1(X) = \{1\}$ can be deduced from $H^1(X, \mathcal{O}) = 0$; see e.g. [269, App. A].

(ii) There are complex surfaces X with the homotopy type of a K3 surface, i.e. simply connected complex surfaces with an intersection pairing on $H^2(X, \mathbb{Z})$ given by (3.4), which, however, are not diffeomorphic to a K3 surface. As proved by Kodaira in [309], these *homotopy K3 surfaces* are all obtained by logarithmic transforms of elliptic K3 surfaces. Note that any complex surface diffeomorphic to a K3 surface is in fact a K3 surface (see [186, VII. Cor. 3.5]) and that due to results of Freedman any homotopy K3 surface is homeomorphic to a K3 surface.

Remark 3.7 Replacing singular cohomology by étale cohomology, similar considerations hold true for arbitrary K3 surfaces. See Milne [405] for basics on étale cohomology.

Indeed, the Kummer sequence $0 \longrightarrow \mu_n \longrightarrow \mathbb{G}_m \longrightarrow \mathbb{G}_m \longrightarrow 0$ for n prime to the characteristic of k and the observation that $\mathrm{Pic}(X) \simeq H^1(X, \mathbb{G}_m)$ is torsion free suffice to show that $H^1_{\text{ét}}(X, \mu_n) \simeq k^*/(k^*)^n$. For k separably closed and using duality, this yields for $\ell \neq \mathrm{char}(k)$:

$$H^1_{\text{ét}}(X, \mathbb{Z}_\ell) = 0 \text{ and } H^3_{\text{ét}}(X, \mathbb{Z}_\ell) = 0.$$

Then, from $c_2(X) = 24$ one can deduce that $H^2_{\text{ét}}(X, \mathbb{Z}_\ell)$ is a free $\mathbb{Z}_\ell$-module of rank 22. However, it is more sensible to consider the Tate twist $H^2_{\text{ét}}(X, \mathbb{Z}_\ell(1))$, which comes with the natural map $c_1 \colon \mathrm{Pic}(X) \longrightarrow H^2_{\text{ét}}(X, \mathbb{Z}_\ell(1))$ and a canonical perfect pairing that takes values in $\mathbb{Z}_\ell$ (and not in $\mathbb{Z}$, as for singular cohomology of a complex K3 surface). In fact, the induced inclusion

$$\mathrm{NS}(X) \otimes \mathbb{Z}_\ell \hookrightarrow H^2_{\text{ét}}(X, \mathbb{Z}_\ell(1))$$

respects the given pairings on both sides and, if $X = X_0 \times_{k_0} k$, also the natural actions of $\mathrm{Gal}(k/k_0)$; see Section **17.2.2**. Note that this proves, as the analogue of (3.3), that

$$\rho(X) \leq 22$$

for all K3 surfaces over arbitrary fields.

4 More Examples

We collect a number of classical construction methods for K3 surfaces. One should, however, keep in mind that most K3 surfaces, especially of high degree, do not admit explicit descriptions. Their existence is solely predicted by deformation theory.

4.1 Smooth hypersurfaces $X \subset \mathbb{P}^2 \times \mathbb{P}^1$ and $X \subset \mathbb{P}^1 \times \mathbb{P}^1 \times \mathbb{P}^1$ of type $(3, 2)$ and $(2, 2, 2)$, respectively, provide K3 surfaces. By choosing polarizations of the form $\mathcal{O}(a, b)|_X$ and $\mathcal{O}(a, b, c)|_X$ one obtains polarized K3 surfaces of various degrees, 8, 10, 12, and many others. Note however that the very general polarized K3 surface of these degrees is not isomorphic to such a hypersurface.

4.2 The following is an example of K3 surfaces of degree 14 that play an important role in the theory of Hilbert schemes of points on K3 surfaces and Fano varieties of lines on cubic fourfolds, as e.g. in the paper [52] by Beauville and Donagi. Consider the Plücker embedding $\mathrm{Gr} := \mathrm{Gr}(2, 6) \hookrightarrow \mathbb{P}^{14}$. It is of codimension six and degree 14. For the latter one may use the general formula that gives the degree of $\mathrm{Gr}(r, n) \subset \mathbb{P}^{\binom{n}{r}-1}$ as (see e.g. [432])

$$(r(n-r))! \prod_{1 \le i \le r < j \le n} (j-i)^{-1}.$$

Then the intersection $X := \mathrm{Gr} \cap \mathbb{P}^8$ with a generic linear subspace $\mathbb{P}^8 \subset \mathbb{P}^{14}$ is a smooth surface with an ample line bundle $L := \mathcal{O}(1)|_X$ of degree $(L)^2 = 14$. Similar to the argument in Example 1.3, (i), one shows that $H^1(X, \mathcal{O}_X) = 0$. The normal bundle sequence

$$0 \longrightarrow \mathcal{T}_X \longrightarrow \mathcal{T}_{\mathrm{Gr}}|_X \longrightarrow L^{\oplus 6} \longrightarrow 0$$

and $\omega_{\mathrm{Gr}} \simeq \mathcal{O}(-6)|_{\mathrm{Gr}}$ immediately show $\omega_X \simeq \mathcal{O}_X$, i.e. X is indeed a K3 surface.

4.3 The more general observation behind the last example is that Fano manifolds of coindex three give rise to K3 surfaces. More precisely, if Y is a Fano manifold such that $\omega_Y \simeq L^{-r}$ with L very ample and $\dim(Y) - r + 1 = 3$, then the generic complete intersection $X := H_1 \cap \cdots \cap H_r$, $H_i \in |L|$, defines a K3 surface. For example, the double cover $\pi : Y \longrightarrow \mathbb{P}^2 \times \mathbb{P}^2$ ramified over a divisor of bidegree $(2, 2)$ is a Fano fourfold of coindex three. Intersecting with $H_1, H_2 \in |\pi^*\mathcal{O}(1, 1)|$, one obtains a K3 surface $X = H_1 \cap H_2$ of degree 12.

Here is one more example of this type. Consider $\mathrm{Gr}(2, 5)$ with its Plücker embedding $\mathrm{Gr}(2, 5) \hookrightarrow \mathbb{P}^9$ and let $\pi : Y \longrightarrow \mathrm{Gr}(2, 5)$ be a double cover branched along a generic section with a quadric. Then $\omega_Y \simeq \pi^*(\omega_{\mathrm{Gr}} \otimes \mathcal{O}(1)) \simeq \pi^*\mathcal{O}(-4)$ and, therefore, Y is a six-dimensional Fano variety of coindex three. The pre-image $X = \pi^{-1}(H_1 \cap \cdots \cap H_4)$ of the intersection of four generic hyperplane sections defines a K3 surface of degree 10.

For more examples, see papers by Mukai, e.g. [431, 433]. In this way one indeed obtains a description of the generic K3 surface of degree $2d = 2, 4, \ldots, 18$ or, equivalently, of genus $g = 2, 3, \ldots, 10$.

4.4 Many interesting K3 surfaces can be described as elliptic surfaces $\pi : X \longrightarrow \mathbb{P}^1$, i.e. π is a surjective morphism with the generic fibre being a smooth elliptic curve. Often such surfaces are given in terms of their Weierstrass normal form. Consider sections $g_2 \in H^0(\mathbb{P}^1, \mathcal{O}(8))$ and $g_3 \in H^0(\mathbb{P}^1, \mathcal{O}(12))$ and let $\overline{X} \subset \mathbb{P}(\mathcal{O}(4) \oplus \mathcal{O}(6) \oplus \mathcal{O})$ be the hypersurface defined by the equation $y^2 z = 4x^3 - g_2 x z^2 - g_3 z^3$. Under certain genericity assumptions on g_2, g_3, e.g. $\Delta := g_2^3 - 27 g_3^2 \neq 0$, the surface $\overline{X}$ has at most ordinary double points and its minimal resolution defines indeed a K3 surface. See for example the books by Friedman et al. [185, 186]. For more on elliptic K3 surfaces, see Chapter 11.

4.5 Classically, a quartic $Y \subset \mathbb{P}^3$ with the maximal number of 16 singular points is also called a Kummer surface. The singular points are all ordinary double points and the minimal resolution $X \longrightarrow Y$ defines a K3 surface. Moreover, such a surface has 16 tropes, i.e. tangent planes that are tangent along a conic in Y. Each trope contains exactly six of the singular points and each singular point is contained in exactly six tropes. The configuration of the 16 points and 16 tropes is called a $(16, 6)$ configuration which in fact determines the quartic uniquely. Due to work of Nikulin [449] (cf. Remark **14**.3.19), it is known that the minimal resolution X is in fact a Kummer surface in the sense of Example 1.3, (iii). In fact, it is the Kummer surface associated with the Jacobian of the genus two curve C given as the double cover $C \longrightarrow \overline{C}$ of the intersection $\overline{C} = Y \cap \mathbb{P}^2$ of a trope $\mathbb{P}^2 \subset \mathbb{P}^3$ with the quartic Y ramified in the six singular points contained in $\overline{C}$. See [151, 211, 248].

References and Further Reading

The standard reference for the theory of complex surfaces is the book [33] by Barth et al., which in particular contains an extra chapter on complex K3 surfaces. Other standard texts on the theory of algebraic surfaces in general are [28, 43, 340, 509, 517]. A very elegant treatment of the Riemann–Roch formula for surfaces can be found in Mumford's lectures [437]. For complex K3 surfaces we strongly recommend the beautiful collection [54] by Beauville et al. We give more specific references in later chapters.

Questions and Open Problems

That $H^0(X, \Omega_X) = 0$ for a K3 surface in positive characteristic is a difficult theorem. It is used in the deformation theory of K3 surfaces and, in particular, in the proof of the important result that K3 surfaces in positive characteristic lift to characteristic zero; see Section **9**.5.3. The result is of course equivalent to $h^1(X, \Omega_X) \leq 20$ and one might wonder whether there is a way to approach the problem from this side. Also, can one show that any K3 surface deforms to a quartic without using this vanishing?

2

Linear Systems

There is a recurrent theme in the theory of algebraic K3 surfaces. The projective geometry of K3 surfaces shows surprising analogies to the theory of linear systems on curves and to a somewhat lesser extent to the theory of line bundles on abelian varieties. This chapter explains the basic aspects of these analogies and in particular Saint-Donat's results on ample linear systems.

We start with a recap of some aspects of the classical theory for curves and state the Kodaira–Ramanujam vanishing theorem in Section 1. The typical features of linear systems on a K3 surface are directly accessible if the linear system is associated with a smooth curve contained in the K3 surface. So, we treat this case first; see Section 2. The general case is then studied in Section 3, where we also give a proof of the Kodaira–Ramanujam vanishing theorem. The last section contains existence results for primitively polarized K3 surfaces of arbitrary even degree.

As we shall not be interested in rationality questions in this section, we assume the ground field k to be algebraically closed. If not mentioned otherwise, its characteristic is arbitrary.

1 General Results: Linear Systems, Curves, Vanishing

We collect standard results on linear systems on curves and explain first the consequences for the geometry of linear systems on K3 surfaces.

1.1 Recall that with any line bundle L on a variety X one associates the complete linear system $|L|$ which by definition is the projectivization of the space $H^0(X, L)$ of global sections or, equivalently, the space of all effective divisors $D \subset X$ linearly equivalent

---

* Thanks to Chenyang Xu for many helpful comments on this chapter.

to L. The *base locus* $\mathrm{Bs}|L|$ of $|L|$ is the maximal closed subscheme of X contained in all $D \in |L|$, i.e. $\mathrm{Bs}|L| = \bigcap_{s \in H^0(X,L)} Z(s)$.

If L has more than one section, i.e. $h^0(X, L) > 1$, then it induces the rational map

$$\varphi_L : X \dashrightarrow \mathbb{P}(H^0(X, L)^*)$$

which is regular on the complement of $\mathrm{Bs}|L|$.

For a surface X the base locus $\mathrm{Bs}|L|$ can have components of dimension zero and one. The *fixed part* of $|L|$ is the one-dimensional part of $\mathrm{Bs}|L|$, and we shall denote it by F. Then $h^0(X, F) = 1$ and the natural inclusion $L(-F) \hookrightarrow L$ yields an isomorphism $H^0(X, L(-F)) \simeq H^0(X, L)$. In this sense, φ_L on $X \setminus \mathrm{Bs}|L|$ can be identified with $\varphi_{L(-F)}$ and the latter can be extended to a morphism defined on $X \setminus \{x_i\}$. Here, $\{x_i\}$ is the finite set of base points of $L(-F)$, which contains the zero-dimensional locus of $\mathrm{Bs}|L|$. The *mobile part* of L, i.e. $M := L(-F)$, is nef (see below) and satisfies $(M)^2 \geq 0$. This observation turns out to be very useful later on.

Decomposing L in its fixed and its mobile part, often written additively as $L = M + F$, is a basic technique in the study of linear systems.

1.2 The case of the canonical linear system on a smooth irreducible curve is understood classically, and many results about linear systems on K3 surfaces are reduced to it. Recall that a smooth irreducible curve C is called *hyperelliptic* if there exists a morphism $C \longrightarrow \mathbb{P}^1$ of degree two. Often, one restricts to the case $g(C) \geq 2$ such that ω_C is base point free and the canonical map φ_{ω_C} is non-constant. Curves of genus two are hyperelliptic. For genus $g > 2$ there exist both non-hyperelliptic and hyperelliptic curves and they are distinguished by the canonical linear system being very ample or not. For the following, see e.g. [236, IV.Prop. 5.2].

Proposition 1.1 *Let C be a smooth irreducible complete curve of genus $g(C) \geq 2$. Then ω_C is very ample if and only if C is not hyperelliptic.*

Recall also that the canonical embedding $C \hookrightarrow \mathbb{P}^{g-1}$ of a non-hyperelliptic curve is projectively normal, i.e. the restriction map

$$H^0(\mathbb{P}^{g-1}, \mathcal{O}(k)) \longrightarrow H^0(C, \omega_C^k) \tag{1.1}$$

is surjective for all k. This is the theorem of Max Noether; see [11, III.2]. If C is an arbitrary curve of genus $g > 2$ and $k \geq 2$ (or $g = 2$ and $k \geq 3$), then ω_C^k is very ample, i.e. $\varphi_{\omega_C^k}$ is a closed embedding; see [236, IV.Cor. 3.2].

1.3 Let us also recall a few standard notions concerning curves on surfaces. Consider a curve $C \subset X$ in a smooth surface X. A priori, C is allowed to be singular, reducible, non-reduced, etc. The *arithmetic genus* of C is by definition

$$p_a(C) := 1 - \chi(C, \mathcal{O}_C).$$

For a curve $C \subset X$ the exact sequence

$$0 \longrightarrow \mathcal{O}(-C) \longrightarrow \mathcal{O}_X \longrightarrow \mathcal{O}_C \longrightarrow 0 \qquad (1.2)$$

shows $p_a(C) = 1 + \chi(X, \mathcal{O}(-C)) - \chi(X, \mathcal{O}_X)$, and the Riemann–Roch formula applied twice turns this into

$$2p_a(C) - 2 = (C.\omega_X \otimes \mathcal{O}(C)). \qquad (1.3)$$

If X is a K3 surface, this becomes

$$2p_a(C) - 2 = (C)^2. \qquad (1.4)$$

The arithmetic genus of a smooth and irreducible curve C coincides with its geometric genus $g(C) := h^0(C, \omega_C)$ and (1.3) confirms the standard adjunction formula

$$\omega_C \simeq (\omega_X \otimes \mathcal{O}(C))|_C.$$

For an arbitrary reduced curve C the *geometric genus* is by definition the genus of the normalization $\nu: \widetilde{C} \longrightarrow C$, i.e. $g(C) := g(\widetilde{C})$. Thus, $p_a(C) = g(C) + h^0(\delta)$, where $\delta = \nu_* \mathcal{O}_{\widetilde{C}}/\mathcal{O}_C$ is concentrated in the singular points of C.

The arithmetic genus $p_a(C)$ of an integral curve C is non-negative by definition. Thus, for an integral curve C on a K3 surface X the formula (1.3) yields $(C)^2 \geq -2$.

Definition 1.2   A (-2)-*curve* on a K3 surface is an irreducible curve C with $(C)^2 = -2$.

Observe that a (-2)-curve C is in fact integral. Moreover, it has arithmetic and geometric genus zero and it is automatically smooth. For the latter, use that $g(C) \leq p_a(C)$ with equality if and only if C is smooth. As we work over an algebraically closed field, all this implies that

$$C \simeq \mathbb{P}^1.$$

In the theory of K3 surfaces, (-2)-curves play a central role and they appear frequently throughout this book.

1.4   Let us state a few immediate consequences of the Riemann–Roch theorem for line bundles on K3 surfaces; see Section 1.2.3 and also the lectures [43, 509] by Beauville and Reid. Assume L is a line bundle on a K3 surface X.

- If $(L)^2 \geq -2$, then $H^0(X, L) \neq 0$ or $H^0(X, L^*) \neq 0$. The converse does not hold.
- If $(L)^2 \geq 0$, then either $L \simeq \mathcal{O}_X$ or $h^0(X, L) \geq 2$ or $h^0(X, L^*) \geq 2$.

- If $h^0(X, L) = 1$ and $D \subset X$ is the effective divisor defined by the unique section of L, then every curve $C \subset D$ satisfies $(C)^2 \leq -2$, and if C is integral, then C is a (-2)-curve and so $C \simeq \mathbb{P}^1$.

Corollary 1.3 *The fixed part F of any line bundle L on a K3 surface is a linear combination of smooth rational curves (with multiplicities), i.e. $F = \sum a_i C_i$ with $a_i \geq 0$ and $C_i \simeq \mathbb{P}^1$.* □

1.5 Recall that a line bundle L on a complete surface is ample if and only if $(L)^2 > 0$ and $(L.C) > 0$ for all closed curves $C \subset X$. This is a special case of the Nakai–Moishezon–Kleiman criterion (cf. Theorem **8**.1.2); see [235, I.Thm. 5.1] or [236, V.Thm. 1.10]. Using the notion of the positive cone $\mathcal{C}_X \subset \mathrm{NS}(X)_\mathbb{R}$ (see Remark 1.2.2), this criterion leads for K3 surfaces to the following result.

Proposition 1.4 *Let L be a line bundle on a K3 surface X. Then L is ample if and only if L is contained in the positive cone $\mathcal{C}_X \subset \mathrm{NS}(X)_\mathbb{R}$ and $(L.C) > 0$ for every smooth rational curve $\mathbb{P}^1 \simeq C \subset X$.*

Proof Only the 'if' needs a proof. First, note that any curve $C \subset X$ with $(C)^2 \geq 0$ is contained in the closure of $\mathcal{C}_X$. For this use that $(C.H) > 0$ for any ample line bundle H. Also, if $L \in \mathcal{C}_X$, then $(L.M) > 0$ for all $M \neq 0$ in the closure of $\mathcal{C}_X$.

Therefore, the hypothesis $L \in \mathcal{C}_X$ alone suffices to conclude that $(L.C) > 0$ for any curve $C \subset X$ with $(C)^2 \geq 0$. However, an integral curve C with $(C)^2 < 0$ is automatically smooth and rational and for those we have $(L.C) > 0$ by assumption. Thus, $(L)^2 > 0$ and $(L.C) > 0$ for all curves, which by the Nakai–Moishezon–Kleiman criterion implies that L is ample. □

A line bundle L on an arbitrary complete variety X is called *nef* if $(L.C) \geq 0$ for all closed curves $C \subset X$.

Corollary 1.5 *Consider a line bundle L on a K3 surface satisfying $(L)^2 \geq 0$ and $(L.C) \geq 0$ for all smooth rational curves $C \simeq \mathbb{P}^1$. Then L is nef unless there exists no such C, in which case L or L^* is nef.*

Proof Use that L is nef if and only if for a fixed ample line bundle H the line bundles $nL + H$ are ample for all $n > 0$. □

Definition 1.6 A line bundle L on a surface is called *big and nef* if $(L)^2 > 0$ and L is nef.[1]

---

[1] *Warning*: This seems to suggest that one should call L big if $(L)^2 > 0$, but this would mean that with L also its dual L^* is big, which we do not want. So 'big' with this definition should be used only together with 'nef'.

Remark 1.7 Often, results proved for ample line bundles in fact also hold true for line bundles which are only big and nef.

As an example of this and as an application of the Hodge index theorem (see Section 1.2.2), let us prove that big and nef curves are 1-connected. Suppose C is a big and nef curve, i.e. $(C)^2 > 0$ and $(C.D) \geq 0$ for any other curve D. Then C is *1-connected*, i.e. for any effective decomposition $C = C_1 + C_2$ one has $(C_1.C_2) \geq 1$.

Note that $C = C_1 + C_2$ can be either read as a decomposition of an effective divisor or as $\mathcal{O}(C) \simeq \mathcal{O}(C_1) \otimes \mathcal{O}(C_2)$. Indeed, 'big and nef' is a numerical property of C and thus depends only on the line bundle $\mathcal{O}(C)$. Note that in the decomposition $C = C_1 + C_2$ the curves C_1 and C_2 are allowed to have common components.

For the proof let $\lambda := (C_1.C)/(C)^2$. Since C is nef and hence $(C_i.C) \geq 0$, $i = 1, 2$, one knows $0 \leq \lambda \leq 1$. As $(C)^2 > 0$, we may assume strict inequality on one side, say $0 \leq \lambda < 1$. If $0 = \lambda$, i.e. $(C_1.C) = 0$, then by the Hodge index theorem $(C_1)^2 < 0$. But then $0 = (C_1.C) = (C_1)^2 + (C_1.C_2)$ proves the assertion. So we can assume $0 < \lambda < 1$, i.e. $(C.C_i) > 0$ for $i = 1, 2$. Then consider $\alpha := \lambda C - C_1 \in \mathrm{NS}(X)_{\mathbb{Q}}$. Then $(\alpha.C) = 0$ and hence, by the Hodge index theorem, either $\alpha = 0$ or $(\alpha)^2 < 0$. In the first case, $(\alpha.C_2) = 0$ and hence $(C_1.C_2) = \lambda(C.C_2) > 0$. If $(\alpha)^2 < 0$, then $(C_1.C_2) = (\lambda C - \alpha.(1 - \lambda)C + \alpha) = \lambda(1 - \lambda)(C)^2 - (\alpha)^2 > 0$.

The next result, a generalization of the classical Kodaira vanishing theorem (valid in full generality only in characteristic zero), is another example that an ampleness assumption can often be weakened to just big and nef. It is at the heart of many geometric results. It holds true for K3 surfaces in positive characteristic (see Section 3.1 for a proof) and for smooth projective varieties in higher dimensions (and is there known as the *Kawamata–Viehweg vanishing* theorem).

Theorem 1.8 (Kodaira–Ramanujam) *Let X be a smooth projective surface over a field k of characteristic zero. If L is a big and nef line bundle, then*

$$H^i(X, L \otimes \omega_X) = 0$$

for $i > 0$.

By Serre duality $H^1(X, L^*) \simeq H^1(X, \omega_X \otimes L)^*$ and so the result can be read as a vanishing for $H^1(X, L^*)$ of a line bundle L satisfying a certain positivity condition.

Remark 1.9 (i) The usual Kodaira vanishing theorem for ample line bundles fails in positive characteristic and so does the stronger Kodaira–Ramanujam theorem above. Pathologies in positive characteristic, i.e. the failure of standard classical facts in characteristic zero, have been studied by Mumford in a series of papers; see [438] and references therein. In particular he constructs a normal projective surface violating the Kodaira vanishing theorem. In [506] Raynaud produces for any algebraically closed field of positive characteristic a smooth projective surface together with an ample line bundle L such that $H^1(X, L \otimes \omega_X) \neq 0$.

(ii) A priori there is no reason why the Kodaira vanishing theorem should hold for line bundles L on K3 surfaces in positive characteristic. But it does and we will see that rather straightforward geometric arguments suffice to prove it; see Proposition 3.1. What is really used in the argument is the vanishing $H^1(X, \mathcal{O}) = 0$ and the fact that a big and nef line bundle L on a K3 surface is effective (but not that ω_X is trivial).

(iii) The deeper reason for the validity of the Kodaira vanishing theorem for K3 surfaces in positive characteristic is revealed by the approach of Deligne and Illusie [144]. Their arguments apply whenever the variety lifts to characteristic zero. And indeed, K3 surfaces do lift to characteristic zero. This is a non-trivial result that relies on work of Rudakov and Šafarevič and Deligne; see Section **9.5.3**.

2 Smooth Curves on K3 Surfaces

We shall study the geometry of K3 surfaces X from the point of view of the smooth curves they contain. The main result is a theorem of Saint-Donat, which we prove in this section assuming the existence of smooth curves; cf. [43, 509].

2.1 The following results give a good first impression of the geometry of K3 surfaces viewed from the curve perspective.

Lemma 2.1 *Let $C \subset X$ be a smooth irreducible curve of genus g on a K3 surface X and $L := \mathcal{O}(C)$. Then $(L)^2 = 2g - 2$ and $h^0(X, L) = g + 1$.*

Proof For the first equality use the adjunction formula (1.4). Then by the Riemann–Roch formula, $\chi(X, L) = g + 1$. Clearly, $h^2(X, L) = h^0(X, L^*) = 0$ and, hence, $h^0(X, L) \geq g + 1$. Using $H^0(C, L|_C) \simeq H^0(C, \omega_C)$ and the exact sequence

$$0 \longrightarrow H^0(X, \mathcal{O}) \longrightarrow H^0(X, L) \longrightarrow H^0(C, L|_C)$$

one deduces $h^0(X, L) = h^0(C, \omega_C) + 1 = g + 1$. $\qquad\qquad\square$

Remark 2.2 (i) The proof shows that $H^0(X, L) \longrightarrow H^0(C, L|_C)$ is surjective. This is an important observation which can also, and in fact more easily, be concluded from $H^1(X, \mathcal{O}) = 0$.

(ii) As $h^0(X, L) = g + 1 = \chi(X, L)$ and $H^2(X, L) \simeq H^0(X, L^*)^* = 0$, the lemma immediately yields $H^1(X, L) = 0$. Alternatively, one can use $0 \longrightarrow \mathcal{O} \longrightarrow L \longrightarrow L|_C \longrightarrow 0$, $H^1(X, \mathcal{O}) = 0$, and the observation that the boundary map $H^1(C, L|_C) \longrightarrow H^2(X, \mathcal{O}_X)$ is Serre dual to the bijective restriction map $H^0(X, \mathcal{O}_X) \longrightarrow H^0(C, \mathcal{O}_C)$.

(iii) A similar argument proves the surjectivity of

$$H^0(X, L^\ell) \longrightarrow\hspace{-1.5em}\longrightarrow H^0(C, L^\ell|_C) \tag{2.1}$$

for all $\ell > 0$. Indeed, on the one hand, the Riemann–Roch formula gives

$$(\ell^2/2)(L)^2 + 2 = \chi(X, L^\ell) \leq h^0(X, L^\ell)$$

and, on the other, the exact sequence $0 \longrightarrow H^0(X, L^{\ell-1}) \longrightarrow H^0(X, L^\ell) \longrightarrow H^0(C, L^\ell|_C)$ shows by induction over ℓ

$$h^0(X, L^\ell) \leq h^0(X, L^{\ell-1}) + h^0(C, L^\ell|_C)$$
$$\leq ((\ell - 1)^2/2)(L)^2 + 2 + \ell \deg(L|_C) + 1 - g(C) = (\ell^2/2)(L)^2 + 2.$$

Hence, equality must hold everywhere, in particular $h^0(X, L^\ell) = h^0(X, L^{\ell-1}) + h^0(C, L^\ell|_C)$, which implies (2.1).

Lemma 2.3 *For a smooth and irreducible curve $C \subset X$ of genus $g \geq 1$ the line bundle $L = \mathcal{O}(C)$ is base point free and the induced morphism $\varphi_L \colon X \longrightarrow \mathbb{P}^g$ restricts to the canonical map $C \longrightarrow \mathbb{P}^{g-1}$.*

Proof Indeed, the surjectivity of $H^0(X, L) \longrightarrow\hspace{-1.5em}\longrightarrow H^0(C, L|_C)$ and the adjunction formula $\omega_C = \mathcal{O}(C)|_C$ yield an embedding $\mathbb{P}^{g-1} = \mathbb{P}(H^0(C, \omega_C)^*) \subset \mathbb{P}(H^0(X, L)^*) = \mathbb{P}^g$. Moreover, L has clearly no base points outside C and ω_C is base point free for $g \geq 1$. Hence, also L is base point free and φ_L restricts to the canonical map on C. □

Remark 2.4 (i) If $g = 2$, then the curve C is hyperelliptic and hence the morphism $\varphi_L \colon X \longrightarrow \mathbb{P}^2$ restricts to a morphism $C \longrightarrow \mathbb{P}^1$ of degree two. Since $(L)^2 = 2$ and $L = \varphi_L^* \mathcal{O}(1)$, φ_L is also of degree two. Thus, in this case X is generically a double cover of $\mathbb{P}^2$ (ramified over a curve of degree six).

(ii) For $g \geq 3$ the morphism φ_L can be of degree two or degree one. More precisely, φ_L is birational, depending on whether the generic curve in $|L|$ is hyperelliptic or not. Accordingly, one calls L hyperelliptic or non-hyperelliptic.

Noether's theorem on the projective normality of non-hyperelliptic curves has a direct analogue for K3 surfaces.

Corollary 2.5 *Suppose C is an irreducible, smooth, non-hyperelliptic curve of genus $g > 2$ on a K3 surface X. Then the linear system $L = \mathcal{O}(C)$ is projectively normal, i.e. the pull-back under φ_L defines for all $k \geq 0$ a surjective map*

$$H^0(\mathbb{P}^g, \mathcal{O}(k)) \longrightarrow\hspace{-1.5em}\longrightarrow H^0(X, L^k).$$

Proof Consider the short exact sequence (1.2) tensored by L^{k+1}:

$$0 \longrightarrow L^k \longrightarrow L^{k+1} \longrightarrow \omega_C^{k+1} \longrightarrow 0.$$

By the Kodaira–Ramanujam vanishing theorem (see Proposition 3.1 or Remark 2.2) the induced maps $H^0(X, L^{k+1}) \longrightarrow H^0(C, \omega_C^{k+1})$ are surjective. Hence, the composition

$$H^0(\mathbb{P}^g, \mathcal{O}(k+1)) \simeq S^{k+1} H^0(X, L) \longrightarrow H^0(X, L^{k+1}) \longrightarrow H^0(C, \omega_C^{k+1})$$

is surjective by the classical Noether theorem (1.1) for non-hyperelliptic curves.

The kernel of $H^0(X, L^{k+1}) \longrightarrow H^0(C, \omega_C^{k+1})$ is spanned by $s \cdot H^0(X, L^k)$, where s is the section defining C. Now use the induction hypothesis $S^k H^0(X, L) \longrightarrow H^0(X, L^k)$ to conclude. $\qquad\square$

Lemma 2.6 *Let $C \subset X$ be a smooth, irreducible curve and $L = \mathcal{O}(C)$. Then, for $k \geq 2, g > 2$, or $k \geq 3, g = 2$, the morphism $\varphi = \varphi_{L^k} \colon X \longrightarrow \mathbb{P}^{k^2(g-1)+1}$ is birational onto its image.*

Proof As we have recalled earlier, ω_C^k for a smooth curve C defines an embedding under the above assumptions on k and g. Since this applies to the generic curve $D \in |L|$ and $\varphi^{-1}\varphi(D) = D$, we conclude that φ is generically an embedding. $\qquad\square$

2.2 The following can be seen as the main result of Saint-Donat's celebrated paper [518, Thm. 8.3]. In characteristic zero it had been proved earlier by Mayer [401]. It is very much in the spirit of the analogous classical result for abelian varieties (see [443]) that for an ample line bundle L on an abelian variety A (in arbitrary characteristic) the line bundle L^k is very ample for $k \geq 3$ (independent of the dimension of A).

Theorem 2.7 *Let L be an ample line bundle on a K3 surface over a field of characteristic $\neq 2$. Then L^k is globally generated for $k \geq 2$ and very ample for $k \geq 3$.*

For a version of the result for big and nef line bundles, see Remark 3.4.

Proof We prove the theorem under the simplifying assumption that the linear systems $|L^k|$ for $k = 1$ and $k = 3$ contain smooth irreducible curves. Then in fact L^k is globally generated for all $k \geq 1$. In Section 3 we discuss some of the crucial arguments that are needed to prove the assertion in general. A posteriori, it turns out that the existence of smooth irreducible curves is equivalent to the existence of just irreducible ones; cf. Remark 3.7.

Suppose there exists a smooth curve $C \in |L|$. By Remark 1.7 it is automatically irreducible. Since L is ample and hence $(L)^2 > 0$, we have $g(C) > 1$ and thus Lemma 2.3 applies. Hence, L is globally generated and so are all powers of it.

In order to prove the second assertion, it suffices to argue that L^k is very ample for $k = 3$. By Lemma 2.6 the line bundles L^k define birational morphisms φ_{L^k} for $k \geq 3$. (If $(L)^2 > 2$, it suffices to assume that $k \geq 2$.) Thus,

$$\varphi := \varphi_{L^3} \colon X \longrightarrow \overline{X} := \varphi(X) \subset \mathbb{P}^{9(g-1)+1}$$

is a birational morphism with $\varphi^*\mathcal{O}(1) = L^3$ ample. If $\overline{X}$ is normal, then $\varphi_*\mathcal{O}_X \simeq \mathcal{O}_{\overline{X}}$ by Zariski's main theorem (see [236, III.Cor. 11.4]) and hence $H^0(\overline{X}, \mathcal{O}(\ell)) = H^0(X, L^{3\ell})$ for all ℓ. As $L^{3\ell}$ is very ample for $\ell \gg 0$, φ is an isomorphism.

As proving the normality of $\overline{X}$ might be tricky, we use instead the existence of a smooth curve $D \in |L^3|$ close to $3C$. Then, by Lemma 2.3, φ is an isomorphism along D, which thus has to be non-hyperelliptic. By Corollary 2.5, $H^0(\overline{X}, \mathcal{O}(\ell)) \twoheadrightarrow H^0(X, L^{3\ell})$ is then surjective. $\qquad\square$

Remark 2.8 Saint-Donat's result can also be seen in the light of *Fujita's conjecture*, which predicts that for an ample line bundle L on a smooth projective variety X (over $\mathbb{C}$) the line bundle $L^k \otimes \omega_X$ is globally generated for $k \geq \dim(X) + 1$ and very ample for $k \geq \dim(X) + 2$; see [191, 357]. Fujita's conjecture is known to hold for surfaces. For K3 surfaces Theorem 2.7 proves a stronger version which does not hold for general surfaces.

Theorem 2.7, which in fact holds true in characteristic two [597], can also be compared with the very general result of Mumford [441, Thm. 3], which for a K3 surface says: if L is an ample and base point free line bundle, then L^k is very ample for $k \geq 3$.

3 Vanishing and Global Generation

We give an idea of some of Saint-Donat's many results in [518]. Some of the proofs below are not presented with all the details and some arguments work only under simplifying assumptions.

3.1 Let X be a K3 surface over an algebraically closed field. The following is the Kodaira–Ramanujam vanishing theorem for K3 surfaces; see Theorem 1.8. Recall that for an irreducible curve C with $(C)^2 > 0$ the associated line bundle $L := \mathcal{O}(C)$ is big and nef.

Proposition 3.1 *Let L be a big and nef line bundle on a K3 surface X. Then*

$$H^1(X, L) = 0.$$

Proof We shall first give the proof under an additional assumption.

(i) Let L be the line bundle $L = \mathcal{O}(C)$ associated with an integral (or just connected and reduced) curve $C \subset X$. The vanishing follows, even without assuming big or nef, from the short exact sequence (1.2), Serre duality $H^1(X, L \otimes \omega_X) \simeq H^1(X, L^*)^*$, and the trivial observation that the restriction $H^0(X, \mathcal{O}_X) \longrightarrow H^0(C, \mathcal{O}_C)$ is an isomorphism in this case. The induced injection

$$H^1(X, \mathcal{O}(-C)) \hookrightarrow H^1(X, \mathcal{O}) = 0$$

yields the assertion.

(ii) Here now is the general proof. First, since L is big and nef, by the Riemann–Roch formula $h^0(X, L) \geq 3$ (see Section 1.2.3). Hence, $L = \mathcal{O}(C)$ for some curve C. Pick a subdivisor $C_1 \subset C$ for which $h^0(C_1, \mathcal{O}_{C_1}) = 1$, e.g. start with an integral component of C. We may assume that C_1 is maximal with $h^0(C_1, \mathcal{O}_{C_1}) = 1$ and then show that $C_1 = C$, which would prove the assertion. If $C_1 \neq C$, then $(C_1.C - C_1) \geq 1$, for C is big and nef and hence 1-connected; see Remark 1.7. But then there exists an integral component C_2 of $C - C_1$ for which $(C_1.C_2) \geq 1$ and hence $H^0(\mathcal{O}_{C_2}(-C_1)) = 0$. The short exact sequence

$$0 \longrightarrow H^0(\mathcal{O}_{C_2}(-C_1)) \longrightarrow H^0(\mathcal{O}_{C_1+C_2}) \longrightarrow H^0(\mathcal{O}_{C_1})$$

yields a contradiction to the maximality of C_1. $\qquad\square$

Remark 3.2 Note that the proof uses only that C is 1-connected and $H^1(X, \mathcal{O}) = 0$. Thus, $H^1(X, \mathcal{O}(-C)) = 0$ for any 1-connected curve C on an arbitrary surface X with $H^1(X, \mathcal{O}) = 0$. This is *Ramanujam's lemma*, which in characteristic zero holds true even without the assumption $H^1(X, \mathcal{O}) = 0$.

Remark 3.3 The vanishing can be used to study linear systems. Let us here give a glimpse of a standard technique which almost shows base point freeness of an ample linear system; see also Proposition 3.5.

Suppose L is a big and nef line bundle on a K3 surface X. Let $L = M + F$ be the decomposition in its mobile part M and its fixed part F (written additively); see Section 1.1. Then M is effective and since mobile, i.e. without fixed part, also nef. Thus, $(M)^2 \geq 0$. Let us assume that the strict inequality $(M)^2 > 0$ holds. So, M is big and nef. One shows that then L has at most isolated base points. Compare this with Corollary 3.15, where it is shown that L is in fact base point free.

Indeed, by Proposition 3.1 one has $H^1(X, M) = 0 = H^1(X, L)$. Hence, $\chi(M) = h^0(M) = h^0(L) = \chi(L)$. Thus, from the Riemann–Roch formula one concludes $(M)^2 = (L)^2$ and hence $2(M.F) + (F)^2 = 0$. Now, L nef yields $(M.F) + (F)^2 = (L.F) \geq 0$ and, therefore, $(M.F) = (F)^2 = 0$. Then the Riemann–Roch formula applied to $F \neq 0$ leads to the contradiction $1 = h^0(F) \geq \chi(F) = 2$. Hence, $F = 0$.

Remark 3.4 Suppose L is a big and nef line bundle. Then by the Base Point Free theorem (see e.g. [121, Lect. 9] or [138, Thm. 7.32]) some positive power L^n is globally generated. (This holds for arbitrary smooth projective varieties in characteristic zero as long as $L \otimes \omega_X^*$ is big and nef.) Then the induced morphism $\varphi_{L^n} : X \longrightarrow \mathbb{P}^N$ is generically injective and, arguing as in the proof of Theorem 2.7, in fact birational. However, φ_{L^n} may contract certain curves, say C_i, (not necessarily irreducible) to points x_i. Then the curves C_i are ADE curves and the points x_i are ADE singularities of $\varphi_{L^n}(X)$; see [33, III, Prop. 2.5] and Section **14**.0.3. Note that also for L only big and nef, L^k is globally generated for $k \geq 2$.

3.2   The next result is the analogue of Lemma 2.3; see also Remark 3.3.

Proposition 3.5   *Suppose L is a line bundle on a K3 surface X with $(L)^2 > 0$ and such that $|L|$ contains an irreducible curve C. Then L is base point free.*

Proof   A complete proof can be found in the collection [517, VIII.3. Lem. 2] by Šafarevič et al. and in [518, Thm. 3.1] by Saint-Donat. Tannenbaum in [587] gives a proof relying more on considerations about multiplicities of base points. Here are the main steps of Saint-Donat's proof. One can assume that C is also reduced. Indeed, if $\mathcal{O}(C)$ is base point free, $\mathcal{O}(rC)$ also is.

Suppose $x \in X$ is a base point of $|L|$ and let $\mathcal{I}_x$ be its ideal sheaf. Then the restriction map $H^0(X, L) \longrightarrow H^0(X, k(x))$ is trivial and hence $H^1(X, L \otimes \mathcal{I}_x) \neq 0$. The latter cohomology can be more easily computed on the blow-up $\tau \colon \widetilde{X} \longrightarrow X$ in $x \in X$. Indeed, if E denotes the exceptional divisor, then $H^1(X, L \otimes \mathcal{I}_x) = H^1(\widetilde{X}, \tau^*L(-E))$. So in order to get a contradiction, it is enough to show the vanishing of the latter which by Serre duality and using $\omega_{\widetilde{X}} \simeq \mathcal{O}(E)$ is isomorphic to $H^1(\widetilde{X}, \tau^*L^*(2E))$.

Ramanujam's lemma (see Remark 3.2) shows $H^1(\widetilde{X}, \tau^*L^*(2E)) = 0$ if $\tau^*L(-2E)$ is 1-connected. Lemma 3.6 in [518] asserts quite generally that if every curve in $|L|$ is 2-connected, then every curve in $|\tau^*L(-2E)|$ is 1-connected. Thus, it suffices to show that under our assumptions every curve in $|L|$ is 2-connected, i.e. that for every decomposition $C_1 + C_2 \in |L|$ one has $(C_1.C_2) \geq 2$.

By the 1-connectedness of $|C|$ (see Remark 1.7) it suffices to exclude $(C_1.C_2) = 1$ for any $C_1 + C_2 \in |C|$. So suppose $(C_1.C_2) = 1$.

If $(C.C_i) \geq 2$, $i = 1, 2$, then $(C_i)^2 \geq 1$, $i = 1, 2$. Since the intersection form is even, in fact $(C_i)^2 \geq 2$. But then $(C_1)^2(C_2)^2 > (C_1.C_2)^2$ violates the Hodge index theorem; see Remark **1**.2.2.

If $(C.C_1) = 1$, or equivalently $(C_1)^2 = 0$, then one obtains a contradiction as follows. By the Riemann–Roch formula $h^0(X, \mathcal{O}(C_1)) \geq 2$. Using the short exact sequence $0 \longrightarrow \mathcal{O}(-C_2) \longrightarrow \mathcal{O}(C_1) \longrightarrow \mathcal{O}_C(C_1) \longrightarrow 0$ and the vanishing of $H^0(X, \mathcal{O}(-C_2))$, one concludes $h^0(C, \mathcal{O}_C(C_1)) \geq 2$. This means that on the irreducible curve C the line bundle $\mathcal{O}_C(C_1)$ of degree one has at least two sections. This implies that C, which is also reduced, is in fact smooth and rational, contradicting $(C)^2 > 0$.

If $(C.C_1) = 0$, then $(C_1)^2 = -1$, violating the evenness of the intersection pairing.

If $(C.C_1) < 0$, then C is an irreducible component of C_1. However, this is absurd, as C would then be linearly equivalent to $C + D$ with $D = C_2 + (C_1 - C)$ effective and intersecting with an ample divisor would show $D = 0$ and hence $C_2 = 0$.   □

It turns out a posteriori that Lemma 2.3 and Proposition 3.5 deal with the same situation.

Corollary 3.6 *Let C be an irreducible curve on a K3 surface X over k with* char(k) $\neq$ 2. *If* $(C)^2 > 0$, *then the generic curve in* $|C|$ *is smooth and irreducible.*

Proof In characteristic zero one can apply Bertini theorem (see [236, III.Cor. 10.9]), which shows that the generic curve in a base point free complete linear system is smooth. Thus, since C itself is irreducible, the generic curve in $|C|$ is smooth and irreducible.

In positive characteristic one can argue as follows. Consider $L = \mathcal{O}(C)$ and the induced regular map $\varphi_L \colon X \longrightarrow \mathbb{P}^g$. Then one can show that φ_L is of degree ≤ 2 (cf. Remark 2.4 and [518, Sec. 5]) and that the image $\varphi_L(X)$ has at most isolated singularities; see [518, 6.5]. By the Bertini theorem [236, II.Thm. 8.18] the generic hyperplane section of $\varphi_L(X)$ is smooth and using $\deg(\varphi_L) \leq 2$ this is true also for its inverse image; see [518, Lem. 5.8.2]. $\square$

Remark 3.7 (i) Thus, the proof of Saint-Donat's theorem (see Theorem 2.7) given in the previous section works in characteristic $\neq 2$ whenever $|L^k|$, $k = 1, 3$, is known to contain an irreducible curve. In [111, Prop. 3.1] one finds comments on the case char $= 2$.

(ii) If L is a line bundle without fixed part (cf. Corollary 3.14) and $(L)^2 > 0$, one can show that $|L|$ contains an irreducible curve and that, therefore, Proposition 3.5 applies.

The reference for this is [518, Prop. 2.6]. The argument there makes use of a version of the Bertini theorem due to Zariski [651] (which is a little difficult to read nowadays). However, it can be replaced by a variant due to Jouanolou [280, Thm. 6.3(4)], which says that if the image of a morphism $\varphi \colon U \longrightarrow \mathbb{P}^n$ is of dimension ≥ 2 and U is irreducible (we work over an algebraically closed field), then the pre-image $\varphi^{-1}(H)$ of the generic hyperplane is irreducible. In our case we work with $U := X \setminus \mathrm{Bs}(L)$. In positive characteristic, $\varphi^{-1}(H)$ might not be reduced. Then Zariski's result essentially says that the multiplicity is a power p^e of the characteristic.

From here on, the argument goes roughly as follows. Let us first assume that $\varphi_L(X)$ is of dimension > 1, i.e. $|L|$ is not composed with a pencil. Since $\mathrm{Bs}(L)$ is empty or of codimension two, Jouanolou's Bertini theorem yields the existence of an irreducible divisor in $|L|$. Thus, it is enough to deal with the case that $|L|$ is composed with a pencil, i.e. that the closure D of $\varphi(X)$ is a curve. We may assume that D is smooth. Suppose L were base point free. Then $(C_x)^2 = 0$ for the fibres C_x of the regular map $\varphi \colon X \longrightarrow D$. On the other hand, $\sum_{x \in H \cap D} C_x \in |L|$ for some hyperplane section H, which contradicts $(L)^2 > 0$. Thus, at least one point of X has to be blown up to extend φ to a regular map $\widetilde{\varphi} \colon \widetilde{X} \longrightarrow D$. The exceptional curve of $\widetilde{X} \longrightarrow X$ maps onto D and hence D is rational. Then $H^0(\mathbb{P}^1, \mathcal{O}(1)) \xrightarrow{\sim} H^0(\widetilde{X}, \widetilde{\varphi}^*\mathcal{O}(1))$. However, $\widetilde{\varphi}^*\mathcal{O}(1)$ corresponds to a complete linear system $|\widetilde{L}|$ with $\widetilde{L}$ a root of L. As $(\widetilde{L})^2 > 0$, the Riemann–Roch formula, implying $h^0(X, \widetilde{L}) > 2$, then yields the contradiction.

Remark 3.8   The above discussion can be summarized as follows: for an irreducible curve C on a K3 surface X in characteristic $\neq 2$ with $(C)^2 > 0$ the linear system $|\mathcal{O}(C)|$ is base point free and its generic member is smooth. One distinguishes the two cases:

- The linear system $|\mathcal{O}(C)|$, i.e. its generic member, is hyperelliptic. Then the morphism $\varphi_{\mathcal{O}(C)}$ is of degree two.
- The linear system $|\mathcal{O}(C)|$, i.e. its generic member, is non-hyperelliptic. Then the morphism $\varphi_{\mathcal{O}(C)}$ is of degree one, i.e. birational.

Example 3.9   As an application, we prove that any K3 surface X with $\mathrm{Pic}(X) = \mathbb{Z} \cdot L$ and such that $(L)^2 = 4$ can be realized as a quartic $X \subset \mathbb{P}^3$ with $\mathcal{O}(\pm 1)|_X \simeq L$.

First, we may assume that L is ample (after passing to its dual if necessary).[2] Then by Riemann–Roch $h^0(X, L) = 4$. Moreover, all curves in $|L|$ are automatically irreducible, as L generates $\mathrm{Pic}(X)$. Thus, Proposition 3.5, Corollary 3.6, and Lemma 2.3 apply. Hence, $\varphi_L \colon X \longrightarrow \mathbb{P}^3$ is a finite morphism, which could be of either degree one or degree two. If φ_L is of degree one, then one argues as in the proof of Theorem 2.7 to show that it is an embedding. If it is degree two, its image $X' := \varphi_L(X)$ is a quadric. If X' is smooth, then $X' \simeq \mathbb{P}^1 \times \mathbb{P}^1$ and, therefore, $\mathrm{Pic}(X') \simeq \mathbb{Z} \times \mathbb{Z}$, which contradicts $\mathrm{Pic}(X) \simeq \mathbb{Z}$. If X' is singular, then $\mathcal{O}(1)|_{X'}$ has a square root as a Weil divisor (see [236, II.Exer. 6.5]), and its pull-back to X would yield a square root of L, which is absurd.

Compare the arguments with Le Potier's more direct proof in [54, Exp. VI].

3.3   For the case of trivial self-intersection one has the following result; see [509, 3.8].

Proposition 3.10   *If a non-trivial nef line bundle L on a K3 surface X satisfies $(L)^2 = 0$, then L is base point free. If char$(k) \neq 2, 3$, then there exists a smooth irreducible elliptic curve E such that $mE \in |L|$ for some $m > 0$*

Proof   Note that $(L)^2 = 0$ implies $h^0(X, L) \geq 2$, as $h^2(X, L) = h^0(X, L^*) = 0$ for the non-trivial nef line bundle L (intersect with an ample curve). Let F be the fixed part of L. Then the mobile part $M := L(-F)$ has at most isolated fixed points. Note that $|M|$ is not trivial, because $h^0(X, L(-F)) = h^0(X, L) \geq 2$. Also recall that M is nef, and so $(M.F) \geq 0$ and $(M)^2 \geq 0$.

Now L nef and $(L)^2 = 0$ imply $(L.M) = (L.F) = 0$. This in turn yields $(M)^2 + (F.M) = 0$, hence $(M)^2 = (F.M) = 0$ and thus $(F)^2 = 0$.

---

[2] Note that this even holds for a complex K3 surface; see note 9 on page 11.

If F is non-trivial, then the Riemann–Roch formula gives $h^0(X, F) \geq 2$, which contradicts F being the fixed part of L. Hence, F is trivial and L can have at most isolated fixed points. But the existence of an isolated fixed point would contradict $(L)^2 = 0$.

Thus, L is base point free. Note that in characteristic zero, the Bertini theorem shows that the generic curve in $|L|$ is smooth but possibly disconnected.

Consider $\varphi_L \colon X \longrightarrow \mathbb{P}^m$, $m := h^0(X, L) - 1$. Since $(L)^2 = 0$, the image of φ_L is a curve $D \subset \mathbb{P}^m$. For the Stein factorization $X \longrightarrow \widetilde{D} \longrightarrow D$ (see [236, III.Cor. 11.5]), we may assume $\widetilde{D}$ smooth. Note that the generic fibre of $X \longrightarrow \widetilde{D}$ is geometrically integral and hence also the closed fibres X_t are for $t \in \widetilde{D}$ in a Zariski open subset; see [28, Ch. 7]. Then by the Leray spectral sequence $H^1(\widetilde{D}, \mathcal{O}) \hookrightarrow H^1(X, \mathcal{O}) = 0$ and therefore $\widetilde{D} \simeq \mathbb{P}^1$.

In characteristic zero, the Bertini theorem shows that the generic fibre E of $\varphi \colon X \longrightarrow \widetilde{D}$ is smooth; the fibres form the pencil $|\varphi^* \mathcal{O}_{\widetilde{D}}(1)|$. Since all fibres are connected, the generic fibre E is also irreducible. As $\widetilde{D} \simeq \mathbb{P}^1$, all fibres are linearly equivalent, which shows $L \simeq \mathcal{O}(mE)$ with $m = h^0(X, L) - 1$.

To deal with the case of positive characteristic observe that the generic fibre of the morphism $\varphi \colon X \longrightarrow \widetilde{D}$ is still a regular curve, but it might not be smooth. In fact, the geometric generic fibre, which is still integral, is either smooth or a rational curve with one cusp and the latter can occur only for char$(k) = 2$ or 3; see Tate's original article [590] or the more recent paper by Schröer [533]. However, if the generic fibre is smooth, then there also exists a smooth elliptic closed fibre; cf. [28, Thm. 7.18]. $\square$

Combined with Corollary 1.5, this shows that, at least in char$(k) \neq 2, 3$, every line bundle L with $(L)^2 = 0$ and $(L.C) \geq 0$ for all $C \simeq \mathbb{P}^1$ is isomorphic to $\mathcal{O}(mE)$ for some integer m and some smooth elliptic curve E.

Example 3.11 Consider the Fermat quartic $X \subset \mathbb{P}^3$, $x_0^4 + \cdots + x_3^4 = 0$, which contains many lines, e.g. the line $\ell \subset X$ given by $x_1 = \xi x_0$, $x_3 = \xi x_2$, with ξ a primitive eighth root of unity. Then the line bundle $L := \mathcal{O}(1) \otimes \mathcal{O}(-\ell)$, with its complete linear system consisting of all planes containing ℓ, satisfies $(L)^2 = 0$ and is clearly nef. Now, an elliptic fibration as predicted by Proposition 3.10 can be described explicitly by projecting X with center ℓ onto a disjoint line in $\mathbb{P}^3$. The fibres are the residual plane cubics of ℓ in the hyperplane intersections of X containing ℓ. Rewriting the Fermat equation as $(x_0^2 + \xi^2 x_1^2)(x_0^2 - \xi^2 x_1^2) + (x_2^2 + \xi^2 x_3^2)(x_2^2 - \xi^2 x_3^2) = 0$ allows one to write an elliptic fibration explicitly as

$$X \longrightarrow \mathbb{P}^1, \ [x_0 : x_1 : x_2 : x_3] \longmapsto [x_0^2 + \xi^2 x_1 : x_2^2 - \xi^2 x_3^2].$$

Combining Proposition 3.10 and Theorem 2.7 (see also Remark 3.4) one obtains the following corollary.

Corollary 3.12   *If L is a nef line bundle on a K3 surface, then L is semiample, i.e. L^n is globally generated for some $n > 0$.*   □

Since a semiample line bundle is obviously also nef, these two concepts coincide on K3 surfaces.

Remark 3.13   Let us mention a few related results.

(o) Assume char$(k) \neq 2, 3$. Suppose L is a non-trivial nef line bundle with $(L)^2 = 0$. Then L is linearly equivalent to a divisor $\sum n_i C_i$ with $n_i > 0$ and C_i (possibly singular) rational curves.

Indeed, by Proposition 3.10 L is linearly equivalent to mE with E a smooth fibre of an elliptic fibration $X \longrightarrow \mathbb{P}^1$. However, any elliptic fibration of a K3 surface has at least one singular fibre and all components of a singular fibre are rational. See Section **11**.1.4 for more on the fibres of elliptic fibrations.

(i) If C is an integral curve of arithmetic genus one, for example a smooth and irreducible elliptic curve, contained in a K3 surface X, then $\mathcal{O}(C)$ is primitive in Pic(X).

Indeed, C then satisfies $(C)^2 = 0$ and hence $h^0(C, \mathcal{O}(C)|_C) \leq 1$. The short exact sequence $0 \longrightarrow \mathcal{O} \longrightarrow \mathcal{O}(C) \longrightarrow \mathcal{O}(C)|_C \longrightarrow 0$ therefore yields $h^0(X, \mathcal{O}(C)) \leq 2$ and by the Riemann–Roch formula $h^0(X, \mathcal{O}(C)) \geq 2$. If $\mathcal{O}(C) \simeq M^\ell$, then $(M)^2 = 0$ and thus $h^0(X, M) \geq 2$. Hence, $h^0(X, M) = h^0(X, \mathcal{O}(C)) = 2$ and for $\ell > 1$ this would show that any curve in $|\mathcal{O}(C)|$ is reducible, which is absurd since the integral curve C is given. Thus, $\ell = 1$. Note that the proposition has actually not been used for this.

(ii) Suppose $X \longrightarrow \mathbb{P}^1$ is an elliptic pencil on a K3 surface X, i.e. the generic curve is an integral curve of arithmetic genus one. Then no fibre is multiple. Just apply (i) to a generic fibre. See Section **11**.1.2 for more on smooth and singular fibres of elliptic fibrations.

(iii) A K3 surface X in char $\neq 2, 3$, is elliptic if and only if there exists a line bundle L on X with $(L)^2 = 0$.[3]

The idea is that if such an L exists then one also finds an L' still satisfying $(L')^2 = 0$ and in addition the condition of the proposition, i.e. L' nef, holds. Roughly this is achieved by passing successively from L to the reflection $L + (L.C)C$ for a (-2)-curve C with $(L.C) < 0$. This process stops. See Example **8**.2.13 for details.

In char $= 2, 3$, the assertion is still true unless X is unirational (and in this case $\rho(X) = 22$ (see Proposition **17**.2.7), and X is supersingular (see Section **18**.3.5)).

3.4   Saint-Donat also observes the following useful fact; see [518, Cor. 3.2].

Corollary 3.14   *A complete linear system $|L|$ on a K3 surface X has no base points outside its fixed part, i.e. Bs$|L| = F$.*

---

[3] *Warning*: This is not saying that L itself comes from an elliptic pencil.

Proof If $\mathcal{O}(F) = L$, then there is nothing to show. So assume $|L(-F)|$ is non-empty. Clearly, this is now a complete linear system which for any curve $D \subset X$ contains one member that intersects D properly. In particular, $(L(-F))^2 \geq 0$. If strict inequality holds, then apply Remark 3.7, (ii), and Proposition 3.5 to show that $L(-F)$ is base point free. If $(L(-F))^2 = 0$, then Proposition 3.10 yields the assertion. $\square$

Corollary 3.15 *Let L be a big and nef line bundle.*

(i) *If the mobile part $M = L(-F)$ is big, then L is base point free. In particular, F is trivial.*

(ii) *Assume char$(k) \neq 2, 3$. If L is not base point free, then $L \simeq \mathcal{O}(mE + C)$ with E smooth elliptic, $C \simeq \mathbb{P}^1$, and $m \geq 2$.*[4]

Proof The first assertion follows from combining Corollary 3.14 with Remark 3.3. Now suppose $F \neq 0$ or, equivalently, $(M)^2 = 0$. Then, by Proposition 3.10, there exists a smooth elliptic curve E with $mE \in |M|$. Note that $m \geq 2$, because by Riemann–Roch $2 < h^0(L) = h^0(mE)$ but $h^0(E) = 2$. Since $0 < (L)^2 = (M+F)^2 = 2(M.F) + (F)^2$ and $(F)^2 < 0$, one has $(M.F) > 0$ and hence $(E.C) > 0$ for at least one integral component C of F, for which we know $C \simeq \mathbb{P}^1$ by Corollary 1.3. As $m \geq 2$, also $M + C$ is big and nef. Applying the arguments to prove (i) to the decomposition $L = (M + C) + F'$, one finds $F' = 0$, i.e. $F = C$. $\square$

4 Existence of K3 Surfaces

In the course of this book we see many examples of K3 surfaces of arbitrary degree, for example by considering Kummer surfaces or explicit equations for elliptic K3 surfaces. It may nevertheless be useful to state the general existence result at this point already.

Definition 4.1 A *polarized K3 surface* of degree $2d$ consists of a projective K3 surface X together with an ample line bundle L such that L is primitive, i.e. indivisible in Pic(X), with $(L)^2 = 2d$.

If 'ample' is replaced by the weaker assumption that L is only 'big and nef' one obtains the notion of *quasi-polarized* (or *pseudo-polarized* or *almost-polarized*) K3 surfaces (X, L) for which L is assumed to be big and nef and primitive.

Note that for any line bundle L the self-intersection $(L)^2$ is even and hence d above is a positive integer. Often, one writes $2g - 2$ for the degree $2d$, because the genus of any smooth curve in $|L|$ is indeed g. One even says that (X, L) is a *polarized K3 surface of genus g* in this case.

---

[4] As recently observed by Ulrike Rieß, in this case one has furthermore $(E.C) = 1$.

4.1   K3 surfaces can be produced by classical methods.

Theorem 4.2   *Let k be an algebraically closed field. For any $g \geq 3$ there exists a K3 surface over k of degree $(2g - 2)$ in $\mathbb{P}^g$.*

Proof   The following is taken from Beauville's book (see [43, Prop. VIII.15]), to which we refer for the complete proof. The primitivity is not addressed there but, at least for generic choices, is easy to check. Just to mention one concrete example: for $g = 3k$ one can consider a generic quartic X in $\mathbb{P}^3$ containing a line $\ell \subset X$. The linear system $H - \ell$, for H the hyperplane section, defines an elliptic pencil $|E|$; cf. Example 3.11. Then consider $L_k := H + (k - 1)E$. It is elementary, e.g. by using [236, II.Rem. 7.8.2], to see that L_k is very ample. For the generic choice of X it is also primitive. (For this, one needs to know a little more.) This yields examples of polarized K3 surfaces of degree $(L_k)^2 = 6k - 2$.   □

4.2   Alternatively, the existence of polarized K3 surfaces of arbitrary degree can be proved by deforming Kummer surfaces. We shall explain the argument in the complex setting. So, let A be a complex abelian surface with a primitive ample line bundle L_0 of degree $(L_0)^2 = 4d$. Any other line bundle in the same numerical equivalence class is then of the form $L = L_0 \otimes M$ for some $M \in \widehat{A} = \operatorname{Pic}^0(A)$. Now, L is symmetric, i.e. $L \simeq \iota^*L$, if and only if $M^2 \simeq \iota^*L_0 \otimes L_0^*$. Such an M always exists, but it is unique only up to the 16 two-torsion points in $\widehat{A}$. Note that a symmetric line bundle L does not necessarily descend to a line bundle $\overline{L}$ on the quotient A/ι, as ι might act non-trivially on the fibre $L(x)$ at one of the fixed point $x = \iota(x)$. However, by changing L by the appropriate two-torsion line bundle, this can be achieved. (As we are working with complex abelian surfaces, one can alternatively argue with the first Chern class in $H^2(A, \mathbb{Z}) \cap H^{1,1}(A)$.)

Now, pulling back $\overline{L}$ under $\pi \colon X \dashrightarrow A/\iota$ to the associated Kummer surface X (see Example 1.1.3), one obtains a line bundle on X, which we again call L, of degree $(L)^2 = 2d$. By construction, L is big and nef, but not ample, as it is trivial along the exceptional curves $E_i \subset X$ contracted by π. Using general deformation theory for the pair (X, L), which is explained in Section 6.2.4, one obtains a deformation (X', L') on which none of the exceptional curves E_i survives or, even stronger, for which $\operatorname{Pic}(X')$ is generated by L'. Using Proposition 1.4, this shows that L' is ample.

For other algebraic closed fields, the argument does not a priori work. For example, for $k = \overline{\mathbb{F}}_p$, the Picard number of X' is always even (see Corollary 17.2.9), and in characteristic zero, e.g. over $\overline{\mathbb{Q}}$, it a priori might happen that the countable many points (X_t, L_t) in the deformation space of (X, L) all come with a (-2)-curve $C_t \subset X_t$ with $(L_t.C_t) = 0$ that does not itself deform (but see Proposition 17.2.15). To conclude the argument for arbitrary algebraically closed field, one works over $\overline{\mathbb{Q}}$ and then reduces modulo primes. As ampleness is an open property; this proves the assertion at least for almost all primes.

If the field is not algebraically closed, then results of this type become more difficult. They are related to questions about rational points in the moduli space of K3 surfaces. In fact, for a fixed finite field $k = \mathbb{F}_q$ the degree of polarized K3 surfaces (X, L) defined over k is bounded; see Proposition **17**.3.8.

References and Further Reading

In [303] Knutsen and Lopez study the vanishing in Proposition 3.1 in the reverse direction. The main result describes geometrically all effective line bundles L with $(L)^2 \geq 0$ for which the vanishing $H^1(X, L) = 0$ holds true.

Tannenbaum [588] proves a criterion that (in characteristic zero) decides for a reduced and connected curve C whether $|C|$ contains an irreducible and smooth curve. It is formulated in terms of the intersection numbers of all possible decompositions of C.

For arbitrary smooth projective surfaces Reider's method (see e.g. [356, 509] for an account) not only yields a proof of the Fujita conjecture but explains the failure of ampleness of adjoint bundles. Even for K3 surfaces these results are interesting. For example, one finds that if an ample line bundle L on a K3 surface satisfies $(L)^2 \geq 5$ and $(L.C) \geq 2$ (resp. $(L)^2 \geq 10$ and $(L.C) \geq 3$) for all curves C, then L is globally generated (resp. very ample); see [356, Cor. 2.6]. Also Mumford's (Kodaira–Ramanujam) vanishing can be approached using Reider's method. We recommend Morrison's lectures [426].

The stronger notion of k-ampleness for line bundles on K3 surfaces has been studied, e.g. by Szemberg et al. in [38, 502]. In particular, Theorem 2.7 has been generalized to the statement that for L ample and $n \geq 2k + 1$ the power L^n is k-ample.

Saint-Donat also discusses equations defining K3 surfaces. More precisely, he considers the natural graded ring homomorphism

$$S^* H^0(X, L) \longrightarrow R(X, L) := \bigoplus H^0(X, L^n)$$

for a linear system $|L|$ containing a smooth irreducible non-hyperelliptic curve with trivial fixed part. Under further rather weak assumptions, he shows that the kernel is generated by elements of degree two; see [518, Thm. 7.2].

By a theorem of Zariski, $R(X, L)$ is finitely generated for semiample line bundles on projective varieties; see [357, I.Ch. 2.7.B]. Since any nef line bundle on a K3 surface is semiample (cf. Corollary 3.12), this shows that for any nef line bundle L on a K3 surface X the section ring $R(X, L)$ is finitely generated.

An effective divisor D can be decomposed not only in its mobile and its fixed part $D = M + F$, but also in its positive and its negative parts. More precisely, any effective divisor D (or any element in $\overline{\mathrm{NE}}(X) \cap \mathrm{NS}(X)$; see Section **8**.3.1) can be decomposed as $D = P + N$, where $P, N \in \mathrm{NS}(X)_{\mathbb{Q}}$ with $P \in \mathrm{Nef}(X)$ (the 'positive part') and $N = \sum a_i C_i$ effective ($a_i \in \mathbb{Q}_{>0}$) such that $(P.C_i) = 0$ and the intersection matrix $((C_i.C_j))$ is negative definite. This is the *Zariski decomposition* of D; see [28] or [357] for more references. The section ring $R(X, L)$ of a big line bundle L on a smooth surface is finitely generated if and only if its positive part P is semiample; see [357, I.Cor. 2.3.23]. Hence, $R(X, L)$ is finitely generated for any big line bundle on a K3 surface.

Seshadri constants on K3 surfaces have been studied in low degree. See for example the paper by Galati and Knutsen [195] where one also finds a survey of known results and further references.

A non-effective version of the Fujita conjecture has been known for a long time. In particular, there is Matsusaka's big theorem (see [357, 363, 395, 507]): for every polynomial $P(t)$ there exists a constant c, depending only on $P(t)$, such that for every smooth projective variety X in characteristic zero and an ample line bundle L on X with Hilbert polynomial $P(t)$ the line bundle L^k is very ample for all $k \geq c$. Matsusaka's theorem is also known in positive characteristic for small dimensions and for varieties with mild singularities. Effective versions of it (in characteristic zero) have been found by Demailly, Siu and others; see [357].

Related to the question for polarized K3 surfaces of prescribed degree is the question of which degree and genus can be realized by a smooth curve on a quartic. This has been addressed by Mori in [422].

Questions and Open Problems

It is natural to wonder how much of the theory generalizes to higher dimensions. For example, if L is an ample line bundle on a projective irreducible symplectic variety X, is then L^2 globally generated and L^3 very ample? Similarly, is X described by quadratic equations? The first case to study would be the Hilbert scheme of length two subschemes of a K3 surface.

3

Hodge Structures

For the reader's convenience we recall the basic definitions and facts concerning (pure) Hodge structures in Section 1. In Section 2 we specialize to Hodge structures of weight one and two and state the global Torelli theorem for curves and K3 surfaces. The latter appears again in subsequent chapters. This section also introduces the transcendental lattice of a Hodge structure of weight two, which we describe explicitly for a few examples. Which lattices can occur as the transcendental lattice of a K3 surface turns out to be essentially a question in lattice theory, to which we return in Chapter 14. In the final Section 3, we study the field of endomorphisms of the transcendental lattice. It turns out to be totally real in most cases. An elementary proof of this result is included. The last section also contains a discussion of the Mumford–Tate group and the conjecture describing it in terms of the algebraic fundamental group.

1 Abstract Notions

We are interested in rational and integral Hodge structures. So, in the following V always stands for either a free $\mathbb{Z}$-module of finite rank or a finite-dimensional vector space over $\mathbb{Q}$.

1.1 By $V_{\mathbb{R}}$ and $V_{\mathbb{C}}$ we denote the real and complex vector spaces obtained by scalar extension. Since V is defined over $\mathbb{Z}$ or $\mathbb{Q}$, respectively, both subrings of $\mathbb{R}$, the complex vector space $V_{\mathbb{C}}$ comes with a real structure, i.e. complex conjugation $v \longmapsto \bar{v}$ is well defined and defines an $\mathbb{R}$-linear isomorphism $V_{\mathbb{C}} \xrightarrow{\sim} V_{\mathbb{C}}$.

Definition 1.1 A *Hodge structure of weight* $n \in \mathbb{Z}$ on V is given by a direct sum decomposition of the complex vector space $V_{\mathbb{C}}$

$$V_{\mathbb{C}} = \bigoplus_{p+q=n} V^{p,q} \tag{1.1}$$

such that $\overline{V^{p,q}} = V^{q,p}$.

Often, one tacitly assumes for $n > 0$ that $V^{p,q} = 0$ for $p < 0$, which is the case, for example, when V is the cohomology of degree n of a projective manifold; see Section 1.2.

One can pass from integral Hodge structures to rational Hodge structures (of the same weight) by simple base change $V \longmapsto V_{\mathbb{Q}}$. Relaxing the obvious notion of isomorphisms between Hodge structures, one calls two integral Hodge structures V and W *isogenous* if the rational Hodge structures $V_{\mathbb{Q}}$ and $W_{\mathbb{Q}}$ are isomorphic.

For a Hodge structure V (integral or rational) of even weight $n = 2k$ the intersection $V \cap V^{k,k}$ is called the space of *Hodge classes* in V. Here, we use the natural inclusion $V \subset V_{\mathbb{C}}$. Note that in general the inclusion $(V \cap V^{k,k})_{\mathbb{C}} \subset V^{k,k}$ is proper. However, $(V_{\mathbb{R}} \cap V^{k,k})_{\mathbb{C}} = V^{k,k}$ and, similarly, $(V_{\mathbb{R}} \cap (V^{p,q} \oplus V^{q,p}))_{\mathbb{C}} = V^{p,q} \oplus V^{q,p}$ (for $p \neq q$), i.e. $V^{p,q} \oplus V^{q,p}$ is defined over $\mathbb{R}$.

Definition 1.2   A *sub-Hodge structure* of a Hodge structure V of weight n is given by a $\mathbb{Q}$-linear subspace (resp. submodule) $V' \subset V$ such that the Hodge structure on V induces a Hodge structure on V', i.e. $V'_{\mathbb{C}} = \bigoplus (V'_{\mathbb{C}} \cap V^{p,q})$.

A sub-Hodge structure $V' \subset V$ of an integral Hodge structure V is called *primitive* if V/V' is torsion free. Any Hodge structure that does not contain any non-trivial proper and primitive (in the integral case) Hodge structure $V' \subset V$ is called *irreducible*.

Note that with this definition, any Hodge class in V spans a one-dimensional sub-Hodge structure and, similarly, the space of all Hodge classes is a sub-Hodge structure.

Example 1.3   The *Tate Hodge structure* $\mathbb{Z}(1)$ is the Hodge structure of weight -2 given by the free $\mathbb{Z}$-module of rank one $(2\pi i)\mathbb{Z}$ (as a submodule of $\mathbb{C}$) such that $\mathbb{Z}(1)^{-1,-1}$ is one-dimensional. Similarly, one defines the rational Tate Hodge structure $\mathbb{Q}(1)$. The reason that $2\pi i$ is put in, instead of just considering the free module $\mathbb{Z}$, is not apparent here. In fact, for many of the applications the difference between the two is not important.

Most of the standard linear algebra constructions have analogues in Hodge theory. We shall be brief, as the details are easy to work out.

(i) The *direct sum* $V \oplus W$ of two Hodge structures V and W of the same weight n is endowed again with a Hodge structure of weight n by setting

$$(V \oplus W)^{p,q} = V^{p,q} \oplus W^{p,q}.$$

(ii) The *tensor product* $V \otimes W$ of Hodge structures V and W of weight n and m, respectively, comes naturally with a Hodge structure of weight $n + m$ by putting

$$(V \otimes W)^{p,q} = \bigoplus V^{p_1,q_1} \otimes W^{p_2,q_2},$$

where the sum is over all pairs of tuples $(p_1, q_1), (p_2, q_2)$ with $p_1 + p_2 = p$.

(iii) For a Hodge structure V of weight n, one defines a Hodge structure of weight $-n$ on the dual $V^* := \operatorname{Hom}_{\mathbb{Z}}(V, \mathbb{Z})$ (or $= \operatorname{Hom}_{\mathbb{Q}}(V, \mathbb{Q})$ if V is rational) by

$$V^{*-p,-q} := \operatorname{Hom}_{\mathbb{C}}(V^{p,q}, \mathbb{C}) \subset \operatorname{Hom}_{\mathbb{C}}(V_{\mathbb{C}}, \mathbb{C}) = V_{\mathbb{C}}^*.$$

Example 1.4 Applied to the Tate Hodge structure, (ii) and (iii) lead to the Hodge structure of weight two $\mathbb{Z}(-1) := \mathbb{Z}(1)^*$ and the Hodge structures $\mathbb{Z}(k) := \mathbb{Z}(1)^{\otimes k}$ for $k > 0$ and $\mathbb{Z}(k) := \mathbb{Z}(-1)^{\otimes -k}$ for $k < 0$, which are of weight $-2k$. Note that the underlying $\mathbb{Z}$-module of $\mathbb{Z}(k)$ is $(2\pi i)^k \mathbb{Z}$. By convention, $\mathbb{Z} = \mathbb{Z}(0)$ is the trivial Hodge structure of weight zero and rank one. The Hodge structures $\mathbb{Q}(k)$, $k \in \mathbb{Z}$, are defined analogously.

For an arbitrary integral or rational Hodge structure V of weight n one defines the Hodge structure of weight $n - 2k$

$$V(k) := V \otimes \mathbb{Z}(k) \quad \text{or} \quad V(k) := V \otimes \mathbb{Q}(k),$$

respectively, for which $V(k)^{p,q} = V^{p+k,q+k}$.

(iv) Let V and W again be Hodge structures of weight n and m. Then a *morphism of weight k* from V to W is a $\mathbb{Z}$-(or $\mathbb{Q}$-)linear map $f\colon V \longrightarrow W$ such that its $\mathbb{C}$-linear extension satisfies $f(V^{p,q}) \subset W^{p+k,q+k}$. Note that for non-trivial f this implies $m = n + 2k$, so that one could actually drop mentioning the weight k. Equivalently, a morphism $f\colon V \longrightarrow W$ of weight k can be thought of as a morphism $f\colon V \longrightarrow W(k)$ of weight zero. Then for the space of morphisms of weight k one has $\operatorname{Hom}_k(V, W) = \operatorname{Hom}_0(V, W(k))$. For $n = m$, one just writes $\operatorname{Hom}(V, W)$. Note also that

$$\operatorname{Hom}_k(V, W) = V^* \otimes W \cap (V^* \otimes W)^{k,k},$$

which is the space of Hodge classes of the Hodge structure $V^* \otimes W$ of weight $2k$. In particular, $\operatorname{Hom}_k(\mathbb{Z}, V) = \operatorname{Hom}(\mathbb{Z}(-k), V) = V \cap V^{k,k}$ is the space of Hodge classes in V.

(v) Similar to the tensor product, one also defines the *exterior product* of Hodge structures. If V is a Hodge structure of weight n, then $\bigwedge^k V$ is the Hodge structure of weight kn with $(\bigwedge^k V)^{p,q}$ being the sum of all $\bigotimes \bigwedge^{k_i}(V^{p_i,q_i})$ with $\sum k_i = k$ and $\sum k_i p_i = p$. The most interesting case for us is the exterior product $\bigwedge^2 V$ of a Hodge structure V of weight one. Here,

$$\left(\bigwedge{}^2 V\right)^{2,0} = \bigwedge{}^2 V^{1,0}, \quad \left(\bigwedge{}^2 V\right)^{1,1} = V^{1,0} \otimes V^{0,1}, \quad \text{and} \quad \left(\bigwedge{}^2 V\right)^{0,2} = \bigwedge{}^2 V^{0,1}.$$

(vi) If V is a Hodge structure, then its *complex conjugate* $\overline{V}$ is a Hodge structure on the same free $\mathbb{Z}$-module or $\mathbb{Q}$-vector space V but with $\overline{V}^{p,q} := V^{q,p}$. For a Hodge structure of weight one given by $V_{\mathbb{C}} = V^{1,0} \oplus V^{0,1}$ it amounts to flipping the two summands.

(vii) With any Hodge structure of weight n one associates the *Hodge filtration*

$$0 \subset F^n V_{\mathbb{C}} \subset F^{n-1} V_{\mathbb{C}} \subset \cdots \subset F^0 V_{\mathbb{C}} \subset V_{\mathbb{C}}, \tag{1.2}$$

where $F^i := \bigoplus_{p \geq i} V^{p,q}$. (For $n > 0$ one often has $F^{n+1} V_{\mathbb{C}} = 0$ or, equivalently, $F^0 V_{\mathbb{C}} = V_{\mathbb{C}}$.) Then

$$F^p V_{\mathbb{C}} \oplus \overline{F^q V_{\mathbb{C}}} = V_{\mathbb{C}}, \tag{1.3}$$

for all $p + q = n + 1$. Conversely, any filtration (1.2) satisfying (1.3) defines a Hodge structure (1.1) by

$$V^{p,n-p} := F^p V_{\mathbb{C}} \cap \overline{F^{n-p} V_{\mathbb{C}}}.$$

Note that the real structure of $V_{\mathbb{C}}$ is needed for this.

Thus, the two notions describe the same mathematical structure. However, the Hodge filtration is more natural when it comes to deforming Hodge structures; see e.g. Lemma **6.2.1**.

1.2 The most important examples of Hodge structures are provided by the cohomology of smooth projective varieties over $\mathbb{C}$ or, more generally, compact Kähler manifolds. For a compact Kähler manifold X the torsion-free part of the singular cohomology $H^n(X, \mathbb{Z})$ comes with a natural Hodge structure of weight n given by the standard Hodge decomposition

$$H^n(X, \mathbb{Z}) \otimes \mathbb{C} = H^n(X, \mathbb{C}) = \bigoplus_{p+q=n} H^{p,q}(X).$$

Here, $H^{p,q}(X)$ could be viewed either as the space of de Rham classes of bidegree (p, q) or as the Dolbeault cohomology $H^q(X, \Omega_X^p)$.

The even part $\bigoplus H^{2k}(X, \mathbb{Q})$ contains all algebraic classes, i.e. classes obtained as fundamental classes $[Z]$ of subvarieties $Z \subset X$. It is not difficult to see that $[Z]$ is an integral class, i.e. that it comes from an element in $H^{2k}(X, \mathbb{Z})$, and that it is contained in $H^{k,k}(X)$. The Hodge conjecture asserts that the space spanned by those is determined entirely by the Hodge structure itself.

Conjecture 1.5 (Hodge conjecture)   *For a smooth projective variety X over $\mathbb{C}$ the subspace of $H^{2k}(X, \mathbb{Q})$ spanned by all algebraic classes $[Z]$ coincides with the space of Hodge classes, i.e. $H^{2k}(X, \mathbb{Q}) \cap H^{k,k}(X) = \langle [Z] \mid Z \subset X \rangle_{\mathbb{Q}}$.*

It is often more appropriate to state the Hodge conjecture in terms of the Tate twist $H^{2k}(X, \mathbb{Q}) \otimes \mathbb{Q}(k)$. It is well known that the Hodge conjecture can fail for Kähler manifolds which are not projective [655] and that the analogous version using the integral Hodge structure does not hold in general [27].

The Hodge conjecture is known for $(1, 1)$-classes (the Lefschetz theorem on $(1, 1)$-classes) and for classes of type $(d - 1, d - 1)$ where $d = \dim X$. Thus, the Hodge conjecture is known to hold for K3 surfaces, but it is open already for self-products $X \times \cdots \times X$ of a K3 surface X.

1.3   The intersection pairing on the middle primitive cohomology or more generally the Hodge–Riemann pairing with respect to a (rational or integral) Kähler class is formalized by the notion of a polarization.

For the following, we shall need the notion of the *Weil operator C*, which acts on $V^{p,q}$ by multiplication with i^{p-q}. It clearly preserves the real vector space $(V^{p,q} \oplus V^{q,p}) \cap V_{\mathbb{R}}$.

Definition 1.6   A *polarization* of a rational Hodge structure V of weight n is a morphism of Hodge structures

$$\psi : V \otimes V \longrightarrow \mathbb{Q}(-n) \tag{1.4}$$

such that its $\mathbb{R}$-linear extension yields a positive definite symmetric form

$$(v, w) \longmapsto \psi(v, Cw)$$

on the real part of $V^{p,q} \oplus V^{q,p}$. Then (V, ψ) is called a *polarized Hodge structure*. A Hodge structure is called *polarizable* if it admits a polarization. An isomorphism $V_1 \xrightarrow{\sim} V_2$ of Hodge structures that is compatible with given polarizations ψ_1 resp. ψ_2 is called a *Hodge isometry*.

Note that ψ as a morphism of Hodge structures is a bilinear pairing on V whose $\mathbb{C}$-linear extension has the property that $\psi(v_1, v_2) = 0$ for $v_i \in V^{p_i,q_i}$ except possibly when $(p_1, q_1) = (q_2, p_2)$ (or, equivalently, $p_1 + p_2 = n$ and $q_1 + q_2 = n$). Integral polarizations are defined analogously.

Here are a few easy consequences of the definition; see e.g. [203, 253]. Assume ψ is a polarization of a Hodge structure V of weight n.

(i) If $n \equiv 1\,(2)$, then ψ is alternating. If $n \equiv 0\,(2)$, then ψ is symmetric. Indeed, working with the $\mathbb{C}$-linear extension, the required symmetry $\psi(v, Cw) = \psi(w, Cv)$ for $v \in V^{p,q}$, $w \in V^{q,p}$ reads $i^{q-p}\psi(v, w) = i^{p-q}\psi(w, v)$. Then use $i^{q-p} = (-1)^n i^{p-q}$.

(ii) The restriction of the $\mathbb{C}$-linear extension of ψ yields a non-degenerate pairing $V^{p,q} \otimes V^{q,p} \longrightarrow \mathbb{C}$.

(iii) For even $n = 2k$, the $\mathbb{R}$-linear extension to $V_{\mathbb{R}}$ yields a positive definite symmetric form $(-1)^{k-q}\psi$ on $(V^{p,q} \oplus V^{q,p}) \cap V_{\mathbb{R}}$. Indeed, for $w \in V^{p,q}$ one computes $C(w + \overline{w}) = (-1)^{k-q}(w + \overline{w})$.

(iv) A polarization of a rational Hodge structure V of weight n leads to an isomorphism of Hodge structures $V \simeq V^*(-n)$.

(v) The restriction of ψ to any sub-Hodge structure $V' \subset V$ defines a polarization of V'. Thus, any sub-Hodge structure of a polarizable Hodge structure is again polarizable.

(vi) For a polarized rational Hodge structure V, any sub-Hodge structure $V' \subset V$ gives rise to a direct sum decomposition (of Hodge structures) $V = V' \oplus V'^{\perp}$, where $V'^{\perp}$ is the orthogonal complement of V' with respect to ψ. If the Hodge structures are only integral, then $V' \oplus V'^{\perp} \subset V$ is a sub-Hodge structure of finite (and often non-trivial) index.

We come back to the Hodge structure $H^n(X, \mathbb{Z})$ (up to torsion) of a compact complex Kähler manifold. On the middle cohomology $H^d(X, \mathbb{Z})$, $d = \dim_{\mathbb{C}} X$, the intersection defines a morphism of Hodge structures as in (1.4). On cohomology groups of smaller degree one can use the Hodge–Riemann pairing, which, however, only for a rational Kähler class defines a morphism of rational Hodge structures and so X needs to be projective. But in both cases the morphism of Hodge structures becomes a polarization only after certain sign changes.

Consider a rational (or integral) Kähler class $\omega \in H^2(X, \mathbb{Q})$ on a smooth projective manifold X. Then define on $H^n(X, \mathbb{Q})$ with $n \leq d = \dim_{\mathbb{C}} X$ the pairing

$$(v, w) \longmapsto (-1)^{n(n-1)/2} \int_X v \wedge w \wedge \omega^{d-n}, \tag{1.5}$$

the *Hodge–Riemann pairing*. One defines the *primitive part* (depending on ω) by

$$H^n(X, \mathbb{Q})_{\mathrm{p}} := \mathrm{Ker}\left(\wedge \omega^{d-n+1} : H^n(X, \mathbb{Q}) \longrightarrow H^{2d-n+2}(X, \mathbb{Q})\right),$$

on which the Hodge–Riemann pairing (1.5) then defines a polarization. This is ensured by the Hodge–Riemann bilinear relation; see e.g. [253, Ch. 3.3]. Strictly speaking one should twist the integral on the right-hand side by $(2\pi i)^{-n}$, so that it really takes values $\mathbb{Q}(-n) = (2\pi i)^{-n}\mathbb{Q}$, but this is often omitted.

Example 1.7 Let us spell this out for Hodge structures of weight one and two.

(i) For degree reasons $H^1(X, \mathbb{Q})_{\mathrm{p}} = H^1(X, \mathbb{Q})$. For any Kähler class $\omega \in H^{1,1}(X) \cap H^2(X, \mathbb{Q})$ the alternating pairing

$$\psi(v, w) = \int_X v \wedge w \wedge \omega^{d-1} \tag{1.6}$$

is indeed a polarization, as $i \int v \wedge \bar{v} \wedge \omega^{d-1} > 0$ for all $0 \neq v \in H^{1,0}(X)$.

(ii) In degree two one has $H^2(X, \mathbb{Q}) = H^2(X, \mathbb{Q})_{\mathrm{p}} \oplus \mathbb{Q} \cdot \omega$. The symmetric pairing

$$\psi(v, w) = -\int_X v \wedge w \wedge \omega^{d-2} \tag{1.7}$$

is positive definite on $H^{1,1}(X) \cap H^2(X, \mathbb{R})_{\mathrm{p}}$ and negative definite on the real part of $(H^{2,0} \oplus H^{0,2})(X)$. In particular, with this definition of the pairing only its restriction to the primitive part actually defines a polarization. However, changing it by a sign on $\mathbb{Q} \cdot \omega$ one obtains a polarization of the full $H^2(X, \mathbb{Q})$. Thus, $H^2(X, \mathbb{Q})$ is polarizable. Note that for $d = 2$, i.e. the case of a surface, the pairing (1.7) is independent of ω and differs from the intersection pairing only by a sign. However, the primitive decomposition clearly depends on ω and so does the modified ψ that gives a polarization on the full $H^2(X, \mathbb{Q})$.

1.4 We shall briefly explain how to interpret Hodge structures as representations of the Deligne torus; see e.g. [203, 253].

Any rational Hodge structure of weight n gives rise to a real representation of $\mathbb{C}^*$, namely the group homomorphism

$$\rho \colon \mathbb{C}^* \longrightarrow \mathrm{GL}(V_\mathbb{R}), \ z \longmapsto \rho(z) \colon v \longmapsto (z^p \bar{z}^q) \cdot v$$

for $v \in V^{p,q}$. In order to check that the representation is indeed real, take $v \in V_\mathbb{R}$ and consider its decomposition $v = \sum v^{p,q}$ according to (1.1) with $\overline{v^{p,q}} = v^{q,p}$. Then $\rho(z)(v) = \sum (z^p \bar{z}^q) \cdot v^{p,q}$ is still real, as $\overline{(z^p \bar{z}^q)} \cdot v^{p,q} = (z^q \bar{z}^p) \cdot v^{q,p}$. Note that the induced representation $\rho|_{\mathbb{R}^*}$ is given by $\rho(t)(v) = t^n \cdot v$. The Weil operator C defined earlier is in this context simply $\rho(i)$.

There is a natural bijection between rational Hodge structures of weight n on a rational vector space V and algebraic representations $\rho \colon \mathbb{C}^* \longrightarrow \mathrm{GL}(V_\mathbb{R})$ with $\mathbb{R}^*$ acting by $\rho(t)(v) = t^n \cdot v$. To see this, we give an inverse construction that associates with an algebraic representation $\rho \colon \mathbb{C}^* \longrightarrow \mathrm{GL}(V_\mathbb{R})$ a Hodge structure.

Let us denote the $\mathbb{C}$-linear extension of ρ by $\rho_\mathbb{C} \colon \mathbb{C}^* \longrightarrow \mathrm{GL}(V_\mathbb{C})$ and let

$$V^{p,q} := \left\{ v \in V_\mathbb{C} \mid \rho_\mathbb{C}(z)(v) = (z^p \bar{z}^q) \cdot v \text{ for all } z \in \mathbb{C}^* \right\}.$$

Then $\rho_\mathbb{C}$ splits into a sum of one-dimensional representations $\lambda_i \colon \mathbb{C}^* \longrightarrow \mathbb{C}^*$ and in order to show that $V_\mathbb{C} = \bigoplus V^{p,q}$ it is enough to show that $\lambda_i(z) = z^p \bar{z}^q$ for some $p + q = n$.

At this point the assumption that ρ is algebraic comes in. As an $\mathbb{R}$-linear algebraic group,

$$\mathbb{C}^* = \left\{ z = \begin{pmatrix} x & -y \\ y & x \end{pmatrix} \right\} \subset \mathrm{GL}(2, \mathbb{R}).$$

Hence, $\rho(z)$ is a matrix whose entries are polynomials in x, y, and the inverse of the determinant $(x^2 + y^2)^{-1}$. So, $\lambda_i(z)$ must be a polynomial in z, $\bar{z}$, and $(z\bar{z})$. Therefore, it is of the form $z^p \bar{z}^q$ for some p, q with $p + q = n$.

Remark 1.8 A better way to say this is in terms of the *Deligne torus*

$$\mathbb{S} := \mathrm{Res}_{\mathbb{C}/\mathbb{R}} \mathbb{G}_{m,\mathbb{C}},$$

which is the real algebraic group described by $\mathbb{S}(A) = (A \otimes_\mathbb{R} \mathbb{C})^*$ for any $\mathbb{R}$-algebra A, so in particular $\mathbb{S}(\mathbb{R}) \simeq \mathbb{C}^*$. Then for any real vector space V there exists a natural bijection

$$\{ \text{ Hodge structures on } V \} \longleftrightarrow \{ \rho: \mathbb{S} \longrightarrow \mathrm{GL}(V_{\mathbb{R}}) \}, \qquad (1.8)$$

where on the right-hand side one considers morphisms of real algebraic groups. To be more precise, if on the left-hand side a Hodge structure of weight n is picked, then its image ρ on the right-hand side satisfies $\rho|_{\mathbb{G}_{m,\mathbb{R}}}: t \longmapsto t^n$.

Example 1.9 (i) If $V_{\mathbb{C}} = \bigoplus V^{p,q}$ is a Hodge structure given by $\rho: \mathbb{C}^* \longrightarrow \mathrm{GL}(V_{\mathbb{R}})$, then the dual Hodge structure V^* defined earlier corresponds to the dual representation $\rho^*: \mathbb{C}^* \longrightarrow \mathrm{GL}(V_{\mathbb{R}}^*)$, which is explicitly given by $\rho^*(z)(f): v \longmapsto f(\rho(z)^{-1}v)$.

(ii) A polarization is in this language described by a bilinear map $\psi: V \otimes V \longrightarrow \mathbb{Q}$ with

$$\psi(\rho(z)v, \rho(z)w) = (z\bar{z})^n \psi(v, w)$$

and such that $\psi(v, \rho(i)w)$ defines a positive definite symmetric form on $V_{\mathbb{R}}$.

(iii) The Tate Hodge structure $\mathbb{Q}(1)$ corresponds to $\mathbb{C}^* \longrightarrow \mathbb{R}^*$, $z \longmapsto (z\bar{z})^{-1}$. Thus, if a Hodge structure on V corresponds to a representation ρ_V, then the Tate twist $V(1)$ corresponds to $\rho_{V(1)}: z \longmapsto (z\bar{z})^{-1}\rho_V(z)$.

2 Geometry of Hodge Structures of Weight One and Two

Only Hodge structures of weight one and two are used in this book, especially those associated with two-dimensional tori and K3 surfaces. The Kummer construction allows one to pass from Hodge structures of weight one of a two-dimensional torus to the Hodge structure of weight two of its associated Kummer surface. The Kuga–Satake construction, to be discussed in Chapter 4, can be seen as a partial converse of this.

2.1 Hodge structures of weight one are all of geometric origin. We shall recall the basic features of this classical theory.

There is a natural bijection between the set of isomorphism classes of integral Hodge structures of weight one and the set of isomorphism classes of complex tori:

$$\{ \text{ complex tori } \} \longleftrightarrow \{ \text{ integral Hodge structures of weight one } \}, \qquad (2.1)$$

which is constructed as follows. For an integral Hodge structure V of weight one, $V_{\mathbb{C}}$ can be projected injectively into $V^{1,0}$. This yields a lattice $V \subset V^{1,0}$ and $V^{1,0}/V$ is a complex torus. Clearly, if V and V' are isomorphic integral Hodge structures of weight one, then $V^{1,0}/V$ and $V'^{1,0}/V'$ are isomorphic complex tori.

Conversely, if $\mathbb{C}^n / \Gamma$ is a complex torus, then $\mathbb{C}^n$ can be regarded as $\Gamma_{\mathbb{R}}$ endowed with an almost complex structure. This yields a decomposition $(\Gamma_{\mathbb{R}})_{\mathbb{C}} = (\Gamma_{\mathbb{R}})^{1,0} \oplus (\Gamma_{\mathbb{R}})^{0,1}$ with $(\Gamma_{\mathbb{R}})^{1,0}$ and $(\Gamma_{\mathbb{R}})^{0,1}$ being the eigenspaces on which $i \in \mathbb{C}$ acts by multiplication by i and $-i$, respectively, defining in this way an integral Hodge structure of weight one.

Using the existence of a $\mathbb{C}$-linear isomorphism $\mathbb{C}^n \simeq \Gamma_{\mathbb{R}} \simeq (\Gamma_{\mathbb{R}})^{1,0}$, the two constructions are seen to be inverse to each other. Finally, one verifies that any isomorphism between two complex tori $\mathbb{C}^n/\Gamma$ and $\mathbb{C}^n/\Gamma'$ is (up to translation) induced by a $\mathbb{C}$-linear isomorphism $\varphi\colon \mathbb{C}^n \xrightarrow{\sim} \mathbb{C}^n$ with $\varphi(\Gamma) = \Gamma'$.

Remark 2.1 If $A = \mathbb{C}^n/\Gamma$ is a complex torus, then the dual torus is

$$\mathrm{Pic}^0(A) \simeq H^1(A, \mathcal{O})/H^1(A, \mathbb{Z}).$$

If as above A is written as $V^{1,0}/V$ for an integral Hodge structure V of weight one, then the dual torus $\mathrm{Pic}^0(A)$ is naturally associated with the dual of the complex conjugate (and not just the dual) Hodge structure, i.e. $\mathrm{Pic}^0(A) \simeq V^{0,1*}/V^*$.

As was mentioned in the context of general Hodge structures, the primitive cohomology of a smooth complex projective variety is polarizable. This in particular applies to abelian varieties, i.e. projective tori. Conversely, the complex torus associated with a polarizable integral Hodge structure of weight one is projective. This yields a bijection

$$\{ \text{ abelian varieties } \} \longleftrightarrow \{ \text{ polarizable integral Hodge structures of weight one } \} \quad (2.2)$$

and, analogously, a bijection between polarized abelian varieties and polarized integral Hodge structures.

The two equivalences (2.1) and (2.2) have analogies for rational Hodge structures. On the geometric side one then considers tori and abelian varieties up to isogeny.

Arguably, the most important application of Hodge structures (of weight one) is the following classical result; see e.g. [221].

Theorem 2.2 (global Torelli theorem) *Two smooth compact complex curves C and C' are isomorphic if and only if there exists an isomorphism $H^1(C, \mathbb{Z}) \simeq H^1(C', \mathbb{Z})$ of integral Hodge structures respecting the intersection pairing (i.e. the polarization).*

Using the above equivalence between polarized abelian varieties and polarized Hodge structures of weight one, the global Torelli theorem for curves can be rephrased in terms of principally polarized Jacobians.

2.2 Let us now turn to Hodge structures of weight two.

Definition 2.3 We call V a Hodge structure of *K3 type* if V is a (rational or integral) Hodge structure of weight two with

$$\dim_{\mathbb{C}}(V^{2,0}) = 1 \quad \text{and} \quad V^{p,q} = 0 \text{ for } |p - q| > 2.$$

The motivation for this definition is, of course, that $H^2(X, \mathbb{Q})$ and $H^2(X, \mathbb{Z})$ of complex K3 surfaces (or a two-dimensional complex torus) X are rational resp. integral Hodge structures of K3 type.

If X is algebraic, then $H^2(X,\mathbb{Z})$ is polarizable. However, as explained above, it is not the (negative of the) intersection pairing that defines a polarization, but the pairing that is obtained from it by changing the sign of the intersection pairing for a rational Kähler (i.e. an ample) class. In fact, there are non-algebraic K3 surfaces for which $H^2(X,\mathbb{Q})$ is not polarizable; see Example 3.2.[1]

The importance of the Hodge structure of a K3 surface becomes apparent by the global Torelli theorem for K3 surfaces, which is to be considered in line with the global Torelli theorem 2.2 for curves and the description of tori and abelian varieties in terms of their Hodge structures of weight one as in (2.1) and (2.2). The *global Torelli theorem*, due to Pjateckiĭ-Šapiro and Šafarevič [493] in the algebraic and to Burns and Rapoport [93] in the non-algebraic case, is the central result in the theory of (complex) K3 surfaces and we come back to it later repeatedly; see Chapters 7 and 16.

Theorem 2.4 (global Torelli theorem) *Two complex K3 surfaces X and X' are isomorphic if and only if there exists an isomorphism $H^2(X,\mathbb{Z}) \simeq H^2(X',\mathbb{Z})$ of integral Hodge structures respecting the intersection pairing.*

Abusively, we also call Hodge isometry an isomorphism $H^2(X,\mathbb{Z}) \simeq H^2(X,\mathbb{Z})$ of Hodge structures that is merely compatible with the intersection pairing (and not necessarily a true polarization); cf. Definition 1.6. Note that a posteriori, one can state the global Torelli theorem for projective K3 surfaces also in terms of polarized Hodge structures which are polarized in the strict sense. This then becomes a Torelli theorem for polarized K3 surfaces. Note that the Hodge isometries in Theorems 2.2 and 2.4 are not necessarily induced by isomorphisms of the varieties themselves.

Any Hodge structure of K3 type contains two natural sub-Hodge structures: first the sub-Hodge structure of all Hodge classes $V^{1,1} \cap V$ and, second the transcendental lattice or transcendental part.

Definition 2.5 For an integral or rational Hodge structure of K3 type V one defines the *transcendental lattice* or *transcendental part T* as the minimal primitive sub-Hodge structure

$$T \subset V \quad \text{with} \quad V^{2,0} = T^{2,0} \subset T_{\mathbb{C}}.$$

The primitivity, i.e. the condition that V/T is torsion free, has to be added for integral Hodge structures, as otherwise minimality cannot be achieved. Clearly, the transcendental lattice T is again of K3 type.

If V is the Hodge structure $H^2(X,\mathbb{Z})$ of a K3 surface X, then

$$V^{1,1} \cap V = H^{1,1}(X) \cap H^2(X,\mathbb{Z}) \simeq \mathrm{NS}(X) \simeq \mathrm{Pic}(X)$$

---

[1] In contrast to the situation for complex tori, the existence of a polarization on $H^2(X,\mathbb{Q})$ does not imply that the K3 surface X is projective (or, equivalently, the existence of a class $\omega \in H^{1,1}(X) \cap H^2(X,\mathbb{Q})$ with $\omega^2 > 0$); cf. Remark **8**.1.3.

(see Section **1**.3.3), and T is called the *transcendental lattice*

$$T(X) \subset H^2(X, \mathbb{Z})$$

of the K3 surface X. It is usually considered as an integral Hodge structure.

Remark 2.6 The transcendental lattice $T(X)$ plays an equally fundamental role in the theory of K3 surfaces as the full cohomology $H^2(X, \mathbb{Z})$. However, there are non-isomorphic (algebraic as well as non-algebraic) K3 surfaces with isometric transcendental lattices. In other words, any Hodge isometry $H^2(X, \mathbb{Z}) \simeq H^2(X', \mathbb{Z})$ induces a Hodge isometry $T(X) \simeq T(X')$, but not vice versa. This is remedied by passing to derived categories. In Chapter 16 we explain that two complex algebraic K3 surfaces X and X' have Hodge isometric transcendental lattices if and only if their bounded derived categories of coherent sheaves are equivalent as $\mathbb{C}$-linear triangulated categories. This result, the *derived global Torelli theorem*, is due to Mukai and Orlov; see Corollary **16**.3.7.

Lemma 2.7 *The transcendental lattice T of a polarizable Hodge structure V of K3 type is a polarizable irreducible Hodge structure of K3 type.*

Proof Suppose $0 \neq T' \subset T$ is a sub-Hodge structure. If T' is not pure of type $(1, 1)$, then the one-dimensional $V^{2,0}$ is contained in $T'_{\mathbb{C}}$. The minimality and primitivity of T implies $T' = T$. If $T'_{\mathbb{C}} = T'^{1,1}$, then the orthogonal complement $T'^{\perp}$ of T' in T (see Section 1.3) satisfies $V^{2,0} \subset T'^{\perp}_{\mathbb{C}}$, contradicting the minimality of T. $\qquad\square$

2.3 Consider a torus $A = \mathbb{C}^n / \Gamma$. Then $H^1(A, \mathbb{Z}) \simeq \Gamma^*$ and

$$H^2(A, \mathbb{Z}) \simeq \bigwedge^2 H^1(A, \mathbb{Z}).$$

The latter can be read as an isomorphism of Hodge structures of weight two.

For $n = 2$, the case of interest to us, $H^1(A, \mathbb{Z})$ is of rank four, and, therefore, $H^2(A, \mathbb{Z})$ is of rank six. Considered with its intersection form one has an isometry

$$H^2(A, \mathbb{Z}) \simeq U^{\oplus 3}.$$

Here, U is the hyperbolic plane; see Section **14**.0.3. Explicitly, if $H^1(A, \mathbb{Z}) = \bigoplus \mathbb{Z} v_i$ with $\int_A v_1 \wedge \cdots \wedge v_4 = 1$, then the three copies of the hyperbolic plane U with the standard bases (e_i, f_i), $i = 1, 2, 3$, are realized by setting $e_1 := v_1 \wedge v_2, f_1 := v_3 \wedge v_4$, $e_2 := v_1 \wedge v_3, f_2 := v_4 \wedge v_2$, and $e_3 := v_1 \wedge v_4, f_3 := v_2 \wedge v_3$.

The Hodge structure on $H^1(A, \mathbb{Z})$ is given by a decomposition into two two-dimensional spaces $H^1(A, \mathbb{C}) \simeq H^{1,0}(A) \oplus H^{0,1}(A)$, e.g. $H^{1,0}(A) \simeq H^0(A, \Omega_A) \simeq T_0^* A \simeq \mathbb{C}^2$. Hence,

$$H^{2,0}(A) \simeq \bigwedge^2 H^{1,0}(A) \simeq \mathbb{C}$$

is one-dimensional and, therefore, $H^2(A, \mathbb{Z})$ is a Hodge structure of K3 type. Note that $H^{1,1}(A) \simeq H^{1,0}(A) \otimes H^{0,1}(A) \simeq \mathbb{C}^4$. In particular, the Néron–Severi group

$$\mathrm{NS}(A) \simeq H^{1,1}(A) \cap H^2(A, \mathbb{Z}) \subset H^2(A, \mathbb{Z}) \simeq U^{\oplus 3}$$

is of rank $0 \leq \rho(A) \leq 4$ and the transcendental lattice is of rank at least two. In fact, if A is an abelian surface, then its transcendental lattice is of rank $6 - \rho(A) \leq 5$; cf. Section 3.1.

2.4 Note that when passing from $H^1(A, \mathbb{Z})$ of a two-dimensional complex torus $A \simeq \mathbb{C}^2/\Gamma$ to $H^2(A, \mathbb{Z})$, information is lost. More precisely,

$$H^1(A, \mathbb{Z}) \longmapsto H^2(A, \mathbb{Z}) \simeq \bigwedge^2 H^1(A, \mathbb{Z}),$$

which maps an integral Hodge structure of weight one and rank four to its second exterior power, is generically two-to-one. In fact, for two two-dimensional tori A and A' there exists a Hodge isometry

$$H^2(A, \mathbb{Z}) \simeq H^2(A', \mathbb{Z})$$

if and only if

$$A \simeq A' \quad \text{or} \quad A' \simeq \mathrm{Pic}^0(A).$$

In particular, $H^2(A, \mathbb{Z})$ does not distinguish between A and its dual. Moreover, $H^2(A, \mathbb{Z})$ of an abelian surface A determines A if and only if A is principally polarized. This was proved by Shioda in [563], where also the surjectivity of the weight two period map for tori was observed. More precisely, if $\sigma \in U^{\oplus 3} \otimes \mathbb{C}$ satisfies

$$(\sigma.\sigma) = 0 \quad \text{and} \quad (\sigma.\bar{\sigma}) > 0, \tag{2.3}$$

then there exist a complex torus $A = \mathbb{C}^2/\Gamma$ and an isometry $U^{\oplus 3} \simeq H^2(A, \mathbb{Z})$ such that the image of σ spans $H^{2,0}(A)$. This is proved by the following elementary computation: if $\sigma = \sum(\alpha_i e_i + \beta_i f_i)$, then the two conditions in (2.3) translate to

$$\sum \alpha_i \beta_i = 0 \quad \text{and} \quad \mathrm{Re}\left(\sum \alpha_i \bar{\beta}_i\right) > 0.$$

After scaling, we may assume $\alpha_1 = 1$. Then let $\Gamma \subset \mathbb{C}^2$ be the lattice spanned by the columns of

$$\begin{pmatrix} 1 & 0 & -\beta_3 & \beta_2 \\ 0 & 1 & \alpha_2 & \alpha_3 \end{pmatrix}.$$

The surjectivity of the period map for K3 surfaces is considerably harder; see Sections **6**.3.3 and **7**.4.1.

2.5 Recall from Examples **1.1.3** and **1.3.3** that with any abelian surface A (here over $\mathbb{C}$) or a complex torus of dimension two, one associates the Kummer surface X as the minimal resolution of the quotient A/ι of A by the natural involution $\iota \colon x \longmapsto -x$:

$$
\begin{array}{ccc}
\widetilde{A} & \longrightarrow & A \\
\pi \downarrow & & \downarrow \\
X & \longrightarrow & A/\iota.
\end{array}
$$

The cohomology of the blow-up $\widetilde{A} \longrightarrow A$ is easily determined: in odd degree nothing changes, but in degree two one finds

$$
H^2(A, \mathbb{Z}) \oplus \bigoplus_{i=1}^{16} \mathbb{Z} \cdot [E_i] \simeq H^2(\widetilde{A}, \mathbb{Z}). \tag{2.4}
$$

Here, the $\mathbb{P}^1 \simeq E_i$ are the exceptional divisors over the 16 fixed points of the involution ι and as before we denote by $\mathbb{P}^1 \simeq \overline{E}_i \subset X$ their images.

As ι^* acts by $-\mathrm{id}$ on $H^1(A, \mathbb{Z})$, it acts as the identity on $H^2(A, \mathbb{Z}) \simeq \bigwedge^2 H^1(A, \mathbb{Z})$. The same holds for the action of the lifted involution $\widetilde{\iota}$ on $H^*(\widetilde{A}, \mathbb{Z})$, i.e. also $\widetilde{\iota}^*[E_i] = [E_i]$. But then any class $\alpha \in H^2(\widetilde{A}, \mathbb{Z})$ is in fact of the form $\alpha = \pi^*\beta$ and hence, by projection formula, $\pi^*\pi_*\alpha = \pi^*\pi_*\pi^*\beta = \pi^*(2\beta) = 2\alpha$. Use this to compare the intersection forms on $\widetilde{A}$ and X:

$$
(\pi_*\alpha . \pi_*\alpha') = 2(\alpha . \alpha'). \tag{2.5}
$$

For example, for $\alpha = [E_i]$ and $[\overline{E}_i] = \pi_*[E_i]$ this gives back $([\overline{E}_i])^2 = -2$, as for any smooth irreducible rational curve on a K3 surface; cf. Section **2.1.3**.

Next, observe that $\pi_* H^2(A, \mathbb{Z}) \subset H^2(X, \mathbb{Z})$ is indeed the orthogonal complement of $\bigoplus_{i=1}^{16} \mathbb{Z} \cdot [\overline{E}_i] \subset H^2(X, \mathbb{Z})$; see [54, Exp. VIII] for a detailed argument. In particular, $\pi_* H^2(A, \mathbb{Z}) \subset H^2(X, \mathbb{Z})$ is a primitive sublattice. As $H^2(A, \mathbb{Z}) \simeq U^{\oplus 3}$, it is abstractly isomorphic to $U(2)^{\oplus 3}$. See Section **14**.0.3 for the notation. However, in contrast to (2.4)

$$
\pi_* H^2(A, \mathbb{Z}) \oplus \bigoplus_{i=1}^{16} \mathbb{Z} \cdot [\overline{E}_i] \subset H^2(X, \mathbb{Z}) \tag{2.6}
$$

is a proper sublattice (of finite index). Indeed, as was noted already in Example **1.1.3**, the line bundle $\mathcal{O}(\sum \overline{E}_i)$ has a square root and, therefore, $\sum [\overline{E}_i]$ is divisible by two in $H^2(X, \mathbb{Z})$. But the situation is even more complicated. The saturation

$$
\bigoplus_{i=1}^{16} \mathbb{Z} \cdot [\overline{E}_i] \subset K \subset H^2(X, \mathbb{Z}),
$$

i.e. the smallest primitive sublattice containing $\bigoplus_{i=1}^{16} \mathbb{Z} \cdot [\overline{E}_i]$, is an overlattice of index 2^5. The lattice K, which is unique up to isomorphism, is called the *Kummer lattice*.

It is an even, negative definite lattice of rank 16 and discriminant 2^6. Thus (2.6) can be refined to

$$\pi_* H^2(A, \mathbb{Z}) \oplus K \subset H^2(X, \mathbb{Z}),$$

with both summands being primitive sublattices. For more information on the lattice theory of this situation, see Section **14**.3.3.

Remark 2.8   It is not difficult to use the surjectivity of the period map for complex tori mentioned above to deduce a similar statement for Kummer surfaces; cf. the proof of Theorem **14**.3.17. More precisely, any Hodge structure of K3 type on $E_8(-1)^{\oplus 2} \oplus U^{\oplus 3}$ with its $(2, 0)$-part contained in $U(2)^{\oplus 3}$ (under the above embedding) is Hodge isometric to the Hodge structure of a Kummer surface.

Also note that for a torus A and its Kummer surface X, there exists an isomorphism of Hodge structures of K3 type

$$T(A) \simeq T(X),$$

which, however, fails to be a Hodge isometry by the factor two in (2.5). In particular, there exists a primitive embedding

$$T(X) \lhook\joinrel\longrightarrow U(2)^{\oplus 3}$$

of the transcendental lattice $T(X)$ of any Kummer surface X. Also,

$$16 \leq \rho(X) \leq 22$$

and, more precisely, $\rho(X) = \rho(A) + 16$. See Corollary **14**.3.20 for a characterization of Kummer surfaces in terms of their transcendental lattice.

2.6   The actual computation of the transcendental lattice $T(X)$ of any particular K3 surface, even such an explicitly described one as the Fermat quartic, can be difficult. Already determining the Picard rank, or equivalently the rank of $T(X)$, or the quadratic form on $T(X) \otimes \mathbb{Q}$, is usually not easy.

For the Fermat quartic $X \subset \mathbb{P}^3$, $x_0^4 + \cdots + x_3^4 = 0$, the computation has been done. The problem is intimately related to the question of whether the lines contained in a Fermat quartic surface generate, rationally or even integrally, $NS(X)$, which turns out to be equivalent to disc $NS(X) = -64$. The answer to this question is affirmative and a modern proof has been given by Schütt, Shioda, and van Luijk in [541], which also contains historical comments (in particular, that disc $NS(X) = -16$ or -64 had already been shown in [493]).

In any case, the final result is that for a Fermat quartic $X \subset \mathbb{P}^3$ one has

$$T(X) \simeq \mathbb{Z}(8) \oplus \mathbb{Z}(8) \text{ and } NS(X) \simeq E_8(-1)^{\oplus 2} \oplus U \oplus \mathbb{Z}(-8) \oplus \mathbb{Z}(-8).$$

See Section **14**.0.3 for the notation. In particular, the discriminant of NS(X) is -64.[2] In Section **17**.1.4 one finds more comments and similar examples.

3 Endomorphism Fields and Mumford–Tate Groups

For any complex K3 surface, one has the two sublattices

$$\text{NS}(X), T(X) \subset H^2(X, \mathbb{Z}).$$

As we shall see, they are each other's orthogonal complement. Thus, there is a natural inclusion $T(X) + \text{NS}(X) \subset H^2(X, \mathbb{Z})$. However, for non-projective X the sum need not be direct nor the inclusion of finite index.

3.1 Recall that the transcendental part, integral or rational, of a Hodge structure of K3 type is the minimal primitive sub-Hodge structure of K3 type; see Definition 2.5. Alternatively, one has the following lemma.

Lemma 3.1 *The transcendental lattice of a complex K3 surface is the orthogonal complement of the Néron–Severi group:*

$$T(X) = \text{NS}(X)^\perp.$$

If X is projective, then $T(X)$ is a polarizable irreducible Hodge structure; cf. Lemma 2.7.

Proof To shorten the notation, we write $T = T(X)$ and $N = N(X)$.

Any integral class orthogonal to T is in particular orthogonal to $H^{2,0}(X)$ and thus of type $(1, 1)$. Then, by Lefschetz theorem on $(1, 1)$-classes (cf. (3.2) in Chapter 1) $T^\perp \subset N$. As $H^{2,0}(X)$ is orthogonal to N and thus contained in $N_\mathbb{C}^\perp$, one has $T \subset N^\perp$ by minimality of T. Taking orthogonal complements yields $N^{\perp\perp} \subset T^\perp$. Combined with the obvious $N \subset N^{\perp\perp}$, this yields

$$T^\perp \subset N \subset N^{\perp\perp} \subset T^\perp.$$

Therefore, equality holds everywhere and thus $T \subset T^{\perp\perp} = N^\perp$. It suffices, therefore, to show $T = T^{\perp\perp}$. To see this, note that $T_\mathbb{R}$ always contains the positive plane $(T^{2,0} \oplus T^{0,2}) \cap T_\mathbb{R}$ and that T is either non-degenerate, and then clearly $T = T^{\perp\perp}$, or has exactly one isotropic direction. In the second case and after diagonalizing the intersection form on $H^2(X, \mathbb{R})$ to $(1, 1, 1, -1, \ldots, -1)$, one may assume that

---

[2] In [564] Shioda shows that in characteristic $p > 0$ one still has disc NS(X) $= -64$ for $p \equiv 1(4)$ (and $\rho(X) = 20$), but disc NS(X) $= -p^2$ for $p \equiv 3(4)$ (and $\rho(X) = 22$); cf. Section **17**.2.3. Are lines still generating NS(X)?

$T_{\mathbb{R}} = \langle e_1, e_2, e_3 + e_4, e_5, \ldots, e_n \rangle$. Then $T_{\mathbb{R}}^{\perp} = \langle e_3 + e_4, e_{n+1}, \ldots, e_{22} \rangle$ and $T_{\mathbb{R}} = T_{\mathbb{R}}^{\perp\perp}$, which implies that the natural inclusion $T \subset T^{\perp\perp}$ is an equality.

If X is projective, the intersection form on $\mathrm{NS}(X)$ is non-degenerate due to the Hodge index theorem. The negative of the intersection pairing (1.7) defines a polarization on the Hodge structure $T(X) = \mathrm{NS}(X)^{\perp}$. For any sub-Hodge structure $T' \subset T(X)$ either $T'_{\mathbb{C}}$ or its orthogonal complement contains $H^{2,0}(X)$. Thus, by the minimality of $T(X)$ either $T' = 0$ or $T' = T(X)$.   □

Example 3.2   For a non projective K3 surface, $T(X) = \mathrm{NS}(X)^{\perp}$ need not be irreducible or polarizable. Suppose X is a K3 surface with $\mathrm{NS}(X)$ spanned by a non-trivial line bundle L of square zero. In particular, X is not algebraic. The existence of such a K3 surface is a consequence of the surjectivity of the period map; see Theorem 7.4.1. In this case $\mathrm{NS}(X)^{\perp}$ is the kernel of $(L. \ ) \colon H^2(X, \mathbb{Z}) \longrightarrow \mathbb{Z}$ and contains the non-trivial sub-Hodge structure spanned by L. Thus, in the non-algebraic case $T(X) = \mathrm{NS}(X)^{\perp}$ is not necessarily irreducible and the intersection $T(X) \cap \mathrm{NS}(X)$ might be non-trivial.

Geometrically, K3 surfaces of this type are provided by elliptic K3 surfaces without any multisection. For a concrete algebraic example consider $V = \mathbb{Q}^4$ with diagonal intersection form $(1, 2, 1, -1)$. Then let $\ell = e_1 + e_4$ and $V^{2,0}$ be spanned by $\sigma := (e_2 + \ell) + i(\sqrt{2}e_3 + \ell)$. One easily checks that $\langle \ell \rangle^{\perp}$ in this case is spanned by e_2, e_3, ℓ. Due to the $\sqrt{2}$ in the definition of σ, the class ℓ is indeed the only Hodge class up to scaling.

In a similar fashion, one constructs examples of complex K3 surfaces with $\mathrm{Pic}(X)$ spanned by a line bundle L with $(L)^2 < 0$.

3.2   Let us next note the following elementary but very useful statement.

Lemma 3.3   *Let T be a rational (or integral) Hodge structure of K3 type, such that there is no proper (primitive) sub-Hodge structure $0 \neq T' \subset T$ of K3 type. If $a \colon T \longrightarrow T$ is any endomorphism of the Hodge structure with $a = 0$ on $T^{2,0}$, then $a = 0$.*

Proof   By assumption, $T' := \mathrm{Ker}(a) \subset T$ is a Hodge structure with $T'^{2,0} \neq 0$ (and T/T' torsion free). Hence, $T' = T$ and so $a = 0$.   □

The result is usually applied to irreducible Hodge structures, e.g. $T(X)_{\mathbb{Q}}$ of a projective K3 surface, but it also applies to $T(X)_{\mathbb{Q}}$ of a non-projective K3 surface. In this case, $T(X)_{\mathbb{Q}}$ may or may not be irreducible, but it still satisfies the assumption of the lemma. The lemma is often used to deduce from $a = \mathrm{id}$ on $T^{2,0}$ that $a = \mathrm{id}$, which is of course equivalent to the above version.

The next result is formulated in the geometric context but it holds for the transcendental lattice of any polarized Hodge structure of K3 type.

Corollary 3.4 *Let $a: T(X) \xrightarrow{\sim} T(X)$ be a Hodge isometry of the transcendental lattice of a complex projective K3 surface. Then there exists an integer $n > 0$ such that $a^n = $ id. In fact, the group of all Hodge isometries of $T(X)$ is a finite cyclic group.*

Proof Consider $V(X) := T(X)_{\mathbb{R}} \cap (H^{2,0} \oplus H^{0,2})(X)$ and its orthogonal complement $V(X)^{\perp}$ in $T(X)_{\mathbb{R}}$. Then the intersection form restricted to $V(X)$ is positive definite and restricted to $V(X)^{\perp}$ negative definite, for X is assumed projective. The decomposition $T(X)_{\mathbb{R}} = V(X) \oplus V(X)^{\perp}$ is preserved by a. Hence, the eigenvalues of $a|_{V(X)}$ and of $a|_{V(X)^{\perp}}$ (and thus also of a itself) are all of absolute value one

On the other hand, a is defined on the integral lattice $T(X)$ and, therefore, its eigenvalues are all algebraic integers. Thus, if λ is the algebraic integer that is the eigenvalue of the action of a on $T(X)^{2,0}$, then $|\lambda_i| = 1$ for all its conjugates λ_i. Hence, by Kronecker's theorem, λ is a root of unity, say $\lambda = \zeta_n$. Then $a^n = $ id on $T(X)^{2,0}$ and, therefore, $a^n = $ id by Lemma 3.3.

For the second statement, one argues that the group of Hodge isometries $T(X) \xrightarrow{\sim} T(X)$ is discrete and a subgroup of the compact $O(V(X)) \times O(V(X)^{\perp})$ and, therefore, necessarily finite. On the other hand, any Hodge isometry of $T(X)$ is determined by its action on $H^{2,0}(X)$. Thus, the group of Hodge isometries of $T(X)$ can be realized as a finite subgroup of $\mathbb{C}^*$ and is, therefore, cyclic. $\qquad\square$

The following has been observed by Oguiso in [470, Lem. 4.1] (see also [386, Lem. 3.7]) and can be used to determine $\mathrm{Aut}(X)$ for a general complex projective K3 surface; see Corollary 15.2.12. It is curious that the same argument comes up when showing that the Tate conjecture implies that the Picard number of any K3 surface over $\overline{\mathbb{F}}_p$ is even; cf. Corollary 17.2.9.

Corollary 3.5 *Let X be a complex projective K3 surface of odd Picard number or, equivalently, with $\mathrm{rk}\, T(X) \equiv 1 \, (2)$. Then the only Hodge isometries of $T(X)$ are $\pm$ id.*

Proof For an isometry a of $T(X)$, λ is an eigenvalue if and only if $\lambda^{-1} = \overline{\lambda}$ is an eigenvalue. Hence, the number of those eigenvalues $\neq \pm 1$ must be even. Therefore, if $T(X)$ is of odd rank, then 1 or -1 occurs as a eigenvalue and a corresponding eigenvector can be chosen in the lattice $T(X)$, i.e. there exists at least one $0 \neq \alpha \in T(X)$ with $a(\alpha) = \pm\alpha$. If now a is a Hodge isometry with $a \neq \pm$id, then $a = \zeta_m \cdot$ id on $H^{2,0}(X)$, $m > 2$, as above. Pairing α with $H^{2,0}(X)$ shows that α has to be orthogonal to $H^{2,0}(X)$, i.e. $\alpha \in H^{1,1}(X, \mathbb{Z})$, which contradicts $\alpha \in T(X)$ (for $T(X)$ is irreducible). $\qquad\square$

The situation changes when one considers rational Hodge isometries. This is discussed next.

3.3 For an arbitrary irreducible rational(!) Hodge structure T of K3 type, one considers its *endomorphism field* $K(T)$ of all morphisms $a: T \longrightarrow T$ of Hodge structures, which is a $\mathbb{Q}$-algebra endowed with a $\mathbb{Q}$-algebra homomorphism

$$\varepsilon \colon K := K(T) := \mathrm{End}_{\mathrm{Hdg}}(T) \longrightarrow \mathbb{C}$$

defined by $a|_{T^{2,0}} = \varepsilon(a) \cdot \mathrm{id}$. Note that at this stage the endomorphisms a are not assumed to be compatible with any polarization and, in fact, T need not even admit a polarization.

Corollary 3.6   *The map ε is injective and K is a number field.*

Proof   The injectivity follows from Lemma 3.3. In particular, K is commutative and obviously finite-dimensional over $\mathbb{Q}$. To show that K is a field, consider a with $\varepsilon(a) \neq 0$. Then $\mathrm{Ker}(a) \subset T$ is a proper sub-Hodge structure, as it does not contain $T^{2,0}$. However, T is irreducible and hence $\mathrm{Ker}(a) = 0$. Therefore, a is an isomorphism and can thus be inverted.   □

What kind of algebraic number fields does one encounter as the endomorphism rings $\mathrm{End}_{\mathrm{Hdg}}(T)$ of irreducible Hodge structures T of K3 type?

Before stating the result, recall that a number field K_0 is called *totally real* if all embeddings $K_0 \hookrightarrow \mathbb{C}$ take image in $\mathbb{R} \subset \mathbb{C}$. An extension $K_0 \subset K$ is a purely imaginary quadratic extension if there exists an element α such that $K = K_0(\sqrt{\alpha})$ and $\rho(\alpha) \in \mathbb{R}_{<0}$ for all embeddings $\rho \colon K_0 \hookrightarrow \mathbb{C}$. If K is a purely imaginary quadratic extension of a totally real field, then K is a *CM field*.

The following result, for which we provide an elementary proof in Section 3.5, is due to Zarhin [648, Thm. 1.5.1]. See also Borcea's [76] for the existence of the Hodge isometry in the CM case.

Theorem 3.7   *Let $K = K(T)$ be the endomorphism field $K(T)$ of a polarizable irreducible rational Hodge structure T of K3 type. Then K is either totally real or a CM field. If K is a CM field, then there exists a Hodge isometry η such that $K = \mathbb{Q}(\eta)$.*

Remark 3.8   In [204] van Geemen shows that any totally real field K is realized as $K(T)$ of a polarized rational Hodge structure T of K3 type. Moreover, $m := \dim_K T \geq 3$ can be prescribed. If $m \cdot [K : \mathbb{Q}] \leq 10$, then K is realized as $K(T(X))$ of the general member X of an $(m - 2)$-dimensional family of K3 surfaces.

Pjateckiĭ-Šapiro and Šafarevič show in [494] that any CM field K with $[K : \mathbb{Q}] \leq 16$ can be realized as $K(T(X))$ of a K3 surface which moreover satisfies $\dim_K T = 1$ or, equivalently, $\dim_{\mathbb{Q}} T = [K : \mathbb{Q}]$; see (3.2). The condition on the degree has recently be weakened by Taelman to $[K : \mathbb{Q}] \leq 20$; see [582]. Note that $\dim_K T = 1$ implies that K is a CM field; see Remark 3.14.

3.4   The endomorphism ring is crucial in determining the Mumford–Tate group. We shall first recall the definition of the Mumford–Tate group and its characterization in terms of Hodge classes.

Consider a polarizable Hodge structure on a rational vector space V in terms of the representation $\rho\colon \mathbb{S} \longrightarrow \mathrm{GL}(V_{\mathbb{R}})$; see Remark 1.8. The *Hodge group* $\mathrm{Hdg}(V)$ and the *Mumford–Tate group* $\mathrm{MT}(V)$ of V are defined as the smallest algebraic subgroups of the linear algebraic group $\mathrm{GL}(V)$ over $\mathbb{Q}$ with

$$\rho(\mathbb{U}(\mathbb{R})) \subset \mathrm{Hdg}(V)(\mathbb{R}) \quad \text{and} \quad \rho(\mathbb{S}(\mathbb{R})) \subset \mathrm{MT}(V)(\mathbb{R}).$$

Here, $\mathbb{U} \subset \mathbb{S}$ is the kernel of the norm $\mathrm{Nm}\colon \mathbb{S} \longrightarrow \mathbb{G}_{m,\mathbb{R}}$, so $\mathbb{U}(\mathbb{R}) = \{z \mid z\bar{z} = 1\} \subset \mathbb{S}(\mathbb{R})$. The two groups can also be related via the surjective and finite morphism

$$\mathrm{Hdg}(V) \times \mathbb{G}_m \longrightarrow \mathrm{MT}(V), \ (g, \mu) \longmapsto g\mu.$$

By definition, Hodge classes in V are invariant under the action of $\mathrm{Hdg}(V)$. More generally:

- A subspace $W \subset \bigoplus V^{\otimes n_i} \otimes V^{*\otimes m_i}$ is a sub-Hodge structure if and only if it is preserved by the natural action of $\mathrm{MT}(V)$.
- In particular, a vector $v \in \bigoplus V^{\otimes n_i} \otimes V^{*\otimes m_i}$ is a Hodge class if and only if it is invariant up to scaling under the natural action of $\mathrm{MT}(V)$.

In fact, the last property characterizes $\mathrm{MT}(V)$. More precisely, if a subgroup of $\mathrm{GL}(V)$ fixes every Hodge class of weight zero in any $\bigoplus V^{\otimes n_i} \otimes V^{*\otimes m_i} \otimes \mathbb{Q}(n)$, then it is contained in $\mathrm{MT}(V)$. This hinges on the fact that $\mathrm{MT}(V)$ is a reductive group for polarizable Hodge structure; see [145, Ch. I] or [203, 420, 529].

Zarhin also proves in [648] the following theorem.

Theorem 3.9 *Let* (T, ψ) *be an irreducible polarized rational Hodge structure of K3 type with endomorphism field* $K = K(T)$. *Then its Hodge group is the subgroup of K-linear special isometries:*

$$\mathrm{Hdg}(T) = \mathrm{SO}_K(T) \subset \mathrm{SO}(T). \tag{3.1}$$

Equivalently, the Hodge group can be described as

$$\mathrm{Hdg}(T) = \begin{cases} \mathrm{SO}(T, \Psi), & \text{if } K \text{ is totally real,} \\ \mathrm{U}(T, \Psi), & \text{if } K \text{ is a CM field.} \end{cases}$$

Here, the pairing $\Psi\colon T \times T \longrightarrow K$ is defined by the condition that for all $a \in K$ one has $\psi(av, w) = \mathrm{Tr}_{K/\mathbb{Q}}(a\Psi(v, w))$. An elementary computation shows that Ψ is symmetric if K is totally real and sesquilinear if K is a CM field. (Complex conjugation on K is described by (3.3) below.)

The inclusion $\mathrm{Hdg}(T) \subset \mathrm{SO}_K(T)$ follows from the minimality of $\mathrm{Hdg}(T)$ and the observation that $\mathbb{U}(\mathbb{R})$ commutes with the action of K and hence $\rho(\mathbb{U}(\mathbb{R})) \subset \mathrm{Hdg}(T)(\mathbb{R}) \subset \mathrm{SO}_K(T)(\mathbb{R})$. Similarly, the inclusion $\mathrm{Hdg}(T) \subset \mathrm{SO}(T, \Psi)$ (resp. $\subset \mathrm{U}(T, \Psi)$) is deduced from the fact that for all $a \in K$ and all $z \in \mathbb{C}^*$ with $z\bar{z} = 1$ one has

$$\psi(a(\rho(z)(v)), \rho(z)(w)) = (z\bar{z}) \cdot \psi(av, w) = \psi(av, w),$$

where one uses that ψ is a polarization. The other inclusion is deduced from a comparison of dimensions. For the description of real points of the Hodge group, see [204, Sec. 2.7].

Remark 3.10 In [494] Pjateckiĭ-Šapiro and Šafarevič define a complex projective K3 surface to have CM if Hdg(T) is commutative. This turns out to be equivalent to $\dim_K T = 1$ (see [648, Rem. 1.5.3]) and implies the weaker property that K is a CM field; cf. Remark 3.14. Moreover, due to [494, Thm. 4] a K3 surface that has CM in the sense of [494] is defined over a number field.

Also note that any complex K3 surface of maximal Picard number $\rho(X) = 20$ has CM. Indeed, in this case Hdg(X) as a subgroup of SO(2) is commutative. Alternatively, observe that

$$\begin{pmatrix} 1 & b/a \\ -b/c & 1 - b^2/ac \end{pmatrix}$$

defines an orthogonal transformation of the rank two lattice with intersection form

$$\begin{pmatrix} 2a & b \\ b & 2c \end{pmatrix}$$

given by $T(X)$, which automatically preserves the Hodge structure; cf. the discussion in Section **14**.3.4. So, $K \neq \mathbb{Q}$ and hence $\dim_K T = 1$.

For the sake of completeness, we mention at this point the *Mumford–Tate conjecture*, which conjecturally relates Hodge theory and Galois theory via a comparison of Mumford–Tate groups and algebraic monodromy groups. Compare the discussion here with the one of the Tate conjecture in Section **17**.3. Here and there, we restrict to degree two. For a geometric version of the following discussion, see Section **6**.4.3.

Let X be a smooth projective variety over a finitely generated field k of characteristic zero with the natural Galois action

$$\rho_\ell \colon \operatorname{Gal}(\bar{k}/k) \longrightarrow \operatorname{GL}(H^2_{\text{ét}}(X, \mathbb{Q}_\ell(1))).$$

Consider the Zariski closure of $\operatorname{Im}(\rho_\ell) \subset \overline{\operatorname{Im}(\rho_\ell)} \subset \operatorname{GL}(H^2_{\text{ét}}(X, \mathbb{Q}_\ell(1)))$, which often is called the *ℓ-adic algebraic monodromy group*.

On the other hand, any embedding $k \hookrightarrow \mathbb{C}$ yields a complex variety $X_{\mathbb{C}}$ whose singular cohomology $H^2(X_{\mathbb{C}}, \mathbb{Q})$ is endowed with the action of the Mumford–Tate group. The Mumford–Tate conjecture then predicts that under the usual comparison isomorphism $H^2_{\text{ét}}(X, \mathbb{Q}_\ell(1)) \simeq H^2(X_{\mathbb{C}}, \mathbb{Q}) \otimes \mathbb{Q}_\ell(1)$ the identity component $\overline{\operatorname{Im}(\rho_\ell)}^0$ coincides with the Mumford–Tate group; cf. [548]. The conjecture has been proved for K3 surfaces over number fields by Tankeev [585, 586] and André [7].

Theorem 3.11 *Let X be a K3 surface over a finitely generated field $k \subset \mathbb{C}$. Then for all ℓ*

$$\overline{\mathrm{Im}(\rho_\ell)}^{\,\circ} = \mathrm{MT}(H^2(X_\mathbb{C}, \mathbb{Q}(1))) \times_\mathbb{Q} \mathbb{Q}_\ell.$$

At least morally, the Mumford–Tate conjecture follows from the conjunction of the Hodge conjecture and the Tate conjecture. The Hodge conjecture for $H^2(X_\mathbb{C}, \mathbb{Z})$, which is nothing but the Lefschetz theorem on $(1, 1)$-classes (see (3.2) in Section **1**.3.3), implies that $\mathrm{NS}(X_\mathbb{C}) \otimes \mathbb{Q}_\ell \simeq H^2(X_\mathbb{C}, \mathbb{Q}_\ell(1))^{\mathrm{MT}}$, and the Tate conjecture for finitely generated fields of characteristic zero, which is proved via the Kuga–Satake construction (see Section **17**.3.2), shows $\mathrm{NS}(X) \otimes \mathbb{Q}_\ell \simeq H^2_{\mathit{ét}}(X, \mathbb{Q}_\ell(1))^{\mathrm{Gal}(\overline{k}/k)}$. So, at least after finite base change:

$$H^2(X_\mathbb{C}, \mathbb{Q}_\ell(1))^{\mathrm{MT}} = H^2_{\mathit{ét}}(X, \mathbb{Q}_\ell(1))^{\overline{\mathrm{Im}(\rho_\ell)}}.$$

For the geometric analogue of the Mumford–Tate conjecture, see Section **6**.4.3.

3.5 Here is a completely elementary proof of Theorem 3.7. Let us consider the embeddings $K \hookrightarrow \mathbb{C}$. We denote by

$$\rho_1, \ldots, \rho_r \colon K \hookrightarrow \mathbb{R} \subset \mathbb{C} \quad \text{and} \quad \sigma_1, \overline{\sigma}_1, \ldots, \sigma_s, \overline{\sigma}_s \colon K \hookrightarrow \mathbb{C}$$

the real and complex embeddings. In particular, $[K : \mathbb{Q}] = r + 2s$. Then $\mathrm{Tr}_{K/\mathbb{Q}}$ of any $a \in K$ can be written as $\mathrm{Tr}_{K/\mathbb{Q}}(a) = \sum \rho_i(a) + \sum \sigma_j(a) + \sum \overline{\sigma}_j(a) = \sum \rho_i(a) + 2 \sum \mathrm{Re}(\sigma_j(a))$ and considering T as a vector space over K yields

$$\dim_\mathbb{Q} T = \dim_K T \cdot [K : \mathbb{Q}] \quad \text{and} \quad \mathrm{Tr}_{T/\mathbb{Q}}(a) = \dim_K T \cdot \mathrm{Tr}_{K/\mathbb{Q}}(a). \tag{3.2}$$

In a first step, we identify a totally real field $K_0 \subset K$ and then show that either $K_0 = K$ or that $K_0 \subset K$ is purely imaginary quadratic. In order to define K_0, we use the polarization ψ and let $\langle \ , \ \rangle := -\psi(\ , \ )$. This is a non-degenerate symmetric bilinear form on T of signature $(2, m)$ such that its $\mathbb{R}$-linear extension is positive definite on $(T^{2,0} \oplus T^{0,2}) \cap T_\mathbb{R}$ and such that the decomposition $T_\mathbb{R} = (T^{1,1} \cap T_\mathbb{R}) \oplus ((T^{2,0} \oplus T^{0,2}) \cap T_\mathbb{R})$ is orthogonal with respect to $\langle \ , \ \rangle$. In particular, the $\mathbb{R}$-linear extension of $\langle \ , \ \rangle$ is negative definite on $(T^{1,1} \cap T_\mathbb{R})$.

Then one defines an involution $K \longrightarrow K$, $a \longmapsto a'$ by the condition

$$\langle av, w \rangle = \langle v, a'w \rangle \tag{3.3}$$

for all $v, w \in T$. In other words, a' is the formal adjoint of a with respect to $\langle \ , \ \rangle$.

Lemma 3.12 *If $a \in K$, then $a' \in K$, i.e. with a also a' preserves the Hodge structure.*

Proof Suppose $w \in T^{1,1}$. Then $\langle av, w \rangle = 0$ for all $v \in T^{2,0} \oplus T^{0,2}$, because av is again of type $(2, 0) + (0, 2)$. Hence $a'(w)$ is orthogonal to $T^{2,0} \oplus T^{0,2}$ and thus again of type $(1, 1)$. The proof that a' preserves $T^{2,0}$ and $T^{0,2}$ is similar. $\qquad \square$

Clearly, $(ab)' = a'b'$, i.e. $a \longmapsto a'$ is an automorphism of K. Also observe that for $a \in K$ and all $v, w \in T$ one has $\langle av, aw \rangle = \langle v, a'aw \rangle$. Hence, $a'a = 1$ if and only if a is an isometry.

Denote by $K_0 \subset K$ the subfield of all $a \in K$ with $a' = a$. Since $a \longmapsto a'$ is an automorphism of K of order two, its fixed field K_0 satisfies $[K : K_0] \leq 2$.

To study K_0, it is more convenient to work with a positive definite symmetric bilinear form, which however is defined only over $\mathbb{R}$. One defines $(\ ,\ )$ on $T_\mathbb{R}$ by setting

$$(\ , \ ) = \langle \ , \ \rangle \ \text{ on } \ (T^{2,0} \oplus T^{0,2}) \cap T_\mathbb{R} \ \text{ and } \ (\ , \ ) = -\langle \ , \ \rangle \ \text{ on } \ T^{1,1} \cap T_\mathbb{R}.$$

As it turns out, for any $a \in K$, the formal adjoint a' with respect to $\langle \ , \ \rangle$ is also the formal adjoint of the $\mathbb{R}$-linear extension of a with respect to $(\ ,\ )$.

For any $0 \neq a \in K$ let

$$\xi_a := a'a = aa' \in K,$$

which satisfies:

(i) $(\xi_a v, w) = (v, \xi_a w)$ for all $v, w \in T_\mathbb{R}$, i.e. ξ_a is self-adjoint.
(ii) $(\xi_a v, v) = (av, av) > 0$ for all $0 \neq v \in T_\mathbb{R}$.
(iii) $\xi_a = a^2$ for $a \in K_0$.

In particular, all eigenvalues of ξ_a are positive and, therefore, $\mathrm{Tr}_{T/\mathbb{Q}}(\xi_a) > 0$ and also $\mathrm{Tr}_{K/\mathbb{Q}}(\xi_a) > 0$.

Lemma 3.13   *Any number field L satisfying $\mathrm{Tr}_{L/\mathbb{Q}}(a^2) > 0$ for all $0 \neq a \in L$ is totally real. In particular, K_0 is totally real.*

Proof   Suppose $s > 0$, i.e. there exists at least one embedding $\sigma_s \colon L \hookrightarrow \mathbb{C}$ which is not real. Using $L \otimes_\mathbb{Q} \mathbb{R} \simeq \mathbb{R}^r \oplus \mathbb{C}^s$, one finds an $a \in L$ such that $\rho_i(a)$ and $\sigma_j(a)$ are all close to zero for all i and all $j < s$, and $\sigma_s(a)$ close to $\sqrt{-1}$.

By assumption $0 < \mathrm{Tr}_{L/\mathbb{Q}}(a^2) = \sum_i \rho_i(a^2) + 2 \sum_{j<s} \mathrm{Re}(\sigma_j(a^2)) + 2\mathrm{Re}(\sigma_s(a^2))$. On the other hand, by construction, $\rho_i(a^2) = \rho_i(a)^2$ and $\sigma_j(a^2) = \sigma_j(a)^2$ for $j < s$ are all close to zero, whereas $\sigma_s(a^2)$ is close to -1. This yields the contradiction $0 < \mathrm{Tr}_{L/\mathbb{Q}}(a^2) < 0$. $\square$

We can now prove the first part of Theorem 3.7.

Proof   We have to show that either $K_0 = K$ or, if not, then K/K_0 is a purely imaginary quadratic extension. As observed earlier, if $K_0 \neq K$, then $[K : K_0] = 2$, and, therefore, we can write $K = K_0(\sqrt{\alpha})$ for some $\alpha \in K_0$.

Fix one real embedding $K_0 \subset \mathbb{R}$ and suppose $\alpha \in \mathbb{R}_{>0}$. The natural inclusion $K_0(\sqrt{\alpha}) \subset \mathbb{R}$ yields one real embedding $\rho_1 \colon K \longrightarrow \mathbb{R}$ and we set $\rho_2 := \rho_1 \circ (\ )'$, which is the identity on K_0 and sends $\sqrt{\alpha}$ to $-\sqrt{\alpha}$.

Let us denote the remaining embeddings of K by $\rho_3, \dots, \rho_d$ (which may be real or complex).

Similar to the argument used in the proof of Lemma 3.13 we choose $a \in K$ such that $\rho_1(a) \sim -1$, $\rho_2(a) \sim 1$, and $\rho_i(a) \sim 0$ for $i \geq 3$. Then, $0 < \mathrm{Tr}_{K/\mathbb{Q}}(\xi_a) = \mathrm{Tr}_{K/\mathbb{Q}}(aa')$ is contradicted by $\mathrm{Tr}_{K/\mathbb{Q}}(aa') = \rho_1(aa') + \rho_2(aa') + \sum_{i \geq 3} \rho_i(aa') \sim \rho_1(aa') + \rho_2(aa') = \rho_1(a)\rho_1(a') + \rho_2(a)\rho_2(a') = 2\rho_1(a)\rho_2(a) \sim -2$. □

Remark 3.14 (i) In the case of complex multiplication, the involution $a \longmapsto a'$ is given by complex conjugation for all complex embeddings.

(ii) If $\dim_{\mathbb{Q}} T$ is odd, then $K_0 = K$, i.e. K is totally real. Indeed, by (3.2) $[K : \mathbb{Q}]$ divides $\dim_{\mathbb{Q}} T$ and $[K : \mathbb{Q}]$ is even for a CM field.

(iii) If $\dim_K T = 1$, i.e. $[K : \mathbb{Q}] = \dim_{\mathbb{Q}} T$, then K is a CM field. To prove this, consider $T \otimes_{\mathbb{Q}} \mathbb{C} = \bigoplus T_\rho$, where the sum runs over all $\rho \colon K \hookrightarrow \mathbb{C}$ and T_ρ is the $\mathbb{C}$-subspace on which the elements $\alpha \in K$ act by multiplication with $\rho(\alpha)$. Clearly $\dim_{\mathbb{C}} T \otimes_{\mathbb{Q}} \mathbb{C} = \dim_{\mathbb{Q}} T = \dim_{\mathbb{Q}} K = [K : \mathbb{Q}]$ and hence $\dim_{\mathbb{C}} T_\rho = 1$. Suppose K were totally real. Then $K = \mathbb{Q}(\alpha)$ with $\varepsilon(\alpha) \in \mathbb{R}$. Hence $T^{2,0}$ and $T^{0,2}$ are both contained in T_ε, which contradicts $\dim_{\mathbb{C}} T_\varepsilon = 1$.[3]

As mentioned above, $a \in K$ is an isometry if and only if $\xi_a = a'a = 1$. For $a \in K_0$, this is possible only if $a = \pm 1$. Thus, in the case of real multiplication, there exist very few Hodge isometries of T. For the CM case, the situation is completely different, as asserted by the second part of Theorem 3.7. This is proved as follows.

Proof Write $K = K_0(\sqrt{-D})$ with $D \in K_0$ positive under each embedding $K_0 \hookrightarrow \mathbb{R}$ and fix a primitive element $\beta \in K_0$, i.e. $K_0 = \mathbb{Q}(\beta)$.

We start out by showing that $\mathbb{Q}(D(\beta + \gamma)) = K_0$ for most $\gamma \in \mathbb{Q}$. To see this, let $\mathbb{Q} \subset M_\gamma := \mathbb{Q}(D(\beta + \gamma)) \subset K_0$. If $M_{\gamma'} \subset M_\gamma$ for $\gamma' \neq \gamma$ in $\mathbb{Q}$, then $M_\gamma = K_0$. Indeed, the inclusion implies $D(\beta + \gamma) - D(\beta + \gamma') \in M_\gamma$ and hence $D \in M_\gamma$. The latter yields $\beta \in M_\gamma$, i.e. $M_\gamma = K_0$. Since K_0 has only finitely many subfields, $M_{\gamma/2} = K_0$ for all but finitely many $\gamma \in \mathbb{Q}$.

Similarly, one defines for $\gamma \in \mathbb{Q}$ the subfield $L_\gamma := \mathbb{Q}(D(\beta + \gamma)^2) \subset K_0$. For an infinite set $S \subset \mathbb{Q}$ the field L_γ is the same for all $\gamma \in S$. Among the infinitely many sums $\gamma + \gamma'$ with $\gamma, \gamma' \in S$ pick one for which $M_{(\gamma+\gamma')/2} = K_0$. Then use $D(2\beta + \gamma + \gamma')(\gamma - \gamma') = D(\beta + \gamma)^2 - D(\beta + \gamma')^2 \in M_\gamma$ to deduce that $K_0 = M_{(\gamma+\gamma')/2} = \mathbb{Q}(D(2\beta + \gamma + \gamma')) = \mathbb{Q}(D(2\beta + \gamma + \gamma')(\gamma - \gamma')) \subset \mathbb{Q}(D(\beta + \gamma)^2, D(\beta + \gamma')^2) = \mathbb{Q}(D(\beta + \gamma)^2) \subset K_0$. Hence, $\mathbb{Q}(D(\beta + \gamma)^2) = K_0$.

From the above discussion we need only that there exists a primitive element of the form D/ξ^2, i.e. $K_0 = \mathbb{Q}(D/\xi^2)$, with $\xi \in K_0$.

---

[3] Thanks to Uli Schlickewei for his help with the argument.

Then let $\eta := (D - \xi^2)/(D + \xi^2) + 2\xi \sqrt{-D}/(D + \xi^2) \in K$ and check that $\eta\eta' = \eta\bar{\eta} = 1$, i.e. η is a Hodge isometry, and $\eta + \bar{\eta} = 2(1 - 2(D/\xi^2 + 1)^{-1})$. The latter shows $D/\xi^2 \in \mathbb{Q}(\eta)$. Since also $\sqrt{-D} \in \mathbb{Q}(\eta)$, this suffices to conclude $K = \mathbb{Q}(\eta)$. □

References and Further Reading

For the theory of Hodge structures, see e.g. [145, 167, 203, 221, 253, 621]. Abelian varieties and complex tori are studied in depth in [64, 65, 139, 443]. In [7, Thm. 1.6.1] André generalizes the results of Zarhin and Tankeev to higher dimensions. Explicit examples of K3 surfaces with real multiplication have been studied by van Geemen in [204] as double planes ramified over the union of six lines. For examples defined over $\mathbb{Q}$, see the work [171] of Elsenhans and Jahnel; cf. Section **17**.2.6.

4

Kuga–Satake Construction

The Kuga–Satake construction associates with any Hodge structure of weight two a Hodge structure of weight one. Geometrically, this allows one to pass from K3 surfaces to complex tori. This chapter introduces the basic ingredients of the construction and collects examples that describe the Kuga–Satake variety for special K3 surfaces explicitly. The Kuga–Satake construction can be performed in families, which is of importance for arithmetic considerations; see Section **6**.4.4 for more details. The appendix contains a brief discussion of Deligne's proof of the Weil conjectures for K3 surfaces which relies on the Kuga–Satake construction.

1 Clifford Algebra and Spin-Group

We begin by recalling some basic facts on Clifford algebras. For background and more information on the subject, see [15, 84, 88, 500].

1.1 Let K be a commutative ring. In all the examples we have in mind, K is either $\mathbb{Z}$ or a subfield of $\mathbb{C}$ like $\mathbb{Q}$ or $\mathbb{R}$. In any case, we shall assume that 2 is not a zero divisor in K.

Consider a free K-module V of finite rank, so a finite-dimensional vector space when K is a field, and a quadratic form q on V. The associated bilinear form is given by $q(v, w) := (1/2)(q(v + w) - q(v) - q(w))$, which takes values in $K[1/2]$. The *tensor algebra*

$$T(V) := \bigoplus_{i \geq 0} V^{\otimes i}$$

with $V^{\otimes 0} := K$ is a graded non-commutative K-algebra. It can also be considered as a K-superalgebra by defining the even part and the odd part of $T(V)$ as

$$T^+(V) := \bigoplus_{i \geq 0} V^{\otimes 2i} \quad \text{and} \quad T^-(V) := \bigoplus_{i \geq 0} V^{\otimes 2i+1}.$$

Let $I := I(q) \subset T(V)$ be the two-sided ideal generated by the even elements $v \otimes v - q(v)$, $v \in V$. Here, $q(v) \in K$ is considered as an element of $K = V^{\otimes 0} \subset T(V)$. The *Clifford algebra* is the quotient K-algebra

$$\mathrm{Cl}(V) := \mathrm{Cl}(V, q) := T(V)/I(q).$$

The Clifford algebra has no longer a natural $\mathbb{Z}$-grading. However, since $I(q)$ is generated by even elements, it still has a natural $\mathbb{Z}/2\mathbb{Z}$-grading and we write:

$$\mathrm{Cl}(V) = \mathrm{Cl}^+(V) \oplus \mathrm{Cl}^-(V).$$

Note that the even part $\mathrm{Cl}^+(V)$, the *even Clifford algebra*, is indeed a K-subalgebra, whereas $\mathrm{Cl}^-(V)$ is only a two-sided $\mathrm{Cl}^+(V)$-submodule of $\mathrm{Cl}(V)$. Also note that $\mathrm{Cl}^-(V)$ naturally contains V as a K-submodule via the projection $V \subset T(V) \twoheadrightarrow \mathrm{Cl}(V)$.

The multiplication in $\mathrm{Cl}(V)$ shall be written as $v \cdot w$ for $v, w \in \mathrm{Cl}(V)$. Then by construction, $v \cdot v = q(v)$ and $v \cdot w + w \cdot v = 2q(v, w)$ in $\mathrm{Cl}(V)$ for all $v, w \in V$. For the latter simply spell out the equality $(v + w) \cdot (v + w) = q(v + w)$. Note that in particular $v \cdot w = -w \cdot v$ if $v, w \in V$ are orthogonal.

If $q = 0$, then $\mathrm{Cl}(V) \simeq \bigwedge^* V$ and for an arbitrary quadratic form q over a field K any choice of an orthogonal basis $v_1, \ldots, v_n$ of V leads to an isomorphism of K-vector spaces

$$\mathrm{Cl}(V) \xrightarrow{\sim} \bigwedge{}^* V$$

mapping $v_{i_1} \cdot \cdots \cdot v_{i_k} \in \mathrm{Cl}(V)$ to $v_{i_1} \wedge \cdots \wedge v_{i_k} \in \bigwedge^k V$. Moreover, the isomorphism is independent of the choice of the orthogonal basis. Thus, using an orthogonal basis $v_1, \ldots, v_n$ of V one can write down a of $\mathrm{Cl}(V)$

$$\mathrm{Cl}(V) = \bigoplus_{a_i \in \{0,1\}} K \cdot v_1^{a_1} \cdot \cdots \cdot v_n^{a_n},$$

which for a K-vector space of dimension n implies

$$\dim \mathrm{Cl}(V) = 2^n.$$

1.2 If $\mathrm{Cl}(V)^*$ denotes the group of units of the Clifford algebra $\mathrm{Cl}(V)$, then

$$\mathrm{CSpin}(V) := \{v \in \mathrm{Cl}(V)^* \mid vVv^{-1} \subset V\}$$

is called the *Clifford group*. Here, as before, we consider V as a submodule of $\mathrm{Cl}(V)$ and define for any $v \in \mathrm{Cl}(V)^*$ the map $V \longrightarrow \mathrm{Cl}(V)$, $w \mapsto v \cdot w \cdot v^{-1}$, the image of which is denoted vVv^{-1}. The *even Clifford group* $\mathrm{CSpin}^+(V)$ is defined similarly as the set of units in $\mathrm{Cl}^+(V)$ mapping V to itself under conjugation.

To study the Clifford group one considers the map $v \longmapsto (w \longmapsto v \cdot w \cdot v^{-1})$, which defines an orthogonal representation

$$\tau : \mathrm{CSpin}(V) \longrightarrow \mathrm{O}(V). \tag{1.1}$$

It is orthogonal, as for any $w \in V$ one has $q(w) = w \cdot w$ and thus $q(v \cdot w \cdot v^{-1}) = (v \cdot w \cdot v^{-1}) \cdot (v \cdot w \cdot v^{-1}) = v \cdot (w \cdot w) \cdot v^{-1} = q(w)(v \cdot v^{-1}) = q(w)$ for $v \in \mathrm{CSpin}(V)$.

Assume now that K is a field. Then, for $v \in V$ with $q(v) \neq 0$ the reflection $s_v \in \mathrm{O}(V)$ is defined as

$$s_v(w) := w - \frac{2q(v, w)}{q(v)} v$$

and characterized by the two properties $s_v(v) = -v$ and $s_v(w) = w$ for any $w \in V$ orthogonal to v. It is straightforward to check that $\tau(v)(v) = v$ and $\tau(v)(w) = -w$ for all $w \in v^\perp \subset V$. Thus, $\tau(v) = -s_v$. Since for a non-degenerate q over a field K the orthogonal group $\mathrm{O}(V)$ is generated by reflections (Cartan–Dieudonné theorem, see e.g. [15, III. Thm. 3.20] or [500, Sec. 5.3.9]), one finds that in this case $\mathrm{SO}(V)$ is contained in $\tau(\mathrm{CSpin}^+(V))$. More precisely, there exists a short exact sequence

$$0 \longrightarrow K^* \longrightarrow \mathrm{CSpin}^+(V) \longrightarrow \mathrm{SO}(V) \longrightarrow 0. \tag{1.2}$$

The *spin group* is the subgroup of the even Clifford group defined by

$$\mathrm{Spin}(V) := \{ v \in \mathrm{CSpin}^+(V) \mid v \cdot v^* = 1 \}. \tag{1.3}$$

Here, $v \longmapsto v^*$ is the anti-automorphism of $\mathrm{Cl}(V)$ defined by $v = v_1 \cdots v_k \longmapsto v_k \cdots v_1$, where $v_i \in V$. In other words, $\mathrm{Spin}(V)$ is defined by the short exact sequence

$$0 \longrightarrow \mathrm{Spin}(V) \longrightarrow \mathrm{CSpin}^+(V) \longrightarrow K^* \longrightarrow 0. \tag{1.4}$$

Note that $v \longmapsto v^*$ does descend from $T(V)$ to $\mathrm{Cl}(V)$, as it preserves the ideal $I(q)$ and clearly satisfies $(v \cdot w)^* = w^* \cdot v^*$. The exactness of (1.2) and the fact that the Clifford norm $v \longmapsto v \cdot v^*$ really takes image in K^* involve the same type of computations.

For q non-degenerate and K an algebraically closed field (the existence of $\sqrt{q(v)} \in K$ for all v suffices) one has

$$\tau(\mathrm{Spin}(V)) = \mathrm{SO}(V).$$

The kernel of the surjection $\tau : \mathrm{Spin}(V) \longrightarrow \mathrm{SO}(V)$ consists of ± 1 and the composition $K^* \longrightarrow \mathrm{CSpin}^+(V) \longrightarrow K^*$ of the maps in (1.2) and (1.4) is $\lambda \longmapsto \lambda^2$.

More conceptually, one defines $\mathrm{Spin}(V)$ as a linear algebraic group defined over K and the above construction describes the group of K-rational points of it. Moreover, the constructions are all natural. For example, for a base change L/K one has natural isomorphisms $\mathrm{Cl}(V_L) \simeq \mathrm{Cl}(V)_L$, $\mathrm{Spin}(V_L) \simeq \mathrm{Spin}(V)_L$, etc., i.e. the set of L-rational points of the algebraic group $\mathrm{Spin}(V)$ defined over K coincides with the group $\mathrm{Spin}(V_L)$ as defined above. Also, the Clifford algebra behaves well under direct sums, i.e. there exists a natural isomorphism of $\mathbb{Z}/2\mathbb{Z}$-graded algebras

$$\mathrm{Cl}(V_1 \oplus V_2) \simeq \mathrm{Cl}(V_1) \otimes \mathrm{Cl}(V_2),$$

where the tensor product is $\mathbb{Z}/2\mathbb{Z}$-graded, i.e. $(v \cdot v') \otimes (w \cdot w') = (-1)^{|v'||w|}(v \otimes w) \cdot (v' \otimes w')$ for $v, v' \in \mathrm{Cl}(V_1)$ and $w, w' \in \mathrm{Cl}(V_2)$ with v' and w homogeneous.

Remark 1.1 The tensor construction $V \longmapsto \mathrm{Cl}^{\pm}(V)$ can be upgraded to a construction for Hodge structures. If V is a Hodge structures of weight zero, then $T^{\pm}(V)$ is endowed with a natural Hodge structure of weight zero. If, in addition, $q \colon V \otimes V \longrightarrow \mathbb{Q}$ is a morphism of Hodge structures, then $\mathrm{Cl}^{\pm}(V) = T^{\pm}/I^{\pm}(q)$ inherits this Hodge structure of weight zero. For example, if V is of type $(1,-1)+(0,0)+(-1,1)$ with $\dim V^{1,-1} = 1$, then also $\mathrm{Cl}^{\pm}(V)$ is of type $(1,-1)+(0,0)+(-1,1)$. This is later applied to $H^2(X, \mathbb{Z}(1))$ of a K3 surface X.

2 From Weight Two to Weight One

We start with a Hodge structure V of K3 type (see Definition 3.2.3) and assume that V is endowed with a polarization $-q$. The first step is to construct a natural complex structure on the real vector space $\mathrm{Cl}^{+}(V_{\mathbb{R}})$ which can be interpreted as a Hodge structure of weight one. The second step consists of constructing a polarization for this new Hodge structure of weight one.

2.1 The Hodge structure of V induces a decomposition of the real vector space

$$V_{\mathbb{R}} = \left(V^{1,1} \cap V_{\mathbb{R}} \right) \oplus \left((V^{2,0} \oplus V^{0,2}) \cap V_{\mathbb{R}} \right).$$

The $\mathbb{C}$-linear hull of the second summand is $V^{2,0} \oplus V^{0,2}$ which by assumption is two-dimensional. Thus, with respect to q the real space $(V^{2,0} \oplus V^{0,2}) \cap V_{\mathbb{R}}$ is a positive plane; see Section 3.1.3.

Pick a generator $\sigma = e_1 + i e_2$ of $V^{2,0}$ with $e_1, e_2 \in V_{\mathbb{R}}$ and $q(e_1) = 1$. Then $q(\sigma) = 0$ implies $q(e_1, e_2) = 0$ and $q(e_2) = 1$, i.e. e_1, e_2 is an orthonormal basis of $(V^{2,0} \oplus V^{0,2}) \cap V_{\mathbb{R}}$. Hence, $e_1 \cdot e_2 = -e_2 \cdot e_1$ in $\mathrm{Cl}(V_{\mathbb{R}})$, which shows (see also (2.6)) that left multiplication with

$$J := e_1 \cdot e_2 \in \mathrm{CSpin}^{+}(V_{\mathbb{R}}) \subset \mathrm{Cl}(V_{\mathbb{R}})$$

induces a complex structure on the real vector space $\mathrm{Cl}(V_{\mathbb{R}})$, i.e. $J^2 \equiv -\mathrm{id}$. Obviously, J preserves $\mathrm{Cl}^{+}(V_{\mathbb{R}})$ and $\mathrm{Cl}^{-}(V_{\mathbb{R}})$.

Let us check that the complex structure J is independent of the choice of the orthonormal basis e_1, e_2. Suppose $\sigma' = e_1' + ie_2'$ is another generator of $V^{2,0}$ with $q(e_1') = 1$. Then $\sigma = \lambda \sigma'$ for some $\lambda \in \mathbb{C}^*$. Writing $\lambda = a + ib$, this yields $e_1 = ae_1' - be_2'$ and $e_2 = ae_2' + be_1'$. A simple calculation in $\mathrm{Cl}(V)$ then reveals that $e_1 \cdot e_2 = (a^2 + b^2)(e_1' \cdot e_2')$. But $q(e_1) = q(e_1') = 1$ implies $a^2 + b^2 = 1$.

Now one defines the *Kuga–Satake Hodge structure* as the Hodge structure of weight one on $\mathrm{Cl}^+(V)$ given by

$$\rho \colon \mathbb{C}^* \longrightarrow \mathrm{GL}(\mathrm{Cl}^+(V)_{\mathbb{R}}), \ x + yi \longmapsto x + yJ. \tag{2.1}$$

Note that by the same procedure one obtains a Hodge structure of weight one on the full Clifford algebra $\mathrm{Cl}(V)$ and on its odd part $\mathrm{Cl}^-(V)$.

Remark 2.1   (i) If the Hodge structure on V is, as in Remark **3**.1.8, interpreted in terms of the morphism of real algebraic groups $\rho \colon \mathbb{S} \longrightarrow \mathrm{GL}(V_{\mathbb{R}})$, then the above construction can be viewed more abstractly as a lift

$$\begin{array}{ccc}
\mathbb{S} & \xrightarrow{\ \rho(1)\ } & \mathrm{SO}(V_{\mathbb{R}}(1)) \subset \mathrm{GL}(V_{\mathbb{R}}) \\
& \searrow{\scriptstyle \widetilde{\rho}} & \ \big\uparrow{\scriptstyle \tau} \\
& & \mathrm{CSpin}^+(V_{\mathbb{R}}),
\end{array}$$

which becomes unique by requiring $\widetilde{\rho}|_{\mathbb{G}_{m,\mathbb{R}}} \colon t \longmapsto t$. Letting $\mathrm{CSpin}^+(V_{\mathbb{R}})$ act by multiplication from the left on $\mathrm{Cl}^+(V_{\mathbb{R}})$ yields the Hodge structure $\widetilde{\rho} \colon \mathbb{S} \longrightarrow \mathrm{GL}(\mathrm{Cl}^+(V_{\mathbb{R}}))$ described by (2.1). Note that the Tate twists $\rho(1)$ and $V_{\mathbb{R}}(1)$, which turn V into a Hodge structure of weight zero, are necessary for the commutativity of the diagram. See also Remark 2.8 and the explicit computation in the proof of Proposition 2.6.

(ii) For $\widetilde{g} \in \mathrm{Spin}(V)$ let $g := \tau(\widetilde{g}) \in \mathrm{SO}(V)$ be its image under $\tau \colon \mathrm{Spin}(V) \longrightarrow\!\!\!\!\!\rightarrow \mathrm{SO}(V)$. Consider a Hodge structure of K3 type on V given by $\sigma = e_1 + ie_2$ and its image under g given by $g(\sigma) = g(e_1) + ig(e_2)$. Denote the two Kuga–Satake Hodge structures on $\mathrm{Cl}^+(V)$ by ρ and ρ', respectively. Then left multiplication with $\widetilde{g}$ defines an isomorphism of Hodge structures

$$(\mathrm{Cl}^+(V), \rho) \xrightarrow{\ \sim\ } (\mathrm{Cl}^+(V), \rho'), \ w \longmapsto \widetilde{g} \cdot w. \tag{2.2}$$

Indeed, $\widetilde{g} \cdot (e_1 \cdot e_2 \cdot w) = g(e_1) \cdot g(e_2) \cdot (\widetilde{g} \cdot w)$ by definition of τ. Note that there is also another isomorphism of Hodge structures

$$(\mathrm{Cl}^+(V), \rho) \xrightarrow{\ \sim\ } (\mathrm{Cl}^+(V), \rho'), \ w \longmapsto \widetilde{g} \cdot w \cdot \widetilde{g}^*, \tag{2.3}$$

which is the natural one induced by $T(g) \colon T(V) \xrightarrow{\ \sim\ } T(V)$.

The distinction between the two isomorphisms plays a role in the family version of the construction to be discussed in Section **6**.4.4. It turns out that (2.2) behaves better with respect to the pairing in Section 2.2.

If the Hodge structure V was defined over $\mathbb{Z}$, then the Kuga–Satake structure is an integral Hodge structure of weight one which is equivalent to giving a complex torus. More precisely, $\mathrm{Cl}^+(V) \subset \mathrm{Cl}^+(V_\mathbb{R})$ is a lattice and the above defines the structure of a complex torus on its quotient.

Definition 2.2 The *Kuga–Satake variety* associated with the integral Hodge structure V of weight two is the complex torus

$$\mathrm{KS}(V) := \mathrm{Cl}^+(V_\mathbb{R})/\mathrm{Cl}^+(V).$$

If the Hodge structure V was defined only over $\mathbb{Q}$, then one obtains an isogeny class of complex tori which for many purposes is already very useful. In fact, in most of the examples below we content ourselves with the description of the Kuga–Satake variety up to isogeny.

Remark 2.3 The Kuga–Satake construction works for any Hodge structure of K3 type with a quadratic form q that is positive definite on the real part of $V^{2,0} \oplus V^{0,2}$ and satisfies $q(\sigma) = 0$ for $\sigma \in V^{2,0}$. This allows one to define the Kuga–Satake variety associated with $H^2(X, \mathbb{Z})$ for an arbitrary (possibly non-projective) K3 surface X and without restricting to the primitive part first.

The dimension of $\mathrm{KS}(V)$ is usually quite high. More precisely, as for $n = \dim_\mathbb{C} V_\mathbb{C}$ the real dimension of $\mathrm{Cl}^+(V_\mathbb{R})$ is 2^{n-1}, the complex dimension of $\mathrm{KS}(V)$ is

$$\dim \mathrm{KS}(V) = 2^{n-2}.$$

This makes it in general very difficult to describe examples explicitly.

Note that the Hodge structures of weight one $\mathrm{Cl}^+(V)$ and $\mathrm{Cl}^-(V)$ are isogenous, although not canonically. Indeed, choosing an orthogonal basis $v_1, \ldots, v_n$ of $V_\mathbb{Q}$ one obtains an isomorphism of complex (with respect to J) vector spaces

$$\mathrm{Cl}^+(V_\mathbb{R}) \xrightarrow{\sim} \mathrm{Cl}^-(V_\mathbb{R}), \quad v \mapsto v \cdot v_n$$

which is defined over $\mathbb{Q}$. This leads to an isogeny of $\mathrm{KS}(V)$ with the analogous torus defined in terms of $\mathrm{Cl}^-(V)$. Since an orthogonal basis of an integral Hodge structure usually does not exist (over $\mathbb{Z}$), an isogeny is the best thing we can hope for even when the given Hodge structure V is integral.

Example 2.4 Consider the case that V can be written as the direct sum of two Hodge structures $V = V_1 \oplus V_2$. As $\dim(V^{2,0}) = 1$, one of the two, say V_2, is pure of type $(1,1)$, i.e. $V_2^{2,0} = 0$. Then $\mathrm{Cl}^+(V)$ as a Hodge structure of weight one is isomorphic to the product of 2^{n_2-1} copies of $\mathrm{Cl}^+(V_1) \times \mathrm{Cl}^-(V_1)$, where $n_2 = \dim(V_2)$. Thus, $\mathrm{KS}(V_1 \oplus V_2)$ is isogenous to the product of 2^{n_2} copies of $\mathrm{KS}(V_1)$, which we shall write as

$$\mathrm{KS}(V_1 \oplus V_2) \sim \mathrm{KS}(V_1)^{2^{n_2}}.$$

2.2 The next step in the Kuga–Satake construction is the definition of a polarization, which makes the complex torus defined above an abelian variety. In order to define a polarization on KS(V), choose two orthogonal vectors $f_1, f_2 \in V$ with $q(f_i) > 0$. Then define a pairing (ignoring the factor $2\pi i$ as usual)

$$Q: \mathrm{Cl}^+(V) \times \mathrm{Cl}^+(V) \longrightarrow \mathbb{Q}(-1), \quad (v, w) \longmapsto \pm \mathrm{tr}(f_1 \cdot f_2 \cdot v^* \cdot w). \tag{2.4}$$

Here, tr denotes the trace of the endomorphism of Cl(V) defined by left multiplication. Since multiplication with f_i interchanges $\mathrm{Cl}^+(V)$ and $\mathrm{Cl}^-(V)$, one has $\mathrm{tr}(f_i) = 0$. Using $\mathrm{tr}(v \cdot w) = \mathrm{tr}(w \cdot v)$, one also finds $\mathrm{tr}(v) = \mathrm{tr}(v^*)$. Observe that Q is preserved under (2.2), but not necessarily under (2.3). The sign in the definition of Q is not given explicitly but is determined in the course of the proof of the following

Proposition 2.5 *Assume V is a Hodge structure of K3 type with a polarization $-q$. Then, with the appropriate sign in (2.4), the pairing Q defines a polarization for the Hodge structure of weight one on $\mathrm{Cl}^+(V)$.*

Proof Let us check that Q is a morphism of Hodge structures. We suppress the sign, as it is of no importance at this point. For $z = x + iy \in \mathbb{C}^*$ one computes

$$
\begin{aligned}
Q(\rho(z)v, \rho(z)w) &= \mathrm{tr}(f_1 \cdot f_2 \cdot (\rho(z) \cdot v)^* \cdot \rho(z) \cdot w) \\
&= \mathrm{tr}(f_1 \cdot f_2 \cdot v^* \cdot (\rho(z)^* \cdot \rho(z)) \cdot w) = (z\bar{z}) Q(v, w),
\end{aligned}
$$

using $J^* = -J$ and hence $\rho(z)^* \cdot \rho(z) = x^2 + y^2$.

It is obvious that Q is non-degenerate. Let us show that $Q(v, \rho(i)w)$ is symmetric.

$$
\begin{aligned}
Q(v, \rho(i)w) &= \mathrm{tr}(f_1 \cdot f_2 \cdot v^* \cdot J \cdot w) = \mathrm{tr}((f_1 \cdot f_2 \cdot v^* \cdot J \cdot w)^*) \\
&= -\mathrm{tr}(w^* \cdot J \cdot v \cdot (f_1 \cdot f_2)^*) = \mathrm{tr}(w^* \cdot J \cdot v \cdot (f_1 \cdot f_2)) \\
&= \mathrm{tr}(f_1 \cdot f_2 \cdot w^* \cdot J \cdot v) = Q(w, \rho(i)v).
\end{aligned}
$$

Here, one uses $(f_1 \cdot f_2)^* = -f_1 \cdot f_2$ and that tr is symmetric.

It is in the last step, when showing that $Q(v, \rho(i)w)$ is positive definite, that the sign has to be chosen correctly and where one uses that $-q$ is a polarization. If e_1, e_2 happen to be rational, which in general they are not, then one can take $f_i = e_i$ and a direct computation yields the result. For the general case one uses that the space of all Hodge structures of K3 type on V has two connected components (see Remark **6.1.6**) and that the property of being positive definite stays constant under deformations of the Hodge structure in one of the two components. Passing from one connected component to the other, to eventually reach the point where one can take $f_i = e_i$, one may have to change the sign. For details we refer to van Geemen [203, Prop. 5.9] or to Satake's original paper [524]. □

2.3 For a rational Hodge structure V of weight two of K3 type (see Definition **3.2.3**) we have defined a Hodge structure $\mathrm{Cl}^+(V)$ of weight one. The tensor product $\mathrm{Cl}^+(V) \otimes \mathrm{Cl}^+(V)$ carries a natural Hodge structure of weight two; see Section **3.1.1**.

Proposition 2.6 *Assume V is a Hodge structure of K3 type with a polarization $-q$. Then there exists an inclusion of Hodge structures of weight two*

$$V \hookrightarrow \mathrm{Cl}^+(V) \otimes \mathrm{Cl}^+(V).$$

Dualizing and using the isogeny between V and V^*, the above construction yields a rational class of type $(2,2)$ in $V \otimes \mathrm{Cl}^+(V) \otimes \mathrm{Cl}^+(V)$ with its natural Hodge structure of weight four. The class is discussed below in the geometric situation.

Proof Choose an element $v_0 \in V$ which is invertible in $\mathrm{Cl}(V)$, i.e. $q(v_0) \neq 0$. Then consider the embedding (ignoring $2\pi i$)

$$V(1) = V \otimes \mathbb{Q}(1) \hookrightarrow \mathrm{End}(\mathrm{Cl}^+(V)), \quad v \mapsto f_v \colon w \mapsto v \cdot w \cdot v_0. \tag{2.5}$$

It is injective, since $f_v(v_1 \cdot v_0) = q(v_0)(v \cdot v_1)$ for all $v_1 \in V$.

We claim that this is a morphism of Hodge structures (of weight zero), which can be checked by the following straightforward computation, but see [203, Prop. 6.3] for a more conceptual proof.

Denote by $\rho_V, \rho_{V(1)}$, and ρ the representations of $\mathbb{C}^*$ corresponding to V, $V(1)$, and $\mathrm{Cl}(V)$, respectively. Then we have to show that $f_{\rho_{V(1)}(z)v}(w) = \rho(z)f_v(\rho(z)^{-1}w)$ for all $w \in V_{\mathbb{R}}$, where

$$f_{\rho_{V(1)}(z)v}(w) = (\rho_{V(1)}(z)v) \cdot w \cdot v_0 \quad \text{and} \quad \rho(z)f_v(\rho(z)^{-1}w) = \rho(z)(v \cdot \rho(z)^{-1}w \cdot v_0).$$

Thus, it suffices to prove

$$\rho_{V(1)}(z)v = \rho(z) \cdot v \cdot \rho(z)^{-1}, \tag{2.6}$$

where on the right-hand side $\rho(z)$ for $z = x + iy$ is viewed as the element $x + yJ = x + y(e_1 \cdot e_2)$ and similarly $\rho(z)^{-1} = (x^2 + y^2)^{-1}(x - y(e_1 \cdot e_2))$. Therefore, the assertion reduces to

$$\rho_{V(1)}(z)v = (x^2 + y^2)^{-1}(x + y(e_1 \cdot e_2)) \cdot v \cdot (x - y(e_1 \cdot e_2)).$$

We can treat the cases $v \in V_{\mathbb{R}}^{1,1}$ and $v \in (V^{2,0} \oplus V^{0,2}) \cap V_{\mathbb{R}}$ separately. In the first case, the assertion follows from $\rho_{V(1)}(z)v = (z\bar{z})^{-1}\rho_V(z)v = v$ (see Example **3.1.9**) and

$$(x + y(e_1 \cdot e_2)) \cdot v \cdot (x - y(e_1 \cdot e_2)) = (x + y(e_1 \cdot e_2)) \cdot (x - y(e_1 \cdot e_2)) \cdot v = (x^2 + y^2)v$$

(use that v is orthogonal to e_1, e_2). For $v = e_1$ one computes $\rho_V(z)v = \mathrm{Re}(z^2)e_1 - \mathrm{Im}(z^2)e_2$ and

$$(x + y(e_1 \cdot e_2)) \cdot v \cdot (x - y(e_1 \cdot e_2)) = (x + y(e_1 \cdot e_2)) \cdot (xe_1 - ye_2)$$
$$= x^2 e_1 - 2xye_2 - y^2 e_1 = (x^2 - y^2)e_1 - 2xye_2.$$

The computation for $v = e_2$ is similar.

The polarization of $\mathrm{Cl}^+(V)$ can be interpreted as an isomorphism $\mathrm{Cl}^+(V)^* \simeq \mathrm{Cl}^+(V) \otimes \mathbb{Q}(1)$ and thus yields $\mathrm{End}(\mathrm{Cl}^+(V)) \simeq \mathrm{Cl}^+(V)^* \otimes \mathrm{Cl}^+(V) \simeq \mathrm{Cl}^+(V) \otimes \mathrm{Cl}^+(V) \otimes \mathbb{Q}(1)$. □

Remark 2.7 Note that the embedding constructed above depends on the choice of the vector v_0. However, for another choice of v_0, say v_0', the two embeddings differ by an automorphism of the Hodge structure of weight one on $\mathrm{Cl}^+(V)$ given by $w \longmapsto (1/q(v_0))w \cdot v_0 \cdot v_0'$. If one wants to avoid the choice of v_0 altogether, then the construction described above naturally yields an injection of Hodge structures

$$ V \hookrightarrow \mathrm{Hom}(\mathrm{Cl}^+(V), \mathrm{Cl}^-(V)). $$

Remark 2.8 By definition, $\rho(z)$ on $\mathrm{Cl}(V)_\mathbb{R}$ acts by left multiplication with an element in $\mathrm{Cl}^+(V)_\mathbb{R}$. Equation 2.6 proves that this element is contained in $\mathrm{CSpin}(V)$ (cf. Remark 2.1), and also shows that for the orthogonal representation (1.1)

$$ \tau \colon \mathrm{CSpin}(V) \longrightarrow O(V) $$

one has $\tau(\rho(z)) = \rho_{V(1)}(z)$. Thus, the Hodge structure of weight zero on $V(1)$, and hence on V, can be recovered from the Hodge structure of weight one on $\mathrm{KS}(V)$ by means of the orthogonal representation $\mathrm{CSpin}(V) \longrightarrow O(V)$.

Thus, the Kuga–Satake construction

$$ \mathrm{KS} \colon \{ \text{ Hodge structures of K3 type } \} \hookrightarrow \{ \text{ Hodge structures of weight one } \} \quad (2.7) $$

is injective.

Let us quickly check that the injectivity also holds on the infinitesimal level. For this, we disturb the $(2, 0)$-form $\sigma = e_1 + ie_2 \in V^{2,0}$ by some $\alpha = \alpha_1 + i\alpha_2 \in V^{1,1}$, so $\sigma + \varepsilon\alpha$ defines a first-order deformation of the $(2, 0)$-part. The induced first-order deformation of $J = e_1 \cdot e_2$ is then $J_\varepsilon = e_1 \cdot e_2 + \varepsilon(\alpha_1 \cdot e_2 + e_1 \cdot \alpha_2)$.

The map

$$ V^{1,1} \longrightarrow \mathrm{Hom}(\mathrm{Cl}^+(V)^{1,0}, \mathrm{Cl}^+(V)^{0,1}), \quad \alpha \longmapsto h_\alpha \colon w \longmapsto (\alpha_1 \cdot e_2 + e_1 \cdot \alpha_2) \cdot w $$

is the differential $d\,\mathrm{KS}$ of (2.7); cf. Proposition 6.2.4. This proves that $d\,\mathrm{KS}$ is injective and also that it is $\mathbb{C}$-linear. Indeed, $h_{i\alpha} = ih_\alpha$, which for $w \in \mathrm{Cl}^+(V)^{1,0}$ follows from $ih_\alpha(w) = -(e_1 \cdot e_2) \cdot h_\alpha(w) = -(e_1 \cdot e_2) \cdot (\alpha_1 \cdot e_2 - \alpha_2 \cdot e_1) \cdot w = (-\alpha_1 \cdot e_1 - \alpha_2 \cdot e_2) \cdot w = h_{i\alpha}(w)$. The latter observation is interpreted as saying that KS is holomorphic; see Section 6.4.4.

2.4 We next explain a version of the above construction which turns out to be important for arithmetic applications. Similarly to (2.5), one constructs a morphism of Hodge structures

$$\mathrm{Cl}^+(V(1)) \hookrightarrow \mathrm{End}(\mathrm{Cl}^+(V)), \quad v \longmapsto f_v \colon w \longmapsto v \cdot w, \tag{2.8}$$

which is injective as $v = f_v(1)$. Note that on the left-hand side, $\mathrm{Cl}^+(V(1))$ is viewed as a Hodge structure of weight zero (see Remark 1.1), whereas on the right-hand side $\mathrm{Cl}^+(V)$ is the Kuga–Satake Hodge structure of weight one.

Let now C be the opposite algebra of $\mathrm{Cl}^+(V)$ (without Hodge structure) which acts on $\mathrm{Cl}^+(V)$ by right multiplication which respects the Hodge structure.

Furthermore, (2.8) is compatible with this action. So, $\mathrm{Cl}^+(V(1)) \hookrightarrow \mathrm{End}_C(\mathrm{Cl}^+(V))$, which is in fact an isomorphism of algebras (and also of Hodge structures)

$$\mathrm{Cl}^+(V(1)) \simeq \mathrm{End}_C(\mathrm{Cl}^+(V)). \tag{2.9}$$

The surjectivity is deduced from computing the dimension of the right-hand side (after passing to a finite extension of $\mathbb{Q}$); see [203, Lem. 6.5]. Moreover, (2.9) is compatible with the action of $\mathrm{CSpin}(V)$ defined by conjugation $w \longmapsto v \cdot w \cdot v^{-1}$ on $\mathrm{Cl}^+(V(1))$ and by $f \longmapsto (w \longmapsto v \cdot f(v^{-1} \cdot w))$ on $\mathrm{End}_C(\mathrm{Cl}^+(V))$, and, in fact, (2.9) is the only algebra isomorphism of $\mathrm{Spin}(V)$-representations; see [141, Prop. 3.5].

Note that (2.9) also holds for the case that V is an integral Hodge structure. Indeed, (2.8) is certainly well defined and becomes the isomorphism (2.9) after tensoring with $\mathbb{Q}$. Then use that the obvious inverse $f \longmapsto f(1)$ is defined over $\mathbb{Z}$.

2.5 Let us apply the abstract Kuga–Satake construction to Hodge structures associated with compact complex surfaces X with $h^{2,0}(X) = 1$, e.g. K3 surfaces or two-dimensional complex tori. The quadratic form q is in this case given by the standard intersection pairing which is positive definite on $(H^{2,0} \oplus H^{0,2})(X) \cap H^2(X, \mathbb{R})$.

Definition 2.9 The *Kuga–Satake variety* $\mathrm{KS}(X)$ of X, for now just a complex torus, is defined as the Kuga–Satake variety associated with $H^2(X, \mathbb{Z})$:

$$\mathrm{KS}(X) := \mathrm{KS}(H^2(X, \mathbb{Z})).$$

There are variants of this construction. The Kuga–Satake construction can also be applied to the transcendental lattice $T(X)$; see Definition 3.2.5. This yields another complex torus $\mathrm{KS}(T(X))$ naturally associated with $H^2(X, \mathbb{Z})$. Note that $\mathrm{KS}(T(X))$ is a complex torus of (complex) dimension $2^{\mathrm{rk}\, T(X)-2}$. For example, if X is a K3 surface of maximal Picard number $\rho(X) = 20$, then $\mathrm{KS}(T(X))$ is an elliptic curve.

Replacing $T(X)$ by any other lattice $T(X) \subset T(X)'$ yields a complex torus $\mathrm{KS}(T(X)')$. If $T(X)_{\mathbb{Q}}$ is a direct summand of $T(X)'_{\mathbb{Q}}$, for example when $T(X)'$ is polarizable, then $\mathrm{KS}(T(X)')$ is isogenous to $\mathrm{KS}(T(X))^{2^d}$ with $d = \mathrm{rk}(T(X)'/T(X))$; see Example 2.4. In particular, if $H^2(X, \mathbb{Z})$ is polarizable, then

$$\mathrm{KS}(X) \sim \mathrm{KS}(T(X))^{2^\rho}$$

where $\rho = \rho(X)$ is the Picard number of X. If X is projective and $\ell \in H^2(X, \mathbb{Z})$ is an ample class, one could also consider $T(X)' := \ell^\perp = H^2(X, \mathbb{Z})_p$, the primitive cohomology of (X, ℓ). Then

$$\mathrm{KS}(X, \ell) := \mathrm{KS}(\ell^\perp)$$

is a complex torus naturally associated with the polarized surface (X, ℓ). Note that the intersection form really is a polarization of $\ell^\perp = H^2(X, \mathbb{Z})_p$ (up to sign), and, therefore, $\mathrm{KS}(X, \ell)$ is an abelian variety. Moreover, there are natural isogenies

$$\mathrm{KS}(X) \sim \mathrm{KS}(X, \ell)^2 \sim \mathrm{KS}(T(X))^{2^\rho}.$$

The relation between the various Kuga–Satake varieties up to isomorphism is more complicated, for in general $H^2(X, \mathbb{Z})$ contains $T(X) \oplus \mathrm{NS}(X)$ as a proper finite index subgroup. However, the description up to isogeny suffices to show for example that $\mathrm{KS}(X)$ is in fact an abelian variety. Note that for a very general (and in particular non-projective) K3 surface or a very general two-dimensional torus, one has $T(X) = H^2(X, \mathbb{Z})$ and the intersection form is not a polarization as it has three positive eigenvalues.

Remark 2.10 The Kuga–Satake construction is of a highly transcendental nature. Essentially, only the transcendental lattice, which encodes algebraic information of X in a very indirect way, really matters for $\mathrm{KS}(X)$. In particular, questions concerning the field of definition of $\mathrm{KS}(X)$, e.g. when X is defined over a number field, are subtle. However, one can show, for example, that the Kuga–Satake variety of a Kummer surface with $\rho = 20$ is defined over some number field. Indeed, X is CM (i.e. $K = \mathrm{End}_{\mathrm{Hdg}}(T(X))$ is a CM field and $\dim_K T(X) = 1$; cf. Remark **3**.3.10), and using the Kuga–Satake correspondence $\mathrm{KS}(X)$ is also shown to be CM [494, Lem. 4].

2.6 Let us revisit Proposition 2.6 in the geometric setting of $H^2(X, \mathbb{Q})$ for X a complex K3 surface or a two-dimensional complex torus (or one of the other natural Hodge structures of weight two considered above). First note that by the Künneth formula there exists an embedding of Hodge structures

$$H^1(\mathrm{KS}(X), \mathbb{Q}) \otimes H^1(\mathrm{KS}(X), \mathbb{Q}) \hookrightarrow H^2(\mathrm{KS}(X) \times \mathrm{KS}(X), \mathbb{Q}),$$

which composed with the inclusion constructed in Proposition 2.6 yields

$$H^2(X, \mathbb{Q}) \subset H^2(\mathrm{KS}(X) \times \mathrm{KS}(X), \mathbb{Q})$$

and thus corresponds to an element

$$\kappa_X \in H^4(X \times \mathrm{KS}(X) \times \mathrm{KS}(X), \mathbb{Q})$$

of type $(2,2)$, the *Kuga–Satake class*. By construction, the embedding and, therefore, the Kuga–Satake class depend on the choice of a non-isotropic vector $v_0 \in H^2(X, \mathbb{Q})$ which we suppress. See the comment after the proof of Proposition 2.6. Of course, if the K3 surface is given together with a polarization, then v_0 could be chosen naturally to be the corresponding class.

As a special case of the Hodge conjecture one has

Conjecture 2.11 (Kuga–Satake–Hodge conjecture) *Suppose X is a smooth complex projective surface with $h^{2,0}(X) = 1$. Then the class κ_X is algebraic.*

The conjecture applies to K3 surfaces as well as to abelian surfaces. Clearly, the above form is equivalent to the analogous one for the transcendental lattices.

The transcendental nature of the Kuga–Satake construction makes it difficult to approach the Kuga–Satake–Hodge conjecture. It is known in a few cases when the Kuga–Satake variety can be described explicitly; see below. Also κ_X is known to be absolute Hodge; cf. Remark 4.6.

3 Kuga–Satake Varieties of Special Surfaces

We outline the description of the Kuga–Satake variety for Kummer surfaces and special double planes.

3.1 Let A be a two-dimensional complex torus. Via the Kuga–Satake construction one associates with A another torus $\mathrm{KS}(A) = \mathrm{KS}(H^2(A, \mathbb{Z}))$ which is of dimension 16. Working with the transcendental lattice or the primitive cohomology yields factors of the latter but they tend to be of rather high dimension as well. What is the (geometric) relation between the tori A and $\mathrm{KS}(A)$? Since a torus is determined by its integral Hodge structure of weight one, this amounts to asking for the relation between the two Hodge structures of weight one $H^1(A, \mathbb{Z})$ and $H^1(\mathrm{KS}(A), \mathbb{Z})$.

We have seen earlier that $H^2(A, \mathbb{Z})$ is indeed isomorphic, as an integral Hodge structure, to $\bigwedge^2 H^1(A, \mathbb{Z})$; cf. Section 3.2.3. This is crucial for the proof of the next result which is due to Morrison [425, Thm. 4.3]. We denote by $\hat{A}$ the complex torus dual to A.

Proposition 3.1 *Let A be a complex torus of dimension two. Then there exists an isogeny*

$$\mathrm{KS}(A) \sim (A \times \hat{A})^4.$$

Proof Here is an outline of the proof, leaving out most of the straightforward but tedious verifications. Compare the arguments below with the more conceptual ones in [113].

To simplify notations, let us denote the Hodge structure of weight one $H^1(A, \mathbb{Q})$ by V. Observe that the $\mathbb{Q}$-vector space $\bigwedge^2 V \simeq H^2(A, \mathbb{Q})$ can be identified with the subspace of $\mathrm{Hom}(V^*, V)$ consisting of all alternating morphisms. Similarly, we view $\bigwedge^2 V^*$ as a subspace of $\mathrm{Hom}(V, V^*)$.

Consider

$$\bigwedge^2 V \longrightarrow \mathrm{End}(V \oplus V^*), \quad u \longmapsto A_u := \begin{pmatrix} 0 & u \\ -u^* & 0 \end{pmatrix},$$

where $u^* \in \bigwedge^2 V^*$ is defined by $u^*(u') = q(u, u')$ with respect to the intersection form q on $\bigwedge^2 V$. Morally, one would like to use Clifford multiplication on the left-hand side and the algebra structure on the right to obtain an algebra morphism $\mathrm{Cl}(\bigwedge^2 V) \longrightarrow \mathrm{End}(V \oplus V^*)$. This can be carried out and, using that $A_u \cdot A_{u'}$ is diagonal, one obtains

$$\mathrm{Cl}^+(\bigwedge^2 V) \longrightarrow \mathrm{End}(V) \oplus \mathrm{End}(V^*). \tag{3.1}$$

As $\dim_{\mathbb{Q}} V = 4$, both sides are of the same dimension, 2^5, and one can indeed check that the morphism is bijective.

Now define a Hodge structure of weight one on $\mathrm{End}(V) \oplus \mathrm{End}(V^*)$ by $(\rho(z)f)(v) = \rho(z)(f(v))$, i.e. only the Hodge structure on the target is used. Clearly, with this Hodge structure $\mathrm{End}(V) \oplus \mathrm{End}(V^*)$ is isomorphic to $(V \oplus V^*)^4$. It remains to show that (3.1) is an isomorphism of Hodge structures.

For this let $\alpha = \alpha_1 + i\alpha_2, \beta = \beta_1 + i\beta_2 \in V^{1,0}$ be a basis such that $e_1 + ie_2 = \alpha \wedge \beta$ satisfies $q(e_1) = 1$. Thus, the complex structure on the left-hand side of (3.1) is given by multiplication with $e_1 \cdot e_2$ which on the right-hand side corresponds to matrix multiplication with

$$-\begin{pmatrix} e_1 \circ e_2^* & 0 \\ 0 & e_1^* \circ e_2 \end{pmatrix}.$$

Using that $e_1 = \alpha_1 \wedge \beta_1 - \alpha_2 \wedge \beta_2$, etc., one checks that, for example, $-(e_1 \circ e_2^*)(\alpha) = i\alpha$. This shows that the morphism (3.1) preserves the given Hodge structures of weight one on the two sides. $\square$

Since an abelian surface A is isogenous to its dual $\hat{A}$, the result of Proposition 3.1 also shows that for abelian surfaces

$$KS(A) \sim A^8. \tag{3.2}$$

For a polarized abelian surface (A, ℓ) the result yields an isogeny $KS(A, \ell) \sim A^4$.

Remark 3.2 The Kuga–Satake–Hodge conjecture 2.11 is known for abelian surfaces. Indeed, by the work of Moonen and Zarhin [421, Thm. 0.1] the Hodge conjecture is known to hold for arbitrary products A^n of any abelian surface A.

3.2 Let us now turn to K3 surfaces. Only in very few cases has the Kuga–Satake variety of a K3 surface been described and in even fewer cases has the Kuga–Satake–Hodge conjecture 2.11 been verified. The latter might not be too surprising as even for the self product $X \times X$ of a K3 surface X the Hodge conjecture has not been proved in general; see Remark **16**.3.11.

Since the Kuga–Satake variety of a two-dimensional complex torus can be described as explained above, it is tempting to approach the case of Kummer surfaces first. The results here are again due to Morrison [425, Cor. 4.6] and Skorobogatov [572]. The case of Kummer surfaces of maximal Picard rank 20 was already discussed by Kuga and Satake in [331].

Proposition 3.3 *Let X be the Kummer surface associated with the complex torus A. Then there exists an isogeny*

$$\mathrm{KS}(X) \sim (A \times \hat{A})^{2^{18}}.$$

For X or, equivalently, A algebraic, one has

$$\mathrm{KS}(X) \sim A^{2^{19}}.$$

Proof We have seen earlier that the rational Hodge structure $H^2(X, \mathbb{Q})$ is isomorphic to the direct sum of $H^2(A, \mathbb{Q})$ and 16 copies of the pure Hodge structure $\mathbb{Q}(-1)$; cf. Section **3**.2.5. In particular, there is an isomorphism of Hodge structures of weight two given by the transcendental lattice $T(X)_\mathbb{Q} \simeq T(A)_\mathbb{Q}$. Note, however, that the polarizations differ by a factor two, i.e. $q_X(\alpha) = 2q_A(\alpha)$. Thus, any orthogonal basis $\{v_i\}$ of $T(A)_\mathbb{Q}$ can also be considered as an orthogonal basis of $T(X)_\mathbb{Q}$. This leads to an isogeny

$$\mathrm{KS}(T(A)) \sim \mathrm{KS}(T(X))$$

and hence

$$\mathrm{KS}(X) \sim \mathrm{KS}(T(X))^{2^{\rho(X)}} \sim \mathrm{KS}(T(A))^{2^{\rho(X)}} \sim \mathrm{KS}(A)^{2^{16}}.$$

Then use Proposition 3.1 or (3.2).   □

The result can be generalized to K3 surfaces that are isogenous to an abelian surface, i.e. such that there exists an isomorphism of Hodge structures $T(X)_\mathbb{Q} \simeq T(A)_\mathbb{Q}$ that is compatible with the intersection forms up to a factor. In this case one finds again $\mathrm{KS}(X) \sim A^{2^{19}}$.

Example 3.4 The latter applies in particular to K3 surfaces with $\rho(X) = 19$ or 20. In fact, as was shown by Shioda and Inose [568] and Morrison [424] (see also Remark **15**.4.1), any K3 surface with $\rho(X) = 19$ or 20 is the double cover of a surface that is birational to a Kummer surface. Moreover, for $\rho(X) = 20$ the Kummer surface is associated with the product of two isogenous elliptic curves $E \sim E'$ and in this case one finds

$$KS(X) \sim E^{2^{20}}.$$

Here, the elliptic curves $E \sim E'$ have complex multiplication and their rational period can be read off directly from the lattice of rank two $T(X)$.[1] See Section **14**.3.4 for more details.

Again, for Kummer surfaces the Kuga–Satake–Hodge conjecture 2.11 is known to hold. Indeed, the correspondence $T(X)_{\mathbb{Q}} \simeq T(A)_{\mathbb{Q}}$ is clearly algebraic and then use again the Hodge conjecture for powers of abelian surfaces; see Remark 3.2.

3.3 Let us now turn to K3 surfaces, that are given as (resolutions of) double covers $X \longrightarrow \mathbb{P}^2$ ramified over six lines; see Example **1**.1.3. Already the description of the transcendental lattice is highly non-trivial in this case; see [394]. Paranjape proves in [486] the following result. The arguments are geometrically more involved.

Proposition 3.5 *Let X be as above. Then*

$$KS(T(X)) \sim B^{2^{18}}$$

for a certain abelian fourfold B which can be described as the Prym variety of a curve C of genus five constructed as a 4 : 1 *cover of an elliptic curve.*

This explicit description, which actually starts with C, allows Paranjape to also prove the Kuga–Satake–Hodge conjecture in this case. For the case that the six lines are tangent to a conic the abelian fourfold B can be replaced by the square of the Jacobian of the natural double cover of the conic.

4 Appendix: Weil Conjectures

The Weil conjectures occupy a very special place in the history of algebraic geometry. They have motivated large parts of modern algebraic geometry. For a short survey with a historical account, see [236, App. C] or [295].

The rationality of the Zeta function and its functional equation had been proved by Dwork by 1960. The analogue of the Riemann hypothesis was eventually proved by Deligne in 1974, who had verified it for K3 surfaces a few years earlier [141]. The arguments in the case of K3 surfaces, in particular the use of the Kuga–Satake construction, have turned out to be powerful for later developments in the theory of K3

---

[1] I am grateful to Matthias Schütt for clarifying comments regarding this point.

surfaces. An independent proof of the Weil conjectures for K3 surfaces also relying on the Kuga–Satake construction is due to Pjateckiĭ-Šapiro and Šafarevič [494].

This appendix gives a rough sketch of the main arguments of Deligne's proof. The techniques he introduced are important for a number of other arithmetic results. In our discussion, we freely use results that are explained only in later chapters and often refer to the original sources for technical details.

4.1 Let us first briefly sketch what the Weil conjectures have to say for K3 surfaces. So, consider a K3 surface X over a finite field $k = \mathbb{F}_q$, $q = p^n$. Let $\overline{X} := X \times_k \overline{k}$ and let $F: X \longrightarrow X$ be the absolute Frobenius acting as the identity on points and by $a \longmapsto a^p$ on $\mathcal{O}_X$. Then $F^n: X \longrightarrow X$ is a morphism of k-varieties and its base change to $\overline{k}/k$ yields the $\overline{k}$-morphism

$$f := F^n \times \mathrm{id} : X \times_k \overline{k} \longrightarrow X \times_k \overline{k} = \overline{X},$$

which in coordinates can alternatively be described by $(a_i) \longmapsto (a_i^q)$. A point $x \in \overline{X}$ has coordinates in $\mathbb{F}_{q^r}$ if and only if it is a fixed point of f^r, i.e. $f^r(x) = x$. If N_r denotes the number of $\mathbb{F}_{q^r}$-points, then the Zeta function of X is defined as

$$Z(X, t) := \exp\left(\sum_{r=1}^{\infty} N_r \frac{t^r}{r}\right).$$

The number of fixed points of f^r can alternatively be expressed by a Lefschetz fixed point formula. For $\ell \neq p$, consider the $\mathbb{Q}_\ell$-linear pull-back map[2]

$$f^{r*} : H_{\acute{e}t}^*(\overline{X}, \mathbb{Q}_\ell) \longrightarrow H_{\acute{e}t}^*(\overline{X}, \mathbb{Q}_\ell).$$

Then $N_r = \sum_i (-1)^i \mathrm{tr}(f^{r*} \mid H_{\acute{e}t}^i(\overline{X}, \mathbb{Q}_\ell))$ and hence

$$Z(X, t) = \prod_i \exp\left(\sum_r \mathrm{tr}\left(f^{r*} \mid H_{\acute{e}t}^i(\overline{X}, \mathbb{Q}_\ell)\right) \frac{t^r}{r}\right)^{(-1)^i}. \qquad (4.1)$$

For a K3 surface, $H_{\acute{e}t}^i(\overline{X}, \mathbb{Q}_\ell) = 0$ for $i = 1, 3$ and $H_{\acute{e}t}^i(\overline{X}, \mathbb{Q}_\ell) \simeq \mathbb{Q}_\ell$ for $i = 0, 4$. Moreover, $f^{*r} = \mathrm{id}$ on $H_{\acute{e}t}^0(\overline{X}, \mathbb{Q}_\ell)$ and $f^{r*} = q^{2r} \cdot \mathrm{id}$ on $H_{\acute{e}t}^4(\overline{X}, \mathbb{Q}_\ell)$, as F^n is a finite morphism of degree q^2. The elementary identity $\exp(\sum_{r=0}^{\infty} t^r/r) = 1/(1-t)$ turns (4.1) for a K3 surface into

$$Z(X, t)^{-1} = (1 - t) \cdot \det\left(1 - f^* t \mid H_{\acute{e}t}^2(\overline{X}, \mathbb{Q}_\ell)\right) \cdot (1 - q^2 t). \qquad (4.2)$$

---

[2] Note that for $F_{\overline{k}}^n : \overline{k} \longrightarrow \overline{k}$, $a \longmapsto a^q$, the morphism $f \circ (\mathrm{id} \times F_{\overline{k}}^n) = F^n \times F_{\overline{k}}^n$ is the absolute Frobenius on $\overline{X}$ which acts trivially on $H_{\acute{e}t}^*(\overline{X}, \mathbb{Q}_\ell)$. Thus, instead of considering the action f^* one could work with the pull-back under $(\mathrm{id} \times F_{\overline{k}}^n)^{-1}$, the geometric Frobenius, as in [141]. See [285] for a discussion of the geometric nature of the geometric Frobenius.

Eventually, one uses the natural non-degenerate symmetric pairing

$$H^2_{\acute{e}t}(\overline{X}, \mathbb{Q}_\ell) \times H^2_{\acute{e}t}(\overline{X}, \mathbb{Q}_\ell) \longrightarrow H^4_{\acute{e}t}(\overline{X}, \mathbb{Q}_\ell),$$

which satisfies $\langle f^*v, f^*w \rangle = q^2 \langle v, w \rangle$. An easy linear algebra argument then shows $\det(f^*) \det(t - f^*) = \det(tf^* - q^2)$ and hence the set of eigenvalues of f^* on $H^2_{\acute{e}t}(\overline{X}, \mathbb{Q}_\ell)$ satisfies (with multiplicities)

$$\{\alpha_1, \ldots, \alpha_{22}\} = \{q^2/\alpha_1, \ldots, q^2/\alpha_{22}\}.$$

Theorem 4.1 (Weil conjectures for K3 surfaces) *The polynomial*

$$P_2(t) := \det\left(1 - f^*t \mid H^2_{\acute{e}t}(\overline{X}, \mathbb{Q}_\ell)\right) = \prod_{i=1}^{22}(1 - \alpha_i t)$$

has integer coefficients, independent of ℓ, and its zeroes $\alpha_i \in \overline{\mathbb{Q}}$ satisfy $|\alpha_i| = q$. Moreover, one may assume $\alpha_i = \pm q$ for $i = 1, \ldots, 2k$ and $\alpha_{i>2k} \neq \pm q$ with $\alpha_{2j-1} \cdot \alpha_{2j} = q^2, j > k$.

After passing to a finite extension, one can in fact assume that an even number of the eigenvalues are just $\alpha_i = q$ and that for all others α_i/q is of absolute value one but not a root of unity.

Note that the statement subsumes the rationality of $Z(X, t)$, its functional equation $Z(X, 1/(q^2t)) = (qt)^{24} \cdot Z(X, t)$, and the analogue of the Riemann hypothesis (saying that the zeroes of $P_2(q^{-s})$ satisfy $\mathrm{Re}(s) = 1$).

Remark 4.2 The proof of the theorem also reveals that f^* is semi-simple, which is not at all obvious from the above and which in fact is known only for very few varieties, such as abelian varieties and K3 surfaces. Explicitly this is stated as [143, Cor. 1.10]. It can also be seen as a consequence of the Tate conjecture; see the proof of Proposition **17.3.5**.

4.2 The following result is the central step in Deligne's proof. It relies heavily on results that are presented in Section **6**.4.

Proposition 4.3 *Let X be a polarized K3 surface over a field K of characteristic zero. Then there exists an abelian variety A defined over a finite extension L/K together with an isomorphism of algebras*

$$\mathrm{Cl}^+(H^2_{\acute{e}t}(X_{\overline{L}}, \mathbb{Z}_\ell(1))_\mathrm{p}) \simeq \mathrm{End}_C(H^1_{\acute{e}t}(A_{\overline{L}}, \mathbb{Z}_\ell)), \tag{4.3}$$

which is invariant under the natural action of $\mathrm{Gal}(\overline{L}/L)$.

Proof One can assume that K is finitely generated. We choose an embedding $K \subset \mathbb{C}$ and consider the Kuga–Satake variety $\mathrm{KS}(X_\mathbb{C})$ associated with the complex K3 surface $X_\mathbb{C} := X \times_K \mathbb{C}$. Ideally, the abelian variety A is obtained by descending $\mathrm{KS}(X_\mathbb{C})$

to a finite extension of K. The arguments below are not quite showing this, but see Remark 4.4.

Here, one uses the primitive cohomology $V := H^2(X_{\mathbb{C}}, \mathbb{Z})_p$ to define $\mathrm{KS}(X_{\mathbb{C}})$. The Kuga–Satake variety $\mathrm{KS}(X_{\mathbb{C}})$ comes with the action $C \hookrightarrow \mathrm{End}(\mathrm{KS}(X_{\mathbb{C}}))$ of the (constant) algebra $C = \mathrm{Cl}^+(V)^{\mathrm{op}}$; see Section 2.4. Moreover, there exists an isomorphism of algebras (see (2.9)),

$$\mathrm{Cl}^+(H^2(X_{\mathbb{C}}, \mathbb{Z})_p(1)) \simeq \mathrm{End}_C(H^1(\mathrm{KS}(X_{\mathbb{C}}), \mathbb{Z})), \qquad (4.4)$$

which in fact is also an isomorphism of Hodge structures (of type $(1, -1) + (0, 0) + (-1, 1)$). It is compatible with the action of $\mathrm{Spin}(V)$ on both sides and it is unique with this property.

The abelian variety $\mathrm{KS}(X_{\mathbb{C}})$ together with the action of C is defined over some finitely generated field extension $K \subset L \subset \mathbb{C}$, i.e. $\mathrm{KS}(X_{\mathbb{C}}) \simeq A \times_L \mathbb{C}$ for some abelian variety A over L. Then, $H^1_{\acute{e}t}(A_{\overline{L}}, \mathbb{Z}_\ell) \simeq H^1_{\acute{e}t}(\mathrm{KS}(X_{\mathbb{C}}), \mathbb{Z}_\ell) \simeq H^1(\mathrm{KS}(X_{\mathbb{C}}), \mathbb{Z}) \otimes \mathbb{Z}_\ell$, and tensoring (4.4) with $\mathbb{Z}_\ell$ this yields an isomorphism of algebras

$$\psi_0 \colon \mathrm{Cl}^+(H^2_{\acute{e}t}(X_{\overline{L}}, \mathbb{Z}_\ell(1))_p) \xrightarrow{\sim} \mathrm{End}_C(H^1_{\acute{e}t}(A_{\overline{L}}, \mathbb{Z}_\ell)). \qquad (4.5)$$

Next, if L/K is not already finite, one views L as a function field of a finite type K-scheme and 'spreads' A with its C action over T. This yields a smooth abelian scheme $b \colon \mathcal{B} \longrightarrow T$ with an action of the constant algebra C. Its generic fibre gives back (A, C). Now, there exists a finite extension K'/K with $T(K') \neq \emptyset$ and specializing (4.5) to $b \in T(K')$ yields $\psi_{0b} \colon \mathrm{Cl}^+(H^2_{\acute{e}t}(X_{\overline{K'}}, \mathbb{Z}_\ell(1))_p) \xrightarrow{\sim} \mathrm{End}_C(H^1_{\acute{e}t}(\mathcal{B}_{b\overline{K'}}, \mathbb{Z}_\ell))$. As it turns out, the family $\mathcal{B} \longrightarrow T$ is in fact isotrivial, i.e. after passing to a finite extension of L (corresponding geometrically to a finite covering of T) the family becomes trivial and so (A, C) itself is defined over a finite extension of K, i.e. $\mathcal{B}_b \times K \simeq A$. However, at this point this is not clear; see Remark 4.4.

Next, consider the natural action of the Galois group $\mathrm{Gal}(\overline{L}/L)$ on both sides of (4.5). In order to show that the isomorphism is compatible with it, which then yields the assertion, one uses that K3 surfaces have 'big monodromy'.[3]

As shown by Corollary 6.4.7, the complex K3 surface $X_{\mathbb{C}}$ sits in a family of polarized complex K3 surfaces with a big monodromy group. The proof of the result reveals that the family is actually defined over K. So, there exists a family of polarized K3 surfaces $f \colon (\mathcal{X}, L) \longrightarrow S$ over K with special fibre $\mathcal{X}_0 \simeq X$ and such that the image of the monodromy representation $\pi_1(S_{\mathbb{C}}, 0) \longrightarrow \mathrm{O}(H^2(X_{\mathbb{C}}, \mathbb{Z})_p)$ is of finite index and, therefore, Zariski dense in $\mathrm{SO}(H^2(X_{\mathbb{C}}, \mathbb{Z})_p)$. By Proposition 6.4.10 and after passing to a finite cover of S, which we suppress, there exists an abelian scheme $a \colon \mathcal{A} \longrightarrow S_{\mathbb{C}}$ with an action of the constant algebra C and an isomorphism of VHS

---

[3] By construction, there exists a natural isomorphism $H^1_{\acute{e}t}(A_{\overline{L}}, \mathbb{Z}_\ell) \simeq \mathrm{Cl}^+(H^2_{\acute{e}t}(X_{\overline{L}}, \mathbb{Z}_\ell(1))_p)$, which, however, might not be compatible with the natural Galois actions.

$$\psi: \mathrm{Cl}^+(P^2 f_{\mathbb{C}*}\mathbb{Z}(1)) \overset{\sim}{\longrightarrow} \underline{\mathrm{End}}_C(R^1 a_*\mathbb{Z}), \tag{4.6}$$

which is the global version of (4.5). (Here, $P^2 f_* \subset R^2 f_*$ denotes the local system that fibrewise corresponds to the primitive cohomology.)

Clearly, the abelian scheme $a: \mathcal{A} \longrightarrow S_{\mathbb{C}}$ is defined over some finitely generated field extension $K \subset L \subset \mathbb{C}$, i.e. a is obtained by base change from some $a_L: \mathcal{A}_L \longrightarrow S_L$. As above, if L/K is not finite already, one spreads the family over some finite type K-scheme T and specializes to a point $b \in T(K')$ for some finite extension K'/L. So, we can assume that L is actually finite. Moreover, (4.6) descends to

$$\psi_{\overline{L}}: \mathrm{Cl}^+(P^2 f_{\overline{L}*}\mathbb{Z}_\ell(1)) \overset{\sim}{\longrightarrow} \underline{\mathrm{End}}_C(R^1 a_{\overline{L}*}\mathbb{Z}_\ell), \tag{4.7}$$

giving back ψ_0 in (4.5) (twisted by $\mathbb{Z}_\ell$) over the distinguished point $0 \in S$. As the local systems in (4.7) are pulled back from S_L, conjugating $\psi_{\overline{L}}$ to $\psi_{\overline{L}}^\sigma$ by an element $\sigma \in \mathrm{Gal}(\overline{L}/L)$ defines an isomorphism between the same local systems. Now, the existence of ψ is equivalent to saying that ψ_0 is invariant under the monodromy action and similarly the fact that $\psi_{\overline{L}}^\sigma$ is an isomorphism between the local systems in (4.7) implies that the conjugate ψ_0^σ is still invariant under the monodromy action. However, as by construction the monodromy group is Zariski dense in $\mathrm{SO}(H^2(X_{\mathbb{C}},\mathbb{Z})_p)$, the conjugate ψ_0^σ is in fact invariant under the $\mathrm{Spin}(V)$-action. But ψ_0 is the unique such isomorphism and, therefore, $\psi_0^\sigma = \psi_0$. Hence, ψ_0 is Galois invariant. $\square$

Remark 4.4 As André in [7, Sec. 1.7] stresses, the above arguments do not directly prove that the abelian variety A over the finite extension L/K in the proposition actually yields the Kuga–Satake variety $\mathrm{KS}(X_{\mathbb{C}})$ when base changed via $K \subset \mathbb{C}$. For this, one needs to verify the isotriviality of $\mathcal{B} \longrightarrow T$ in the above proof. This is achieved by observing that the monodromy action of the family induces the trivial action on $\mathrm{End}(H^1(\mathcal{B}_t,\mathbb{Z}))$. Following [7, Lem. 5.5.1], one argues that $\underline{\mathrm{End}}(R^1 b_{\mathbb{C}*}\mathbb{Z})$ is isomorphic to the constant system associated with $\mathrm{Cl}^+(H^2(X_{\mathbb{C}},\mathbb{Z})_p)$ (so, roughly, that the specialization ψ_{0b} is canonical), for which one again uses the fact that K3 surfaces have big monodromy. Thus, $\pi_1(T_{\mathbb{C}})$ acts by scalars and, as $\mathcal{B} \longrightarrow T$ is a polarized family, necessarily via a finite group.

André also shows that A is independent of the embedding $K \subset \mathbb{C}$ and explains how to control the finite extension L/K.

4.3 We can now outline the rest of Deligne's argument to prove the Weil conjectures. First, one needs to lift any given K3 surface Y_0 over the finite field $\mathbb{F}_q$ to characteristic zero. Following the discussion in Section **9.5.3**, there exists a polarized family $Y \longrightarrow \mathrm{Spec}(R)$ of K3 surfaces over a complete DVR R of mixed characteristic with residue field a finite extension $k/\mathbb{F}_q$ such that the closed fibre is $Y_0 \times k$. Let K denote the fraction field of R.

Thus, the generic fibre X of $Y \longrightarrow \mathrm{Spec}(R)$ is a K3 surface over the field K of characteristic zero to which one can apply Proposition 4.3. Hence, after passing to a finite extension of K, which we suppress, there exist an abelian variety A over K with an action of the algebra C and a Galois invariant isomorphism (4.3). The inertia subgroup $I_K \subset \mathrm{Gal}(\overline{K}/K)$, i.e. the kernel of the natural surjection $\mathrm{Gal}(\overline{K}/K) \longrightarrow\!\!\!\!\!\rightarrow \mathrm{Gal}(\overline{k}/k)$, acts trivially on $H^2_{\acute{e}t}(X_{\overline{K}}, \mathbb{Z}_\ell(1))_\mathrm{p}$, as the polarized surface X reduces to the smooth $Y_0 \times k$. Thus, I_K acts trivially on the left-hand side of (4.3) and hence on the right-hand side as well. The latter implies that, after finite base change, I_K acts trivially on $H^1_{\acute{e}t}(A_{\overline{K}}, \mathbb{Z}_\ell)$; for the details of this part of the argument, see [141].

Now, specialization yields $H^2_{\acute{e}t}(Y_0 \times \overline{k}, \mathbb{Z}_\ell(1))_\mathrm{p} \simeq H^2_{\acute{e}t}(X_{\overline{K}}, \mathbb{Z}_\ell(1))_\mathrm{p}$ (see [405, VI.Cor. 4.2]). By Néron–Ogg–Šafarevič theory (see [550]), the inertia group acts trivially on $H^1_{\acute{e}t}(A_{\overline{K}}, \mathbb{Z}_\ell)$ if and only if A reduces to a smooth abelian variety A_0 over k for which there then exists an isomorphism

$$\mathrm{Cl}^+(H^2_{\acute{e}t}(Y_0 \times \overline{k}, \mathbb{Z}_\ell(1))_\mathrm{p}) \simeq \mathrm{End}_C(H^1_{\acute{e}t}(A_0 \times \overline{k}, \mathbb{Z}_\ell)) \qquad (4.8)$$

of $\mathrm{Gal}(\overline{k}/k)$-modules, i.e. compatible with the action of the Frobenius.

The Weil conjectures for abelian varieties had been proved by Weil himself already in [633] and so all eigenvalues of the Frobenius action on $H^1_{\acute{e}t}(A_0 \times \overline{k}, \mathbb{Q}_\ell)$ have absolute value $\sqrt{q}$ and, therefore, absolute value one on the right-hand side of (4.8). Using a Galois invariant embedding $\bigwedge^2 H^2_{\acute{e}t}(Y_0 \times \overline{k}, \mathbb{Q}_\ell(1))_\mathrm{p} \subset \mathrm{Cl}^+(H^2_{\acute{e}t}(Y_0 \times \overline{k}, \mathbb{Q}_\ell(1))_\mathrm{p}) \simeq \mathrm{End}_C(H^1_{\acute{e}t}(A_0 \times \overline{k}, \mathbb{Q}_\ell))$, one concludes that the eigenvalues of the Frobenius action on $H^2_{\acute{e}t}(Y_0 \times \overline{k}, \mathbb{Q}_\ell(1))_\mathrm{p}$ all have absolute value one. Note that at this point one uses that $b_2(Y_0 \times \overline{k}) \geq 3$.

Remark 4.5 Proposition 4.3 holds as well for the Kuga–Satake variety associated with the full cohomology and so $\mathrm{Cl}^+(H^2_{\acute{e}t}(X_{\overline{L}}, \mathbb{Q}_\ell(1))) \simeq \mathrm{End}_C(H^1_{\acute{e}t}(A_{\overline{L}}, \mathbb{Q}_\ell))$ for some abelian variety A, which is isogenous to a power of the original one. Now, the polarization of $X_{\overline{L}}$ is Galois invariant, and, therefore, right multiplication by its class defines a Galois invariant embedding $H^2_{\acute{e}t}(X_{\overline{L}}, \mathbb{Q}_\ell(1)) \hookrightarrow \mathrm{Cl}^+(H^2_{\acute{e}t}(X_{\overline{L}}, \mathbb{Q}_\ell(1)))$; see Proposition 2.6. Therefore, for any K3 surface X over a field K of characteristic zero there exists a Galois invariant embedding

$$H^2_{\acute{e}t}(X_{\overline{L}}, \mathbb{Q}_\ell) \hookrightarrow H^1_{\acute{e}t}(A_{\overline{L}}, \mathbb{Q}_\ell) \otimes H^1_{\acute{e}t}(A_{\overline{L}}, \mathbb{Q}_\ell), \qquad (4.9)$$

where A is an abelian variety defined over a finite extension L/K. This is an essential ingredient for the proof of the Tate conjecture in characteristic zero (see Section **17**.3.2), and the analogous statement holds in positive characteristic (but there not implying the Tate conjecture, due to the lack of a Lefschetz theorem on $(1, 1)$-classes).

Remark 4.6 Recall that (4.9) is conjectured to be algebraic and that this has been verified for Kummer surfaces; see Conjecture 2.11 and Remark 3.2. In particular, it is known to be 'absolute' for Kummer surfaces, cf. [113, 145] for the notion of absolute classes. If one now puts an arbitrary K3 surface in a family connecting it to a Kummer surface, then Deligne's Principle B (see [145]), applied to the family version of the Kuga–Satake construction (see Proposition **6**.4.10), implies that the Kuga–Satake construction is absolute for all K3 surfaces. See the lectures of Charles and Schnell [113] for more details.

Remark 4.7 As Deligne explains in the introduction to [141], the motivic nature of the Kuga–Satake construction served as a guiding principle in his proof of the Weil conjectures for K3 surfaces. Its motivic nature was discussed further by André in [7, 8], who moreover showed that the Kuga–Satake construction is 'motivated'. As a consequence, one concludes that the motive of any K3 surface, as an object in André's category of motives, is contained in the Tannaka subcategory generated by abelian varieties.

***

References and Further Reading

The Kuga–Satake construction is also used for the study of projective hyperkähler (or irreducible holomorphic symplectic) manifolds (see Section **10**.3.4), which are higher-dimensional analogues of K3 surfaces for which the second cohomology is also a Hodge structure of K3 type. See, for example, [7, 111].

The Kuga–Satake variety of a K3 surface with real multiplication, i.e. such that the ring $\text{End}_{\text{Hdg}}(T(X))$ is a totally real field, has been studied by van Geemen [204] and Schlickewei [528].

In [196, 377] one finds a detailed discussion of the Kuga–Satake variety associated with a sub-Hodge structure of weight two of certain abelian fourfolds. Double covers of $\mathbb{P}^1 \times \mathbb{P}^1$ and their Kuga–Satake varieties have been studied in [311].

In [622] Voisin observed that one can gain some flexibility in the above construction by splitting it into two steps. One can first define a Hodge structure of weight two on $\text{Cl}(V)$ by setting $\text{Cl}(V)^{2,0} := V^{2,0} \otimes \bigoplus_k \bigwedge^{k-1}(V^{1,1})$, which is in a certain sense compatible with the algebra structure; cf. Remark 1.1. Then one associates with any Hodge structure of weight two on $\text{Cl}(V)$ compatible with the algebra structure a Hodge structure of weight one.

The polynomial $P_2(t)$ for a K3 surface over a finite field splits into an algebraic part, and a transcendental part, $P_2(t) = P_2(t)_{\text{alg}} \cdot P_2(t)_{\text{tr}}$, according to whether α_i/q is a root of unity or not. The transcendental part enjoys remarkable properties. For example, it has a unique irreducible factor. For a review of some of the properties of $P_2(t)_{\text{tr}}$, see Taelman [582] in which he also proves that any polynomial satisfying these properties can actually be realized.

Questions and Open Problem

Clearly, the main open problem here is Conjecture 2.11. It is known that the Kuga–Satake class is an absolute Hodge class; cf. Remark 4.6. This had been implicitly proved already by Deligne [141], which predates the notion of absolute Hodge classes, and explicitly by André [7] and Charles and Schnell [113]. It would be very interesting to find other examples of K3 surfaces for which the Kuga–Satake–Hodge conjecture can be proved and detect any general pattern behind those.

Is there a more explicit proof of the algebraicity of the Kuga–Satake correspondence for abelian surfaces than the one that uses the full Hodge conjecture for self-products of abelian surfaces? (See Remark 3.2.)

At this point it seems unlikely that the transcendental construction of the Kuga–Satake variety can be replaced by an algebro-geometric one.

Grothendieck at some point wondered whether maybe every variety is dominated by a product of curves (DPC). If that were true, one could prove the Weil conjectures (in fact also the Tate conjecture; see Remark 17.3.3) by reducing to the case of curves. In a letter to Grothendieck in 1964, Serre produced a counterexample in dimension two [127]. His surface is realized as a subvariety of an abelian variety and is in particular not a K3 surface. In fact, it is not known whether there exist K3 surfaces that are not DPC. More examples of varieties that are not DPC were constructed by Schoen in [531].

5

Moduli Spaces of Polarized K3 Surfaces

It is often preferable not to study individual K3 surfaces but to consider all (of a certain degree or with a certain projective embedding, etc.) simultaneously. This leads to the concept of moduli spaces of K3 surfaces, and this chapter is devoted to the various existence results for moduli spaces of polarized K3 surfaces as quasi-projective varieties, algebraic spaces, or Deligne–Mumford stacks.

In Section 1 the moduli functor is introduced and three existence results are stated. They are discussed in greater detail in Sections 2 and 4. In Section 3 we study the local structure of the moduli spaces and prove finiteness results for automorphism groups of polarized K3 surfaces.

Moduli spaces of polarized complex K3 surfaces of different degrees are all contained in the larger, but badly behaved, moduli space of complex (not necessarily algebraic) K3 surfaces, for which we refer to Chapter 6.

1 Moduli Functor

We shall work over a Noetherian base S. The cases we are most interested in are $S = \mathrm{Spec}(\mathbb{C})$, $S = \mathrm{Spec}(\overline{\mathbb{Q}})$, $S = \mathrm{Spec}(K)$ for a number field K, $S = \mathrm{Spec}(\mathbb{F}_q)$, and $S = \mathrm{Spec}(\mathcal{O})$ with $\mathcal{O}$ the ring of integers in a number field, e.g. $\mathcal{O} = \mathbb{Z}$.

For a given positive integer d one considers the *moduli functor*

$$\mathcal{M}_d \colon (Sch/S)^o \longrightarrow (Sets), \ T \longmapsto \{(f \colon X \longrightarrow T, L)\} \tag{1.1}$$

that sends a scheme T of finite type over S to the set $\mathcal{M}_d(T)$ of equivalence classes of pairs (f, L) with $f \colon X \longrightarrow T$ a smooth proper morphism and $L \in \mathrm{Pic}_{X/T}(T)$[1] such that for

---

[1] Note that by definition $\mathrm{Pic}_{X/T}(T) = H^0(T, R^1 f_* \mathbb{G}_m)$, which is obtained by étale sheafification of the functor $T' \longmapsto \mathrm{Pic}(X_{T'})/\mathrm{Pic}(T)$. We often (over) simplify by thinking of L as an actual line bundle L modulo line bundles coming from T, although it possibly only exists after passing to an étale cover.

all geometric points $\text{Spec}(k) \longrightarrow T$, i.e. k an algebraically closed field, the base change yields a K3 surface X_k with a primitive ample line bundle L_{X_k} such that $(L_{X_k})^2 = 2d$, i.e. (X_k, L_k) is a polarized K3 surface of degree $2d$; cf. Definition **2.4.1**.[2]

By definition, $(f, L) \sim (f', L')$ if there exists a T-isomorphism $\psi \colon X \xrightarrow{\sim} X'$ and a line bundle L_0 on T such that $\psi^* L' \simeq L \otimes f^* L_0$. For an S-morphism $g \colon T' \longrightarrow T$ one defines $\mathcal{M}_d(T) \longrightarrow \mathcal{M}_d(T')$ as the pull-back $(f \colon X \longrightarrow T, L) \longmapsto (f_{T'} \colon X \times_T T' \longrightarrow T', g_X^* L)$. Here, $g_X \colon X \times_T T' \longrightarrow X$ is the base change of g.

Recall that a (fine) *moduli space* for $\mathcal{M}_d$ is an S-scheme M_d together with an isomorphism of functors

$$\mathcal{M}_d \simeq \underline{M}_d := h_{M_d}$$

(the functor of points associated with the scheme M), i.e. M_d represents $\mathcal{M}_d$.

Due to the existence of automorphisms, a fine moduli space does not exist, so one can only hope for a *coarse moduli space*. A coarse moduli space is by definition an S-scheme M together with a functor transformation

$$\Psi \colon \mathcal{M}_d \longrightarrow \underline{M}_d$$

with the following two properties:

(i) For any algebraically closed field k the induced map $\mathcal{M}_d(\text{Spec}(k)) \longrightarrow \underline{M}_d(\text{Spec}(k))$ is bijective. (By definition, $\underline{M}_d(\text{Spec}(k))$ coincides with the set $M_d(k)$ of k-rational points of M_d.)

(ii) For any S-scheme N and any natural transformation $\Phi \colon \mathcal{M}_d \longrightarrow \underline{N}$ there exists a unique S-morphism $\pi \colon M \longrightarrow N$ such that $\Phi = \underline{\pi} \circ \Psi$.

1.1   The following result is due to Pjateckiĭ-Šapiro and Šafarevič [493]. Their proof relies on the global Torelli theorem and the quasi-projectivity of arithmetic quotients of the period domain due to Baily and Borel [29]. We come back to this later; see Corollary **6.4.3**.

An alternative proof was given by Viehweg in [615]. His arguments rely on geometric invariant theory (GIT), but without actually proving that points in the appropriate Hilbert scheme corresponding to K3 surfaces are stable. In [427] Ian Morrison proves that the generic K3 surface defines a GIT stable point if $d \geq 6$.

---

[2] By our definition, a K3 surface is a surface over a field that is a K3 surface over the algebraic closure. So in fact, for any point $\text{Spec}(k) \longrightarrow T$ the fibre is a K3 surface. However, in principle an ample line bundle could acquire a root after base field extension. So we have to require L to be primitive, i.e. not the power of any other line bundle, over the algebraic closure. See Chapter 17 for more on the behavior of the Picard group under base change.

Theorem 1.1 *For $S = \text{Spec}(\mathbb{C})$ the moduli functor $\mathcal{M}_d$ can be coarsely represented by a quasi-projective variety M_d.*[3]

See Section 2.3 for more details and comments on the case of positive characteristic.

1.2 The existence of a quasi-projective coarse moduli space is far from being trivial, using periods or GIT. However, the existence of the coarse moduli space as an algebraic space is much easier and follows from very general existence results for group quotients in the category of algebraic spaces.

Theorem 1.2 *The moduli functor $\mathcal{M}_d$ can be coarsely represented by a separated algebraic space M_d which is locally of finite type over S.*

The existence of the coarse moduli space as an algebraic space was intensively studied by Popp for $S = \text{Spec}(\mathbb{C})$; see [496, 497, 498]. His construction relies on the existence of certain group quotients in the category of analytic spaces which he then endows with an algebraic structure. For the existence of the coarse moduli space as an algebraic space over other fields, see Viehweg's book [616, Ch. 9]. The existence of the coarse moduli space as an algebraic space in much broader generality can be deduced from a more recent result by Keel and Mori [289]; we also recommend Lieblich [364] for an account of their result. See also the related result by Kollár in [313] and Section 2.3 for more details.

1.3 Instead of considering $\mathcal{M}_d$ as a contravariant functor $(Sch/S)^o \longrightarrow (Sets)$ one can view it as a groupoid over (Sch/S). More precisely, one can consider the category $\mathcal{M}_d$ of all $(f : X \longrightarrow T, L)$ as before. The projection $(f : X \longrightarrow T, L) \longmapsto T$ then defines a functor $\mathcal{M}_d \longrightarrow (Sch/S)$. A morphism in $\mathcal{M}_d$ is defined to be a fibre product diagram

$$\begin{array}{ccc} X' & \xrightarrow{\tilde{g}} & X \\ \downarrow & & \downarrow \\ T' & \xrightarrow{g} & T \end{array}$$

with $\tilde{g}^* L \simeq L'$. The isomorphism is not part of the datum.

In fact, in shifting the point of view like this one takes into account automorphisms of K3 surfaces, which are responsible for the non-existence of a fine moduli space as

---

[3] From this and Theorem 1.2 one can conclude that $\mathcal{M}_d$ admits a quasi-projective coarse moduli space over any field of characteristic zero. The only thing that needs checking is that the algebraic space coarsely representing $\mathcal{M}_d$ over k can be completed to a complete algebraic space. This is provided by the Nagata compactification theorem; see [129]. Then use that a complete algebraic space that becomes projective after base field extension is itself projective.

well as for singularities of the coarse moduli space. Indeed, the fibre of $\mathcal{M}_d \longrightarrow (Sch/S)$ over an S-scheme T, which consists of all $(f\colon X \longrightarrow T, L)$ as before, is not merely a set but in fact a groupoid, i.e. a category in which all morphisms are isomorphisms and in particular the endomorphisms of $(f\colon X \longrightarrow T, L)$ are precisely the automorphisms of the polarized K3 surface (X, L) (over T).

Instead of representing the functor $\mathcal{M}_d$ by a quasi-projective scheme or an algebraic space, one now studies it in the realm of stacks. This approach goes back to Deligne and Mumford in [146], where it was successfully applied to the moduli functor of curves. In analogy to their result, one has the following result; see [510] and Section 4.

Theorem 1.3 *The groupoid $\mathcal{M}_d \longrightarrow (Sch/S)$ is a separated Deligne–Mumford stack of finite type.*

The result of Keel and Mori [289] in fact shows that any separated Deligne–Mumford stack of finite type has a coarse moduli space in the category of algebraic spaces. This gives back Theorem 1.2 (however, relying on essentially the same techniques).

As explained by Rizov in [510], taking into account automorphisms of polarized K3 surfaces resolves the singularities of the moduli stack. More precisely, one can show that $\mathcal{M}_d$ is a smooth Deligne–Mumford stack over $\mathrm{Spec}(\mathbb{Z}[1/(2d)])$; cf. Corollary 3.6 and Remark 3.2.

1.4   To conclude the introduction to this chapter, we mention the possibility of partially compactifying the moduli space by adding quasi-polarized K3 surfaces. Recall that a quasi-polarized (or pseudo-polarized) K3 surface (X, L) of degree $2d$ consists of a smooth K3 surface X together with a primitive big and nef line bundle L such that $(L)^2 = 2d$.

The corresponding moduli functor $\mathcal{M}'_d$ can be defined analogously to $\mathcal{M}_d$ in (1.1). Many of the arguments and constructions that are explained below work in this more general context. However, the moduli functor defined in this way has a disadvantage over $\mathcal{M}_d$: it is not separated. More precisely, there exist families

$$(X, L), (Y, M) \longrightarrow \mathrm{Spec}(R) \tag{1.2}$$

of quasi-polarized K3 surfaces over a DVR R such that the generic fibres are isomorphic, $(X_\eta, L_\eta) \simeq (Y_\eta, M_\eta)$, but the isomorphism does not extend over the closed fibres to an isomorphism of the families (although the special fibres themselves are again isomorphic to each other); cf. the remarks in Section 2.3, the proof of Proposition 7.5.5, and [128] for an explicit algebraic example.

The way out is to add not quasi-polarized K3 surfaces (X, L) but polarized 'singular K3 surfaces' $(\overline{X}, \overline{L})$. For any big and nef line bundle L on a K3 surface, L^3 is base point free and the induced morphism $\varphi_{L^3}\colon X \longrightarrow \overline{X}$ contracts only ADE curves and so $\overline{X}$ has only rational double points; cf. Remark **2**.3.4 and Section **11**.2.2. Applied to (1.2), one obtains isomorphic polarized families $(\overline{X}, \overline{L}) \simeq (\overline{Y}, \overline{L}) \longrightarrow \mathrm{Spec}(R)$ with a singular central fibre. So, if the moduli functor $\overline{\mathcal{M}}_d$ is defined accordingly, it is a smooth Deligne–Mumford stack with a quasi-projective coarse moduli space $\overline{M}_d$. As the moduli space M_d, $\overline{M}_d$ also admits a description via periods in characteristic zero; see Section **6**.4.1.

2 Via Hilbert Schemes

In all approaches to the moduli space of polarized K3 surfaces the Hilbert scheme plays the central role. Due to general results of Grothendieck [225], the Hilbert scheme is always represented by a projective scheme and the part of it that parametrizes K3 surfaces defines a quasi-projective subscheme. The quotient by the action of a certain PGL, identifying the various projective embeddings, yields the desired moduli space. So, the question of whether $\mathcal{M}_d$ has a coarse moduli space becomes a question about the nature of this quotient. For the shortest outline of the construction (for the more general class of symplectic varieties), see André [7, Sec. 2.3] and for a detailed discussion of the general theory see the monographs [444, 616]. We shall discuss the various steps in this process in the case of $S = \mathrm{Spec}(k)$ and shall often, for simplicity, assume that k is algebraically closed. The result, however, is used for general Noetherian base S, e.g. when proving that $\mathcal{M}_d$ is a Deligne–Mumford stack; see Proposition 4.10.

2.1 Consider the Hilbert polynomial P of a polarized K3 surface (X, L) of degree $2d = (L)^2$. By the Riemann–Roch theorem $P(t) = dt^2 + 2$; see Section **1**.2.3. Let $N := P(3) - 1$. This choice is prompted by the theorem of Saint-Donat saying that for any ample L the line bundle L^3 is very ample; see Theorem **2**.2.7. Hence, any (X, L) with L ample and $(L)^2 = 2d$ can be embedded into $\mathbb{P}^N$ such that $\mathcal{O}(1)|_X \simeq L^3$. Finally, the Hilbert polynomial of $X \subset \mathbb{P}^n$ with respect to $\mathcal{O}(1)$ is $P(3t)$.

Consider the Hilbert scheme

$$\mathrm{Hilb} := \mathrm{Hilb}_{\mathbb{P}^N}^{P(3t)}$$

of all closed subschemes $Z \subset \mathbb{P}^N$ with Hilbert polynomial $P(3t)$. Then Hilb is a projective scheme representing the Hilbert functor,

$$\underline{\mathrm{Hilb}}: (Sch/k)^o \longrightarrow (Sets),$$

mapping a k-scheme T to the set of T-flat closed subschemes $Z \subset T \times \mathbb{P}^N$ such that all geometric fibres $Z_s \subset \mathbb{P}^N_{k(s)}$ have Hilbert polynomial $P(3t)$. In particular, the Hilbert scheme comes with a universal family $\mathcal{Z} \subset \mathrm{Hilb} \times \mathbb{P}^N$ with flat projection $\mathcal{Z} \longrightarrow \mathrm{Hilb}$.

Proposition 2.1   *There exists a subscheme $H \subset \mathrm{Hilb}$ with the following universal property: a morphism $T \longrightarrow \mathrm{Hilb}$ factors through $H \subset \mathrm{Hilb}$ if and only if the pull-back*

$$f \colon \mathcal{Z}_T \longrightarrow T$$

of the universal family $\mathcal{Z} \longrightarrow \mathrm{Hilb}$ satisfies:

  (i) *The morphism $f \colon \mathcal{Z}_T \longrightarrow T$ is a smooth family with all fibres being K3 surfaces.*
  (ii) *If $p \colon \mathcal{Z}_T \longrightarrow \mathbb{P}^N$ is the natural projection, then*

$$p^*\mathcal{O}(1) \simeq L^3 \otimes f^*L_0$$

     for some $L \in \mathrm{Pic}(\mathcal{Z}_T)$ and $L_0 \in \mathrm{Pic}(T)$.
  (iii) *The line bundle L in (ii) is primitive on each geometric fibre.*
  (iv) *For all fibres $\mathcal{Z}_s$ of $f \colon \mathcal{Z}_T \longrightarrow T$, restriction yields isomorphisms*

$$H^0(\mathbb{P}^N_{k(s)}, \mathcal{O}(1)) \xrightarrow{\ \sim\ } H^0(\mathcal{Z}_s, L_s^3).$$

Proof   The subscheme H is in fact an open subscheme of Hilb, but it is slightly easier to prove its existence as a locally closed subscheme.

Smoothness, irreducibility, and vanishing of $H^1(\mathcal{O})$ are all open properties and, therefore, define an open subset of the Hilbert scheme. Since triviality of a line bundle is a closed condition, $\omega \simeq \mathcal{O}$ is a priori a closed condition, but, in fact, it is also open. Indeed, if one fibre is a K3 surface, then $\chi(\mathcal{O}) = 2$ and hence $h^0(\omega) = h^2(\mathcal{O}) \geq 1$ for all fibres in an open neighbourhood. Also, $h^0(\omega^2) \leq 1$ and $h^1(\omega^*) = 0$ are open conditions. Since by Riemann–Roch and Serre duality $h^0(\omega^*) = 2 + h^1(\omega^*) - h^0(\omega^2)$, one finds that $h^0(\omega^*) \neq 0$ for all fibres in an open neighbourhood. However, $h^0(\omega) \neq 0 \neq h^0(\omega^*)$ if and only if ω is trivial.

Thus, (i) describes an open subscheme of Hilb. It is straightforward to see that (iv) is also an open condition. Thus, (i) and (iv) together define an open subscheme H' to which we restrict. Next, one shows that there exists a universal subscheme $H \subset H'$ defined by (ii) and (iii).

The relative Picard scheme $\mathrm{Pic}_{\mathcal{Z}'/H'} \longrightarrow H'$ of the restriction $\mathcal{Z}' \longrightarrow H'$ of the universal family is a disjoint union of H'-projective schemes $\mathrm{Pic}^{Q(t)}_{\mathcal{Z}'/H'} \longrightarrow H'$ parametrized by the possible Hilbert polynomials $Q(t)$; see [82, 176]. Clearly, there are only finitely many possibilities to write $P(3t) = Q(nt)$ with $Q(t)$ the Hilbert polynomial of an actual line bundle on a K3 surface (i.e. with integral coefficients). Since the map

$$\mathrm{Pic}^{Q(t)}_{\mathcal{Z}'/H'} \longrightarrow \mathrm{Pic}^{Q(nt)}_{\mathcal{Z}'/H'}, \ M \longmapsto M^n$$

is an H'-morphism and all schemes are projective over H', its image is closed. In fact, as can easily be seen, the morphism is a closed embedding. This yields a universal locally closed subscheme $Y \subset \mathrm{Pic}^{P(3t)}_{\mathcal{Z}'/H'}$ parametrizing line bundles M which can fibrewise be written as $M \simeq L^3$ for some primitive L.

The line bundle $\mathcal{O}(1)$ can be viewed as a section of $\mathrm{Pic}_{\mathcal{Z}'/H'}^{P(3t)} \longrightarrow H'$ and one defines H as the pre-image of $Y \subset \mathrm{Pic}_{\mathcal{Z}'/H'}^{P(3t)}$ under this section.[4] Note that under our assumptions, there exists a Poincaré bundle on $\mathrm{Pic}_{\mathcal{Z}_H/H} \times_H \mathcal{Z}_H$ so that L and L_0 as in (ii) exist for $\mathcal{Z}_H \longrightarrow H$.

For later use note that (ii) and (iv) in particular show $f_*(L^3) \simeq L_0^* \otimes \mathcal{O}_T^{N+1}$. $\qquad\square$

Thus, H together with the restriction of the universal family $\mathcal{X} := \mathcal{Z}_H \longrightarrow H$ represents the functor

$$\underline{H} \colon (Sch/k)^o \longrightarrow (Sets)$$

that maps T to the set of all T-flat closed subschemes $Z \subset T \times \mathbb{P}^N$ satisfying (i)–(iv).

Clearly, mapping $Z \subset T \times \mathbb{P}^N$ to $(f = p_1 \colon Z \longrightarrow T, L)$ with L as in (ii) defines a functor transform $\underline{H} \longrightarrow \mathcal{M}_d$. The only thing that needs checking at this point is whether L is uniquely determined (up to pull-back of line bundles on T). This is due to the fact mentioned in the proof above that $L \mapsto L^3$ defines a closed embedding $\mathrm{Pic}^{P(t)} \longrightarrow \mathrm{Pic}^{P(3t)}$. The injectivity on the level of sets is implied by the torsion freeness of the Picard group of any K3 surface; see Remark 1.2.5.

2.2 The Hilbert scheme $\mathrm{Hilb} = \mathrm{Hilb}_{\mathbb{P}^N}^{P(3t)}$ comes with a natural $\mathrm{PGL} := \mathrm{PGL}(N+1)$-action. It can functorially be defined as the functor transformation

$$\underline{\mathrm{PGL}} \times \underline{\mathrm{Hilb}}_{\mathbb{P}^N}^{P(3t)} \longrightarrow \underline{\mathrm{Hilb}}_{\mathbb{P}^N}^{P(3t)}$$

that sends $(A \in \mathrm{PGL}(T), Z \subset T \times \mathbb{P}^N)$ to $(\varphi_A(Z) \subset T \times \mathbb{P}^N)$. Here, the isomorphism $\varphi_A \colon T \times \mathbb{P}^N \xrightarrow{\sim} T \times \mathbb{P}^N$ is obtained by viewing A as a family of automorphisms of $\mathbb{P}^N$ varying over T.

Clearly, the conditions (i)–(iv) above are invariant under the PGL-action. Hence, H is preserved and we obtain an action

$$\mathrm{PGL} \times H \longrightarrow H.$$

Moreover, the natural transformation $\underline{H} \longrightarrow \mathcal{M}_d$, which just forgets the projective embedding, is equivariant and hence yields a functor

$$\Theta \colon \underline{H}/\mathrm{PGL} \longrightarrow \mathcal{M}_d.$$

Proposition 2.2 *The natural transformation* $\Theta \colon \underline{H}/\mathrm{PGL} \longrightarrow \mathcal{M}_d$ *is injective and locally surjective.*

---

[4] In fact, $H \subset H'$ is also open, but this needs some deformation theory. Indeed, the obstruction to deform a line bundle L in a given family sideways lies in $H^2(X, \mathcal{O}_X)$. If $M \simeq L^3$, then the two obstruction classes differ by a factor 3 and so L deforms whenever M does (at least if $\mathrm{char}(k) \neq 3$).

Proof Local surjectivity means that for any $(f : X \longrightarrow T, L)$ there exists an étale open covering $T := \bigcup T_i$ such that the restrictions $(f_i : X_{T_i} \longrightarrow T_i, L|_{X_{T_i}})$ are in the image of $\Theta(T_i) : (\underline{H}/\underline{PGL})(T_i) \longrightarrow \mathcal{M}_d(T_i)$. This is shown as follows: the direct image $f_*(L^3)$ is locally free of rank $N + 1$ and the higher direct images are trivial; cf. Proposition 2.3.1. After passing to an open cover of T, we may assume that $f_*(L^3)$ is in fact free, i.e. $f_*(L^3) \simeq \mathcal{O}_T^{N+1}$. Moreover, since L^3 is fibrewise very ample, the surjection $\mathcal{O}_X^{N+1} \longrightarrow L^3$ obtained by pull-back defines a closed embedding $X \hookrightarrow T \times \mathbb{P}^N$. Then (i)–(iv) are satisfied by construction.[5]

For the injectivity we have to show that two $Z, Z' \subset T \times \mathbb{P}^N$ in $\underline{H}(T)$ are isomorphic as polarized families if and only if their projective embeddings differ by an automorphism of $\mathbb{P}^N$. If $(f : Z \longrightarrow T, L)$ and $(f' : Z' \longrightarrow T, L')$ define the same element in $\mathcal{M}_d$, then there exists an isomorphism $\psi : Z \xrightarrow{\sim} Z'$ with $\psi^* L' \simeq L \otimes f^* L_0$ for some $L_0 \in \text{Pic}(T)$. The given embeddings induce trivializations of $f'_*(L'^3)$ and $f_*(L^3)$. The induced isomorphism

$$\mathcal{O}_T^{N+1} \simeq f'_*(L'^3) \simeq f_*(L^3) \otimes L_0^3 \simeq \mathcal{O}_T^{N+1} \otimes L_0^3$$

corresponds to an $A \in \underline{PGL}(T)$ and the closed embeddings $Z, Z' \subset T \times \mathbb{P}^N$ differ by the automorphism φ_A of $T \times \mathbb{P}^N$. $\square$

Note that Θ in particular induces a bijection $[\underline{H}/\underline{PGL}](k) \xrightarrow{\sim} \mathcal{M}_d(k)$.[6] By the following result the question of whether $\mathcal{M}_d$ has a coarse moduli space is reduced to the existence of a categorical quotient of the action $PGL \times H \longrightarrow H$.

Proposition 2.3 *Suppose there exists a categorical quotient $\pi : H \longrightarrow Q = H/PGL$ whose k-rational points parametrize the orbits of the action. Then Q is a coarse moduli space for $\mathcal{M}_d$.*

Proof Recall that by definition a *categorical quotient* is a morphism $\pi : H \longrightarrow Q$ such that the two morphisms $PGL \times H \longrightarrow H$, obtained by projection and group action, composed with π coincide and such that any other $\pi' : H \longrightarrow Q'$ with this property factors uniquely through π:

$$
\begin{array}{ccc}
H & \xrightarrow{\ \pi\ } & Q \\
& \pi' \searrow & \big\downarrow {\scriptstyle q} \\
& & Q'.
\end{array}
$$

Here is a sketch of the argument: to construct $\mathcal{M}_d \longrightarrow Q$ use the local surjectivity of $\underline{H}/\underline{PGL} \longrightarrow \mathcal{M}_d$. Then any $(f, L) \in \mathcal{M}_d(T)$ can locally over the open sets of some

---

[5] It looks as if the T_i could be chosen Zariski open. Remember, however, that L itself may only exist on an étale cover.

[6] Under our assumption that k is algebraically closed, the set of k-rational points of the quotient stack (see Examples 4.4 and 4.5) $[\underline{H}/\underline{PGL}]$ is indeed just $H(k)/PGL(k)$.

covering $T = \bigcup T_i$ first be lifted to $\underline{H}(T_i)$ and then mapped to $\underline{Q}(T_i)$. Due to the PGL-invariance of $\pi : H \longrightarrow Q$ and the injectivity of $\underline{H}/\underline{\text{PGL}} \longrightarrow \mathcal{M}_d$, the image does not depend on the lift to $\underline{H}(T_i)$. Since $\underline{Q}$ is a sheaf (in the Zariski topology), the images in $\underline{Q}$ of the lifts to the $\underline{H}(T_i)$ glue. Eventually, this yields the functorial $\mathcal{M}_d \longrightarrow \underline{Q}$. Since $\underline{Q}(k)$ parametrizes the orbits of the PGL-action by assumption and since by Proposition 2.2 the same holds for $\mathcal{M}_d(k)$, one has $\mathcal{M}_d(k) \overset{\sim}{\longrightarrow} \underline{Q}(k)$.

It remains to prove the minimality of $\mathcal{M}_d \longrightarrow \underline{Q}$, which is proved similarly. Any $\mathcal{M}_d \longrightarrow \underline{N}$ can be composed with $\underline{H} \longrightarrow \mathcal{M}_d$, which yields an invariant $\underline{H} \longrightarrow \underline{N}$. The latter corresponds to an invariant $H \longrightarrow N$ which by the universality property of the categorical quotient $\pi : H \longrightarrow Q$ factors uniquely through a morphism $Q \longrightarrow N$. It is not difficult to see that $\mathcal{M}_d \longrightarrow \underline{Q} \longrightarrow \underline{N}$ is the original transformation. □

On purpose, we were vague about the geometric nature of the quotient and in fact the proof is so general that it works in many settings. The best possible case is that Q is a quasi-projective scheme. This can in fact be achieved for $k = \mathbb{C}$, a result due to Viehweg [615], and yields Theorem 1.1 which we state again as the following.

Theorem 2.4 *For $k = \mathbb{C}$, there exists a categorical quotient $\pi : H \longrightarrow Q$ with Q a quasi-projective scheme. Its k-rational points parametrize the orbits of the action, i.e. $[\underline{H}/\underline{\text{PGL}}](k) \overset{\sim}{\longrightarrow} \underline{Q}(k)$. So, $M_d := Q$ is a quasi-projective coarse moduli space for $\mathcal{M}_d$.*

Usually a quasi-projective quotient would be constructed by GIT methods, i.e. by showing that a smooth K3 surface yields a point in H that is stable with respect to the action of PGL and an appropriate linearization; see Mumford's original [444] for the foundations of GIT. However, this direct approach works only in low dimensions, e.g. for curves. Viehweg's techniques avoid a direct check of GIT stability. They do not seem to generalize to positive characteristic, and, therefore, the existence of quasi-projective coarse moduli spaces in positive characteristic had been an open problem for a long time. Recently, the quasi-projectivity has been proved by Maulik [398] for $p \geq 5$ and $p \nmid d$ and by Madapusi Pera [387] for any $p > 2$.

Example 2.5 There is, however, one case where the standard GIT techniques do work and really yield the moduli space as a quasi-projective variety. This is the classical case of hypersurfaces in projective spaces. For K3 surfaces one is looking at quartics $X \subset \mathbb{P}^3$. In particular, in this case we do not have to pass from the ample line bundle $L := \mathcal{O}(1)$ to its power L^3, as L itself is already very ample.

Let $H \subset |\mathcal{O}(4)|$ be the open subscheme parametrizing smooth quartics. Thus, if $x \in H(k)$ corresponds to the hypersurface $X \subset \mathbb{P}^3$, then X is a K3 surface. Clearly, H is invariant under the natural action of PGL $:=$ PGL(4) and the quotient $H(k)/\text{PGL}(k)$ parametrizes all polarized K3 surfaces which are isomorphic to a quartic in $\mathbb{P}^3$.

Now, GIT shows that the quotient H/PGL exists as a categorical quotient such that its k-rational points correspond to the orbits of the PGL(k)-action on $H(k)$ if every point in $H(k)$ is stable. Recall that a point $x \in H(k)$ is GIT-*stable* if the stabilizer of x is a finite subgroup of PGL(k) and there exists an invariant section $s \in H^0(H, \mathcal{L}^n)$ for some $n > 0$ satisfying: (i) $s(x) \neq 0$, (ii) $H_s := H \setminus Z(s)$ is affine, and (iii) the action of PGL on H_s is closed. Here, $\mathcal{L}$ is an SL-linearized ample line bundle on H.

It is now a classical fact that smooth hypersurfaces of degree $d \geq 3$ in $\mathbb{P}^n$ do correspond to stable points. The line bundle $\mathcal{L}$ is in this case $\mathcal{O}(1)$ on the projective space $|\mathcal{O}(d)|$. An invariant polynomial not vanishing in a point x corresponding to a hypersurface $X \subset \mathbb{P}^n$ is provided by the discriminant. See the textbooks by Mumford et al. [444, Ch. 4] or Mukai [434, Ch. 5.2] for details.

For results dealing with the stability of complete intersection K3 surface, see the more recent article by Li and Tian [361].

2.3   The existence of a coarse moduli space as an algebraic space is much easier or at least can be deduced from very general principles. According to a result of Keel and Mori [289] one has[7] the following theorem.

Theorem 2.6   *If G is a linear algebraic group acting properly on a scheme of finite type H (over, say, a Noetherian base), then a categorical quotient $\pi : H \longrightarrow Q = H/G$ exists as a separated algebraic space. Moreover, for any algebraically closed field k it induces a bijection $H(k)/G(k) \xrightarrow{\sim} Q(k)$.*

In order to deduce Theorem 1.2 from this, it remains to show that PGL $\times H \longrightarrow H$ is a *proper action*, i.e. that the graph morphism PGL $\times H \longrightarrow H \times H$, $(g, x) \mapsto (gx, x)$ is proper. Working over an algebraically closed field and with a linear algebraic group, this is equivalent to the following two statements:

(i)   The orbit PGL $\cdot x$ of any $x = (X \subset \mathbb{P}_k^N) \in H(k)$ is closed in H.
(ii) The stabilizer Stab(x), i.e. the fibre of PGL $\longrightarrow$ PGL $\cdot x$, is finite.

There are various approaches to the properness. One uses a famous theorem of Matsusaka and Mumford [396] and proves the properness, i.e. (i) and (ii), in one go.[8] The argument applies to geometrically non-ruled varieties and, therefore, in particular to K3 surfaces (in arbitrary characteristic!). The Matsusaka–Mumford theorem for those is the following statement: suppose $(f : X \longrightarrow \mathrm{Spec}(R), L)$ and $(f' : X' \longrightarrow \mathrm{Spec}(R), L')$ are two smooth projective families of polarized varieties over a discrete valuation ring R. Then any polarized isomorphism $(X_\eta, L_\eta) \simeq (X'_\eta, L'_\eta)$ over the generic point $\mathrm{Spec}(\eta)$ can be extended to a polarized isomorphism $(X, L) \simeq (X', L')$ over R. This proves that

---

[7] This can be compared with a result of Artin [17, Cor. 6.3] saying that the quotient of an algebraic space by a flat equivalence relation is again an algebraic space. It is not applicable to our situation, as the equivalence relation induced by a group action is often not flat.

[8] I am grateful to Max Lieblich for a discussion of this point.

the moduli functor is separated, which together with the valuative criterion for proper morphisms then proves the properness of the group action; see [616, Lem. 7.6] for the complete argument.[9]

A direct proof for the finiteness of stabilizers can be given rather easily; see Proposition 3.3. Moreover, for complex K3 surfaces $\mathrm{Aut}(X, L) \longrightarrow \mathrm{Aut}(H^2(X, \mathbb{Z}))$ is injective (cf. Proposition **15**.2.1), which allows one to pass to an étale cover of H on which the action becomes free; see Section **6**.4.2.[10] This second approach is closer to the construction of the moduli space via periods.

Remark 2.7 Suppose $(X, L) \longrightarrow S$ is a family of polarized K3 surfaces. The Matsusaka–Mumford result can also be used to show that the automorphism groups $\mathrm{Aut}(X_t, L_t)$ of the fibres (X_t, L_t) form a proper and in fact a finite S-scheme $\mathrm{Aut}(X, L) \longrightarrow S$. In particular, $\{t \in S \mid \mathrm{Aut}(X_t, L_t) \neq \{\mathrm{id}\}\}$ is a proper closed subscheme of S.

Remark 2.8 Moduli spaces of polarized projective varieties have been intensively studied, e.g. by Viehweg in [616]. As shown by Kollár in [314], moduli spaces need not always be quasi-projective even when they can be represented by algebraic spaces.

3 Local Structure

We continue to denote by $H \subset \mathrm{Hilb} = \mathrm{Hilb}_{\mathbb{P}^N}^{P(3t)}$ the open subscheme parametrizing K3 surfaces as in Proposition 2.1. Recall that $P(t) = dt^2 + 2$ and $N = P(3) - 1$.

3.1 As it turns out, the Hilbert scheme parametrizing K3 surfaces is smooth, which later leads to the fact that the moduli space of polarized K3 surfaces is nearly smooth.

Proposition 3.1 *Suppose that the characteristic of k is prime to $6d$. Then the scheme H is smooth of dimension $19 + N^2 + 2N = 18 + (9d + 2)^2$.*

Proof Consider a point $x \in H$ corresponding to an embedded K3 surface $X \subset \mathbb{P}_k^N$. As H is an open subscheme of Hilb, the tangent space $T_x H$ of H at x is naturally isomorphic to $\mathrm{Hom}(\mathcal{I}_X, \mathcal{O}_X)$ and the obstruction space is $\mathrm{Ext}^1(\mathcal{I}_X, \mathcal{O}_X)$. Since X is smooth, the tangent and obstruction spaces can therefore be computed as $H^0(X, \mathcal{N})$ and $H^1(X, \mathcal{N})$,

---

[9] In particular, the argument shows that the group of automorphisms of a polarized geometrically non-ruled smooth projective variety (X, L) is finite. Note that Matsusaka–Mumford theorem really works over $\mathbb{Z}$.

[10] The faithfulness in finite characteristic goes back to Ogus [478, Cor. 2.5]; see [510, Prop. 3.4.2]. More precisely, Ogus shows faithfulness of the action on crystalline cohomology and Rizov uses this to prove faithfulness of the action on étale cohomology $H_{\text{ét}}^2(X, \mathbb{Z}_\ell)$ for $\ell \neq p \neq 2$; cf. Remark **15**.2.2

where $\mathcal{N} := \mathcal{N}_{X/\mathbb{P}^N}$ is the normal bundle of $X \subset \mathbb{P}^N_k$. See the books by Hartshorne, Kollár, and Sernesi [237, 312, 544] for general accounts of deformation theory.

Both spaces can be computed by means of the normal bundle sequence

$$0 \longrightarrow \mathcal{T}_X \longrightarrow \mathcal{T}_{\mathbb{P}^N}|_X \longrightarrow \mathcal{N} \longrightarrow 0,$$

which combined with $H^0(X, \mathcal{T}_X) = H^2(X, \mathcal{T}_X) = 0$ (see Section 1.2.4) leads to

$$0 \longrightarrow H^0(X, \mathcal{T}_{\mathbb{P}^N}|_X) \longrightarrow H^0(X, \mathcal{N}) \longrightarrow H^1(X, \mathcal{T}_X) \longrightarrow H^1(X, \mathcal{T}_{\mathbb{P}^N}|_X) \longrightarrow H^1(X, \mathcal{N}) \longrightarrow 0.$$

From the Euler sequence restricted to X and the vanishing of $H^i(X, \mathcal{O}(1)|_X)$ for $i = 1, 2$ (cf. Proposition 2.3.1) and of $H^1(X, \mathcal{O}_X)$ one then deduces

$$H^1(X, \mathcal{T}_{\mathbb{P}^N}|_X) \simeq H^2(X, \mathcal{O}_X) \simeq k$$

and a short exact sequence

$$0 \longrightarrow k \longrightarrow H^0(X, \mathcal{O}(1)|_X)^{N+1} \longrightarrow H^0(X, \mathcal{T}_{\mathbb{P}^N}|_X) \longrightarrow 0.$$

Thus, if

$$H^1(X, \mathcal{T}_X) \longrightarrow H^1(X, \mathcal{T}_{\mathbb{P}^N}|_X) \tag{3.1}$$

is non-trivial, then the obstruction space $H^1(X, \mathcal{N})$ is trivial and the dimension of the tangent space is obtained by a straightforward computation. To check the non-triviality of (3.1), consider its Serre dual

$$H^1(X, \Omega_{\mathbb{P}^N}|_X) \longrightarrow H^1(X, \Omega_X).$$

The first Chern class $c_1(\mathcal{O}(1)) \in H^1(\mathbb{P}^N, \Omega_{\mathbb{P}^N})$ restricts to the first Chern class $c_1(L^3) \in H^1(X, \Omega_X)$, which is shown to be non-trivial as follows. The image of the intersection number $18d = (L^3)^2 \in \mathbb{Z}$ under $\mathbb{Z} \longrightarrow k$ can be computed as the residue of $c_1^2(L^3) \in H^2(X, \Omega_X^2)$. Hence, if $6d$ or, equivalently, $(L^3)^2$ is prime to the characteristic of k, then $c_1^2(L^3) \neq 0$ and, hence, $c_1(L^3) \neq 0$. $\qquad\square$

Remark 3.2 The above arguments are valid in broad generality and even for the universal construction over $\mathbb{Z}$. One obtains a smooth Hilbert scheme over $\mathbb{Z}[1/(6d)]$; see [7, 510]. This result in particular implies the smoothness of the moduli space as a Deligne–Mumford stack; see Remark 4.11. However, by a more direct argument avoiding the Hilbert scheme, smoothness can be shown over $\mathbb{Z}[1/(2d)]$.

3.2 The finiteness of the automorphism group of polarized K3 surfaces, to be proven next, is subsequently used to show that the moduli space is étale locally the quotient of a smooth scheme by the action of a finite group.

Proposition 3.3 *Let X be a K3 surface over a field k and L an ample line bundle on X. Then the group of automorphisms $f: X \xrightarrow{\sim} X$ (over k) with $f^*L \simeq L$ is finite.*

Proof We freely use that $H^0(X, \mathcal{T}_X) = 0$, which is easy for $k = \mathbb{C}$ and substantially more difficult for a field of positive characteristic; see Sections **1**.2.4 and Sections **9**.5.1.

Let $(L)^2 = 2d$ and $P(t) := dt^2 + 2$. The graph of an automorphism $f: X \xrightarrow{\sim} X$ is a closed subscheme $\Gamma_f \subset X \times X$ and thus corresponds to a k-rational point of the Hilbert scheme $\mathrm{Hilb}_{X \times X}$ of closed subschemes of $X \times X$. Clearly, f is uniquely determined by its graph.

If $f^*L \simeq L$, then the Hilbert polynomial of Γ_f with respect to the ample line bundle $L \boxtimes L$ on $X \times X$ is given by $P(2t)$. Indeed,

$$\chi(\Gamma_f, (L \boxtimes L)^n|_{\Gamma_f}) = \chi(X, (L \otimes f^*L)^n) = \chi(X, L^{2n}) = P(2n).$$

Thus, the graph of f defines a k-rational point of $\mathrm{Hilb}_{X \times X, L \boxtimes L}^{P(2t)}$, which by general results due to Grothendieck is a projective scheme; cf. [225, 176].

The tangent space of a k-rational point of $\mathrm{Hilb}_{X \times X}$ corresponding to a closed subscheme $Z \subset X \times X$ is naturally isomorphic to the k-vector space $\mathrm{Hom}(\mathcal{I}_Z, \mathcal{O}_Z)$. If Z is a smooth subscheme, then $\mathrm{Hom}(\mathcal{I}_Z, \mathcal{O}_Z) \simeq H^0(Z, \mathcal{N})$, where $\mathcal{N} := \mathcal{N}_{Z/X \times X}$ denotes the normal bundle of Z in $X \times X$. Use the normal bundle sequence

$$0 \longrightarrow \mathcal{T}_Z \longrightarrow \mathcal{T}_{X \times X}|_Z \longrightarrow \mathcal{N} \longrightarrow 0$$

to see that $\mathcal{N} \simeq \mathcal{T}_X$ for a graph $Z = \Gamma_f$. Indeed, identifying X with the graph of f via $(\mathrm{id}, f): X \xrightarrow{\sim} \Gamma_f \subset X \times X$ and writing $\mathcal{T}_{X \times X}|_{\Gamma_f} \simeq \mathcal{T}_X \oplus f^*\mathcal{T}_X \simeq \mathcal{T}_X \oplus \mathcal{T}_X$ yields the isomorphism. The embedding $\mathcal{T}_Z \longrightarrow \mathcal{T}_{X \times X}|_Z$ is given by $v \longmapsto v \oplus df(v)$, which can be split by $(v, w) \longmapsto v$.

Hence, the tangent space of $\mathrm{Hilb}_{X \times X}$ at the point $[\Gamma_f]$ is naturally isomorphic to $H^0(X, \mathcal{T}_X)$, which is trivial. Thus, the k-rational points of $\mathrm{Hilb}_{X \times X}$ corresponding to the graph of automorphisms of (X, L) are (reduced) isolated points of $\mathrm{Hilb}_{X \times X, L \boxtimes L}^{P(2t)}$. Since a projective scheme can have only finitely many irreducible components, the set of those k-rational points that correspond to $[\Gamma_f]$ of automorphisms with $f^*L \simeq L$ is finite. $\quad\square$

Remark 3.4 For later use note that the proof in fact shows that $\mathrm{Isom}((X, L), (X', L'))$ for two polarized K3 surfaces over a field k is a finite set of reduced points. Both properties are needed to show that the moduli space of polarized K3 surfaces is a Deligne–Mumford stack; see Proposition 4.10.

Corollary 3.5 *Any automorphism $f: X \xrightarrow{\sim} X$ of a K3 surface X with $f^*L \simeq L$ for some ample line bundle is of finite order.* $\quad\square$

3.3 The local description of the Hilbert scheme and the finiteness of the group of automorphisms leads to the following result on the local structure of the coarse moduli space. As it turns out, M_d is no longer smooth but not far from it either. For simplicity, we state the result for the case of a quasi-projective moduli space over $\mathbb{C}$, for which it can also be deduced from the period description explained in Section **6**.4.1.

Corollary 3.6 *Let M_d be the coarse moduli space of polarized complex K3 surfaces of degree $2d$. Then étale locally M_d is the quotient of a smooth scheme by a finite group.*

Proof This is an immediate consequence of Luna's étale slice theorem, which more generally asserts the following: if a reductive group G acts on a variety Y over k such that a good quotient $Y \longrightarrow Y/G$ exists, then through any point $x \in Y$ with closed orbit $G \cdot x$ there exists a locally closed Stab(x)-invariant subscheme S (the *slice* through x) such that $S \times^{\mathrm{Stab}(x)} G \longrightarrow Y$ and $S/\mathrm{Stab}(x) \longrightarrow Y/G$ are étale. Moreover, if Y is smooth, then S can be chosen smooth as well. See [328] or [444].

In our case, $Y = H$ and $G = \mathrm{PGL}$. The quotient exists by Theorem 2.4 and all orbits are closed due to the properness of the action; see Section 2.3. The stabilizer Stab(x) of a point $x \in H$ corresponding to some polarized K3 surface $X \subset \mathbb{P}^N$, $L^3 \simeq \mathcal{O}(1)|_X$, is isomorphic to $\mathrm{Aut}(X, L)$, which is finite by Proposition 3.3. $\square$

A version of Luna's étale slice theorem valid in positive characteristic has been proved in [32] and in fact the corollary remains valid in positive characteristic. This is the statement that the moduli functor $\mathcal{M}_d$ is a smooth Deligne–Mumford stack, which shall be explained next.

4 As Deligne–Mumford Stack

For many purposes it is enough to know that the moduli space of polarized K3 surfaces exists as a Deligne–Mumford stack. In fact, it is even preferable to view the moduli space as a stack, for the stack keeps track of the automorphism groups of the K3 surfaces. As a consequence, for example, the moduli stack is smooth but the coarse moduli space is not.

4.1 In the introduction we have exhibit $\mathcal{M}_d$ also as a groupoid over (Sch/S), more precisely as a category over (Sch/S) fibred in groupoids (CFG over S). For the definition of a CFG, see the introduction by Deligne and Mumford in [146, Sec. 4] or [1, Tag 04SE]. The conditions are easily verified in our situation.

Recall that a CFG $\mathcal{N}$ over (Sch/S) is *representable* if there exists an S-scheme U such that $\underline{U} \simeq \mathcal{N}$. Here, $\underline{U}$ is the CFG with the set $\mathrm{Mor}_S(T, U)$ as the groupoid over T.

Let $(X_1 \longrightarrow T, L_1), (X_2 \longrightarrow T, L_2) \in \mathcal{M}_d$ and define

$$\mathrm{Isom}_T((X_1, L_1), (X_2, L_2)) \colon (Sch/T)^o \longrightarrow (Sets)$$

as the functor that maps $T' \longrightarrow T$ to the set of isomorphisms $\psi \colon X_{1T'} \xrightarrow{\sim} X_{2T'}$ over T' with $\psi^* L_{2T'} \simeq L_{1T'}$ up to tensoring with the pull-back of a line bundle on T'. (The isomorphism between the line bundles is not part of the datum.)

Proposition 4.1 *The functor*

$$\mathrm{Isom}_T((X_1, L_1), (X_2, L_2)) \colon (Sch/T)^o \longrightarrow (Sets)$$

is a sheaf in the étale topology.

Proof In fact, and this is what is needed later, the functor is representable and thus in particular a sheaf. This follows from the representability of the Hilbert scheme by embedding $\mathrm{Isom}_T(X_1, X_2)$ into $\mathrm{Hilb}_{X_1 \times_T X_2}$ as an open subscheme. Considering only isomorphisms that respect the polarizations ensures that the image is contained in the part of the Hilbert scheme for which the Hilbert polynomial with respect to the product ample line bundle $L_1 \boxtimes L_2$ equals $\chi(L_1^{2n}) = 4dn^2 + 2$. More precisely, using properness of the relative Picard scheme of the universal family $\mathcal{Z}$ over Hilb, one finds that $\mathrm{Isom}((X_1, L_1), (X_2, L_2))$ is a locally closed subscheme of $\mathrm{Isom}(X_1, X_2)$; cf. the proof of Proposition 2.1. □

Proposition 4.2 *Every descent datum in $\mathcal{M}_d$ is effective.*

Proof We have to show the following. Suppose $T' \longrightarrow T$ is an étale (or just fpqc) covering in (Sch/S). Denote the natural projections by

$$p_i \colon T'' := T' \times_T T' \longrightarrow T' \quad \text{and} \quad p_{ij} \colon T' \times_T T' \times_T T' \longrightarrow T''.$$

If for $(f' \colon X' \longrightarrow T', L') \in \mathcal{M}_d(T')$ an isomorphism $\varphi \colon p_1^*(X', L') \xrightarrow{\sim} p_2^*(X', L')$ satisfies the cocycle condition $p_{23}^* \varphi \circ p_{12}^* \varphi = p_{13}^* \varphi$, then there exist $(X, L) \in \mathcal{M}_d(T)$ and an isomorphism $\lambda \colon (X, L)_{T'} \xrightarrow{\sim} (X', L')$ inducing φ. Moreover, (X, L) and λ are unique up to canonical isomorphism.

The idea of the proof is to use effective descent for quasi-coherent sheaves and morphisms of quasi-coherent sheaves. Indeed, by assumption X' is isomorphic to the relative $\mathrm{Proj}(\mathcal{S}')$, where $\mathcal{S}'$ is the quasi-coherent graded sheaf of algebras $\bigoplus f_*'(L'^k)$. The descent datum given by φ translates immediately into a descent datum for $\mathcal{S}'$. Note that the algebra structure is encoded by morphisms between quasi-coherent sheaves. Effective descent for quasi-coherent sheaves then yields a quasi-coherent graded sheaf of algebras $\mathcal{S}$ on T and X is defined as its relative Proj. The argument does not use any particular properties of K3 surfaces and so the result holds true in broad generality. See e.g. [55] for details on descent theory.[11] □

Corollary 4.3 *The groupoid $\mathcal{M}_d$ of primitively polarized K3 surfaces is a stack.*

---

[11] As before, in the proof we have assumed L' to be an actual line bundle, although it may exist only étale locally. This has no effect on the descent of $\mathrm{Proj}(\mathcal{S}')$, but $\mathcal{S}$ and L may exist only étale locally.

Proof By definition, a CFG $\mathcal{M}$ is a stack if Propositions 4.1 and 4.2 hold.   □

Example 4.4 Clearly, the CFG $\underline{U}$ associated with an S-scheme U is a stack. The other source for examples is group quotients. If H is a scheme with a group scheme G acting on H (everything over S), then $[H/G]$ is the CFG with sections over T consisting of all principal G-bundles P over T together with a G-equivariant morphism $P \longrightarrow H$. Morphisms in $[H/G]$ are pull-back diagrams. Then $[H/G]$ is a stack.

Example 4.5 The most important example of a quotient stack in the present context is the one given by the action of PGL on the open subscheme H of the Hilbert scheme Hilb studied in the previous sections. There exists a natural isomorphism of stacks

$$[H/\text{PGL}] \overset{\sim}{\longrightarrow} \mathcal{M}_d.$$

All the main ideas for the construction of this isomorphism have been explained already. Consider a section of $[H/\text{PGL}]$ over T given by a principal PGL-bundle $P \longrightarrow T$ and a G-equivariant morphism $P \longrightarrow H$. The latter is given by a polarized K3 surface $X \longrightarrow P$ together with an embedding $X \subset \mathbb{P}_P^N$. The PGL-action produces a descent datum, and effective descent for $\mathcal{M}_d$ (see Proposition 4.2) yields a section of $\mathcal{M}_d$ over T.

To show that this yields an isomorphism of stacks start with $(f \colon X \longrightarrow T, L)$ in $\mathcal{M}_d(T)$ and consider the locally free sheaf $f_*(L^3)$ on T. Each choice of a basis in the fibre of $f_*(L^3)$ yields an embedding of the fibre of X into $\mathbb{P}^N$. Thus, the associated PGL-bundle (of frames in the fibres of $f_*(L^3)$) comes with a natural morphism $P \longrightarrow H$. The verification that the functor is fully faithful is straighforward.

4.2 It turns out that $\mathcal{M}_d$ is much more than just a stack; it is a Deligne–Mumford stack. We shall need to find an étale or at least a smooth atlas for it.

Remark 4.6 Recall that a morphism of CFG $\mathcal{M} \longrightarrow \mathcal{N}$ is *representable* if for any $\underline{U} \longrightarrow \mathcal{N}$ the fibre product $\underline{U} \times_{\mathcal{N}} \mathcal{M}$ is representable. By [146, Prop. 4.4] (see also [55]) the diagonal morphism

$$\Delta \colon \mathcal{M} \longrightarrow \mathcal{M} \times_{(Sch/S)} \mathcal{M}$$

of a stack $\mathcal{M}$ is representable if and only if for all $\underline{T} \longrightarrow \mathcal{M} \longleftarrow \underline{T'}$ the fibre product $\underline{T} \times_{\mathcal{M}} \underline{T'}$ is representable. In fact, it is enough to consider the case $T = T'$.

Thus, the diagonal of the stack $\mathcal{M}_d$ is representable if for all $(X_1, L_1), (X_2, L_2) \in \mathcal{M}_d(T)$ the sheaf $\text{Isom}_T((X_1, L_1), (X_2, L_2))$ is representable. This we have noted already in the proof of Proposition 4.1. Hence, in our situation the diagonal is representable.

As usual, a morphism of CFG $\mathcal{M} \longrightarrow \mathcal{N}$ is said to have a certain scheme theoretic property (e.g. quasi-compact, separated, étale, etc.) if it is representable and if for every $\underline{U} \longrightarrow \mathcal{N}$ the morphism of schemes representing $\underline{U} \times_{\mathcal{N}} \mathcal{M} \longrightarrow \underline{U}$ has this property.

Definition 4.7 A stack $\mathcal{M}$ over (Sch/S) is called a *Deligne–Mumford stack* if in addition the following two conditions are satisfied:

(i) The stack $\mathcal{M}$ is quasi-separated, i.e. diagonal morphism $\Delta\colon \mathcal{M} \longrightarrow \mathcal{M} \times_{(Sch/S)} \mathcal{M}$ is representable, quasi-compact, and separated.[12]

(ii) There exists a scheme U and an étale surjective morphism $\underline{U} \longrightarrow \mathcal{M}$ (over S).

Remark 4.8 In our geometric situation and, in particular, for the construction of the moduli space of K3 surfaces, one could try to rigidify the situation by introducing additional structures (level structures); e.g. one would consider every K3 surface together with all isomorphisms $H^2(X,\mathbb{Z}/\ell\mathbb{Z}) \simeq \Lambda/\ell\Lambda$. This produces a finite étale covering $\widetilde{H} \longrightarrow H$ with Galois group, say Γ, and such that the PGL-action on H lifts naturally to a free action on $\widetilde{H}$. Then the existence of $\widetilde{H}/\mathrm{PGL}$ as a scheme is easier and this quotient can be used as an étale cover of $\mathcal{M}_d$ as required in **(ii)**. Another advantage of $\widetilde{H}/\mathrm{PGL}$ over H/PGL is the existence of a universal family. See Section 6.4.2 for more details on this approach (in the complex setting).

However, as an alternative to the approach sketched in the last remark one can use the following result; see [146, Thm. 4.21] and also [162] or [55, Ch. 5].

Theorem 4.9 *Let $\mathcal{M}$ be a quasi-separated stack over a Noetherian scheme S. Then $\mathcal{M}$ is a Deligne–Mumford stack if*

(iii) *The diagonal $\Delta\colon \mathcal{M} \longrightarrow \mathcal{M} \times_{(Sch/S)} \mathcal{M}$ is unramified and*

(iv) *There exists a scheme U of finite type over S and a smooth surjective morphism $\underline{U} \longrightarrow \mathcal{M}$ (over S).*[13]

This can be used to prove Theorem 1.3, which we state again as the following.

Proposition 4.10 *The stack $\mathcal{M}_d$ of primitively polarized K3 surfaces of degree $2d$ over a Noetherian base S is a Deligne–Mumford stack.*

Proof (i) By Remark 4.6, the diagonal of $\mathcal{M}_d$ is representable. We show that it is actually finite and hence quasi-compact and separated. Consider two polarized K3 surfaces $(X_1, L_1), (X_2, L_2) \in \mathcal{M}_d(\mathrm{Spec}(R))$ over a discrete valuation ring R. Then by the theorem of Matsusaka–Mumford (see Section 2.3), any isomorphism over the generic point of $\mathrm{Spec}(R)$ extends uniquely to an isomorphism over R. Thus, by the valuative criterion $\mathrm{Isom}_T((X_1, L_1), (X_2, L_2))$ is proper over T. Together with the finiteness of

---

[12] Often, the separatedness of the diagonal is added as an additional condition and not seen as part of the definition.

[13] In a certain sense, the existence of U in **(iv)** is an analogue (less precise) of Luna's étale slice theorem mentioned in the proof of Corollary 3.6. The existence of U as in **(iv)** suggests taking étale sections to produce the étale covering in **(ii)**. Note however, that in general one cannot find an étale slice through every point of a smooth atlas $\underline{U} \longrightarrow \mathcal{M}$; see [55, Ex. 5.7]. Also note that if U in **(iv)** is smooth, then also the étale atlas in **(ii)** can be chosen smooth.

Aut(X, L) of a polarized K3 surface over a field (see Proposition 3.3), this proves the finiteness of the diagonal.

(iii) Recall that a morphism, locally of finite type, is unramified if all geometric fibres are discrete and reduced; see [226, Ch. 17]. Thus, it suffices to show that for two families $(X, L) \longrightarrow T \longleftarrow (X', L')$ the fibres of $\text{Isom}_T((X, L), (X', L')) \longrightarrow T$ over geometric points consist of reduced isolated points. However, the fibre over a geometric point $t \in T$ is $\text{Isom}_{k(t)}((X_t, L_t), (X'_t, L'_t))$, which has this property; see Remark 3.4.

(iv) We use the PGL-action on the (open) subscheme $H \subset \text{Hilb}$. As was mentioned in Example 4.5, $[H/\text{PGL}] \overset{\sim}{\longrightarrow} \mathcal{M}_d$. Thus, one has to show that $\underline{H} \longrightarrow [H/\text{PGL}]$ is (formally) smooth. So, consider $(X, L) \in \mathcal{M}_d(\text{Spec}(A))$ and an ideal $I \subset A$ with $I^2 = 0$. Let $A_0 := A/I$ and suppose that (X, L) lifts to $\underline{H}$ over $\text{Spec}(A_0)$. Thus, there exists a principal PGL-bundle $P \longrightarrow \text{Spec}(A)$ together with an equivariant morphism $P \longrightarrow H$. The restriction to $\text{Spec}(A_0) \subset \text{Spec}(A)$ yields a principal bundle $P_0 \longrightarrow \text{Spec}(A_0)$. The existence of the lift over $\text{Spec}(A_0)$ to $\underline{H}$ implies the existence of a morphism $\text{Spec}(A_0) \longrightarrow H$ which via pull-back yields $P_0 \longrightarrow \text{Spec}(A_0)$. In particular, the latter is a trivial PGL-bundle. In other words, one has $\text{Spec}(A_0) \longrightarrow P_0 \subset P \longrightarrow H$. In order to show that the projection $\underline{H} \longrightarrow [H/\text{PGL}]$ is formally smooth, one needs to extend the composition $\text{Spec}(A_0) \longrightarrow H$ to a morphism $\text{Spec}(A) \longrightarrow H$. But this can be obtained by simply passing to the closure of $\text{Spec}(A_0)$ in P and by composing with $P \longrightarrow H$. Equivalently, if $P_0 \longrightarrow \text{Spec}(A_0)$ is trivial, then so is $P \longrightarrow \text{Spec}(A)$.

Underlying the above arguments is the following general observation: if a smooth group scheme (over S) acts on H (of finite type over S) with finite and reduced stabilizers, then the quotient $\underline{H}/\underline{G}$ is a Deligne–Mumford stack.[14]   □

Remark 4.11 Using Section 3, one finds that $\mathcal{M}_d$ over $\mathbb{Z}$ is smooth over $\mathbb{Z}[1/(2d)]$ (see [510] or the footnote to Theorem 4.9). In particular, over a field of characteristic zero all $\mathcal{M}_d$ are smooth Deligne–Mumford stacks. Note however that the coarse moduli space M_d is singular, due to the existence of K3 surfaces with non-trivial automorphisms.

References and Further Reading

As pointed out to me by Chenyang Xu, the construction of the moduli space of polarized complex K3 surfaces (X, L) can also be based on Donaldson's result in [158], which states that for sufficiently high n the pair (X, L^n) defines a Chow stable point. Details of this construction and in particular a comparison of the various ample line bundles on the moduli space have not been addressed in the literature.

---

[14] The properness, e.g. in Theorem 2.6, is needed to ensure that the geometric points of the quotient parametrize orbits. A priori this is not an issue for the stack, but it becomes one when one wants to pass to its coarse moduli space. In fact, in [289] Keel and Mori also show that any separated Deligne–Mumford stack of finite type has a coarse moduli space in the category of algebraic spaces. So, a fortiori, the assumption that the stabilizers are finite and reduced implies the properness of the action.

For questions related to compactification of the moduli space, of K3 surfaces, see the papers by Friedman [183], Olsson [481], and Scattone [526].

The geometry of the moduli spaces of K3 surfaces, for example their Kodaira dimensions, has recently been studied intensively; see the original article of Gritsenko et al. [223] or Voisin's survey [623]. For high degree they tend to be of general type. In contrast, moduli spaces of K3 surfaces of small degree may be unirational (more concretely for $d \leq 12$ and $d = 17, 19$); see Mukai's papers, e.g. [435] or [623]. Ultimately, this is related to the fact that general K3 surfaces of small degree can be described as complete intersections in Fano varieties; cf. Section 1.4.3. In Kirwan's article [294] one finds among other things a computation of the intersection cohomology of the moduli space of stable quartics.

In [206] van der Geer and Katsura show that the maximal dimension of a complete subvariety of the 19-dimensional moduli space M_d in characteristic zero is at most 17 (it equals 17 in the moduli space $\overline{M}_d$ of quasi-polarized K3 surfaces). The paper also proves some cycle class relations. See also [154], [201], [205], [436] for other recent topics related to moduli spaces of K3 surfaces.

Instead of fixing a polarization, which can be thought of as a primitive sublattice of the Picard group generated by an ample line bundle, it is interesting to study moduli spaces of K3 surfaces with a fixed lattice of higher rank primitively contained in the Picard group. Moduli spaces of lattice polarized K3 surfaces have been introduced by Dolgachev in [150] in the context of mirror symmetry and using period domains as in Chapter 6. For a brief discussion of the algebraic approach, see Beauville [48].

Most Deligne–Mumford stacks are in fact quotient stacks; see Kresch's article [330] for the precise statement and further references.

Questions and Open Problems

As mentioned in the text, it is more difficult to prove the quasi-projectivity of the coarse moduli space of polarized K3 surfaces in positive (or mixed) characteristic. This has been achieved by Maulik and Madapusi Pera in [387, 398]; see also Benoist's Bourbaki survey [58]. Later we shall see that in characteristic zero the moduli space of polarized K3 surfaces of fixed degree is connected and in fact irreducible. Again, this is much harder in positive characteristic, but has been proved recently in [387] for the case that $p^2 \nmid d$. It would be interesting to have proofs of both statements that do not rely on the Kuga–Satake construction.

6

Periods

Hodge structures (of complex K3 surfaces) are parametrized by period domains. The first section recalls three descriptions of the period domain of Hodge structures of K3 type: as an open subset of a smooth quadric, in terms of positive oriented planes, and as a tube domain. In Section 2 we review the basic deformation theory relevant for our purposes and introduce the local period map associated with any local family of K3 surfaces. The local Torelli theorem 2.8, a key result for K3 surfaces but valid for a much broader class of varieties, is explained. In Section 3 we state two cornerstone results in the theory of complex K3 surfaces: the surjectivity of the period map, Theorem 3.1, and the global Torelli theorem 3.4. Their proofs, however, are postponed to Chapter 7. The last section shows how these results can be used to give an alternative construction of the moduli space of polarized complex K3 surfaces, which allows one to derive global information. We also return to the Kuga–Satake construction and the Mumford–Tate group and discuss how they behave in families. The Appendix summarizes results concerning Kulikov models for degenerations of K3 surfaces.

1 Period Domains

In the following, Λ is a non-degenerate lattice with its bilinear form $(\ .\ )$. For its signature (n_+, n_-) we assume $n_+ \geq 2$. In fact, only the three cases $n_+ = 2, 3, 4$ are of importance for us and often Λ will be the K3 lattice $E_8(-1)^{\oplus 2} \oplus U^{\oplus 3}$ or the orthogonal complement $\Lambda_d := \ell^\perp$ of a primitive vector $\ell \in E_8(-1)^{\oplus 2} \oplus U^{\oplus 3}$ of positive square $(\ell)^2 = 2d$. Note that $\Lambda_d \simeq E_8(-1)^{\oplus 2} \oplus U^{\oplus 2} \oplus \mathbb{Z}(-2d)$; see Example **14**.1.11. An example with $n_+ = 4$ is provided by the full cohomology $H^*(X, \mathbb{Z})$ of a complex K3 surface X. For the necessary lattice theory, in particular all the notations, we refer to Chapter 14.

1.1   Consider the associated complex vector space $\Lambda_{\mathbb{C}} := \Lambda \otimes_{\mathbb{Z}} \mathbb{C}$ endowed with the $\mathbb{C}$-linear extension of (.) which corresponds to a homogenous quadratic polynomial. Its zero locus in $\mathbb{P}(\Lambda_{\mathbb{C}})$ is a quadric which is smooth due to the assumption that (.) is non-degenerate. Consider the open (in the classical topology) subset of this quadric

$$D := \{x \in \mathbb{P}(\Lambda_{\mathbb{C}}) \mid (x)^2 = 0, \ (x.\overline{x}) > 0\} \subset \mathbb{P}(\Lambda_{\mathbb{C}}),$$

to which we refer as the *period domain* associated with Λ and which is considered as a complex manifold. Note that the second condition really is well posed, as $(\lambda x.\overline{\lambda x}) = (\lambda \overline{\lambda})(x.\overline{x})$ and $\lambda \overline{\lambda} \in \mathbb{R}_{>0}$ for all $\lambda \in \mathbb{C}^*$. Also note that D itself depends only on the real vector space $\Lambda_{\mathbb{R}}$ together with the real linear extension of (.).

Remark 1.1   Denote by $\ell_x \subset \Lambda_{\mathbb{C}}$ the line corresponding to a point $x \in D$. Then for the tangent space of D at x there exists a natural isomorphism

$$T_x D \simeq \mathrm{Hom}(\ell_x, \ell_x^{\perp}/\ell_x).$$

Indeed, $T_x\mathbb{P}(\Lambda_{\mathbb{C}}) \simeq \mathrm{Hom}(\ell_x, \Lambda_{\mathbb{C}}/\ell_x)$ and writing out the infinitesimal version of $(x)^2 = 0$ shows that $T_x D \subset T_x\mathbb{P}(\Lambda_{\mathbb{C}})$ consists of all linear maps $\ell_x \longrightarrow \Lambda_{\mathbb{C}}/\ell_x$ with image orthogonal to ℓ_x.

Proposition 1.2   *There exists a natural bijection between D and the set of Hodge structures of K3 type on Λ such that for all non-zero $(2,0)$-classes σ:*

(i) $(\sigma)^2 = 0.$
(ii) $(\sigma.\overline{\sigma}) > 0.$
(iii) $\Lambda^{1,1} \perp \sigma.$

Proof   The $(2,0)$-part of any Hodge structure of K3 type on Λ defines a line in $\Lambda_{\mathbb{C}}$. If the Hodge structure satisfies (i) and (ii), then the line defines a point in D. Conversely, if a point x in D is given, then there exists a Hodge structure with ℓ_x as its $(2,0)$-part satisfying (i) and (ii). Adding condition (iii) makes it unique. Indeed, $\Lambda^{1,1}$ is the complexification of the real vector subspace of $\Lambda_{\mathbb{R}}$ defined as the orthogonal complement of the plane spanned by $\mathrm{Re}(\sigma)$ and $\mathrm{Im}(\sigma)$. (The plane is non-degenerate and in fact positive definite; see below for a related discussion.)   □

Note that in the above proposition, (.) does not necessarily polarize the Hodge structure (see Definition 3.1.6) as we do not assume that it is definite on $\Lambda^{1,1} \cap \Lambda_{\mathbb{R}}$.

Example 1.3   (i) If X is a complex K3 surface, then the natural Hodge structure on $\Lambda = H^2(X, \mathbb{Z})$ is of the above type.

(ii) Suppose $n_+ > 2$ and fix $\ell \in \Lambda$ with $(\ell)^2 > 0$. Then $\ell^{\perp} \subset \Lambda$ induces a linear embedding $\mathbb{P}(\ell_{\mathbb{C}}^{\perp}) \subset \mathbb{P}(\Lambda_{\mathbb{C}})$ and the period domain associated with $\ell^{\perp}$ is obtained as the intersection of the period domain $D \subset \mathbb{P}(\Lambda_{\mathbb{C}})$ with the hyperplane $\mathbb{P}(\ell_{\mathbb{C}}^{\perp})$.

(iii) As a special case of (ii), consider the K3 lattice $\Lambda = E_8(-1)^{\oplus 2} \oplus U^{\oplus 3}$ and let $\ell = e_1 + df_1$, where e_1, f_1 is the standard basis of the first copy of U. Let $\Lambda_d := \ell^\perp$ and

$$D_d \subset \mathbb{P}(\Lambda_{d\mathbb{C}}) \subset \mathbb{P}(\Lambda_{\mathbb{C}})$$

be the associated period domain. In fact, any primitive $\ell \in \Lambda$ with $(\ell)^2 = 2d$ is of this form after applying a suitable orthogonal transformation of Λ; see Corollary **14**.1.10.

Remark 1.4 The period domain introduced above is a special case of Griffiths's period domains parametrizing (polarized) Hodge structures of arbitrary weight. See [220, 103, 167, 621].

1.2 The period domain D associated with a lattice Λ as above has two other realizations, as a Grassmannian and as a tube domain; see e.g. [525, App. Sec. 6].

Let us first consider the real vector space $\Lambda_{\mathbb{R}}$ with the $\mathbb{R}$-linear extension of $(\,.\,)$. The Grassmannian $\mathrm{Gr}(2, \Lambda_{\mathbb{R}})$ of planes in $\Lambda_{\mathbb{R}}$ is a real manifold of dimension $2(\dim \Lambda_{\mathbb{R}} - 2) = 2(n_+ + n_- - 2)$. Let $\mathrm{Gr}^{\mathrm{p}}(2, \Lambda_{\mathbb{R}}) \subset \mathrm{Gr}(2, \Lambda_{\mathbb{R}})$ be the open set of all planes $P \subset \Lambda_{\mathbb{R}}$ for which the restriction $(\,.\,)|_P$ is positive definite and let $\mathrm{Gr}^{\mathrm{po}}(2, \Lambda_{\mathbb{R}})$ be the manifold of all such positive planes together with the choice of an orientation. Thus, $\mathrm{Gr}^{\mathrm{po}}(2, \Lambda_{\mathbb{R}})$ can be realized as a natural covering of degree two

$$\mathrm{Gr}^{\mathrm{po}}(2, \Lambda_{\mathbb{R}}) \longrightarrow\!\!\!\!\!\rightarrow \mathrm{Gr}^{\mathrm{p}}(2, \Lambda_{\mathbb{R}}). \tag{1.1}$$

Proposition 1.5 *There exist diffeomorphisms*

$$D \xrightarrow{\sim} \mathrm{Gr}^{\mathrm{po}}(2, \Lambda_{\mathbb{R}}) \xrightarrow{\sim} O(n_+, n_-)/SO(2) \times O(n_+ - 2, n_-).$$

Proof Here, $O(n_+, n_-)$ denotes the orthogonal group of $\mathbb{R}^{n_+ + n_-}$ endowed with the diagonal quadratic form $\mathrm{diag}(1, \ldots, 1, -1, \ldots, -1)$ of signature (n_+, n_-). By choosing an identification of $\Lambda_{\mathbb{R}}$ with $\mathbb{R}^{n_+ + n_-}$, one obtains a natural transitive action of $O(n_+, n_-)$ on $\mathrm{Gr}^{\mathrm{po}}(2, \Lambda_{\mathbb{R}})$. The stabilizer of the plane spanned by the first two unit vectors (with the natural orientation) v_1, v_2 is the subgroup $SO(2) \times O(\langle v_1, v_2 \rangle^\perp) \simeq SO(2) \times O(n_+ - 2, n_-)$.

For the first diffeomorphism consider the map

$$D \longrightarrow \mathrm{Gr}^{\mathrm{po}}(2, \Lambda_{\mathbb{R}}), \quad x \longmapsto \mathbb{R} \cdot \mathrm{Re}(x) \oplus \mathbb{R} \cdot \mathrm{Im}(x).$$

Note that $(x)^2 = 0$ and $(x.\bar{x}) > 0$ imply that $e_1 := \mathrm{Re}(x)$ and $e_2 := \mathrm{Im}(x)$ are orthogonal to each other and $(e_1)^2 = (e_2)^2 > 0$. The map is well defined as λx with $\lambda \in \mathbb{C}^*$ defines the same oriented plane. Conversely, an oriented positive plane P with a chosen oriented orthonormal basis e_1, e_2 can be mapped to $x = e_1 + i e_2$. $\qquad\square$

Remark 1.6 Using the description of D in terms of positive planes, it is not difficult to see that D is connected for $n_+ > 2$ and that it has two connected components for

$n_+ = 2$. In the second case, the orientation of two positive planes can be compared via orthogonal projections. In the description of D as a subset of $\mathbb{P}(\Lambda_{\mathbb{C}})$ the two components in the decomposition

$$D = D^+ \sqcup D^-$$

can be interchanged by complex conjugation $x \longmapsto \bar{x}$. Equivalently, for $n_+ = 2$ the covering (1.1) is trivial; i.e. in this case $\mathrm{Gr}^{\mathrm{po}}(2, \Lambda_{\mathbb{R}})$ consists of two disjoint copies of the target.

For the tube domain realization choose an orthogonal decomposition $\Lambda_{\mathbb{R}} = U_{\mathbb{R}} \oplus W$ for which we have to assume $n_- > 0$. (Shortly we also assume $n_+ = 2$.) Consider the coordinates on $U_{\mathbb{R}}$ corresponding to the standard basis e, f with $(e)^2 = (f)^2 = 0$ and $(e.f) = 1$. A point $x \in \mathbb{P}(\Lambda_{\mathbb{C}})$ corresponding to $\alpha e + \beta f + z \in U_{\mathbb{C}} \oplus W_{\mathbb{C}}$ shall be denoted $[\alpha : \beta : z]$ and the associated *tube domain* is defined as

$$\mathcal{H} := \{z \in W_{\mathbb{C}} \mid (\mathrm{Im}(z))^2 > 0\}$$

and comes with the structure of a complex manifold in the obvious way.

Proposition 1.7 *Assume $n_+ = 2$. Then the map $z \longmapsto [1 : -(z)^2 : \sqrt{2}z]$ defines a biholomorphic map*

$$\mathcal{H} \xrightarrow{\sim} D.$$

Proof Let us verify that the map takes values in D. For $x = (1, -(z)^2, \sqrt{2}z)$ one finds $(x)^2 = (e - (z)^2 f.e - (z)^2 f) + 2(z)^2 = 0$ and $(x.\bar{x}) = (e - (z)^2 f.e - \overline{(z)}^2 f) + 2(z.\bar{z}) = -2\mathrm{Re}(z)^2 + 2(z.\bar{z}) = 4(\mathrm{Im}(z))^2 > 0$ for $z \in \mathcal{H}$.

For the inverse map consider $x = [\alpha : \beta : \gamma] \in D$. Suppose $\alpha = 0$. Then (β, γ) corresponds to a positive plane in $\mathbb{R} \cdot f \oplus W$, but the latter has only one positive direction. Hence, $\alpha \neq 0$ and we may thus assume $\alpha = 1$. But then the quadratic equation for D implies $2\beta + (\gamma)^2 = 0$ and the inequality $(x.\bar{x}) > 0$ yields $(\mathrm{Im}(\gamma))^2 > 0$. $\qquad \square$

Example 1.8 For $n_+ = 2$ and $n_- = 1$ one finds the following familiar picture. In this case we may assume $W_{\mathbb{C}} = \mathbb{C}$ with the standard quadratic form and then

$$\mathcal{H} = \mathbb{H} \sqcup (-\mathbb{H}).$$

The above biholomorphic map is then simply $\mathbb{H} \xrightarrow{\sim} D^+, z \longmapsto [1 : -z^2 : \sqrt{2}z]$.

The isomorphism becomes more interesting when the two sides are considered with their natural actions of $\mathrm{SL}(2, \mathbb{Z})$ and $O(\Lambda)$, respectively; see below.

Recall that $\mathbb{H}$ is biholomorphic to the unit disk in $\mathbb{C}$. In the same vein, the period domain D^+ for $n_+ = 2$ is always a bounded symmetric domain (of type IV).

1.3   The period domain D associated with the lattice Λ comes with a natural action of the discrete group $O(\Lambda)$. The action is well behaved only for $n_+ = 2$. More precisely, the group $O(\Lambda)$ is not expected to act properly discontinuously on D for $n_+ > 2$. In particular, the quotient $O(\Lambda)\backslash D$ is not expected to be Hausdorff for $n_+ > 2$.

Example 1.9   For a geometric-inspired example for this phenomenon consider a complex K3 surface X with an infinite automorphism group $\mathrm{Aut}(X)$. (An example can be constructed by considering an elliptic K3 surface with a non-torsion section; see Section **15**.4.2.) Then the infinite subgroup $\mathrm{Aut}(X) \subset O(H^2(X,\mathbb{Z}))$ (see Proposition **15**.2.1) is contained in the stabilizer of the point $x \in D \subset \mathbb{P}(H^2(X,\mathbb{C}))$ corresponding to the Hodge structure of X. See also the proof of Proposition **7**.1.3 to see how bad the action can really be.

Remark 1.10   The action of $O(\Lambda)$ on D for $n_+ = 2$ is properly discontinuous. This can be seen as a consequence of the following general result: if $K \subset G$ is a compact subgroup of a locally compact topological group which is Hausdorff, then the action of a subgroup $H \subset G$ on G/K is properly discontinuous if and only if H is discrete in G; see e.g. [638, Lem. 3.1.1] for the elementary proof. The example applies to our case, as for $n_+ = 2$ the group $SO(2) \times O(n_+ - 2, n_-) \simeq SO(2) \times O(n_-)$ is compact (use Proposition 1.5).

Thus, for the rest of this section we shall restrict to the case $n_+ = 2$ and will consider, slightly more generally, the action of subgroups $\Gamma \subset O(\Lambda)$ of finite index, so of *arithmetic subgroups*. Recall that two subgroups $\Gamma_1, \Gamma_2 \subset H$ are *commensurable* if their intersection $\Gamma_1 \cap \Gamma_2$ has finite index in Γ_1 and Γ_2. For an algebraic group $G \subset GL(n, \mathbb{Q})$ a subgroup $\Gamma \subset G(\mathbb{Q})$ is arithmetic if it is commensurable with $G(\mathbb{Q}) \cap GL(n, \mathbb{Z})$.

We use without proof the following classical results, due to Borel and Baily and Borel. See [80, Prop. 17.4], Milne's lecture notes [408, Sec. 3] for a short review, or Satake's book [525, IV. Lem. 4.2].

Proposition 1.11   *Let $\Gamma \subset G(\mathbb{Q})$ be an arithmetic subgroup. Then there exists a (normal) subgroup of finite index $\Gamma' \subset \Gamma$ which is torsion free.*

The subgroup Γ' can be given by a congruence condition, i.e. $\Gamma' := \{g \in \Gamma \mid g \equiv \mathrm{id}\,(\ell)\}$ for some large ℓ. A typical example of a torsion-free arithmetic group is the congruence subgroup $\Gamma(p) \subset SL(2, \mathbb{Z})$, $p \geq 3$, of matrices $A \equiv \mathrm{id}\,(p)$.

The proposition can be applied to $G = O(\Lambda_\mathbb{Q})$ and any $\Gamma \subset O(\Lambda)$ of finite index.

Proposition 1.12   *If $\Gamma \subset O(\Lambda)$ is a torsion-free subgroup of finite index, then the natural action of Γ on the period domain D is free and the quotient $\Gamma\backslash D$ is a complex manifold.*

Proof We use that Γ acts properly discontinuous; see Remark 1.10. Therefore, the stabilizer of any point is finite and thus trivial if Γ is torsion free. Hence, the action is free. The open neighbourhoods U of a point $x \in D$ for which $gU \cap U = \emptyset$ for all id $\neq g \in \Gamma$ can be used as holomorphic charts for the image of x in the quotient. See e.g. [408, Prop. 3.1] for a detailed proof. $\qquad\square$

The main result in this context is, however, the following theorem of Baily–Borel for which the original paper [29] seems to be the only source. We apply the theorem to the period domain D associated with a lattice Λ of signature $(2, n_-)$. It is, however, valid for arithmetic groups acting on arbitrary bounded symmetric domains.

Theorem 1.13 (Baily–Borel) *If $\Gamma \subset O(\Lambda)$ is torsion free, then $\Gamma \backslash D$ is a smooth quasi-projective variety.*

For a subgroup $\Gamma \subset O(\Lambda)$ which is not torsion free the quotient $\Gamma \backslash D$ still exists as a quasi-projective variety, but it is only normal in general. Indeed, passing to a finite index torsion-free subgroup $\Gamma' \subset \Gamma$ first (cf. Proposition 1.11), one can construct the smooth quasi-projective quotient $\Gamma' \backslash D$. Then view $\Gamma \backslash D$ as a finite quotient of the smooth $\Gamma' \backslash D$.

2 Local Period Map and Noether–Lefschetz Locus

Small deformations of K3 surfaces are faithfully measured by the induced deformations of their associated Hodge structures. This is the content of the local Torelli theorem, which can be phrased by saying that the local period map identifies the universal deformation $\mathrm{Def}(X)$ of a K3 surface with an open subset of the period domain D introduced above. (The existence of the universal deformation is a completely general fact, which is only stated.) We shall introduce the local period map and explain why it is holomorphic. The locus, in $\mathrm{Def}(X)$ or D, of those deformations that have non-trivial Picard group, the so-called Noether–Lefschetz locus, consists of countably many smooth codimension one subsets and we prove that it is dense. We come back to more algebraic aspects of the Noether–Lefschetz locus in Section **17.2**.

2.1 Let $f \colon X \longrightarrow S$ be a smooth proper family of complex K3 surfaces $X_t := f^{-1}(t)$. For simplicity we shall mainly consider the case that S is a complex manifold (and then X is). Usually we also assume that S (and hence X) is connected and we shall fix a distinguished point $0 \in S$. Such a family is called *non-isotrivial* if the fibres X_t are not all isomorphic.

The locally constant system $R^2 f_* \mathbb{Z}$ with fibre $H^2(X_t, \mathbb{Z})$ at $t \in S$ corresponds to a representation of $\pi_1(S)$ on $H^2(X_0, \mathbb{Z})$. In particular, if S is simply connected, e.g. S a disk in $\mathbb{C}^n$, then $R^2f_*\mathbb{Z}$ is canonically isomorphic to the constant system $\underline{H}^2(X_0,\mathbb{Z})$. The same arguments apply to $R^2f_*\mathbb{Q}$ and $R^2f_*\mathbb{C}$. Clearly, $R^2f_*\mathbb{Z} \otimes_{\mathbb{Z}} \mathbb{C} \simeq R^2f_*\mathbb{C}$.

These local systems induce a flat holomorphic vector bundle $R^2f_*\mathbb{Z} \otimes_{\mathbb{Z}} \mathcal{O}_S \simeq R^2f_*\mathbb{C} \otimes_{\mathbb{C}} \mathcal{O}_S$. Its fibre at a point $t \in S$ is naturally isomorphic to $H^2(X_t,\mathbb{C})$ and thus contains the line $H^{2,0}(X_t)$. These lines glue to a holomorphic subline bundle due to the following.

Lemma 2.1 *There is a natural injection* $f_*\Omega^2_{X/S} \subset R^2f_*\mathbb{C} \otimes_{\mathbb{C}} \mathcal{O}_S$ *of holomorphic bundles which in each fibre yields the natural inclusion* $H^{2,0}(X_t) \subset H^2(X_t,\mathbb{C})$.

Proof This can be proved by an explicit computation (see e.g. [54, Exp. V]) or by a more conceptual argument as follows (see e.g. [60, Ch. 3]).

First recall that on a complex K3 surface X the constant sheaf $\underline{\mathbb{C}}$ has a resolution

$$\underline{\mathbb{C}} \longrightarrow \mathcal{O}_X \longrightarrow \Omega^1_X \longrightarrow \Omega^2_X.$$

In other words, $\underline{\mathbb{C}}$ is quasi-isomorphic to the holomorphic de Rham complex

$$\Omega^\bullet_X : \mathcal{O}_X \longrightarrow \Omega^1_X \longrightarrow \Omega^2_X.$$

Thus, singular cohomology $H^i(X,\mathbb{C})$ can also be computed as the hypercohomology $H^i(X,\Omega^\bullet_X)$ of the de Rham complex $\Omega^\bullet_X$. (Note that although the sheaves Ω^i_X are coherent, the de Rham complex is not a complex of coherent sheaves; the differential is only $\mathbb{C}$-linear, but not $\mathcal{O}_X$-linear.)

The natural $\Omega^2_X[-2] \longrightarrow \Omega^\bullet_X$ (the shift simply puts Ω^2_X in degree two) is a morphism of complexes. The induced map $H^0(X,\Omega^2_X) \simeq H^2(X,\Omega^2_X[-2]) \longrightarrow H^2(X,\mathbb{C})$ is the inclusion given by the Hodge decomposition.

Similarly, in the relative context of a smooth proper family of K3 surfaces $f: X \longrightarrow S$ the relative de Rham complex $\Omega^\bullet_{X/S} : \mathcal{O}_X \longrightarrow \Omega^1_{X/S} \longrightarrow \Omega^2_{X/S}$ is $f^{-1}\mathcal{O}_S$-linear and quasi-isomorphic to $f^{-1}\mathcal{O}_S$. Thus, using the projection formula,

$$R^if_*\mathbb{C} \otimes_{\mathbb{C}} \mathcal{O}_S \simeq R^if_*(f^{-1}\mathcal{O}_S) \simeq R^if_*(\Omega^\bullet_{X/S}).$$

Again, the natural $\Omega^2_{X/S}[-2] \longrightarrow \Omega^\bullet_{X/S}$ is a morphism of complexes of $f^{-1}\mathcal{O}_S$-sheaves and the induced $f_*\Omega^2_{X/S} \simeq R^2f_*(\Omega^2_{X/S}[-2]) \longrightarrow R^2f_*\Omega^\bullet_{X/S} \simeq R^2f_*\mathbb{C}\otimes_{\mathbb{C}}\mathcal{O}_S$ is the desired inclusion of coherent sheaves. $\square$

Remark 2.2 (i) The above remarks apply more generally to smooth proper families of complex surfaces or compact Kähler manifolds. This leads to the notion of *variations of Hodge structures* (VHS) of arbitrary weight. The case of Hodge structures of K3 type and of Hodge structures of weight one are the only cases of interest to us.

(ii) Note that the lines $H^{0,2}(X_t) \subset H^2(X_t,\mathbb{C})$ also glue to a subbundle of $R^2f_*\mathbb{C}\otimes_{\mathbb{C}}\mathcal{O}_S$ of rank one. However, this inclusion does not define a holomorphic subbundle. In fact, it is rather the quotients $H^2(X_t,\mathbb{C}) \longrightarrow H^{0,2}(X_t)$ or, more globally, the natural

$\Omega^{\bullet}_{X/S} \longrightarrow \mathcal{O}_X$ that should be considered. They yield a holomorphic quotient bundle $R^2 f_* \mathbb{C} \otimes_{\mathbb{C}} \mathcal{O}_S \longrightarrow\!\!\!\!\to R^2 f_* \mathcal{O}_X$.

It is possible to replace the smooth S by an arbitrary complex space, the arguments showing that $f_* \Omega^2_{X/S} \subset R^2 f_* \mathbb{C} \otimes_{\mathbb{C}} \mathcal{O}$ forms a holomorphic subbundle (or, equivalently, a coherent locally free subsheaf with locally free cokernel) can be modified to cover this case.

2.2 Let us step back and consider the more general situation of a holomorphic subbundle $E \subset \mathcal{O}_S^{N+1}$ of rank r. The universality property of the Grassmannian says that a subbundle of this type is obtained as the pull-back of the universal subbundle on $\mathrm{Gr}(r, N+1)$ under a uniquely determined holomorphic map $S \longrightarrow \mathrm{Gr}(r, N+1)$. For $r = 1$, the classifying map is a morphism $S \longrightarrow \mathbb{P}^N$ and the universal subbundle on $\mathbb{P}^N$ is $\mathcal{O}(-1) \subset \mathcal{O}^{N+1}$ (the dual of the evaluation map). Explicitly, the image of $t \in S$ in $\mathbb{P}^N$ is the line given by the fibre $E(t) \subset \mathcal{O}^{N+1}(t) \simeq \mathbb{C}^{N+1}$.

In our situation of a family of K3 surfaces $X \longrightarrow S$, the holomorphic map $S \longrightarrow \mathbb{P}^N$ becomes the *period map*. For this we have to assume that S is simply connected. To simplify notations, we also fix a *marking* of X_0, i.e. an isomorphism of lattices $\varphi \colon H^2(X_0, \mathbb{Z}) \xrightarrow{\sim} \Lambda$ with the K3 lattice $\Lambda := E_8(-1)^{\oplus 2} \oplus U^{\oplus 3}$. Using that S is simply connected, this yields canonical markings for all fibres $H^2(X_t, \mathbb{Z}) \simeq H^2(X_0, \mathbb{Z}) \simeq \Lambda$.

Proposition 2.3 *The period map defined by*

$$\mathcal{P} \colon S \longrightarrow \mathbb{P}(\Lambda_{\mathbb{C}}), \ t \longmapsto [\varphi(H^{2,0}(X_t))]$$

is a holomorphic map that takes values in the period domain $D \subset \mathbb{P}(\Lambda_{\mathbb{C}})$. It depends on the distinguished point $0 \in S$ and the marking φ.

Proof After the discussion above, one needs only to verify that $\mathcal{P}(t) \in D$. But this follows from $\int \sigma \wedge \sigma = 0$ and $\int \sigma \wedge \overline{\sigma} > 0$ for any $0 \neq \sigma \in H^{2,0}(X_t)$, as was observed already in Example 1.3. $\qquad\square$

The differential of the period map can be described cohomologically. It is, however, geometrically more instructive to state the result without appealing to the chosen marking $\varphi \colon H^2(X_0, \mathbb{Z}) \xrightarrow{\sim} \Lambda$.

Proposition 2.4 (Griffiths transversality) *Under the above assumptions, the differential*

$$d\mathcal{P}_0 \colon T_0 S \longrightarrow T_{\mathcal{P}(0)} D \xrightarrow{\sim} \mathrm{Hom}(H^{2,0}(X_0), H^{2,0}(X_0)^{\perp}/H^{2,0}(X_0))$$

can be described as the composition of the Kodaira–Spencer map $T_0 S \longrightarrow H^1(X_0, \mathcal{T}_{X_0})$ and the natural map $H^1(X_0, \mathcal{T}_{X_0}) \xrightarrow{\sim} H^1(X_0, \Omega_{X_0})$ given by contraction with a chosen $0 \neq \sigma \in H^{2,0}(X_0)$.

Proof For the description of $T_{\mathcal{P}(0)}D$, use Remark 1.1. The inclusion $H^1(X_0, \Omega_{X_0}) \subset H^2(X_0, \mathbb{C})$, given by the Hodge decomposition, yields a natural identification

$$H^1(X_0, \Omega_{X_0}) \xrightarrow{\sim} H^{2,0}(X_0)^{\perp}/H^{2,0}(X_0)$$

implicitily used in the statement.

Recall that the Kodaira–Spencer map is the boundary map of the obvious short exact sequence $0 \longrightarrow T_{X_0} \longrightarrow T_X|_{X_0} \longrightarrow f^*T_S|_{X_0} \longrightarrow 0$, which is the restriction of the dual of

$$0 \longrightarrow f^*\Omega_S \longrightarrow \Omega_X \longrightarrow \Omega_{X/S} \longrightarrow 0, \tag{2.1}$$

where one uses $f^*T_S|_{X_0} \simeq T_0 S \otimes_{\mathbb{C}} \mathcal{O}_{X_0}$.

The result is a special case of Griffiths transversality (cf. [621, Ch. 12] or [60]) describing the differential of arbitrary variations of Hodge structures. If $R^2 f_* \mathbb{C} \otimes_{\mathbb{C}} \mathcal{O}_S$ is viewed with its natural flat (Gauss–Manin) connection ∇, then Griffiths transversality is the statement that $\nabla(F^p) \subset F^{p-1} \otimes \Omega_S$, which we apply to $p = 2$ and so $\nabla(f_* \Omega^2_{X/S}) \subset F^1 f_*(\Omega^{\bullet}_{X/S}) \otimes \Omega_S$. The proof in this case using spectral sequences relies on the exact sequence

$$0 \longrightarrow f^*\Omega^1_S \otimes \Omega^{\bullet}_{X/S}[-1] \longrightarrow \Omega^{\bullet}_X/(f^*\Omega^2_S \otimes \Omega^{\bullet}_X) \longrightarrow \Omega^{\bullet}_{X/S} \longrightarrow 0, \tag{2.2}$$

which is a version of (2.1) for complexes.   □

2.3 Next we need to recall a few general concepts from deformation theory. Let $X \longrightarrow S$ be a smooth proper family and X_0 the fibre over a distinguished point $0 \in S$. For the general theory we have to allow singular and even non-reduced base S. In the following only the germ of the family in $0 \in S$ plays a role and all statements have to be read in this sense.

If $S' \longrightarrow S$ is a holomorphic map sending a distinguished point $0' \in S'$ to $0 \in S$, then the pull-back family is obtained as the fibre product

$$
\begin{array}{ccc}
X' := X \times_S S' & \longrightarrow & X \\
\downarrow & & \downarrow \\
S' & \longrightarrow & S.
\end{array}
$$

The family $X \longrightarrow S$ is *complete* (for the distinguished fibre X_0) if any other family $X' \longrightarrow S'$ with $X'_0 \simeq X_0$ is isomorphic to the pull-back under some $S' \longrightarrow S$. If, moreover, the map $S' \longrightarrow S$ is unique, then $X \longrightarrow S$ is called the *universal deformation*. Clearly, the universal deformation is unique up to unique isomorphism.

The ultimate aim of deformation theory for a manifold X_0 is to produce a universal deformation $X \longrightarrow S$ with special fibre X_0. But this cannot always be achieved. If $X \longrightarrow S$ is complete, but only the tangent of the map $S' \longrightarrow S$ is unique, then $X \longrightarrow S$

is called *versal*. Note that a (uni)versal family $X \longrightarrow S$ might not be (uni)versal for the nearby fibres X_t. The (uni)versal deformation of a manifold X_0, if it exists, shall be denoted

$$X \longrightarrow \mathrm{Def}(X_0).$$

The main results concerning deformation of compact complex manifolds are summarized by the following results, mostly due to Kuranishi and Kodaira; see [310].

Theorem 2.5 *Every compact complex manifold X_0 has a versal deformation. Moreover, there exists an isomorphism $T_0\mathrm{Def}(X_0) \simeq H^1(X_0, \mathcal{T}_{X_0})$.*

(i) *If $H^2(X_0, \mathcal{T}_{X_0}) = 0$, then a smooth(!) versal deformation exists.*
(ii) *If $H^0(X_0, \mathcal{T}_{X_0}) = 0$, then a universal deformation exists.[1]*
(iii) *The versal deformation $X \longrightarrow S$ of X_0 is versal and complete for any of its fibres X_t if $h^1(X_t, \mathcal{T}_{X_t}) \equiv$ const.*

Remark 2.6 Note that the isomorphism of the given manifold X_0 with the distinguished fibre of $X \longrightarrow S$ is part of the datum. In particular, even when $H^0(X_0, \mathcal{T}_{X_0}) = 0$, so a universal deformation exists, the group $\mathrm{Aut}(X_0)$ acts on the base of the universal deformation $\mathrm{Def}(X_0)$. In particular, if X_0 admits non-trivial automorphisms, then there might exist different fibres $X_t, X_{t'}$ which are isomorphic to each other.

It is not difficult to see (cf. the proof of Proposition 5.2.1) that the nearby fibres X_t in a deformation of a K3 surface X_0 are again K3 surfaces.

Corollary 2.7 *Let X_0 be a complex K3 surface. Then X_0 admits a smooth universal deformation $X \longrightarrow \mathrm{Def}(X_0)$ with $\mathrm{Def}(X_0)$ smooth of dimension 20.*

Proof This follows immediately from the vanishing $H^0(X_0, \mathcal{T}_{X_0}) = H^2(X_0, \mathcal{T}_{X_0}) = 0$ and $h^1(X_0, \mathcal{T}_{X_0}) = 20$; see Section 1.2.4. □

The following marks the beginning of the theory of complex K3 surfaces.[2]

Proposition 2.8 (local Torelli theorem) *Let $X \longrightarrow S := \mathrm{Def}(X_0)$ be the universal deformation of a complex K3 surface X_0. Then the period map*

$$\mathcal{P}: S \longrightarrow D \subset \mathbb{P}(H^2(X_0, \mathbb{C}))$$

is a local isomorphism.

---

[1] The conditions in (i) and (ii) are sufficient but not necessary. For example, a Calabi–Yau manifold can have $H^2(X, \mathcal{T}_X) \neq 0$ but still the versal deformation is smooth, due to a result of Tian and Todorov.

[2] Grauert in [215] for Kummer surfaces and later Kodaira in [308] attribute this result to Andreotti and Weil (see also Weil's report [634], where it is attributed to Nierenberg), and Pjateckiĭ-Šapiro and Šafarevič in [493] refer to [517, Ch. IX] and attribute it to Tjurina.

Proof Implicitly in the statement, the base S of the universal deformation X_0 is thought of as a small open disk in $\mathbb{C}^{20}$. In particular, S is contractible and thus simply connected. So the period map is indeed well defined.

Since $h^1(X_t, \mathcal{T}_{X_t}) \equiv 20$, the deformation is universal for all fibres X_t. Moreover, after identifying $H^2(X_t, \mathbb{Z}) \simeq H^2(X_0, \mathbb{Z})$ the period map $\mathcal{P}$ (with respect to X_0) can also be considered as the period map for the nearby fibres. As D and S are smooth of dimension 20, it thus suffices to show that $d\mathcal{P}_0$ is bijective. By Proposition 2.4, $d\mathcal{P}_0 \colon T_0 S \simeq H^1(X_0, \mathcal{T}_{X_0}) \longrightarrow H^1(X_0, \Omega_{X_0})$ is given by contraction with $\sigma \colon \mathcal{T}_{X_0} \xrightarrow{\sim} \Omega_{X_0}$ and hence indeed bijective.   □

2.4 A deformation theory for polarized manifolds, i.e. manifolds together with an ample line bundle, exists and yields results similar to Theorem 2.5; cf. [237, 544]. For our purpose a more ad hoc approach is sufficient. Suppose $X \longrightarrow S$ is the universal deformation of a K3 surface X_0 and L_0 is a non-trivial line bundle on X_0. Let ℓ be its cohomology class in $H^2(X_0, \mathbb{Z})$. Clearly, ℓ is a $(1, 1)$-class on X_0 and thus orthogonal to the period $H^{2,0}(X_0)$ of X_0. In other words, $\mathcal{P}(0) \in D \cap \mathbb{P}(\ell_{\mathbb{C}}^{\perp})$. In fact, an arbitrary class $0 \neq \ell \in H^2(X_0, \mathbb{Z})$ is a $(1, 1)$-class (and hence corresponds to a unique line bundle L_0 on X_0) if and only if $\mathcal{P}(0) \in D \cap \mathbb{P}(\ell_{\mathbb{C}}^{\perp})$.

If for S a small open disk as before natural identifications $H^2(X_t, \mathbb{Z}) \simeq H^2(X_0, \mathbb{Z})$ are chosen, then the same reasoning applies to all fibres X_t: the class $\ell \in H^2(X_t, \mathbb{Z})$ is a $(1, 1)$-class on X_t (and hence corresponds to a unique line bundle L_t on X_t) if and only if $\mathcal{P}(t) \in D \cap \mathbb{P}(\ell_{\mathbb{C}}^{\perp})$. Using the local Torelli theorem (see Proposition 2.8), one finds that the set of points $t \in S$ in which ℓ is of type $(1, 1)$ is a smooth hypersurface[3]

$$S_\ell \subset S.$$

Over S_ℓ the class ℓ can be viewed as a section of $R^2 f_* \mathbb{Z}|_{S_\ell}$ that vanishes under the projection $R^2 f_* \mathbb{Z}|_{S_\ell} \longrightarrow R^2 f_* \mathcal{O}_X|_{S_\ell}$.

Observe that for S as above, there are natural isomorphisms $H^2(X, \mathbb{Z}) \simeq \Gamma(S, R^2 f_* \mathbb{Z})$, $H^2(X, \mathcal{O}_X) \simeq \Gamma(S, R^2 f_* \mathcal{O}_X)$, and similarly for the restricted family $X|_{S_\ell} \longrightarrow S_\ell$. Using the exponential sequence on $X|_{S_\ell}$, one finds that over S_ℓ the class ℓ gives rise to a uniquely determined line bundle L on X. Below the discussion is applied to K3 surfaces with a polarization.

2.5 Let $f \colon X \longrightarrow S$ be a smooth proper family of complex K3 surfaces over a connected base and let $\rho_0 := \min\{\rho(X_t) \mid t \in S\}$. The *Noether–Lefschetz locus* of the family is the set

$$\mathrm{NL}(X/S) := \{t \in S \mid \rho(X_t) > \rho_0\}.$$

---

[3] For $(\ell)^2 = 0$ the quadric in $\mathbb{P}(\ell_{\mathbb{C}}^{\perp})$ is singular, but S_ℓ is nevertheless smooth.

The following result is usually attributed to Green (see [621, Prop. 17.20]), and Oguiso [471].

Proposition 2.9 *If* $f: X \longrightarrow S$ *is a non-isotrivial smooth proper family of K3 surfaces over a connected base, then* $\mathrm{NL}(X/S) \subset S$ *is dense.*

Proof It is clearly enough to consider the case that S is a one-dimensional disk. Furthermore, we may assume that the Picard number of the special fibre is minimal, i.e. $\rho(X_0) = \rho_0$. The assumption that the family is non-isotrivial is saying that the period map $\mathcal{P}: S \longrightarrow \mathbb{P}(H^2(X_0, \mathbb{C}))$ is non-constant. The assertion is now equivalent to the density of $\mathcal{P}(S) \cap \bigcup \ell^{\perp}$ in $\mathcal{P}(S)$, where the union runs over all classes $\ell \in H^2(X_0, \mathbb{Z}) \setminus \mathrm{NS}(X_0)$. Note that $\mathcal{P}(S) \subset \ell^{\perp}$ for all $\ell \in \mathrm{NS}(X_0)$, i.e. $\mathcal{P}(S) \subset D \cap \mathrm{NS}(X_0)^{\perp}$. It is not difficult to see that indeed $(D \cap \mathrm{NS}(X_0)^{\perp}) \cap \bigcup \ell^{\perp}$ is dense in $D \cap \mathrm{NS}(X_0)^{\perp}$; see Proposition 7.1.3. However, the assertion here is slightly stronger. For this consider the total space of the Hodge bundle

$$H^{1,1} := \{(\alpha, t) \mid \alpha \in H^{1,1}(X_t)\} \subset H^2(X_0, \mathbb{C}) \times S$$

and the projection $p: H^{1,1} \longrightarrow H^2(X_0, \mathbb{C})$. As $\mathcal{P}$ is non-constant, the holomorphic map p is open. Hence, the image of $H_{\mathbb{R}}^{1,1} := H^{1,1} \cap (H^2(X_0, \mathbb{R}) \times S)$ contains an open subset of $H^2(X_0, \mathbb{R})$ as a dense subset. Eventually, use the density of $H^2(X_0, \mathbb{Q}) \subset H^2(X_0, \mathbb{R})$ (or rather of the complements of $\mathrm{NS}(X_0)_{\mathbb{Q}}$ and $\mathrm{NS}(X_0)_{\mathbb{R}}$) to conclude that the locus of points $(\alpha, t) \in H^{1,1}$ with $\alpha \in H^{1,1}(X_t, \mathbb{Q}) \setminus \mathrm{NS}(X_0)_{\mathbb{Q}}$ is dense in $H^{1,1}$. Therefore, also its image in S, which is nothing but $\mathrm{NL}(X/S)$, is dense in S. For technical details, see [621, Sec. 17.3.4]. $\square$

Remark 2.10 In the algebraic setting the result is often stated as follows: if $f: X \longrightarrow S$ is a smooth proper family of complex K3 surfaces over a quasi-projective base S with constant Picard number $\rho(X_t)$, then the family is isotrivial. A weaker version, assuming the base to be projective, was proved by means of automorphic forms in [78].

In scheme-theoretic terms the result asserts that the natural specialization map

$$\mathrm{sp}: \mathrm{NS}(X_{\bar{\eta}}) \hookrightarrow \mathrm{NS}(X_t)$$

which is injective for all $t \in S$ (see Proposition 17.2.10) fails to be surjective (even after tensoring with $\mathbb{Q}$) for a dense set of closed points $t \in S$. Here, $\eta \in S$ is the generic point of S. Note that in positive characteristic the result does not hold; there exist non-isotrivial families of supersingular K3 surfaces (see Section 18.3.4).

The Noether–Lefschetz locus is further discussed in Section 17.1.3.

3 Global Period Map

The approach of the previous section can be globalized, in particular allowing non-simply connected base S. This leads to a global version of the above local Torelli theorem, to be discussed in Chapter 7, and eventually to an alternative construction of the moduli space of polarized complex K3 surfaces.

3.1 Consider a smooth proper family $f\colon X \longrightarrow S$ of K3 surfaces over an arbitrary base S. The locally constant system $R^2 f_* \mathbb{Z}$ on S has fibres (non-canonically) isomorphic to $\Lambda := E_8(-1)^{\oplus 2} \oplus U^{\oplus 3}$. Consider the infinite étale covering

$$\widetilde{S} := \mathrm{Isom}(R^2 f_* \mathbb{Z}, \Lambda) \longrightarrow S$$

with fibres being the set of isometries $H^2(X_t, \mathbb{Z}) \xrightarrow{\sim} \Lambda$. In other words, $\widetilde{S} \longrightarrow S$ is the natural $O(\Lambda)$-principal bundle associated with $R^2 f_* \mathbb{Z}$. In particular, $\widetilde{S}$ comes with a natural action of $O(\Lambda)$, the quotient of which gives back S.

The pull-back of $f\colon X \longrightarrow S$ under $\widetilde{S} \longrightarrow S$ yields a smooth proper family $\widetilde{f}\colon \widetilde{X} \longrightarrow \widetilde{S}$ of K3 surfaces for which $R^2 \widetilde{f}_* \mathbb{Z}$ is a constant local system. Indeed, in $(t, \varphi) \in \widetilde{S}$ with $t \in S$ and $\varphi\colon H^2(X_t, \mathbb{Z}) \xrightarrow{\sim} \Lambda$ the fibre of $R^2 \widetilde{f}_* \mathbb{Z}$ is canonically isomorphic to Λ. These identifications glue to an isomorphism $R^2 \widetilde{f}_* \mathbb{Z} \xrightarrow{\sim} \underline{\Lambda}$.

The period map for $\widetilde{f}\colon \widetilde{X} \longrightarrow \widetilde{S}$ is thus well defined and yields a holomorphic map

$$\mathcal{P}\colon \widetilde{S} \longrightarrow D \subset \mathbb{P}(\Lambda_\mathbb{C}).$$

Clearly, the period map is equivariant with respect to the natural actions of $O(\Lambda)$ on the two sides. This yields a commutative diagram

$$
\begin{array}{ccc}
\widetilde{S} & \xrightarrow{\ \mathcal{P}\ } & D \\
\downarrow & & \downarrow \\
S & \xrightarrow{\ \overline{\mathcal{P}}\ } & O(\Lambda) \backslash D.
\end{array}
$$

As was explained in Example 1.9, the action of $O(\Lambda)$ on the period domain D associated with a lattice of signature (n_+, n_-) with $n_+ > 2$ is not properly discontinuous and hence the quotient $O(\Lambda) \backslash D$ is not Hausdorff. For this reason, the resulting map $\overline{\mathcal{P}}\colon S \longrightarrow O(\Lambda) \backslash D$ is difficult to use in practice. Working with polarizations improves the situation; this shall be explained next.

3.2 Consider a smooth proper family $f\colon X \longrightarrow S$ of K3 surfaces and assume there exists a relatively ample line bundle L on X. Eventually, we work with algebraic families, i.e. X and S are schemes of finite type over $\mathbb{C}$ and f is regular, but the following construction works equally well in the setting of complex spaces. Via its first Chern class, the line bundle L induces a global section $\ell \in \Gamma(S, R^2 f_* \mathbb{Z})$. Consider the locally constant system $\ell^\perp \subset R^2 f_* \mathbb{Z}$, the orthogonal complement of ℓ with respect to the fibrewise intersection product. Then the fibres of $\ell^\perp$ are lattices of signature $(2, 19)$ and if in addition L is (fibrewise) primitive, then as abstract lattices they are isomorphic to Λ_d where $2d \equiv (L_t)^2$; see Example 1.3, (iii). For simplicity we add this assumption.

Similar to the construction above, one passes from S to the étale cover $\widetilde{S}' \longrightarrow S$ parametrizing isometries $\ell_t^\perp \simeq \Lambda_d$ that extend to $H^2(X_t, \mathbb{Z}) \simeq \Lambda$. Thus, $\widetilde{S}' \longrightarrow S$ is a principal $\widetilde{O}(\Lambda_d)$-bundle, where

$$\widetilde{O}(\Lambda_d) := \{g|_{\Lambda_d} \mid g \in O(\Lambda),\ g(e_1 + df_1) = e_1 + df_1\}.^4$$

Extending $\ell_t^\perp \simeq \Lambda_d$ to $H^2(X_t, \mathbb{Z}) \simeq \Lambda$ by sending L to $e_1 + df_1$ defines an embedding $\widetilde{S}' \hookrightarrow \widetilde{S}$. The composition $\mathcal{P}_d$ with the period map $\mathcal{P} \colon \widetilde{S} \longrightarrow D$ takes values in $D_d = D \cap \mathbb{P}(\Lambda_d \mathbb{C})$, so

$$\mathcal{P}_d \colon \widetilde{S}' \longrightarrow D_d \subset \mathbb{P}(\Lambda_d \mathbb{C}).$$

Moreover, $\mathcal{P}_d$ is equivariant with respect to the action of $\widetilde{O}(\Lambda_d)$ and thus yields the commutative diagram

$$\begin{array}{ccccc}
\widetilde{S}' & \xrightarrow{\ \mathcal{P}_d\ } & D_d & \lhook\joinrel\longrightarrow & D \\
\downarrow & & \downarrow & & \downarrow \\
S & \xrightarrow{\ \overline{\mathcal{P}}_d\ } & \widetilde{O}(\Lambda_d)\backslash D_d & \longrightarrow & O(\Lambda)\backslash D.
\end{array} \qquad (3.1)$$

Now, $\widetilde{O}(\Lambda_d)$ is an arithmetic subgroup of $O(\Lambda_d)$ and by Baily–Borel (see Theorem 1.13) the quotient $\widetilde{O}(\Lambda_d)\backslash D_d$ is a normal quasi-projective variety.

3.3 Two of the main results in the theory of K3 surfaces, the surjectivity of the period map and the global Torelli theorem (cf. Theorems **3.2.4**, **7.5.3**), can be formulated in terms of the period maps discussed above. We shall here only state these results and come back to their proofs in Chapter 7. Both results can be best phrased in terms of moduli spaces of marked (polarized) K3 surfaces, which shall be introduced first.

The *moduli space of marked K3 surfaces N* can be constructed in a rather ad hoc manner. Maybe the most surprising aspect of the following construction, apart from its simplicity, is that the moduli space of marked K3 surfaces turns out to be a fine(!) moduli space; cf. Section **7.2.1**.

As a set, N consists of all isomorphism classes of pairs (X, φ) with X a K3 surface and $\varphi \colon H^2(X, \mathbb{Z}) \xrightarrow{\sim} \Lambda$ a marking (i.e. an isomorphism of lattices). To introduce the

---

[4] Equivalently, $\widetilde{O}(\Lambda_d)$ is the subgroup of $O(\Lambda_d)$ of all isometries acting trivially on the discriminant; see Section **14.2.2**.

structure of a complex manifold on N, one glues the universal deformation spaces of the various K3 surfaces as follows. For any K3 surface X_0 consider its universal deformation $X \longrightarrow \mathrm{Def}(X_0)$. A given marking $\varphi \colon H^2(X_0, \mathbb{Z}) \overset{\sim}{\longrightarrow} \Lambda$ induces canonically markings of all fibres. Note that by the local Torelli theorem (see Proposition 2.8), the induced map

$$\mathrm{Def}(X_0) \lhook\joinrel\longrightarrow D$$

is injective. Since $X \longrightarrow \mathrm{Def}(X_0)$ is universal for each of the fibres, the pairs $(\mathrm{Def}(X_0), \varphi)$ can be glued along the intersections $\mathrm{Def}(X_0) \cap \mathrm{Def}(Y_0)$ in D. Thus, the complex structures of the universal deformation spaces $\mathrm{Def}(X_0)$ for all K3 surfaces (together with a marking) define a global complex structure on N. Moreover, since the natural map $\mathrm{Aut}(X) \longrightarrow \mathrm{O}(H^2(X, \mathbb{Z}))$ is injective for K3 surfaces (see Proposition **15**.2.1), the universal families $X \longrightarrow \mathrm{Def}(X_0)$ glue to a global universal family

$$f \colon X \longrightarrow N$$

together with a marking $R^2 f_* \mathbb{Z} \overset{\sim}{\longrightarrow} \underline{\Lambda}$. For more details, see [54, Exp. XIII].[5]

Warning: The moduli space N of marked K3 surface exists as a (20-dimensional) complex manifold, but it is not Hausdorff.

Using the universal marking $\varphi \colon R^2 f_* \mathbb{Z} \overset{\sim}{\longrightarrow} \underline{\Lambda}$ of the universal family $f \colon X \longrightarrow N$ one obtains a global period map

$$\mathcal{P} \colon N \longrightarrow D \subset \mathbb{P}(\Lambda_{\mathbb{C}}),$$

which due to the local Torelli theorem is a local isomorphism. The following theorem relies on the existence of (hyper)kähler metrics on K3 surfaces (cf. Section **7**.3.2), which is discussed in Theorem **7**.4.1.

Theorem 3.1 (Surjectivity of the period map)   *The global period map*

$$\mathcal{P} \colon N \longrightarrow\!\!\!\!\!\rightarrow D$$

is surjective.

Remark 3.2   In this general form, the surjectivity of the period map is due to Todorov [601]. His argument relies on the proof of Calabi's conjecture by Yau and previous work of Kulikov [335] and Persson and Pinkham [492] for algebraic K3 surfaces. An alternative argument, still using the existence of hyperkähler metrics on K3 surfaces

---

[5] As in the algebraic context, N represents a moduli functor, namely $\mathcal{N} \colon (Compl)^o \longrightarrow (Sets)$, $S \longmapsto \{(f \colon X \longrightarrow S, \varphi)\}/\sim$. Here, $(Compl)$ is the category of complex spaces and $f \colon X \longrightarrow S$ is a smooth and proper family of K3 surfaces with marking $\varphi \colon R^2 f_* \mathbb{Z} \overset{\sim}{\longrightarrow} \underline{\Lambda}$ (as usual, compatible with the intersection pairing). One defines $(f \colon X \longrightarrow S, \varphi) \sim (f' \colon X' \longrightarrow S, \varphi')$ if there exists an isomorphism $g \colon X \overset{\sim}{\longrightarrow} X'$ with $f' \circ g = f$ and $\varphi' = \varphi \circ g^*$.

(of Kähler type), was later given by Looijenga [378]. A slightly shorter proof of the surjectivity using a less precise description of the Kähler cone (that generalizes to higher dimensions) can be found in [250]. See Chapter 7 for more details.

Remark 3.3 In concrete terms, the surjectivity of the period map asserts that for any Hodge structure of K3 type on the lattice $\Lambda = E_8(-1)^{\oplus 2} \oplus U^{\oplus 3}$, which is signed (i.e. such that the pairing is positive definite on $(\Lambda^{2,0} \oplus \Lambda^{0,2})_\mathbb{R}$), there exists a K3 surface X together with a Hodge isometry $H^2(X, \mathbb{Z}) \simeq \Lambda$; i.e. if the Hodge structure is given by $x \in \Lambda_\mathbb{C}$, then there exists a K3 surface X and an isometry $\varphi \colon H^2(X, \mathbb{Z}) \xrightarrow{\sim} \Lambda$ such that $\varphi^{-1}(x)$ spans $H^{2,0}(X)$.

Note that for Hodge structures of K3 type on the lattice $U^{\oplus 3} \simeq H^2(\mathbb{C}^2/\Gamma, \mathbb{Z})$ this is much easier to achieve. Indeed, any signed Hodge structure of K3 type on the lattice $U^{\oplus 3}$ is realized by a Hodge structure on $H^2(T, \mathbb{Z})$ for some two-dimensional complex torus $T = \mathbb{C}^2/\Gamma$ which is in fact unique up to taking its dual. This was studied by Shioda in [563]; cf. Section 3.2.4.

3.4 Similarly to the above, one can construct the moduli space N_d of primitively polarized marked K3 surfaces of degree $2d$. Points of N_d parametrize triples (X, L, φ) with L an ample line bundle on the complex K3 surface X and $\varphi \colon H^2(X, \mathbb{Z}) \xrightarrow{\sim} \Lambda$ an isometry mapping L to the distinguished class $e_1 + df_1$ (which, in particular, makes L a primitive line bundle).[6]

The arguments to construct N_d use in addition that for the universal deformation $f \colon X \longrightarrow S := \mathrm{Def}(X_0)$ the restriction $X|_{S_\ell} \longrightarrow S_\ell$ for the class ℓ induced by some ample line bundle L_0 on X_0 form the universal deformation of the pair (X_0, L_0); cf. Section 2.4.

The fine moduli space N_d obtained in this way is a complex manifold which turns out to be Hausdorff; see Section 5.2.3. Any $g \in \widetilde{O}(\Lambda_d)$ induces a natural orthogonal transformation of Λ by mapping ℓ to itself. This defines an action

$$\widetilde{O}(\Lambda_d) \times N_d \longrightarrow N_d, \ (g, (X, \varphi)) \longmapsto (X, g \circ \varphi).$$

Clearly, the quotient $\widetilde{O}(\Lambda_d) \backslash N_d$ parametrizes all primitively polarized K3 surfaces (X, L) of degree $2d$. The relation with the algebraic moduli spaces M_d is discussed in Section 4.1 below.

The global period map

$$\mathcal{P}_d \colon N_d \longrightarrow D_d \subset \mathbb{P}(\Lambda_{d\mathbb{C}})$$

---

[6] More formally, one may consider the functor $\mathcal{N}_d \colon (Compl)^o \longrightarrow (Sets)$, $S \longmapsto \{(f \colon X \longrightarrow S, L, \varphi)\}/\sim$ with $f \colon X \longrightarrow S$ a smooth, proper family of K3 surfaces, $L \in \mathrm{Pic}(X)$ ample on all fibres, and the marking $\varphi \colon R^2 f_* \mathbb{Z} \xrightarrow{\sim} \underline{\Lambda}$ mapping the section ℓ corresponding to L to the distinguished constant section $e_1 + df_1$ of $\underline{\Lambda}$. The equivalence relation $\sim$ is induced by the natural notion of isomorphisms of such triples.

is a local isomorphism which is compatible with the action of $\widetilde{O}(\Lambda_d)$. Thus, one also has a holomorphic map $\overline{\mathcal{P}}_d \colon \widetilde{O}(\Lambda_d) \backslash N_d \longrightarrow \widetilde{O}(\Lambda_d) \backslash D_d$ of complex spaces. Due to a result of Pjateckiĭ-Šapiro and Šafarevič [493], one has the following theorem.

Theorem 3.4 (global Torelli theorem)   *The period maps*

$$\mathcal{P}_d \colon N_d \lhook\joinrel\longrightarrow D_d \quad and \quad \overline{\mathcal{P}}_d \colon \widetilde{O}(\Lambda_d) \backslash N_d \lhook\joinrel\longrightarrow \widetilde{O}(\Lambda_d) \backslash D_d$$

are injective.

More explicitly, the global Torelli theorem can be rephrased as follows.

Corollary 3.5   *Let (X, L) and (X', L') be two polarized complex K3 surfaces. Then $(X, L) \simeq (X', L')$ if and only if there exists a Hodge isometry $H^2(X, \mathbb{Z}) \simeq H^2(X', \mathbb{Z})$ mapping ℓ to ℓ'.*

As before, ℓ is the cohomology class of L and similarly for ℓ'.

Remark 3.6   (i) The rough idea of the proof of the global Torelli theorem is as follows: using the local Torelli theorem one easily shows that $\mathcal{P}_d$ is locally an open embedding. In order to show that it is injective, it suffices to show that all fibres over points of a dense subset of the image consist of a single point. In other words, it suffices to prove the corollary for a dense set of K3 surfaces. In the original [493] and in later work, Kummer surfaces were used to provide this set.

(ii) An alternative proof was later given by Friedman [183]. He deduces the global Torelli theorem for N_d from the properness of the period map and the global Torelli theorem for N_{d-1}. Eventually he proves it for $d = 1$, i.e. for double planes, which had also been studied in [244, 551, 600].

(iii) Burns and Rapoport generalized in [93] the global Torelli theorem from polarized K3 surfaces to arbitrary complex K3 surfaces (of Kähler type). It turns out that although $\mathcal{P} \colon N \longrightarrow D$ is no longer injective, due to the non-Hausdorffness of N, the period map between the quotients is indeed a bijection

$$\overline{\mathcal{P}} \colon O(\Lambda) \backslash N \xrightarrow{\sim} O(\Lambda) \backslash D. \tag{3.2}$$

We emphasize again that the quotient on the right-hand side of (3.2) (and hence on the left-hand side as well) is badly behaved. In some sense, by passing to the quotient by the action of $O(\Lambda)$ one gets rid of the non-Hausdorffness of N but creates it anew on the quotient of D.

More in the spirit of Corollary 3.5, the global Torelli theorem for unpolarized K3 surfaces can also be stated as follows. Two complex K3 surfaces X and X' are isomorphic if and only if there exists a Hodge isometry $H^2(X, \mathbb{Z}) \simeq H^2(X', \mathbb{Z})$. The proof of this will be discussed Section 7.5.5.

Warning: The period map $\mathcal{P}\colon N \longrightarrow D$ is surjective but not injective (it is injective only over general points) and the polarized period map $\mathcal{P}_d\colon N_d \longrightarrow D_d$ is injective but not surjective. For the quotients one has

$$\overline{\mathcal{P}}\colon O(\Lambda)\backslash N \xrightarrow{\sim} O(\Lambda)\backslash D,$$

but

$$\overline{\mathcal{P}}_d\colon \widetilde{O}(\Lambda_d)\backslash N_d \lhook\joinrel\longrightarrow \widetilde{O}(\Lambda_d)\backslash D_d$$

is still only an immersion.

Remark 3.7 The complement $(\widetilde{O}(\Lambda_d)\backslash D_d) \setminus \mathrm{Im}(\overline{\mathcal{P}}_d)$ of the image can be described explicitly as the union of all hyperplane sections $\bigcup \delta^{\perp}$, where $\delta \in \Lambda_d$ with $(\delta)^2 = -2$:

$$(\widetilde{O}(\Lambda_d)\backslash D_d) \setminus \mathrm{Im}(\overline{\mathcal{P}}_d) = \bigcup_{\delta \in \Delta(\Lambda_d)} \delta^{\perp}.$$

Indeed, for $x = \mathcal{P}(X,\varphi) \in D_d$, the class $\varphi^{-1}(e_1 + df_1)$ corresponds to a primitive line bundle L on X with $(L)^2 = 2d$. By Corollary 8.2.9, there exist (-2)-curves $C_1, \ldots, C_n \subset X$ such that $\ell' := \pm(s_{[C_1]} \circ \cdots \circ s_{[C_n]})(\ell)$ is nef. In fact, it is ample unless there exists a (-2)-class $\delta' \in \mathrm{NS}(X)$ with $(\ell'.\delta') = 0$. The latter is equivalent to $(\ell.\delta) = 0$ for $\delta := (s_{[C_n]} \circ \cdots \circ s_{[C_1]})(\delta')$. As $x = \mathcal{P}(X,\varphi) = \mathcal{P}(X, \varphi \circ s_{[C_1]} \circ \cdots \circ s_{[C_n]})$ this shows that any class $x \in D_d$ not contained in $\bigcup \delta^{\perp}$ is contained in the image of $\mathcal{P}_d$. Conversely, if the period $x \in D_d$ of a polarized marked K3 surface (X, φ, L) were contained in $\delta^{\perp} \subset D_d$ for some (-2)-class $\delta \in \Lambda_d$, then the line bundle L would be orthogonal to the class $\varphi^{-1}(\delta)$. However, by the Riemann–Roch formula $\varphi^{-1}(\delta)$ is of the form $\pm[C]$ for an effective curve $C \subset X$, contradicting the ampleness of L. For the notations $s_{[C_i]}$, etc., we refer to Chapter 8.

4 Moduli Spaces of K3 Surfaces via Periods and Applications

Let us return to the construction of the moduli space of polarized K3 surfaces. For the definition of the moduli functor $\mathcal{M}_d$ and its moduli space M_d, see Section 5.1. We work over $\mathbb{C}$. A construction of the moduli space using the global Torelli theorem and the period map has been initiated by Pjateckiĭ-Šapiro and Šafarevič in [493, 494].[7] The idea in this setting is to construct M_d as an open subvariety of the quasi-projective variety $\widetilde{O}(\Lambda_d)\backslash D_d$. Let us explain this approach briefly.

---

[7] However, as far as I can see, the result is not actually stated as such in either of these two papers.

4.1   We use the notation of Section **5.2.1**, $P(t) := dt^2 + 2$ and $N := P(3) - 1$, and consider the corresponding Hilbert scheme Hilb $:= \mathrm{Hilb}_{\mathbb{P}^N}^{P(3t)}$. The universal (open) subscheme $H \subset$ Hilb (see Proposition **5.2.1**) parametrizes polarized K3 surfaces (X, L) with $X \hookrightarrow \mathbb{P}^N$ and $L^3 \simeq \mathcal{O}(1)|_X$. The coarse moduli space M_d of interest would be a categorical quotient of H by the natural action of PGL $:=$ PGL$(N + 1)$ such that the closed points parametrize the orbits.

Consider the universal family $f: X \longrightarrow H$ and apply the construction of Section 3.2 to the underlying complex manifolds. Thus, we obtain a complex manifold $\widetilde{H}$ with an étale map $\widetilde{H} \longrightarrow H$ which is the principal $\widetilde{O}(\Lambda_d)$-bundle associated with $\ell^\perp \subset R^2 f_* \mathbb{Z}$. Here, ℓ is the global section of $R^2 f_* \mathbb{Z}$ induced by the first Chern class of the global L on $X \longrightarrow H$. The pull-back family $\widetilde{f}: \widetilde{X} \longrightarrow \widetilde{H}$ comes with a natural marking $R^2 \widetilde{f}_* \mathbb{Z} \simeq \underline{\Lambda}$ that maps ℓ to the constant section $e_1 + df_1$. Similarly to (3.1), one obtains a commutative diagram

$$
\begin{array}{ccc}
\widetilde{H} & \xrightarrow{\ \mathcal{P}_d\ } & D_d \\
\downarrow & & \downarrow \\
H & \xrightarrow{\ \overline{\mathcal{P}}_d\ } & \widetilde{O}(\Lambda_d) \backslash D_d.
\end{array}
$$

Clearly, the period map $\overline{\mathcal{P}}_d: H \longrightarrow \widetilde{O}(\Lambda_d) \backslash D_d$ is PGL-equivariant and due to the global Torelli theorem 3.4 the set of orbits H/PGL, which is nothing but $\widetilde{O}(\Lambda_d) \backslash N_d$, injects into $\widetilde{O}(\Lambda_d) \backslash D_d$. Due to the local Torelli theorem (cf. Proposition 2.8), this describes the set of orbits H/PGL as an open (in the classical topology) subset of the algebraic variety $\widetilde{O}(\Lambda_d) \backslash D_d$ (cf. Theorem 1.13). In order to give H/PGL itself the structure of an algebraic variety one needs the following result due to Borel [79].[8]

Theorem 4.1 (Borel)   *If Y is a non-singular complex variety and $\varphi: Y \longrightarrow \widetilde{O}(\Lambda_d) \backslash D_d$ is a holomorphic map, then φ is algebraic.*

Remark 4.2   Usually the theorem is stated with $\widetilde{O}(\Lambda_d)$ replaced by a finite index torsion-free subgroup $\Gamma \subset \widetilde{O}(\Lambda_d)$ such that the quotient $\Gamma \backslash D_d$ is smooth; see Propositions 1.11 and 1.12. Replacing $H \longrightarrow \widetilde{O}(\Lambda_d) \backslash D_d$ by $H' = \Gamma \backslash \widetilde{H} \longrightarrow \Gamma \backslash D_d$, one can easily reduce to this case.

Corollary 4.3   *The orbit space H/PGL exists as quasi-projective variety M_d which is a coarse moduli space for the moduli functor $\mathcal{M}_d$ on $(\mathrm{Sch}/\mathbb{C})^\circ$ of primitively polarized K3 surfaces of degree $2d$. Moreover, M_d can be realized as a Zariski open subscheme of the quasi-projective variety $\overline{M}_d := \widetilde{O}(\Lambda_d) \backslash D_d$.*

---

[8] In [494] Pjateckiĭ-Šapiro and Šafarevič attribute the result also to Kobayashi but I could not trace the reference.

Proof Since $\varphi \colon H \longrightarrow \overline{M}_d = \widetilde{O}(\Lambda_d) \backslash D_d$ is algebraic, its image is constructible. On the other hand, it is analytically open by the local Torelli theorem. Hence, it is open in the Zariski topology. Thus, $M_d := \varphi(H)$ has a natural algebraic structure and its closed points parametrize effectively all primitively polarized K3 surfaces of degree $2d$.

In order to prove that M_d with this definition is a coarse moduli space for $\mathcal{M}_d$ (see Section 5.1), one needs to construct a natural $\mathcal{M}_d \longrightarrow \underline{M}_d$ inducing the above bijection of $\mathcal{M}_d(\mathbb{C})$ with (the closed points) of M_d. If $(X, L) \longrightarrow S$ is an algebraic family of primitively polarized K3 surfaces, first pass to the induced family of complex spaces and then use $S \longrightarrow \widetilde{O}(\Lambda_d) \backslash D_d$ as constructed in (3.1), which takes image in M_d. This provides a holomorphic map $S \longrightarrow M_d$ which, again by Theorem 4.1, is algebraic. $\qquad\square$

The description of M_d as an open subset of the arithmetic quotient $\widetilde{O}(\Lambda_d) \backslash D_d$ allows one to derive global information about the moduli space M_d. For example, M_d can be proved to be irreducible by observing that the two connected components $D_d^{\pm} \subset D_d$ are interchanged by $\widetilde{O}(\Lambda_d)$. More precisely, if Λ_d is written as $E_8(-1)^{\oplus 2} \oplus U^{\oplus 2} \oplus \mathbb{Z}(-2d)$, then the isometry that interchanges the two summands of $U^{\oplus 2}$ has the required effect.

Corollary 4.4 *For each $d > 0$ the moduli space M_d of polarized complex K3 surfaces of degree $2d$ is an irreducible quasi-projective variety of dimension 19.* $\qquad\square$

The moduli space M_d is not smooth, but as a consequence of the above discussion, it can be viewed as a smooth orbifold. In the algebraic terminology, M_d is the coarse moduli space of a smooth Deligne–Mumford stack; see Remark **5.4.11**.

Remark 4.5 At least pointwise, it is easy to see that the quotient $\overline{M}_d = \widetilde{O}(\Lambda_d) \backslash D_d$ can also be viewed as a coarse moduli space, namely as the moduli space of quasi-polarized (also called pseudo-polarized) K3 surfaces (X, L); i.e. L in this case is only big and nef. To see this, just repeat the discussion in Remark 3.7, which can be rephrased as

$$M_d = \overline{M}_d \setminus \bigcup \delta^{\perp}.$$

As explained in Section **5.1.4**, the corresponding moduli functor is not separated, and, for this reason, it is preferable to regard $\overline{M}_d$ as the moduli space of polarized 'singular K3 surfaces' (surfaces with rational double point singularities whose minimal resolution are K3 surfaces; see Section **14**.0.3). And then indeed, the quasi-projective variety $\overline{M}_d = \widetilde{O}(\Lambda_d) \backslash D_d$ coarsely represents the corresponding moduli functor $\overline{\mathcal{M}}_d$.

4.2 As mentioned earlier, the moduli space M_d of polarized complex K3 surfaces, constructed either algebraically as described in Section **5.2.2** or as a Zariski open subset of the quotient $\widetilde{O}(\Lambda_d) \backslash D_d$ as above, is a coarse moduli space only; i.e. it comes without a universal family

In Remark **5**.4.8 it was alluded to already that one can, however, find a finite cover

$$\pi : M_d^{\mathrm{lev}} \longrightarrow M_d$$

over which a 'universal family' exists. In other words, there exists a quasi-projective
variety M_d^{lev} and a polarized family

$$(X, L) \longrightarrow M_d^{\mathrm{lev}}$$

of K3 surfaces of degree $2d$ such that the classifying morphism $\pi : M_d^{\mathrm{lev}} \longrightarrow M_d$ is finite
and surjective. Note that M_d^{lev} is not unique but depends on the choice of a level, which
is usually given in form of an integer $\ell \gg 0$. Moreover, M_d^{lev} can be constructed as a
smooth(!) quasi-projective variety. Recall that M_d itself is not smooth; see Section **5**.3.3.

The construction can be performed in the algebraic setting but is most easily
explained in the complex setting using periods. First recall that in Theorem 1.13 one
had to choose a torsion-free subgroup $\Gamma \subset \widetilde{O}(\Lambda_d)$ to get a smooth quasi-projective
variety $\Gamma \backslash D_d$. The choice of Γ can be made explicit as follows. Let

$$\Gamma_\ell := \{g \in \widetilde{O}(\Lambda_d) \mid g \equiv \mathrm{id}\,(\ell)\},$$

and then indeed for $\ell \gg 0$, the group Γ_ℓ is torsion free and, therefore, acts freely on D_d.

Next, the restriction of the universal family of marked K3 surfaces $(X, \varphi) \longrightarrow N$ (see
Section 3.3) to the moduli space of marked polarized K3 surfaces N_d (see Section 3.4)
yields a universal family of marked polarized K3 surfaces $(X, L, \varphi) \longrightarrow N_d$. By Theorem
3.4 the period map $N_d \hookrightarrow D_d$ is an open embedding. Clearly, N_d is preserved by the
action of $\widetilde{O}(\Lambda_d)$, but, more importantly, this action can be lifted to an action of $\widetilde{O}(\Lambda_d)$
on the universal family (X, L, φ). For this one has to use another part of the global Torelli
theorem (see Section **7**.5.2), saying that any Hodge isometry $H^2(X, \mathbb{Z}) \xrightarrow{\sim} H^2(X', \mathbb{Z})$
mapping a polarization to a polarization can be lifted uniquely to an isomorphism
$X \xrightarrow{\sim} X'$. Restricting the action of $\widetilde{O}(\Lambda_d)$ to the subgroup $\Gamma_\ell \subset \widetilde{O}(\Lambda_d)$ yields a
free action

$$\Gamma_\ell \times (X, L) \longrightarrow (X, L)$$

on the universal family $(X, L, \varphi) \longrightarrow N_d$ that lifts the action of Γ_ℓ on N_d. The action
is free, simply because it is free on N_d already. Taking the quotient yields a universal
family over the quasi-projective variety

$$M_d^{\mathrm{lev}} := \Gamma_\ell \backslash N_d.$$

Unravelling the construction shows that M_d^{lev} parametrizes polarized K3 surfaces
(X, L) together with an isomorphism $H^2(X, \mathbb{Z}/\ell\mathbb{Z}) \xrightarrow{\sim} \Lambda \otimes \mathbb{Z}/\ell\mathbb{Z}$ compatible with the
pairing and mapping L to the class of the distinguished class $e_1 + df_1$ or, equivalently,

with an isometry of the primitive cohomology $H^2(X, \mathbb{Z}/\ell\mathbb{Z})_p \simeq \Lambda_d \otimes \mathbb{Z}/\ell\mathbb{Z}$. So, M_d^{lev} is the moduli space of polarized K3 surfaces of degree $2d$ with a $\Lambda/\ell\Lambda$-*level structure*.

Remark 4.6   By the discussion in Section 5.4.2, it is natural to view $\mathcal{M}_d$ as a smooth Deligne–Mumford stack with its coarse moduli space given by M_d. Now, viewing M_d as the quotient of M_d^{lev} by the action of the finite group $G := \widetilde{O}(\Lambda_d)/\Gamma_\ell$ makes M_d into the coarse moduli space of another smooth Deligne–Mumford stack $[M_d^{\mathrm{lev}}/G]$. The two stacks are in fact isomorphic. Indeed, a point in $[M_d^{\mathrm{lev}}/G]$ is principal G-bundle $P \longrightarrow T$ with a G-equivariant morphism $P \longrightarrow M_d^{\mathrm{lev}}$. The universal family over M_d^{lev} can be pulled back to yield a family $X_P \longrightarrow P$. The G-action on P can be lifted to an action on X_P and its quotient yields a family over T. This defines a morphism $[M_d^{\mathrm{lev}}/G] \longrightarrow \mathcal{M}_d$ which turns out to be an isomorphism. Compare also the comments in Example 5.4.5 and Remark 5.4.8.

4.3   Consider a family of polarized K3 surfaces $f : X \longrightarrow S$ over a smooth connected algebraic variety S and the induced variation of polarized Hodge structures $R^2f_*\mathbb{Z}$. The image $\mathrm{Im}(\rho)$ of the monodromy representation

$$\rho : \pi_1(S, t) \longrightarrow O(H^2(X_t, \mathbb{Q}))$$

is the *monodromy group* (see as well the discussion in Section 7.5.3), and its Zariski closure

$$\overline{\mathrm{Im}(\rho)} \subset O(H^2(X_t, \mathbb{Q}))$$

is called the *algebraic monodromy group*.

As a consequence of the discussion in the previous section, we mention the following corollary.

Corollary 4.7   *Any polarized complex K3 surface (X_0, L_0) of degree $2d$ sits in a smooth family of polarized K3 surfaces $(X, L) \longrightarrow S$ over a smooth, connected, complex algebraic variety S such that $\mathrm{Im}(\rho) \subset O(H^2(X_0, \mathbb{Z}))$ is a finite index subgroup of the subgroup $O(H^2(X_0, \mathbb{Z}))$ fixing L_0 (which is isomorphic to $\widetilde{O}(\Lambda_d)$).*

Proof   Indeed, take the universal family $(X, L) \longrightarrow S := M_d^{\mathrm{lev}}$ as above and consider the induced period map $M_d^{\mathrm{lev}} \longrightarrow \Gamma_\ell \backslash D_d$ which is algebraic due to Theorem 4.1. Now use the general fact that for any dominant morphism between smooth, connected, complex varieties the induced map between their fundamental groups has finite cokernel. In our situation it shows that the image of $\pi_1(M_d^{\mathrm{lev}}) \longrightarrow \pi_1(\Gamma_\ell \backslash D_d)$ has finite index. As $\pi_1(\Gamma_\ell \backslash D_d) \simeq \Gamma_\ell$, which is of finite index in $\widetilde{O}(\Lambda_d)$, this proves the claim.   □

Remark 4.8   The assertion can be improved. Indeed, as shown by Pjateckiĭ-Šapiro and Šafarevič in [494, Cor. 2], there exists a family $(X, L) \longrightarrow S$ for which $\mathrm{Im}(\rho)$

equals $\widetilde{O}(\Lambda_d) \subset O(H^2(X_0, \mathbb{Z}))$. The family is realized by the standard Hilbert scheme construction outlined in Section 5.2.1, so $S \subset$ Hilb is an open set. The assertion then follows from the observation that the natural $\widetilde{O}(\Lambda_d)$-bundle associated with $R^2 f_* \mathbb{Z}$ is connected.

Due to Deligne's theorem (see [621, Ch. 15]), the invariant part $H^2(X_t, \mathbb{Q})^{\mathrm{Im}(\rho)}$ is the image of $H^2(\overline{X}, \mathbb{Q}) \longrightarrow H^2(X_t, \mathbb{Q})$, where $X \subset \overline{X}$ is an arbitrary smooth projective compactification. The arithmetic analogue of the algebraic monodromy group is the group $\overline{\mathrm{Im}(\rho_\ell)}$. See Section 3.3.4 for its relation to the Mumford–Tate group $\mathrm{MT}(H^2(X_t, \mathbb{Q}))$. The following result, a special case of a completely general fact due to Deligne [141, Prop. 7.5] (see also André [6]), relates the Mumford–Tate group to the algebraic fundamental group.

Theorem 4.9 *For any family of polarized K3 surfaces $X \longrightarrow S$, there exists a countable union of proper closed subvarieties $S' \subset S$ such that for $t \in S \setminus S'$ the Mumford–Tate group $\mathrm{MT}(H^2(X_t, \mathbb{Q}))$ is constant and contains a finite index subgroup of the algebraic monodromy group $\overline{\mathrm{Im}(\rho)}$ and, in particular, its identity component.*

Proof Here is a sketch of the argument. As mentioned in Section 3.3.4, the Mumford–Tate group is the subgroup of $\mathrm{GL}(H^2(X_t, \mathbb{Q}))$ that fixes all Hodge classes in $H^2(X_t, \mathbb{Q})^{\otimes m}$, for all $m > 0$. Thus, it stays constant for points $t \in S$ for which the space of Hodge classes on all powers X_t^m is minimal. Due to the result of Cattani, Deligne, and Kaplan [108] this is the complement of a countable union $S' \subset S$ of proper, closed, algebraic subvarieties. For $t \in S \setminus S'$ the space of Hodge classes in $H^2(X_t, \mathbb{Q})^{\otimes m}$ comes with a polarization that is preserved by the discrete monodromy group. Thus, for each fixed m the induced action of $\pi_1(S, t)$ on the space of Hodge classes factors over a finite group. Equivalently, a finite index subgroup of $\pi_1(S, t)$ acts trivially on the space of Hodge classes in $H^2(X_t, \mathbb{Q})^{\otimes m}$ and its image under ρ is therefore contained in $\mathrm{MT}(H^2(X_t, \mathbb{Q}))$. Applying this to the finite number of m needed to determine the Mumford–Tate group proves that there exists a finite index subgroup of $\pi_1(S, t)$ such that its image under ρ is contained in $\mathrm{MT}(H^2(X_t, \mathbb{Q}))$. □

The result confirms Zarhin's description (see Theorem 3.3.9) of the Mumford–Tate group of the very general polarized K3 surface as $\mathrm{MT}(H^2(X, \mathbb{Q})) \simeq O(T(X) \otimes \mathbb{Q})$.

4.4 We shall now explain relative versions of the Kuga–Satake construction of Chapter 4. To start out, consider a variation of Hodge structures (VHS) of K3 type V over S, e.g. $V = R^2 f_* \mathbb{Z}$ for $f : X \longrightarrow S$ a smooth family of K3 surfaces. Applying fibrewise the Kuga–Satake construction yields a family of Hodge structures $\mathrm{Cl}^+(V_t)$ of weight one parametrized by points $t \in S$. The first-order computation in Remark 4.2.8 shows that this indeed yields a VHS of weight one, which is denoted $\mathrm{Cl}^+(V)$. Associated with it, there is a family of complex tori. Note however that this construction is transcendental, and so even for a family of polarized K3 surfaces $(X, L) \longrightarrow S$ and $V := \ell^{\perp} \subset R^2 f_* \mathbb{Z}$ the VHS $Cl^+(V)$ is in general not algebraic. One reason is that the fibrewise polarization Q in Section **4.2.2** depends on the choice of positive orthogonal vectors f_1, f_2 which might not exist globally.

Consider now the universal family $f \colon (X, L) \longrightarrow N_d$ of polarized marked K3 surfaces of degree $2d$ over the open set $N_d \subset D_d$ of the period domain $D_d \subset \mathbb{P}(\Lambda_{d\mathbb{C}})$; see Theorem 3.4. It gives rise to the constant systems $R^2 f_* \mathbb{Z} \simeq \underline{\Lambda}$ and $\ell^{\perp} \simeq \underline{\Lambda}_d$, which come with natural VHS of weight two given by

$$f_* \Omega^2_{X/N_d} \subset \ell^{\perp} \otimes \mathcal{O}_{N_d} \subset R^2 f_* \mathbb{Z} \otimes \mathcal{O}_{N_d}.$$

The action of any torsion-free subgroup $\Gamma \subset \widetilde{O}(\Lambda_d)$ on N_d naturally lifts to an action on $\ell^{\perp}$. Passing to the quotient defines a locally constant system also called V on $\Gamma \backslash N_d$ which still carries a VHS. As explained before, the action of Γ lifts naturally to an action on the total space of the family $f \colon X \longrightarrow N_d$. Hence, there exists also a universal family over $\Gamma \backslash N_d$ inducing V. In any case, with V over $\Gamma \backslash N_d$ one associates the VHS of weight one $Cl^+(V)$ as above. Alternatively, one can obtain $Cl^+(V)$ as the quotient of the VHS $Cl^+(\underline{\Lambda}_d)$ on N_d by the natural action of Γ. If instead of V one uses the VHS $V(1)$ of weight zero, one obtains a VHS of weight zero $Cl^+(V(1))$.

However, if we now assume that there exists a torsion-free subgroup $\widetilde{\Gamma} \subset \mathrm{Spin}(\Lambda_{d\mathbb{Q}}) \cap Cl^+(\Lambda_d)$ with $\tau \colon \widetilde{\Gamma} \xrightarrow{\sim} \Gamma$ under $\tau \colon \mathrm{Spin}(V) \longrightarrow SO(V)$ (see Remark **4.2.1**), then there exists another VHS of weight one on $\Gamma \backslash N_d$ that is algebraic. Indeed, instead of taking the quotient of $Cl^+(\underline{\Lambda}_d)$ by Γ one takes its quotient by $\widetilde{\Gamma}$ acting by left multiplication. As the polarization Q (depending on the choice of positive orthogonal vectors $f_1, f_2 \in \Lambda_d$) is preserved under left multiplication, the quotient by this action defines a polarized VHS of weight one on $\widetilde{\Gamma} \backslash N_d = \Gamma \backslash N_d$ which corresponds to a family of polarized abelian varieties

$$a \colon A \longrightarrow \Gamma \backslash N_d, \tag{4.1}$$

i.e. $R^1 a_* \mathbb{Z} \simeq \widetilde{\Gamma} \backslash Cl^+(\underline{\Lambda}_d)$. Note that this family of polarized abelian varieties is not only holomorphic, but due to Theorem 4.1 (or rather the analogous version for the moduli space of abelian varieties) in fact algebraic.

Observe that the two VHS $R^1 a_* \mathbb{Z}$ and $Cl^+(V)$ are fibrewise (non-canonically) isomorphic. Note also, that the right action of $C := Cl^+(\Lambda_d)^{\mathrm{op}}$ (see Section **4.2.4**), commutes with the left action of $\widetilde{\Gamma}$ and, therefore, descends to an action of the (constant) $\mathbb{Z}$-algebra C on the VHS $R^1 a_* \mathbb{Z}$ or, equivalently, on the abelian scheme (4.1).

The second reason for using the lift $\widetilde{\Gamma}$ to define a is that only with this definition one obtains the family version of (2.9) in Section **4.2.4**:

$$Cl^+(V(1)) \simeq \underline{\mathrm{End}}_C(R^1 a_* \mathbb{Z}), \tag{4.2}$$

which should be read as an isomorphism of VHS of weight zero and in particular of the underlying locally constant systems.

Let us, for a moment, think of the local systems in (4.2) as representations of $\pi_1(\Gamma\backslash N_d)^9$ on the fibres $\mathrm{Cl}^+(H^2(X_t,\mathbb{Z}(1))_p)$ and $\mathrm{End}_C(\mathrm{Cl}^+(H^2(X_t,\mathbb{Z})_p))$ for some fixed $t \in N$. Then the observation in Section 4.2.4 that (2.9) there is compatible with the action of CSpin and the fact that by construction the monodromy action factorizes over CSpin imply that (2.9) is indeed invariant under the monodromy action. It therefore corresponds to an isomorphism of the local systems in (4.2) as claimed.

Proposition 4.10 *Let V be a polarized VHS of K3 type over a smooth complex variety S, e.g. $V = \ell^\perp \subset R^2 f_* \mathbb{Z}$ for a polarized family of K3 surfaces $(X, L) \longrightarrow S$.*

Then there exist a finite étale cover $S' \longrightarrow S$, an abelian scheme $a\colon A \longrightarrow S'$, and an isomorphism of VHS of weight zero over S'

$$\mathrm{Cl}^+(V_{S'}(1)) \simeq \underline{\mathrm{End}}_C(R^1 a_* \mathbb{Z}),$$

where C is the constant $\mathbb{Z}$-algebra $\mathrm{Cl}^+(V_t)^{\mathrm{op}}$.

Proof We shall give the proof in the geometric situation, so that we can use the notation introduced before, and follow the approach in Section 4.2.

We have to choose a finite index subgroup Γ of $\Gamma_\ell := \{g \in \widetilde{O}(\Lambda_d) \mid g \equiv \mathrm{id}\,(\ell)\}$. Such a Γ is torsion free for $\ell \gg 0$ and we can consider the finite étale cover $S' \longrightarrow S$ obtained as the quotient by Γ of the étale cover of S that over $t \in S$ parametrizes all isometries $H^2(X_t,\mathbb{Z})_p \simeq \Lambda_d$. Then the classifying morphism $S' \longrightarrow \Gamma \backslash N_d$ is well defined and algebraic due to Borel's theorem 4.1.

Now let $\widetilde{\Gamma} \subset \mathrm{Spin}(\Lambda_{d\mathbb{Q}}) \cap \mathrm{Cl}^+(\Lambda_d)$ be the subgroup of all $\widetilde{g}$ with $\overline{\widetilde{g}} \equiv \mathrm{id}\,(\ell)$. Then $\tau\colon \widetilde{\Gamma} \xrightarrow{\sim} \Gamma$ and so we can pull back (4.1) and the isomorphism (4.2). $\qquad\square$

In general, we cannot expect A to exist over S; passing to the finite étale cover $S' \longrightarrow S$ really is necessary. Also note that the abelian scheme $a\colon A \longrightarrow S'$ is very special as it comes with complex multiplication provided by the action of C.

5 Appendix: Kulikov Models

The moduli space of polarized (or just quasi-polarized) K3 surfaces is not proper and it is an important and interesting problem to compactify it naturally, for example by allowing singular degenerations of K3 surfaces. The easiest case are one-dimensional degenerations. This leads to the notion of Kulikov models. In this Appendix we survey the theory of Kulikov models and mention a few additional results.

---

[9] Which factors via Γ, as the construction could have been performed over D_d.

5.1 Consider a one-dimensional degeneration of K3 surfaces. More precisely, let

$$X \longrightarrow \Delta \subset \mathbb{C}$$

be a proper, flat, surjective morphism from a smooth threefold X to a one-dimensional disk Δ such that all fibres X_t, $t \neq 0$, are K3 surfaces, and so in particular are smooth. The central fibre X_0 can be arbitrarily singular, reducible, and even non-reduced.

A *modification* of $X \longrightarrow \Delta$ is a family of the same type $X' \longrightarrow \Delta$ (so in particular X' is smooth), such that there exists a birational morphism $X \dashrightarrow X'$, which is compatible with the projections to Δ and an isomorphism over Δ^*. The degeneration is called *semistable* if the special fibre X_0 is reduced with local normal crossings.

After base change $\Delta \longrightarrow \Delta$, $z \longmapsto z^m$, every degeneration $X \longrightarrow \Delta$ admits a modification that is semistable. This is a consequence of a general result for degenerations of smooth varieties due to Mumford [290, Ch. II]. Most of it follows from Hironaka's resolution of singularities; the hard part is to get the central fibre reduced. Note that Mumford's result actually yields a semistable degeneration such that the irreducible components of $X_0 = \bigcup Y_i$ are smooth, i.e. X_0 has strict normal crossing.

If $X \longrightarrow \Delta$ is already semistable and we assume in addition that the irreducible components Y_i of the central fibre $X_0 = \bigcup Y_i$ are algebraic or, at least, Kähler, then, according to Kulikov [337] and Persson–Pinkham [492], the situation can be improved further.

Theorem 5.1 *Any semistable degeneration $X \longrightarrow \Delta$ of K3 surfaces admits a modification $f' : X' \longrightarrow \Delta$ with trivial canonical bundle $\omega_{X'}$.*

The new family $X' \longrightarrow \Delta$ is called a *Kulikov model* of the original degeneration and one would like to describe its central fibre.

5.2 The special fibre of a Kulikov model can be classified according to the following result due to Kulikov [335], Persson [490], and Friedman–Morrison [187].

Theorem 5.2 *Let $X \longrightarrow \Delta$ be a Kulikov degeneration. Then the central fibre X_0 is of one of the following three types:*

I X_0 *is a smooth K3 surface.*

II X_0 *is a chain of elliptic ruled surfaces Y_i with rational surfaces on either end and such that the curves $D_{ij} := Y_i \cap Y_j$, $i \neq j$, are (at worse nodal) elliptic, or*

III X_0 *is a union of rational surfaces Y_i such that for fixed i the curves $D_{ij} := Y_i \cap Y_j \subset Y_i$, $j \neq i$, form a cycle of (at most nodal) rational curves on Y_i and such that the dual graph of $X_0 = \bigcup Y_i$ is a triangulation of S^2.*

Without the assumption that the components are Kähler, other types of surfaces can occur as irreducible components of the central fibre X_0; see [458]. Also note that already for type I the Kulikov model need not be unique. For example, any (-2)-curve in the smooth central fibre X_0 with normal bundle $\mathcal{O}(-1)^{\oplus 2}$ in X can be flopped (Atiyah flop), which results in a non-isomorphic Kulikov model. The same phenomenon may cause the Kulikov model to be non-algebraic; cf. [231] for another example. See the article by Miranda and Morrison [415] for more on normal forms of the special fibre for types II and III.

Types I and II are viewed as the easy cases, (see [187]), so attention has focused on type III degenerations. For example in [182, 188], Friedman, together with Scattone, addresses the question of whether any X_0 as in type III with additional combinatorial requirements on the self-intersections $(D_{ij})^2_{\bar{Y}_i}$ can be smoothened to a degeneration of K3 surfaces. See also [351]. Note that after allowing a further base change, one can arrange all components Y_i to be smooth.

5.3   Each degeneration $X \longrightarrow \Delta$ induces a monodromy representation

$$\rho \colon \mathbb{Z} \simeq \pi_1(\Delta^*, t) \longrightarrow \mathrm{O}(H^2(X_t, \mathbb{Z}));$$

see Section 4.3. It is described by the action of a simple loop around the origin

$$T \colon H^2(X_t, \mathbb{Z}) \xrightarrow{\sim} H^2(X_t, \mathbb{Z}),$$

the *monodromy operator*, which is known to be quasi-unipotent [219, Thm. 3.1], i.e.

$$(T^m - \mathrm{id})^n = 0$$

for some $m, n > 0$, which are chosen minimal. For example, for the family $X \longrightarrow \Delta$ obtained by smoothening an A_1-singularity (see the proof of Proposition 7.5.5), the monodromy T is the reflection s_δ with δ the cohomology class corresponding to the exceptional (-2)-curve over the singularity. In this case, $m = 2$ and $n = 1$. In general, if the degeneration is semistable, then $m = 1$; i.e. T is unipotent.

Observe that, if T is the monodromy operator for a family $X \longrightarrow \Delta$, then T^m is the monodromy operator for the family obtained by base change $\Delta \longrightarrow \Delta, z \longmapsto z^m$. Hence, after base change, the monodromy becomes unipotent.

The next result determines the type of the Kulikov model in terms of the integers m and n; cf. [242, 338]. We let $X \longrightarrow \Delta$ be a one-dimensional degeneration of K3 surfaces with monodromy satisfying $(T^m - \mathrm{id})^n = 0$ such that $m, n > 0$ are minimal.

Theorem 5.3   *There exists a modification which is a Kulikov model if and only if the monodromy is unipotent, i.e. $m = 1$. For a Kulikov model $X \longrightarrow \Delta$, so $m = 1$, the central fibre X_0 is of type I if $n = 1$, of type II if $n = 2$, and of type III if $n = 3$.*

Note that it is not true in general that degenerations with trivial monodromy can be filled in smoothly; see [181] for a counterexample using quintics in $\mathbb{P}^3$. So, from this perspective, already the first assertion is not trivial. In fact, the historically first approach, due to Kulikov [336], to prove the surjectivity of the period map (cf. Theorem 7.4.1) was based on degenerations and relied on this fact. Roughly, it is enough to show that any smooth family of K3 surfaces $X^* \longrightarrow \Delta^*$ with trivial monodromy can be filled in smoothly. The approach via twistor lines presented in Chapter 7, although using the deep fact that K3 surfaces admit Ricci flat Kähler metrics, seems conceptually cleaner.

5.4 For the compactification of the moduli space of polarized K3 surfaces, however, a version of the above is needed that takes polarizations of the smooth fibres X_t, $t \in \Delta^*$, into account. This was initiated by Shepherd-Barron [553] (see also [352, Thms. 2.8 and 2.11]), who showed that a Kulikov model $X \longrightarrow \Delta$ with a line bundle L on X that induces a quasi-polarization L_t on all $X_{t \neq 0}$ can be modified to yield a family $X' \longrightarrow \Delta$ together with a line bundle L' that is also nef on the special fibre. However, the new family may not be a Kulkov model any longer. The singularities of the special fibre X'_0 have been studied in [315] for arbitrary surfaces. In the language of the minimal model program one strives for the relative log canonical model of the pair (X, L) and so the central fibre will have semi–log canonical surface singularities.

For low degree this has been investigated further in [183, 352, 526, 551, 552, 578, 599]. Global aspects leading to partial compactifications of the moduli space of polarized K3 surfaces by log smooth K3 surfaces have been studied by Olsson [481]. The comparison between the various compactifications using GIT, periods, etc., is very intricate even in low degrees.

The semistable MMP in positive characteristic plays a crucial role in Maulik's proof of the Tate conjecture [398]. The paper by Liedtke and Matsumoto [374] studies the situation in mixed characteristic, in particular proving an arithmetic analogue of Theorem 5.3 detecting whether a K3 surface has good reduction.

References and Further Reading

O'Grady in [465] proves that the rank of the Picard group (or more precisely the rank of its image in H^2) of $\overline{M}_d := \widetilde{O}(\Lambda_d) \setminus D_d$, of which one should think as the moduli space of quasi-polarized K3 surfaces (see Remark 4.5), can be arbitrarily large. The more recent article of Maulik and Pandharipande [399] investigates so-called Noether–Lefschetz loci which produce explicit divisors. Li and Tian in [361] and, more generally, Bergeron et al. in [59] prove that $\mathrm{Pic}(\overline{M}_d) \simeq H^2(\overline{M}_d, \mathbb{Q})$ for all d and that the Noether–Lefschetz divisors generate the cohomology.

Dolgachev in [150] gives a description of the moduli space of lattice polarized K3 surfaces as an arithmetic quotient of an appropriate period domain. We also highly recommend the lectures by Dolgachev and Kondō [154].

7

Surjectivity of the Period Map and Global Torelli

We present a proof of the global Torelli theorem that is not quite standard. It is inspired by the approach used in higher dimensions; see the Bourbaki talk [259] on Verbitsky's paper [612]. On the way, we prove the surjectivity of the period map, also deviating slightly from the classical arguments.

Starting with Section 3, we use that every complex K3 surface is Kähler, a deep result due to Siu and Todorov. The reader may also simply add this as a condition to the definition of a complex K3 surface. More importantly, we make use of the existence of Ricci-flat metrics on K3 surfaces (see Theorems 3.6 and 9.4.11), which is a consequence of the Calabi conjecture proved by Yau.

1 Deformation Equivalence of K3 Surfaces

Using the local Torelli theorem (see Proposition 6.2.8), one proves that all complex K3 surfaces are deformation equivalent, and hence diffeomorphic, to each other.

1.1 Two compact complex manifolds X_1 and X_2 are *deformation equivalent* if there exists a smooth proper holomorphic morphism $X \longrightarrow B$ such that

(i) The (possibly singular) base B is connected.
(ii) There exist points $t_1, t_2 \in B$ and isomorphisms $X_{t_1} \simeq X_1$ and $X_{t_2} \simeq X_2$.

By the theorem of Ehresmann, any two fibres of $X \longrightarrow B$ are diffeomorphic. Thus, deformation-equivalent complex manifolds are in particular diffeomorphic. The converse is in general not true; i.e. there exist diffeomorphic compact complex manifolds which are not deformation equivalent. However, producing explicit examples is not easy.[1]

For K3 surfaces, all topological invariants like Betti numbers and intersection form are independent of the particular K3 surface. In fact, even the Hodge numbers $h^{p,q}(X)$ do not depend on X. This may be seen as evidence for the following theorem of Kodaira [308, Thm. 13], which in particular shows that all complex K3 surfaces can be realized by some complex structure I on a fixed differentiable manifold M of dimension four, for example the differentiable manifold underlying a smooth quartic in $\mathbb{P}^3$.

Theorem 1.1 *Any two complex K3 surfaces are deformation equivalent.*

The theorem can also be seen as a consequence of the much harder global Torelli theorem; see Remark **6.3.6**. Moreover, the description of the moduli space of polarized K3 surfaces as in Corollary **6.4.3** shows that any two polarized K3 surfaces $(X_1, L_1), (X_2, L_2)$ are deformation equivalent in the sense that they are isomorphic to fibres of a polarized smooth family over a connected base.

Remark 1.2 Note that the theorem does not immediately yield the analogous result for algebraic K3 surfaces in positive characteristic, because smooth families might acquire singularities under reduction modulo p. It is known, however, that the moduli space of polarized K3 surfaces of fixed degree $2d$ is irreducible for $p^2 \nmid d$, due to work of Madapusi Pera [387]. In order to connect K3 surfaces of different degrees, one would need to prove the existence of K3 surfaces (over a fixed algebraically closed field) admitting polarizations L_1, L_2 with given $2d_1 = (L_1)^2$ and $2d_2 = (L_2)^2$.

1.2 Before presenting a proof of Theorem 1.1, we need a general density result which is useful in many other situations as well. It is closely related to the density of the Noether–Lefschetz locus; see Proposition **6.2.9**.

In the following, Λ denotes the K3 lattice $\Lambda := E_8(-1)^{\oplus 2} \oplus U^{\oplus 3}$ and $D \subset \mathbb{P}(\Lambda_\mathbb{C})$ is the period domain as introduced in Section **6.1.1**.

Proposition 1.3 *Let $0 \neq \alpha \in \Lambda$. Then the set*

$$\bigcup_{g \in O(\Lambda)} g(\alpha^\perp \cap D) = \bigcup_{g \in O(\Lambda)} g(\alpha)^\perp \cap D$$

is dense in D.

Proof We use the following observation: let $\Lambda = \Lambda' \oplus U$ be any orthogonal decomposition and let (e, f) be a standard basis of the hyperbolic plane U. For $B \in \Lambda'$

---

[1] In fact, two diffeomorphic algebraic surfaces which are not of general type are deformation equivalent; see [186]. However, this does not hold any longer for surfaces of general type. For simply connected counterexamples, see [107].

with $(B)^2 \neq 0$ we define the 'B-field shift' $\varphi_B \in O(\Lambda)$ by $\varphi_B(f) = f$, $\varphi_B(e) = e + B - ((B)^2/2) \cdot f$, and $\varphi_B(x) = x - (B.x)f$ for $x \in \Lambda'$. It is easy to see that indeed with this definition $\varphi_B \in O(\Lambda)$. This corresponds to multiplication by $\exp(B)$ as introduced in Section **14**.2.3.

Observe that for any $y \in \Lambda_{\mathbb{R}}$ one has

$$\lim_{k \to \infty} \varphi_B^k[y] = [f] \in \mathbb{P}(\Lambda_{\mathbb{R}})$$

whenever the Λ'-component y' of y is either trivial or satisfies $(B.y') \neq 0$. Hence, the closure $\overline{O}$ of the orbit $O(\Lambda) \cdot [\alpha] \subset \mathbb{P}(\Lambda_{\mathbb{R}})$ contains an isotropic vector. In other words, for the closed set $Z \subset \mathbb{P}(\Lambda_{\mathbb{R}})$ of all isotropic vectors up to scaling, one has $\overline{O} \cap Z \neq \emptyset$.

It is enough to show that for any point in D corresponding to a positive plane $P \subset \Lambda_{\mathbb{R}}$ there exists an automorphism $g \in O(\Lambda)$ such that $g(\alpha)$ is arbitrarily close to $P^{\perp}$. Indeed, in this case there exists a subspace $W \subset \Lambda_{\mathbb{R}}$ of codimension two and signature $(1, 19)$, which is close to $P^{\perp}$ and contains $g(\alpha)$, and, therefore, the point in D corresponding to $W^{\perp}$ is contained in $g(\alpha)^{\perp}$ and close to the point corresponding to P.

Since $P^{\perp}$ contains some isotropic vector $v \in P^{\perp}$, so $[v] \in Z$, it suffices to show not only that Z intersects the closure $\overline{O}$ of the orbit $O(\Lambda) \cdot [\alpha]$, but that in fact $Z \subset \overline{O}$ and so in particular $[v] \in \overline{O}$. However, $\overline{O} \cap Z$ is closed and $O(\Lambda)$-invariant. It therefore is enough to prove that for any (one would be enough) $[y] \in Z$ the $O(\Lambda)$-orbit $O_y := O(\Lambda) \cdot [y]$ is dense in Z. This is proved in two steps.

(i) The closure $\overline{O}_y$ contains the subset $\{[x] \in Z \mid x \in \Lambda\}$. Indeed, for any primitive $x \in \Lambda$ with $x^2 = 0$ one finds an orthogonal decomposition $\Lambda = \Lambda' \oplus U$ with $x = f$, where (e, f) is a standard basis of the hyperbolic plane U; cf. Corollary **14**.1.14. As we have observed above, $\lim \varphi_B^k[y] = [f] = [x]$ if $B \in \Lambda'$ is chosen such that $(B)^2 \neq 0$ and $(B.y') \neq 0$. Hence, $[x] \in \overline{O}_y$.

(ii) The set $\{[x] \in Z \mid x \in \Lambda\}$ is dense in Z. Indeed, if we write $\Lambda = \Lambda' \oplus U$ as before, then the dense open subset $V \subset Z$ of points of the form $[x' + \lambda e + f]$ with $\lambda \in \mathbb{R}$, $x' \in \Lambda'_{\mathbb{R}}$, can be identified with the affine quadric $\{(x', \lambda) \mid 2\lambda + (x')^2 = 0\} \subset \Lambda_{\mathbb{R}} \times \mathbb{R}$ and thus is given as the graph of the rational polynomial map $\Lambda'_{\mathbb{R}} \longrightarrow \mathbb{R}$, $x' \longmapsto -(x')^2/2$. Therefore, the rational points are dense in V.

Combining both steps yields the assertion.   □

1.3   We are now ready to give the proof of the theorem.

PROOF OF THEOREM 1.1. We follow the arguments given by Le Potier in [54, Exp. VI]. The first step consists of showing that a K3 surface X with $\mathrm{Pic}(X)$ generated by a line bundle L of square $(L)^2 = 4$ is a quartic surface; see Example **2**.3.9.

Next, one shows that for any K3 surface X_0 one of the fibres of its universal deformation $X \longrightarrow \mathrm{Def}(X_0)$ has a Picard group of the above form. Since the base

Def(X_0) is by definition only the germ of a complex manifold, this result actually says more, namely that any K3 surface is arbitrarily close to a smooth quartic in $\mathbb{P}^3$.

Fix an isometry $H^2(X_0, \mathbb{Z}) \simeq \Lambda$. Then, by the local Torelli theorem, Proposition **6.2.8**, the period map $\mathcal{P} \colon \mathrm{Def}(X_0) \longrightarrow D \subset \mathbb{P}(\Lambda_{\mathbb{C}})$ is a local isomorphism. Now apply Proposition 1.3 to a primitive $\alpha \in \Lambda$ with $(\alpha)^2 = 4$ to conclude that there exists a primitive class $\ell \in \Lambda$ with $(\ell)^2 = 4$ and such that $\mathcal{P}(\mathrm{Def}(X_0)) \cap \ell^\perp \neq \emptyset$. As $\mathcal{P}(\mathrm{Def}(X_0)) \subset D$ is open, the Picard group of the fibre X_t over the very general point $t \in \mathcal{P}(\mathrm{Def}(X_0)) \cap \ell^\perp$ is generated by ℓ (using the natural isometry $H^2(X_t, \mathbb{Z}) \simeq H^2(X_0, \mathbb{Z}) \simeq \Lambda$). Hence, X_t is a quartic.

To conclude the proof, it suffices to observe that any two smooth quartics in $\mathbb{P}^3$ are deformation equivalent, as smooth quartics are all parametrized by an open and hence connected subset of $|\mathcal{O}_{\mathbb{P}^3}(4)|$. □

As explained in Remark **1.3.6**, it is easy to show that the profinite completion of $\pi_1(X)$ of a complex K3 surface (or the algebraic fundamental group of an algebraic K3 surface over a separably closed field) is trivial. This is strengthened by the following corollary.

Corollary 1.4 *Every complex K3 surface is simply connected.*

Proof This is a consequence of standard Lefschetz theory. Morse theory can be used to show that the relative homotopy groups $\pi_i(M, X)$ are trivial for $i < \dim(M)$, when M is a compact complex manifold and $X \subset M$ is the zero set of a regular section of an ample line bundle on M; see e.g. Le Potier's talk [54, Exp. VI]. This can then be applied to the case of a quartic $X \subset \mathbb{P}^3$ to obtain $\pi_2(\mathbb{P}^3, X) = 0$. The latter implies the injectivity of the natural map $\pi_1(X) \hookrightarrow \pi_1(\mathbb{P}^3)$. Since $\mathbb{P}^3$ is simply connected, this proves the result. □

2 Moduli Space of Marked K3 Surfaces

We first recall notions from the previous chapter, most importantly the notion of the moduli space of marked K3 surfaces. This space is not Hausdorff and we discuss its 'Hausdorff reduction' in Section 2.2.

2.1 By definition (see Section **6.3.3**), the *moduli space of marked K3 surfaces*

$$N = \{(X, \varphi)\}/_\sim$$

parametrizes marked K3 surfaces (X, φ) up to equivalence. A marking $\varphi \colon H^2(X, \mathbb{Z}) \xrightarrow{\sim} \Lambda$ is an isometry between $H^2(X, \mathbb{Z})$ with its intersection form and the K3 lattice $\Lambda = E_8(-1)^{\oplus 2} \oplus U^{\oplus 3}$. Two marked K3 surfaces, (X, φ) and (X', φ'), are equivalent,

$(X, \varphi) \sim (X', \varphi')$, if there exists a biholomorphic map $g \colon X \xrightarrow{\sim} X'$ such that $\varphi \circ g^* = \varphi'$.

The moduli space N has the structure of a 20-dimensional complex manifold, obtained by gluing the bases of the universal deformations $X \longrightarrow \mathrm{Def}(X_0)$ for all K3 surfaces X_0. In particular, the $\mathrm{Def}(X_0)$ form a basis of open sets in N.

Also recall that the local period maps glue to the global period map

$$\mathcal{P} \colon N \longrightarrow D \subset \mathbb{P}(\Lambda_{\mathbb{C}}),$$

where $D = \{x \in \mathbb{P}(\Lambda_{\mathbb{C}}) \mid (x)^2 = 0,\ (x.\bar{x}) > 0\}$ is the period domain. The period map is a local isomorphism, which turns out to be surjective on each connected component of N.

Two words of warning. First, N is a complex manifold but it is not Hausdorff. Second, it is a priori not clear that the universal families $X \longrightarrow \mathrm{Def}(X_0)$ and $Y \longrightarrow \mathrm{Def}(Y_0)$ glue over the intersection $\mathrm{Def}(X_0) \cap \mathrm{Def}(Y_0)$ in N. Of course, one would like to have a universal family $(X, \varphi) \longrightarrow N$ and it does exist, but in order to glue the local families one uses that non-trivial automorphisms of K3 surfaces always act non-trivially on the cohomology; cf. Section **6**.3.3 and Proposition **15**.2.1.

2.2 Using the period map, the moduli space N can be 'made Hausdorff'. This slightly technical procedure was introduced by Verbitsky in the context of general compact hyperkähler manifolds. The result can be phrased as follows.

Proposition 2.1 *The period map $\mathcal{P} \colon N \longrightarrow D \subset \mathbb{P}(\Lambda_{\mathbb{C}})$ factorizes uniquely over a Hausdorff space $\overline{N}$, i.e. there exists a complex Hausdorff manifold $\overline{N}$ and locally biholomorphic maps factorizing the period map:*

$$\mathcal{P} \colon N \longrightarrow\!\!\!\!\twoheadrightarrow \overline{N} \longrightarrow D,$$

such that $(X, \varphi), (X', \varphi') \in N$ map to the same point in $\overline{N}$ if and only if they are inseparable points of N.

A complete and elementary proof can be found in the Bourbaki talk [259]. Note however that the general notion of a 'Hausdorff reduction' of a non-Hausdorff manifold is not well defined. In fact, Proposition 2.1 relies on a result that describes the consequence of $(X, \varphi), (X', \varphi')$ being non-separated geometrically.

Proposition 2.2 *Suppose $(X, \varphi), (X', \varphi') \in N$ are distinct inseparable points. Then $X \simeq X'$ and $\mathcal{P}(X, \varphi) = \mathcal{P}(X', \varphi')$ is contained in $\alpha^{\perp}$ for some $0 \neq \alpha \in \Lambda$.*

Proof The second assertion is equivalent to $\rho(X) = \rho(X') \geq 1$; see Remark 3.8. The composition $\psi := \varphi^{-1} \circ \varphi' \colon H^2(X', \mathbb{Z}) \xrightarrow{\sim} H^2(X, \mathbb{Z})$ is a Hodge isometry that is not induced by an isomorphism $X' \xrightarrow{\sim} X$, as otherwise $(X, \varphi) = (X', \varphi')$ as points in N.

Moreover, ψ preserves the positive cone, i.e. $\psi(\mathcal{C}_{X'}) = \mathcal{C}_X$, as the only other possibility $\psi(\mathcal{C}_{X'}) = -\mathcal{C}_X$ contradicts the assumption that the two points are inseparable; see the arguments below. For the definition of the positive cone, see Remark **1.2.2** and Section **8.1.1**.

If one had proved already that any Hodge isometry that preserves the Kähler cone, i.e. $\psi(\mathcal{K}_{X'}) = \mathcal{K}_X$, is induced by an isomorphism (see Theorem 5.3), then one could conclude that $\mathcal{K}_X$ is strictly smaller than $\mathcal{C}_X$. By Theorem **8.5.2** this would imply the existence of a (smooth rational) curve $C \subset X$ and, therefore, $\rho(X) \geq 1$. Alternatively, one could use Proposition 3.7.

However, the description of isomorphisms between X and X' in terms of Hodge isometries preserving the Kähler cone is better seen as a part of the global Torelli theorem we want to prove here.

So, in order not to turn in circles, one rather proves the assertion directly by a degeneration argument that was first used by Burns and Rapoport in [93]. The following sketch is copied from [250, Thm. 4.3] and [259, Prop. 4.7]. One first constructs a bimeromorphic correspondence between X and X' roughly as follows. By assumption, there exists a sequence $t_i \in N$ converging simultaneously to (X, φ) and to (X', φ'). For the universal deformations of X and X' this corresponds to isomorphisms $g_i \colon X_{t_i} \longrightarrow X'_{t_i}$ compatible with the induced markings of X_{t_i} and X'_{t_i}. The graphs Γ_{g_i} converge to a cycle

$$\Gamma_\infty = Z + \sum Y_i \subset X \times X'$$

of which the component Z defines a bimeromorphic correspondence and the components Y_i do not dominate X or X'. As X and X' are minimal surfaces of non-negative Kodaira dimension, the bimeromorphic correspondence Z is in fact the graph of an isomorphism $X \simeq X'$. Now, either the Y_i do not occur, and then $(X, \varphi) \simeq (X', \varphi')$ via Z, or their images in X and X' yield non-trivial curves and hence $\rho(X) = \rho(X') \geq 1$. □

3 Twistor Lines

We start with a discussion of twistor lines in the period domain D and show later how they can be lifted to curves in the moduli space N.

3.1 For the following, Λ can be any lattice of signature $(3, b - 3)$.

A subspace $W \subset \Lambda_{\mathbb{R}}$ of dimension three is called a *positive three-space* if the restriction of $(\, . \,)$ to W is positive definite. To such a W one associates its *twistor line*

$$T_W := D \cap \mathbb{P}(W_{\mathbb{C}}),$$

which is a smooth quadric in $\mathbb{P}(W_{\mathbb{C}}) \simeq \mathbb{P}^2$. Hence, as a complex manifold $T_W \simeq \mathbb{P}^1$.

Two distinct points $x, y \in D$ are contained in one twistor line if and only if their associated positive planes $P(x)$ and $P(y)$ span a positive three-space $\langle P(x), P(y) \rangle \subset \Lambda_{\mathbb{R}}$. Here we use the interpretation of D as the Grassmannian of oriented positive planes; see Proposition **6**.1.5.

A twistor line T_W is called *generic* if $W^{\perp} \cap \Lambda = 0$ or, equivalently, if there exists a vector $w \in W$ with $w^{\perp} \cap \Lambda = 0$ or, still equivalently, if there exists a point $x \in T_W$ with $x^{\perp} \cap \Lambda = 0$. In fact, if W is generic, then $x^{\perp} \cap \Lambda = 0$ for all except countably many points $x \in T_W$.

Definition 3.1   Two points $x, y \in D$ are called *equivalent* if there exists a chain of generic twistor lines $T_{W_1}, \ldots, T_{W_k}$ and points $x = x_1, \ldots, x_{k+1} = y$ with $x_i, x_{i+1} \in T_{W_i}$.

The following rather easy observation suffices to prove the global surjectivity of the period map; see Section 4.1.

Proposition 3.2   *Any two points $x, y \in D$ are equivalent.*

Proof   We follow Beauville's account in [54]. Since D is connected, it suffices to show that every equivalence class is open.

Consider $x \in D$ and choose an oriented basis a, b for the corresponding positive plane, i.e. $P(x) = \langle a, b \rangle$. Pick c such that $\langle a, b, c \rangle$ is a positive three-space. Then for any (a', b') in an open neighbourhood of (a, b) the spaces $\langle a, b', c \rangle$ and $\langle a', b', c \rangle$ are still positive three-spaces. Let T_1, T_2, and T_3 be the twistor lines associated with $\langle a, b, c \rangle$, $\langle a, b', c \rangle$ and $\langle a', b', c \rangle$, respectively. Then $P(x) = \langle a, b \rangle, \langle a, c \rangle \in T_1, \langle a, c \rangle, \langle b', c \rangle \in T_2$, and $\langle b', c \rangle, \langle a', b' \rangle \in T_3$. Thus, x and $\langle a', b' \rangle$ are connected via the chain of the three twistor lines T_1, T_2, and T_3.

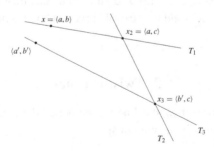

If we choose in the above argument c such that $c^{\perp} \cap \Lambda = 0$, then the twistor lines associated with the positive three-spaces $\langle a, b, c \rangle$, $\langle a, b', c \rangle$, and $\langle a', b', c \rangle$ are all generic. Hence, x and the period corresponding to $\langle a', b' \rangle$ are equivalent.   □

In order to prove that the period map is a covering map, which eventually leads to the global Torelli theorem, one also needs a local version of the surjectivity (cf. Section 4.2) which in turn relies on a local version of Proposition 3.2.

In the following, we consider *balls in D* and write $B \subset \overline{B} \subset D$ when $\overline{B}$ is a closed ball in a differentiable chart in D. In particular, B is the open set of interior points in $\overline{B}$.

Definition 3.3 Two points $x, y \in B \subset \overline{B} \subset D$ are called *equivalent as points in B* if there exist a chain of generic twistor lines $T_{W_1}, \ldots, T_{W_k}$ and points $x = x_1, \ldots, x_{k+1} = y \in B$ such that x_i, x_{i+1} are contained in the same connected component of $T_{W_i} \cap B$.

The proof of Proposition 3.2 can be adapted to prove the following local version, only this time nearby points are connected by a chain of four twistor lines. We refer to [259, Prop. 3.10] for details.

Proposition 3.4 *For a given ball $B \subset \overline{B} \subset D$ any two points $x, y \in B$ are equivalent as points in B.*

A much easier and intuitively rather obvious result is the following [259, Lem. 3.11].

Lemma 3.5 *Consider a ball $B \subset \overline{B} \subset D$. Then, for any point $x \in \partial B := \overline{B} \setminus B$ there exists a generic twistor line $x \in T_W \subset D$ with $x \in \partial(B \cap T_W)$. In other words, the boundary of B can be connected to its interior by means of generic twistor lines.*

3.2 To use twistor lines geometrically, the existence of hyperkähler metrics is crucial. The next result is a consequence of Yau's solution to the Calabi conjecture (cf. Theorem 9.4.11) for which no easier or more direct proof for the case of K3 surfaces is known. There are however various proofs available for the fact that K3 surfaces are always Kähler. The gaps in the original proof of Todorov [601] were filled by Siu in [571]. The result completed the proof of a conjecture by Kodaira that any compact complex surface with even Betti number is Kähler. A more direct proof of Kodaira's conjecture, not relying on the Kodaira classification of surfaces, was later given by Buchdahl [90] and independently by Lamari [346], both relying on results of Demailly on smoothing of currents. Since $b_1(X) = 0$ for complex K3 surfaces (see Section 1.3.2), their result provides a new proof for the fact that K3 surfaces are Kähler.

For the following it is useful to think of a K3 surface X as a differentiable manifold M endowed with a complex structure $I \in \mathrm{End}(TM)$.

Theorem 3.6 *For any Kähler class $\alpha \in H^2(X, \mathbb{R})$ there exists a Kähler metric g and complex structures J and K such that:*

(i) *The metric g is Kähler with respect to the three complex structures I, J, and K.*
(ii) *The Kähler form $\omega_I := g(I\,,\,)$ represents α, i.e. $[\omega_I] = \alpha$.*
(iii) *The complex structures I, J, and K satisfy the usual relations $K = I \circ J = -J \circ I$.*

In fact for each $(a, b, c) \in S^2$ also $\lambda = aI + bJ + cK$ is a complex structure on M with respect to which g is Kähler. The corresponding Kähler form is $\omega_\lambda := g(\lambda\,,\,)$ and there exists a natural non-degenerate holomorphic two-form σ_λ on (M, λ), e.g. $\sigma_J = \omega_K + i\omega_I$. Thus, one obtains a family of K3 surfaces (M, λ) parametrized by points $(a, b, c) \in S^2 \simeq \mathbb{P}^1$. This can indeed be put together to form a holomorphic family as follows. Let $\mathbb{I}$ be the endomorphism of the tangent bundle of $M \times \mathbb{P}^1$ defined by

$$\mathbb{I}: T_m M \oplus T_\lambda \mathbb{P}^1 \longrightarrow T_m M \oplus T_\lambda \mathbb{P}^1, \ (v, w) \longmapsto (\lambda(v), I_{\mathbb{P}^1}(w)).$$

Then $\mathbb{I}$ is an almost complex structure which can be shown to be integrable; see [243]. The complex threefold obtained in this way is denoted $\mathcal{X}(\alpha)$. The second projection defines a smooth holomorphic map

$$\mathcal{X}(\alpha) \longrightarrow T(\alpha) := \mathbb{P}^1,$$

which is called the *twistor space* associated with the Kähler class α. By construction, the fibre over λ is the K3 surface described by (M, λ).

Since $\mathcal{X}(\alpha)$ as a differentiable manifold is simply $M \times \mathbb{P}^1$, any marking $\varphi: H^2(X, \mathbb{Z}) \xrightarrow{\sim} \Lambda$ of $X = (M, I)$ extends naturally to a marking of all fibres (M, λ). Thus, the period map defines a holomorphic map $\mathcal{P}: T(\alpha) \longrightarrow D \subset \mathbb{P}(\Lambda_\mathbb{C})$. In fact, the period map identifies $T(\alpha) = \mathbb{P}^1$ with the twistor line $T_{W_\alpha} \subset D$ associated with the positive three-space $W_\alpha := \varphi\langle[\omega_I], [\mathrm{Re}(\sigma_I)], [\mathrm{Im}(\sigma_I)]\rangle = \varphi(\mathbb{R} \cdot \alpha \oplus (H^{2,0}(X) \oplus H^{0,2}(X))_\mathbb{R})$, i.e.

$$\mathcal{P}: \mathbb{P}^1 \simeq T(\alpha) \xrightarrow{\sim} T_{W_\alpha} \subset D.$$

We shall need a converse to this observation, i.e. a result that describes twistor lines in D that can be realized in this way. In particular, one wants to know which classes in $H^2(X, \mathbb{R})$ are Kähler. A complete answer to this question is known and is discussed in Chapter 8, but for our purposes here the following result is sufficient.

Suppose X is a K3 surface which we assume to be Kähler. Then the *Kähler cone* $\mathcal{K}_X \subset H^{1,1}(X, \mathbb{R})$ of all Kähler classes is contained in the *positive cone* $\mathcal{C}_X$:

$$\mathcal{K}_X \subset \mathcal{C}_X \subset H^{1,1}(X, \mathbb{R}),$$

which is the distinguished component of the set $\{\alpha \in H^{1,1}(X, \mathbb{R}) \mid (\alpha)^2 > 0\}$ that contains one Kähler class (and hence in fact all); cf. Section **8.5.1**. The algebraic analogue of the positive cone was defined in Remark **1.2.2**. The Kähler cone plays the role of the ample cone in our setting.

Proposition 3.7 *If* $\mathrm{Pic}(X) = 0$*, then any class α in the positive cone $\mathcal{C}_X$ is Kähler:*

$$\rho(X) = 0 \implies \mathcal{K}_X = \mathcal{C}_X.$$

Proof This can be seen as a consequence of a deep theorem due to Demailly and Paun [147]: for any compact Kähler manifold X the Kähler cone $\mathcal{K}_X \subset H^{1,1}(X, \mathbb{R})$ is a connected component of the cone of classes $\alpha \in H^{1,1}(X, \mathbb{R})$ defined by the condition $\int_Z \alpha^d > 0$ for all subvarieties $Z \subset X$ of dimension $d > 0$.

Clearly, if X is a K3 surface, only curves and X itself have to be tested. However, for $\mathrm{Pic}(X) = 0$, there are no curves in X and only the integral for $Z = X$ has to be computed. But this condition reads $(\alpha)^2 > 0$, i.e. $\pm\alpha \in \mathcal{C}_X$. □

There are more direct approaches to the above due to Buchdahl [90] and Lamari [346] as well as more complete results describing the Kähler cone for any K3 surface; see Theorem **8.5.2**.

Remark 3.8 If the period of a K3 surface X is known, then it is easy to decide whether $\mathrm{Pic}(X) = 0$. In fact, if $x = \mathcal{P}(X, \varphi)$, then $\mathrm{Pic}(X) \neq 0$ if and only if there exists a class $0 \neq \alpha \in \Lambda$ with $x \in \alpha^\perp$. Indeed, $\varphi^{-1}(\alpha) \in H^2(X, \mathbb{Z})$ is a $(1, 1)$-class (or, equivalently, is contained in $\mathrm{NS}(X) = H^{1,1}(X) \cap H^2(X, \mathbb{Z})$) if and only if it is orthogonal to $H^{2,0}(X)$. But by definition of the period map, x spans $\varphi(H^{2,0}(X))$.

Proposition 3.9 *Consider a marked K3 surface $(X, \varphi) \in N$ and assume that its period $\mathcal{P}(X, \varphi)$ is contained in a generic twistor line $T_W \subset D$. Then there exists a unique lift of T_W to a curve in $\overline{N}$ through (X, φ), i.e. there exists a commutative diagram*

with (X, φ) in the image of $\tilde{\imath}$.

Proof Since $\mathcal{P}\colon \overline{N} \longrightarrow D$ is locally biholomorphic, the inclusion $i\colon \Delta \subset T_W \hookrightarrow D$ of a small open one-dimensional disk containing $0 = \mathcal{P}(X, \varphi) \in \Delta$ can be lifted to $\tilde{\imath}\colon \Delta \hookrightarrow \overline{N}, t \mapsto (X_t, \varphi_t)$ with $\tilde{\imath}(0) = (X, \varphi)$. By construction, the space $\overline{N}$ is Hausdorff (see Proposition 2.1) and hence $\tilde{\imath}\colon \Delta \hookrightarrow \overline{N}$ is unique.

As T_W is a generic twistor line, the set $T_W \cap \bigcup_{0 \neq \alpha \in \Lambda} \alpha^\perp$ is countable and thus for general $t \in \Delta$ one has $\mathrm{Pic}(X_t) = 0$; see Remark 3.8. Let us fix such a general t and denote by σ_t a generator of $H^{2,0}(X_t)$.

The natural marking of X_t induced by φ shall be denoted $\varphi_t\colon H^2(X_t, \mathbb{Z}) \xrightarrow{\sim} \Lambda$ and so, by construction, $\varphi_t(\sigma_t) \in W_{\mathbb{C}}$. Therefore, there exists a class $\alpha_t \in H^2(X_t, \mathbb{R})$ such that $\varphi_t(\alpha_t)$ is orthogonal to $\varphi_t\langle\mathrm{Re}(\sigma_t), \mathrm{Im}(\sigma_t)\rangle \subset W$ and contained in W. Hence, α_t is of type $(1, 1)$ on X_t and $\pm\alpha_t \in \mathcal{C}_{X_t}$, as W is a positive three-space. Due to Proposition 3.7 and using $\mathrm{Pic}(X_t) = 0$ for our fixed generic t, this implies $\pm\alpha_t \in \mathcal{K}_{X_t}$.

Now consider the twistor space $\mathcal{X}(\alpha_t) \longrightarrow T(\alpha_t)$ for X_t endowed with the Kähler class $\pm\alpha_t$. Since $\varphi_t\langle\alpha_t, \mathrm{Re}(\sigma_t), \mathrm{Im}(\sigma_t)\rangle = W$, the period map yields a natural identification $\mathcal{P}\colon T(\alpha_t) \xrightarrow{\sim} T_W$.

Both $T(\alpha_t)$ and $\tilde{i}(\Delta)$ contain the point t and map locally isomorphically to T_W. Again by the uniqueness of lifts for a local homeomorphism between Hausdorff spaces, this proves $0 \in T(\alpha_t)$, which yields the assertion. $\square$

4 Local and Global Surjectivity of the Period Map

The surjectivity of the period map is a direct consequence of the description of the Kähler cone of a general K3 surface as provided by Proposition 3.9. Section 4.2 presents a local version of it, which is crucial for the proof of the global Torelli theorem given in Section 5.

4.1 The surjectivity of the period map was first proved by Todorov [601]. The algebraic case had earlier been studied by Kulikov [335]. The proof given here is closer to Siu's [570] as presented by Beauville in [54]. However, our approach differs in one crucial step. Whereas classically the period map is shown to be surjective by using the Nakai criterion for ampleness on a projective K3 surface (see Proposition **2.1.4**), we here use that by Proposition 3.7 all classes in the positive cone of a K3 surface of Picard number zero are Kähler. This is again inspired by the higher-dimensional theory in [250]. The following is a slightly refined version of Theorem **6.3.1**.

Theorem 4.1 (Surjectivity of the period map)   *Let N^o be a connected component of the moduli space N of marked K3 surfaces. Then the restriction of the period map*

$$\mathcal{P}\colon N^o \longrightarrow D \subset \mathbb{P}(\Lambda_{\mathbb{C}})$$

is surjective.

Proof   Since by Proposition 3.2 any two points $x, y \in D$ are equivalent, it is enough to show that $x \in \mathcal{P}(N^o)$ if and only if $y \in \mathcal{P}(N^o)$ for any two points $x, y \in T_W \subset D$ contained in a generic twistor line T_W. This is an immediate consequence of Proposition 3.9 which shows that the generic twistor line T_W can be lifted through any given pre-image (X, φ) of x. Indeed, then y is also contained in the image of the lift of T_W. $\square$

4.2   Recall that a continuous map $\pi\colon T \longrightarrow D$ is a *covering space* if every point in D admits an open neighbourhood $U \subset D$ such that $\pi^{-1}(U)$ is the disjoint union $\coprod U_i$ of open subsets $U_i \subset T$ such that the maps $\pi\colon U_i \xrightarrow{\sim} U$ are homeomorphisms.

The local version of the surjectivity relies on the following technical but intuitively quite obvious criterion for a local homeomorphism to have the covering property. A proof relying on arguments due to Markman in [612] can be found in [259, Prop. 5.6].

Lemma 4.2 *Suppose a continuous map* $\pi: T \longrightarrow D$ *between topological Hausdorff manifolds is a local homeomorphism. Then* $\pi: T \longrightarrow D$ *is a covering space if for any ball* $B \subset \overline{B} \subset D$ *and any connected component* C *of the closed subset* $\pi^{-1}(\overline{B})$ *one has* $\pi(C) = \overline{B}$.

The surjectivity of $\mathcal{P}: N \longrightarrow D$ is implied by the covering property. Thus, the next assertion can be seen as a local and stronger version of Theorem 4.1, very much like Proposition 3.4 is a local and stronger version of Proposition 3.2.

Proposition 4.3 *The period map induces a covering space* $\mathcal{P}: \overline{N} \longrightarrow D$.

Proof In order to apply Lemma 4.2, one first adapts the arguments of the proof of Theorem 4.1 to show $B \subset \mathcal{P}(C)$. Clearly, $\mathcal{P}(C)$ contains at least one point of B, because $\mathcal{P}$ is a local homeomorphism.

Next, due to Proposition 3.4, any two points $x, y \in B$ are equivalent as points in B. Thus, it suffices to show that $x \in \mathcal{P}(C)$ if and only if $y \in \mathcal{P}(C)$ for any two points $x, y \in B$ contained in the same connected component of the intersection $T_W \cap B$ with T_W a generic twistor line. If $x = \mathcal{P}(X, \varphi)$ with $(X, \varphi) \in C$, choose a local lift of the inclusion $x \in \Delta \subset T_W$ according to Proposition 3.9 and then argue literally as in the proof of Theorem 4.1. The assumption that x, y are contained in the same connected component of $T_W \cap B$ ensures that the twistor deformation $T(\alpha_t)$ constructed in the proof of Proposition 3.9 connects (X, φ) to a point over y that is indeed still contained in C.

It remains to prove that also the boundary $\overline{B} \setminus B$ is contained in $\mathcal{P}(C)$. For this apply Lemma 3.5 to any point $x \in \overline{B} \setminus B$ and lift the generic twistor line connecting x with a point in B to a twistor deformation as before. □

5 Global Torelli Theorem

We deviate from the traditional approach and discuss how the global properties of the period map established above can be used to prove the global Torelli theorem. In the end, it is reduced to a statement about monodromy groups.

5.1 An immediate consequence of the fact that the period map on the Hausdorff reduction $\overline{N}$ of N is a covering map is the following.

Corollary 5.1 *The period map* $\mathcal{P}: N \longrightarrow D$ *is generically injective on each connected component* N^o *of* N.

Proof   Here, 'generically injective' means injective on the complement of a countable union of proper analytically closed subsets.

If N^o is a connected component, then its Hausdorff reduction $\overline{N}^o$, which is a connected component of $\overline{N}$, is a covering space of the period domain D. However, D is simply connected, which can be deduced from the description $D \simeq O(3,19)/SO(2) \times O(1,19)$ given in Proposition 6.1.5; cf. [252, Sec. 4.7]. As any connected covering space of a simply connected target must be a homeomorphism, this proves

$$\mathcal{P}\colon \overline{N}^o \xrightarrow{\sim} D.$$

Hence, by Propositions 2.1, 2.2, and 3.7, $N^o \longrightarrow \overline{N}^o \xrightarrow{\sim} D$ is generically injective.   □

Remark 5.2   If $(X,\varphi),(X',\varphi') \in N$ are contained in the same connected component with $\mathcal{P}(X,\varphi) = \mathcal{P}(X',\varphi')$, then they coincide as points in N, and hence $X \simeq X'$, or are at least inseparable. In the second case, one argues as in the proof of Proposition 2.2 to see that X and X' are bimeromorphic. Hence, also in this case, $X \simeq X'$.

5.2   Let us now relate the above result to the classical formulation of the global Torelli theorem which for convenience we recall from Section 6.3.4.

Theorem 5.3 (global Torelli theorem)   *Two complex K3 surfaces X and X' are isomorphic if and only if there exists a Hodge isometry $H^2(X,\mathbb{Z}) \simeq H^2(X',\mathbb{Z})$.*

Moreover, for any Hodge isometry $\psi\colon H^2(X,\mathbb{Z}) \xrightarrow{\sim} H^2(X',\mathbb{Z})$ with $\psi(\mathcal{K}_X) \cap \mathcal{K}_{X'} \neq \emptyset$ there exists a (unique) isomorphism $f\colon X' \xrightarrow{\sim} X$ with $f^* = \psi$.

The 'only if' direction is obvious, as any biholomorphic map induces a Hodge isometry. So, let us assume a Hodge isometry $\psi\colon H^2(X,\mathbb{Z}) \xrightarrow{\sim} H^2(X',\mathbb{Z})$ is given. Pick any marking $\varphi\colon H^2(X,\mathbb{Z}) \xrightarrow{\sim} \Lambda$ and let $\varphi' := \varphi \circ \psi^{-1}\colon H^2(X',\mathbb{Z}) \xrightarrow{\sim} \Lambda$ be the induced marking of X'. Then $\mathcal{P}(X,\varphi) = \mathcal{P}(X',\varphi')$. Thus, if one knew already that (X,φ) and (X',φ') are contained in the same connected component N^o of N, then $X \simeq X'$ by Corollary 5.1 and Remark 5.2.

The remaining question now is whether (X,φ) and (X',φ') are always in the same connected component. First of all, we may change the sign of φ' without affecting $\mathcal{P}(X,\varphi) = \mathcal{P}(X',\varphi')$, for $-\mathrm{id} \in O(\Lambda)$ acts trivially on the period domain D. In fact, Corollary 5.1 shows that $-\mathrm{id}$ does indeed not(!) preserve connected components.

5.3   Before really going into the proof of the global Torelli theorem in Section 5.5, we need to discuss the monodromy group. By Theorem 1.1 any two complex K3 surfaces are deformation equivalent. Thus, if X and X' are K3 surfaces, then they admit markings φ and φ', respectively, such that (X,φ) and (X',φ') are contained in the same connected component N^o of N. In fact, we may even pick an arbitrary

φ and choose φ' accordingly. Therefore, in order to show that N has at most two connected components (interchanged by $(X, \varphi) \mapsto (X, -\varphi)$), it suffices to show that for one (possibly very special) K3 surface X and two arbitrary markings φ_1, φ_2 of X the marked K3 surfaces (X, φ_1) and $(X, \pm\varphi_2)$ are contained in same connected component.

This can be phrased in terms of the monodromy of K3 surfaces. To state the result, fix a K3 surface X of which we think as (M, I), i.e. a differentiable manifold M with a complex structure I. If $\mathcal{X} \longrightarrow S$ is a smooth proper morphism over a connected (possibly singular) base S with $X = \mathcal{X}_t$ for some fixed point $t \in S$, then the *monodromy* of this family is the image of the monodromy representation (see also Section **6.4.3**)

$$\pi_1(S, t) \longrightarrow \mathrm{O}(H^2(X, \mathbb{Z})).$$

By $\mathrm{Mon}(X)$ we denote the subgroup of $\mathrm{O}(H^2(X, \mathbb{Z}))$ generated by the monodromies of all possible families $\mathcal{X} \longrightarrow S$ with central fibre $X \simeq \mathcal{X}_t$. By the construction of the monodromy representation, which relies on the theorem of Ehresmann, $\mathrm{Mon}(X)$ is a subgroup of the image of the natural action $\mathrm{Diff}(X) \longrightarrow \mathrm{O}(H^2(X, \mathbb{Z}))$. Note that the action on $H^2(X, \mathbb{Z})$ induced by any automorphism of X is contained in $\mathrm{Mon}(X)$, e.g. by gluing the trivial deformation $X \times \mathbb{C}$ via the automorphism viewed as an isomorphism of two fibres X_{t_1}, X_{t_2}, to a family over the singular base $S := \mathbb{C}/(t_1 = t_2)$.

Proposition 5.4 *If* $\mathrm{O}(H^2(X, \mathbb{Z}))/\mathrm{Mon}(X)$ *is generated by* $-\mathrm{id}$, *then the moduli space* N *has at most two connected components and* $\mathcal{P}\colon N \longrightarrow D$ *is generically injective on each of them.*

Proof Write $X = (M, I)$ and identify $H^2(X, \mathbb{Z}) = H^2(M, \mathbb{Z})$. Clearly, any two markings φ_1 and φ_2 differ by an orthogonal transformation. Thus, under the assumption of the proposition, we can choose ψ in $\mathrm{Mon}(X)$ such that $\psi = \pm\varphi_1^{-1} \circ \varphi_2$.

To conclude, we only have to show that for $\psi \in \mathrm{Mon}(X)$ the marked K3 surfaces (X, φ_1) and $(X, \varphi_1 \circ \psi)$ are in the same connected component of N. For this, write $\psi = \psi_1 \circ \cdots \circ \psi_n$ with $\psi_i \in \mathrm{Im}(\pi_1(S_i, t_i) \longrightarrow \mathrm{O}(H^2(X, \mathbb{Z})))$. Here $\mathcal{X}_i \longrightarrow S_i$ are smooth proper connected families with fibre over $t_i \in S_i$ identified with X. Clearly, it suffices to show the assertion for $\psi = \psi_i$. The local system $R^2\pi_*\mathbb{Z}$ on S can be trivialized locally, i.e. identified with $\underline{H}^2(X, \mathbb{Z})$. The monodromy ψ is then obtained by following these trivializations along a closed path in S beginning in $t \in S$. The classifiying maps to N necessarily stay in the same connected component.

The injectivity follows from Corollary 5.1. $\qquad\square$

Note that due to the deformation equivalence of any two K3 surfaces, the assumption on the monodromy group in the proposition is independent of the specific K3 surface, and, as we see next, it can be proved using some standard lattice theory.

5.4 Let, as before, Λ be the K3 lattice $E_8(-1)^{\oplus 2} \oplus U^{\oplus 3}$, which is unimodular of signature $(3, 19)$. The *spinor norm* of an orthogonal transformation $g \in O(\Lambda_{\mathbb{R}})$ of the underlying real vector space is defined as follows. For a reflection

$$s_\delta : v \longmapsto v - 2\frac{(v.\delta)}{(\delta)^2}\delta,$$

where $\delta \in \Lambda_{\mathbb{R}}$ with $(\delta)^2 \neq 0$, the spinor norm is $+1$ if $(\delta)^2 < 0$ and -1 otherwise. This is extended multiplicatively to $O(\Lambda_{\mathbb{R}})$ by writing any g as a composition of reflections s_δ, which is possible due to the Cartan–Dieudonné theorem; see e.g. [15, 500]. We let

$$O^+(\Lambda) \subset O(\Lambda)$$

be the index two subgroup of orthogonal transformations of spinor norm $+1$ or, equivalently, the subgroup of all orthogonal transformations preserving any given orientation of the three positive directions in $\Lambda_{\mathbb{R}}$. A classical result essentially due to Wall and later improved by Ebeling and Kneser (see Theorem **14**.2.2) asserts that any $g \in O^+(\Lambda)$ can be written as a product $\prod s_{\delta_i}$ with $\delta_i \in \Lambda$ such that $(\delta_i)^2 = -2$.

As a consequence of this abstract result for orthogonal transformations of Λ, one obtains a description of the monodromy group.

Proposition 5.5 *The monodromy group* $\mathrm{Mon}(X) \subset O(H^2(X, \mathbb{Z}))$ *of a K3 surface X coincides with the index two subgroup* $O^+(H^2(X, \mathbb{Z}))$ *of all transformations with trivial spinor norm. In particular,* $O(H^2(X, \mathbb{Z}))/\mathrm{Mon}(X) = \{\pm 1\}$.

Proof It suffices to show that any reflection s_δ associated with a (-2)-class $\delta \in \Lambda$ can be realized by a monodromy transformation.[2] The argument goes roughly as follows: fix M and a class $\delta \in H^2(M, \mathbb{Z})$ with $(\delta)^2 = -2$. We use the surjectivity of the period map (see Theorem 4.1) to show that there exists a complex structure I on M such that for $X = (M, I)$ one has $H^{1,1}(X, \mathbb{Z}) = \mathbb{Z}\delta$. Under these conditions there exists a unique smooth rational curve $\mathbb{P}^1 \simeq C \subset X$ with $[C] = \pm\delta$. Indeed, the Riemann–Roch formula for the line bundle L corresponding to δ shows $\chi(L) = 1$ and so $H^0(X, L) \neq 0$ or $H^0(X, L^*) \neq 0$. But any curve $C \in |L|$ is necessarily integral and satisfies $(C)^2 = -2$ and hence $C \simeq \mathbb{P}^1$; see Section **2**.1.3.

Then by the Grauert–Mumford contraction theorem (see e.g. [33, Thm. III.2.1]) there exists a contraction $X \longrightarrow \overline{X}$ of C to a singular point $0 \in \overline{X}$ with $X \setminus C = \overline{X} \setminus \{0\}$. The surface $\overline{X}$ is compact but singular and locally described as the zero set of $x^2 + y^2 + z^2 = 0$ (a rational normal double point or A_1-singularity; cf. Section **14**.0.3).

The surface $\overline{X}$ can be smoothened, i.e. there exists a proper flat family $\mathcal{X} \longrightarrow \Delta$ over a one-dimensional disk Δ such that $\mathcal{X}_0 = \overline{X}$ and the restriction $\mathcal{X}^* \longrightarrow \Delta^*$ to the punctured disk is smooth; see [94]. Locally this is given by the equation $x^2 + y^2 + z^2 = t$.

---

[2] This is a folklore result; see e.g. [184] or [397], where the argument is presumably explained in detail.

Consider the base change $\Delta \longrightarrow \Delta$, $u \longmapsto u^2$, and the pull-back $\mathcal{X} \times_{\Delta^*} \Delta^*$, which can be completed to a smooth proper family $\mathcal{Y} \longrightarrow \Delta$ with a natural identification $\mathcal{Y}_0 = X$.[3] The fibres of the two families $\mathcal{Y}^*, \mathcal{X}^* \longrightarrow \Delta^*$ are all diffeomorphic to M by parallel transport. Eventually, one observes that the monodromy of the family $\mathcal{X}^* \longrightarrow \Delta^*$ on $H^2(\mathcal{X}_{t \neq 0}, \mathbb{Z})$ under this identification is nothing but s_δ (Picard–Lefschetz formula); cf. e.g. [621, Ch. 15].

This proves $O^+(H^2(X, \mathbb{Z})) \subset \mathrm{Mon}(X)$. Since $-\mathrm{id}$ is not contained in $\mathrm{Mon}(X)$ (cf. Corollary 5.1 and the discussion in Section 5.2), equality holds. □

5.5 Proof of the Global Torelli Theorem 5.3. As explained already, to prove the first part of the assertion, it suffices to combine Propositions 5.4 and 5.5 with Remark 5.2.

To prove the second part, i.e. that any Hodge isometry with $\psi(\mathcal{K}_X) \cap \mathcal{K}_{X'} \neq \emptyset$ is induced by a unique isomorphism, use again the arguments in the proof of Proposition 2.2 showing that there exists a correspondence $\Gamma_\infty = Z + \sum Y_i$ with $[\Gamma_\infty]_* = \psi$ such that Z is the graph of an isomorphism $f \colon X \xrightarrow{\sim} X'$ and the components Y_i project onto curves $C_i \subset X$ and $C'_i \subset X'$.

Now, for $\alpha \in \mathcal{K}_X$ let $\alpha' := f_* \alpha \in \mathcal{K}_{X'}$ and $\beta := \psi_* \alpha = \alpha' + \sum (C_i.\alpha)[C'_i] = \alpha' + [D]$. Note that D is an effective curve, for α as a Kähler class is positive on all curves C_i. If $D = 0$, then f lifts ψ, which proves the claim. If not, use that ψ is an isometry to deduce $(\alpha)^2 = (\beta)^2 = (\alpha')^2 + 2(\alpha'.D) + (D)^2$. However, as also $(\alpha)^2 = (\alpha')^2$, this implies $2(\alpha'.D) + (D)^2 = 0$. Now, since α' is Kähler as well, $0 \leq (\alpha'.D)$ and hence

$$(\beta.D) = (\alpha'.D) + (D)^2 \leq 2(\alpha'.D) + (D)^2 = 0,$$

excluding β from being contained in $\mathcal{K}_{X'}$.

For later use in the proof of Theorem **8.5.2** we remark that it is enough to know that $(\alpha'.D) \geq 0$ for all $\mathbb{P}^1 \simeq D \subset X'$. Indeed, all other integral curves would have $(D)^2 \geq 0$ and $(\alpha'.D) \geq 0$ would follow from $\alpha' \in \mathcal{C}_{X'}$.

The uniqueness of f follows from Proposition 15.2.1. □

Remark 5.6 (i) Conversely, our discussion in Section 5.3 shows that the global Torelli theorem can be used to prove the assertion on the monodromy. This was remarked by Borcea in [75].[4]

(ii) Similarly, in [494] Pjateckiĭ-Šapiro and Šafarevič explain how to use the global Torelli theorem to construct for any $d > 0$ a family of polarized K3 surfaces $X \longrightarrow S$ of degree $2d$ over a smooth connected algebraic variety S such that the image of

---

[3] In fact, there are two ways to do this, which gives rise to the 'Atiyah flop', but this is not essential at this point.

[4] Thanks to Robert Friedman and Kieran O'Grady for insisting that the description of the monodromy group of a K3 surface should rather be derived directly and independently from the global Torelli theorem.

the monodromy presentation $\pi_1(S,t) \longrightarrow O(H^2(X_t, \mathbb{Z}))$ is the subgroup $\widetilde{O}(\Lambda_d)$ of all orthogonal transformations of $H^2(X_t, \mathbb{Z})$ fixing the polarization; cf. Section **6.3.2**, Corollary **6.4.7**, and Remark **6.4.8**.

5.6   Maybe this is a good point to say a few words about the diffeomorphism group of K3 surfaces. This is not relevant for the global Torelli theorem, but clearly linked to the monodromy group and very interesting in itself.

Obviously, Mon$(X) \subset \rho(\text{Diff}(X))$, where $\rho \colon \text{Diff}(X) \longrightarrow O(H^2(X, \mathbb{Z}))$ is the natural action. But in fact equality holds; see [157].

Theorem 5.7 (Donaldson)   *For any K3 surface X one has*

$$\text{Mon}(X) = \rho(\text{Diff}(X)) = O^+(H^2(X, \mathbb{Z})),$$

which is an index two subgroup of $O(H^2(X, \mathbb{Z}))$.

Writing $H^2(X, \mathbb{Z}) = E_8(-1)^{\oplus 2} \oplus U^{\oplus 3}$, it suffices to show that $\text{id}_{E_8(-1)^{\oplus 2} \oplus U^{\oplus 2}} \oplus (-\text{id}_U)$ is not contained in the image of ρ or, equivalently, that all diffeomorphisms respect any given orientation of three positive directions in Λ.

Donaldson's proof relies on SU(2) gauge theory, but it seems likely that the easier Seiberg–Witten invariants could also be used.

Once $\rho(\text{Diff}(X))$ is described completely, one turns to the kernel of ρ.

Question 5.8   *Is the kernel of* $\rho \colon \text{Diff}(X) \longrightarrow O(H^2(X, \mathbb{Z}))$ *connected? In other words, is any diffeomorphism of a K3 surface that acts trivially on cohomology isotopic to the identity?*

Work of Ruberman seems to suggest that the answer to this question might be negative; see [209, Rem. 4.1]. The question is also linked to the following stronger version of the global Torelli theorem conjectured by Weil [634], which has in fact been addressed recently by Buchdahl in [91].

Question 5.9   *Suppose two complex structures* I, I' *on M define K3 surfaces* $X = (M, I)$ *and* $X = (M, I')$ *with* $H^{2,0}(X) = H^{2,0}(X')$ *as subspaces inside the fixed* $H^2(M, \mathbb{C})$. *Does there exist a diffeomorphism g isotopic to the identity such that* $g^*I = I'$?

If the Kähler cones of X and X' coincide (or at least intersect), then the global Torelli theorem does indeed show that there exists $g \in \text{Ker}(\rho)$ with $g^*I = I'$. Since, however, it seems doubtful that $\text{Ker}(\rho)$ consists of isotopies only, this does not quite answer Weil's original question (in the above modified form).

Remark 5.10 In fact, an affirmative answer to Question 5.9 would also imply an affirmative answer to Question 5.8.[5] Indeed, fix an arbitrary complex structure I on M and consider for $g \in \text{Ker}(\rho)$ the induced complex structure $I' := g^*I$. Then $H^{2,0}(X) = H^{2,0}(X')$. Hence, there exists a diffeomorphism h isotopic to the identity such that $h^*I = I'$. But then $(g \circ h^{-1})^*I = I$ and so $g \circ h^{-1}$ is an automorphism of $X = (M, I)$ acting trivially on $H^*(X, \mathbb{Z})$. Therefore, by Proposition 15.2.1, $g = h$ and, in particular, g is isotopic to the identity, too.

To conclude the interlude on the diffeomorphism group, let us mention the following result, which is called the *Nielsen realization problem* for K3 surfaces; see [209].

Proposition 5.11 *The natural projection* $\text{Diff}(X) \longrightarrow \rho(\text{Diff}(X))$ *has no section (as a group homomorphism).*

The analogous statement holds for $\text{Diff}(X) \longrightarrow \pi_0(\text{Diff}(X))$, though at this point we cannot exclude that in fact $\pi_0(\text{Diff}(X)) \xrightarrow{\sim} \rho(\text{Diff}(X))$.

In Section 16.3.3 we study a rather similar looking picture:

$$\text{Aut}(\text{D}^b(X)) \longrightarrow \text{Aut}(\widetilde{H}(X, \mathbb{Z})), \quad \Phi \longmapsto \Phi^H,$$

which is the 'mirror dual' of the above, as was explained by Szendrői in [581].

6 Other Approaches

Compare the following comments with the discussion in Remark 6.3.6.

6.1 Originally, the global Torelli theorem was proved by Pjateckiĭ-Šapiro and Šafarevič [493] for algebraic K3 surfaces and by Burns and Rapoport [93] for K3 surfaces of Kähler type (but, as shown later by Todorov and Siu, every K3 surface is Kähler). In both proofs the global Torelli theorem is first and rather directly shown for Kummer surfaces.

The idea is that a complex torus $A = \mathbb{C}^2/\Gamma$ is determined by its Hodge structure of weight one $H^1(A, \mathbb{Z})$, and when passing to the Hodge structure of weight two $H^2(A, \mathbb{Z}) = \bigwedge^2 H^1(A, \mathbb{Z})$ not much of the information is lost (in fact, only A and its dual torus have isomorphic Hodge structures of weight two; see Section 3.2.4). The associated Kummer surface contains $H^2(A, \mathbb{Z})$ as the orthogonal complement of the additional 16 (-2)-classes (see Section 3.2.5) and one needs to control those in order to conclude.

---

[5] As was pointed out to me by Andrey Soldatenkov.

The difficult part is to pass from Kummer surfaces, which are dense in the moduli space (see Remark **14**.3.24), to arbitrary K3 surfaces. We refrain from presenting the classical proof, as several very detailed accounts of it exist in the literature; see e.g. Looijenga's article [378] or the expositions in [33, 54].

6.2 In principle one could try to replace in the classical approach Kummer surfaces by any other class of K3 surfaces which are dense in the moduli space and for which a direct proof of the global Torelli theorem is available. Elliptic K3 surfaces, quartics, or double covers of $\mathbb{P}^2$ come to mind, but unfortunately direct proofs are difficult to come by even for these special classes of K3 surfaces.

Note that instead of determining the monodromy group $\mathrm{Mon}(X)$ completely as done in Section 5.4, one could have concluded the proof of the global Torelli theorem in Section 5.2 by proving it for one single(!) K3 surface. More precisely, if there exists a marked K3 surface $(X, \varphi) \in N$ such that $\mathcal{P}^{-1}(\mathcal{P}(X, \varphi)) = \{(X, \varphi), (X, -\varphi)\}$ (up to non-Hausdorff issues), then N has only two connected components and thus Theorem 5.3 follows.

Is there any K3 surface for which a proof of the global Torelli theorem can be given directly (and more easily than for general Kummer surfaces)? A good candidate would be a K3 surface with $\rho(X) = 20$ and $T(X) \simeq T(2)$. Then X is a very special Kummer surface associated with $E_1 \times E_2$. Here E_i are elliptic curves isogenous to the elliptic curve determined by $H^{0,2}(X)/H^2(X, \mathbb{Z})$, which is determined by the period $\mathcal{P}(X, \varphi)$. See Remark **14**.3.24.

6.3 There are other approaches towards the global Torelli theorem that also rely on global information. Notably, Friedman's proof [183] for algebraic K3 surfaces uses partial compactifications of the moduli space of polarized K3 surfaces. The properness of the extended period map allows him to reduce by induction to the degree two case, i.e. to double covers of $\mathbb{P}^2$, which is then dealt with by global arguments (and not directly). The paper also contains a proof of the surjectivity of the period map, which in degree two had also been proved by Shah in [551].

6.4 Let us also mention Buchdahl's article [91] again which not only starts with a clear account of the history of this part of the theory of K3 surfaces, but also gives easier and more streamlined proofs of the main results: deformation equivalence, global Torelli theorem, and surjectivity of the period map. His approach does not use the fact that the period map is a covering (after passing to the Hausdorff reduction), but instead relies on more analytical methods originated by Demailly. Of course, also in our approach ultimately complex analysis is used (for example in the description of the Kähler cone and for the existence of Kähler metrics on K3 surfaces).

References and Further Reading

There are very interesting aspects of Section 1 for real K3 surfaces; see [54, Exp. XIV] or the more recent survey [293] by Kharlamov.

The monodromy group of special families of K3 surfaces can be very interesting, see for example [614].

The presentation given here is very much inspired by the approach to higher-dimensional generalizations of K3 surfaces. For a survey of the basic aspects of compact hyperkähler manifolds, see [251].

Questions and Open Problems

In [219, Sec. 7] Griffiths suggested to prove a global Torelli theorem for K3 surfaces by directly reconstructing the function field of an algebraic K3 surface X from its period $\mathcal{P}(X, \varphi)$. Unfortunately, nothing along this line has ever been worked out.

Concerning the differential topology of K3 surfaces, the two main open problems are the description of the kernel of $\mathrm{Diff}(X) \longrightarrow \mathrm{O}(H^2(X, \mathbb{Z}))$ (see Question 5.8) and the description of the group of symplectomorphisms (up to isotopy) of a K3 surface X viewed as a manifold with a real symplectic structure given by a Kähler form; cf. Seidel's article [543]. The latter is related to the group of autoequivalences of the derived category $\mathrm{D}^b(\mathrm{Coh}(X))$ via mirror symmetry; see Chapter 16.

8

Ample Cone and Kähler Cone

Here we shall discuss the structure of various cones, ample, nef, effective, that are important for the study of K3 surfaces. We concentrate on the case of algebraic K3 surfaces and their cones in the Néron–Severi lattice, but include in Section 5 a brief discussion of the Kähler cone of complex K3 surfaces.

0.5 To fix conventions, let us recall some basic notions concerning cones. By definition, a subset $\mathcal{C} \subset V$ of a real vector space V is a *cone* if $\mathbb{R}_{>0} \cdot \mathcal{C} = \mathcal{C}$. It is *convex* if in addition for $x, y \in \mathcal{C}$ also $x + y \in \mathcal{C}$. A ray $\mathbb{R}_{>0} \cdot x \subset \mathcal{C}$ of a closed cone $\mathcal{C}$ is called *extremal* if for all $y, z \in \mathcal{C}$ with $y + z \in \mathbb{R}_{>0} \cdot x$ also $y, z \in \mathbb{R}_{>0} \cdot x$.

A closed cone is *polyhedral* if $\mathcal{C}$ is the convex hull of finitely(!) many $x_1, \ldots, x_k \in V$, i.e. $\mathcal{C} = \sum_{i=1}^{k} \mathbb{R}_{\geq 0} \cdot x_i$. If $V = \Gamma_{\mathbb{R}} := \Gamma \otimes_{\mathbb{Z}} \mathbb{R}$ for some free $\mathbb{Z}$-module Γ, the cone is *rational polyhedral* if in addition the x_i can be chosen such that $x_i \in \Gamma_{\mathbb{Q}}$ (or, equivalently, $x_i \in \Gamma$). A closed cone $\mathcal{C}$ is *locally polyhedral* at $x \in \mathcal{C}$ if there exists a neighbourhood of x in $\mathcal{C}$ which is polyhedral. A cone $\mathcal{C}$ is *circular* at $x \in \partial\mathcal{C}$ if there exists an open subset $x \in U \subset \partial\mathcal{C}$ of the boundary such that the closure $\overline{\mathcal{C}}$ is not locally polyhedral at any point of U.

A *fundamental domain* for the action of a discrete group G acting continuously on a topological manifold M is (usually) defined as the closure $\overline{U}$ of an open subset $U \subset M$ such that M can be covered by $g\overline{U}$, $g \in G$, and such that for $g \neq h \in G$ the intersection $g\overline{U} \cap h\overline{U}$ does not contain interior points of gU or hU.

1 Ample and Nef Cone

We shall start with a few general facts on ample and nef classes on projective surfaces. Then, we explain how they can be made more precise for K3 surfaces. This section should also serve as a motivation for the more abstract discussion in Section 2.

1.1 Most of the following results hold for arbitrary smooth projective varieties, but for simplicity we shall restrict to the two-dimensional case. For an account of the general theory we recommend Lazarsfeld's comprehensive monograph [357].

First, recall the Hodge index theorem (see Proposition **1.2.4**): if $H \in \mathrm{NS}(X)$ is ample (or, weaker, if $(H)^2 > 0$), then the intersection form is negative definite on its orthogonal complement $H^\perp \subset \mathrm{NS}(X)$.

Definition 1.1 The *positive cone*

$$\mathcal{C}_X \subset \mathrm{NS}(X)_\mathbb{R}$$

is the connected component of the set $\{\alpha \in \mathrm{NS}(X) \mid (\alpha)^2 > 0\}$ that contains one ample class (or, equivalently, all of them).

The *ample cone*

$$\mathrm{Amp}(X) \subset \mathrm{NS}(X)_\mathbb{R}$$

is the set of all finite sums $\sum a_i L_i$ with $L_i \in \mathrm{NS}(X)$ ample and $a_i \in \mathbb{R}_{>0}$. The *nef cone*

$$\mathrm{Nef}(X) \subset \mathrm{NS}(X)_\mathbb{R}$$

is the set of all classes $\alpha \in \mathrm{NS}(X)_\mathbb{R}$ with $(\alpha.C) \geq 0$ for all curves $C \subset X$.

Note that $\mathcal{C}_X$, $\mathrm{Amp}(X)$, and $\mathrm{Nef}(X)$ are all convex cones. The ample cone $\mathrm{Amp}(X)$ is by definition the convex cone spanned (over $\mathbb{R}$) by ample line bundles, but the nef cone $\mathrm{Nef}(X)$ is in general not spanned by nef line bundles. Indeed, the effective nef cone $\mathrm{Nef}^{\mathrm{e}}(X)$ (see Section 4.1), i.e. the set of all finite sums $\sum a_i L_i$ with $L_i \in \mathrm{NS}(X)$ nef and $a_i \in \mathbb{R}_{\geq 0}$, can be strictly smaller than $\mathrm{Nef}(X)$ (its closure, however, always gives back $\mathrm{Nef}(X)$). A famous example for this phenomenon, a particular ruled surface, goes back to Mumford and Ramanujam; see [235, I.10], [357, I.1.5], and also Section 3.2, **(i)** and **(v)**.

In order to understand the relation between $\mathrm{Amp}(X)$ and $\mathrm{Nef}(X)$ one needs the following classical result; see [235, I.Thm. 5.1] or [236, V.Thm. 1.10].

Theorem 1.2 (Nakai–Moishezon–Kleiman) *A line bundle L on a smooth projective surface X over an arbitrary field k is ample if and only if*

$$(L)^2 > 0 \quad \textit{and} \quad (L.C) > 0 \tag{1.1}$$

for all curves $C \subset X$.

Remark 1.3 (i) Due to Grauert's ampleness criterion (see [33, Thm. 6.1]), the theorem still holds for complex (a priori not necessarily projective) surfaces. Another consequence of Grauert's criterion is that a smooth compact complex surface is projective if and only if there exists a line bundle L with $(L)^2 > 0$; cf. page 11.

(ii) Note that the weaker inequality $(\alpha.C) \geq 0$ for all curves C already implies $(\alpha)^2 \geq 0$, i.e.

$$\text{Nef}(X) \subset \overline{C}_X. \tag{1.2}$$

Indeed, otherwise by the Hodge index theorem $\alpha^\perp$ would cut C_X into two parts, in one of which one would find a line bundle $L \in C_X$, so in particular $(L)^2 > 0$, and such that $(L.\alpha) < 0$. The first inequality (together with the Riemann–Roch theorem) implies that some positive multiple of L is effective, whereas the latter would then contradict the nefness of α.

(iii) The two inequalities $(\alpha)^2 > 0$ and $(\alpha.C) > 0$ in (1.1) also describe the ample cone $\text{Amp}(X)$ and not only the integral classes in $\text{Amp}(X)$. Every class $\alpha \in \text{Amp}(X)$ satisfies these inequalities, but the converse is not obvious. Any class α satisfying (1.1) is contained in a polyhedral cone spanned by rational classes α' arbitrarily close to α. We may assume that they all satisfy $(\alpha')^2 > 0$. However, a priori there could be a sequence of curves C_i with $(\alpha.C_i) \longrightarrow 0$, which leaves open the possibility that $(\alpha'.C_i) < 0$ for one of the α'. This does not happen for α contained in the interior of $\text{Nef}(X)$ or, equivalently, if all $(\alpha.C)$ can be bounded by some $\varepsilon_0 > 0$. For K3 surfaces X the following discussion excludes this scenario.

If a sequence of such curves C_i existed, then they could be assumed integral and, since $C_i^\perp$ approaches α, such that $(C_i)^2 < 0$. But then $(C_i)^2 = -2$; cf. Section **2.1.3**. However, as is explained below (see Section 2.2 or Lemma 3.5), hyperplanes of the form $\delta^\perp$ with $(\delta)^2 = -2$ do not accumulate in the interior of C_X.

The proof that the ample cone of an arbitrary projective surface is indeed described by the two inequalities is more involved.

(iv) As a consequence of (ii) and (iii), one can deduce the ampleness of every class $\alpha + \varepsilon H$ with $\alpha \in \text{Nef}(X)$, H ample, and $\varepsilon > 0$. Indeed, $(\alpha + \varepsilon H.C) \geq \varepsilon(H.C) \geq \varepsilon > 0$ and $(\alpha + \varepsilon H)^2 = (\alpha)^2 + 2\varepsilon(\alpha.H) + \varepsilon^2(H)^2 > 0$.

Corollary 1.4 *Let X be a smooth projective surface over an arbitrary field. Then the ample cone $\text{Amp}(X)$ is the interior of the nef cone $\text{Nef}(X)$. The latter is the closure of the former:*

$$\text{Amp}(X) = \text{Int}\,\text{Nef}(X) \subset \text{Nef}(X) = \overline{\text{Amp}}(X).$$

Proof The nef cone is closed by definition. If $H \in \text{NS}(X)$ is ample and $L_1, \ldots, L_\rho$ form a basis of $\text{NS}(X)$, then for $|\varepsilon_1|, \ldots, |\varepsilon_\rho| \ll 1$ the class $H_\varepsilon := H + \sum \varepsilon_j L_j$ still satisfies $(H_\varepsilon)^2 > 0$ and $(H_\varepsilon.C) > 0$ for all curves C (use that for an integral class H actually $(H.C) \geq 1$). Hence, for all small $\varepsilon_j \in \mathbb{Q}$ the class H_ε is ample (rational), and, therefore, a small open neighbourhood of H is still contained in $\text{Amp}(X)$. If H is replaced by a (real) class $\alpha \in \text{Amp}(X)$, then write $\alpha = \sum a_i M_i$ with M_i ample and $a_i \in \mathbb{R}_{>0}$. The above argument goes through, by choosing the ε_j small enough for all of the finitely many M_i. Hence, $\text{Amp}(X)$ is open.

The inclusion $\text{Amp}(X) \subset \text{Nef}(X)$ is obvious and hence $\text{Amp}(X) \subset \text{Int Nef}(X)$. On the other hand, if $\alpha \in \text{Int Nef}(X)$, then for any ample H and $|\varepsilon| \ll 1$ the class $\alpha - \varepsilon H$ is still nef. Then write $\alpha = (\alpha - \varepsilon H) + \varepsilon H$, which is a sum of a nef and an ample class and hence itself ample. (Note that here one uses only (iii) and (iv) in Remark 1.3 in the easier case that the nef class is contained in the interior of the nef cone.) □

Corollary 1.5 *For every class α in the boundary $\partial \text{Nef}(X)$ of the nef cone one has $(\alpha)^2 = 0$ or there exists a curve C with $(\alpha.C) = 0$.*

Proof Suppose there is no curve C with $(\alpha.C) = 0$ and $(\alpha)^2 > 0$. Then α would be ample and hence contained in the interior of $\text{Nef}(X)$. (Note that here we use the stronger version of (iii) in Remark 1.3.) □

1.2 Let now X be a projective K3 surface over an algebraically closed field k.[1] The following more precise version of Theorem 1.2 for K3 surfaces was stated in a slightly different form already as Proposition 2.1.4. Use that if an integral curve C is not isomorphic to $\mathbb{P}^1$, then $(C)^2 \geq 0$ (cf. Section 2.1.3) and that then $(\alpha.C) > 0$ holds automatically for any $\alpha \in \mathcal{C}_X$.

Corollary 1.6 *A line bundle L on a projective K3 surface X is ample if and only if*

(i) $(L)^2 > 0$,
(ii) $(L.C) > 0$ *for every smooth rational curve $\mathbb{P}^1 \simeq C \subset X$, and*
(iii) $(L.H) > 0$ *for one ample divisor H (or, equivalently, for all of them).* □

An (almost) equivalent reformulation is the following corollary (use Remark 1.3, (iii)).

Corollary 1.7 *Let X be a projective K3 surface. Then*

$$\text{Amp}(X) = \{\alpha \in \mathcal{C}_X \mid (\alpha.C) > 0 \text{ for all } \mathbb{P}^1 \simeq C \subset X\}.$$ □

Note that if X does not contain any smooth rational curve at all, then being ample is a purely numerical property; i.e. it can be read off from just the lattice $\text{NS}(X)$ with its intersection form and one ample (or just effective) divisor $H \in \text{NS}(X)$. The latter is only needed to single out the positive cone as one of the two connected components of the set of all classes α with $(\alpha)^2 > 0$. Corollary 1.5 for K3 surfaces now reads as the following.

Corollary 1.8 *Let $\alpha \in \partial \text{Nef}(X)$. Then $(\alpha)^2 = 0$ or there exists a smooth rational curve $\mathbb{P}^1 \simeq C \subset X$ with $(\alpha.C) = 0$.* □

---

[1] The only reason why k has to be assumed algebraically closed here is that otherwise smooth curves of genus zero may not be isomorphic to $\mathbb{P}^1$. Note that the characteristic can be arbitrary.

2 Chambers and Walls

We recall some standard facts concerning hyperbolic reflection groups that are used to describe the ample and the Kähler cone of a K3 surface. The subject itself is vast; the classical references are [83, 617]. The Weyl group of a K3 surface is a special case of a Coxeter group, although usually an infinite one, and most of the literature deals only with finite ones. The arguments given below are deliberately ad hoc and the reader familiar with the general theory of Coxeter groups should rather use it to conclude.[2]

2.1   Consider a real vector space V of dimension $n+1$ endowed with a non-degenerate quadratic form $(\, . \,)$ of signature $(1, n)$. Thus, abstractly $(V, (\, . \,))$ is isomorphic to $\mathbb{R}^{n+1}$ with the quadratic form $x_0^2 - x_1^2 - \cdots - x_n^2$.

The set $\{x \in V \mid (x)^2 > 0\}$ has two connected components which are interchanged by $x \longmapsto -x$. We usually distinguish one of the two connected components, say $\mathcal{C} \subset V$, and call it the *positive cone*. Thus,

$$\{x \in V \mid (x)^2 > 0\} = \mathcal{C} \sqcup (-\mathcal{C}).$$

Note that x, y with $(x)^2 > 0$ and $(y)^2 > 0$ are in the same connected component if and only if $(x.y) > 0$.

The subset $\mathcal{C}(1)$ of all $x \in \mathcal{C}$ with $(x)^2 = 1$ is isometric to the hyperbolic n-space $\mathbb{H}^n := \{x \in \mathbb{R}^{n+1} \mid x_0^2 - x_1^2 - \cdots - x_n^2 = 1, \, x_0 > 0\}$. By writing

$$\mathcal{C} \simeq \mathcal{C}(1) \times \mathbb{R}_{>0} \simeq \mathbb{H}^n \times \mathbb{R}_{>0},$$

questions concerning the geometry of $\mathcal{C}$ can be reduced to analogous ones for $\mathbb{H}^n$.

We write $\mathrm{O}(V)$ for the orthogonal group $\mathrm{O}(V; (\, . \,))$, which is abstractly isomorphic to $\mathrm{O}(1, n)$. By $\mathrm{O}^+(V) \subset \mathrm{O}(V)$ we denote the index two subgroup of transformations preserving the positive cone $\mathcal{C}$. The induced action $\mathrm{O}^+(V) \times \mathcal{C}(1) \longrightarrow \mathcal{C}(1)$ is transitive and the stabilizer of $x \in \mathcal{C}(1)$ is the orthogonal group $\mathrm{O}(x^\perp)$ of the negative definite space $x^\perp \subset V$. Thus,

$$\mathcal{C}(1) \simeq \mathrm{O}^+(V)/\mathrm{O}(x^\perp) \simeq \mathrm{O}^+(1, n)/\mathrm{O}(n).$$

Remark 2.1   Any discrete subgroup $H \subset \mathrm{O}(V)$ acts properly discontinuously from the left on $\mathcal{C}(1) \simeq \mathrm{O}^+(V)/\mathrm{O}(x^\perp)$; i.e. for every $x \in \mathcal{C}(1)$ there exists an open neighbourhood $x \in U \subset \mathcal{C}(1)$ such that $g(U) \cap U = \emptyset$ for all $g \in H$ except for the finitely many ones in $\mathrm{Stab}(x)$. See Remark 6.1.10 for the general statement and [638, Lem. 3.1.1] for an elementary proof.

---

[2] See also Ogus's complete and detailed account in [479, Prop. 1.10]. The standard sources focus on complex K3 surfaces and he checks that all the arguments are indeed valid for K3 surfaces in positive characteristic and in particular for supersingular K3 surfaces.

2.2 Let now Γ be a lattice of signature $(1, n)$ (think of $NS(X)$ of an algebraic K3 surface X) and consider $V := \Gamma_{\mathbb{R}}$ with the induced quadratic form. Then Remark 2.1 applies to the discrete subgroup $O^+(\Gamma) := O(\Gamma) \cap O^+(V)$, which leads to the following observation.

Remark 2.2 For any subset $I \subset O^+(\Gamma)$ the set

$$\text{Fix}_I := \bigcup_{g \in I} \text{Fix}(g) \subset \mathcal{C}$$

is closed and so is $\text{Fix}_I \cap \mathcal{C}(1)$.

To see this, let $x \in \mathcal{C}(1) \setminus \text{Fix}_I$. Then $\text{Stab}(x) \cap I = \emptyset$. Since the action of $O^+(\Gamma)$ on $\mathcal{C}(1)$ is properly discontinuous, there exists an open neighbourhood $x \in U \subset \mathcal{C}(1)$ with $g(U) \cap U = \emptyset$ for all $g \in I$. Hence, $U \cap \text{Fix}_I = \emptyset$, i.e. U is an open neighbourhood of x contained in $\mathcal{C}(1) \setminus \text{Fix}_I$.

Next consider the set of *roots*

$$\Delta := \{\delta \in \Gamma \mid (\delta)^2 = -2\}.$$

It is often convenient to distinguish a subset of *positive roots* $\Delta_+ \subset \Delta$, i.e. a subset with the property that $\Delta = \Delta_+ \sqcup (-\Delta_+)$.[3]

With any $\delta \in \Delta$ one associates the *reflection* $s_\delta \in O^+(\Gamma) \subset O^+(\Gamma_{\mathbb{R}})$ defined by

$$s_\delta : x \longmapsto x + (x.\delta)\delta.$$

Thus, $s_\delta(\delta) = -\delta$ and $s_\delta = \text{id}$ on $\delta^\perp$. To see that s_δ really preserves $\mathcal{C}$ use $(s_\delta(x).x) = (x)^2 + (x.\delta)^2 > 0$ for $x \in \mathcal{C}$. Frequently we use the observation that for arbitrary $g \in O(\Gamma_{\mathbb{R}})$

$$g \circ s_\delta = s_{g(\delta)} \circ g. \tag{2.1}$$

We call $\text{Fix}(s_\delta) = \mathcal{C} \cap \delta^\perp$ the *wall* associated with $\delta \in \Delta$. Although walls could and often do accumulate towards the boundary $\partial \mathcal{C} \subset \{x \mid (x)^2 = 0\}$, Remark 2.2 applied to $I = \{s_\delta \mid \delta \in \Delta\} \subset O^+(\Gamma)$ shows that the union of all walls

$$\bigcup_{\delta \in \Delta} \delta^\perp \subset \mathcal{C}$$

is closed and hence locally finite in $\mathcal{C}$.

Remark 2.3 Here is a more ad hoc argument for the same fact, taken from Ogus [479]. Pick a class $h \in \mathcal{C} \cap \Gamma$ and complete $h/\sqrt{(h)^2}$ to an orthogonal basis of $\Gamma_{\mathbb{R}} \simeq \mathbb{R}^{n+1}$

---

[3] In the geometric situation, when the δ correspond to divisor classes on a K3 surface X, a natural choice is given by the effective divisors. Indeed, if a (-2)-class $\delta \in NS(X)$ corresponds to a line bundle L, then by the Riemann–Roch theorem either L or L^* is effective.

such that (.) on $\Gamma_{\mathbb{R}}$ is given by $x_0^2 - x_1^2 - \cdots - x_n^2$. We write accordingly any $x \in \Gamma_{\mathbb{R}}$ as $x = x_0 + x'$ with $x' \in h_{\mathbb{R}}^{\perp}$.

Now let $\langle\,,\,\rangle$ be the inner product $-(\,.\,)$ on $h_{\mathbb{R}}^{\perp}$ and let $\|\,\|$ be the associated norm. Then for $\delta \in \Delta$ one has $|\delta_0|^2 - \|\delta'\|^2 = -2$ and hence $\|\delta'\| \le |\delta_0| + 2$. For $x \in \delta^{\perp}$ one finds $x_0 \cdot \delta_0 = \langle x', \delta' \rangle$ and, by Cauchy–Schwarz, $|x_0 \cdot \delta_0| \le \|\delta'\| \cdot \|x'\| \le (|\delta_0| + 2) \cdot \|x'\|$ and, therefore, $|\delta_0| \cdot (|x_0| - \|x'\|) \le 2\|x'\|$.

Now, when x approaches h, then $\|x'\| \longrightarrow 0$ and $x_0 \longrightarrow \sqrt{(h)^2}$. But then also $|\delta_0| \longrightarrow 0$ and hence $\|\delta'\|$ is bounded. Since Δ is discrete, this suffices to conclude that a small open neighbourhood U of h is met by only finitely many walls, i.e. $\delta^{\perp} \cap U \ne \emptyset$.

See also the proof of Lemma 3.5 where the argument is repeated in a slightly different form.

The connected components of the open complement $\mathcal{C} \setminus \bigcup_{\delta \in \Delta} \delta^{\perp}$ are called *chambers* and shall be denoted

$$\mathcal{C}_0, \mathcal{C}_1, \ldots \subset \mathcal{C}.$$

The discussion above leads to the following proposition.

Proposition 2.4   *The chamber structure of $\mathcal{C}$ induced by the roots Δ is locally polyhedral in the interior of $\mathcal{C}$; i.e. for every chamber $\mathcal{C}_i \subset \mathcal{C}$ the cone $\overline{\mathcal{C}}_i$ is locally polyhedral in the interior of $\mathcal{C}$.* $\square$

2.3   Two elements $x, y \in \mathcal{C}$ are in the same chamber if and only if $(x.\delta) \cdot (y.\delta) > 0$ for all $\delta \in \Delta$. Moreover, a chamber $\mathcal{C}_0 \subset \mathcal{C}$ is uniquely determined by the sequence of signs of $(\delta.\mathcal{C}_0)$, $\delta \in \Delta$. Equivalently, the choice of a chamber $\mathcal{C}_0 \subset \mathcal{C}$ is determined by the choice of a set of positive roots $\Delta_+ := \{\delta \mid (\delta.\,)|_{\mathcal{C}_0} > 0\} \subset \Delta$. One also defines the smaller subset $\Delta_{\mathcal{C}_0} \subset \Delta_+$ of all δ such that $\delta^{\perp}$ defines a wall of $\mathcal{C}_0$ of codimension one, i.e. $\delta^{\perp}$ intersects the closure of $\mathcal{C}_0$ in codimension one.

The *Weyl group* W is the subgroup of $O^+(\Gamma)$ generated by all s_{δ}, i.e.

$$W := \langle s_{\delta} \mid \delta \in \Delta \rangle \subset O^+(\Gamma),$$

which due to (2.1) is a normal subgroup of $O^+(\Gamma)$.

Note that for $\delta, \delta' \in \Delta$ and $x \in \delta^{\perp}$ one has $(s_{\delta'}(x).s_{\delta'}(\delta)) = (x.\delta) = 0$, i.e. $s_{\delta'}(x) \in s_{\delta'}(\delta)^{\perp}$. Hence, W preserves the union of walls $\bigcup_{\delta \in \Delta} \delta^{\perp}$ and, thus, acts on the set of chambers.

Remark 2.5   Suppose $\mathcal{C}_0 \subset \mathcal{C}$ is a chamber and let $\Delta_{\mathcal{C}_0} \subset \Delta_+ \subset \Delta$ be as above.
(i) Then

$$W_{\mathcal{C}_0} := \langle s_{\delta} \mid \delta \in \Delta_{\mathcal{C}_0} \rangle$$

acts transitively on the set of chambers. Indeed, one can connect C_0 with any other chamber by a path $\gamma : [0,1] \longrightarrow C$ that passes through just one wall of codimension one at a time. Using compactness, this yields a finite sequence of chambers $C_0, \dots, C_n$ such that C_i and C_{i+1} are separated by one wall $\delta_i^\perp$ with $\delta_i \in \Delta_{C_i} \cap (-\Delta_{C_{i+1}})$. So, in particular $s_{\delta_i}(C_i) = C_{i+1}$. For simplicity we assume $n = 2$ and leave the general case as an exercise.

Use $s_\delta(\Delta_{C_0}) = \Delta_{s_\delta(C_0)}$, which is straightforward to check, to show that $\delta_1 = s_{\delta_0}(\delta_1')$ for some $\delta_1' \in \Delta_{C_0}$. Then, $(s_{\delta_1} \circ s_{\delta_0})(C_0) = s_{\delta_1}(C_1) = C_2$, but by (2.1)

$$s_{\delta_1} \circ s_{\delta_0} = s_{s_{\delta_0}(\delta_1')} \circ s_{\delta_0} = s_{\delta_0} \circ s_{\delta_1'} \in W_{C_0}.$$

Hence, there exists an element in W_{C_0} that maps C_0 to C_n.

(ii) We claim that the transitivity of the action of W_{C_0} implies that

$$W_{C_0} = W. \tag{2.2}$$

For this it suffices to show that $s_\delta \in W_{C_0}$ for all $\delta \in \Delta$. By the local finiteness of walls, $\delta^\perp$ defines a wall of codimension one of some chamber C_1. Then by (i) there exists $g \in W_{C_0}$ with $g(C_0) = C_1$ and, therefore, a $\delta' \in \Delta_{C_0}$ with $g(\delta') = \delta$. Now use (2.1) to conclude that $s_\delta = s_{g(\delta')} = g \circ s_{\delta'} \circ g^{-1}$, which is contained in W_{C_0}.

Proposition 2.6 *The Weyl group W acts simply transitively on the set of chambers.*

Proof It remains to exclude the existence of some id $\ne g \in W$ leaving invariant a chamber $C_0 \subset C$, i.e. $g(C_0) = C_0$. Let $\Delta_{C_0} \subset \Delta_+ \subset \Delta$ be as before. Then by (2.2) one can write $g = s_{\delta_0} \circ \cdots \circ s_{\delta_\ell}$ with $\delta_i \in \Delta_{C_0}$. Choose ℓ minimal. Use the δ_i to define a generic closed path γ as in Remark 2.5, (i), with $\gamma(0), \gamma(1) \in C_0$. The walls that are crossed by γ are $(s_{\delta_0} \circ \cdots \circ s_{\delta_i})(\delta_{i+1}^\perp)$. Since γ is closed, all hyperplanes crossed by γ occur twice. So, for example, $\delta_0^\perp = (s_{\delta_0} \circ \cdots \circ s_{\delta_i})(\delta_{i+1}^\perp)$ for some i, and hence, using (2.1) again, one finds for $g' := s_{\delta_0} \circ \cdots \circ s_{\delta_i}$ that

$$g' \circ s_{\delta_{i+1}} = s_{g'(\delta_{i+1})} \circ g' = s_{\delta_0} \circ g' = s_{\delta_1} \circ \cdots \circ s_{\delta_i}.$$

This contradicts the minimality of ℓ. $\qquad \square$

Remark 2.7 The above can be used for the description of the ample cone. However, for the description of the Kähler cone one needs to modify the setting slightly. Instead of a lattice Γ of signature $(1, n)$ one starts with a lattice Λ of signature $(3, n)$ (think of the K3 lattice). Then for any positive plane $P \subset \Lambda_{\mathbb{R}}$ the orthogonal complement $V := P^\perp$ has signature $(1, n)$. The set of (-2)-classes to be considered in this situation is

$$\Delta_P := \{\delta \in \Lambda \mid (\delta)^2 = -2, \; \delta \in P^\perp\},$$

which depends on P. The associated reflections generate the Weyl group

$$W_P := \langle s_\delta \mid \delta \in \Delta_P \rangle,$$

which is a discrete subgroup of $\{g \in O(\Lambda_{\mathbb{R}}) \mid g|_P = \mathrm{id}\} \simeq O(P^\perp)$.

Note that often $\Lambda \cap P^\perp$ does not span $P^\perp$ (it is actually trivial most of the time) and that the natural map $W_P \longrightarrow O(\Lambda \cap P^\perp)$ is not necessarily injective. Nevertheless, Remark 2.5 and Propositions 2.4 and 2.6 still hold true and are proved by the same arguments.

2.4   Let us now assume that Γ is indeed $NS(X)$ of a projective K3 surface X over an algebraically closed field. Then $\Delta \subset NS(X)$ is the set of line bundles L with $(L)^2 = -2$. By the Riemann–Roch theorem, such a line bundle L or its dual L^* is effective; cf. Section 1.2.3. If $(L)^2 = -2$ and L is effective, then the fixed part of $|L|$ contains a (-2)-curve. For example, the union $C = C_1 + C_2$ of two (-2)-curves intersecting transversally in one point also satisfies $(C)^2 = -2$. Or, if $C_1 \subset X$ is an integral curve with $(C_1)^2 = 6$ not intersecting a (-2)-curve C_2, then $(C_1 + 2C_2)^2 = -2$ with Bs $|C_1 + 2C_2| = 2C_2$.

By Corollary 1.7, the ample cone $\mathrm{Amp}(X) \subset \mathcal{C}_X$; is one of the chambers defined by Δ. Moreover, to verify whether a class $\alpha \in \mathcal{C}_X$ is in fact contained in $\mathrm{Amp}(X)$ it suffices to check $(\alpha.C) > 0$ for all $\mathbb{P}^1 \simeq C \subset X$. Thus, by Remark 2.5, the Weyl group W is generated by reflections $s_{[C]}$ for all (-2)-curves C, i.e.

$$W = \langle s_{[C]} \mid \mathbb{P}^1 \simeq C \subset X \rangle.$$

Remark 2.8   Every $\mathbb{P}^1 \simeq C \subset X$ defines a codimension one wall of $\mathrm{Amp}(X)$; in particular no (-2)-class is superfluous for cutting out $\mathrm{Amp}(X)$. Indeed, for any ample H the class $x := H + (1/2)(H.C)[C]$ is contained in $[C]^\perp$, but for any other $\mathbb{P}^1 \simeq C' \subset X$ one has $(x.C') \geq (H.C') > 0$, because $(C.C') \geq 0$. The argument is taken from Sterk's article [577].

Corollary 2.9   *For any $\alpha \in \mathcal{C}_X$ there exist smooth rational integral curves $C_1, \ldots, C_n \subset X$ such that $(s_{[C_1]} \circ \cdots \circ s_{[C_n]})(\alpha)$ is nef. If, moreover, $(\alpha.\delta) \neq 0$ for all $\delta \in NS(X)$ with $(\delta)^2 = -2$, then*

$$(s_{[C_1]} \circ \cdots \circ s_{[C_n]})(\alpha) \in \mathrm{Amp}(X).$$
□

Remark 2.10   The result can be rephrased as follows: under the stated assumptions on α, there exist finitely many smooth rational curves $\mathbb{P}^1 \simeq C_i \subset X$ such that for the cycle $\Gamma := \Delta + \sum C_i \times C_i$ on $X \times X$ viewed as a correspondence the image $[\Gamma]_*(\alpha)$ under the induced map $[\Gamma]_* : H^2(X, \mathbb{Z}) \xrightarrow{\sim} H^2(X, \mathbb{Z})$ is a Kähler (or at least nef) class. This version is more natural in the context of Fourier–Mukai transforms; see Section **16**.3.1 and higher-dimensional generalizations.

The discussion above is summarized by the following.

Corollary 2.11 *For a projective K3 surface X, the cone* $\text{Nef}(X) \cap \mathcal{C}_X$ *is a fundamental domain for the action of the Weyl group* $W \subset O^+(\text{NS}(X))$ *on the positive cone* $\mathcal{C}_X$. *Moreover, W is generated by reflections* $s_{[C]}$ *with* $\mathbb{P}^1 \simeq C \subset X$ *and* $\text{Nef}(X)$ *is locally polyhedral in the interior* $\mathcal{C}_X$. □

Note that in general $\text{Nef}(X) \cap \partial \mathcal{C}_X$ might consist of a single ray which for $\rho(X) \geq 3$ could not be a fundamental domain for the action of W on $\partial \mathcal{C}_X$. Whether one wants to consider $\text{Nef}(X)$ as a fundamental domain for the action of W on $\overline{\mathcal{C}}_X$ is largely a matter of convention.

Also note that, although $\text{Nef}(X) \subset \overline{\mathcal{C}}_X$ is cut out by the inequalities $(C. \ ) \geq 0$ for all $\mathbb{P}^1 \simeq C \subset X$, it need not be (locally) polyhedral in the closed cone $\overline{\mathcal{C}}_X$, as extremal rays of $\text{Nef}(X)$ may accumulate towards $\partial \mathcal{C}_X$. However, the failure of $\text{Nef}(X)$ being (locally) polyhedral is essentially due to the action of $\text{Aut}(X)$ only; see Theorem 4.2.

For later reference, let us state also the following easy consequence.

Corollary 2.12 *If* $\text{NS}(X)$ *contains a class* α *with* $(\alpha)^2 = 2d > 0$, *then X admits a quasi-polarization L, i.e. a big and nef line bundle, with* $(L)^2 = 2d$. □

Remark 2.13 We come back to Remark **2**.3.13, (iii): a K3 surface X in characteristic zero (in fact, char $\neq 2, 3$ suffices) is elliptic if and only if there exists a non-trivial line bundle L with $(L)^2 = 0$.

Using Proposition **2**.3.10, it suffices to show that there exists a nef line bundle L' with $(L')^2 = 0$. Passing to its dual if necessary, we may assume $L \in \overline{\mathcal{C}}_X$, and therefore, by Riemann–Roch, L is effective. If L is not nef, then there exists a (-2)-curve C with $(L.C) < 0$. Clearly, $s_{[C]}(L)$ is still in $\overline{\mathcal{C}}_X$, and hence effective. Moreover,

$$0 < (s_{[C]}(L).H) = (L.H) + (L.C)(C.H) < (L.H)$$

for a fixed ample class H. If the new $s_{[C]}(L)$ is still not nef, continue. Since the degree with respect to the fixed H has to be positive but decreases at every step, this process stops. Thus, one finds a sequence of (-2)-curves $C_1, C_2, \ldots, C_k$ such that $(s_{[C_k]} \circ \cdots \circ s_{[C_1]})(L)$ is nef.[4] This has the surprising consequence that as soon as $\partial \mathcal{C}_X \cap \text{NS}(X) \neq \{0\}$, there also exists a nef class in $\partial \mathcal{C}_X$. See also Remark **3**.7.

3 Effective Cone

The cone of curves is by definition dual to the nef cone. It might a priori have a round part and a locally polyhedral part generated by smooth rational curves. However, due to

---

[4] The argument follows Barth et al. [33, VIII.Lem. 17.4]. Ideally one would like to argue with the chamber structure directly and say that there must exist an element $g \in W$ with $g(L)$ being contained in the nef cone. However, a priori the chambers might accumulate towards the boundary $\partial \mathcal{C}_X$.

a result by Kovács, to be explained in this section, in most cases the closure of the cone of curves is either completely round or locally polyhedral everywhere. We start with a general discussion of the cone of curves. Then in Section 3.2 we motivate Kovács's result by explaining a number of particular cases by drawing pictures and finally state his result saying that these special cases exhaust all possibilities. The proof is given in Section 3.3.

3.1   Dually to the nef cone one defines the effective cone, which plays a fundamental role in the minimal model program for higher-dimensional algebraic varieties. In dimension two, curves and divisor are the same thing, so the duality between them has a slightly different flavor compared with the higher-dimensional case.

Definition 3.1   Let X be a smooth projective surface. The *effective cone*

$$NE(X) \subset NS(X)_{\mathbb{R}},$$

also called the *cone of curves*, is the set of all finite sums $\beta = \sum a_i[C_i]$ with $C_i \subset X$ irreducible (or integral) curves and $a_i \in \mathbb{R}_{\geq 0}$.

As we shall see, $NE(X)$ is in general neither open nor closed. Its closure $\overline{NE}(X)$ is called the *Mori cone*. The following is a special case of the duality between effective curves and nef divisors on arbitrary smooth projective varieties; see e.g. [357, I.Prop. 1.4.28].

Theorem 3.2   *On a smooth projective surface X the Mori cone and the nef cone are dual to each other, i.e.*

$$\overline{NE}(X) = \{\beta \mid (\alpha.\beta) \geq 0 \, \text{for all } \alpha \in Nef(X)\}$$

and

$$Nef(X) = \{\alpha \mid (\alpha.\beta) \geq 0 \, \text{for all } \beta \in NE(X)\}.$$

Proof   The right-hand sides are by definition the dual cones $Nef(X)^*$ and $\overline{NE}(X)^*$, respectively. Since $Nef(X)^{**} = Nef(X)$ and $\overline{NE}(X)^{**} = \overline{NE}(X)$, it suffices to prove one of the two assertions. But the second is just the definition of the nef cone. For general facts on duality between cones, see [511].   □

As by Corollary 1.4 $Amp(X)$ is the interior of $Nef(X)$, one obtains the following description of the ample cone, which, again, is a general fact; see [357, I.Thm. 1.4.29] or, in the case of surfaces, [33, Prop. 7.5].

Corollary 3.3   *For the ample cone one has*

$$Amp(X) = \{\alpha \in NS(X)_{\mathbb{R}} \mid (\alpha.\beta) > 0 \, \text{for all } \beta \in \overline{NE}(X) \setminus \{0\}\}.$$

Some of the following remarks are already more specific to K3 surfaces. So from now on we shall assume that X is a projective K3 surface over an algebraically closed field.

Remark 3.4   (i) One knows that

$$\text{NE}(X) \subset \overline{C}_X + \sum \mathbb{R}_{\geq 0} \cdot [C], \tag{3.1}$$

where the curves C are smooth, integral, and rational; see Section 2.1.3. On the other hand, by the Riemann–Roch theorem every integral class in C_X is effective. Hence

$$\overline{\text{NE}}(X) = \overline{C}_X + \sum_{C \simeq \mathbb{P}^1} \mathbb{R}_{\geq 0} \cdot [C]. \tag{3.2}$$

(ii) Also,

$$\text{Nef}(X) \subset \overline{\text{NE}}(X),$$

for $\text{Nef}(X)$ is the closure of $\text{Amp}(X)$ and the latter is clearly contained in $\text{NE}(X)$. Or use (3.2) combined with $\text{Nef}(X) \subset \overline{C}_X$; see (1.2).

(iii) If $C \subset X$ is an integral curve with $(C)^2 \leq 0$, then $\text{NE}(X)$ is spanned by $[C]$ and all $\beta \in \text{NE}(X)$ with $(\beta.C) \geq 0$. Indeed, any curve C' not containing C satisfies $(C'.C) \geq 0$. Note that in particular $[C] \in \partial\text{NE}(X)$.

(iv) The class $[C]$ of any $\mathbb{P}^1 \simeq C \subset X$ defines an extremal ray of $\overline{\text{NE}}(X)$. Indeed, if $[C] = \beta + \beta'$ with $\beta, \beta' \in \overline{\text{NE}}(X)$, then using (iii) one finds that in fact $\beta, \beta' \in \mathbb{R}_{\geq 0} \cdot [C]$.

Lemma 3.5   *Let H be an ample divisor on a K3 surface X. Then for any N there are at most finitely many curves $\mathbb{P}^1 \simeq C \subset X$ with $(C.H) \leq N$. The same holds for H replaced by any real ample class $\alpha \in \text{Amp}(X)$.*

Proof   This is in fact an abstract result that has nothing to do with K3 surfaces. It has essentially been proved already in Section 2.2; see Remark 2.3. Nevertheless, we prove it again, and, moreover, in two different ways.

(i) Since $(C)^2 = -2$ for any $\mathbb{P}^1 \simeq C \subset X$, fixing $(C.H)$ is equivalent to fixing the Hilbert polynomial of $C \subset X$. Now, the Hilbert scheme of all subvarieties of X with fixed Hilbert polynomial is a projective scheme. As smooth integral rational curves do not deform, they correspond to connected components of the projective Hilbert scheme, of which there exist only finitely many ones. Hence, there are only finitely many $\mathbb{P}^1 \simeq C \subset X$ with fixed $(C.H)$.

(ii) Alternatively, one could use the following purely numerical argument. The given ample class H can be completed to an orthogonal basis of $\text{NS}(X)_{\mathbb{R}} \simeq \mathbb{R}^\rho$ such that the quadratic form is $(H)^2 x_1^2 - x_2^2 - \cdots - x_\rho^2$. For fixed $(C.H) = N$ the classes $[C]$ are all in the compact set $\{(N/(H)^2, x_2, \ldots, x_\rho) \mid x_2^2 + \cdots + x_\rho^2 = 2 + N^2/(H)^2\}$ which intersects the discrete $NS(X)$ in only finitely many points.

(iii) Finally, if only a real class $\alpha \in \text{Amp}(X)$ is fixed, then the argument in (ii) shows that there are at most finitely many $\mathbb{P}^1 \simeq C \subset X$ with $(C.\alpha)$ fixed. To exclude that $(C.\alpha)$ gets arbitrarily small, use the arguments in Section 2.2.   □

The next immediate consequence, at least its second part, has also been proved abstractly already in Proposition 2.4.

Corollary 3.6   *Outside* $\overline{C}_X$ *the cone* $\overline{NE}(X)$ *is locally polyhedral. Dually, the cone* $\text{Nef}(X) \cap C_X$ *is locally polyhedral in the open cone* C_X.

Proof   Indeed, the intersection of the closed cone $\overline{NE}(X)$ with the cone $\{x \mid (H.x)^2 \le k|(x)^2|\}$ is polyhedral for all $k > 0$. Clearly, any $x \in \overline{NE}(X) \setminus \overline{C}_X$ is contained in such an intersection for k large enough.   □

Kawamata proves in [287, Thm. 1.9] a more general statement covering in particular all surfaces with trivial canonical bundle.

3.2   The structure of the effective cone of a K3 surface can be intricate. There is one result however that shows that not everything that in principle is possible also occurs. This result is due to Kovács [327]. We prepare the ground by first looking at some pictures.

Since for $\rho(X) = 1$ effective and ample cone coincide and form just one ray spanned by an ample class, the first instructive examples can be found for $\rho(X) = 2$. There are four cases that can occur:

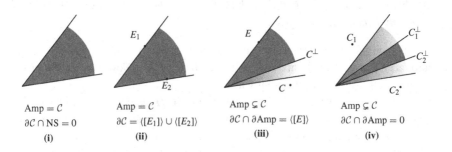

$$
\begin{array}{cccc}
\text{Amp} = C & \text{Amp} = C & \text{Amp} \subsetneq C & \text{Amp} \subsetneq C \\
\partial C \cap NS = 0 & \partial C = \langle [E_1] \rangle \cup \langle [E_2] \rangle & \partial C \cap \partial \text{Amp} = \langle [E] \rangle & \partial C \cap \partial \text{Amp} = 0 \\
\text{(i)} & \text{(ii)} & \text{(iii)} & \text{(iv)}
\end{array}
$$

(i) $\text{Amp}(X) = C_X$ and $\partial C_X \cap NS(X) = \{0\}$. Then by Theorem 3.2[5]

$$
NE(X) = \text{Amp}(X).
$$

---

[5] This is an example where the cone of curves $NE(X)$ is not closed and the nef cone $\text{Nef}(X)$ is not spanned by nef line bundles.

(ii) $\text{Amp}(X) = \mathcal{C}_X$ and there exist two smooth elliptic curves E_1 and E_2 such that $\partial \mathcal{C}_X = \mathbb{R}_{\geq 0} \cdot [E_1] \sqcup \mathbb{R}_{\geq 0} \cdot [E_2]$. Then

$$\text{NE}(X) = \mathbb{R}_{\geq 0} \cdot [E_1] + \mathbb{R}_{\geq 0} \cdot [E_2].$$

(iii) $\text{Amp}(X) \subsetneq \mathcal{C}_X$ and there exist smooth integral curves E and C of genus one and zero, respectively, such that the two boundaries of $\text{Nef}(X) = \overline{\text{Amp}}(X)$ are $\mathbb{R}_{\geq 0} \cdot [E]$ and the ray orthogonal to $[C]$. Then

$$\text{NE}(X) = \mathbb{R}_{\geq 0} \cdot [E] + \mathbb{R}_{\geq 0} \cdot [C].$$

(iv) $\text{Amp}(X) \subsetneq \mathcal{C}_X$ and there exist smooth integral rational curves C_1 and C_2 such that the boundaries of $\overline{\text{Amp}}(X)$ are the two rays orthogonal to $[C_1]$ and $[C_2]$. In particular, $\partial \text{Amp}(X)$ is contained in the interior of $\mathcal{C}_X$ and

$$\text{NE}(X) = \mathbb{R}_{\geq 0} \cdot [C_1] + \mathbb{R}_{\geq 0} \cdot [C_2].$$

Note that in (i) $\text{Nef}(X) = \overline{\mathcal{C}}_X$ is polyhedral but not rational polyhedral. In the remaining cases $\text{Nef}(X)$ and $\text{NE}(X)$ are in fact both rational polyhedral.

Remark 3.7 For purely numerical reasons the case that only one of the two rays of $\partial \text{Amp}(X)$ is spanned by a class in $\text{NS}(X)$ cannot occur. For example, if $\text{Amp}(X) = \mathcal{C}_X$ and E is smooth elliptic (and thus spans one ray of $\partial \text{Amp}(X)$) and H is ample, then $2(H.E)H - (H)^2 E$ spans the other ray; cf. Lemma 3.13.

Also, as a consequence of Remark 2.13 or by a purely numerical argument, one finds that in the case (iv) none of the two boundaries $\partial \mathcal{C}_X$ is rational.

Remark 3.8 Here are a few more comments concerning elliptic and rational curves. In case (i) there exist neither smooth rational nor smooth elliptic curves. In case (ii) no smooth rational curve can exist, as its orthogonal complement would cut $\mathcal{C}_X$ and hence $\text{Amp}(X)$ could not be maximal.

All smooth elliptic curves E_1, E_2, and E in (ii) and (iii) can be replaced by integral rational (but singular) curves. Indeed, any nef line bundle L with $(L)^2 = 0$ defines an elliptic fibration $\pi \colon X \longrightarrow \mathbb{P}^1$; see Proposition 2.3.10. The elliptic curves are smooth fibres of the corresponding fibration. Then take a singular and hence rational fibre of π, which has to be irreducible due to $\rho(X) = 2$; cf. Remark 2.3.13 and Corollary 11.1.7.

Let us now look at the case $\rho(X) > 2$. We shall try to visualize this by assuming $\rho(X) = 3$ and by taking a cut with $(H.\ ) = 1$, for a fixed ample class H.

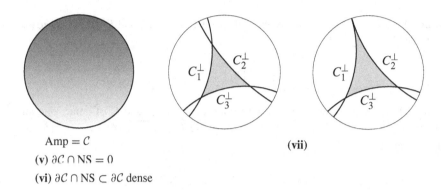

$\text{Amp} = \mathcal{C}$

(v) $\partial\mathcal{C} \cap \text{NS} = 0$

(vi) $\partial\mathcal{C} \cap \text{NS} \subset \partial\mathcal{C}$ dense

(vii)

(v) As in **(i)**, $\text{Amp}(X) = \mathcal{C}_X$ and $\partial\mathcal{C}_X \cap \text{NS}(X) = \{0\}$. Then

$$\text{NE}(X) = \text{Amp}(X).$$

(vi) As in **(ii)**, $\text{Amp}(X) = \mathcal{C}_X$ and its closure $\text{Nef}(X) = \overline{\text{Amp}}(X) = \overline{\mathcal{C}}_X$ is the closure of the cone spanned by all classes $[E]$ of smooth elliptic curves E:

$$\mathcal{C}_X = \text{Amp}(X) \subsetneq \text{NE}(X) \subsetneq \overline{\text{NE}}(X) = \overline{\mathcal{C}}_X.$$

(vii) As in **(iv)**, $\text{Amp}(X) \subsetneq \mathcal{C}_X$ and $\overline{\text{NE}}(X)$ is the closure of the cone spanned by all classes $[C]$ of smooth rational curves C:

$$\overline{\text{NE}}(X) = \sum \mathbb{R}_{\geq 0} \cdot [C_i],$$

where the C_i are (-2)-curves.

Note that we have not drawn the analogue of **(iii)** for $\rho(X) > 2$. Clearly, in **(v)** and **(vi)** the nef cone is not polyhedral, not even locally. In **(vii)** it is at least locally polyhedral.

Remark 3.9 As in Remark 3.8, instead of the smooth elliptic curves E in **(vi)** one could use singular rational curves. As $\text{Amp}(X)$ is maximal in this case, the singular fibres of the elliptic fibration associated with E are irreducible (of type I_1 or II), for no smooth rational curve can exist; see Section 11.1.4.

The last picture in **(vii)** is realized e.g. by an elliptic K3 surface with a section C_3 and a reducible fibre $C_1 + C_2$ (type I_2 or III). In this case, one can assume $(C_3.C_1) = 1$ and $(C_3.C_2) = 0$. By Hodge index theorem and an easy computation, $C_3^\perp$ meets $C_1^\perp$ and $C_2^\perp$ in the interior of $\mathcal{C}_X$.

Remark 3.10 If X does not admit any (-2)-class, then any ray $\mathbb{R}_{>0} \cdot L \subset \partial\mathcal{C}_X$ is spanned by a smooth elliptic curve (assuming $\text{char}(k) \neq 2, 3$; see Proposition 2.3.10).

The following is the main result of Kovács [327].

Theorem 3.11 *Let X be a projective K3 surface of Picard number $\rho(X) \geq 2$. Then Amp(X) and NE(X) are as in one of the cases (i)–(vii). Moreover, $\rho(X) \leq 4$ in (v) and $\rho(X) \leq 11$ in (vi).*

The main steps of the proof are sketched below. This theorem has a series of important and actually quite surprising consequences and we begin with those.

Corollary 3.12 *Let X be a projective K3 surface.*

(i) *If $\rho(X) = 2$, then NE(X) (or Nef(X) = $\overline{\text{Amp}}(X)$) is rational(!) polyhedral if and only if X contains a smooth elliptic or a smooth rational curve.*

(ii) *For $\rho(X) \geq 3$, either X does not contain any smooth rational curves at all or $\overline{\text{NE}}(X)$ is the closure of the cone spanned by all smooth rational curves $C \subset X$.*

(iii) *Either $\overline{\text{NE}}(X)$ is completely circular or has no circular parts at all. For $\rho(X) \geq 3$ the former case is equivalent to $\overline{\text{NE}}(X) = \overline{\text{Amp}}(X) = \overline{C}_X$.*

Proof The first assertion follows from an inspection of the cases (i)–(iv).

If X does not contain any (-2)-curve at all, then $\text{Amp}(X) = C_X$ by Corollary 1.7 and $\overline{\text{NE}}(X) = \text{Nef}(X) = \overline{C}_X$ by Theorem 3.2. Hence, $\overline{\text{NE}}(X)$ is completely circular for $\rho(X) \geq 3$.

If X contains a (-2)-curve, then (iii), (iv) (for $\rho(X) = 2$), or (vii) describe $\overline{\text{NE}}(X)$. In particular, for $\rho(X) \geq 3$ it is the closure of the cone spanned by smooth integral rational curves. As this cone is locally polyhedral outside ∂C_X by Corollary 3.6, $\overline{\text{NE}}(X)$ has no circular parts at all. □

As Kovács explains in detail in [327], all cases allowed by the theorem do in fact occur.

3.3 The proof of Theorem 3.11 starts with the following elementary observation. We follow the original [327] quite closely. Suppose $\alpha, \beta \in \text{NS}(X)$ with $\alpha \in C_X$ and $0 \neq \beta \in \partial C_X$. Then $\gamma := 2(\alpha.\beta)\alpha - (\alpha)^2\beta \in \mathbb{R} \cdot \alpha \oplus \mathbb{R} \cdot \beta$ is also contained in ∂C_X and β and γ are on 'opposite sides' of $\mathbb{R}_{>0} \cdot \alpha$ (i.e. α is contained in the convex cone spanned by β and γ); cf. Remark 3.7.

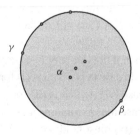

By varying $\alpha \in C_X \cap \text{NS}(X)_{\mathbb{Q}}$, this immediately yields

Lemma 3.13   *If $\partial \mathcal{C}_X \cap \mathrm{NS}(X) \neq \{0\}$, then $\partial \mathcal{C}_X \cap \mathrm{NS}(X)_\mathbb{Q}$ is dense in $\partial \mathcal{C}_X$.*   □

Corollary 3.14   *Assume that X does not contain any (-2)-curve and that $\rho(X) \geq 5$. Then $\mathrm{Amp}(X)$ is described as in* (**vi**).

Proof   Indeed, $\mathrm{Amp}(X) = \mathcal{C}_X$ by Corollary 1.7 and $\partial \mathcal{C}_X \cap \mathrm{NS}(X) \neq \{0\}$ by Hasse–Minkowski; cf. [547, IV.3.2]. Now combine Lemma 3.13 and Remark 3.10 to show that indeed (**vi**) describes $\overline{\mathcal{C}}_X$ as the closure of the cone spanned by smooth elliptic curves.   □

Remark 3.15   Similar arguments show the following. If $\alpha \in \mathcal{C}_X \cap \mathrm{NS}(X)$ and $\beta \in \mathrm{NE}(X)$ with $(\beta)^2 = -2$, then there exists an effective class $\gamma \in (\mathbb{R} \cdot \alpha \oplus \mathbb{R} \cdot \beta) \cap \mathrm{NS}(X)$ with either $(\gamma)^2 = 0$ or $= -2$ and again β and γ on opposite sides of $\mathbb{R}_{>0} \cdot \alpha$.

Moreover, $(\gamma)^2 = 0$ can be achieved if and only if $2(\alpha)^2 + (\alpha.\beta)^2$ is a square in $\mathbb{Q}$. Indeed, then $x^2(2(\alpha)^2 + (\alpha.\beta)^2) - y^2 = 0$ has a positive integral solution and one can set

$$\gamma := 2x\alpha - (y - (\alpha.\beta)x)\beta \tag{3.3}$$

(and use $y > (\alpha.\beta)x$, due to $(\alpha)^2 > 0$, to see that γ is effective). If $2(\alpha)^2 + (\alpha.\beta)^2$ is not a square in $\mathbb{Q}$, then the (infinitely many) solutions to Pell's equation $x^2(2(\alpha)^2 + (\alpha.\beta)^2) - y^2 = -1$ yield γ defined by (3.3) with $(\gamma)^2 = -2$. Also, one can arrange things such that γ is effective and on the opposite side of $\mathbb{R} \cdot \alpha$.

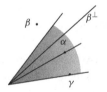   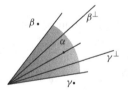

PROOF OF THEOREM 3.11.   Note that by purely lattice theoretic considerations (cf. Corollary **14**.3.8) any X with $\rho(X) \geq 12$ in fact contains a (-2)-curve and then Corollary 3.14 does not apply.

On the other hand, if X does not contain any (-2)-curve but $\rho(X) < 5$, then either $\partial \mathcal{C}_X \cap \mathrm{NS}(X) = \{0\}$, and we are in case (**v**) (which is (**i**) for $\rho(X) = 2$), or $\partial \mathcal{C}_X \cap \mathrm{NS}(X) \neq \{0\}$, and then we are again in case (**vi**) (which is (**ii**) for $\rho(X) = 2$).

It remains to deal with the case that X contains a (-2)-curve $C \subset X$. For $\rho(X) = 2$, Remark 3.15 shows that (**iii**) or (**iv**) must hold; i.e. if $\partial \mathrm{Amp}(X)$ contains one of the rays of $\partial \mathcal{C}_X$, then this ray is spanned by an integral class.

It remains to show that **(vii)** holds if X contains a (-2)-curve C and satisfies $\rho(X) \geq 3$. More precisely, we have to show that in this case $\overline{\text{NE}}(X)$ is the closure of the cone spanned by (-2)-curves or, equivalently, that $\partial\text{NE}(X)$ has no circular part; see the argument in the proof of Corollary 3.12.

Suppose $\partial\text{NE}(X)$ has a circular part. By Corollary 3.6, this can happen only when there exists an open subset $U \subset \partial\mathcal{C}_X$ that is at the same time contained in $\partial\text{NE}(X)$. We can assume that $U = \mathbb{R}_{>0} \cdot U$. Now choose an integral class $\alpha \in \mathcal{C}_X$ arbitrarily close to U. By Remark 3.15 one finds an effective class $\gamma \in (\mathbb{R}\cdot\alpha \oplus \mathbb{R}\cdot[C]) \cap \text{NS}(X)$ with $(\gamma)^2 = 0$ or $= -2$ such that γ and $[C]$ are on opposite sides of $\mathbb{R}_{>0}\cdot\alpha$. As α approaches $U \subset \partial\text{NE}(X)$, only $(\gamma)^2 = 0$ is possible and, moreover, $\gamma \in U$. In other words, such a $U \subset \partial\mathcal{C}_X$ always contains an integral effective class $\gamma \in U \cap \text{NS}(X)$.

Consider $\gamma_\varepsilon := (1 - \varepsilon)\gamma + \varepsilon[C]$, which is an effective rational class for sufficiently small $\varepsilon \in \mathbb{Q}_{>0}$. Then $(\gamma.C) > 0$, as otherwise $(\gamma_\varepsilon)^2 = -2\varepsilon^2((\gamma.C)+1)+2\varepsilon(\gamma.C) < 0$, which would contradict $\gamma \in \partial\text{NE}(X) \cap \partial\mathcal{C}_X$. Since $\rho(X) \geq 3$, there exists an integral class $\gamma' \in (\mathbb{R}\cdot\gamma \oplus \mathbb{R}\cdot[C])^\perp$ with $(\gamma')^2 < 0$. Then define

$$\gamma_n := -2n^2(\gamma')^2(\gamma.C)^3\gamma - 2n(\gamma.C)^2\gamma' + [C]$$

and check $(\gamma_n)^2 = -2$. Moreover, $\mathbb{R}_{>0} \cdot \gamma_n$ converges to $\mathbb{R}_{>0} \cdot \gamma$. By Riemann–Roch and using $(\gamma.H) > 0$ and hence $(\gamma_n.H) > 0$ for $n \gg 0$ and a fixed ample class H, one concludes that the classes γ_n are effective. Hence, γ is contained in the closure of the cone spanned by (-2)-curves contradicting the assumption that $\overline{\text{NE}}(X)$ is circular in γ. This concludes the proof of Theorem 3.11. $\square$

The last step in the proof illustrates the phenomenon that $\overline{\text{NE}}(X)$ is locally polyhedral outside $\overline{\mathcal{C}}_X$ but not necessarily in points of the boundary $\partial\mathcal{C}_X$.

4 Cone Conjecture

As explained in Section 2.4, the action of the Weyl group W on the positive cone $\mathcal{C}_X$ admits a fundamental domain of the form $\text{Nef}(X) \cap \mathcal{C}_X$. Moreover, $\text{Nef}(X) \subset \overline{\mathcal{C}}_X$ is locally polyhedral in the interior of $\mathcal{C}_X$, but not necessarily at points in $\partial\mathcal{C}_X$. However, the only reason for not being locally polyhedral in $\partial\mathcal{C}_X$ and for not being polyhedral (and not only locally) altogether is the possibly infinite automorphism group $\text{Aut}(X)$. To make this precise, we have to replace the nef cone by the *effective nef cone* $\text{Nef}^e(X)$.

The main result Theorem 4.2 of this section, due to Sterk (following suggestions by Looijenga [577]), is a particular case of the Kawamata–Morrison cone conjecture. Totaro's survey [604] is highly recommended; see also his paper [603] for the technical details and Lazić's [358, Sec. 6] for a discussion of the conjecture for general Calabi–Yau varieties as part of the minimal model program.

4.1   In order to phrase the cone theorem properly, one needs to introduce

$$\mathrm{Nef}^e(X) \subset \mathrm{Nef}(X)$$

as the real convex hull of $\mathrm{Nef}(X) \cap \mathrm{NS}(X)$. Note that $\mathrm{Nef}^e(X)$ need not be closed, e.g. in the cases **(i)** and **(v)** in Section 3.2 the nef cone is maximal $\mathrm{Nef}(X) = \overline{C}_X$, but $\mathrm{Nef}^e(X)$ is the open cone C_X. Or in case **(vi)**, again $\mathrm{Nef}(X) = \overline{C}_X$, but of course only the rational rays in ∂C_X can be contained in $\mathrm{Nef}^e(X)$. In any case, the closure $\overline{\mathrm{Nef}}^e(X)$ always gives back $\mathrm{Nef}(X)$.

Note that $\mathrm{Nef}^e(X)$ is rational polyhedral if and only if $\mathrm{Nef}(X)$ is. Of course, in this case the two cones coincide, but they might coincide without being rational polyhedral. Recall that a convex cone that is rationally polyhedron is by definition spanned by finitely many rational rays.

Example 4.1   (i) Neither of the two cones $\mathrm{Nef}(X)$ or $\mathrm{Nef}^e(X)$ is in general locally polyhedral. Again case **(vi)** is an example, in which case $\mathrm{Nef}^e(X) \cap \partial C_X$ is dense in ∂C_X. Every open neighbourhood of an arbitrary point in ∂C_X, rational or not, intersects infinitely many walls of $\mathrm{Nef}^e(X)$.

(ii) In [604, Sec. 4] Totaro describes an example of a K3 surface X with $\rho(X) = 3$ for which $\mathrm{Nef}(X) = \mathrm{Nef}^e(X)$ intersect ∂C_X in only one ray (corresponding to the fibre class of an elliptic fibration), but along this ray the cones are not locally polyhedral; i.e. infinitely many walls corresponding to infinitely many sections of the elliptic fibration accumulate towards this ray.[6] See also Example 4.3.

4.2   Recall that a rational polyhedral fundamental domain for the action of a group G on a cone C_0 (often not closed) is a rational polyhedral (and hence automatically closed) cone $\Pi \subset C_0$ such that $C_0 = \bigcup_{g \in G} g(\Pi)$ and $g(\Pi) \cap h(\Pi)$ does not contain interior points for $g \neq h$.

---

[6] The dark region in the picture represents a fundamental domain for the action of $\mathrm{Aut}(X)$ on $\mathrm{Nef}^e(X)$, as shall be explained shortly.

Theorem 4.2 *Let X be a projective K3 surface over an algebraically closed field k of characteristic $\neq 2$. The action of $\operatorname{Aut}(X)$ on the effective nef cone $\operatorname{Nef}^e(X) \subset \operatorname{NS}(X)_{\mathbb{R}}$ admits a rational polyhedral fundamental domain Π.*

Proof Here is an outline of the argument for complex K3 surfaces following Kawamata [287]. We shall use that $\operatorname{Aut}(X)$ acts faithfully on $H^2(X, \mathbb{Z})$; see Proposition **15.2.1**.

The first thing to note is that the subgroup $\operatorname{Aut}_s(X) \subset \operatorname{Aut}(X)$ of symplectic automorphisms (i.e. those that act trivially on $T(X)$ (see Section **15.1**) and the Weyl group W can both be seen as subgroups of $O^+(\operatorname{NS}(X))$. In fact,

$$\operatorname{Aut}_s(X) \ltimes W \subset O^+(\operatorname{NS}(X)) \tag{4.1}$$

is a finite index subgroup. This is also stated as Theorem **15.2.6** and here is the proof.

One first remarks that the kernel of the natural map $O(\operatorname{NS}(X)) \longrightarrow O(\operatorname{NS}(X)^*/\operatorname{NS}(X))$ is clearly of finite index. Next, by Corollary 2.11 any $g_0 \in O^+(\operatorname{NS}(X))$ can be modified by an element of $h \in W \subset O^+(\operatorname{NS}(X))$ such that the new element $g := h \circ g_0$ preserves the chamber $\operatorname{Amp}(X) \subset \mathcal{C}_X$. If in addition, $g = \operatorname{id}$ on the discriminant lattice $\operatorname{NS}(X)^*/\operatorname{NS}(X)$, then g can be extended by id on the transcendental lattice $T(X)$ to a Hodge isometry of $H^2(X, \mathbb{Z})$; see Proposition **14.2.6**. Moreover, this Hodge isometry respects the ample cone and, by the global Torelli theorem **7.5.3**, is therefore induced by an automorphism $f \in \operatorname{Aut}_s(X)$.

Eventually one uses that the action of $O^+(\operatorname{NS}(X))$ on $\mathcal{C}_X$ admits a rational polyhedral fundamental domain $\Pi \subset \mathcal{C}_X$. This is a very general statement on lattices of signature $(1, \rho - 1)$, which has nothing to do with the geometry of K3 surfaces. The standard reference for this is the book by Ash et al [23, Ch. II.4], but see also [379, Sec. 4] and the survey in [499, Thm. 2.5]. In [577, Sec. 2] one finds a rather detailed explanation. The precise statement is as follows: there exists a rational polyhedral domain $\Pi \subset \mathcal{C}_X^e$ for the action of $O^+(\operatorname{NS}(X))$ on the effective positive cone $\mathcal{C}_X^e$ which is defined as the real convex hull of $\overline{C}_X \cap \operatorname{NS}(X)$. In fact, in [577] one finds the explicit description

$$\Pi = \{x \in \mathcal{C}_X^e \mid (H.g(x) - x) \geq 0 \quad \text{for all } g \in O^+(\operatorname{NS}(X))\}, \tag{4.2}$$

where H is a fixed ample class.

For $g = s_{[C]}$ the inequality defining Π reads $(H.x + (x.C)[C]) \geq (H.x)$, which is equivalent to $(x.C) \geq 0$, as $(H.C) > 0$. Therefore, $\Pi \subset \operatorname{Nef}^e(X)$, and, moreover, Π is up to finite index a rational polyhedral fundamental domain for the action of $\operatorname{Aut}(X)$ on $\operatorname{Nef}^e(X)$. In fact, a rational polyhedral fundamental domain for the action of $\operatorname{Aut}(X)$ can be described similarly to (4.2), where H is chosen such that its stabilizer in $\operatorname{Aut}(X)$ is trivial. See Totaro's [603, Thm. 3.1, Lem. 2.2 arxiv version] or Looijenga's [379, Prop. 4.1, Appl. 4.14].

The assertion for projective K3 surfaces over algebraically closed fields of characteristic zero follows from the complex case. For positive characteristic the arguments above were adapted by Lieblich and Maulik in [369]. Very roughly, if X is not supersingular, then NS(X) and Aut(X) lift to characteristic zero. In positive characteristic $\neq 2$ one applies Ogus's crystalline theory [479]. □

Example 4.3 If $\pi : X \longrightarrow \mathbb{P}^1$ is an elliptic K3 surface with rk MW$(X) > 0$ (see Section **11.3.2**), then the induced natural inclusion MW$(X) \hookrightarrow$ Aut(X) (see Remark **15**.4.5) yields an infinite subgroup of (symplectic) automorphisms all fixing the class $[X_t] \in \partial C_X \cap$ NS(X). Hence, any neighbourhood of $[X_t]$ intersects infinitely many copies of the fundamental domain. Compare this to Example 4.1, (ii).

Corollary 4.4 *The effective nef cone* Nef$^e(X)$ *is rational polyhedral if and only if* Aut(X) *is finite.* □

Remark 4.5 It is instructive to compare the above results with the case of complex abelian surfaces. The nef cone of an abelian surface A can be understood more explicitly, mainly because $1 \le \rho(A) \le 4$.

If $\rho(A) = 2$, then Nef(A) is obviously polyhedral, but not necessarily rational polyhedral; cf. Section 3.2, case (**i**). In fact, it is rational polyhedral if and only if $A = E_1 \times E_2$ with E_1 and E_2 non-isogenous elliptic curves without complex multiplication. For $\rho(A) = 3, 4$, the nef cone is not rational polyhedral. Bauer in [37, Thm. 7.2] gives a detailed description. The analogue of Theorem 4.2 for abelian surfaces was proved by Kawamata in [287].

4.3 Here are a few consequences of the theorem that show that the convex geometry of the natural cones, ample, nef, etc., has strong implications for the geometry of a K3 surface, which we continue to assume to be over an algebraically closed field.

For the first one, see the article [577] by Sterk. Lieblich and Maulik [369] checked the case char$(k) > 0$.

Corollary 4.6 *The set of (-2)-curves up to automorphisms*

$$\{C \subset X \mid C \simeq \mathbb{P}^1\}/\text{Aut}(X)$$

is finite. More generally, for any d there are only finitely many orbits of the action of Aut(X) *on the set of classes $\alpha \in$ NS(X) of the form $\alpha = [C]$ with $C \subset X$ irreducible[7] and $(\alpha)^2 = (C)^2 = 2d$.*

Proof Throughout the proof one uses that every $\mathbb{P}^1 \simeq C \subset X$ defines a wall of codimension one of Nef(X); see Remark 2.8.

Let now Π be a rational polyhedral fundamental domain for the action of Aut(X) on Nef$^e(X)$. Then Nef$^e(X)$ and Π share finitely many walls $[C_1]^\perp, \ldots, [C_n]^\perp$ for certain

---

[7] For $d \ge 0$ it suffices to require that C is nef.

$\mathbb{P}^1 \simeq C_1, \ldots, C_n \subset X$. Now, consider another $\mathbb{P}^1 \simeq C \subset X$. Then there exist an $f \in \mathrm{Aut}(X)$ such that $f^*[C]^\perp$ is one of the $[C_i]^\perp$. But two (-2)-curves that define the same wall coincide. Hence, every $\mathrm{Aut}(X)$-orbit on the set of (-2)-curves meets the finite set $\{C_1, \ldots, C_n\}$.

For the second part, i.e. $(C)^2 = 2d \geq 0$, one first observes that the (rational) polyhedral fundamental domain $\Pi \subset \mathrm{Nef}^e(X)$ contains only finitely many integral classes with $(\alpha)^2 = 2d \geq 0$. (This needs an extra but still elementary argument when Π contains a class in $\partial \mathcal{C}_X$.) Now, if $C \subset X$ is an irreducible curve with $(C)^2 = 2d \geq 0$, then C is nef, i.e. $[C] \in \mathrm{Nef}^e(X)$. Hence, there exists $f \in \mathrm{Aut}(X)$ such that $f^*[C] \in \Pi$, which is then one of these finitely many classes. $\qquad\square$

The next result is due to Pjateckiĭ-Šapiro and Šafarevič [493, §7 Thm. 1] and Sterk [577].

Corollary 4.7 *Consider the following conditions:*

(i) *The effective cone $\mathrm{NE}(X)$ is rational polyhedral.*
(ii) *The quotient $O(\mathrm{NS}(X))/W$ is finite.*
(iii) *The group $\mathrm{Aut}(X)$ is finite.*
(iv) *There are only finitely many smooth rational curves contained in X.*

Then (i) $\Leftrightarrow$ (ii) $\Leftrightarrow$ (iii) $\Rightarrow$ (iv). *If X contains at least one (-2)-curve, then also* (iv) $\Rightarrow$ (i).

Proof Assume (i). Thus, $\mathrm{NE}(X)$ and hence $\overline{\mathrm{NE}}(X)$ are rational polyhedral. But then also the dual $\mathrm{Nef}(X)$ and $\mathrm{Nef}^e(X)$ are rational polyhedral. Hence, by Corollary 4.4, $\mathrm{Aut}(X)$ is finite and, therefore, (i) implies (iii).

Next, (ii) and (iii) are equivalent, because $\mathrm{Aut}_s(X) \ltimes W \subset O(\mathrm{NS}(X))$ is of finite index; see the proof of Theorem 4.2 or Theorem 15.2.6.

Now assume (iii). Then, by Corollary 4.4, $\mathrm{Nef}^e(X)$ is rational polyhedral, and, therefore, also $\mathrm{Nef}(X) = \overline{\mathrm{Nef}}^e(X)$ is. This in turn implies that $\mathrm{NE}(X)$ is rational polyhedral and hence (i) holds, e.g. by going through **(i)**–**(vii)**. As all smooth rational curves define a wall of codimension one of $\mathrm{Nef}(X)$ (see Remark 2.8), (ii) and (iii) imply (iv).

For $\rho(X) \geq 3$, Corollary 3.12 shows that either X does not contain any smooth rational curves or $\overline{\mathrm{NE}}(X)$ is the closure of the cone spanned by smooth rational curves. So, if X contains a (-2)-curve and (iv) is assumed, then $\mathrm{NE}(X) = \overline{\mathrm{NE}}(X)$ is rational polyhedral. (Note that for $\mathrm{Nef}(X) = \overline{\mathcal{C}}_X$, the existence of a rational polyhedral domain for $\mathrm{Nef}^e(X)$ requires $\mathrm{Aut}(X)$ to be infinite.)

If $\rho(X) = 2$ and X contains a (-2)-curve, then only **(iii)** and **(iv)** can occur, and in both cases $\mathrm{NE}(X)$ is rational polyhedral.

See again [369] for the case of positive characteristic. $\qquad\square$

Thus, whether NE(X) is rational polyhedral can be read off from NS(X) alone. In fact, there are only finitely many choices for the hyperbolic lattice NS(X) such that NE(X) is rational polyhedral. An explicit (but still quite involved) complete classification of these lattices is known; cf. [13, Thm. 2.12] or [149, Cor. 4.2.4, Thm. 2.2.2] and see Section **15**.2.4.

Rephrasing the above in terms of the nef cone yields the following corollary.

Corollary 4.8   *If* Nef(X) *is not rationally polyhedral, then* Aut(X) *is infinite.*   □

In practice, it is often quite mysterious where these infinitely many automorphisms come from, e.g. in the cases (**i**) and (**v**) in Section **3**.2. See Section **15**.2.5 for examples with infinite Aut(X) but with no (-2)-curve.

Example 4.9   The following, which can be deduced by closed inspection of (**i**)–(**iv**), was observed in [493, Sec. 7]: if $\rho(X) = 2$, then Aut(X) is finite if and only if there exists a class $\alpha \in$ NS(X) with $(\alpha)^2 = -2$ or $= 0$. See also Section **14**.2 and Example **15**.2.11.

Another immediate, and partially more geometric, consequence of the cone conjecture can be spelled out as follows.

Corollary 4.10   *Up to the action of* Aut(X), *there exist only finitely many ample line bundles L on X with* $(L)^2$ *fixed. Equivalently, for a fixed K3 surface X_0 the set*

$$\{(X, L) \in M_d \mid X \simeq X_0\}$$

is finite.   □

Here, M_d is the moduli space of polarized K3 surfaces (X, L) with $(L)^2 = 2d$; see Chapter **5**.

5 Kähler Cone

We quickly explain what happens if instead of the ample cone in NS(X)$_\mathbb{R}$ one considers the Kähler cone $\mathcal{K}_X$ inside $H^{1,1}(X, \mathbb{R})$.

5.1   For this let X be a complex K3 surface. It is known that any complex K3 surface is Kähler (see comments in Section **7**.3.2) and we shall assume this here. A Kähler metric g on X with induced Kähler form ω defines a class $[\omega] \in H^{1,1}(X, \mathbb{R})$. A class in $H^{1,1}(X, \mathbb{R})$ is called a *Kähler class* if it can be represented by a Kähler form. Since a positive linear combination of Kähler forms is again a Kähler form, the set of all Kähler classes in $H^{1,1}(X, \mathbb{R})$ describes a convex cone.

Definition 5.1 The Kähler cone

$$\mathcal{K}_X \subset H^{1,1}(X,\mathbb{R})$$

is the open convex cone of all Kähler classes $[\omega] \in H^{1,1}(X,\mathbb{R})$. The *positive cone* of the complex K3 surface X is the connected component

$$\mathcal{C}_X \subset H^{1,1}(X,\mathbb{R})$$

of the open set $\{\alpha \in H^{1,1}(X,\mathbb{R}) \mid (\alpha)^2 > 0\}$ that contains $\mathcal{K}_X$.

Note that the inclusion

$$\mathrm{NS}(X)_{\mathbb{R}} = H^{1,1}(X,\mathbb{Z})_{\mathbb{R}} \subset H^{1,1}(X,\mathbb{R}) \tag{5.1}$$

is in general strict. In fact, for very general X one has $\mathrm{NS}(X) = 0$. Equality in (5.1) holds only for K3 surfaces of Picard number $\rho(X) = 20$. Observe that $\mathcal{K}_X$ and $\mathcal{C}_X$ are open convex cones of real dimension 20 independent of X, whereas the dimension of $\mathrm{Amp}(X)$ depends on $\rho(X)$.

Under the inclusion (5.1), one has $\mathrm{Amp}(X) = \mathcal{K}_X \cap \mathrm{NS}(X)_{\mathbb{R}}$ by Kodaira's characterization of positive classes; (cf. [221, 253, 621]) and $\mathcal{C}_X \cap \mathrm{NS}(X)_{\mathbb{R}}$ gives back the positive cone in $\mathrm{NS}(X)_{\mathbb{R}}$ (for which we used the same notation). Recall that X is projective if and only if $\mathcal{K}_X$ contains an integral class or, equivalently, if $\mathcal{C}_X$ contains an integral class; see Remark 1.3.

The following is the Kähler analogue of Corollary 1.7.

Theorem 5.2 *The Kähler cone $\mathcal{K}_X$ is described by*

$$\mathcal{K}_X = \{\alpha \in H^{1,1}(X,\mathbb{R}) \mid \alpha \in \mathcal{C}_X \quad \text{and} \quad (\alpha.C) > 0 \quad \text{for all} \quad \mathbb{P}^1 \simeq C \subset X\}. \tag{5.2}$$

Moreover, if α is in the boundary of $\mathcal{K}_X$, then $(\alpha)^2 = 0$ or there exists a $\mathbb{P}^1 \simeq C \subset X$ with $(\alpha.C) = 0$.

Note that the cone described by the right-hand side of (5.2) is open by Remark 1.3, which remains true in the Kähler setting due to Remark 2.7. In Section 7.3.2 we stated a special case of the theorem as the following corollary.

Corollary 5.3 *If $\mathrm{Pic}(X) = 0$, then $\mathcal{K}_X = \mathcal{C}_X$.* □

In fact, this corollary was there used to deduce the global Torelli theorem, on which the classical proof of Theorem 5.2, to be sketched below, relies. So there is a certain (but no actual) circular touch to the argument.

PROOF OF THEOREM 5.2. The theorem can nowadays be seen as a consequence of the much more general result by Demailly and Paun [147], which in dimension two had been proved by Buchdahl [90] and Lamari [346, 347]. However historically, its first proof relied on the surjectivity of the period map (cf. Section 7.4.1) and the global Torelli theorem 7.5.3.

Let us sketch the classical approach following [54, Exp. X]. Let $\mathcal{K}_X^o$ denote the set of classes $\alpha \in \mathcal{C}_X$ with $(\alpha.C) > 0$ for all $\mathbb{P}^1 \simeq C \subset X$ and such that the positive three-space $W := \langle \mathrm{Re}(\sigma), \mathrm{Im}(\sigma), \alpha \rangle$ contains a class $0 \neq \alpha_0 \in W \cap H^2(X, \mathbb{Q})$. Here, $0 \neq \sigma \in H^{2,0}(X)$. Note that if $\alpha \in \mathcal{K}_X^o$, then $(\alpha.\delta) \neq 0$ for all (-2)-classes $\delta \in \mathrm{NS}(X)$. Now write $W = \alpha_0^\perp \oplus \mathbb{R} \cdot \alpha_0$ and use the surjectivity of the period map (cf. Section 7.4.1) to realize $\alpha_0^\perp$ as the period of a K3 surface X_0. Passing from X to X_0 has the advantage that due to the existence of the rational class $\alpha_0 \in H^{1,1}(X_0, \mathbb{Q})$ with $(\alpha_0)^2 > 0$ one now has an algebraic K3 surface X_0; see Remark 1.3. Note that α_0 has still the property that $(\alpha_0.\delta_0) \neq 0$ for all (-2)-classes $\delta_0 \in \mathrm{NS}(X_0)$. Indeed, otherwise $\delta_0 \in W^\perp$ and hence $\delta_0 \in \mathrm{NS}(X)$ and $(\alpha.\delta_0) = 0$.

By Corollary 2.9, there exists a Hodge isometry g of $H^2(X_0, \mathbb{Z})$ such that $g(\alpha_0)$ becomes Kähler. Using the twistor space construction for $g(\alpha_0)$ (see Section 7.3.2), one finds a K3 surface X' together with a Hodge isometry $\tilde{g} \colon H^2(X, \mathbb{Z}) \xrightarrow{\sim} H^2(X', \mathbb{Z})$ mapping α to a Kähler class $\alpha' := g(\alpha)$. (Tacitly, we are using the natural identification $H^2(X, \mathbb{Z}) \simeq H^2(X_0, \mathbb{Z})$.) By the global Torelli theorem 7.5.3 or rather its proof in Section 7.5.5 one can lift $\tilde{g}$ to an isomorphism $X \xrightarrow{\sim} X'$, for α is positive on all $\mathbb{P}^1 \simeq C \subset X$. In particular, all $\alpha \in \mathcal{K}_X^o$ are Kähler classes. As $\mathcal{K}_X^o$ is dense in the open cone described by the right-hand side of (5.2), this is enough to conclude.   □

Remark 5.4 In the Kähler setting too, the notions of nef classes and of the nef cone exist. Since in general there are too few curves (sometimes none) to measure positivity of classes, one rather uses an analytic definition. A posteriori it turns out that the corresponding nef cone is again the closure $\overline{\mathcal{K}}_X$ of the Kähler cone; see [147] for further references.

5.2 The definition of the Weyl group of a complex K3 surface remains unchanged,

$$W := \langle s_\delta \mid \delta \in \mathrm{NS}(X) \text{ with } (\delta)^2 = -2 \rangle,$$

but it is now considered as a subgroup of $O(H^2(X, \mathbb{Z}))$. Note that all s_δ act as id on $H^{2,0} \oplus H^{0,2}$ and in fact on $T(X) = \mathrm{NS}(X)^\perp$ (which, however, might not complement $\mathrm{NS}(X)$); see Example 3.3.2). The set of roots $\Delta := \{\delta \in \mathrm{NS}(X) \mid (\delta)^2 = -2\}$ induces a chamber decomposition of $\mathcal{C}_X \subset H^{1,1}(X, \mathbb{R})$. Due to Remark 2.7, the main results of Sections 2.2 and 2.4 still hold true.

Proposition 5.5 *The Weyl group W of any complex K3 surface acts simply transitively on the set of chambers of the positive cone $\mathcal{C}_X \subset H^{1,1}(X, \mathbb{R})$. Moreover, $\overline{\mathcal{K}}_X \cap \mathcal{C}_X$ is a fundamental domain for this action. It is locally polyhedral in the interior of $\mathcal{C}_X$.*   □

5.3 The ample cone and the positive cone in $\mathrm{NS}(X)_\mathbb{R}$ of an algebraic K3 surface behave badly under deformation of the surface. Since the Picard number usually jumps on dense subsets (see Sections **6.2.5** and **17**.1.3), even their dimension cannot be expected to be locally constant. The situation is different for the Kähler cone. Already the positive cone $\mathcal{C}_X \subset H^{1,1}(X, \mathbb{R})$ behaves well under deformation; it is always a real open cone of dimension 20, and the family of positive cones $\mathcal{C}_{X_t}$ for a deformation $X \longrightarrow S$ forms a real manifold $\mathcal{C}_{X/S} \longrightarrow S$ of relative real dimension 20 over S. This still holds true for the family of Kähler cones $\mathcal{K}_{X/S} \longrightarrow S$ but takes a bit more effort to prove. For the following we refer to [33, Ch. VIII.9] and [54, Exp. IX]. The result can also be deduced from the general fact that for a family of $(1, 1)$-classes, being Kähler is an open condition.

Proposition 5.6 *For a smooth family of K3 surfaces $f : X \longrightarrow S$, the family of Kähler cones $\mathcal{K}_{X_t} \subset \mathcal{C}_{X_t} \subset H^{1,1}(X_t, \mathbb{R})$ forms an open subset*

$$\mathcal{K}_{X/S} \subset (R^1 f_* \Omega_{X/S})_{\mathbb{R}}$$

of the real vector bundle on the right-hand side with fibres $H^{1,1}(X_t, \mathbb{R})$.

5.4   Some of the results for $\mathrm{Amp}(X)$ do not carry over to $\mathcal{K}_X$. Here are two examples.

Lemma 3.13 fails in two ways. If X is not algebraic, then $\mathrm{NS}(X)$ might very well be of rank one generated by a line bundle L with $(L)^2 = 0$ and, in particular, up to sign no other primitive integral class (of square zero) exists. But even when X is algebraic Lemma 3.13 yields only density of rational classes in the boundary of the positive cone in $\mathrm{NS}(X)_{\mathbb{R}}$. For example, if $\rho(X) < 20$, then the rational classes in the positive cone $\mathcal{C}_X \subset H^{1,1}(X, \mathbb{R})$ can certainly not be dense.

Theorem 3.11 is false in the general Kähler case and so are Theorem 4.2 and Corollaries 3.12 and 4.7. For example, using the surjectivity of the period map (see Theorem 7.4.1), one ensures the existence of a K3 surface X such that $\mathrm{NS}(X)$ is generated by three pairwise orthogonal (-2)-classes $[C_i]$, $i = 1, 2, 3$. So, $\rho(X) = 3$, but $\mathcal{K}_X$ has a large circular part. Only within the interior of $\mathcal{C}_X$ is it cut out by the three hyperplanes $[C_i]^\perp$.

References and Further Reading

It can be quite difficult to describe the ample cone $\mathrm{Amp}(X)$ of a particular K3 surface X explicitly. See e.g. Nikulin's article [454] or the more recent [30] by Baragar for $\rho(X) = 3, 4$. The latter examines certain fractals associated with the ample cone.

The Cox ring of a K3 surface is by definition

$$\mathrm{Cox}(X) := \bigoplus_{L \in \mathrm{NS}(X)} H^0(X, L).$$

It contains the usual section rings $\bigoplus H^0(X, L^n)$ for any L, which are finitely generated subrings for all nef L; see comments at the end of Chapter 2. However, the Cox ring is rarely finitely generated. In fact, as shown by Artebani et al. in [13] the Cox ring of a projective complex K3 surface is finitely generated if and only if NE(X) is rational polyhedral. Another proof for $\rho(X) = 2$ was given by Ottem in [484], which also contains explicit descriptions of generators of Cox(X) in this case; see also [13, Thm. 3.2]. The more general case of klt Calabi–Yau pairs in dimension two is the subject of Totaro [603].

Coming back to Corollary 4.10: in [556, Cor. 1.7] Shimada shows that (under certain assumptions on X) up to automorphisms any H with $(H)^2 = 2d$ also satisfies an upper bound $(H.H_0) \leq \phi(d)$ with $\phi(d)$ a explicit linear function in d. This should allow one to bound the cardinality of the finite set.

The interior of $\overline{\mathrm{NE}}(X)$ is the *big cone* Big(X), i.e. the set of positive real linear combinations $\sum a_i L_i$ with L_i big, i.e. $h^0(L_i^n) \sim n^2$. See [357, Ch. 2.2]. Also the big cone admits a chamber decomposition. In fact, it admits two. First, one can look at the *Weyl chambers*, which by definition are the connected components of Big(X) $\setminus \bigcup [C]^\perp$, where the union is over all $\mathbb{P}^1 \simeq C \subset X$. Clearly, on $\mathcal{C}_X \subset$ Big(X) this gives back the classical chamber decomposition. Second, one can consider the decomposition into *Zariski chambers*. By definition, two big divisors L_1, L_2 are contained in the same Zariski chamber if the negative parts in the Zariski decomposition of L_1 are the (-2)-curves with $(L_2.C) = 0$ (see comments at the end of Chapter 2). These two notions have been compared carefully in [39]. In particular, it is shown that the two decompositions coincide if and only if one cannot find two (-2)-curves C_1, C_2 with $(C_1.C_2) = 1$.

The cone conjecture for deformations of Hilb$^n(X)$ with X a K3 surface has recently been established by Markman and Yoshioka in [390] and in general by Amerik and Verbitsky in [4].

Questions and Open Problems

Presumably, (4.1) and Corollary 4.6 may fail for general complex K3 surfaces; i.e. Aut$_s(X) \ltimes W \subset$ O(NS(X)) might not be of finite index and there may be infinitely many (-2)-curves and yet finite Aut(X). It could be interesting to describe explicit examples.

Using the surjectivity of the period map, it is easy to show that every class $\alpha \in H^2(M, \mathbb{R})$ (with M the differentiable manifold underlying a K3 surface) with $(\alpha)^2 > 0$ is a Kähler class $[\omega]$ with respect to some complex structure $X = (M, I)$ (automatically defining a K3 surface). This implies that every symplectic structure on M compatible with the standard orientation of M can cohomologically be realized by a (hyper)Kähler structure. It is however unknown whether this remains true on the level of forms or whether the space of symplectic structure in one cohomology class is connected. This is related to the discussion in Section 7.5.6 and at the end of Chapter 7. For further references, see [520].

9

Vector Bundles on K3 Surfaces

In this chapter, a few explicit and geometrically relevant bundles on K3 surfaces and their properties are studied in detail. In particular, stability of the tangent bundle and of bundles naturally associated with line bundles and curves is discussed. Stability of the tangent bundle can be seen as a strengthening of the non-existence of vector fields on K3 surfaces and is known only in characteristic zero. We mention the algebraic approach due to Miyaoka and the analytic one that uses the existence of Ricci-flat Kähler metrics; see Section 4. Vector bundle techniques developed by Lazarsfeld to prove that generic curves in integral linear systems on K3 surfaces are Brill–Noether general are outlined in Section 2. In the Appendix we outline how the vanishing $H^0(X, \mathcal{T}_X) = 0$ in general is used to lift K3 surfaces from positive characteristic to characteristic zero.

1 Basic Techniques and First Examples

This section introduces the Mukai pairing, proves that the tangent bundle of a very general complex K3 surface is simple and studies the rigid bundle obtained as the kernel of the evaluation map of a big and nef line bundle.

1.1 For the convenience of the reader we collect a few standard results on coherent sheaves on a smooth surface X. For proofs and more results, see Friedman's book [185] or [266].

(o) A coherent sheaf F on a smooth surface X is *torsion free* if its torsion $T(F)$ is trivial.

(i) If F is torsion free, then F is locally free on the complement $X \setminus \{x_1, \ldots, x_n\}$ of a finite set of closed points.

(ii) The dual $F^* := \mathcal{H}om(F, \mathcal{O}_X)$ of any coherent sheaf F is locally free.

(iii) The double dual $F^{**} := \mathcal{H}om(\mathcal{H}om(F, \mathcal{O}_X), \mathcal{O}_X)$ of a coherent sheaf F is called its *reflexive hull*. There is a natural morphism $F \longrightarrow F^{**}$ which is injective if and only if F is torsion free. Its cokernel is a sheaf with zero-dimensional support.

(iv) The rank of a coherent sheaf F can be defined as $\dim_{k(x)} F(x)$ with $x \in X$ generic (or even the generic point) or as the rank of the locally free sheaf F^*. Then, $\text{rk}(F) = 0$ if and only if F is a torsion sheaf.

(v) Any torsion-free sheaf of rank one is isomorphic to a sheaf $M \otimes I_Z$ with $M \in \text{Pic}(X)$ and I_Z the ideal sheaf of a subscheme $Z \subset X$ of dimension zero.

1.2 Arguably, the two most important techniques in the study of the geometry of a K3 surface X (over an arbitrary field k) are Serre duality and the Riemann–Roch formula. Serre duality for two coherent sheaves $E, F \in \text{Coh}(X)$ or, more generally, for two bounded complexes of coherent sheaves $E, F \in D^b(X)$ asserts that there exist natural isomorphisms

$$\text{Ext}^i(E, F) \simeq \text{Ext}^{2-i}(F, E)^*.$$

In a more categorical language this can be phrased by saying that the derived category

$$D^b(X) := D^b(\text{Coh}(X))$$

of the abelian category $\text{Coh}(X)$ is endowed with a *Serre functor* $S \colon D^b(X) \xrightarrow{\sim} D^b(X)$ which is isomorphic to the double shift $E \longmapsto E[2]$; cf. Section **16**.1.3. For a coherent sheaf E one can use a locally free resolution of E to compute $\text{Ext}^i(E, F)$. For complexes, the description

$$\text{Ext}^i(E, F) = \text{Hom}_{D^b(X)}(E, F[i])$$

might be more useful.

Also note that for coherent sheaves E and F one has $\text{Ext}^i(E, F) = 0$ for $i > 2$ and $i < 0$, which fails for arbitrary complexes. Thus, Serre duality for sheaves reduces to the two isomorphisms $\text{Hom}(E, F) \simeq \text{Ext}^2(F, E)^*$ and $\text{Ext}^1(E, F) \simeq \text{Ext}^1(F, E)^*$.

Assume $E = F$ (sheaves or complexes). Then Serre duality $\text{Ext}^1(E, E) \simeq \text{Ext}^1(E, E)^*$ can be seen as the existence of a non-degenerate quadratic form on $\text{Ext}^1(E, E)$. The duality pairing

$$\text{Ext}^1(E, E) \times \text{Ext}^1(E, E) \longrightarrow k \qquad (1.1)$$

is in fact obtained by composing $\alpha \in \text{Hom}_{D^b(X)}(E, E[1])$ with $\beta \in \text{Hom}_{D^b(X)}(E, E[1]) \simeq \text{Hom}_{D^b(X)}(E[1], E[2])$ followed by the trace giving $\text{tr}(\beta \circ \alpha) \in H^2(X, \mathcal{O}) \simeq k$.[1]

Proposition 1.1 *The Serre duality pairing (1.1) is non-degenerate and alternating.*

Proof There are various ways of proving this. See [266, Ch. 10] for a proof using Čech cohomology. Here is an argument using Dolbeault cohomology. It works only

---

[1] The last isomorphism is not canonical. Viewing $H^2(X, \mathcal{O})$ as $H^0(X, \omega_X)^*$, it depends on the choice of a trivializing section of ω_X, i.e. a non-trivial regular two-form.

for locally free sheaves on complex K3 surface, but it shows clearly how the sign comes up.

One uses that for E locally free $\operatorname{Ext}^1(E, E) = H^1(X, \mathcal{E}nd(E))$ can be computed as the first cohomology of the $\bar{\partial}$-complex

$$\mathcal{A}^0(\mathcal{E}nd(E)) \longrightarrow \mathcal{A}^{0,1}(\mathcal{E}nd(E)) \longrightarrow \mathcal{A}^{0,2}(\mathcal{E}nd(E)).$$

Classes $\alpha, \beta \in \operatorname{Ext}^1(E, E)$ can be represented by $\sum \alpha_i \otimes \gamma_i$ resp. $\sum \beta_j \otimes \delta_j$ with α_i, β_j differentiable endomorphisms of the complex bundle and γ_i, δ_j forms of type $(0, 1)$. Then the Serre duality pairing of α and β yields $\sum \operatorname{tr}(\beta_j \circ \alpha_i) \otimes (\delta_j \wedge \gamma_i)$. Clearly, $\operatorname{tr}(\beta_j \circ \alpha_i) = \operatorname{tr}(\alpha_i \circ \beta_j)$ but $\delta_j \wedge \gamma_i = -\gamma_i \wedge \delta_j$. □

Recall from Section 1.2.4 that the Hirzebruch–Riemann–Roch formula for arbitrary coherent sheaves (or bounded complexes of such) on a K3 surface takes the form

$$\chi(F) = \int \operatorname{ch}(F)\operatorname{td}(X) = \operatorname{ch}_2(F) + 2\operatorname{rk}(F). \tag{1.2}$$

This is generalized to an expression for a quadratic form as follows. Define for E and F the Euler pairing

$$\chi(E, F) := \sum (-1)^i \dim \operatorname{Ext}^i(E, F).$$

Then, Serre duality implies $\chi(E, F) = \chi(F, E)$; i.e. the Euler pairing is symmetric. Note that for $E = \mathcal{O}_X$ one finds $\chi(\mathcal{O}_X, F) = \chi(F)$ and, more generally, for E locally free $\chi(E, F) = \chi(E^* \otimes F)$. Then (1.2) generalizes to

$$\chi(E, F) = \int \operatorname{ch}^*(E)\operatorname{ch}(F)\operatorname{td}(X) = \int (\operatorname{ch}^*(E) \sqrt{\operatorname{td}(X)})(\operatorname{ch}(F) \sqrt{\operatorname{td}(X)}). \tag{1.3}$$

Here, ch^* is defined by $\operatorname{ch}^*_i = (-1)^i \operatorname{ch}_i$, which for a locally free sheaf E yields $\operatorname{ch}^*(E) = \operatorname{ch}(E^*)$, and $\sqrt{\operatorname{td}(X)} = 1 + (1/24)c_2(X)$.

Definition 1.2 The *Mukai vector* for (complexes of) sheaves is defined by

$$v(E) := \operatorname{ch}(E) \sqrt{\operatorname{td}(X)} = (\operatorname{rk}(E), c_1(E), \operatorname{ch}_2(E) + \operatorname{rk}(E))$$
$$= (\operatorname{rk}(E), c_1(E), \chi(E) - \operatorname{rk}(E)).$$

The Mukai vector can be considered in cohomology (étale, singular, crystalline, de Rham), in the Chow ring $\operatorname{CH}^*(X)$ (see Section 12.1.4), or in the numerical Grothendieck group (see Section 16.2.4). As at this point, this is of no importance for our discussion, we shall be vague about it.

Example 1.3 For later reference, we record the special cases:

$$v(k(x)) = (0, 0, 1), \quad v(\mathcal{O}_X) = (1, 0, 1), \quad \text{and} \quad v(L) = (1, c_1(L), c_1^2(L)/2 + 1)$$

for $L \in \operatorname{Pic}(X)$.

In order to express $\chi(E,F)$ as an intersection of Mukai vectors, one introduces the *Mukai pairing*. This can be done for arbitrary K3 surfaces (see Section **16**.2.4), but for complex K3 surfaces it can be most conveniently phrased using singular cohomology.

Definition 1.4 For a complex K3 surface X the *Mukai pairing* on $H^*(X, \mathbb{Z})$ is

$$\langle \alpha, \beta \rangle = (\alpha_2.\beta_2) - (\alpha_0.\beta_4) - (\alpha_4.\beta_0),$$

where $(\,.\,)$ denotes the usual intersection form on $H^*(X, \mathbb{Z})$.

In other words, the Mukai pairing differs from the intersection form only by a sign in the pairing on $H^0 \oplus H^4$. See also Section **1**.3.3, where $(\,.\,)$ was considered only on $H^2(X, \mathbb{Z})$.

With this definition, the Hirzebruch–Riemann–Roch formula (1.3) becomes

$$\chi(E, F) = -\langle v(E), v(F) \rangle. \tag{1.4}$$

1.3 Let us henceforth assume that we work over an algebraically closed field. Then a sheaf E is *simple* if $\mathrm{End}(E) = k.$[2] Then

$$\chi(E, E) = 2 - \dim \mathrm{Ext}^1(E, E) \leq 2$$

and hence $\langle v(E), v(E) \rangle \geq -2$. Typical examples with $\langle v(E), v(E) \rangle = -2$ are provided by line bundles $E = L \in \mathrm{Pic}(X)$ and the structure sheaf $E = \mathcal{O}_C$ of a smooth rational curve $\mathbb{P}^1 \simeq C \subset X$; cf. Section **16**.2.3.

Remark 1.5 The inequality $\langle v(E), v(E) \rangle \geq -2$ for a simple sheaf E can be spelled out as $(\mathrm{rk}(E) - 1)c_1^2(E) - 2\mathrm{rk}(E)c_2(E) + 2\mathrm{rk}(E)^2 \leq 2$ or, equivalently,

$$\Delta(E) := 2\mathrm{rk}(E)c_2(E) - (\mathrm{rk}(E) - 1)c_1^2(E) \geq 2(\mathrm{rk}(E)^2 - 1).$$

For $\mathrm{rk}(E) \geq 1$, this yields the weak *Bogomolov inequality* $\Delta(E) \geq 0$, which holds for semistable(!) sheaves on arbitrary surfaces; cf. Section 3.1.

It is surprisingly difficult to find simple bundles explicitly. In fact, there are very few naturally given bundles on a K3 surface. The tangent bundle $\mathcal{T}_X$ and bundles derived from it by linear algebra operations, such as tensor and symmetric powers, are the only natural non-trivial bundles. Proving simplicity of $\mathcal{T}_X$ is not trivial and is usually seen as a consequence of stability; see Section 3.1. For complex K3 surfaces X with $\mathrm{Pic}(X) = 0$, and then for a Zariski open subset of all K3 surfaces (including many projective ones), the simplicity of $\mathcal{T}_X$ can be proved by an elementary argument, which we shall explain next.

---

[2] Over an arbitrary field a sheaf is called simple if $\mathrm{End}(E)$ is a division algebra.

Example 1.6 (i) Suppose $\mathcal{T}_X$ were not simple. Then there exists an endomorphism $0 \neq \varphi \colon \mathcal{T}_X \longrightarrow \mathcal{T}_X$ which is not an isomorphism. Indeed, for any $\psi \colon \mathcal{T}_X \longrightarrow \mathcal{T}_X$ which is not of the form $\lambda \cdot \mathrm{id}$ pick a point $x \in X$ and an eigenvalue λ of $\psi_x \colon \mathcal{T}_X \otimes k(x) \longrightarrow \mathcal{T}_X \otimes k(x)$. Then, $\varphi := \psi - \lambda \cdot \mathrm{id}$ is not an isomorphism at the point x and not trivial either. As a morphism of sheaves, φ cannot be injective, for otherwise $\mathrm{Coker}(\varphi)$ would be a non-trivial torsion sheaf with trivial Chern classes. Hence, $\mathrm{Ker}(\varphi) \neq 0$. Since $\mathrm{Im}(\varphi)$ is torsion free and hence of homological dimension ≤ 1 (see e.g. [266, Ch. 1]), $\mathrm{Ker}(\varphi)$ has homological dimension zero. Thus, $\mathrm{Ker}(\varphi)$ is a line bundle and $\mathrm{Ker}(\varphi) \simeq \mathcal{O}_X$ if $\mathrm{Pic}(X) = 0$. This, however, contradicts $H^0(X, \mathcal{T}_X) = 0$ (see Section 1.2.4 and the Appendix).

(ii) Note that the arguments work for arbitrary sheaves and show: a locally free sheaf E that is not simple admits a non-trivial endomorphism $\varphi \colon E \longrightarrow E$ with non-trivial kernel.

A straightforward computation yields $\langle v(\mathcal{T}_X), v(\mathcal{T}_X) \rangle = 88$, which together with the fact that $\mathcal{T}_X$ is simple (at least over $\mathbb{C}$) yields

$$\dim \mathrm{Ext}^1(\mathcal{T}_X, \mathcal{T}_X) = 90.^3$$

1.4 In the case of a polarized K3 surface X, or more precisely of an embedded K3 surface $X \subset \mathbb{P}^N$, the restricted tangent bundle $\mathcal{T}_{\mathbb{P}^N}|_X$ also appears naturally and one might ask whether it has particular properties. By the Euler sequence, the twisted dual $\Omega_{\mathbb{P}^N}(1)|_X$ can often be described as the vector bundle M_L associated naturally with the line bundle $L = \mathcal{O}(1)|_X$, which we now define.

Definition 1.7 For a globally generated, big and nef line bundle L let M_L be the kernel of the evaluation map:

$$0 \longrightarrow M_L \longrightarrow H^0(X, L) \otimes \mathcal{O}_X \overset{\mathrm{ev}}{\longrightarrow} L \longrightarrow 0. \tag{1.5}$$

Note that M_L is of rank $h^0(L) - 1$ and satisfies $H^0(X, M_L) = H^1(X, M_L) = 0$; for the latter use $H^1(X, \mathcal{O}) = 0$. The following result is taken from Camere's thesis [97].

Example 1.8 For any globally generated, big and nef line bundle L the vector bundle M_L is simple. Indeed, tensoring (1.5) with $E_L := M_L^*$ yields the short exact sequence

$$0 \longrightarrow M_L \otimes E_L \longrightarrow H^0(X, L) \otimes E_L \longrightarrow L \otimes E_L \longrightarrow 0,$$

the long cohomology sequence of which has the form

$$\cdots \longrightarrow H^1(X, L \otimes E_L) \longrightarrow H^2(X, M_L \otimes E_L) \longrightarrow H^0(X, L) \otimes H^2(X, E_L) \longrightarrow 0.$$

---

[3] A geometric interpretation of this number would be interesting.

By Serre duality and definition of M_L, $H^2(X, E_L) \simeq H^0(X, M_L)^* = 0$. To compute $H^1(X, L \otimes E_L)$, dualize (1.5) and tensor with L. The long cohomology sequence reads

$$\cdots \longrightarrow H^0(X, L)^* \otimes H^1(X, L) \longrightarrow H^1(X, L \otimes E_L) \longrightarrow H^2(X, \mathcal{O}_X) \simeq k \longrightarrow \cdots .$$

However, $H^1(X, L) = 0$ by Proposition **2**.3.1, and, therefore, $H^2(X, M_L \otimes E_L)$ is at most one-dimensional. Thus, M_L is indeed simple, since $H^2(X, M_L \otimes E_L) \simeq \mathrm{End}(M_L)^*$.

Also note that M_L is in fact simple and *rigid*, i.e. also $\mathrm{Ext}^1(M_L, M_L) = 0$. This follows from $\langle v(M_L), v(M_L) \rangle = -2$, using $v(M_L) = h^0(X, L) \cdot v(\mathcal{O}_X) - v(L)$.

Restricting the Euler sequence $0 \longrightarrow \Omega_{\mathbb{P}^N} \otimes \mathcal{O}(1) \longrightarrow H^0(X, L) \otimes \mathcal{O} \longrightarrow \mathcal{O}(1) \longrightarrow 0$ to the projective embedding $X \hookrightarrow \mathbb{P}^N$ induced by a very ample linear system $|L|$ shows

$$\Omega_{\mathbb{P}^N}|_X \simeq M_L \otimes L^*,$$

which is therefore simple.

2 Simple Vector Bundles and Brill–Noether General Curves

The evaluation map for a line bundle on a curve viewed as a sheaf on the ambient surface is another source of interesting examples of bundles. The construction provides a link between Brill–Noether theory on curves in K3 surfaces and the theory of bundles on K3 surfaces. This has led to many important results.

2.1 If we allow ourselves to use more of the geometry of the K3 surface X, in particular curves contained in the surface, then more vector bundles can be exhibited. A standard technique in this context uses elementary transformations along curves. Here is an outline of the construction; for more details, see [266, Ch. 5] or the survey by Lazarsfeld [356, Sec. 3].

Let $C \subset X$ be a curve and A a line bundle on C, simultaneously viewed as a torsion sheaf on X. If E is a vector bundle on X and $E|_C \longrightarrow\!\!\!\!\!\rightarrow A$ a surjection on C, then the kernel F of the composition $E \longrightarrow E|_C \longrightarrow\!\!\!\!\!\rightarrow A$, which is a sheaf on X, is called the *elementary transformation* of E along C (but it clearly depends also on A and the surjection). Thus, there exists a short exact sequence on X,

$$0 \longrightarrow F \longrightarrow E \longrightarrow A \longrightarrow 0. \tag{2.1}$$

Lemma 2.1 *The elementary transformation F is locally free and satisfies*

$$\det(F) \simeq \det(E) \otimes \mathcal{O}(-C) \text{ and } c_2(F) = c_2(E) - (C.c_1(E)) + \deg(A).$$

Proof The first assertion can be checked locally and so we may can assume $A \simeq \mathcal{O}_C$. Using the locally free resolution $0 \longrightarrow \mathcal{O}(-C) \longrightarrow \mathcal{O}_X \longrightarrow \mathcal{O}_C \longrightarrow 0$, one finds for the homological dimension that $\mathrm{dh}(\mathcal{O}_C) = 1$. Since F is the kernel of $E \twoheadrightarrow \mathcal{O}_C$ with E locally free, this is enough to conclude $\mathrm{dh}(F) = 0$; i.e. F is locally free.

The line bundle A is trivial on the complement of finitely many points $x_1, \ldots, x_n \in C$. Therefore, as a vector bundle on X is uniquely determined by its restriction to $X \setminus \{x_i\}$, to compute $\det(F)$, we may assume $A \simeq \mathcal{O}_C$. Then conclude by using $\det(\mathcal{O}_C) \simeq \mathcal{O}(C)$. To compute $c_2(F)$, use the Riemann–Roch formula on C; cf. [266, Prop. 5.2.2]. $\square$

Dualizing the exact sequence (2.1) yields a short exact sequence of the form

$$0 \longrightarrow E^* \longrightarrow F^* \longrightarrow A^* \otimes \mathcal{O}_C(C) \longrightarrow 0. \tag{2.2}$$

Here, A^* denotes the dual of the line bundle A on C (and not the dual on X, which is trivial). Indeed, the injection of locally free sheaves $F \longrightarrow E$, which generically is an isomorphism, dualizes to an injection $E^* \longrightarrow F^*$. So the only thing to check is that indeed $\mathcal{E}xt^1_X(A, \mathcal{O}_X) \simeq A^* \otimes \mathcal{O}_C(C)$. If $A = L|_C$ for some line bundle L on X, then dualizing $0 \longrightarrow L(-C) \longrightarrow L \longrightarrow A \longrightarrow 0$ yields this isomorphism. The general case can be reduced to this by writing $A = L|_C \otimes \mathcal{O}_C(-x_1 - \cdots - x_n)$ and the fact that the computation of $\mathcal{E}xt^i_X(k(x_j), \mathcal{O}_X)$ is purely local.

2.2 Let us apply this general construction to the following special situation. Consider a globally generated line bundle A on a curve $C \subset X$ and let $r := h^0(C, A) - 1$. For $E := \mathcal{O}_X^{r+1}$ and the evaluation map $E|_C \simeq \mathcal{O}_C^{r+1} \twoheadrightarrow A$, the elementary transformation of E along C is in this case described by

$$0 \longrightarrow F \longrightarrow \mathcal{O}_X^{r+1} \longrightarrow A \longrightarrow 0. \tag{2.3}$$

The following result is due to Lazarsfeld [354].

Proposition 2.2 *Assume in addition that $A^* \otimes \mathcal{O}_C(C)$ is globally generated and that every curve in the linear system $|C|$ is reduced and irreducible. Then the elementary transformation F in (2.3) is locally free and simple. (As it turns out, F is in fact μ-stable; see Corollary 3.3.)*

Proof Clearly, F is simple if and only if its dual $G := F^*$ is simple. By (2.2) the bundle G sits in a short exact sequence $0 \longrightarrow \mathcal{O}_X^{r+1} \longrightarrow G \longrightarrow A^* \otimes \mathcal{O}_C(C) \longrightarrow 0$. Using $H^1(X, \mathcal{O}_X) = 0$, this shows that G is globally generated.

If G is not simple, then there exists a non-trivial endomorphism $\varphi \colon G \longrightarrow G$ with non-trivial kernel; see Example 1.6. For $K := \mathrm{Im}(\varphi)$, one has a short exact sequence

$$0 \longrightarrow K \longrightarrow G \longrightarrow G/K \longrightarrow 0$$

with K torsion free of rank $0 < s < r + 1$.

Since G is globally generated and K and G/K are both quotients of G, their determinants are also globally generated and hence of the form $\det(K) \simeq \mathcal{O}(C_1)$

and $\det(G/K) \simeq \mathcal{O}(C_2)$ for some effective curves C_1, C_2. They are both non-trivial, which can be proved as follows. The surjectivity of $G \twoheadrightarrow K$ and the vanishing $\operatorname{Hom}(G, \mathcal{O}_X) = H^0(X, F) = 0$ imply that K is globally generated with $\operatorname{Hom}(K, \mathcal{O}_X) = 0$. The restriction $K|_D$ to a generic ample curve D is locally free and globally generated. Thus, there exists a short exact sequence

$$0 \longrightarrow (K|_D)^* \longrightarrow \mathcal{O}_D^{s+1} \longrightarrow \det(K)|_D \longrightarrow 0$$

of vector bundles on D; see [266, Ch. 5].

For sufficiently positive D, the restriction map $\operatorname{Hom}(K, \mathcal{O}_X) \longrightarrow \operatorname{Hom}(K|_D, \mathcal{O}_D)$ is surjective and thus $H^0(D, (K|_D)^*) = \operatorname{Hom}(K|_D, \mathcal{O}_D) = 0$. Hence, $h^0(D, \det(K)|_D) \geq s + 1$. This clearly implies $\deg(K|_D) > 0$ and hence $C_1 \neq 0$. For G/K, which is not necessarily torsion free, one applies the argument to $G \twoheadrightarrow (G/K)/T(G/K)$. Note that the torsion part $T(G/K)$ has an effective (but possibly trivial) determinant as well.

On the other hand, $\det(G) \simeq \mathcal{O}(C)$, which leads to

$$\mathcal{O}(C_1 + C_2) \simeq \det(K) \otimes \det(G/K) \simeq \det(G) \simeq \mathcal{O}(C),$$

i.e. $C_1 + C_2 \in |C|$. This contradicts the assumption on $|C|$. ☐

Remark 2.3 The proposition is typically applied to the case that $\mathcal{O}(C)$ generates $\operatorname{Pic}(X)$, as in this case the assumption on $|C|$ is automatically satisfied. However, the proof shows that it is enough to assume that C is ample with $(C)^2$ minimal among all intersection numbers $(C.D)$ with D effective.

2.3 The above construction was used in Lazarsfeld's influential paper [354] to deduce properties of curves on K3 surfaces from the geometry of the ambient K3 surfaces or, more precisely, from the Riemann–Roch formula $\chi(F, F) = -\langle v(F), v(F) \rangle$. For an alternative proof of the following fact, see [355].

Corollary 2.4 *Let C be a smooth curve on a K3 surface X such that all curves in $|C|$ are reduced and irreducible. Then every line bundle $A \in \operatorname{Pic}(C)$ satisfies*

$$\rho(A) := g(C) - h^0(A) \cdot h^1(A) \geq 0.$$

Proof Assume first that A and $A^* \otimes \mathcal{O}_C(C)$ are both globally generated. The constructions in Section 2.2 and Proposition 2.2 yield the simple bundle F, which satisfies

$$-\langle v(F), v(F) \rangle = \chi(F, F) \leq 2.$$

On the other hand, a simple computation using Lemma 2.1 shows

$$\langle v(F), v(F) \rangle = 2\rho(A) - 2,$$

which immediately gives the assertion $\rho(A) \geq 0$.

It remains to reduce the general case to the case that A and $A^* \otimes \omega_C \simeq A^* \otimes \mathcal{O}_C(C)$ are globally generated. If $h^0(A) = 0$ or $h^1(A) = 0$, then $\rho(A) = g(C) \geq 0$ and thus the assertion holds. Suppose $h^0(A) \neq 0$, but A is not globally generated. Let D be the fixed locus of A. Hence, $A(-D)$ is globally generated, $h^0(A) = h^0(A(-D))$, and

$$h^1(A) = h^0(A^* \otimes \omega_C) \leq h^0(A^*(D) \otimes \omega_C) = h^1(A(-D)).$$

Therefore, $\rho(A) \geq \rho(A(-D))$. Thus, it suffices to prove the assertion for A globally generated. One argues similarly to reduce to the case that $A^* \otimes \omega_C$ is globally generated without introducing base points for A. This is left as an exercise. $\qquad\square$

A few words putting the corollary in perspective; see also [11] or the surveys in [9, 355, 356]. Brill–Noether theory for smooth projective curves C studies the *Brill–Noether loci*

$$W^r_d(C) \subset \operatorname{Pic}^d(C)$$

of all line bundles A on C of degree d with $h^0(A) \geq r+1$. The $W^r_d(C)$ are determinantal subvarieties of $\operatorname{Pic}^d(C)$ given locally by the vanishing of certain minors; see e.g. [11]. To study the $W^r_d(C)$, one introduces the *Brill–Noether number*

$$\rho(g, r, d) := g - (r+1)(g-d+r).$$

If $\rho(g, r, d) \geq 0$, then the Brill–Noether locus $W^r_d(C)$ is non-empty (Kempf and Kleiman–Laksov) and, if $\rho(g, r, d) \geq 1$, it is also connected (Fulton–Lazarsfeld); cf. [11].

Moreover, the Brill–Noether number is the *expected dimension* of $W^r_d(C)$. More precisely, if $W^r_d(C)$ is non-empty, then $\dim W^r_d(C) \geq \rho(g, r, d)$ and equality holds for generic curves C. The latter is a result due to Griffiths and Harris [222], which was proved using degeneration techniques that do not allow the explicit description of Brill–Noether general curves. Part of this statement is that $W^r_d(C)$ is empty if $\rho(g, r, d) < 0$. In this sense, smooth curves on K3 surfaces defining integral linear systems are Brill–Noether general. This is made precise by the following result.

Corollary 2.5 *Let C be a smooth curve on a K3 surface X such that all curves in $|C|$ are reduced and irreducible. If $\rho(g, r, d) < 0$, then $W^r_d(C) = \emptyset$.*

Proof Suppose $A \in W^r_d(C)$. Then $h^0(A) \geq r+1$ and $\deg(A) = d$. By Riemann–Roch, $h^1(A) = g - 1 - d + h^0(A) \geq g - d + r$. Hence, by Corollary 2.4,

$$\rho(g, r, d) = g - (r+1)(g-d+r) \geq g - h^0(A) \cdot h^1(A) = \rho(A) \geq 0. \qquad\square$$

It is worth pointing out that the proof of Corollary 2.5 does not involve any degeneration techniques, unlike the original in [222].

Remark 2.6 (i) In fact, Lazarsfeld shows in [354] that the generic(!) curve C in an integral linear system $|C_0|$ is Brill–Noether general in the broader sense that all $W_d^r(C)$ are of dimension $\rho(g, r, d)$ and smooth away from $W_d^{r+1}(C) \subset W_d^r(C)$. See also Pareschi's variation of the argument in [488].

(ii) The literature on the generic behavior of curves on K3 surfaces is vast. For example, in [218] Green and Lazarsfeld show that all smooth curves in a linear system on a K3 surface have the same Clifford index. Recall that the *Clifford index* of a line bundle A is $\deg(A) - 2(h^0(A) - 1)$ and the Clifford index of C is the minimum of those over all A with $h^0(A), h^1(A) \geq 2$. A conjecture of Green relates the Clifford index of a curve to the properties of the minimal resolution of the canonical ring. It turns out that again curves on K3 surfaces are more accessible. See Beauville's Bourbaki talk [49] for a survey and for further references.

(iii) In the same spirit, Harris and Mumford asked whether all smooth curves in an ample linear system $|L|$ on a K3 surface have the same gonality. And indeed, as Ciliberto and Pareschi show in [120], this is the case, unless the K3 surface is a double plane and $L = \pi^* \mathcal{O}(3)$, which were known to be counterexamples [155].

3 Stability of Special Bundles

The section is devoted to the stability of the bundle F in (2.3) associated with a line bundle on a curve by relating it to the kernel of the evaluation map on the curve itself. For simplicity we work over an algebraically closed field (using in particular that under this assumption a simple bundle has only scalar endomorphisms).

3.1 We start with the definition of μ-stability on K3 surfaces. So, let X be an algebraic K3 surface over a field k with an ample line bundle H or a complex K3 surface with a Kähler class $\omega \in H^{1,1}(X)$. The *degree* of a coherent sheaf E on X with respect to H or ω is defined as

$$\deg_H(E) := (c_1(E).H) \text{ resp. } \deg_\omega(E) := (c_1(E).\omega).$$

Recall that $c_1(E)$ of an arbitrary coherent sheaf E is $c_1(\det(E))$, where $\det(E)$ can be computed by means of a locally free resolution[4] $0 \longrightarrow E_n \longrightarrow \cdots \longrightarrow E_0 \longrightarrow E \longrightarrow 0$ as $\det(E) = \prod \det(E_i)^{(-1)^i}$. In the following, we shall often write $\deg(E)$ in both situations while keeping in mind the dependence on H resp. ω.

If E is not a torsion sheaf, one defines its *slope* (again depending on H or ω) as

$$\mu(E) := \frac{\deg(E)}{\text{rk}(E)}.$$

---

[4] Locally free resolutions of coherent sheaves exist on non-projective complex (K3) surfaces too; see [534].

Definition 3.1 A torsion-free sheaf E is called μ-*stable* (or *slope stable*) if for all subsheaves $F \subset E$ with $0 < \operatorname{rk}(F) < \operatorname{rk}(E)$ one has

$$\mu(F) < \mu(E).$$

Similarly, a torsion-free sheaf E is called μ-*semistable* if only the weaker inequality $\mu(F) \leq \mu(E)$ is required. Note that a non-trivial subsheaf of a torsion-free sheaf is itself torsion free and hence its slope is well defined.

Here are a few standard facts concerning slope stability of sheaves on smooth surfaces.

(i) Any line bundle is μ-stable. The sum $E_1 \oplus E_2$ of two μ-stable sheaves E_1, E_2 is never μ-stable, and it is μ-semistable if and only if $\mu(E_1) = \mu(E_2)$.

(ii) For a short exact sequence

$$0 \longrightarrow F \longrightarrow E \longrightarrow G \longrightarrow 0$$

with $\operatorname{rk}(F) \neq 0 \neq \operatorname{rk}(G)$, one has

$$\mu(F) < \mu(E) \quad \text{if and only if} \quad \mu(E) < \mu(G).$$

Indeed, $\deg(E) = \deg(F) + \deg(G)$, $\operatorname{rk}(E) = \operatorname{rk}(F) + \operatorname{rk}(G)$, and hence $\mu(E) - \mu(F) = (\operatorname{rk}(G)/\operatorname{rk}(F))(\mu(G) - \mu(E))$, which yields the assertion. Alternatively, draw a picture of the ranks and degrees of the involved sheaves.

Thus, a torsion-free sheaf E is μ-stable if $\mu(E) < \mu(G)$ for all quotients $E \twoheadrightarrow G$ with $0 < \operatorname{rk}(G) < \operatorname{rk}(E)$. Since the degree of a torsion sheaf is always non-negative, one has $\mu(G/T(G)) \leq \mu(G)$, and thus only torsion-free quotients need to be tested. If E itself is locally free, then the torsion freeness of G translates into the local freeness of F. Therefore, to check μ-stability of a locally free E only locally free subsheaves $F \subset E$ need to be tested. A similar result holds for μ-semistability.

(iii) Any μ-stable sheaf E is simple. Indeed, otherwise there is a non-trivial $\varphi \colon E \longrightarrow E$ with a non-trivial kernel (see Example 1.6) and in particular $0 < \operatorname{rk}(\operatorname{Im}(\varphi)) < \operatorname{rk}(E)$. Now use μ-stability for $E \twoheadrightarrow \operatorname{Im}(\varphi)$ and $\operatorname{Im}(\varphi) \subset E$ to derive the contradiction $\mu(E) < \mu(\operatorname{Im}(\varphi)) < \mu(E)$.

Remark 3.2 There is some kind of converse to this statement proved by Mukai in [429, Prop. 3.14]: if $\operatorname{Pic}(X) \simeq \mathbb{Z}$ and the Mukai vector of a simple sheaf E is primitive with $\langle v(E), v(E) \rangle = -2$ or 0, then E is μ-semistable (and in fact stable, in the sense of Definition 10.1.4).

(iv) A torsion-free sheaf E is μ-stable if and only if its dual sheaf E^* is μ-stable. In particular, the μ-stability of a torsion-free sheaf E is equivalent to the μ-stability of its reflexive hull E^{**}, which is locally free. Moreover, $\mu(E) = \mu(E^{**})$.

(v) Any μ-semistable torsion-free sheaf E satisfies the *Bogomolov inequality*

$$\Delta(E) = 2\operatorname{rk}(E)c_2(E) - (\operatorname{rk}(E) - 1)c_1^2(E) \geq 0. \tag{3.1}$$

Note that the μ-semistability of E depends on the choice of the polarization, but the Bogomolov inequality does not. For K3 surfaces this is not surprising, as we have shown in Remark 1.5 that it holds for arbitrary simple torsion-free sheaves. See e.g. [266, Thm. 3.4.1] for a proof of the Bogomolov inequality in general.

3.2   As in Section 2, we consider the elementary transformation

$$0 \longrightarrow F \longrightarrow \mathcal{O}_X^{r+1} \longrightarrow A \longrightarrow 0$$

for a globally generated line bundle A on a curve $C \subset X$ with $r + 1 = h^0(A)$. Then $\mu(F) = -\deg \mathcal{O}(C)/(r+1)$.

As a strengthening of Proposition 2.2, one has the following corollary (as in Remark 2.3, the assumption on C can be weakened).

Corollary 3.3   *If $\mathcal{O}(C)$ generates $\mathrm{Pic}(X)$ and $A^* \otimes \omega_C$ is globally generated as well, then the elementary transformation F is μ-stable.*

Proof   First note that if $F' \subset F$ is a locally free subsheaf of rank s, then $\det(F') \subset \bigwedge^s F \subset \bigwedge^s \mathcal{O}_C^{r+1} = \mathcal{O}_X^n$. Thus, $\mathcal{O}_X \subset \det(F')^*$. As in the proof of Proposition 2.2, one argues that if also $A^* \otimes \omega_C$ and hence F^* are globally generated, then $\det(F')^* \simeq \mathcal{O}(C_1)$ with $C_1 \subset X$ a non-trivial curve.

Under the assumption that $\rho(X) = 1$, the line bundle $\mathcal{O}(C)$ is automatically ample and the slope is taken with respect to it. If $F' \subset F$ is as above, then $\det(F')^* \simeq \mathcal{O}(C_1) \simeq \mathcal{O}(kC)$ for some $k > 0$. Hence, $\deg(F') = k \deg(F) < 0$, which for $\mathrm{rk}(F') < \mathrm{rk}(F)$ shows $\mu(F') < \mu(F)$. $\square$

3.3   Recall from Section 1.4 the definition of M_L associated with any globally generated, big and nef line bundle L on a K3 surface X as the kernel of the evaluation map $H^0(X, L) \otimes \mathcal{O}_X \longrightarrow L$. In Example 1.8 we have seen that M_L or, equivalently, its dual E_L is always simple. A result of Camere [97] shows that M_L is in fact μ-stable with respect to L. (Note that stability can be formally defined with respect to any line bundle H, although later in the theory ampleness becomes crucial.) Let us start by recalling the analogous statement for curves.

Theorem 3.4   *Let C be a smooth projective curve and $L \in \mathrm{Pic}^d(C)$ be a globally generated line bundle. The kernel M_L of the evaluation map $H^0(C, L) \otimes \mathcal{O}_C \longrightarrow L$ is stable if one of the following conditions holds:*

  (i) *$L \simeq \omega_C$ and C is non-hyperelliptic or*
 (ii) *$d > 2g$ or*
(iii) *$d = 2g$, C is non-hyperelliptic, and L is general.*

A proof for (i) can be found in the article [487] by Paranjape and Ramanan. A short argument for (ii) is given by Ein and Lazarsfeld in [164], and Beauville treats in [47] the case (iii).

As a consequence of this theorem, or rather of a technical lemma proved by Paranjape in this context, the following result is proved in [97].

Corollary 3.5 *Let L be a globally generated ample line bundle on a K3 surface X. Then M_L is μ_L-stable.*

Note that for a smooth $C \in |L|$ the restriction $L|_C$ is isomorphic to the canonical bundle ω_C. Moreover, $M_L|_C \simeq M_{\omega_C} \oplus \mathcal{O}_C$, and by Theorem 3.4 the bundle M_{ω_C} is stable if C is not hyperelliptic. However, this does not quite prove the assertion of the corollary.[5] Instead of going into the details of the proof, which would require a discussion of [487] and a special discussion of the hyperelliptic case, we shall link the bundle M_L to the elementary transformation F discussed in Section 3.2.

Lemma 3.6 *Let L be a globally generated line bundle on a K3 surface X and $C \in |L|$. Then the elementary transformation F of $H^0(C, L|_C) \otimes \mathcal{O}_X$ along $L|_C$, i.e. the kernel of the evaluation map $H^0(C, L|_C) \otimes \mathcal{O}_X \twoheadrightarrow L|_C$, is isomorphic to the bundle M_L, which is the kernel of the evaluation map $H^0(X, L) \otimes \mathcal{O}_X \twoheadrightarrow L$.*

Proof Use the commutative diagram

$$
\begin{array}{ccc}
\mathcal{O}_X & \xrightarrow{\;=\;} & \mathcal{O}_X \\
\downarrow & & \downarrow \\
M_L \longrightarrow H^0(L) \otimes \mathcal{O}_X & \longrightarrow & L \\
\downarrow{\scriptstyle\simeq} \qquad\qquad \downarrow & & \downarrow \\
F \longrightarrow H^0(L|_C) \otimes \mathcal{O}_X & \longrightarrow & L|_C.
\end{array}
$$
□

Since $L|_C$ and $L^*|_C \otimes \omega_C \simeq \mathcal{O}_C$ are globally generated, Corollary 3.3 immediately leads to the following special case of Corollary 3.5.

Corollary 3.7 *If L is a globally generated line bundle on a K3 surface X that generates $\mathrm{Pic}(X)$, then M_L is μ_L-stable.* □

4 Stability of the Tangent Bundle

The tangent bundle $\mathcal{T}_X$ of a K3 surface is μ-stable if and only if all line bundles $L \subset \mathcal{T}_X$ are of negative degree $\deg(L)$ (with respect to a fixed polarization or a Kähler class).

---

[5] Camere also notes that the ampleness is not really essential. If L is just globally generated and satisfies $(L)^2 \geq 2$, then M_L is μ_L-semistable. For the stability only the case $g(C) = 6$ poses a problem.

Example 4.1 If X is a complex K3 surface with $\text{Pic}(X) = 0$, then the only line bundle $\mathcal{T}_X$ could contain is $\mathcal{O}_X$. Since $H^0(X, \mathcal{T}_X) = 0$ by Hodge theory, this is excluded as well. Thus, for the generic complex K3 surface the μ-stability of $\mathcal{T}_X$ follows from Hodge theory.[6] The weaker assertion that $\mathcal{T}_X$ is simple in this case has been explained already in Example 1.6.

There are two approaches to the stability of $\mathcal{T}_X$. Both are limited to the case of characteristic zero, but for different reasons.

(i) The algebraic approach relies on general results of Miyaoka and Mori about the existence of foliations and rational curves. Working in characteristic zero allows one to reduce to large(!) finite characteristic p. It is worth pointing out that Miyaoka's techniques prove only that $\mathcal{T}_X$ does not contain a line bundle of positive degree. The vanishing $H^0(X, \mathcal{T}_X) = 0$ has to be dealt with separately, using Hodge theory, to exclude the case of degree zero line bundles. Also note that although characteristic p methods are applied, the stability of the tangent bundle in positive characteristic is not known. The algebraic approach proves μ-stability of the tangent bundle for algebraic complex K3 surfaces.

(ii) The analytic approach uses the existence of a Kähler–Einstein metric on any complex K3 surface. This makes use of the fact that K3 surfaces are Kähler (a result due to Siu and Todorov) and of Yau's solution to the Calabi conjecture; see Section 7.3.2. Since a Kähler–Einstein metric describes in particular a Hermite–Einstein metric on the tangent bundle, slope stability follows immediately from the easy direction of the Kobayashi–Hitchin correspondence. Eventually, this approach proves the μ-stability of the tangent bundle for all complex K3 surfaces.

4.1 The following statement is a consequence of a general theorem due to Miyaoka applied to K3 surfaces. The original result is [417, Thm. 8.4]; see also [418]. A simplified proof was given by Shepherd-Barron in [316].[7]

Theorem 4.2 *Suppose (X, H) is a polarized K3 surface over an algebraically closed field of characteristic zero. If $L \subset \mathcal{T}_X$ is a line bundle with torsion-free quotient and such that $(H.L) > 0$, then through a generic closed point $x \in X$ there exists a rational curve $x \in C \subset X$ with $\mathcal{T}_C(x) \subset L(x) \subset \mathcal{T}_X(x)$.*

Remark 4.3 In [316] the result is phrased for normal complex projective varieties, but then L has to be a part of the Harder–Narasimhan filtration of $\mathcal{T}_X$, which is automatic for K3 surfaces. In addition, the degree of the curves can be bounded: $(C.H) \leq 4(H)^2/(H.L)$.

---

[6] The existence of a Kähler metric is not needed, as the Hodge decomposition holds for all compact complex surfaces; see Section 1.3.3.

[7] Thanks to Nick Shepherd-Barron for helpful discussions on topics touched upon in this section.

Corollary 4.4 *Let (X, H) be a polarized K3 surface over an algebraically closed field of characteristic zero. Then $\mathcal{T}_X$ does not contain any line bundle of positive degree.*

Proof Suppose there exists a line bundle $L \subset \mathcal{T}_X$ with $(H.L) > 0$. By base change to a larger field, we may assume that the base field is uncountable. Then, by the theorem and a standard Hilbert scheme argument, the surface X must be uniruled.

Indeed, $\mathrm{Pic}(X)$ is countable, but for any non-empty open subscheme $U \subset X$ the set $U(k)$ cannot be covered by a countable union of curves. Hence, by the theorem there exists a linear system $|L|$ that contains uncountably many rational curves. As being rational is a (closed) algebraic condition, one finds a curve $D \subset |L|$ parametrizing only rational curves. The restriction $\mathcal{C} \longrightarrow D$ of the universal family comes with a dominant map $\mathcal{C} \longrightarrow X$. Resolving singularities eventually yields a rational dominant map $D \times \mathbb{P}^1 \dashrightarrow X$; i.e. X is uniruled. Resolving indeterminacies one obtains a surjective morphism $Y \longrightarrow X$ with Y a smooth surface birational to $D \times \mathbb{P}^1$.

In characteristic zero the morphism $Y \longrightarrow X$ is generically étale (see [236, III.Cor. 10.7]), and hence $H^0(X, \omega_X) \longrightarrow H^0(Y, \omega_Y)$ is injective. On the one hand, $H^0(X, \omega_X) \neq 0$, as X is a K3 surface, and on the other hand $H^0(Y, \omega_Y) = 0$, as Y is birational to $D \times \mathbb{P}^1$ and $h^0(\omega)$ is a birational invariant; see [236, II.Thm. 8.19].[8] This gives a contradiction and thus proves the assertion. $\qquad\square$

Proposition 4.5 *The tangent bundle $\mathcal{T}_X$ of a polarized K3 surface (X, H) in characteristic zero is μ-stable.*

Proof Let $L \subset \mathcal{T}_X$ be a line bundle. By Corollary 4.4 $(H.L) \leq 0$. If $(H.L) = 0$ but L is not trivial, then $(H'.L) > 0$ with respect to some other polarization H', contradicting Corollary 4.4. Hence, either $(H.L) < 0$ or $L \simeq \mathcal{O}_X$. The latter case can be excluded in characteristic zero by Hodge theory: $H^0(X, \mathcal{T}_X) \simeq H^0(X, \Omega_X) = 0$; see Section 1.2.4. $\qquad\square$

Similar techniques can be used to approach the non-existence of global vector fields on K3 surfaces in positive characteristic. Currently, there are three proofs known [349, 462, 513]. The first step in two of them consists of showing that the existence of a non-trivial vector field would imply that the K3 surface is unirational. This was shown by Rudakov and Šafarevič in [513]. The short proof given by Miyaoka in [418, Cor. III.1.13] can be extended to prove the following result.

Proposition 4.6 *Let X be a K3 surface defined over an algebraically closed field k of characteristic $p > 0$. If $\mathcal{T}_X$ is not μ-stable, e.g. if $H^0(X, \mathcal{T}_X) \neq 0$, then X is unirational.*

---

[8] See [312, IV.Cor. 1.11] for the following general result: if X is a smooth, proper and separably uniruled variety, then $H^0(X, \omega_X^m) = 0$ for all $m > 0$.

Proof Here is an outline of the argument. Suppose $L \subset \mathcal{T}_X$ is a subsheaf of rank one with $(H.L) \geq 0$. We may assume that L is saturated, i.e. that $\mathcal{T}_X/L$ is torsion free. As in the arguments in characteristic zero, one would like to view L as the tangent directions of a certain foliation. A local calculation and $\mathrm{rk}\, L = 1$ show that $L \subset \mathcal{T}_X$ is preserved by the Lie bracket, i.e. $[L, L] \subset L$. Next one needs to show that L is p-closed, i.e. that with ξ a local section of $L \subset \mathcal{T}_X$ also $\xi^p \in \mathcal{T}_X$ lies in L. Here, ξ^p is the pth power of the derivation ξ.

Assume first that $(H.L) > 0$. Using [165, Lem. 4.2], it suffices to show that the $\mathcal{O}_X$-linear homomorphism $L \longrightarrow \mathcal{T}_X/L$, $\xi \longmapsto \overline{\xi^p}$ is trivial, which follows from $\mathcal{T}_X/L$ being torsion free of degree $-(H.L) < 0$. Thus, $L \subset \mathcal{T}_X$ indeed defines a foliation and its quotient $\pi : X \longrightarrow Y$ is obtained by endowing X with the structure sheaf

$$\mathcal{O}_Y := \mathrm{Ann}(L) := \{a \in \mathcal{O}_X \mid \xi(a) = 0, \ \forall \xi \in L\} \subset \mathcal{O}_X.$$

By construction, Y is normal. Indeed, any rational function t on Y integral over $\mathcal{O}_Y$ is also integral over $\mathcal{O}_X$ and hence regular on X. However, t is regular on a dense open subset $U \subset Y$ and as a rational function on Y annihilated by L. But then t as a regular function on X is annihilated by L everywhere. Moreover, Y is smooth if and only if $\mathcal{T}_X/L$ is locally free; see [418, I.Prop. 1.9].

Since $\mathcal{O}_X^p \subset \mathrm{Ann}(L)$, the absolute Frobenius factors through $X \longrightarrow Y \longrightarrow X^{(1)} \longrightarrow X$. It is thus enough to show that Y is rational. The canonical bundle formula [418, I.Cor. 1.11] yields in the present situation $\pi^*\omega_Y \simeq L^{-(p-1)}$ over the smooth locus of Y. As $(H.L) > 0$, this implies $H^0(Y, \omega_Y^n) = 0$ for all $n > 0$. Hence, by the Bombieri–Mumford–Enriques classification, Y is a ruled surface. However, the base of the ruling $Y \longrightarrow C$ has to be rational, for $g(C) \leq h^1(Y, \mathcal{O}_Y)$ and by Leray spectral sequence $h^1(Y, \mathcal{O}_Y) \leq h^1(X, \mathcal{O}_X) = 0$. Hence, Y is rational.

If $(H.L) = 0$, then either there exists a polarization H' with $(H'.L) > 0$, in which case one argues as before, or $L = \mathcal{O}_X$. As above, $\mathcal{O}_X \simeq L \subset \mathcal{T}_X$ is involutive, and to show p-closedness Miyaoka argues as follows: any local section of L is in this case of the form $f\xi$, where $\xi \in H^0(X, \mathcal{T}_X)$ spans L. Thus, it suffices to show that ξ^p is still in L. If not, then $\xi \wedge \xi^p$ would be a non-trivial global section of $\Lambda^2 \mathcal{T}_X \simeq \mathcal{O}_X$, and, therefore, ξ would have no zeroes. The latter, however, contradicts $c_2(X) = 24 > 0$. Alternatively, one could argue that the map $\mathcal{O}_X \longrightarrow \mathcal{T}_X/\mathcal{O}_X$, $\xi \longmapsto \overline{\xi^p}$ defines a global section of $\mathcal{T}_X/\mathcal{O}_X$ which is isomorphic to some ideal sheaf I_Z. So, either this section vanishes or Z is empty. However, the latter would say that $\mathcal{T}_X$ is an extension of $\mathcal{O}_X$ by itself, which would contradict $c_2(X) = 24$.

Now, as before, we use the quotient $\pi : X \longrightarrow Y$. The canonical bundle formula shows this time that ω_Y is trivial on the smooth locus of Y. For the minimal desingularization $\widetilde{Y} \longrightarrow Y$ one has $\omega_{\widetilde{Y}} \simeq \mathcal{O}(\Sigma a_i E_i)$ with $a_i \leq 0$; see [393, Thm. 4.6.2]. If Y were not ruled, then $H^0(\widetilde{Y}, \omega_{\widetilde{Y}}^n) \neq 0$ for some $n > 0$. Thus, only $a_i = 0$ can occur, and using $H^1(X, \mathcal{O}_X) = 0$ one finds that $\widetilde{Y}$ is either a K3 or an Enriques surface. On the other hand, since $\pi : X \longrightarrow Y$ is a homeomorphism and X is a K3 surface,

$$22 = b_2(X) = b_2(Y) \le b_2(\widetilde{Y}) \le 22.$$

Hence, $\widetilde{Y} \simeq Y$ is a smooth(!) K3 surface, and, therefore, $\mathcal{T}_X/L$ is locally free (and in fact $\simeq \mathcal{O}_X$). But ξ must have zeroes. Contradiction. $\qquad\square$

Note that in the proof one actually shows that X is dominated by a rational variety via a purely inseparable morphism. Of course, $H^0(X, \mathcal{T}_X) = 0$ is known even when X is unirational, but it seems to be an open question whether $\mathcal{T}_X$ is always stable.

Later we shall see that a unirational K3 surface X has maximal Picard number $\rho(X) = 22$ (see Proposition **17**.2.7) and vice versa (see Section **18**.3.5).

4.2   The standard reference for the differential geometry of complex vector bundles is Kobayashi's book [305]. In condensed form some of the following results can also be found in [253].

Let h be a hermitian metric on a holomorphic vector bundle E on a compact complex manifold X. The *Chern connection* on E is the unique hermitian connection ∇ on E with $\nabla^{0,1} = \overline{\partial}_E$. Let F_h denote its curvature, which is a global section of $\mathcal{A}^{1,1}(\mathcal{E}nd(E))$. If X is endowed with a Kähler form ω, then the form part of $F_h = \nabla \circ \nabla$ can be contracted with respect to ω to yield a global differentiable section $\Lambda_\omega F_h$ of the complex bundle $\mathcal{E}nd(E)$.

Definition 4.7   A hermitian structure h on E is called *Hermite–Einstein* (HE) if

$$i \cdot \Lambda_\omega F_h = \lambda \cdot \mathrm{id}_E \tag{4.1}$$

for some $\lambda \in \mathbb{R}$.

It is important to note that the HE condition depends not only on the hermitian structure h of E but also on the choice of the Kähler structure on X.

Let us assume for simplicity that X is a surface. Then the scalar λ in the HE condition (4.1) is uniquely determined by the slope $\mu_\omega(E)$; see Section 3.1. In fact,

$$\lambda = 4\pi \frac{\mu_\omega(E)}{(\omega)^2}.$$

Remark 4.8   For the following two results, see e.g. [253, App. 4.B].

(i) It is not difficult to produce an HE metric on a line bundle. The curvature is the unique harmonic representative of $c_1(E)$ (up to scaling).

(ii) A holomorphic bundle E that admits an HE structure satisfies the *Bogomolov–Lübke inequality*

$$\Delta(E) = 2\mathrm{rk}(E)c_2(E) - (\mathrm{rk}(E) - 1)c_1^2(E) \ge 0. \tag{4.2}$$

Line bundles are always μ-stable. Moreover, (4.2) is (3.1) for μ-semistable sheaves. This might serve as a motivation for the following deep result due to Donaldson, Uhlenbeck, and Yau. The difficult direction is the 'if' part, as it requires the construction of a special metric. For the proof one has to consult the original sources, [305] or the more recent account [381] by Lübke and Teleman.

Theorem 4.9 (Kobayashi–Hitchin correspondence)	*A holomorphic vector bundle on a compact Kähler manifold X admits a Hermite–Einstein metric if and only if E is μ-polystable.*

A bundle is μ-*polystable* if it is isomorphic to a direct sum $\bigoplus E_i$ with all E_i μ-stable of the same slope $\mu_\omega(E_i)$. Clearly, μ-polystable bundles are automatically μ-semistable, but the converse does not hold.

If E is the holomorphic tangent bundle $\mathcal{T}_X$, then the two metric structures, h on E and ω on X, can be related to each other. Requiring that they are equal, the HE condition becomes the following notion.

Definition 4.10	A Kähler structure on X is called *Kähler–Einstein* (KE) if the underlying hermitian structure on $\mathcal{T}_X$ is Hermite–Einstein.

This condition is stronger than just saying that $\mathcal{T}_X$ admits an HE metric with respect to ω. In fact, if a KE metric on X exists, then the *Miyaoka–Yau inequality* holds, which is stronger than (4.2). For surfaces the Miyaoka–Yau inequality reads

$$3c_2(X) - c_1^2(X) \geq 0,$$

instead of the Bogomolov inequality, $4c_2(X) - c_1^2(X) \geq 0$.

The KE condition can equivalently be expressed as the Einstein condition for the underlying Kähler metric; see [253, Cor. 4.B.13]. In particular, a Ricci-flat Kähler metric is automatically KE. Since the cohomology class of the Ricci curvature equals $2\pi c_1(X)$, this happens only for compact Kähler manifolds with trivial $c_1(X) \in H^2(X, \mathbb{R})$. The following is an immediate consequence of Yau's solution to the Calabi conjecture. For the special case of K3 surfaces, see Theorem 7.3.6.

Theorem 4.11 (Calabi–Yau)	*Let X be a compact Kähler manifold with $c_1(X) = 0$ in $H^2(X, \mathbb{R})$. Then any Kähler class in $H^2(X, \mathbb{R})$ can be uniquely represented by a Kähler form that defines a Kähler–Einstein structure on X.*

Since a KE structure on X is in particular an HE structure on $\mathcal{T}_X$, the theorem implies the following.

Corollary 4.12	*Let X be a complex K3 surface which is Kähler. Then $\mathcal{T}_X$ is μ-stable with respect to any Kähler class.*

Proof	By Theorems 4.9 and 4.11, the tangent bundle $\mathcal{T}_X$ is μ-polystable with respect to any Kähler class ω on X. Thus, $\mathcal{T}_X$ is μ-stable or a direct sum of line bundles $L \oplus M$ of the same degree with respect to any Kähler class, i.e. $\deg_\omega(L) = \deg_\omega(M)$ for all Kähler classes. Since $\deg_\omega(\mathcal{T}_X) = 0$, one has in the second case $\deg_\omega(L) = \deg_\omega(M) = 0$ for all ω, which implies $L \simeq M \simeq \mathcal{O}_X$. But this contradicts $c_2(X) = 24$. □

It is known that any complex K3 surface is Kähler, a highly non-trivial statement due to Todorov and Siu, and, therefore, the corollary holds in fact for all complex K3 surfaces; cf. Section **7.3.2**.

The differential geometric approach yields more. Due to a general result of Kobayashi [304] one also knows the following.[9]

Corollary 4.13 *Let X be a complex K3 surface. Then $H^0(X, S^m\mathcal{T}_X) = 0$ for all $m > 0$.*

Note that tensor powers of the tangent bundle might very well have global sections, for example $\mathcal{T}_X \otimes \mathcal{T}_X \simeq S^2\mathcal{T}_X \oplus \mathcal{O}_X$.

5 Appendix: Lifting K3 Surfaces

The fact that K3 surfaces do not admit any non-trivial vector fields is a central result in the theory. The proof is easy in characteristic zero and technically involved in general. All the existing proofs in positive characteristic are either rather lengthy or use techniques beyond the scope of this book. So, we only state the result (again) and say a few words about the strategy of the three existing proofs.

The most important consequence is the smoothness of the deformation space of a K3 surface and the liftability of any K3 surface in characteristic $p > 0$ to characteristic zero. The latter is the key to many results in positive characteristic, as it unleashes the power of Hodge theory for arithmetic considerations.

5.1 The following result in complete generality is due to Rudakov and Šafarevič [513, Thm. 7]; see also their survey [514].

Theorem 5.1 *Let X be a K3 surface over an arbitrary field k. Then*

$$H^0(X, \mathcal{T}_X) = 0.$$

It is enough to verify the assertion for K3 surfaces over algebraically closed fields. In characteristic zero, the theorem follows from Hodge theory, as

$$H^0(X, \mathcal{T}_X) \simeq H^0(X, \Omega_X) \simeq H^{1,0}(X) \simeq \overline{H^{0,1}(X)} = 0$$

(cf. Section **1.3.3**) and can also be seen as a shadow of μ-stability of $\mathcal{T}_X$ (cf. Corollary 4.12).

---

[9] Thanks to John Ottem for reminding me of Kobayashi's article.

In positive characteristic, Proposition 4.6 proves the assertion in the case that X is not unirational. To treat unirational K3 surfaces (over a field of characteristic $p > 0$) one first evokes the relatively easy Proposition **17**.2.7, showing that any unirational K3 surface has maximal Picard number $\rho(X) = 22$. But K3 surfaces of Picard number $\rho(X) \geq 5$ are all elliptic[10] (see Proposition **11**.1.3), and, therefore, the theorem is reduced to the case of unirational elliptic K3 surfaces. Those are dealt with by using the fairly technical [513, Thm. 6].

In [462] Nygaard provides an alternative proof of Theorem 5.1. He first reduces to the unirational case as above, and so can assume that $\rho(X) = 22$ and hence $NS(X) \otimes \mathbb{Z}_\ell \simeq H^2_{et}(X, \mathbb{Z}_\ell(1))$. Then, combining the fact that K3 surfaces with maximal Picard number $\rho(X) = 22$ are automatically supersingular (cf. Corollary **18**.3.9) with the slope spectral sequence (see Section **18**.3.3), he concludes that $H^2(X, \Omega_X) = 0$. By Serre duality the latter is equivalent to the assertion.

Another proof can be found in the article [349] by Lang and Nygaard. Their arguments do not require the reduction to the case of unirational K3 surfaces as a first step and roughly proceed as follows. First, one proves that the $d\colon H^0(X, \Omega_X) \longrightarrow H^0(X, \Omega_X^2) \simeq k$ is trivial; i.e. all one-forms are closed. In the second step, a result of Oda is applied to show that the space of (infinitely) closed forms in $H^0(X, \Omega_X)$ is a quotient of the dual of the Dieudonné module associated with the p-torsion in $NS(X)$. But $NS(X)$ is torsion free and, therefore, $H^0(X, \Omega_X) = 0$.

5.2   Deformations of K3 surfaces, both with or without polarizations, have been discussed twice already. In Section **5**.3, the local structure of the moduli space of polarized K3 surfaces was approached by first embedding all K3 surfaces in question into a projective space and then applying the deformation theory for Hilbert schemes. In Section **6**.2.3, deformation theory of general compact complex manifolds was reviewed and then applied to complex K3 surfaces without polarization. Building up on this, Section **6**.2.4 explained how to deal with the polarized case.

Let us now first rephrase the local deformation theory from a more functorial point of view, applying Schlessinger's theory [527].

Start with a K3 surface X over an arbitrary perfect field k. Let W be a fixed complete Noetherian local ring with residue field $W/\mathfrak{m} \simeq k$. The two examples relevant for us are $W = k$ and $W = W(k)$, the ring of Witt vectors. Recall that $W(k) = \varprojlim W_n(k)$, which for $k = \mathbb{F}_p$ becomes $W(k) = \varprojlim \mathbb{Z}/p^n\mathbb{Z} \simeq \mathbb{Z}_p$. Under our assumption that k is perfect, $W(k)$ is a DVR of characteristic zero.

Next consider the category (Art/W) of Artinian local W-algebras A with residue field k and let

---

[10] For simplicity, we assume char$(k) > 3$, otherwise one would also have to deal with quasi-elliptic fibrations.

$$\text{Def}_X\colon (Art/W) \longrightarrow (Sets)$$

be the deformation functor that maps A to the set of all $(\mathcal{X}, \varphi)$, where $\mathcal{X} \longrightarrow \text{Spec}(A)$ is flat and proper and $\varphi\colon \mathcal{X}_0 \xrightarrow{\sim} X$ is an isomorphism of k-schemes. Here, $\mathcal{X}_0$ is the fibre over the closed point of $\text{Spec}(A)$ which has residue field k. General deformation theory [527], briefly outlined in Section **18**.1.3, combined with Theorem 5.1 yields the following fundamental result.

Proposition 5.2 *The functor* Def_X *is pro-representable by a smooth formal W-scheme of dimension* 20

$$\text{Def}(X) \simeq \text{Spf}(W[[x_1,\ldots,x_{20}]]).$$ $\square$

For $W = W(k)$ this setting mixes the deformation theory for X as a k-variety with the liftability of X to characteristic zero. On the one hand, the deformation theory of X as a k-scheme is controlled by the closed formal subscheme

$$\text{Def}(X/k) \simeq \text{Spf}(k[[x_1,\ldots,x_{20}]]) \subset \text{Def}(X) \simeq \text{Spf}(W(k)[[x_1,\ldots,x_{20}]]).$$

On the other hand, the question of whether X can be lifted to characteristic zero asks for a scheme $\mathcal{X}$ with a flat and proper morphism $\mathcal{X} \longrightarrow \text{Spf}(W(k))$ with closed fibre $\mathcal{X}_0 \simeq X$. In both contexts, the obstructions are classes in $H^2(X, \mathcal{T}_X)$. For example, to extend X to first order to $\mathcal{X}_1 \longrightarrow \text{Spec}(k[x]/(x^2))$ or $\mathcal{X}_1 \longrightarrow \text{Spec}(W_2(k))$, respectively, defines a class in $H^2(X, \mathcal{T}_X)$. Similarly, the obstructions to extend a flat and proper scheme $\mathcal{X}_{n-1}$ over $\text{Spec}(k[x]/(x^n))$ or $\text{Spec}(W_n(k))$ further to a flat and proper scheme $\mathcal{X}_n$ over $\text{Spec}(k[x]/(x^{n+1}))$ or $\text{Spec}(W_{n+1}(k))$ is again a class in $H^2(X, \mathcal{T}_X)$. Now, due to Theorem 5.1 and Serre duality, $H^2(X, \mathcal{T}_X) \simeq H^0(X, \mathcal{T}_X)^* = 0$ and so the deformations of X are unobstructed.

Remark 5.3 Deligne and Illusie in [144] show that for an arbitrary smooth projective variety X over k the existence of a flat extension $\mathcal{X}_1 \longrightarrow \text{Spec}(W_2(k))$ implies the degeneration of the Hodge–Frölicher spectral sequence as well as Kodaira vanishing $H^i(X, L^*) = 0$ for $i > 0$ and any ample line bundle L; cf. Remark **2**.1.9 and Proposition **2**.3.1.

Note that the proposition in particular implies that for any K3 surface X over a perfect field k of characteristic $p > 0$ there exists a smooth formal scheme

$$\mathcal{X} \longrightarrow \text{Spf}(W(k))$$

with special fibre $\mathcal{X}_0 \simeq X$. Whether this formal lift is algebraizable, i.e. whether it can be extended to a smooth proper scheme over $\text{Spec}(W(k))$, is a priori not clear. The only general method to approach this question is Grothendieck's existence theorem; see for example Illusie's account of it [176, Thm 8.4.10]. It asserts that $\mathcal{X} \longrightarrow \text{Spf}(W(k))$ is algebraizable to a smooth and proper scheme $\widetilde{\mathcal{X}} \longrightarrow \text{Spec}(W(k))$ if there exists a line bundle $\mathcal{L}$ on the formal scheme $\mathcal{X}$ with ample restriction $\mathcal{L}|_{\mathcal{X}_0}$ to the closed fibre. This naturally leads to a deformation theory for K3 surfaces endowed with an additional line bundle, which we discuss next.

5.3   Indeed, the deformation theory becomes more subtle if a polarization of the K3 surface X or just a non-trivial line bundle L on it is taken into account. We consider the deformation functor

$$\mathrm{Def}_{(X,L)} : (Art/W) \longrightarrow (Sets) \tag{5.1}$$

parametrizing $\mathcal{X} \longrightarrow \mathrm{Spec}(A)$ and $\varphi := \mathcal{X}_0 \xrightarrow{\sim} X$ as above and, additionally, line bundles $\mathcal{L}$ on $\mathcal{X}$ such that $\varphi^* L \simeq \mathcal{L}|_{\mathcal{X}_0}$. This new deformation functor is obstructed by classes in $H^2(X, \mathcal{O}_X) \simeq k$ which are indeed non-trivial in general. This has been observed already for $k = \mathbb{C}$ in Section **6.2.4**. For other fields k, the situation is similar when one restricts to deformations of (X, L) over k. Note, however, that for $\mathrm{char}(k) = p > 0$ the space $H^2(X, \mathcal{O}_X)$ is p-torsion and so for each order n a possible obstruction to deform $\mathcal{L}_n$ on $\mathcal{X}_n \longrightarrow \mathrm{Spec}(k[x]/(x^{n+1}))$ to order $n + 1$ is annihilated by passing to $\mathcal{L}_n^p$.

The more interesting question concerns the lifting of (X, L). This is addressed by the following result due to Deligne [144, Thm. 1.6].

Theorem 5.4   *Let X be a K3 surface over a perfect field of characteristic $p > 0$ with a non-trivial line bundle L. Then the deformation functor (5.1) with $W = W(k)$ is pro-representable by a formal Cartier divisor*

$$\mathrm{Def}(X, L) \subset \mathrm{Def}(X),$$

which is flat over $\mathrm{Spf}(W(k))$ of relative dimension 19.

Thus, $\mathrm{Def}(X, L) \subset \mathrm{Def}(X)$ is defined by one equation and, by flatness, this equation is not divisible by p, so that the closed fibre $\mathrm{Def}(X, L)_0$ is still of dimension 19. To prove the flatness, one has to ensure that the one equation cutting out $\mathrm{Def}(X, L)$, which corresponds to the one-dimensional obstruction space $H^2(X, \mathcal{O}_X)$, is not the one describing $\mathrm{Spf}(k[[x_1, \ldots, x_{20}]]) \subset \mathrm{Spf}(W(k)[[x_1, \ldots, x_{20}]])$. In other words, one needs to show that L cannot be extended to a line bundle on the universal deformation of X as a k-scheme which lives over $\mathrm{Spf}(k[[x_1, \ldots, x_{20}]])$.

However, as $\mathrm{Def}(X, L) \longrightarrow \mathrm{Spf}(W(k))$ is a priori not smooth (but see Remark 5.7 below), a lift of L to any given lift $\mathcal{X} \longrightarrow \mathrm{Spf}(W(k))$ of X might not exist. Instead, Deligne proves the following corollary.

Corollary 5.5   *Let $\mathcal{X} \longrightarrow \mathrm{Spf}(W(k))$ be a formal lift of a K3 surface $X \simeq \mathcal{X}_0$ and let L be a line bundle on X. Then there exists a complete DVR $W(k) \subset W'$, finite over $W(k)$, such that L extends to a line bundle on the formal scheme $\mathcal{X} \times_{W(k)} \mathrm{Spf}(W') \longrightarrow \mathrm{Spf}(W')$.*

The aforementioned existence result of Grothendieck allows one to conclude immediately the following liftability result.

Corollary 5.6 *Let X be a K3 surface over a perfect field endowed with an ample line bundle L. Then there exists a complete DVR $W(k) \subset W'$, finite over $W(k)$, and a smooth proper scheme*

$$\widetilde{\mathcal{X}} \longrightarrow \mathrm{Spec}(W')$$

together with a line bundle $\widetilde{\mathcal{L}}$ on $\widetilde{\mathcal{X}}$ such that $\widetilde{\mathcal{X}}_0 \simeq X$ and $\widetilde{\mathcal{L}}|_{\mathcal{X}_0} \simeq L$. □

Note that due to the vanishing $H^1(X, \mathcal{O}_X)$ or, alternatively, due to the injectivity of the specialization map $\mathrm{sp} \colon \mathrm{Pic}(\widetilde{\mathcal{X}}_\eta) \hookrightarrow \mathrm{Pic}(X)$ (see Proposition **17**.2.10), there exists at most one extension of L to any given lift $\widetilde{\mathcal{X}}$.

Remark 5.7 Ogus [478, Cor. 2.3] proves that in fact most K3 surfaces admit projective lifts to $\mathrm{Spec}(W(k))$. Combined with the Tate conjecture, which has been proven since then (see Section **17**.3), one finds that for any K3 surface X over an algebraically closed field k of characteristic $p > 2$ there exists a smooth and projective scheme $\widetilde{\mathcal{X}} \longrightarrow \mathrm{Spec}(W(k))$ with closed fibre $\mathcal{X}_0 \simeq X$.

More precisely, Ogus proves [478, Prop. 2.2] that $\mathrm{Def}(X, L) \longrightarrow \mathrm{Spf}(W(k))$ is smooth whenever L is not a pth power and X is not 'superspecial'. The superspecial case is dealt with separately; see [478, Rem. 2.4] and also [372, Thm. 2.9]. This applies to all K3 surfaces of finite height.

5.4 For any K3 surface X over a field of characteristic $p > 0$ the first Chern class induces an injection

$$\mathrm{NS}(X)/p \cdot \mathrm{NS}(X) \hookrightarrow \mathrm{NS}(X) \otimes k \hookrightarrow H^2_{\mathrm{dR}}(X),$$

which, moreover, leads to an injection $\mathrm{NS}(X)/p \cdot \mathrm{NS}(X) \hookrightarrow \mathrm{NS}(X) \otimes k \hookrightarrow H^1(X, \Omega_X)$ unless X is supersingular; cf. Proposition **17**.2.1. In [369] Lieblich and Maulik observed that this is enough to lift the entire Picard group. See Section **18**.3 for the notion of the height of a K3 surface.

Proposition 5.8 *Let X be a K3 surface over a perfect field k of characteristic $p > 0$. Assume that X is of finite height. Then there exists a projective lift $\mathcal{X} \longrightarrow \mathrm{Spec}(W(k))$, $\mathcal{X}_0 \simeq X$, such that the specialization defines an isomorphism*

$$\mathrm{NS}(\mathcal{X}_\eta) \xrightarrow{\sim} \mathrm{NS}(X).$$

Remark 5.9 Charles [111, Prop. 1.5] and Lieblich and Olsson [371, Prop. A.1] prove a version that covers supersingular K3 surfaces as well: for a K3 surface X over a perfect field k of characteristic $p > 0$ and line bundles $L_1, \ldots, L_\rho$ with $\rho \leq 10$ and

L_1 ample there exists a complete DVR $W(k) \subset W'$, finite over $W(k)$, and a projective lift $\mathcal{X} \longrightarrow \mathrm{Spec}(W')$ such that the image of the specialization map $\mathrm{NS}(\mathcal{X}_\eta) \longrightarrow \mathrm{NS}(X)$ contains $L_1, \ldots, L_\rho$.

References and Further Reading

We have not discussed bundles on special K3 surfaces like elliptic K3 surfaces. We recommend Friedman's book [185]. Spherical bundles, in particular on double planes, have been studied by Kuleshov in [332, 333]. The existence of stable bundles on K3 surfaces has been treated by Kuleshov and Yoshioka in [334, 645]; cf. Section **10**.3.1.

Questions and Open Problems

What is known about stability and simplicity of the tangent bundle for char$(k) > 0$? I am not aware of any result in this direction apart from $H^0(X, \mathcal{T}_X) = 0$.

The result of Rudakov and Šafarevič (see Proposition 4.6) shows that in positive characteristic the existence of a non-trivial vector field implies unirationality of the K3 surface. I wonder if this can be turned into a completely algebraic (not using Hodge theory) proof of $H^0(X, \mathcal{T}_X) = 0$ in characteristic zero. In fact, if $H^0(X, \mathcal{T}_X) \neq 0$, then the reduction in all primes would be unirational and this might show that X itself is unirational, which is absurd in characteristic zero.[11]

It would be interesting to find a purely algebraic proof of Corollary 4.13 relying only on the stability of $\mathcal{T}_X$ and $H^0(X, \mathcal{T}_X)$.

---

[11] As Matthias Schütt points out, one could maybe refer to Bogomolov–Zarhin, who show that ordinary reductions have density one. One would need to check that they do not use $H^0(X, \mathcal{T}_X) = 0$ and that there is no problem with deforming to a K3 surface defined over a number field.

10

Moduli Spaces of Sheaves on K3 Surfaces

After having studied special sheaves and bundles on K3 surfaces in Chapter 9, we now pass to the study of all sheaves on a given K3 surface. This naturally leads to moduli spaces of (stable) sheaves. A brief outline of the general theory can be found in Section 1. In Section 2 the tangent space and the symplectic structure of the moduli space of sheaves on K3 surfaces is discussed. Low-dimensional moduli spaces and the Hilbert scheme, viewed as a moduli space of sheaves, are dealt with in Section 3.

1 General Theory

Most of the material recalled in this first section is covered by [266].

1.1 Let X be a smooth projective variety over a field k which for simplicity we assume algebraically closed. The moduli space of sheaves that is best understood is the Picard scheme Pic_X representing the functor $(Sch/k)^o \longrightarrow (Sets)$ mapping a k-scheme S to the set $\{L \in \mathrm{Pic}(S \times X)\}/\sim$, where $L \sim L \otimes p^*M$ for all $M \in \mathrm{Pic}(S)$; see e.g. [82, 176]. In particular, the k-rational points of Pic_X form the Picard group $\mathrm{Pic}(X)$.

The Picard scheme itself is neither projective nor of finite type, but it decomposes as

$$\mathrm{Pic}_X = \bigsqcup \mathrm{Pic}_X^P$$

with projective components. Here, Pic_X^P parametrizes line bundles on X with fixed Hilbert polynomial $P \in \mathbb{Q}[t]$ with respect to a chosen ample line bundle $\mathcal{O}(1)$ on X.

Note that the Hilbert polynomial $P(L, m) := \chi(X, L(m))$ of a line bundle on X can be computed via the Hirzebruch–Riemann–Roch formula as

$$\chi(X, L(m)) = \int \mathrm{ch}(L)\mathrm{ch}(\mathcal{O}(m))\mathrm{td}(X).$$

Obviously, the expression depends only on $c_1(L)$ and $(X, \mathcal{O}(1))$.

The naive question this theory raises is the following: if one generalizes the Picard functor as above to the functor of higher rank vector bundles or arbitrary coherent sheaves, is the resulting functor again representable?

The following two examples immediately show that care is needed when leaving the realm of line bundles.

Example 1.1   Consider on $\mathbb{P}^1$ the bundles $E_n := \mathcal{O}(n) \oplus \mathcal{O}(-n)$, $n > 0$. First observe that $h^0(E_n) = n + 1$. In particular the bundles E_n are pairwise non-isomorphic. On the other hand, they are all of rank two with trivial first Chern class $c_1(E_n) = 0$. All higher Chern classes of E_n are trivial for dimension reasons.

Suppose there exists a moduli space parametrizing in particular all bundles E_n. Since $h^0(E) \geq m$ is a closed condition, the infinitely many bundles E_n would lead to a strictly descending chain of closed subschemes, which obviously excludes M from being of finite type. Thus, fixing the Hilbert polynomial or even all numerical invariants does not ensure that a moduli space, if it exists at all, is of finite type.

Example 1.2   (i) On $\mathbb{P}^1$ the extension group $\mathrm{Ext}^1(\mathcal{O}(1), \mathcal{O}(-1)) = H^1(\mathbb{P}^1, \mathcal{O}(-2))$ is one-dimensional. Hence, there exists for any $\lambda \in k$ a unique (up to scaling) non-trivial extension $0 \longrightarrow \mathcal{O}(-1) \longrightarrow E_\lambda \longrightarrow \mathcal{O}(1) \longrightarrow 0$. In fact, these bundles together form a vector bundle E on $\mathbb{A}^1 \times \mathbb{P}^1$ such that the restriction to $\{\lambda\} \times \mathbb{P}^1$ is isomorphic to E_λ.

It is easy to see that for $\lambda \neq 0$ all bundles are isomorphic to each other, in fact $E_\lambda \simeq \mathcal{O} \oplus \mathcal{O}$. On the other hand, $E_0 \simeq \mathcal{O}(-1) \oplus \mathcal{O}(1)$.

Thus, if the moduli functor of higher rank bundles on $\mathbb{P}^1$ were represented by a scheme M, then the universality property would induce a morphism $\mathbb{A}^1 \longrightarrow M$ mapping all closed points $\lambda \neq 0$ to the point $x \in M$ corresponding to $\mathcal{O} \oplus \mathcal{O}$ and the origin $0 \in \mathbb{A}^1$ to the point $y \in M$ given by $\mathcal{O}(-1) \oplus \mathcal{O}(1)$. Thus, if $x \neq y$, which would be the case if M really represented the functor, then M cannot be separated.

(ii) A similar example can be produced on an elliptic curve C by considering extensions of the form $0 \longrightarrow \mathcal{O} \longrightarrow E_\lambda \longrightarrow \mathcal{O} \longrightarrow 0$ with $\lambda \in H^1(C, \mathcal{O}) \simeq k$.

1.2   From the above examples it is clear that for sheaves other than invertible ones, extra conditions need to be added in order to construct a well-behaved moduli space. This condition is stability. As there are several notions of stability, let us for now call the extra condition just $(*)$. Then we are interested in

$$
\begin{aligned}
\mathcal{M}: (Sch/k)^o &\longrightarrow (Sets), \\
S &\longmapsto \{E \in \mathrm{Coh}(S \times X) \mid E \ S\text{-flat}, \forall s \in S : P(E_s) = P, (*) \text{ for } E_s\}/\!\sim.
\end{aligned}
$$

Here, E_s denotes the restriction of E to the fibre $\{s\} \times X$, P is a fixed Hilbert polynomial (see Section 1.3 below), and $\sim$ is as before defined by the action of $\mathrm{Pic}(S)$.

Definition 1.3 The functor $\mathcal{M}$ is *corepresented* by a scheme M if there exists a transformation $\mathcal{M} \longrightarrow \underline{M} = h_M$ (functor of points) with the universal property that any other transformation $\mathcal{M} \longrightarrow \underline{N}$ with $N \in (Sch/k)$ factorizes over a uniquely determined $M \longrightarrow N$, i.e.

We say that M is a *moduli space* for $\mathcal{M}$.

Recall from Section 5.1 that a coarse moduli space satisfies the additional requirement that the induced $\mathcal{M}(k) \longrightarrow M(k)$ is a bijection. For a fine moduli space one needs the even stronger condition $\mathcal{M} \xrightarrow{\sim} \underline{M}$, which is equivalent to the existence of a universal family $\mathcal{E}$ on $M \times X$. Allowing the map $\mathcal{M}(k) \longrightarrow M(k)$ to contract certain sets, i.e. to map different sheaves on X to the same point in M, eventually solves the non-separation problem hinted at in Example 1.2.

1.3 In Section 9.3.1 we have already encountered μ-stability. So one could try to define the additional condition $(*)$ as μ-(semi)stability. This works perfectly well for smooth curves, which was the starting point of the theory; see e.g. Mumford et al.'s classic book [444]. However, in higher dimensions the better notion is (Gieseker) stability. It has (at least) three advantages over μ-stability: (i) fewer objects are identified under $\mathcal{M}(k) \longrightarrow M(k)$, (ii) the translation to GIT-stability is more direct, and (iii) stability for torsion sheaves makes sense.

Before we can properly define stability, let us recall some facts on Hilbert polynomials. For an arbitrary projective scheme X with an ample line bundle $\mathcal{O}(1)$ the Hilbert polynomial of a sheaf E is

$$P(E, m) := \chi(E(m)) = \sum_{i=0}^{d} \alpha_i(E) \frac{m^i}{i!}.$$

Here, $d := \dim(E) := \dim \operatorname{supp}(E)$ and $\alpha_0, \ldots, \alpha_d \in \mathbb{Z}$. For a sheaf E of rank r and Chern classes c_1, c_2 on a smooth surface X this becomes

$$P(E, m) = \int \left(r + c_1 + \frac{c_1^2 - 2c_2}{2} \right) \left(1 + mH + \frac{m^2 (H)^2}{2} \right) \operatorname{td}(X)$$

$$= \frac{r(H)^2}{2} m^2 + m((H.c_1) + r(H.c_1(X))) + \text{const},$$

where we write H for the first Chern class of $\mathcal{O}(1)$.

Note that $\mathrm{rk}(E) = \alpha_d(E)/\alpha_d(\mathcal{O}_X)$ for a sheaf E of maximal dimension $d = \dim(X)$. For sheaves of smaller dimension the role of torsion-free sheaves is played by *pure sheaves*. A coherent sheaf E of dimension d is called pure if $\dim(F) = \dim(E)$ for every non-trivial subsheaf $F \subset E$. Thus, a sheaf of maximal dimension is pure if and only if it is torsion free.

Definition 1.4   The *reduced Hilbert polynomial* of a sheaf E is defined as

$$p(E, m) := \frac{P(E, m)}{\alpha_d(E)}.$$

A coherent sheaf E is called *stable* if E is pure and

$$p(F, m) < p(E, m), \quad m \gg 0,$$

for all proper non-trivial subsheaves $F \subset E$.

A sheaf is called *semistable* if only the weak inequality is required. Recall that the inequality of polynomials $f(m) < g(m)$ for $m \gg 0$ is equivalent to the inequality of their coefficients with respect to the lexicographic order.

Let us spell out what stability means for sheaves on surfaces.

(i) Suppose E is a sheaf of dimension zero; i.e. its support consists of finitely many closed points. Then $P(E, m) \equiv \mathrm{const}$ and hence $p(E, m) = 1$. Clearly, such a sheaf is always pure and semistable. It is stable if and only if $E \simeq k(x)$ for some closed point $x \in X$.

(ii) For a vector bundle E supported on an integral curve $C \subset X$ one easily computes that μ-stability of E on C is equivalent to stability of E viewed as a torsion sheaf on X.

(iii) If E is of maximal dimension two, then

$$p(E, m) = \frac{m^2}{2} + m\left(\frac{(H.c_1(E))}{\mathrm{rk}(E)(H)^2} + \frac{(H.c_1(X))}{(H)^2} \right) + \frac{\alpha_0(E)}{\mathrm{rk}(E)(H)^2}.$$

Hence, E is stable if and only if E is torsion free and for all non-trivial proper subsheaves $F \subset E$ one of the two conditions hold:

$$\frac{(H.c_1(F))}{\mathrm{rk}(F)} < \frac{(H.c_1(E))}{\mathrm{rk}(E)} \tag{1.1}$$

or

$$\frac{(H.c_1(F))}{\mathrm{rk}(F)} = \frac{(H.c_1(E))}{\mathrm{rk}(E)} \quad \text{and} \quad \frac{\alpha_0(F)}{\mathrm{rk}(F)} < \frac{\alpha_0(E)}{\mathrm{rk}(E)}. \tag{1.2}$$

The slope of a torsion-free sheaf E is by definition $\mu(E) = (H.c_1(E))/\mathrm{rk}(E)$. As μ-stability is defined in terms of inequality (1.1) in Section **9**.3.1, this immediately yields

Corollary 1.5   *The following implications hold for any torsion-free sheaf:*

$$\mu\text{-stable} \Rightarrow \text{stable} \Rightarrow \text{semistable} \Rightarrow \mu\text{-semistable}. \qquad \square$$

1.4 Analogously to results on μ-stability observed in Section **9**.3.1, one can show that stability of a pure sheaf E of dimension d is equivalent to either of the two conditions:

- $p(F, m) < p(E, m)$, $m \gg 0$, for all non-trivial proper subsheaves $F \subset E$ with pure quotient E/F of dimension d or
- $p(E, m) < p(G, m)$, $m \gg 0$, for all non-trivial proper quotients $E \twoheadrightarrow G$.

Moreover, $\mathrm{Hom}(E_1, E_2) = 0$ if E_1, E_2 are semistable with $p(E_1, m) > p(E_2, m)$, $m \gg 0$. If E is stable, then $\mathrm{End}(E)$ is a division algebra and hence isomorphic to k. (Recall, we are assuming $k = \bar{k}$.)

Proposition 1.6 *Let E be a semistable sheaf. Then there exists a filtration*

$$0 \subset E_0 \subset \cdots \subset E_n = E$$

such that all quotients E_{i+1}/E_i are stable with reduced Hilbert polynomial $p(E, m)$. The isomorphism type of the graded object

$$\mathrm{JH}(E) := \bigoplus E_{i+1}/E_i$$

is independent of the filtration.

The filtration itself is called the (or rather a) *Jordan–Hölder filtration* and is not unique in general.

Definition 1.7 Two semistable sheaves E and F are called *S-equivalent* if $\mathrm{JH}(E) \simeq \mathrm{JH}(F)$.

1.5 The following result is needed only for K3 surfaces or families of K3 surfaces, but it holds for arbitrary projective varieties.

Theorem 1.8 *For fixed Hilbert polynomial P the functor*

$$\mathcal{M} \colon (Sch/k)^o \longrightarrow (Sets),$$
$$S \longmapsto \{E \in \mathrm{Coh}(S \times X) \mid E \text{ } S\text{-flat}, P(E_s) = P, E_s \text{ semistable}\}/\sim$$

is corepresented by a projective k-scheme M. The closed points of M parametrize the S-equivalence classes of semistable sheaves with Hilbert polynomial P.[1]

The result in this generality is due to Maruyama and Simpson. The boundedness in positive characteristic was proved by Langer. For more on the history of this result, see [266].

---

[1] There is the following relative version of this result: if $X \longrightarrow S$ is a projective morphism of k-schemes of finite type and $\mathcal{O}(1)$ is a relative ample line bundle on X, then the analogously defined moduli functor $\mathcal{M} \colon (Sch/S)^o \longrightarrow (Sets)$ is corepresented by a projective S-scheme $M \longrightarrow S$. See [266, Thm. 4.3.7].

Example 1.9 We have described already the stable sheaves of dimension zero. This immediately yields the following explicit description of a moduli space. Let P be the constant polynomial n. Then the moduli space M corepresenting $\mathcal{M}$ is naturally isomorphic to the symmetric product $S^n(X)$. See [266, Ex. 4.3.6].

The first step in the construction of the moduli space in general consists of showing that adding stability produces a bounded family. (Note that in Example 1.1 the bundles are indeed not semistable except for E_0.) In particular there exists n_0 such that for all $E \in \mathcal{M}(k)$ and all $n \geq n_0$ the sheaf $E(n)$ is globally generated with trivial higher cohomology. Thus, for any $E \in \mathcal{M}(k)$ there exists a point in the Quot-scheme $\mathrm{Quot}^P_{X/V\otimes\mathcal{O}(-n)}$ of the form $[V\otimes\mathcal{O}(-n) \twoheadrightarrow E]$. Here, V is a vector space of dimension $P(n)$. For the notion of the Quot-scheme, see [176, 225, 266].

The next step involves the construction of the Quot-scheme as a projective scheme. For this, one chooses a high twist and maps $[V \otimes \mathcal{O}(-n) \twoheadrightarrow E]$ to the point $[V \otimes H^0(X, \mathcal{O}(m - n)) \twoheadrightarrow H^0(X, E(m))]$ in $\mathrm{Gr} := \mathrm{Gr}(V \otimes H^0(X, \mathcal{O}(m - n)), P(m))$.

Next, one has to prove, using the Hilbert–Mumford criterion, that the (semi)stability of the sheaf E is equivalent to the GIT-(semi)stability of the point $[V\otimes\mathcal{O}(-n) \twoheadrightarrow E]$ in $\mathrm{Quot}^P_{X/V\otimes\mathcal{O}(-n)}$ with respect to the natural $\mathrm{GL}(V)$-action and the standard polarization on Gr. In the last step, one has to twist all sheaves once more in order to control also all potentially destabilizing quotients.

Remark 1.10 Note that the general result of Keel and Mori (see [289] and Section 5.2.3) on quotients of proper linear group actions can be applied here as well, but only to the stable locus. Their result proves that there exists a separated algebraic space that is a coarse moduli space for the subfunctor $\mathcal{M}^s \subset \mathcal{M}$ of stable sheaves. The properness of the group action can be deduced from a result of Langton [266, Thm. 2.B.1].

1.6 If $\mathcal{M}$ has a fine moduli space M as in Theorem 1.8, then applying the isomorphism $\mathcal{M} \xrightarrow{\sim} \underline{M}$ to $\mathrm{Spec}(k[x]/x^2)$ shows that the tangent space T_tM at a point $t \in M$ corresponding to a stable sheaf $E \in \mathcal{M}(k)$ is naturally isomorphic to $\mathrm{Ext}^1(E, E)$. If M is only a coarse moduli space (in a neighbourhood of t), this is still true but one needs to argue via the Quot-scheme.

Proposition 1.11 *Let M be the moduli space of $\mathcal{M}$ and let $t \in M$ be a point corresponding to a stable sheaf $E \in \mathcal{M}(k)$.*

(i) *Then there exists a natural isomorphism*

$$T_tM \simeq \mathrm{Ext}^1(E, E).$$

(ii) *If $\mathrm{Ext}^2(E, E) = 0$, then M is smooth at $t \in M$.*

(iii) *If the trace map* $\mathrm{Ext}^2(E, E) \longrightarrow H^2(X, \mathcal{O})$ *is injective and* Pic_X *is smooth at the point corresponding to the determinant* $\det(E)$, *then* M *is smooth at* $t \in M$.

Proof The moduli space M is constructed as a $\mathrm{PGL}(V)$-quotient of an open subscheme

$$\mathcal{R} \subset Q := \mathrm{Quot}^P_{X/V \otimes \mathcal{O}(-n)}.$$

Moreover, over the stable part $M^s \subset M$, which is also open, the quotient morphism

$$\mathcal{R}^s \longrightarrow M^s$$

is a principal bundle.

The tangent space $T_q Q$ at a quotient $q = [V \otimes \mathcal{O}(-n) \longrightarrow E] \in Q(k)$ is naturally isomorphic to $\mathrm{Hom}(K, E)$,[2] where K is the kernel, and the obstruction space is $\mathrm{Ext}^1(K, E)$. Now apply $\mathrm{Hom}(\ , E)$ to the exact sequence

$$0 \longrightarrow K \longrightarrow V \otimes \mathcal{O}(-n) \longrightarrow E \longrightarrow 0.$$

Using the vanishing $H^i(X, E(n)) = 0$, $i > 0$, one immediately obtains an isomorphism $\mathrm{Ext}^1(K, E) \xrightarrow{\sim} \mathrm{Ext}^2(E, E)$, proving (ii), and an exact sequence

$$0 \longrightarrow \mathrm{End}(E) \longrightarrow \mathrm{Hom}(V \otimes \mathcal{O}(-n), E) \xrightarrow{\alpha} \mathrm{Hom}(K, E) \longrightarrow \mathrm{Ext}^1(E, E) \longrightarrow 0.$$

Since the image of α describes the tangent space of the $\mathrm{PGL}(V)$-orbit, this yields (i).

(iii) The trace map for locally free sheaves can be defined in terms of a Čech covering. For arbitrary sheaves one first passes to a locally free resolution. Consider the obstruction class

$$o(\mathcal{E}, A) \in \mathrm{Ext}^2_X(\mathcal{E}, \mathcal{E} \otimes I) \simeq \mathrm{Ext}^2_X(E, E \otimes_k I)$$

to lift an $\overline{A} = A/I$-flat deformation $\mathcal{E}$ of $\mathcal{E} \otimes A/\mathfrak{m} \simeq E$ to an A-flat deformation, where, as usual, we assume $\mathfrak{m} \cdot I = 0$. Then according to Mukai [428, (1.13)] the image of $o(\mathcal{E}, A)$ under the trace map is $o(\det(\mathcal{E}), A)$. If Pic_X is smooth at $\det(E)$, the latter vanishes. $\square$

2 On K3 Surfaces

From now on X is a K3 surface over a field k and for simplicity we continue to assume that k is algebraically closed. Then the Picard scheme Pic_X consists of reduced isolated points. Indeed, if $\ell \in \mathrm{Pic}_X(k)$ corresponds to a line bundle L on X, then

$$\mathrm{Ext}^1(L, L) \simeq H^1(X, \mathcal{O}) = 0.\text{[3]}$$

---

[2] This is the sheaf analogue of the classical fact that the tangent space of the Grassmannian at a point corresponding to a subspace $U \subset V$ is isomorphic to $\mathrm{Hom}(U, V/U)$.

[3] The behaviour of the Picard group under base field extension is interesting (cf. Chapter 17). For example, if a line bundle L lives only on $X_{k'}$ for some field extension k'/k, the argument still works, for then $\mathrm{Ext}^1(L, L) \simeq H^1(X_{k'}, \mathcal{O}) = 0$. Here we may even allow k not algebraically closed.

Moduli spaces of sheaves other than line bundles are more interesting. The most influential paper on the subject is Mukai's [429], which contains a wealth of interesting results. To start, let us fix a Mukai vector instead of the Hilbert polynomial. Recall from Section 9.1.2 that the Mukai vector $v(E)$ of a sheaf E is

$$v(E) := (\text{rk}(E), c_1(E), \text{ch}_2(E) + \text{rk}(E)) = (\text{rk}(E), c_1(E), \chi(E) - \text{rk}(E)).$$

For $k = \mathbb{C}$ the Mukai vector is usually considered as an element in $H^*(X, \mathbb{Z})$ and otherwise in the numerical Grothendieck group (see Sections 12.1.3, 16.1.2, and 16.2.4):

$$N(X) := K(X)/\sim.$$

By definition $E_1 \sim E_2$ if $\chi(E_1, F) = \chi(E_2, F)$ for all $F \in \text{Coh}(X)$. Note that for $k = \mathbb{C}$, the numerical Grothendieck group $N(X)$ is naturally isomorphic to $H^*(X, \mathbb{Z}) \cap (H^0(X) \oplus H^{1,1}(X) \oplus H^4(X))$; see Section 16.3.1.

Since $P(E, m) = \chi(E(m)) = -\langle v(E), v(\mathcal{O}(-m)) \rangle$ (see (1.4) in Section 9.1.2), the Mukai vector determines the Hilbert polynomial. Conversely, if E is an S-flat sheaf on $S \times X$ with S connected, then $v(E_s)$ is constant. Indeed, $\chi(E_s, F)$ is constant for all F on X.

Thus, instead of fixing the Hilbert polynomial, it is more convenient, at least for K3 surfaces, to fix the Mukai vector. So we shall fix $v = (r, l, s) \in N(X)$ and consider the moduli functor $\mathcal{M}(v)$ of semistable sheaves with its moduli space $M(v)$. The open (possibly empty) subscheme parametrizing stable sheaves shall be denoted

$$M(v)^s \subset M(v),$$

which is a coarse moduli space for $\mathcal{M}(v)^s$. Note that although we are not using the Hilbert polynomial to fix the numerical invariants of the sheaves, the polarization still enters the picture via the stability condition. Thus, implicitly $M(v)$ depends on H. When we want to stress this dependence, we write $M_H(v)$.

2.1 For the following discussion, see also Section 9.1.2. Due to Serre duality, the local structure of moduli spaces of sheaves on K3 surfaces is particularly accessible. Indeed, for any sheaf E on a K3 surface X one has $\text{Ext}^2(E, E) \simeq \text{End}(E)^*$. Thus, if E is a stable (and hence a simple) sheaf, one finds $\text{Ext}^2(E, E) \simeq k$. In fact, Serre duality in degree two is described by composition and the trace, i.e. the pairing

$$\text{Ext}^2(E, E) \times \text{End}(E) \longrightarrow \text{Ext}^2(E, E) \xrightarrow{\text{tr}} H^2(X, \mathcal{O}) \simeq k$$

is non-degenerate. In particular, the trace $\text{tr}: \text{Ext}^2(E, E) \longrightarrow H^2(X, \mathcal{O})$ is Serre dual to the natural inclusion $H^0(X, \mathcal{O}) \longrightarrow \text{End}(E)$, $\lambda \longmapsto \lambda \cdot \text{id}$ and hence non-trivial;

see [428, Sec. 1].[4] Thus, Proposition 1.11 applies and shows that at a point $t \in M(v)^s$ corresponding to a sheaf E the moduli space $M(v)^s$ is smooth of dimension

$$\dim M(v)^s = \dim \operatorname{Ext}^1(E, E).$$

As remarked before, the Picard scheme of a K3 surface consists of isolated reduced points, for $H^1(X, \mathcal{O}) = 0$, and is thus in particular smooth. For higher rank sheaves the moduli spaces are often not discrete anymore. Since for $E \in M(v)^s$

$$\chi(E, E) = \sum (-1)^i \dim \operatorname{Ext}^i(E, E) = 2 - \dim \operatorname{Ext}^1(E, E)$$

and, on the other hand, $\chi(E, E) = -\langle v, v \rangle$, one finds the following corollary.

Corollary 2.1 *Either $M(v)^s$ is empty or a smooth, quasi-projective variety of dimension $2 + \langle v, v \rangle$.* □

2.2 As an interlude, let us state a few observations on the existence of a universal sheaf. The existence usually simplifies the arguments but is often not essential for any particular result one wants to prove for the moduli space.

(i) If there exists a vector $v' \in N(X)$ with $\langle v, v' \rangle = 1$, then the moduli space $M(v)^s$ is fine; i.e. there exists a universal family $\mathcal{E}$ on $M(v)^s \times X$. We briefly indicate the arguments needed to prove this assertion, but refer to [266, Sec. 4.6] for the details. Indeed, on $\operatorname{Quot}^P_{X/V \otimes \mathcal{O}(-n)} \times X$ and hence on $\mathcal{R}^s \times X$ there always exists a universal quotient $V \otimes (\mathcal{O}_{\mathcal{R}^s} \boxtimes \mathcal{O}_X(-n)) \twoheadrightarrow \mathcal{E}$. The naturally GL($V$)-linearized sheaf $\mathcal{E}$ descends under the action of GL(V) on $\mathcal{R}^s$ to a sheaf on $M(v)^s \times X$ if and only if the kernel of GL(V) $\longrightarrow$ PGL(V) (i.e. the center of GL(V)) acts trivially on $\mathcal{E}$, which it does not, as it is actually of weight one. If now there exists a complex of coherent sheaves F with $v(F) = -v'$, then the natural linearization of the line bundle $\mathcal{L}(F) := \det p_*(\mathcal{E} \otimes q^* F^*)$ on $\mathcal{R}^s$ is of weight $1 = \langle v, v' \rangle$. Hence, the center of GL(V) acts trivially on the linearized sheaf $p^* \mathcal{L}(F)^* \otimes \mathcal{E}$, which therefore descends to a universal sheaf on $M^s(v) \times X$.

(ii) However, even when no $v' \in N(X)$ with $\langle v, v' \rangle = 1$ is available, a *twisted*(!) universal sheaf on $M \times X := M(v)^s \times X$ always exists. To explain this notion, recall that by Luna's étale slice theorem (cf. Section 5.3.3), there exists an étale (or analytic, in the complex setting) covering $\bigcup U_i \subset \mathcal{R}^s$ of M. Denote by $\mathcal{E}_i$ the restriction of $\mathcal{E}$ to $U_i \times X$. On the intersection $(U_i \times_M U_j) \times X$ the two sheaves $\mathcal{E}_i$ and $\mathcal{E}_j$ differ by the line bundle $\mathcal{L}_{ij} := p_* \mathcal{H}om(\mathcal{E}_i, \mathcal{E}_j)$ so that

$$\mathcal{E}_j|_{(U_i \times_M U_j) \times X} \simeq \mathcal{E}_i|_{(U_i \times_M U_j) \times X} \otimes p^* \mathcal{L}_{ij}.$$

---

[4] We remark again (see Section 1.2) that implicitly ones fixes a trivializing section of ω_X in all of this. A priori, the pairing gives only natural isomorphisms $\operatorname{Ext}^i(E, E) \otimes H^0(X, \omega_X) \simeq \operatorname{Ext}^{2-i}(E, E)^*$ and $\operatorname{Ext}^i(E, E) \simeq \operatorname{Ext}^{2-i}(E, E \otimes \omega_X)^*$.

By refining the covering $\bigcup U_i$, we may assume that there exist $\xi_{ij} \colon \mathcal{L}_{ij} \xrightarrow{\sim} \mathcal{O}_{U_{ij}}$, which together with the natural isomorphisms $\mathcal{L}_{ik} \simeq \mathcal{L}_{ij} \otimes \mathcal{L}_{jk}$ over the triple intersections U_{ijk} give rise to $\alpha_{ijk} := (\xi_{ij} \otimes \xi_{jk}) \circ \xi_{ik}^{-1} \in \Gamma(U_{ijk}, \mathcal{O}^*)$ defining a Brauer class $\alpha \in \mathrm{Br}(M)$. By construction, the sheaves $\mathcal{E}_i$ on $U_i \times X$ descend to an $\{\alpha_{ijk}\} \boxtimes 1$-twisted sheaf on $M \times X$. See Section **16**.5 for some comments on twisted sheaves.

In Căldăraru's thesis [96, Prop. 3.3.2], using arguments of Mukai in [429, Thm. A.6], the reasoning is closer to (i) by changing the sheaves $\mathcal{E}_i$ by linearized line bundles of weight one. Via Hodge theory, so in the complex setting, the obstruction class $\alpha \in \mathrm{Br}(M)$ is by [96, Thm. 5.4.3] identified as a generator of the kernel of a natural surjection $\mathrm{Br}(M) \longrightarrow \mathrm{Br}(X)$, i.e.

$$0 \longrightarrow \langle \alpha \rangle \longrightarrow \mathrm{Br}(M) \longrightarrow \mathrm{Br}(X) \longrightarrow 0. \tag{2.1}$$

Here, $M = M_H(v)^s$ is assumed to be smooth, projective, and two-dimensional. See (5.6) in Remark **11**.5.9 for a special case.

Note that neither the universal nor the twisted universal sheaf is unique, e.g. they can always be modified by line bundles on $M(v)^s$. More precisely, if $\mathcal{E}$ is a (twisted) universal sheaf on $M(v)^s \times X$, then so is $\mathcal{E} \otimes p^*\mathcal{L}$ for any line bundle $\mathcal{L}$ on $M(v)^s$. In **(ii)** this ambiguity is already contained in the choice of the trivializations ξ_{ij}.

2.3　We can now come back to the tangent bundle of the moduli space. The description of the tangent space of the moduli space at stable points provided by Proposition 1.11 generalizes to the following.

Corollary 2.2　*Suppose there exists a universal family $\mathcal{E}$ over $M(v)^s \times X$. Then there is a natural isomorphism $\mathcal{T}_{M(v)^s} \xrightarrow{\sim} \mathcal{E}xt_p^1(\mathcal{E}, \mathcal{E})$.*

Proof　Here, $p \colon M(v)^s \times X \longrightarrow M(v)^s$ denotes the projection and $\mathcal{E}xt_p^1(\mathcal{E}, \mathcal{E})$ denotes the relative Ext-sheaf. For example, if $\mathcal{E}$ is locally free, then $\mathcal{E}xt_p^1(\mathcal{E}, \mathcal{E}) \simeq R^1p_*(\mathcal{E}^* \otimes \mathcal{E})$. To shorten notation, we write $M = M(v)$ and $M^s = M(v)^s$.

Roughly, one should think of the isomorphism $\mathcal{T}_{M^s} \xrightarrow{\sim} \mathcal{E}xt_p^1(\mathcal{E}, \mathcal{E})$ as obtained by gluing the isomorphisms $T_t M^s \xrightarrow{\sim} \mathrm{Ext}^1(\mathcal{E}_t, \mathcal{E}_t)$. Since the dimension of $\mathrm{Ext}^1(\mathcal{E}_t, \mathcal{E}_t)$ stays constant over M^s, these spaces are indeed given as the fibres $\mathcal{E}xt_p^1(\mathcal{E}, \mathcal{E}) \otimes k(t)$.

A better way to define the *Kodaira–Spencer map* inducing the isomorphism is to use the Atiyah class $A(\mathcal{E}) \in \mathrm{Ext}^1(\mathcal{E}, \mathcal{E} \otimes \Omega_{M^s \times X})$. Its image in $H^0(M^s, \mathcal{E}xt_p^1(\mathcal{E}, \mathcal{E}) \otimes \Omega_{M^s})$ can be interpreted as a natural map $\mathcal{T}_{M^s} \longrightarrow \mathcal{E}xt_p^1(\mathcal{E}, \mathcal{E})$, which fibrewise yields the isomorphism $T_t M^s \xrightarrow{\sim} \mathrm{Ext}^1(\mathcal{E}_t, \mathcal{E}_t)$. For details, see [266, 10.1.8].　$\square$

Remark 2.3　The existence of the universal family is not needed for the corollary. Although $\mathcal{E}$ might not exist (or exists only étale locally), the relative Ext-sheaves $\mathcal{E}xt_p^i(\mathcal{E}, \mathcal{E})$ always do. For example, if all sheaves parametrized by $M(v)^s$ are locally free, then $\mathcal{E}^* \otimes \mathcal{E}$ on the Quot-scheme exists and descends to $M(v)^s \times X$.

Alternatively, one can work with a twisted universal sheaf, as introduced before. The twists of the two factors in the relative Ext-sheaf cancel each other out (similarly to twists by a line bundle coming from the moduli space), so that $\mathcal{E}xt^i_p(\mathcal{E}, \mathcal{E})$ becomes a well-defined untwisted sheaf.

Once $\mathcal{T}_{M(v)^s} \xrightarrow{\sim} \mathcal{E}xt^1_p(\mathcal{E}, \mathcal{E})$ is constructed, one can globalize Serre duality (cf. Proposition 9.1.1) to a non-degenerate alternating pairing

$$\mathcal{T}_{M(v)^s} \times \mathcal{T}_{M(v)^s} \xrightarrow{\sim} \mathcal{E}xt^1_p(\mathcal{E}, \mathcal{E}) \times \mathcal{E}xt^1_p(\mathcal{E}, \mathcal{E}) \longrightarrow \mathcal{O}_{M(v)^s}.$$

Corollary 2.4 *The moduli space of stable sheaves $M(v)^s$ is endowed with a natural regular two-form $\sigma \in H^0(M(v)^s, \Omega^2_{M(v)^s})$ which is everywhere non-degenerate.* □

This was first observed by Mukai in [428]. In fact, the two-form exists more generally on the moduli space of simple sheaves on X (which, however, is in general not separated). Moreover, the two-form is closed, which can be deduced from an explicit description using the Atiyah class; cf. [266, Ch. 10].

2.4 The condition that the moduli space is smooth and parametrizes isomorphism classes of sheaves (and not merely S-equivalence classes) is essentially equivalent to the non-existence of properly semistable sheaves. Thus, it is important to understand under which conditions on the Mukai vector v and the polarization H semistability is equivalent to stability.

A torsion-free semistable sheaf E fails to be stable if there exists a proper saturated subsheaf $F \subset E$ with

$$p(F, m) \equiv p(E, m)$$

or, in other words, if $\langle \mathrm{rk}(F)v(E) - \mathrm{rk}(E)v(F), v(\mathcal{O}(m)) \rangle = 0$ for all m. The latter is equivalent to

$$\text{(i) } (\xi_{E,F}.H) = 0 \quad \text{and} \quad \text{(ii) } \mathrm{rk}(F)(\chi(E) - \mathrm{rk}(E)) = \mathrm{rk}(E)(\chi(F) - \mathrm{rk}(F)), \quad (2.2)$$

where $\xi_{E,F} := \mathrm{rk}(F)c_1(E) - \mathrm{rk}(E)c_1(F)$.

Let us now fix the Mukai vector $v \in N(X)$. It can be uniquely written as $v = mv_0$ with $m \in \mathbb{Z}_{>0}$ and $v_0 \in N(X)$ primitive; i.e. v_0 cannot be divided further or, equivalently, m is maximal. Let us first consider the case $m = 1$, i.e. v itself is primitive. Then if (i) and (ii) hold for all H, then $(\mathrm{rk}(E)/\mathrm{rk}(F))v(F) = v(E) = v = v_0$, which is absurd, as $\mathrm{rk}(F) < \mathrm{rk}(E)$. This leads to the following result.

Proposition 2.5 *Assume $v = (r, \ell, s) \in N(X)$ is primitive. Then, with respect to a generic choice of H, any semistable sheaf E with $v(E) = v$ is stable. Hence, $M_H(v) = M_H(v)^s$, which is smooth and projective of dimension $\langle v, v \rangle + 2$ if not empty.*

The polarization is *generic* if it is contained in the complement of a locally finite union of hyperplanes in $NS(X)_\mathbb{R}$.

Proof For the case of torsion-free sheaves, i.e. $rk > 0$, the proof is in [266, App. 4.C]. The argument given above only shows that for a generic choice of H equalities (i) and (ii) in (2.2) can be excluded for subsheaves $F \subset E$ with a fixed Mukai vector $v(F)$. To complete the proof, one shows the following two things.

First, if E is μ_H-semistable and $F \subset E$ is μ_H-destabilizing, then

$$(\xi_{E,F}.H) = 0 \quad \text{and either} \quad \xi_{E,F} = 0 \quad \text{or} \quad (-rk(E)^2/4)\Delta(v) \le \xi_{E,F}^2 < 0. \qquad (2.3)$$

Here, $\Delta(v) = \Delta(E) = 2rk(E)c_2(E) - (rk(E) - 1)c_1(E)^2$. See [266, Thm. 4.C.3]. The proof uses the Hodge index theorem and the Bogomolov inequality for F.

Second, the union of walls

$$W_\xi := \{H \in NS(X)_\mathbb{R} \text{ ample} \mid \xi, H \text{ satisfying (2.3)}\}$$

is locally finite. See [266, Lem. 4.C.2].

The case of torsion sheaves has been dealt with by Yoshioka in [645, Sec. 1.4], but see also the thesis [654] by Zowislok. Let us show the existence of a generic H in this case. If $v(E) = (0, \ell, s)$ with $\ell \ne 0$ and hence $(\ell.H) > 0$ for all ample H (the case $\ell = 0$ being trivial), then $s = \chi(E)$ and $p(E, m) = m + \chi(E)/(\ell.H)$.

As $M_H(v) \xrightarrow{\sim} M_H(v.ch(H))$ via $E \longmapsto E(H)$, one can assume that $s \ne 0$. Thus, E is semistable if $\chi(F)/(\ell'.H) \le \chi(E)/(\ell.H)$ for all $F \subset E$ with $v(F) = (0, \ell', \chi(F))$. If equality holds for a proper subsheaf F and all H, then $\ell' = (\chi(F)/\chi(E)) \cdot \ell$. If $supp(E) = \sum n_i C_i$ and $supp(F) = \sum m_i C_i$, then clearly $m_i \le n_i$ and hence $\chi(F)/\chi(E) \le 1$. But if v is primitive, then this implies $\chi(F) = \chi(E)$, $\ell' = \ell$, and hence $m_i = n_i$. From the latter one deduces that E/F is a torsion sheaf and hence $\chi(E) - \chi(F) = \chi(E/F) > 0$. Contradiction. The local finiteness of the wall structure is proven as in the case $r > 0$. $\qquad\square$

Remark 2.6 If $v = mv_0$ with v_0 primitive and $m > 1$, then $M(v)$ is expected neither to coincide with $M(v)^s$ nor to be smooth. In [281] Kaledin, Lehn, and Sorger show that for $\langle v_0, v_0 \rangle > 2$ the moduli space $M(v)$ is still locally factorial; i.e. all local rings are UFD, and in particular normal. The same result holds for $\langle v_0, v_0 \rangle = 2$ and $m > 2$.

2.5 The theory is not void; i.e. these moduli spaces are not all empty. But this is a highly non-trivial statement. For general results on the existence of stable sheaves on algebraic surfaces, see references in [266, Ch. 5]. Roughly, for arbitrary surfaces one can prescribe $rk(E)$ and $det(E)$ and prove existence of μ-(semi)stable vector bundles for large $c_2(E) \gg 0$. For K3 surfaces the situation is better due to the following result.

Theorem 2.7 *Let X be a complex projective K3 surface with an ample line bundle H. For any $v = (r, \ell, s) \in \mathbb{Z} \oplus \mathrm{NS}(X) \oplus \mathbb{Z}$ with $\langle v, v \rangle \geq -2$ and such that $r > 0$ or ℓ ample (or, weaker, $(\ell)^2 \geq -2$ and $(\ell.H) > 0$), there exists a semistable sheaf E with $v(E) = v$.*

(i) Mukai in [429, Thms. 5.1 and 5.4] showed that for primitive $v = (r, \ell, s)$ with $r > 0$ and $\langle v, v \rangle = 0$ there exists a μ_H-semistable sheaf E with $v(E) = v$. If $H = \ell$, then E can be chosen to be μ_H-stable. These sheaves are first constructed on special (so-called monogonal) K3 surfaces by the methods explained in Section **9**.3 and then deformed to sheaves on arbitrary K3 surfaces.

Note that once a μ-(semi)stable sheaf E has been found, taking the kernel of surjections

$$E \longrightarrow \bigoplus_{i=1}^{n} k(x_i)$$

yields μ-(semi)stable sheaves with Mukai vector $(r, \ell, s - n)$. However, this does not produce μ-semistable sheaves for all possible Mukai vectors, as $\langle (r, \ell, s - n), (r, \ell, s - n) \rangle = \langle v, v \rangle - 2rn = -2rn$. Also note that the primitivity of v is not essential for the construction of μ-semistable sheaves, as μ-semistability of E implies μ-semistability of $E^{\oplus n}$.

(ii) The existence of stable sheaves with $\langle v, v \rangle = -2$ is due to Kuleshov; see Remark 3.3.

(iii) The result in the above form is implicitly part of a result by Yoshioka [645, Thm. 8.1] which asserts that $M_H(v)^s$ and $\mathrm{Hilb}^{\langle v, v \rangle/2+1}(X)$ are deformation equivalent for primitive v. Once more, the actual construction is done on special elliptic surfaces. General deformation theory yields the result for all K3 surfaces. Originally, for $r = 0$ it was assumed that ℓ is ample, but in [647, Cor. 3.5] the hypothesis was weakened to just $(\ell)^2 \geq -2$ and $(\ell.H) > 0$, which implies that ℓ is effective.

In principle, the arguments should go through for arbitrary algebraically closed fields. However, since global deformations to elliptic K3 surfaces are used, details would need to be checked in positive characteristic, but the irreducibility of the moduli space proved by Madapusi Pera in [387] should be enough. It would be highly desirable to produce stable bundles and sheaves more directly, but no techniques seem to exist that would reproduce the theorem in full generality.

3 Some Moduli Spaces

Let us consider moduli spaces of dimension zero and two and those that are provided by Hilbert schemes of points on K3 surfaces.

3.1 In analogy to the case of Pic_X, one can study *rigid sheaves*. Suppose $t \in M(v)^s$ corresponds to a rigid sheaf E, i.e. $\mathrm{Ext}^1(E, E) = 0$. Then t is a reduced isolated point of $M(v)$. Note that in this case, $\langle v, v \rangle = -2$. Indeed,

$$\langle v, v \rangle = -\chi(E, E) = -\dim \mathrm{Ext}^0(E, E) - \dim \mathrm{Ext}^2(E, E)$$

and $\mathrm{Ext}^2(E, E) \simeq \mathrm{Ext}^0(E, E)^* \simeq k$, by Serre duality and stability of E.[5]

The moduli space $M(v)^s$ for a (-2)-vector v is not only discrete, it in fact consists of at most one point. Note however that in general $M_H(v)^s$ may parametrize different sheaves for different polarizations H. The beautiful argument goes back to Mukai; see [266, Thm. 6.16].

Proposition 3.1 *If $\langle v, v \rangle = -2$, then $M(v)^s$ consists of at most one reduced point. If $M(v)^s \neq \emptyset$, then $M(v)^s = M(v)$.*

Proof Suppose $E, F \in M(v)^s$. Then $\chi(E, F) = -\langle v(E), v(F) \rangle = -\langle v, v \rangle = 2$. Hence, $\mathrm{Hom}(E, F) \neq 0$ or $\mathrm{Hom}(F, E) \simeq \mathrm{Ext}^2(E, F)^* \neq 0$. Since E and F are both stable with the same Hilbert polynomial, this yields $E \simeq F$. The same argument also applies when only E is stable, which proves the second assertion. $\qquad\square$

Remark 3.2 If $r > 0$, then a stable rigid sheaf is automatically locally free. Indeed such a sheaf E is torsion free and isomorphic to the kernel of a surjection $E^{**} \twoheadrightarrow S$ with S of dimension zero. A quotient of this form can be deformed such that the support of S, which is the singularity set of E, actually changes. Hence, E itself deforms non-trivially, which contradicts $\mathrm{Ext}^1(E, E) = 0$; see [266, Thm. 6.16] for details.

Remark 3.3 The existence of simple rigid bundles is non-trivial. Suppose v is a Mukai vector with $r > 0$ and $\langle v, v \rangle = -2$. Then there indeed exists a (usually non-unique) sheaf E with $v(E) = v$ such that E is rigid and simple. This result is due to Kuleshov [332]. The sheaf is automatically locally free and for $\rho(X) = 1$ even stable, as explained by Mukai in [429, Prop. 3.14]. The existence is proved by first constructing such a bundle explicitly on a special (elliptic) K3 surface and then deforming it to any K3 surface. For generic polarization the bundle can even be assumed to be stable. See Section 2.5 for the general existence statement.

3.2 Let $v \in N(X)$ with $\langle v, v \rangle = 0$. Then $M(v)^s$ is empty or smooth and two-dimensional. The analogue of Proposition 3.1 is the following result, again due to Mukai.

Proposition 3.4 *Let $\langle v, v \rangle = 0$ and let $M_1 \subset M(v)^s$ be a complete connected (or, equivalently, irreducible) component. Then $M_1 = M(v)^s = M(v)$.*

---

[5] Note that for numerical considerations of this sort $k = \overline{k}$ is crucial.

Proof For the detailed proof, see [266, Thm. 6.1.8]. Let us here just outline the main steps under the simplifying assumption that there exists a locally free universal family $\mathcal{E}$ on $M_1 \times X$. The reader should have no problem modifying the arguments to also cover the general case.

Consider a semistable sheaf F on X with $v(F) = v$, and let us compute $R^i p_*$ $(q^* F \otimes \mathcal{E}^*)$, where p and q denote the two projections from $M_1 \times X$. Fibrewise one has $H^i(X, F \otimes \mathcal{E}_t^*) = \mathrm{Ext}^i(\mathcal{E}_t, F) = 0$ for $i = 0, 2$ and $F \not\cong \mathcal{E}_t$. Since $\chi(\mathcal{E}_t, F) = -\langle v(\mathcal{E}_t), v(F) \rangle = -\langle v, v \rangle = 0$, in fact $H^i(X, F \otimes \mathcal{E}_t^*) = 0$ for all i whenever $F \not\cong \mathcal{E}_t$. Thus, if the point $[F] \in M(v)$ corresponding to F is not contained in M_1, then $R^i p_*(q^* F \otimes \mathcal{E}^*) = 0$ for all i.

On the other hand, one shows that $R^i p_*(q^* F \otimes \mathcal{E}^*) = 0$ for $i = 0, 1$, even when $t = [F] \in M_1$, and that

$$R^2 p_*(q^* F \otimes \mathcal{E}^*) \otimes k([F]) \longrightarrow H^2(X, F \otimes \mathcal{E}_t^*) \simeq \mathrm{Ext}^2(F, F) \tag{3.1}$$

is surjective. The latter assertion follows from standard base change theorem [236, III.Thm. 12.11]. For the vanishing of $R^i p_*(q^* F \otimes \mathcal{E}^*)$, $i = 0, 1$, one however needs to use more of the proof of the base change theorem, which shows the existence (locally) of a complex of locally free sheaves $\mathcal{K}^\bullet : 0 \longrightarrow \mathcal{K}^0 \longrightarrow \mathcal{K}^1 \longrightarrow \mathcal{K}^2 \longrightarrow \cdots$ with $\mathcal{H}^i(\mathcal{K}^\bullet) \simeq R^i p_*(q^* F \otimes \mathcal{E}^*)$. This can be combined with the observation above that the support of the sheaves $R^i p_*(q^* F \otimes \mathcal{E}^*)$ is contained in the point $[F]$ as follows. For $i = 0$ one uses the inclusion $R^0 p_*(q^* F \otimes \mathcal{E}^*) \subset \mathcal{K}^0$ into the torsion-free sheaf $\mathcal{K}^0$ to conclude vanishing. From this, one deduces a short exact sequence

$$0 \longrightarrow \mathcal{K}^0 \longrightarrow \mathrm{Ker}(d^1) \longrightarrow R^1 p_*(q^* F \otimes \mathcal{E}^*) \longrightarrow 0.$$

However, since in the exact sequence

$$0 \longrightarrow \mathrm{Ker}(d^1) \longrightarrow \mathcal{K}^1 \longrightarrow \mathrm{Im}(d^1) \longrightarrow 0$$

the sheaf $\mathrm{Im}(d^1)$ is torsion free and $\mathcal{K}^1$ is locally free, the sheaf $\mathrm{Ker}(d^1)$ must be locally free as well. Now use that the quotient of the locally free sheaves $\mathrm{Ker}(d^1)$ by $\mathcal{K}^0$ is either trivial or concentrated in codimension ≤ 1. Thus, $R^1 p_*(q^* F \otimes \mathcal{E}^*) = 0$.

Finally, use the Grothendieck–Riemann–Roch formula (see [236, App. A])

$$\mathrm{ch}(Rp_*(q^* F \otimes \mathcal{E}^*)) = p_* \left\{ \mathrm{ch}(q^* F \otimes \mathcal{E}^*) q^* \mathrm{td}(X) \right\} = p_* \left\{ \mathrm{ch}(\mathcal{E}^*) q^* v(F) q^* \sqrt{\mathrm{td}(X)} \right\}.$$

The right-hand side depends only on $\mathcal{E}$ and $v(F)$, whereas the left-hand side is trivial for $[F] \notin M_1$ and equals $\mathrm{ch}(R^2 p_*(q^* F \otimes \mathcal{E}^*)) \neq 0$ otherwise (use (3.1)). This contradiction shows $M(v) \setminus M_1 = \emptyset$. $\qquad\square$

Corollary 3.5 *Assume $v = (r, \ell, s)$, $r \geq 0$, is primitive with $\langle v, v \rangle = 0$. Assume that ℓ is effective if $r = 0$ or, equivalently, that $(\ell.H) > 0$. Then for generic H, the moduli space $M(v)$ is a K3 surface.*

Proof   Indeed, by Mukai's result (see Corollary 2.4 and Proposition 2.5) and Theorem 2.7 one knows that $M := M(v)$ is a smooth projective irreducible surface endowed with an everywhere non-degenerate regular two-form $\sigma \in H^0(M, \Omega_M^2)$. Hence, $\omega_M \simeq \mathcal{O}_M$.

Thus, it remains to prove that $H^1(M, \mathcal{O}) = 0$. If one works over $k = \mathbb{C}$, then one could study the correspondence $H^*(X, \mathbb{Q}) \xrightarrow{\sim} H^*(M, \mathbb{Q})$ which is given by the cohomology class $\mathrm{ch}(\mathcal{E}) \sqrt{\mathrm{td}(X \times M)}$; cf. Proposition **16**.3.2. Or one uses the Leray spectral sequence

$$E_2^{i,j} = H^i(M, \mathcal{E}xt_p^j(\mathcal{E}, \mathcal{E})) \Rightarrow \mathrm{Ext}^{i+j}(\mathcal{E}, \mathcal{E}),$$

which immediately yields $H^1(M, \mathcal{O}) \hookrightarrow \mathrm{Ext}^1(\mathcal{E}, \mathcal{E})$. The vanishing of the latter is however not so easy to prove. Eventually, both arguments reduce to a statement about the composition of certain Fourier–Mukai equivalences; see [266, Lem. 6.1.10].

From a derived category point of view one could argue as follows; cf. the proof of Proposition **16**.2.1. The functor $\Phi_{\mathcal{E}} \colon \mathrm{D}^b(M) \xrightarrow{\sim} \mathrm{D}^b(X)$, $G \mapsto q_*(p^*G \otimes \mathcal{E})$ is fully faithful, which can be shown using a criterion of Bondal and Orlov; see [254, Prop. 7.1] or Proposition **16**.1.6. In fact, it is an equivalence, for X and M are smooth surfaces with trivial canonical bundle (use another criterion of Bondal and Orlov; see [254, Prop. 7.6] or Lemma **16**.1.7). See Proposition **16**.2.3 for more details. This then allows one to reverse the role of X and M and compute $\mathrm{Ext}^1(\mathcal{E}, \mathcal{E})$ via the spectral sequence $E_2^{i,j} = H^i(X, \mathcal{E}xt_q^j(\mathcal{E}, \mathcal{E})) \Rightarrow \mathrm{Ext}^{i+j}(\mathcal{E}, \mathcal{E})$. Since $H^1(X, \mathcal{O}) = H^0(X, \mathcal{T}_X) = 0$ and $\mathcal{T}_X \simeq \mathcal{E}xt_q^1(\mathcal{E}, \mathcal{E})$, this immediately shows $\mathrm{Ext}^1(\mathcal{E}, \mathcal{E}) = 0$ as required; see Proposition **16**.2.1 for more details. If a universal family $\mathcal{E}$ is not available, replace it in the above arguments by a twisted universal sheaf. $\square$

Example 3.6   In [266, Ex. 5.3.7] one finds the following example. Let $X \subset \mathbb{P}^3$ be a general quartic and let $v = (2, \mathcal{O}_X(-1), 1)$. Then $M(v) \simeq X$, where the isomorphism is given by mapping a sheaf F to the point $x \in X$, the ideal sheaf I_x of which is the quotient of a uniquely determined (up to the natural GL(3)-action) injection $F \hookrightarrow \mathcal{O}_X^{\oplus 3}$. In other words, the isomorphism is given by the shift of the spherical twist $I_x \mapsto T_{\mathcal{O}}(I_x)[1]$ (cf. Section **16**.2.3). There are other examples where the two-dimensional moduli space is isomorphic to the original K3 surface, but this is not typical.

Remark 3.7   Let us mention another result which may convey an idea of how a K3 surface X is related to the K3 surface that is given by some moduli space $M(v)$. Assume that X is a complex projective K3 surface and $M(v) = M(v)^s$ is two-dimensional. Then there exists a natural isomorphism

$$H^2(M(v), \mathbb{Z}) \simeq v^\perp / \mathbb{Z} \cdot v \tag{3.2}$$

respecting the Hodge structures of weight two and the quadratic forms. Here, $v^\perp \subset \widetilde{H}(X, \mathbb{Z})$ is endowed with the Mukai pairing and the Hodge structure of K3 type given by $H^{2,0}(X) \subset v_{\mathbb{C}}^{\perp}$. The result is due to Mukai; see [266, Thm. 6.1.14]. It is remarkable that it holds without assuming the existence of a universal sheaf.[6]

For a fine two-dimensional moduli space $M(v) = M(v)^s$ with a universal family $\mathcal{E}$ one deduces (3.2) from the Hodge isometry

$$\widetilde{H}(M(v), \mathbb{Z}) \xrightarrow{\sim} \widetilde{H}(X, \mathbb{Z}), \quad \alpha \longmapsto q_*(p^*\alpha.v(\mathcal{E})) \tag{3.3}$$

between the full cohomologies endowed with the Mukai pairing. See also Proposition **16**.3.2 for an analogous result for derived equivalent K3 surfaces. Indeed, once (3.3) has been proved, use that under this isomorphism v is the image of $[\text{pt}] \in H^4(M(v), \mathbb{Z})$, for the image of the skyscraper sheaf of a point $[E] \in M(v)$ under $D^b(M(v)) \simeq D^b(X)$ is E, and then use that $v^{\perp}/\mathbb{Z} \cdot v \simeq [\text{pt}]^{\perp}/\mathbb{Z} \cdot [\text{pt}] \simeq H^2(M(v), \mathbb{Z})$.

A concrete realization of (3.2) is described by Corollary **11**.4.7.

3.3 After moduli spaces of dimension zero and two, Hilbert schemes are the most accessible ones. Recall that for $P \equiv n$ or equivalently $v = (0, 0, n)$, the moduli space $M(v)$ is isomorphic to the symmetric product $S^n(X)$; see Example 1.9. If $n > 1$, then $M(v)^s = \varnothing$.

Let us now consider $v = (1, 0, 1 - n)$. Then any $E \in M(v)$ is a torsion-free sheaf of rank one and hence of the form $I_Z \otimes L$ with $Z \subset X$ a subscheme of dimension zero and $L \in \text{Pic}(X)$; see Section **9**.1.1. Note that a torsion-free sheaf of rank one does not contain any non-trivial subsheaf with torsion-free quotient and that it is therefore automatically stable with respect to any polarization. Using the exact sequence

$$0 \longrightarrow I_Z \otimes L \longrightarrow L \longrightarrow \mathcal{O}_Z \longrightarrow 0,$$

one finds

$$v = v(E) = v(I_Z \otimes L) = v(L) - v(\mathcal{O}_Z) = (1, c_1(L), c_1^2(L)/2 + 1 - h^0(\mathcal{O}_Z)).$$

Hence, $L \simeq \mathcal{O}_X$ and $h^0(\mathcal{O}_Z) = n$. This observation links the moduli space $M(v) = M(1, 0, 1 - n)$ to the Hilbert scheme $\text{Hilb}^n(X)$ of subschemes $Z \subset X$ of length n. Recall that $\text{Hilb}^n(X)$ represents the functor

$$\text{Hilb}_X^n : (Sch/k)^o \longrightarrow (Sets)$$

that maps a k-scheme S to the set of all S-flat closed subschemes $\mathcal{Z} \subset S \times X$ with geometric fibres $\mathcal{Z}_t \subset X$ of length n; cf. Section **5**.2.1.

---

[6] Over an arbitrary algebraically closed field k, when a priori one cannot speak of the period of X and $M(v)$, the same proof at least shows that the two lattices $\text{NS}(M(v))$ and $v^{\perp}/\mathbb{Z} \cdot v$ (with $v^{\perp}$ the orthogonal complement in $N(X)$) are isometric.

Proposition 3.8   *Mapping a subscheme $Z \subset X$ to its ideal sheaf I_Z induces an isomorphism*

$$\mathrm{Hilb}^n(X) \xrightarrow{\sim} M(1,0,1-n).$$

Proof   Sending $\mathcal{Z} \subset S \times X$ in $\mathrm{Hilb}^n_X(S)$ to its ideal sheaf $\mathcal{I}_{\mathcal{Z}}$, which defines an element in $\mathcal{M}(1,0,1-n)(S)$, defines a functor transformation

$$\mathrm{Hilb}^n_X \longrightarrow \mathcal{M}(1,0,1-n).$$

Conversely, if $E \in \mathcal{M}(1,0,1-n)(S)$, then the spaces $\mathrm{Hom}(E_t, \mathcal{O}_{\{t\} \times X})$ glue to an invertible sheaf $M := \mathcal{H}om_p(E, \mathcal{O}_{S \times X})$ on S, and the naturally induced map $E \otimes p^*M \longrightarrow \mathcal{O}_{S \times X}$ is fibrewise an embedding and hence $E \otimes p^*M \simeq \mathcal{I}_{\mathcal{Z}}$ for some $\mathcal{Z} \in \mathrm{Hilb}^n_X(S)$. Since $E \sim E \otimes p^*M$, this yields an inverse map.   □

Remark 3.9   It can be shown that $\mathrm{Hilb}^n(X)$ admits a unique (up to scaling) regular, everywhere non-degenerate two-form σ; i.e. the one given by Corollary 2.4 is unique up to scaling. Moreover, $H^0(\mathrm{Hilb}^n(X), \Omega^{2i})$ is spanned by $\sigma^{\wedge i}$ and $H^0(\mathrm{Hilb}^n(X), \Omega^{2i+1})$ $= 0$. For $k = \mathbb{C}$ this is equivalent to $\mathrm{Hilb}^n(X)$ being simply connected. See [269] or, for a direct argument, [266, Thm. 6.24]. The result is due to Beauville [45] and to Fujiki for $n = 2$.

3.4   Moduli spaces of stable sheaves on K3 surfaces were for a long time hoped to produce higher-dimensional irreducible symplectic manifolds in abundance. A smooth projective variety M (or a compact Kähler manifold) is called *irreducible symplectic* if $H^0(M, \Omega^2_M)$ is spanned by an everywhere non-degenerate two form σ; i.e. σ induces an isomorphism $\mathcal{T}_M \xrightarrow{\sim} \Omega_M$ and is unique up to scaling, and M does not admit any non-trivial finite étale covering. The Hodge structure of weight two $H^2(M, \mathbb{Z})$ of an irreducible symplectic (projective) manifold M plays the same central role in higher dimensions as $H^2(X, \mathbb{Z})$ for K3 surfaces. It can be endowed with a natural quadratic form, the *Beauville–Bogomolov form*. For a survey on irreducible symplectic manifolds, see [251] and Lehn [359].

   The existence of the symplectic structure on the moduli space, due to Mukai, was a very promising first step. However, the irreducibility, i.e. the uniqueness of the two-form and the simply connectedness, was difficult to establish. In some cases, this could be shown by relating a higher rank moduli space to the Hilbert scheme. For example, for a Mukai vector $v = (r, \ell, s)$ with ℓ primitive it was shown (in [214] for $r = 2$ and by O'Grady in [466] for $r \geq 2$) that for generic H the moduli space $M_H(v)$ is an irreducible symplectic projective manifold whose Hodge numbers equal those of the Hilbert scheme of X of the same dimension. This was later generalized by Yoshioka in [645] to the case that only v is primitive and allowing torsion sheaves.

However, moduli spaces of stable sheaves do not provide really new examples, which was first observed in [250], where birational irreducible symplectic manifolds are shown to be always deformation equivalent.

Theorem 3.10 *Suppose $v = (r, \ell, s)$ is a primitive Mukai vector and H is generic. Assume $r > 0$ (or $r = 0$ and $(\ell.H) > 0$) and $\langle v, v \rangle \geq -2$. Then $M_H(v)$ is an irreducible symplectic projective manifold deformation equivalent to $\mathrm{Hilb}^{\langle v,v \rangle+2}(X)$. Moreover, if $\langle v, v \rangle > 0$, then there exists a Hodge isometry $H^2(M_H(v), \mathbb{Z}) \simeq v^\perp$.*

The deformation equivalence to the Hilbert scheme was proved in [250, Cor. 4.8] for v_1 primitive and later by Yoshioka in [645, App. 8] in general. For $r = 0$, see [647, Cor. 3.5]. For primitive v and generic H it was computed explicitly for $M_H(v)$ by O'Grady [466] and Yoshioka [644].

We cannot resist stating here Göttsche's formulae expressing the Betti and Euler numbers of all $\mathrm{Hilb}^n(X)$ (and then for all moduli spaces $M_H(v)$ covered by Theorem 3.10) simultaneously, using the corresponding generating series:

$$\sum_{n=0}^{\infty} \sum_{i=0}^{4n} b_i(\mathrm{Hilb}^n(X)) t^{i-2n} q^n = \left(\prod_{m=1}^{\infty} (1 - t^{-2}q^m)(1 - q^m)^{22}(1 - t^2 q^m) \right)^{-1}$$

$$\sum_{n=0}^{\infty} e(\mathrm{Hilb}^n(X)) q^n = \left(\prod_{m=1}^{\infty} (1 - q^m)^{24} \right)^{-1}.$$

A similar formula exists for the Hodge numbers. We refer to Göttsche's original article [212] for comments and proofs.

References and Further Reading

The Picard scheme and moduli spaces of stable sheaves also exist for varieties over non-algebraically closed fields (after étale sheafification if one wants to have a chance to represent the functor). The construction is compatible with base field extensions.

There is one other series of examples of irreducible symplectic projective manifolds, also discovered by Beauville. They are obtained as the fibre of the summation $\mathrm{Hilb}^{n+1}(A) \longrightarrow A$, where A is any abelian surface. For $n = 1$ this gives back the Kummer surface associated with A.

O'Grady studied moduli spaces $M_H(v)$ with non-primitive v. In one case he could show that although the moduli space is singular, it can be resolved symplectically. This indeed leads to an example of an irreducible symplectic projective manifold of dimension 10, which is truly new, i.e. topologically different from $\mathrm{Hilb}^5(X)$ and generalized Kummer varieties. By work of Kaledin, Lehn, and Sorger [281] and by Choy and Kiem [118] we know that O'Grady's example is the only one that admits a symplectic resolution. This follows almost immediately from the result mentioned in Remark 2.6. In [281] one also finds a generalization of Mukai's irreducibility result to the case of non-primitive vectors: if $M(v)$ has a connected component parametrizing purely stable sheaves only, then it equals this component. Singular two-dimensional moduli spaces are studied in great detail in [482].

Two-dimensional fine moduli spaces of stable sheaves on a K3 surface X are also called Fourier–Mukai partners of X, because the Fourier–Mukai transform

$$\mathrm{D}^b(X) \longrightarrow \mathrm{D}^b(M), \quad F \longmapsto p_*(q^*F \otimes \mathcal{E}),$$

defines an equivalence of triangulated categories. In fact, all Fourier–Mukai partners are of this form. The number of isomorphism classes of Fourier–Mukai partners of a fixed K3 surface X is finite, but with X varying it is unbounded. See the articles by Hosono et al. [247] and Stellari [575] and the discussion in Chapter 16.

A finer study of moduli spaces of (Bridgeland stable) complexes on K3 surfaces has just begun. See [266] for some comments and references and the recent preprints [41, 42, 641].

Questions and Open Problems

It is still an open question whether moduli spaces of stable sheaves on a fixed K3 surface, maybe with additional conditions on the prescribed Mukai vectors, are derived equivalent as soon as their dimensions coincide. This is even open for cases where the moduli spaces are known to be birational (except for dimension two and four).

The link between the Brauer groups of a K3 surface and of a non-fine moduli space as expressed by (2.1) has been proved using only Hodge theory and so is a priori valid only for complex projective K3 surfaces. Is there a purely algebraic argument for it?

11

Elliptic K3 Surfaces

The literature on elliptic surfaces is vast. Elliptic surfaces play a central role both in complex geometry and in arithmetic. We restrict ourselves to the case of elliptic K3 surfaces and do not hesitate to take short cuts whenever possible. The discussion of the Jacobian fibration of an elliptic K3 surface from the point of view of moduli spaces of sheaves on K3 surfaces is not quite standard. Popular sources for elliptic surfaces include, among many others, [33, 133, 186, 375, 414].

1 Singular Fibres

We shall begin with the definition of an elliptic K3 surface and a classical existence result. The main part of this section reviews Kodaira's classification of singular fibres of elliptic fibrations of K3 surfaces.

1.1   In the following X is an algebraic K3 surface over an arbitrary algebraically closed field k. To simplify, we shall exclude the cases char$(k) = 2, 3$ from the start. Most of what is said holds verbatim for non-projective complex K3 surfaces. We will explicitly state when this is not the case.

Definition 1.1   An *elliptic K3 surface* is a K3 surface X together with a surjective morphism $\pi : X \longrightarrow \mathbb{P}^1$ such that the geometric generic fibre is a smooth integral curve of genus one or, equivalently, if there exists a closed point $t \in \mathbb{P}^1$ such that X_t is a smooth integral curve of genus one.

The morphism $\pi : X \longrightarrow \mathbb{P}^1$ itself is called an *elliptic fibration* of the K3 surface X and it is automatically flat (use [236, II.Prop. 9.7]). In particular, for the arithmetic genus of all fibres $X_t \subset X$ one has $p_a(X_t) = 1 - \chi(\mathcal{O}_{X_t}) = 1$. We shall abusively speak of the smooth fibres of $\pi : X \longrightarrow \mathbb{P}^1$ as elliptic curves, although we do not assume the existence

---

[*] Thanks to Matthias Schütt for detailed comments on this chapter.

219

of a (distinguished) section, and the fibres, therefore, come without a distinguished origin.

Example 1.2 (i) Let X be the Kummer surface associated with the product of two elliptic curves $E_1 \times E_2$, see Example **1**.1.3. Then the two projections induce two natural elliptic fibrations $\pi_i \colon X \longrightarrow E_i/\pm \, \simeq \, \mathbb{P}^1$. In fact, X has many more, as explained by Shioda and Inose [568, Thm. 1].

(ii) Consider a general elliptic pencil of cubics in $\mathbb{P}^2$ as an elliptic fibration $\widetilde{\pi} \colon \widetilde{\mathbb{P}}^2 \longrightarrow \mathbb{P}^1$ of the blow-up of the nine fixed points $\widetilde{\mathbb{P}}^2 \longrightarrow \mathbb{P}^2$. To have a concrete example in mind, consider the *Hesse* (or *Dwork*) *pencil* $x_0^3 + x_1^3 + x_2^3 - 3\lambda \cdot x_0 x_1 x_2$.

A double cover $X \longrightarrow \widetilde{\mathbb{P}}^2$ branched over the union $F_{t_0} \sqcup F_{t_1}$ of two smooth fibres of $\widetilde{\pi}$ describes a K3 surface, which can also be obtained via base change with respect to a double cover $\mathbb{P}^1 \longrightarrow \mathbb{P}^1$ branched over $t_0, t_1 \in \mathbb{P}^1$ or as the minimal resolution of the double plane branched along the sextic described by the union $F_{t_0} \cup F_{t_1} \subset \mathbb{P}^2$.

(iii) The Fermat quartic $X \subset \mathbb{P}^3$, $x_0^3 + \cdots + x_3^4 = 0$, admits many elliptic fibrations; see Example **2**.3.11.

The generic K3 surface is not elliptic but elliptic ones are rather frequent. In fact, the following result for $k = \mathbb{C}$ combined with Proposition 7.1.3 implies that elliptic K3 surfaces are parametrized by a dense codimension one subset in the moduli space of all K3 surfaces.

Proposition 1.3 *Let X be a K3 surface over an algebraically closed field k with* char$(k) \neq 2, 3$.

(i) *Then X admits an elliptic fibration if and only if there exists a non-trivial line bundle L with $(L)^2 = 0$.*

(ii) *If $\rho(X) \geq 5$, then X admits an elliptic fibration.*

(iii) *The surface X admits at most finitely many non-isomorphic elliptic fibrations.*

Proof For the first part, see Remark **8**.2.13. In order to prove the second, use the consequence of the Hasse–Minkowski theorem saying that any indefinite form of rank at least five represents zero; see [547, IV.3.2]. Then apply (i). For (iii), see Corollary **8**.4.6.   □

As we are considering only algebraic K3 surfaces in this chapter, an elliptic K3 surface X satisfies $\rho(X) \geq 2$. A K3 surface with $2 \leq \rho(X) < 5$ may or may not admit an elliptic fibration. For non-projective complex K3 surfaces (ii) and (iii) above may fail.

Remark 1.4 Similarly, a K3 surface X admits an elliptic fibration with a section if there exists an embedding $U \lhook\joinrel\longrightarrow \mathrm{NS}(X)$. For complex projective K3 surfaces this is the case, e.g. if $\rho(X) \geq 12$; see Corollary **14**.3.8. As it turns out, elliptic K3 surfaces (with a section) are dense in the moduli space of all complex (not necessarily algebraic) K3 surfaces and also in the moduli space of polarized K3 surfaces; see Remark **14.3.9**.

1.2 Before passing to the classification of singular fibres of elliptic fibrations of K3 surfaces, let us state a few general observations.

Remark 1.5 (i) If $\pi : X \longrightarrow C$ is a surjective morphism from a K3 surface X onto a curve C, then C is rational. So $C \simeq \mathbb{P}^1$ if C is smooth. Indeed, after Stein factorization, we may assume $\pi_* \mathcal{O}_X \simeq \mathcal{O}_C$ and C smooth. Then the Leray spectral sequence yields an injection $H^1(C, \mathcal{O}_C) \hookrightarrow H^1(X, \mathcal{O}_X) = 0$ and hence $C \simeq \mathbb{P}^1$.

(ii) If $\pi : X \longrightarrow \mathbb{P}^1$ is an elliptic fibration of a K3 surface, then not all fibres of π can be smooth. Indeed, if π were smooth, then $R^1\pi_*\mathbb{Z} \simeq \mathbb{Z}^2$ (as $\mathbb{P}^1$ is simply connected) and hence the Leray spectral sequence would yield the contradiction $H^1(X, \mathbb{Z}) \simeq \mathbb{Z}^2$. Here we assume that X is a complex K3 surface, but the same argument works in the algebraic context using étale cohomology. Alternatively, one could use the Leray spectral sequence to deduce the contradiction $e(X) = e(\mathbb{P}^1) \cdot e(X_t) = 0$.

(iii) Any smooth irreducible fibre of an arbitrary surjective morphism $\pi : X \longrightarrow \mathbb{P}^1$ from a K3 surface X is automatically an elliptic curve. Indeed, for a smooth fibre X_t, $t \in \mathbb{P}^1$, one has $\mathcal{O}(X_t) \simeq \pi^*\mathcal{O}(1)$ and hence $\mathcal{O}(X_t)|_{X_t} \simeq \mathcal{O}_{X_t}$. Therefore, by adjunction formula $\omega_{X_t} \simeq \mathcal{O}_{X_t}$.

(iv) An arbitrary surjective morphism $\pi : X \longrightarrow \mathbb{P}^1$ has a geometrically integral generic fibre if and only if $\mathcal{O}_{\mathbb{P}^1} \simeq \pi_*\mathcal{O}_X$. In this case, also the closed fibres X_t are integral for $t \in \mathbb{P}^1$ in a non-empty Zariski open subset; see [28, Ch. 7]. In characteristic zero, the generic fibre is smooth by Bertini (see [236, III.Cor. 10.9]), and, therefore, $\pi : X \longrightarrow \mathbb{P}^1$ is an elliptic fibration. This is still true in positive characteristic $\neq 2, 3$. Indeed, by a result of Tate the geometric generic fibre is either smooth or a rational curve with one cusp. Moreover, the latter cannot occur for char$(k) \neq 2, 3$. See [28, Thm. 7.18] or [440, 533, 590] and the proof of Proposition **2.3.10**.

The next result holds true more generally (modified appropriately) for arbitrary (even non-projective complex) elliptic surfaces. The first part is known as the *canonical bundle formula* (see e.g. [33, V.12] or [185, Thm. 7.15]), and an important consequence of it should in the general context be read as $\chi(\mathcal{O}_X) = \deg(R^1\pi_*\mathcal{O}_X)^*$.

Proposition 1.6 *Let* $\pi : X \longrightarrow \mathbb{P}^1$ *be an elliptic K3 surface.*

(i) *Then* $\pi_*\mathcal{O}_X \simeq \mathcal{O}_{\mathbb{P}^1}$ *and* $R^1\pi_*\mathcal{O}_X \simeq \mathcal{O}_{\mathbb{P}^1}(-2)$.

(ii) *All fibres* X_t, $t \in \mathbb{P}^1$, *are connected.*

(iii) *No fibre is multiple (but possibly non-reduced).*

(iv) *(Zariski's lemma) If* $X_t = \sum_{i=1}^{\ell} m_i C_i$ *with* C_i *integral, then* $(C_i.X_t) = 0$. *Moreover,* $(\sum n_i C_i)^2 \leq 0$ *for all choices* $n_i \in \mathbb{Z}$ *and equality holds if and only if* $n_1/m_1 = \cdots = n_\ell/m_\ell$.

Proof Consider the short exact sequence $0 \longrightarrow \mathcal{O}(-X_t) \longrightarrow \mathcal{O}_X \longrightarrow \mathcal{O}_{X_t} \longrightarrow 0$ and the induced exact sequences

$$0 \longrightarrow k \simeq H^0(X, \mathcal{O}_X) \longrightarrow H^0(X_t, \mathcal{O}_{X_t}) \longrightarrow H^1(X, \mathcal{O}(-X_t)) \longrightarrow H^1(X, \mathcal{O}_X) = 0$$

$$0 = H^1(X, \mathcal{O}_X) \longrightarrow H^1(X_t, \mathcal{O}_{X_t}) \longrightarrow H^2(X, \mathcal{O}(-X_t)) \longrightarrow H^2(X, \mathcal{O}_X) \simeq k \longrightarrow 0.$$

The generic fibre X_t is an elliptic curve and thus $h^0(X_t, \mathcal{O}_{X_t}) = h^1(X_t, \mathcal{O}_{X_t}) = 1$. Therefore, $H^1(X, \mathcal{O}(-X_t)) = 0$ and $H^2(X, \mathcal{O}(-X_t)) \simeq k^2$. But $H^i(X, \mathcal{O}(-X_t))$ is independent of t which in turn implies $h^0(X_t, \mathcal{O}_{X_t}) = h^1(X_t, \mathcal{O}_{X_t}) = 1$ for all $t \in \mathbb{P}^1$. This is enough to conclude that the natural map $\mathcal{O}_{\mathbb{P}^1} \longrightarrow \pi_* \mathcal{O}_X$ is an isomorphism and, using [236, III.Thm. 12.11], that $R^1 \pi_* \mathcal{O}_X$ is invertible. Write $R^1 \pi_* \mathcal{O}_X \simeq \mathcal{O}(d)$ and use the Leray spectral sequence to show $k \simeq H^2(X, \mathcal{O}_X) \simeq H^1(\mathbb{P}^1, R^1 \pi_* \mathcal{O}_X)$ and hence $d = -2$.[1]

Stein factorization [236, III.Cor. 11.5] yields connectedness of all fibres. This could also be concluded directly from Zariski's connectedness principle (see [236, III.Exer. 11.4] or [375, Thm. 5.3.15]) or even more directly from $h^0(X_t, \mathcal{O}_{X_t}) = 1$.

Suppose a fibre X_t is of the form mC, $m \geq 2$ (with C possibly reducible). Then use

$$0 \longrightarrow \mathcal{O}((m-1)C) \longrightarrow \mathcal{O}(X_t) \longrightarrow \mathcal{O}(X_t)|_C \longrightarrow 0,$$

$\mathcal{O}(X_t)|_C \simeq \mathcal{O}_C$, $h^0(X, \mathcal{O}(X_t)) = h^0(\mathbb{P}^1, \mathcal{O}(1)) = 2$, and the fact that the composition $H^0(\mathbb{P}^1, \mathcal{O}(1)) \longrightarrow H^0(X, \mathcal{O}(X_t)) \longrightarrow H^0(C, \mathcal{O})$ is not zero, to deduce $h^0(X, \mathcal{O}((m-1)C)) = 1$. But by Riemann–Roch $\chi(X, \mathcal{O}((m-1)C)) = 2$, as $(C)^2 = (1/m)^2 (X_t)^2 = 0$, and hence $h^0(X, \mathcal{O}((m-1)C)) \geq 2$, for $h^2(X, \mathcal{O}((m-1)C)) = 0$ for $m \geq 2$. Contradiction. (See also Remark 2.3.13 for a slightly different proof.)

For Zariski's lemma use that for any fibre $\mathcal{O}(X_t)|_{X_t} \simeq \mathcal{O}_{X_t}$. So, $\mathcal{O}(X_t)|_{C_i} \simeq \mathcal{O}_{C_i}$ and thus $(C_i.X_t) = 0$. The second assertion can be rephrased by saying that the intersection form is semi-negative definite on $\bigoplus \mathbb{Z}[C_i]$ with radical $\sum m_i[C_i]$. The first part can be seen as a consequence of the Hodge index theorem; see Section 1.2.3. To prove the full statement, we introduce the notation $c_{ij} := (C_i.C_j)$ and $a_{ij} := m_i m_j c_{ij}$. Clearly, $a_{ij} \geq 0$ for $i \neq j$ and $\sum_j a_{ij} = (m_i C_i.X_t) = 0$ for fixed i and similarly $\sum_i a_{ij} = 0$ for fixed j. If now $C = \sum n_i C_i$ and $\overline{n}_i := n_i/m_i$, then

$$(C)^2 = \sum n_i n_j c_{ij} = \sum \overline{n}_i \overline{n}_j a_{ij} = - \sum_{i<j} (\overline{n}_i - \overline{n}_j)^2 a_{ij} \leq 0.$$

This shows again semi-negativity of the intersection form and, moreover, using connectivity of X_t that $(C)^2 < 0$ except for $\overline{n}_1 = \cdots = \overline{n}_\ell$.   □

Combined with Section 2.1.3 this yields the following corollary.

---

[1] Alternatively, one can use relative (Grothendieck–Verdier) duality to conclude. Indeed, taking cohomology in degree -2 of $R\pi_* \omega_X[\dim(X)] \simeq (R\pi_* \mathcal{O})^* \otimes \omega_{\mathbb{P}^1}[\dim(\mathbb{P}^1)]$ yields $\pi_* \mathcal{O}_X \simeq (R^1 \pi_* \mathcal{O}_X)^* \otimes \mathcal{O}_{\mathbb{P}^1}(-2)$. More concretely, the fibrewise trace map $H^1(X_t, \omega_{X_t}) \xrightarrow{\sim} k$ glues to an isomorphism $R^1 \pi_* \omega_\pi \xrightarrow{\sim} \mathcal{O}_{\mathbb{P}^1}$, but $\omega_\pi \simeq \omega_X \otimes \pi^* \omega_{\mathbb{P}^1}^* \simeq \pi^* \mathcal{O}(2)$. This approach is applicable to general elliptic surfaces.

Corollary 1.7 *Either a fibre X_t of an elliptic fibration $X \longrightarrow \mathbb{P}^1$ is irreducible (and in fact integral) or a curve of the form $X_t = \sum_{i=1}^{\ell} m_i C_i$ with (-2)-curves $C_i \simeq \mathbb{P}^1$ and $(m_1, \ldots, m_\ell) = (1)$.*
Moreover, if $\ell > 2$, then $(C_i.C_j) \leq 1$ for all i,j, whereas $(C_1.C_2) = 2$ for $\ell = 2$.

Proof The corollary can also be deduced directly from Section **2.1.4**. Indeed, if $X_t = C + C'$, then $0 \longrightarrow \mathcal{O}(C) \longrightarrow \mathcal{O}(X_t) \longrightarrow \mathcal{O}(X_t)|_{C'} \longrightarrow 0$ shows $h^0(X, \mathcal{O}(C)) = 1$.

For the last part observe that $(C_i + C_j)^2 < 0$ for $\ell > 2$, which yields the result. For $\ell = 2$, the equation $(m_1 C_1 + m_2 C_2)^2 = 0$ has up to scaling only one solution: $m_1 = m_2 = 1$ and $(C_1.C_2) = 2$. $\qquad\qquad\square$

One can similarly show that for $\ell > 3$ there are no cycles of length three as then $(C_1 + C_2 + C_3)^2 < 0$. But this can also be read off from the description of the dual graphs to be discussed next.

1.3 The following table lists all possible fibres of an elliptic fibration of a K3 surface. See Theorem 1.9 for the precise statement.

I_0	smooth elliptic		$\tilde{A}_0$	$e = 0$
I_1	rational curve with DP		$\tilde{A}_0$	$e = 1$
II	rational curve with cusp		$\tilde{A}_0$	$e = 2$
III			$\tilde{A}_1$	$e = 3$
I_2			$\tilde{A}_1$	$e = 2$
IV			$\tilde{A}_2$	$e = 4$
$I_{n \geq 3}$			$\tilde{A}_{n-1}$	$e = n$
$I^*_{n \geq 0}$			$\tilde{D}_{n+4}$	$e = n + 6$
II^*			$\tilde{E}_8$	$e = 10$
III^*			$\tilde{E}_7$	$e = 9$
IV^*			$\tilde{E}_6$	$e = 8$

The first step in the classification of the singular fibres is pure lattice theory and describes the dual graph of the singular fibres. Recall that the vertices of the *dual graph* of a curve $\sum m_i C_i$ correspond to the irreducible components C_i and two vertices C_i, C_j are connected by $(C_i.C_j)$ edges. Note that the dual graph a priori does not take into account the multiplicities m_i.

For the following recall the definition of extended Dynkin diagrams and, in particular, that a graph of type $\tilde{A}_n$, $\tilde{D}_n$ and $\tilde{E}_n$ has $n+1$ vertices. See Section **14.0.3** and the table above.

Corollary 1.8 *Let $X_t = \sum m_i C_i$ be a fibre of an elliptic fibration $\pi: X \longrightarrow \mathbb{P}^1$ of a K3 surface. Then the dual graph of X_t is one of the extended Dynkin diagrams $\tilde{A}_n$, $\tilde{D}_n$, $\tilde{E}_6$, $\tilde{E}_7$, or $\tilde{E}_8$.*

Proof This is a consequence of a general classification result; see [33, I.2] for the precise statement and references, [414, I.6], or Section **14.0.3**. It can be applied once the fibre X_t is known to be connected and two components C_i, C_j intersect in at most one point (and there transversally). The case of two irreducible components, i.e. $X_t = C_1 + C_2$ with $(C_1.C_2) = 2$, has to be dealt with separately. $\square$

The dual graph of X_t does not always determine the isomorphism type of the curve X_t. But this happens only for fibres with up to five ireducible components. Those can be described explicitly.

- *Irreducible fibres* X_t are either smooth elliptic or rational curves with one singular point which can be an ordinary double point or a cusp. This follows immediately from $p_a(X_t) = 1$. Note that the dual graph is in all three cases $\tilde{A}_0$ (although this notation is not commonly used). The three cases are called I_0, I_1, resp. II.

- *Two irreducible components*: If X_t has two irreducible components, then by Corollary 1.7 one knows $X_t = C_1 + C_2$ with C_1 and C_2 meeting either in one point with multiplicity two (type III) or in two points with multiplicity one (type I_2). The two types are not distinguished by their dual graph, which is $\tilde{A}_1$ in both cases.

- *Three irreducible components*: If X_t has three irreducible components, then $X_t = C_1 + C_2 + C_3$ and any two of the curves C_i meet transversally in one point. Moreover, either all three curves meet in the same point (type IV) or the intersection points are all different (type I_3). Note that again the dual graph, which is $\tilde{A}_2$, cannot distinguish between the two types.

- *Five irreducible components*: For type I_0^* the isomorphism type is defined by the choice of the four points on the central component. The graph for this one-parameter family of choices is always $\tilde{D}_4$.

In particular, if $X_t = \sum m_i C_i$ is a singular fibre, then any two of the curves C_i meet transversally in at most one point and no three meet in a single point with the exception of the fibres of type I_2, III, and IV.

1.4 The next result was proved by Kodaira for complex (also non-projective) surfaces in [307, Thm. 6.2] and was later confirmed by work of Néron and Tate in positive characteristic; see [447, 593].

Theorem 1.9 *The isomorphism types of the singular fibres $X_t = \sum m_i C_i$ of an elliptic K3 surface $\pi : X \longrightarrow \mathbb{P}^1$ are classified by the table above. The vertices are labelled by the coefficients m_i and the last column gives the topological Euler number.*

Proof It remains to determine the multiplicities m_i of the components of a singular fibre $X_t = \sum m_i C_i$. Since the radical of the intersection form on $\bigoplus \mathbb{Z}[C_i]$ is spanned by $\sum m_i[C_i]$ and no fibre is multiple, it suffices to prove that with the multiplicities m_i as given in the theorem one indeed has $(\sum m_i C_i)^2 = 0$. This is straightforward to check. □

Remark 1.10 Observe that if one removes a vertex of multiplicity one in one of the extended Dynkin graphs, one obtains a usual Dynkin graph of type A_n, D_n, or E_n, respectively. Also, using Zariski's lemma and the known bounds for the Picard number $\rho(X) \leq 20$ and ≤ 22 (depending on the characteristic), one can bound n in the fibres of type I_n and I_n^*.

Example 1.11 (i) Consider the natural elliptic fibration $\pi : X \longrightarrow E_1/\pm \simeq \mathbb{P}^1$ of the Kummer surface X associated with the product of two elliptic curves $E_1 \times E_2$. The only singular fibres are over images of the four two-torsion points. The fibre of $(E_1 \times E_2)/\pm \longrightarrow E_1/\pm$ over such a point is $E_2/\pm \simeq \mathbb{P}^1$ and contains four of the 16 singular points of $(E_1 \times E_2)/\pm$, which have to be blown up when passing to X. Hence, $X \longrightarrow \mathbb{P}^1$ has four singular fibres; all four are of type I_0^*. Shioda and Inose [568, Thm. 1] describe elliptic fibrations on X with other types of elliptic fibres, e.g. II^* and IV^*.

(ii) Let us come back to Example 1.2, (ii), and consider a pencil of cubics in $\mathbb{P}^2$ spanned by a smooth cubic C and another cubic D. The isomorphism type of the fibres X_t of the elliptic fibration $\widetilde{\mathbb{P}}^2 \longrightarrow \mathbb{P}^1$ (and then also of $X \longrightarrow \mathbb{P}^1$) corresponding to D depends on the singularities of D and how D and C intersect. Suppose D intersects C transversally (in nine points), then for D integral the fibre F is of type I_0, I_1, or II. If D is allowed to have two or three components, one can realize I_2 and III or I_3 and IV, respectively.

For the Hesse pencil $x_0^3 + x_1^3 + x_2^3 - 3\lambda \cdot x_0 x_1 x_2$ the fibration $\widetilde{\mathbb{P}}^2 \longrightarrow \mathbb{P}^1$ has four singular fibres; in particular for $\lambda = \infty$ the fibre is easily seen to be the union of three lines. The same holds for $\lambda = 1, \zeta_3, \zeta_3^2$. All other fibres are smooth. Hence, $X \longrightarrow \mathbb{P}^1$ is an elliptic K3 surface with eight singular fibres, all of type I_3.

Remark 1.12 The possible configurations of singular fibres is restricted by the global topology of the K3 surface. Let us demonstrate this for K3 surfaces over $\mathbb{C}$ in terms of the classical topology. Using the additivity and multiplicativity of the topological Euler number one finds

$$24 = e(X) = \sum e(X_t), \tag{1.1}$$

where the sum is over all singular fibres (in fact $e(X_t) = 0$ for smooth X_t). The Euler number $e(X_t)$ depends only on the type I_1, II, etc., which is recorded in the above table. For example, if X_t is of type III*, then $e(X_t) = e(\text{III}^*) = 9$. In particular, there can be at most two fibres of type III*.

The generic elliptic K3 surface has only singular fibres of type I_1 (see [186, I.1.4]) and exactly 24 of those: $24 = 24 \cdot e(I_1)$. In Example 1.11, (i), we found $24 = 4 \cdot e(I_0^*)$.

In [416] Miranda and Persson study elliptic K3 surfaces (with a section) with fibres of type I_{n_i} only. (These fibres are the only *semistable fibres* and so elliptic K3 surfaces of this type are called semistable.) Now, combine the Shioda–Tate formula (see Corollary 3.4), with Equation (1.1) which in this case reads $\sum n_i = 24$. One finds that there are at least six singular fibres. It turns out that there are 1242 tuples $(n_1, \ldots, n_s)$, $s \geq 6$, with $\sum n_i = 24$ and all except for 135 can indeed be realized. See also [414, X.4].

2 Weierstrass Equation

Every elliptic surface with a section can be reconstructed from its Weierstrass model. We present the beautiful theory for K3 surfaces only. For details and the general situation the reader is referred to [282, 413].

2.1 Let us briefly recall some of the standard facts on elliptic curves. An elliptic curve is by definition a smooth connected projective curve E of genus one over a (not necessarily algebraically closed) field K with $E(K) \neq \emptyset$ or, equivalently, a smooth projective curve E over K which is isomorphic to a curve in $\mathbb{P}^2_K$ defined by a cubic equation which on the affine chart $z \neq 0$ takes the form

$$y^2 + a_1 xy + a_3 y = x^3 + a_2 x^2 + a_4 x + a_6. \tag{2.1}$$

See e.g. [375, Cor. 7.4.5]. Note that E given by (2.1) has a distinguished K-rational point $[0 : 1 : 0]$ which is taken as the origin. Conversely, for any $p \in E(K)$ the line bundle $\mathcal{O}(3p)$ is very ample and defines an embedding $E \hookrightarrow \mathbb{P}^2_K$.

If $\text{char}(K) \neq 2, 3$, which is assumed throughout, a linear coordinate change transforms (2.1) into the *Weierstrass equation*

$$y^2 = x^3 + a_4 x + a_6 \tag{2.2}$$

or more commonly into

$$y^2 = 4x^3 - g_2 x - g_3. \tag{2.3}$$

The pair (g_2, g_3) is well defined up to the action of (λ^4, λ^6), $\lambda \in K^*$. The *discriminant* is in this notation defined as

$$\Delta(E) := -16(4a_4^3 + 27a_6^2) = g_2^3 - 27g_3^2.$$

It is independent up to scaling by 12th powers in K. An equation of the form (2.2) defines a smooth curve if and only if $\Delta(E) \neq 0$. For example, $y^2 = x^3$ and $y^2 = x^3 - 3x + 2$ define rational curves with one cusp and one ordinary double point, respectively.

The *j-invariant* of E is defined as

$$j(E) = -1728 \cdot \frac{(4a_4)^3}{\Delta(E)} = 1728 \cdot \frac{g_2^3}{\Delta(E)},$$

which indeed depends only on E and not on the particular Weierstrass equation. Conversely, if K is algebraically closed or, more precisely, if $(a_4/a_4')^{1/4}, (a_6/a_6')^{1/6} \in K$, then $j(E) = j(E')$ implies $E \simeq E'$; see [569, III.1].

Consider a field K with a discrete valuation v. Then (2.1) for an elliptic curve E over K is called *minimal* if $v(a_i) \geq 0$ for all i and $v(\Delta(E))$ is minimal. Thus, if $v(a_i) \geq 0$ and $v(\Delta(E)) < 12$ (or $v(g_2) < 4$ or $v(g_3) < 6$), then the equation is minimal. See [569, VII.1] for a converse. For $K = k(t)$ and an equation $y^2 = 4x^3 - g_2 x - g_3$ these assumptions say that $g_2, g_3 \in k[t]$ and $\Delta(E) \in k[t]$ has order of vanishing < 12 (or the analogous condition for g_2 or g_3). The minimality condition eventually translates into the Weierstrass fibration associated with an elliptic K3 surface having only simple singularities; see below.

2.2 Consider an elliptic K3 surface $\pi \colon X \longrightarrow \mathbb{P}^1$. A *section* of π is a curve $C_0 \subset X$ with $\pi|_{C_0} \colon C_0 \xrightarrow{\sim} \mathbb{P}^1$. Thus, a section meets every fibre transversally. In particular, it meets a singular fibre $X_t = \sum m_i C_i$ in exactly one of the irreducible components C_{i_0} (and in a smooth point!) for which, moreover, $m_{i_0} = 1$. Note that then by Kodaira's classification of singular fibres the reduced curve $\sum_{i \neq i_0} C_i$ is an ADE curve (i.e. its dual graph is of ADE type) which therefore can be contracted to a simple surface singularity.[2]

Let us assume that a section of π exists and let us fix one $C_0 \subset X$. The exact sequence $0 \longrightarrow \mathcal{O}_X \longrightarrow \mathcal{O}(C_0) \longrightarrow \mathcal{O}_{C_0}(-2) \longrightarrow 0$ induces the long exact sequence

$$0 \longrightarrow \mathcal{O}_{\mathbb{P}^1} \longrightarrow \pi_* \mathcal{O}(C_0) \longrightarrow \mathcal{O}_{\mathbb{P}^1}(-2) \longrightarrow R^1 \pi_* \mathcal{O}_X \longrightarrow 0,$$

where the vanishing $R^1 \pi_* \mathcal{O}(C_0) = 0$ is deduced from the corresponding vanishing on the fibres. Note that $\pi_* \mathcal{O}(C_0)$ is a line bundle, as $h^0(X_t, \mathcal{O}(p)) = 1$ for any point $p \in X_t$ in an arbitrary fibre X_t. (It is in fact enough to test smooth fibres, as $\pi_* \mathcal{O}(C_0)$ is clearly torsion free.) Thus, the cokernel of $\mathcal{O}_{\mathbb{P}^1} \longrightarrow \pi_* \mathcal{O}(C_0)$ is a torsion sheaf, but also contained in the torsion-free $\mathcal{O}_{\mathbb{P}^1}(-2)$. Hence,

---

[2] Also called rational double point, du Val, canonical, or Kleinian singularity; cf. Section **14**.0.3.

$$\mathcal{O}_{\mathbb{P}^1} \simeq \pi_* \mathcal{O}(C_0) \quad \text{and} \quad \mathcal{O}_{\mathbb{P}^1}(-2) \simeq R^1 \pi_* \mathcal{O}_X$$

(the latter confirming (i) in Proposition 1.6). Similarly, using the short exact sequences

$$0 \longrightarrow \mathcal{O}(C_0) \longrightarrow \mathcal{O}(2C_0) \longrightarrow \mathcal{O}_{C_0}(-4) \longrightarrow 0$$

and

$$0 \longrightarrow \mathcal{O}(2C_0) \longrightarrow \mathcal{O}(3C_0) \longrightarrow \mathcal{O}_{C_0}(-6) \longrightarrow 0$$

one proves

$$\pi_* \mathcal{O}(2C_0) \simeq \mathcal{O}_{\mathbb{P}^1}(-4) \oplus \mathcal{O}_{\mathbb{P}^1} \text{ and } \pi_* \mathcal{O}(3C_0) \simeq \mathcal{O}_{\mathbb{P}^1}(-4) \oplus \mathcal{O}_{\mathbb{P}^1}(-6) \oplus \mathcal{O}_{\mathbb{P}^1} =: F.$$

Thus, the linear system $\mathcal{O}(3C_0)|_{X_t}$ on the fibres X_t (or rather the natural surjection $\pi^* F = \pi^* \pi_* \mathcal{O}(3C_0) \longrightarrow\!\!\!\!\!\!\rightarrow \mathcal{O}(3C_0)$) defines a morphism

$$\varphi \colon X \longrightarrow \mathbb{P}(F^*)$$

with $\varphi^* \mathcal{O}(1) \simeq \mathcal{O}_p(3C_0)$, which is a closed embedding of the smooth fibres and contracts all components of singular fibres X_t that are not met by C_0. (It is not difficult to see that $\mathcal{O}(3C_0)$ is indeed base point free on all fibres.) The image $\overline{X}$ is the *Weierstrass model* of the elliptic surface X:

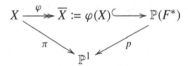

The surface $\overline{X}$ has at most simple singularities and is indeed just the contraction of all ADE curves obtained by removing the unique component in every singular fibre met by the section C_0. The fibres of $\overline{X} \longrightarrow \mathbb{P}^1$ are either smooth elliptic or irreducible rational with one cusp or one ordinary double point. Fibrations of this type with a section are called *Weierstrass fibrations*.

In order to determine $\mathcal{O}(\overline{X})$ of the hypersurface $\overline{X} \subset \mathbb{P}(F^*)$, use the adjunction formula $\mathcal{O}_{\overline{X}} \simeq \omega_{\overline{X}} \simeq (\omega_{\mathbb{P}(F^*)} \otimes \mathcal{O}(\overline{X}))|_{\overline{X}}$ and the relative Euler sequence expressing $\omega_{\mathbb{P}(F^*)}$ to show $\mathcal{O}(\overline{X}) \simeq \mathcal{O}_p(3) \otimes p^* \mathcal{O}_{\mathbb{P}^1}(12)$. Thus, $\overline{X} \subset \mathbb{P}(F^*)$ is described by one equation, which can be viewed as a section:

$$f \in H^0(\mathbb{P}(F^*), \mathcal{O}_p(3) \otimes p^* \mathcal{O}_{\mathbb{P}^1}(12)).$$

Now use $H^0(\mathbb{P}(F^*), \mathcal{O}_p(3) \otimes p^* \mathcal{O}_{\mathbb{P}^1}(12)) \simeq H^0(\mathbb{P}^1, S^3(F) \otimes \mathcal{O}_{\mathbb{P}^1}(12))$ and view x, y, and z as the local coordinates of the direct summands $\mathcal{O}_{\mathbb{P}^1}(-4)$, $\mathcal{O}_{\mathbb{P}^1}(-6)$, and $\mathcal{O}_{\mathbb{P}^1}$ of F.

In this sense, a Weierstrass equation (2.3) $y^2 z = 4x^3 - g_2 x z^2 - g_3 z^3$ with coefficients

$$g_2 \in H^0(\mathbb{P}^1, \mathcal{O}_{\mathbb{P}^1}(8)) \quad \text{and} \quad g_3 \in H^0(\mathbb{P}^1, \mathcal{O}_{\mathbb{P}^1}(12)) \tag{2.4}$$

can be seen as a section of $p_*\mathcal{O}_p(3) \otimes \mathcal{O}_{\mathbb{P}^1}(12)$. For example, g_2 is here interpreted as a section of $\mathcal{O}_{\mathbb{P}^1}(8) = [xz^2]\mathcal{O}(-4)_{\mathbb{P}^1} \otimes \mathcal{O}_{\mathbb{P}^1}(12)$. The discriminant in this situation is the non-trivial section

$$\Delta := g_2^3 - 27g_3^2 \in H^0(\mathbb{P}^1, \mathcal{O}_{\mathbb{P}^1}(24)).$$

Applying the standard coordinate changes, one can always reduce to the situation that f has this form and (g_2, g_3) is unique up to passing to $(\lambda^4 g_2, \lambda^6 g_3)$, $\lambda \in k^*$. Moreover, since $\overline{X}$ has at worst simple singularities, there is no point of $\mathbb{P}^1$ in which g_2 vanishes of order ≥ 4 and g_3 vanishes of order ≥ 6. See e.g. [414, III.3] or [282] for details. The proof relies on a detailed analysis of singularities of the double cover in Remark 2.2 below.

Remark 2.1 The construction can be reversed. Suppose g_2 and g_3 as in (2.4) with $\Delta \neq 0$ are given such that for the vanishing order one has $\min\{3\nu(g_2), 2\nu(g_3)\} < 12$ in every point of $\mathbb{P}^1$. Then the Weierstrass equation (2.3) defines a surface $\overline{X} \subset \mathbb{P}(F^*)$ with at worst simple singularities. Its minimal desingularization $X \longrightarrow \overline{X}$ is a K3 surface which admits an elliptic fibration given by $\pi: X \longrightarrow \overline{X} \longrightarrow \mathbb{P}^1$. It comes with a section C_0 obtained by the intersection with $\mathbb{P}(\mathcal{O}_{\mathbb{P}^1}(6)) \subset \mathbb{P}(F^*)$. See [580, II.4] for more details.

Remark 2.2 Similarly one studies the surjection $\pi^*\pi_*\mathcal{O}(2C_0) \twoheadrightarrow \mathcal{O}(2C_0)$ which gives rise to a cover $X \longrightarrow \mathbb{F}_4 = \mathbb{P}(\mathcal{O}_{\mathbb{P}^1}(4) \oplus \mathcal{O}_{\mathbb{P}^1})$ of degree two. See [414, III.2] for the description of the branching curve.

It is sometimes convenient to describe an elliptic K3 surface not in terms of its Weierstrass equation but, for example, by an equation of the form $y^2 = x(x^2 + ax + b)$. In this case $a \in H^0(\mathbb{P}^1, \mathcal{O}(4))$ and $b \in H^0(\mathbb{P}^1, \mathcal{O}(8))$.

2.3 To relate the discussions in the previous two sections, let us now consider the generic fibre of an elliptic fibration $\pi: X \longrightarrow \mathbb{P}^1$:

$$E := X_\eta = \pi^{-1}(\eta).$$

Then E is a smooth curve of genus one over $K := k(\eta) \simeq k(t)$. Moreover, π admits a section if and only if $E(K) \neq \emptyset$. More precisely, $E(K)$ is naturally identified with the set of sections of π; cf. Section 3.2 below.

Suppose a section C_0 of π is fixed and $f \in H^0(\mathbb{P}(F^*), \mathcal{O}_p(3) \otimes p^*\mathcal{O}_{\mathbb{P}^1}(12))$ is an induced Weierstrass equation. In order to use f to describe the generic fibre E one needs first to fix a point in $\mathbb{P}^1$ or rather a section $0 \neq s \in H^0(\mathbb{P}^1, \mathcal{O}_{\mathbb{P}^1}(1))$ defining a local parameter at this point. Then $s^{-8}g_2$, $s^{-12}g_3$, and $s^{-24}\Delta$ are contained in the function field K. Moreover, the Weierstrass equation for E obtained in this way is minimal for the valuation of K given by the chosen point in $\mathbb{P}^1$. Note that $j(E)$ can be directly read off from $g_2 \in H^0(\mathbb{P}^1, \mathcal{O}_{\mathbb{P}^1}(8))$ and $\Delta \in H^0(\mathbb{P}^1, \mathcal{O}_{\mathbb{P}^1}(24))$ by interpreting g_2^3/Δ as a rational function.

Typically the isomorphism type of the smooth fibres of an elliptic K3 surface $X \longrightarrow \mathbb{P}^1$ varies. Whether this is indeed the case is encoded by the j-invariant. The smooth fibres of π are all isomorphic if and only if $j(E) \in k$. In this case, $\pi \colon X \longrightarrow \mathbb{P}^1$ is called *isotrivial*.

Example 2.3 (i) The Kummer surface associated with the product of two elliptic curves $E_1 \times E_2$ comes with a natural elliptic fibration $\pi \colon X \longrightarrow \mathbb{P}^1 = E_1/\pm$; see Example 1.11, (i). Clearly, this family is isotrivial, as all smooth fibres are isomorphic to E_2. In particular, the j-invariant of the generic fibre E over $K = k(\eta) \simeq k(t)$ is $j(E) = j(E_2) \in k$, and so the two elliptic curves E and $E_2 \times_k K$ over K have the same j-invariant but become isomorphic only after passage to an appropriate extension of K, namely, $k(E_1)/K$.

After a coordinate change on $\mathbb{P}^1$, one can assume that $E_1 \longrightarrow E_1/\pm \simeq \mathbb{P}^1$ ramifies over $0, 1, \infty$, and λ. Up to scaling, the discriminant function is the polynomial $\Delta = (z_0 z_1 (z_0 - z_1)(z_0 - \lambda z_1))^6$.

(ii) Consider an elliptic K3 surface associated with the Hesse pencil $x_0^3 + x_1^3 + x_2^3 - 3\lambda \cdot x_0 x_1 x_2$ as in Example 1.2, (ii). Then depending on the quadratic base change $\mathbb{P}^1 \longrightarrow \mathbb{P}^1$, $t = [t_0 : t_1] \longmapsto [\lambda(t) : 1]$ the j-invariant is given by

$$j(t) = 27 \left(\frac{\lambda(t)(\lambda(t)^3 + 8)}{\lambda(t)^3 - 1} \right)^3.$$

Here, $[\lambda(t){:}1]$ has to be chosen such that ramification does not take place over $\infty, 1, \zeta_3$, or ζ_3^2. Computing the Weierstrass equation for this example is actually a little cumbersome, but knowing that there are eight fibres of type I_3 (see Example 1.11, (ii)) is enough to write Δ (depending on the quadratic base change). In any case, the Hesse pencil induces an elliptic K3 surface that is not isotrivial.

2.4 Since $\Delta \in H^0(\mathbb{P}^1, \mathcal{O}_{\mathbb{P}^1}(24))$, an elliptic fibration $\pi \colon X \longrightarrow \mathbb{P}^1$ of a K3 surface X has 24 singular fibres if counted properly. For example, for the Kummer surface associated with $E_1 \times E_2$ we have seen that there are only four singular fibres, which are all of type I_0^*, but each one has to be counted with multiplicity six. This is a general phenomenon. If X_t is a singular fibre of type I_0^*, then Δ vanishes of order six at t. More generally, the order of vanishing δ of Δ at t is determined by the type of X_t:

	I_0	I_1	II	III	IV	$I_{n \geq 2}$	I_n^*	II*	III*	IV*
δ	0	1	2	3	4	n	$n + 6$	10	9	8

See [414, IV.3] for proofs. One obviously has

$$\sum_{t \in \mathbb{P}^1} \delta_t = 24, \tag{2.5}$$

which should be seen as the analogue of (1.1). In fact, comparing the above table with the one in Theorem 1.9 reveals that

$$e(X_t) = \delta(t).$$

This can be seen as a consequence of the local versions of the formulae $\chi(\mathcal{O}_X) = -\deg(R^1\pi_*\mathcal{O}_X)$ and $e(X) = 12 \cdot \chi(\mathcal{O}_X) = \deg \mathcal{O}(24)$.

It turns out that the type of fibre X_{t_0} together with the value $j(t_0)$ of the j-function and the multiplicity with which it is attained determine the germ of the family $X \longrightarrow \mathbb{P}^1$ near X_{t_0}; see [414, VI.1], where one also finds a description of the monodromy groups around these fibres.

2.5 Let $\pi : X \longrightarrow \mathbb{P}^1$ be an elliptic fibration with a section $C_0 \subset X$. Then all smooth fibres X_t come with a distinguished point and hence a group structure. This still holds for the smooth part of a singular fibre. More precisely, if $X_t = \sum m_i C_i$ is an arbitrary fibre and $X_t' \subset X_t$ denotes the open set of smooth points, then X_t' has a natural group structure. Note that all components C_i with $m_i > 1$ have to be completely discarded when passing to X_t'. The idea for the construction is to use local (complex, étale, etc., depending on the situation) sections through smooth points x, y of a fibre X_t and add the two sections on the generic smooth fibre. The resulting section then intersects X_t in a point $x + y$, which has to be smooth again. If $X' \subset X$ is the open set of π-smooth points, then the fibres of $X' \longrightarrow \mathbb{P}^1$ are the X_t' and the construction gives rise to a group scheme structure on X' over $\mathbb{P}^1$.

For example, for a fibre of type I_1, i.e. a nodal rational curve, or a fibre of type II, i.e. a rational curve with a cusp, one finds the multiplicative group $\mathbb{G}_m$ and the additive group $\mathbb{G}_a$, respectively. It coincides with the classical group structure on the smooth points of a singular cubic in $\mathbb{P}^2$.

Note that the fibre X_t' might be disconnected. Its connected component $(X_t')^o$, containing the intersection with C_0 and the group of components $X_t'/(X_t')^o$, can be read off from Kodaira's table in Theorem 1.9.

Corollary 2.4 *For the smooth fibres of the π-smooth part $X' \longrightarrow \mathbb{P}^1$ one of the following holds:*

(i) *$(X_t')^o = X_t$ is a smooth elliptic curve for type I_0 or*
(ii) *$(X_t)^o \simeq \mathbb{G}_m$ and $X_t'/(X_t')^o \simeq \mathbb{Z}/n\mathbb{Z}$ for type $I_{n \geq 1}$ or*
(iii) *$(X_t)^o \simeq \mathbb{G}_a$ for all other types with $X_t'/(X_t')^o \simeq \mathbb{Z}/2\mathbb{Z}$ for type III and III*, $X_t'/(X_t')^o \simeq (\mathbb{Z}/2\mathbb{Z})^2$ for type I_n^*, $X_t'/(X_t')^o \simeq \mathbb{Z}/3\mathbb{Z}$ for type IV and IV*, and $X_t'/(X_t')^o \simeq \{1\}$ for type II and II*.* $\qquad\square$

Remark 2.5 (i) One could similarly treat the Weierstrass model $\overline{X} \longrightarrow \mathbb{P}^1$. The fibres of the resulting smooth fibration $\overline{X}' \longrightarrow \mathbb{P}^1$ are either smooth elliptic or of isomorphism type $\mathbb{G}_m$ or $\mathbb{G}_a$. In fact, $\overline{X}'_t \simeq (X'_t)^o$.

(ii) The π-smooth part $X' \longrightarrow \mathbb{P}^1$ of the elliptic K3 surface $\pi\colon X \longrightarrow \mathbb{P}^1$ is the *Néron model* of the generic fibre $E = X_\eta$ and has a distinguished universal property: any rational map $Y \longrightarrow X'$ over $\mathbb{P}^1$ of a scheme Y smooth over $\mathbb{P}^1$ extends to a morphism; see e.g. [18].

3 Mordell–Weil Group

For this section we recommend the articles by Cox [134] and Shioda [565]; see also Shioda's survey article [566]. Throughout we consider an elliptic K3 surface $\pi\colon X \longrightarrow \mathbb{P}^1$ with a section $C_0 \subset X$. In particular, $C_0 \simeq \mathbb{P}^1$ and $(C_0)^2 = -2$. The generic fibre of π is denoted as before,

$$E := X_\eta = \pi^{-1}(\eta).$$

It is an elliptic curve over the function field $K = k(\eta) \simeq k(t)$ of $\mathbb{P}^1$. The origin $o_E \in E(K)$ is chosen to be the point of intersection with C_0.

Recall that the set of sections of $\pi\colon X \longrightarrow \mathbb{P}^1$ is naturally identified with the set $E(K)$ of K-rational points of E by mapping a section C to its intersection with E. Conversely, the closure $\overline{p} \subset X$ of a point $p \in E(K)$ defines a section of π, for example, $\overline{o}_E = C_0$:

$$E(K) \longleftrightarrow \{ C \hookrightarrow X \mid \pi|_C \colon C \xrightarrow{\sim} \mathbb{P}^1 \}.$$

3.1 A divisor $D = \sum n_i C_i$ on X is called *vertical* if all components are supported on some fibres; i.e. the images $\pi(C_i)$ are closed points. In particular, if D is vertical, then $(D.X_t) = 0$. An irreducible curve C is *horizontal* if $\pi|_C \colon C \longrightarrow \mathbb{P}^1$ is surjective, e.g. any section is horizontal.

Restriction to the generic fibre yields a group homomorphism

$$\mathrm{Div}(X) \longrightarrow \mathrm{Div}(E),$$

the kernel of which consists of vertical divisors. As the function fields of X and E coincide, one also obtains a homomorphism

$$\mathrm{Pic}(X) \longrightarrow \mathrm{Pic}(E),$$

the kernel of which is the subgroup of all line bundles linearly equivalent to vertical divisors.

For an arbitrary $L \in \mathrm{Pic}(X)$, let $d_L := (L.X_t)$ be its *fibre degree*. Then the line bundle $L|_E \otimes \mathcal{O}(-d_L \cdot o_E)$ is a degree zero line bundle on the elliptic curve E and thus, by Abel's theorem, isomorphic to $\mathcal{O}(p_L - o_E)$ for a unique point $p_L \in E(K)$. For example, if $L = \mathcal{O}(C_0)$, then $p_L = o_E$. More generally, if $L = \mathcal{O}(C)$ for a section $C \subset X$ of $\pi : X \longrightarrow \mathbb{P}^1$, then p_L is the point of intersection of C and E.

For arbitrary L we shall denote by $\overline{p}_L \subset X$ the section corresponding to the point $p_L \in E$. Then

$$L \simeq \mathcal{O}(\overline{p}_L) \otimes \mathcal{O}((d_L - 1)C_0) \otimes \mathcal{O}(nX_t) \otimes \mathcal{O}\left(\sum n_i C_i\right), \qquad (3.1)$$

where $n = (L.C_0) + 2(d_L - 1) - (\overline{p}_L.C_0)$ and C_i are certain irreducible fibre components not met by the section C_0.

3.2 The set of sections of π is endowed with the group structure of $E(K)$ (with the origin o_E). This gives rise to the following definition.

Definition 3.1 The *Mordell–Weil group* $\mathrm{MW}(X)$ of an elliptic K3 surface $\pi : X \longrightarrow \mathbb{P}^1$ is the set of sections or, alternatively,

$$\mathrm{MW}(X) \simeq E(K).$$

Although not reflected by the notation, $\mathrm{MW}(X)$ depends of course on the elliptic fibration $\pi : X \longrightarrow \mathbb{P}^1$.

With this definition, the map $L \longmapsto p_L$ is easily seen to define a group homomorphism

$$\mathrm{NS}(X) \simeq \mathrm{Pic}(X) \longrightarrow \mathrm{MW}(X),$$

which is surjective, because $p_L = p$ for $L = \mathcal{O}(\overline{p})$.

The following result holds more generally, e.g. for minimal regular elliptic surfaces, and is an immediate consequence of the above discussion.

Proposition 3.2 *There exists a short exact sequence*

$$0 \longrightarrow A \longrightarrow \mathrm{NS}(X) \longrightarrow \mathrm{MW}(X) \longrightarrow 0,$$

where A is the subgroup generated by vertical divisors and the section C_0. In particular, the Mordell–Weil group is a finitely generated abelian group. □

Remark 3.3 That the Mordell–Weil group is finitely generated is reminiscent of the Mordell–Weil theorem asserting that $E(K)$ is a finitely generated group for any number (or, more generally, finitely generated) field K; see [175, Ch. VI]. But the result also shows that the elliptic curves E over K, e.g. for $K = \mathbb{C}(t)$, appearing as generic fibres of an elliptic K3 (or regular elliptic) surface are rather special.

Corollary 3.4 (Shioda–Tate) *Let $\pi : X \longrightarrow \mathbb{P}^1$ be an elliptic K3 surface with a section $C_0 \subset X$ and let r_t denote the number of irreducible components of a fibre X_t. Then*

$$\rho(X) = 2 + \sum_t (r_t - 1) + \mathrm{rk}\,\mathrm{MW}(X). \qquad (3.2)$$

Proof It suffices to prove $\operatorname{rk} A = 2 + \sum_t (r_t - 1)$, which is a consequence of Zariski's lemma; see Proposition 1.6. Indeed, the intersection form restricted to the part generated by the components of the singular fibres not met by the section has rank $\sum_t (r_t - 1)$. The hyperbolic plane generated by the fibre class X_t and the section C_0 is orthogonal to it and, of course, of rank two. $\square$

Example 3.5 If X is the Kummer surface associated with the product $E_1 \times E_2$ of two elliptic curves viewed with its elliptic fibration $X \longrightarrow \mathbb{P}^1 = E_1/\pm$, then the Shioda–Tate formula yields $\rho(X) = 18 + \operatorname{rk} \operatorname{MW}(X)$. In particular, if E_1 and E_2 are isogenous, then $\operatorname{rk} \operatorname{MW}(X) > 0$ (which can also be verified more explicitly by looking at the section induced by the graph of an isogeny).

Remark 3.6 If $\operatorname{char}(k) = 0$, then $\rho(X) \leq 20$ and the Shioda–Tate formula shows that for an elliptic K3 surface with section fibres of type I_n and I_m^* can occur only for $n \leq 19$ and $m \leq 14$.

That the same bound still holds in positive characteristic $\neq 2, 3$, although one has only $\rho(X) \leq 22$, was shown in by Schütt and Shioda in [535, 567]. There one also finds a classification of elliptic K3 surfaces realizing the maximal singular fibres I_{19} and I_{14}^*.

Remark 3.7 Assume $\operatorname{char}(k) = 0$. An elliptic K3 surface $\pi : X \longrightarrow \mathbb{P}^1$ with a section is called *extremal* if $\operatorname{rk} A = 20$ or, equivalently, if $\rho(X) = 20$ and $\operatorname{MW}(X)$ is a finite group. Extremal elliptic K3 surfaces have been classified in terms of lattices by Shimada and Zhang in [557]. There are 325 cases. For many of them explicit descriptions have been found by Schütt in [536]. Note that every K3 surface with $\rho(X) = 20$ admits a *Shioda–Inose structure* (see Remark **15**.4.1), i.e. a rational map of degree two onto a Kummer surface which in this case is associated with the product of two isogenous elliptic curves $E \times E'$ and hence is itself elliptic (but not extremal). See also [491] where Persson studies the analogy between extremal elliptic K3 surfaces and maximizing double planes; cf. Section **17**.1.4.

In the proof of the Shioda–Tate formula we have seen already that A can be written as a direct orthogonal sum (cf. Example **14**.0.3):

$$A = U \oplus R.$$

The hyperbolic lattice U is spanned by the classes of a fibre X_t and the section C_0 and the orthogonal R is the negative definite lattice spanned by all fibre components not met by C_0. Note that by Kodaira's classification R is a direct sum of lattices of ADE type. For example, for the canonical elliptic fibration of a Kummer surface $X \longrightarrow E_1/\pm$ associated with $E_1 \times E_2$ the lattice R is isomorphic to $D_4^{\oplus 4}$.

Remark 3.8 (i) Specialization yields for every closed fibre X_t a natural map

$$\operatorname{MW}(X) \simeq E(K) \longrightarrow X_t,$$

which in geometric terms is obtained by intersecting a section with X_t. Its image is contained in the smooth part of X_t. For smooth fibres X_t, this is a group homomorphism. Moreover, in characteristic zero the map induces an injection on torsion points

$$\mathrm{MW}(X)_{\mathrm{tors}} \hookrightarrow X_t,$$

or, in other words, two distinct sections whose difference is torsion on the generic fibre never meet. This can be proved by associating with any torsion section C_1 a symplectic automorphism $f_{C_1}: X \xrightarrow{\sim} X$: see Remark **15.4.6**. Alternatively, one can use the following argument which proves that a torsion section C_1 cannot intersect C_0 in any smooth fibre. Locally around a smooth fibre X_{t_0}, a complex elliptic K3 surface $X \longrightarrow \mathbb{P}^1$ can be written as $\mathbb{C}/(\mathbb{Z} + \tau(t)\mathbb{Z}) \times \Delta$. The point of intersection $x(t)$ of C_1 with X_t is contained in $\mathbb{Q} + \tau(t)\mathbb{Q}$ and if C_0 and C_1 intersect in X_{t_0} in the same point, then $x(t_0) = 0$ and hence $x(t) \equiv 0$ locally if C_1 is a torsion section. A similar argument applies for the intersection with singular fibres; see [414, VII.Prop. 3.2].

If $\mathrm{char}(k) = p > 0$, specialization is still injective on the torsion part prime to p, but may fail in general; cf. [476].

(ii) The torsion subgroup can also be controlled via the restriction to the group of connected components of all fibres, which yields an injective group homomorphism

$$\mathrm{MW}(X)_{\mathrm{tors}} \hookrightarrow \prod_t X'_t/(X'_t)^o.$$

Indeed, if C_1 is a torsion section of order k, then $kC_1 \in A = U \oplus R$ can be written as $kC_1 = aC_0 + b[X_t] + \alpha$. Intersecting with $[X_t]$ yields $a = k$, from which one concludes $b = k((C_0.C_1) + 2)$ by further intersecting with C_0. Finally, if C_1 and C_0 intersects the same connected component of each fibre, then $(C_1.\alpha) = 0$. This leads to the contradiction $-2k = 2k((C_0.C_1) + 1)$, unless $C_0 = C_1$. For an alternative proof, see Lemma 3.10.

3.3 The map that sends a point $p \in E(K)$ to the line bundle $\mathcal{O}(\overline{p})$, which was used to prove the surjectivity of $\mathrm{NS}(X) \twoheadrightarrow \mathrm{MW}(X)$, is in general not additive. In fact, the quotient $\mathrm{MW}(X) = \mathrm{NS}(X)/A$ is in general not even torsion free; i.e. $A \subset \mathrm{NS}(X)$ might not be primitive. Since $A \subset \mathrm{NS}(X)$ is a non-degenerate sublattice, the orthogonal direct sum $A \oplus A^\perp \subset \mathrm{NS}(X)$ is of finite index. Moreover, $A^\perp$ is a negative definite lattice with $\mathrm{rk}\, A^\perp = \mathrm{rk}\, \mathrm{MW}(X)$ given by (3.2) and discriminant (see **(14.0.2)**)

$$\mathrm{disc}\, A^\perp = \frac{\mathrm{disc}\, \mathrm{NS}(X)}{\mathrm{disc}\, A} \cdot (\mathrm{NS}(X) : A \oplus A^\perp)^2.$$

The projection $\mathrm{NS}(X) \twoheadrightarrow \mathrm{MW}(X)$ yields a surjection $A_\mathbb{Q} \oplus A^\perp_\mathbb{Q} \simeq \mathrm{NS}(X)_\mathbb{Q} \twoheadrightarrow \mathrm{MW}(X)_\mathbb{Q}$ with kernel $A_\mathbb{Q}$ and, therefore, a natural injection of groups

$$\mathrm{MW}(X)/\mathrm{MW}(X)_{\mathrm{tors}} \hookrightarrow \mathrm{MW}(X)_\mathbb{Q} \xrightarrow{\sim} A^\perp_\mathbb{Q} \subset \mathrm{NS}(X)_\mathbb{Q}.$$

The intersection form $(\ .\ )$ on $NS(X)_{\mathbb{Q}}$ induces a non-degenerate quadratic form on $MW(X)/MW(X)_{\text{tors}}$.

Definition 3.9 The *Mordell–Weil lattice* of an elliptic K3 surface $\pi : X \longrightarrow \mathbb{P}^1$ is the group $MW(X)/MW(X)_{\text{tors}}$ endowed with $\langle\ .\ \rangle := -(\ .\ )$.

Warning: The quadratic form $\langle\ .\ \rangle$ on $MW(X)$ really takes values in $\mathbb{Q}$ and not necessarily in $\mathbb{Z}$.

So calling $MW(X)/MW(X)_{\text{tors}}$ a lattice is slightly abusive. But it can easily be turned into a positive definite lattice in the traditional sense by passing to an appropriate positive multiple of $\langle\ .\ \rangle$ (this has been made explicit by Shioda [565, Lem. 8.3]) or by restricting to a distinguished finite index subgroup. More precisely, let

$$MW(X)^0 \subset MW(X)$$

be the subgroup of sections intersecting every fibre in the same component as the given section C_0. Since by the discussion in Section 2.5 the component of a singular fibre met by C_0 (or rather its smooth part) is a subgroup, this really defines a subgroup.

Lemma 3.10 *The subgroup $MW(X)^0$ is torsion free and of finite index in $MW(X)$. Moreover, $\langle\ .\ \rangle$ restricted to $MW(X)^0$ is integral, even, and positive definite.*

Proof Suppose $C \in MW(X)^0$ is n-torsion. Then $L := \mathcal{O}(n(C - C_0))$ is linearly equivalent to a vertical divisor. As C and C_0 intersect the same component of every fibre, $(L.D) = 0$ for every vertical curve D. But then $(L)^2 = 0$ and by Zariski's lemma, Proposition 1.6, (iv), $L = \mathcal{O}(\ell X_t)$. Now use $(L.C) = \ell = (L.C_0)$ to deduce $n(-2 - (C_0.C)) = n((C.C_0) + 2)$. Thus, if n were non-zero, then $(C.C_0) = -2$, which is absurd unless $C = C_0$.

To prove finite index, observe first that $MW(X)^0$ is contained in the kernel of the map

$$MW(X) \longrightarrow A_R = R^*/R \qquad (3.3)$$

which is induced by $NS(X) \longrightarrow R^*/R$ (restriction of the intersection form). If $C \in MW(X)$ does not meet a singular fibre $X_t = \sum m_i C_i$ in the same component as C_0, say $(C.C_i) = 1$ but $(C_0.C_i) = 0$, then $m_i = 1$ and by using that R is an orthogonal direct sum of lattices of ADE type one proves $(C.\ ) \neq 0$ in R^*/R; see [414, VII.2]. Hence, $MW(X)^0$ is the kernel of (3.3) and, thus, of finite index. Alternatively, use that if $N := \prod |X_t'/(X_t')^0|$, then $N \cdot C \in MW(X)^0$ for all $C \in MW(X)$.

For $C \in MW(X)^0$, one finds

$$C = (C_0 - (2 + (C.C_0))X_t) \oplus \alpha_C \in A \oplus A^\perp \subset NS(X).$$

Thus, the intersection form restricted to $MW(X)^0$ is indeed integral, as for $C, C' \in MW(X)^0$ one has $\langle C.C' \rangle = -(\alpha_C.\alpha_{C'}) \in \mathbb{Z}$ and $\langle C, C \rangle = -(\alpha_C)^2 \in 2\mathbb{Z}$. By the Hodge index theorem, $(\,,\,)$ has signature $(1, \rho(X) - 1)$ on $NS(X)$ and as A contains a hyperbolic plane, it is negative definite on $A^\perp$. □

The bilinear form on $MW(X)^0$ can be written explicitly as

$$\langle C.C' \rangle = 2 + (C.C_0) + (C_0.C') - (C.C').$$

The induced quadratic form

$$MW(X)^0 \longrightarrow \mathbb{Z}, \quad C \longmapsto \langle C.C \rangle = 4 + 2(C.C_0)$$

is (up to a factor two) the restriction of the canonical height (as introduced by Manin [389] and Tate [594])

$$\hat{h}_{E/K} : E(K)/E(K)_{\text{tors}} \longrightarrow \mathbb{Q}.$$

The *regulator* of E over K (or of the elliptic surface X) is by definition

$$R_{E/K} = R_X := \text{disc}\,(MW(X)/MW(X)_{\text{tors}}) = \det \langle\,.\,\rangle \in \mathbb{Q}.$$

A more Hodge theoretic approach to the Mordell–Weil group goes back to Cox and Zucker in [135]. In particular, they observe

$$|\text{disc}\,NS(X)| = \frac{R_X}{|MW(X)_{\text{tors}}|^2} \cdot \prod n_t = \frac{R_X}{|MW(X)_{\text{tors}}|^2} \cdot |\text{disc}\,A|,$$

where n_t is the number of components of the fibre X_t appearing with multiplicity one (which in fact equals $|\text{disc}\,R_t|$ due to [559, Lem. 1.3]). See also [568, Lem. 1.3] for the case $MW(X)_{\text{tors}} = MW(X)$.

The Mordell–Weil lattice is a positive definite lattice of rank

$$0 \le \text{rk}\,MW(X) \le \text{rk}\,NS(X) - 2.$$

Hence, $\text{rk}\,MW(X) \le 18$ in characteristic zero and ≤ 20 in positive characteristic.

Remark 3.11 (i) In characteristic zero, all values $0 \le \text{rk}\,MW(X) \le 18$ are realized. This has been proved by Cox in [134] using surjectivity of the period map. Explicit equations for $\text{rk} \ne 15$ have been given by Kuwata [342] and for $\text{rk} = 15$ by Kloosterman [298] and Top and De Zeeuw [602].

(ii) The torsion group $MW(X)_{\text{tors}}$ is a finite group of the form $\mathbb{Z}/n\mathbb{Z} \times \mathbb{Z}/m\mathbb{Z}$; use e.g. Remark 3.8. As every element in $MW(X)$ defines a symplectic automorphism of X, one knows $n, m \le 8$ in characteristic zero; cf. Remark **15**.4.5. Moreover, as Cox shows in [134, Thm. 2.2] whenever $MW(X)_{\text{tors}} \ne 0$ for a (non-isotrivial) elliptic fibration, then $\text{rk}\,MW(X) \le 10$. More precisely, he gives a complete and finite list for non-trivial $MW(X)_{\text{tors}}$ and for the possible Mordell–Weil ranks in each case. For example, if $MW(X)_{\text{tors}} \simeq \mathbb{Z}/8\mathbb{Z}$, then $\text{rk}\,MW(X) = 0$.

The computation hinges on the observation that $MW(X)_{tors}$ is a subgroup of A_R, which is a direct sum of discriminant groups of lattices of ADE type. See also [414, VII.3].[3]

(iii) The opposite of extremal elliptic K3 surfaces (see Remark 3.7) in characteristic zero are elliptic K3 surfaces $\pi: X \longrightarrow \mathbb{P}^1$ with $\text{rk } MW(X) = 18$. Those have been studied by Nishiyama [459] and Oguiso [471].

(iv) In [461] Nishiyama proves that the minimal $\langle v.v \rangle$, for non-torsion $v \in MW(X)$, for all elliptic K3 surfaces in characteristic zero is $\frac{11}{420}$. The minimum is attained for certain very special K3 surfaces of Picard rank 19 and 20.

4 Jacobian Fibration

The previous section dealt with elliptic K3 surfaces with a section. What about those that do not admit a section? They do exist and here we shall explain how one can pass from an elliptic K3 surface $\pi: X \longrightarrow \mathbb{P}^1$ without a section to one with a section. This is achieved by looking at the relative Jacobian $J(X) \longrightarrow \mathbb{P}^1$. In Section 5 we shall discuss the Tate–Šafarevič group which conversely controls all elliptic K3 surfaces with the same Jacobian.

In Sections 4.1 and 4.2 $\pi: X \longrightarrow \mathbb{P}^1$ is assumed to be an algebraic elliptic K3 surface over an algebraically closed field k. In Section 4.3 we add comments on the non-algebraic case.

4.1 We consider an elliptic K3 surface $\pi: X \longrightarrow \mathbb{P}^1$ and introduce the relative (compactified) Jacobian $J(X) \longrightarrow \mathbb{P}^1$. Jacobians of elliptic surfaces are usually studied via Picard functors and Néron models; see [3, 82, 176]. Let us sketch this first, before later viewing the construction from the point of view of moduli spaces of stable sheaves. The latter approach in particular shows that $J(X)$ is again a K3 surface, without first analysing the fibres of $J(X)$, and allows one to control the period of $J(X)$.

Start with an elliptic K3 surface $\pi: X \longrightarrow \mathbb{P}^1$ and consider its generic fibre $E = X_\eta$ as a smooth genus one curve over the function field $K \simeq k(t)$ of $\mathbb{P}^1$. The Jacobian $\text{Jac}(E)$ is again a smooth genus one curve over K representing the étale sheafification of the functor

$$\text{Pic}_E^0 : (Sch/K)^o \longrightarrow (Ab), \quad S \longmapsto \text{Pic}^0(E \times S)/\sim,$$

where $\text{Pic}^0(E \times S)$ is the group of line bundles which are of degree zero on the fibres of $E \times S \longrightarrow S$ and the equivalence relation $\sim$ is generated by the natural action of

---

[3] Is there a deeper relation between Cox's finite list of possible torsion groups $MW(X)_{tors}$ and Mazur's result that for an elliptic curve over $\mathbb{Q}$ the torsion group $E(\mathbb{Q})_{tors}$ is isomorphic to $\mathbb{Z}/n\mathbb{Z}$, $n = 1, \ldots, 10, 12$ or $\mathbb{Z}/2\mathbb{Z} \times \mathbb{Z}/n\mathbb{Z}$, $n = 1, \ldots, 4$?

Pic(S). The existence of Jac(E) is a fundamental fact, which we take for granted. See [236, Ch. IV.4] for the assertion over an algebraically closed field and [176, Part 5] for a general discussion. In particular, there exists a natural functor transformation $\mathrm{Pic}^0_E \longrightarrow \underline{\mathrm{Jac}}(E)$ which, moreover, yields isomorphisms $\mathrm{Pic}^0_E(S) \xrightarrow{\sim} \mathrm{Mor}_K(S, \mathrm{Jac}(E))$ as soon as $\mathrm{Pic}^0_E(S) \neq \emptyset$. So, for example, $\mathrm{Jac}(E)(\overline{K}) = \mathrm{Pic}^0(E_{\overline{K}})$ and Jac(E) coarsely represents Pic^0_E. See Section **5**.1 for the notion of coarse moduli spaces.

Note that E might not have any K-rational points. However, Jac(E) always has, as $[\mathcal{O}_E] \in \mathrm{Jac}(E)(K)$. So, Jac($E$) is a smooth elliptic curve over K. Also recall that in general a Poincaré bundle on $E \times \mathrm{Jac}(E)$ may exist only after an appropriate base field extension.

The residue field $k(\xi)$ of the generic point $\xi \in \mathrm{Jac}(E)$ is a finitely generated field extension of K of transcendence degree one and thus of transcendence degree two over k. It can therefore be realized as the function field of a surface over k. The natural inclusion $K \hookrightarrow k(\xi)$ corresponds to a dominant rational map of such a surface to $\mathbb{P}^1$. The *Jacobian*

$$\mathrm{J}(X) \longrightarrow \mathbb{P}^1$$

of the original elliptic fibration $X \longrightarrow \mathbb{P}^1$ is then defined as the unique relatively minimal smooth model. In particular, its function field $K(\mathrm{J}(X))$ is just $k(\xi) = K(\xi)$, where ξ is regarded as the generic point of the surface J(X) and, simultaneously, of the elliptic curve Jac(E) over K. The uniqueness is important and is used throughout.

Alternatively, one could first look at the relative Jacobian fibration

$$\mathrm{Jac}(X/\mathbb{P}^1) \longrightarrow \mathbb{P}^1$$

coarsely representing (or representing the étale sheafification of) the functor

$$(Sch/\mathbb{P}^1)^o \longrightarrow (Sets), \quad T \longmapsto \mathrm{Pic}^0(X \times_{\mathbb{P}^1} T)/\sim.$$

Due to the existence of reducible fibres, Jac($X/\mathbb{P}^1$) is in general not separated. The existence as an algebraic space is due to a general result by Artin (see [82, Sec. 8.3]), but presumably, in the situation at hand it is in fact a (non-separated) scheme.

The fibre of Jac($X/\mathbb{P}^1$) $\longrightarrow \mathbb{P}^1$ over a closed point $t \in \mathbb{P}^1$ is Jac(X_t) and over the generic fibre one recovers Jac(E).[4] So, J(X) $\longrightarrow \mathbb{P}^1$ can also be seen as a relatively minimal smooth model of Jac($X/\mathbb{P}^1$) $\longrightarrow \mathbb{P}^1$. Note that in both descriptions one sees

---

[4] As it turns out, $\hat{\pi} \colon \mathrm{Jac}(X/\mathbb{P}^1) \longrightarrow \mathbb{P}^1$ is locally smooth (but not separated). This can be proved by a standard argument from deformation theory. For L a line bundle on X_t and the inclusion $i \colon X_t \hookrightarrow X$ applying $\mathrm{Hom}_{X_t}(\ ,L)$ to the exact triangle $L[1] \longrightarrow i^* i_* L \longrightarrow L$ and using adjunction yields the exact sequence

$$\mathrm{Ext}^1_{X_t}(L,L) \longrightarrow \mathrm{Ext}^1_X(i_*L, i_*L) \longrightarrow \mathrm{Ext}^0_{X_t}(L,L) \longrightarrow \mathrm{Ext}^2_{X_t}(L,L).$$

The morphism in the middle can be interpreted as the differential of $\hat{\pi}$ at $[L]$. However, for the locally free L on the curve X_t, one has $\mathrm{Ext}^2_{X_t}(L,L) \simeq H^2(X_t, \mathcal{O}_{X_t}) = 0$, i.e. $\hat{\pi}$ is smooth at $[L]$.

that $J(X) \longrightarrow \mathbb{P}^1$ admits a natural section, given either as the closure of the point $[\mathcal{O}_E] \in$ Jac(E) in $J(X)$ or by interpreting $\mathcal{O}_X \in \operatorname{Pic}^0(X)$ as a section of $\operatorname{Jac}(X/\mathbb{P}^1) \longrightarrow \mathbb{P}^1$.

Remark 4.1   Similarly, one defines elliptic surfaces

$$J^d(X) \longrightarrow \mathbb{P}^1$$

for arbitrary d as the unique relatively minimal smooth compactification of $\operatorname{Jac}^d(X/\mathbb{P}^1)$ coarsely representing $T \longmapsto \operatorname{Pic}^d(X \times_{\mathbb{P}^1} T)/\sim$ or of $\operatorname{Jac}^d(E)$ coarsely representing the functor $S \longmapsto \operatorname{Pic}^d(E \times S)/\sim$. However, for $d \neq 0$ it comes without a (natural) section.

Remark 4.2   Note that the rational map that sends $x \in X_t$ to $\mathcal{O}_{X_t}(x)$ extends by minimality to an isomorphism

$$X \xrightarrow{\ \sim\ } J^1(X).$$

To be more precise, the ideal sheaf of $x \in X_t$ is locally free if and only if $x \in X_t$ is a smooth point of the fibre X_t (and in particular not contained in a multiple fibre component). Hence, $x \longmapsto \mathcal{O}_{X_t}(x)$ is regular on the π-smooth part $X' \subset X$.

Definition 4.3   The *index* d_0 of an elliptic fibration $\pi : X \longrightarrow \mathbb{P}^1$ is the minimal fibre degree $d_0 = (C.[X_t])$ of a curve $C \subset X$ such that $\pi : C \longrightarrow \mathbb{P}^1$ is finite.

Equivalently, the index d_0 is the smallest positive $\deg(L|_{X_t}) = (L.X_t)$, $L \in \operatorname{Pic}(X)$. Indeed, if $(L.X_t) > 0$, then $H^0(X, L(nX_t)) \neq 0$ for $n \gg 0$.

Thus, $\pi : X \longrightarrow \mathbb{P}^1$ admits a section if and only if $d_0 = 1$. Also note that a complex elliptic K3 surface is projective if and only if $d_0 < \infty$. Indeed, if D is a divisor of positive fibre degree, then $(D + nX_t)^2 > 0$ for $n \gg 0$; cf. Remark **1**.3.4 or Remark **8**.1.3. See Section 4.3 for more on the non-projective case.

Remark 4.4   Let $C \subset X$ be of degree d_0 over $\mathbb{P}^1$. Then the rational map that sends $L \in \operatorname{Pic}^d(X_t)$ to $L \otimes \mathcal{O}(C)|_{X_t} \in \operatorname{Pic}^{d+d_0}(X_t)$ extends to an isomorphism of elliptic surfaces

$$J^d(X) \xrightarrow{\ \sim\ } J^{d+d_0}(X).$$

In particular, if $\pi : X \longrightarrow \mathbb{P}^1$ admits a section, then

$$X \xrightarrow{\ \sim\ } J^d(X)$$

for all d.

In this sense, the above construction yields at most d_0 different elliptic surfaces

$$J^0(X) = J(X), \ J^1(X) \simeq X, \ldots, \ J^{d_0-1}(X).$$

Although these surfaces are usually different, one can nevertheless show that

$$J^d(X)_t \simeq X_t \tag{4.1}$$

for all d and all closed $t \in \mathbb{P}^1$. This is clear for smooth fibres, because then $J^d(X)_t \simeq \text{Pic}^d(X_t) \simeq X_t$ (over the algebraically closed field k), and for elliptic surfaces with a section, because then in fact $J^d(X) \simeq X$ for all d. For singular fibres, one reduces to the case with a section by constructing a local (analytic, étale, etc.) section through a smooth point of the fibre X_t. Here one uses that an elliptic fibration of a K3 surface does not admit multiple fibres. So in fact there exists an open (analytic or étale) covering of $\mathbb{P}^1$ over which $X \longrightarrow \mathbb{P}^1$ and $J^d(X) \longrightarrow \mathbb{P}^1$ become isomorphic.

Also note that

$$J(J^d(X)) \simeq J(X)$$

for all d. Indeed, both elliptic surfaces come with a section and all their fibres are isomorphic. More algebraically, if $E = X_\eta$, then all $E_d := \text{Jac}^d(E)$ are torsors under the elliptic curve $E_0 := \text{Pic}^0(E)$ by $E_0 \times E_d \longrightarrow E_d$, $(M, L) \longmapsto M \otimes L$. However, the curve E_d is a torsor under only one elliptic curve, namely its dual, and hence $E_0 \xrightarrow{\sim} \text{Pic}^0(E_d)$; see [409, Thm. 7.19] or [569, X.Thm. 3.8] and also Section 5.1.

Finally we remark that there are natural rational maps from X to all $J(X), \ldots, J^{d_0-1}(X)$:

$$X - - \succ J^d(X)$$

obtained by mapping a line bundle L of degree one on a smooth fibre X_t to L^d. Use $J(X) \simeq J^{d_0}(X)$ to get $X - - \succ J(X)$.

4.2 Let us now explain how to use moduli spaces of stable sheaves to give a modular construction for the compactification $J(X)$ and not only for the open part $\text{Jac}(X/\mathbb{P}^1)$. This directly proves that $J(X)$ and in fact all $J^d(X)$ are K3 surfaces. Moreover, it allows one to control their Picard groups and their periods.

Recall from Chapter 10 the notation $M_H(v)$ for the moduli space of semistable sheaves F on X with Mukai vector $v(F) = v$, where semistability is measured with respect to the polarization H. If one thinks of a line bundle $L \in \text{Pic}^d(X_t)$ on a fibre X_t as a sheaf on X, i.e. as $F := i_*L$, one is led to choose

$$v = v_d := v(i_*L) = (0, [X_t], d).$$

Here, $[X_t]$ is the class of the fibre X_t. It is not hard to see that any line bundle L on a smooth fibre X_t gives rise to a sheaf i_*L on X that is stable with respect to any H. This yields an open immersion

$$\text{Jac}^d((X/\mathbb{P}^1)_{\text{reg}}) \hookrightarrow M_H(v_d), \qquad (4.2)$$

where $(X/\mathbb{P}^1)_{\text{reg}}$ is the union of all smooth fibres.

For all choices of d the Mukai vector $v_d = (0, [X_t], d)$ is primitive, and hence for generic H the moduli space $M_H(v_d)$ is a K3 surface; cf. Proposition **10.2.5** and Corollary **10.3.5**. But then $M_H(v)$ can be taken as the unique minimal model of $J^d(X)$.

Summarizing, one obtains the following which can also be deduced from [133, Thm. 5.3.1]; cf. Proposition 5.4.

Proposition 4.5 *For all d, the Jacobian fibration of degree d associated with an elliptic K3 surface $\pi : X \longrightarrow \mathbb{P}^1$ defines an elliptic K3 surface $J^d(X) \longrightarrow \mathbb{P}^1$.* $\qquad\square$

Remark 4.6 (i) The embedding (4.2) can in general not be extended to a modular embedding of $\mathrm{Jac}(X/\mathbb{P}^1)$ into $M_H(v)$. Indeed, a line bundle L on a reducible fibre X_t does not necessarily define a stable sheaf i_*L on X.

(ii) The choice of a generic polarization is essential. For example, if X_t of type I_2 and H restricts to $\mathcal{O}(1)$ on each component, then the fibre of $M_H(v_0) \longrightarrow \mathbb{P}^1$ over t is of type I_1; see [96, Thm. 6.3.11]. Choosing a generic polarization blows up the node which results in a I_2-fibre as needed to ensure $J(X)_t \simeq X_t$.

Combining $J^d(X) \simeq M_H(v_d)$ with Remark **10.3.7** yields the following corollary.

Corollary 4.7 *Let $X \longrightarrow \mathbb{P}^1$ be a complex projective elliptic K3 surface and let $v_d :=$ $(0, [X_t], d)$, $d \neq 0$. Then there exists an isometry of Hodge structures*

$$H^2(J^d(X), \mathbb{Z}) \simeq v_d^\perp / \mathbb{Z} \cdot v_d. \tag{4.3}$$

In particular, $\rho(J^d(X)) = \rho(X)$ for all d. $\qquad\square$

Although $d = 0$ is explicitly excluded, the result nevertheless describes also the Hodge structure and Picard group in this case. Indeed, $J(X) \simeq J^{d_0}(X)$ for d_0, the index of the elliptic fibration, and hence

$$H^2(J(X), \mathbb{Z}) \simeq v_{d_0}^\perp / \mathbb{Z} \cdot v_{d_0}. \tag{4.4}$$

In [291] Keum used the Hodge isometry described by (4.4) to define a lattice embedding $\mathrm{NS}(X) \hookrightarrow \mathrm{NS}(J(X))$, $\alpha \mapsto ((\alpha.[X_t])/d_0, \alpha, 0) \in v_{d_0}^\perp / \mathbb{Z} \cdot v_{d_0}$, the cokernel of which is generated by $(0, 0, 1)$. Since $d_0 \cdot (0, 0, 1) \in \mathrm{NS}(X)$ under this identification and d_0 is minimal with this property, one finds (cf. Section **14.0.1**)

$$\mathrm{disc}\,\mathrm{NS}(X) = d_0^2 \cdot \mathrm{disc}\,\mathrm{NS}(J(X)). \tag{4.5}$$

This immediately yields the following result due to Keum [291].

Corollary 4.8 *If $\mathrm{disc}\,\mathrm{NS}(X)$ of an elliptic K3 surface $\pi : X \longrightarrow \mathbb{P}^1$ is square free, then π admits a section.* $\qquad\square$

It is not difficult to generalize (4.5) to arbitrary $d \neq 0$ by describing a common overlattice for $\mathrm{NS}(X)$ and $\mathrm{NS}(J^d(X))$. Eventually, one finds

$$\mathrm{disc}\,\mathrm{NS}(X) = \mathrm{g.c.d.}(d, d_0)^2 \cdot \mathrm{disc}\,\mathrm{NS}(J^d(X)), \tag{4.6}$$

which sometimes excludes two Jacobians of different degrees $J^{d_1}(X)$ and $J^{d_2}(X)$ from being isomorphic K3 surfaces. See also Example **16.2.11**.

Remark 4.9 As explained above, $J^d(X) \longrightarrow \mathbb{P}^1$ can be described as moduli spaces. However, the moduli space is not always fine; i.e. there does not always exist a universal sheaf $\mathcal{P}$ on $X \times J^d(X)$. In fact, general existence results (see [266, Cor. 4.6.7] or Section 10.2.2) assert that a universal family exists whenever the index d_0 and d are coprime. In this case there exists an equivalence of derived categories

$$D^b(X) \simeq D^b(J^d(X)); \tag{4.7}$$

see Example 16.2.4.

A universal sheaf always exists étale locally and the obstruction to glue these locally defined universal sheaves yields a class α_d in the Brauer group of $J^d(X)$; see Section 10.2.2. We shall get back to this shortly; see Remark 5.9. In any case (see Section 16.4.1), this then yields an equivalence

$$D^b(X) \simeq D^b(J^d(X), \alpha_d). \tag{4.8}$$

The equivalence can also be interpreted as saying that X is a fine moduli space of α_d-twisted sheaves on $J^d(X)$.

All these observations hold true over any algebraically closed field k, which leads to the next remark.

Remark 4.10 The assertions on the Néron–Severi group still hold for elliptic K3 surfaces over arbitrary algebraically closed fields. For d and d_0 coprime, this can be deduced directly from the equivalence $D^b(X) \simeq D^b(J^d(X))$ and the induced isometry between their extended Néron–Severi groups

$$N(X) \simeq N(J^d(X));$$

see Example 16.2.11. This in particular shows that (4.6) continues to hold for elliptic K3 surfaces over arbitrary algebraically closed fields at least for g.c.d. $(d, d_0) = 1$.[5]

4.3 The above algebraic approach is problematic when it comes to non-projective complex elliptic K3 surfaces $\pi \colon X \longrightarrow \mathbb{P}^1$. Nevertheless, it is possible to define the relative Jacobian $J(X) \longrightarrow \mathbb{P}^1$ and all its relatives $J^d(X) \longrightarrow \mathbb{P}^1$ also in the complex setting.

Plainly, we cannot work with the Jacobian of the generic fibre and so cannot define $J(X)$ as the relatively minimal model of it. One can however restrict to the open part of all smooth fibres $\pi \colon (X/\mathbb{P}^1)_{\mathrm{reg}} \longrightarrow \mathbb{P}^1$ and define $J((X/\mathbb{P}^1)_{\mathrm{reg}})$ as the total space of $R^1\pi_*\mathcal{O}/R^1\pi_*\mathbb{Z}$. Then one still needs to argue that a (relatively minimal) compactification yielding $J(X) \longrightarrow \mathbb{P}^1$ exists, which is not granted on general grounds.

---

[5] And there is little doubt that using the twisted version (4.8) it can be proved for all $d \neq 0$ and hence proving (4.5) for arbitrary k. Equality of their Picard numbers for arbitrary $d \neq 0$ is easier to show.

Kodaira's approach is to say that locally analytically $X \longrightarrow \mathbb{P}^1$ has sections, as elliptic K3 surfaces come without multiple fibres, and every (local) elliptic fibration with a section is its own Jacobian fibration. The remaining bit is to glue the local families according to the cocycle for $R^1\pi_*\mathbb{Z}$. We refer to [33, 186] and the original paper by Kodaira [307].

A priori, it should be possible to define the $J^d(X) \longrightarrow \mathbb{P}^1$ as a moduli space of sheaves on the fibres of $X \longrightarrow \mathbb{P}^1$ as explained in Section 4.2. However, the theory of moduli spaces of sheaves (especially of torsion sheaves as in our case) on non-projective manifolds has not yet been sufficiently developed to be used here.

5 Tate–Šafarevič Group

The Tate–Šafarevič group of an elliptic K3 surface $X_0 \longrightarrow \mathbb{P}^1$ with a section parametrizes all elliptic (K3) surfaces $X \longrightarrow \mathbb{P}^1$ for which $X_0 \longrightarrow \mathbb{P}^1$ is isomorphic to the Jacobian fibration $J(X) \longrightarrow \mathbb{P}^1$. The notion is modeled on the Tate–Šafarevič and the Weil–Châtelet group of an elliptic curve, which shall be briefly recalled. The difference between the two groups, however, disappears for K3 surfaces. The algebraic and the complex approach to the Tate–Šafarevič group are quite similar, but in the latter context also non-algebraic surfaces are parametrized. Both versions are presented, using the occasion to practice the different languages.

5.1 Let us start with the Weil–Châtelet group of an elliptic curve E_0 over a field K, where K later is the function field $k(t)$ of $\mathbb{P}^1_k$ with k algebraically closed (or a finite field).

A *torsor* under E_0 is a smooth projective curve of genus one over K together with a simply transitive action

$$E_0 \times E \longrightarrow E, \ (p, x) \longrightarrow p + x.$$

Isomorphisms of E_0-torsors are defined in the obvious manner. In particular, E is isomorphic to the trivial torsor E_0 if and only if $E(K) \neq \emptyset$. Then the *Weil–Châtelet group* is the set

$$\mathrm{WC}(E_0) := \{E = E_0\text{-torsor}\}/_{\simeq}.$$

This groups contains the Tate–Šafarevič group $\mathrm{III}(E_0)$ as a subgroup, which is in general a proper subgroup. As in our applications both groups coincide, we do not go into the definition of $\mathrm{III}(E_0)$.

If E is an E_0-torsor, then $E_0 \simeq \mathrm{Jac}(E)$ as algebraic groups. Indeed, choose any $x \in E(K')$ for some finite extension K'/K and define $E_0 \xrightarrow{\sim} \mathrm{Jac}(E), p \longmapsto \mathcal{O}_E((p+x)-x)$, which is an isomorphism over K. For an isomorphism $f \colon E \xrightarrow{\sim} E'$ of E_0-torsors this isomorphism changes by f_*. Conversely, if a group isomorphism $E_0 \xrightarrow{\sim} \text{Jac}(E)$, $p \longmapsto L_p$, is given, then define $E_0 \times E \longrightarrow E$, $(p, x) \longmapsto y$ with y the unique point satisfying $\mathcal{O}(x) \otimes L_p \simeq \mathcal{O}(y)$. This turns E into an E_0-torsor. As a result, one obtains a natural bijection

$$\text{WC}(E_0) \simeq \{(E, \varphi) \mid \varphi \colon E_0 \xrightarrow{\sim} \text{Jac}(E)\}/_{\simeq}, \tag{5.1}$$

where $(E, \varphi) \simeq (E', \varphi')$ if there exists an isomorphism $f \colon E \xrightarrow{\sim} E'$ with $f_* \circ \varphi = \varphi'$. A cohomological description of the Weil–Châtelet group is provided by the next result.

Proposition 5.1 *There exists a natural bijection*

$$\text{WC}(E_0) \simeq H^1(K, E_0). \tag{5.2}$$

Here, $H^1(K, E_0) := H^1(\text{Gal}(\overline{K}/K), E_0(\overline{K}))$ is the Galois cohomology of E_0, where $\overline{K}$ denotes the separable closure of K.

Proof For any E_0-torsor E the Galois group $G := \text{Gal}(\overline{K}/K)$ acts naturally on $E(\overline{K})$. If $x \in E(\overline{K})$, then $G \longrightarrow E_0(\overline{K})$, $g \longmapsto p_g$ with $g \cdot x = p_g + x$, defines a continuous crossed homomorphism and thus an element in $H^1(G, E_0(\overline{K}))$. Choosing a different point $x' \in E(\overline{K})$ the crossed homomorphism differs by a boundary. Hence, there is a well-defined map $\text{WC}(E_0) \longrightarrow H^1(G, E_0(\overline{K}))$.

Conversely, a crossed homomorphism $G \longrightarrow E_0(\overline{K})$, $g \longmapsto p_g$, yields for all g an isomorphism $\varphi_g \colon E_0 \times \overline{K} \longrightarrow E_0 \times \overline{K}$, $\varphi_g(x) = p_g + x$. The isomorphisms φ_g satisfy the cocycle condition $\varphi_g \cdot (g \cdot \varphi_h) = \varphi_{gh}$ and thus define a descent datum, which in turn yields E over K that splits the descent datum after extension to $\overline{K}$. Note that one does not really need the full machinery of descent at this point, as smooth curves are uniquely determined by their function field and it suffices to construct $K(E)$ by descent. See [409, Ch. IV.7] or [569, Ch. X] for details and references. □

The right-hand side of (5.2) has the structure of an abelian group. The induced group structure on the left-hand side has the property that the product E_3 of two E_0-torsors E_1, E_2 comes with a morphism $\psi \colon E_1 \times E_2 \longrightarrow E_3$ satisfying $\psi(p + x_1, q + x_2) \longrightarrow (p + q) + \psi(x_1, x_2)$; see [569, p. 355].

Remark 5.2 Let E be an E_0-torsor and so $\text{Jac}(E) \simeq E_0$. Then, $\text{Jac}^d(E)$ (see Section 4.1) admits canonically the structure of an E_0-torsor, i.e.

$$\text{Jac}^d(E) \in \text{WC}(\text{Jac}(E)).$$

Using the group structure of $\text{WC}(E_0)$ and writing $E \simeq \text{Jac}^1(E)$, one finds that $[\text{Jac}^d(E)] = d \cdot [E] \in \text{WC}(E_0)$. Indeed, $\text{Jac}^{d_1}(E) \times \text{Jac}^{d_2}(E) \longrightarrow \text{Jac}^{d_1 + d_2}(E)$, $(L, M) \longmapsto L \otimes M$ satisfies the above property with respect to the natural action of $\text{Jac}(E)$.

The Weil–Châtelet group can be defined more generally for any group scheme, e.g. for the smooth part X'_t of a singular fibre X_t of an elliptic K3 surface $\pi \colon X \longrightarrow \mathbb{P}^1$; see

Section 2.5. So, in particular, one can speak of WC(X'_t). However, over an algebraically closed field k this group is trivial and only when put in families does it become interesting. It leads to the notion of the Weil–Châtelet group of an elliptic surface; see below.

5.2 Let $\pi: X_0 \longrightarrow \mathbb{P}^1$ be an elliptic K3 surface with a section $C_0 \subset X_0$. We shall work over an algebraically closed field k, but the arguments remain valid for $k = \mathbb{F}_q$.

Consider the open set $X'_0 \subset X_0$ of π-smooth points as a group scheme $X'_0 \longrightarrow \mathbb{P}^1$; see Section 2.5. A torsor under X'_0 is defined similarly to the absolute case as a smooth fibration $X' \longrightarrow \mathbb{P}^1$ with a group action

$$X'_0 \times_{\mathbb{P}^1} X' \longrightarrow X',$$

making the fibres X'_t torsors under $(X'_0)_t$. The set of all X'_0-torsors is the Weil–Châtelet group of X_0, but due to the absence of multiple fibres, it equals the Tate–Šafarevič group of X_0 and we shall, therefore, not distinguish between the two.

Definition 5.3 The *Tate–Šafarevič group*[6] of an elliptic K3 surface $\pi: X_0 \longrightarrow \mathbb{P}^1$ with a section is the set of all torsors under the group scheme $X'_0 \longrightarrow \mathbb{P}^1$:

$$\text{Ш}(X_0) := \{(X' \longrightarrow \mathbb{P}^1) = (X'_0 \longrightarrow \mathbb{P}^1)\text{-torsor}\}/_{\simeq}.$$

The next result is the converse of Proposition 4.5. It allows us to interpret $\text{Ш}(X_0)$ in terms of Jacobian fibrations.

Proposition 5.4 *Let $X \longrightarrow \mathbb{P}^1$ be a relatively minimal elliptic surface such that its Jacobian $X_0 := J(X) \longrightarrow \mathbb{P}^1$ is a K3 surface. Then X is a K3 surface itself.*

Proof There are various approaches to the assertion. The first one is to view X as a moduli space of twisted sheaves on X_0; cf. Remarks 4.9 and 5.9. This, in the spirit of the arguments used to prove Proposition 4.5, would essentially immediately show that X is a K3 surface.

The second approach uses the observation that étale locally $J(X)$ and X are isomorphic if in addition one assumes the existence of local sections also for X, i.e. that $X \longrightarrow \mathbb{P}^1$ has no multiple fibres. This then implies $e(J(X)) = e(X) = 24$ and hence $R^1\pi_*\mathcal{O}_X \simeq \mathcal{O}(-2)$. (In [133, Prop. 5.3.6] this is proved without the additional assumption on X.) The general canonical bundle formula for elliptic fibrations yields $\omega_X \simeq \mathcal{O}_X$. Moreover, $\chi(\mathcal{O}_X) = 2$ and hence $h^1(X, \mathcal{O}_X) = 0$. Altogether this indeed proves that X is a K3 surface.

For yet another approach in the case of complex K3 surfaces, see Remark 5.15. $\quad\square$

---

[6] The name respects the correct alphabetic order in Cyrillic.

For the following note that by definition the generic fibre E of a torsor $X' \longrightarrow \mathbb{P}^1$ under $X'_0 \longrightarrow \mathbb{P}^1$ is a torsor under the generic fibre E_0 of $X_0 \longrightarrow \mathbb{P}^1$ and so $\mathrm{Jac}(E) \simeq E_0$.

Corollary 5.5 *Let $\pi : X_0 \longrightarrow \mathbb{P}^1$ be an elliptic K3 surface with a section and let E_0 be its generic fibre.*

(i) *Taking the generic fibre of a torsor under $X'_0 \longrightarrow \mathbb{P}^1$ (the open part of π-smooth points) defines an isomorphism*

$$\mathrm{III}(X_0) \xrightarrow{\sim} \mathrm{WC}(E_0).$$

(ii) *The Tate–Šafarevič group $\mathrm{III}(X_0)$ can be naturally identified with the set of pairs (X, φ) with $X \longrightarrow \mathbb{P}^1$ an elliptic K3 surface and an isomorphism $\varphi : X_0 \xrightarrow{\sim} \mathrm{J}(X)$ over $\mathbb{P}^1$ respecting the group scheme structures on X'_0 and $\mathrm{J}(X)'$.*

Proof (i) An E_0-torsor E is trivial if and only if it admits a K-rational point. Similarly, a $(X'_0 \longrightarrow \mathbb{P}^1)$-torsor $X' \longrightarrow \mathbb{P}^1$ is trivial if and only if it admits a section. Now, if the generic fibre E of X' admits a K-rational point, then its closure in X' defines a section. Hence, $X' \longmapsto E$ defines an injection $\mathrm{III}(X_0) \longrightarrow \mathrm{WC}(E_0)$.

To prove surjectivity, one considers the relatively minimal model $X \longrightarrow \mathbb{P}^1$ of an E_0-torsor E. Then $\mathrm{J}(X) \simeq X_0$ and, hence, X is a K3 surface. However, for an elliptic K3 surface $\pi : X \longrightarrow \mathbb{P}^1$ the π-smooth part X' is a torsor under $\mathrm{J}(X)'$. This is the global version of the observation (4.1) that $X_t \simeq \mathrm{J}(X)_t$ for all t.

(ii) Combine (5.1) with (i) and the proposition. Once more, one uses the uniqueness of the minimal model to extend $E_0 \simeq \mathrm{Jac}(E)$ to an isomorphism $X_0 \simeq \mathrm{J}(X)$. $\square$

The reason behind the isomorphism between the Weil–Châtelet groups of the surface and the generic fibre is the absence of multiple fibres in an elliptic fibration of a K3 surface. The situation is more involved for arbitrary elliptic surfaces.

Proposition 5.6 *Let $\pi : X_0 \longrightarrow \mathbb{P}^1$ be an elliptic K3 surface with a section $C_0 \subset X_0$. Then there exist natural isomorphisms*

$$\mathrm{III}(X_0) \simeq H^2(X_0, \mathbb{G}_m) \simeq \mathrm{Br}(X_0).$$

Proof The proof, inspired by the discussion of Friedman and Morgan in [186], is split in three parts. See also [606, Ch. 5.3]. We work in the étale topology.

(i) There exists a natural isomorphism

$$H^2(X_0, \mathbb{G}_m) \simeq H^1(\mathbb{P}^1, R^1\pi_*\mathbb{G}_m). \tag{5.3}$$

This holds without the assumption that there exists a section and follows from the Leray spectral sequence $E_2^{p,q} = H^p(\mathbb{P}^1, R^q\pi_*\mathbb{G}_m) \Rightarrow H^{p+q}(X_0, \mathbb{G}_m)$ and the following facts: $\pi_*\mathbb{G}_m \simeq \mathbb{G}_m$ (which is proved similarly to $\pi_*\mathcal{O}_{X_0} \simeq \mathcal{O}_{\mathbb{P}^1}$), $R^2\pi_*\mathbb{G}_m = 0$ (see [227, III.Cor. 3.2]), and $H^q(\mathbb{P}^1, \mathbb{G}_m) = 0$ for $q \geq 2$ (see [142, III.Prop. 3.1]).

(ii) Denote by $\mathcal{X}_0$ the sheaf of étale local sections $\sigma : U \longrightarrow X_0$ of $X_0 \longrightarrow \mathbb{P}^1$ or, equivalently, of $X_0' \longrightarrow \mathbb{P}^1$. Since the latter is a group scheme, $\mathcal{X}_0$ is indeed a sheaf of abelian groups on $\mathbb{P}^1$. Then

$$\text{III}(X_0) \simeq H^1(\mathbb{P}^1, \mathcal{X}_0). \tag{5.4}$$

This is the relative version of Proposition 5.1 and is proved analogously.

(iii) The sheaf $R^1\pi_*\mathbb{G}_m$ is associated with the presheaf $U \longmapsto \text{Pic}(X_0 \times_{\mathbb{P}^1} U)$. Using this, one defines a natural sheaf homomorphism

$$\mathcal{X}_0 \longrightarrow R^1\pi_*\mathbb{G}_m, \quad \sigma \longmapsto \mathcal{O}(\sigma(U) - C_0|_U).$$

Here, σ is a section of $X_0' \longrightarrow \mathbb{P}^1$ over the étale open set U and $\sigma(U)$ and $C_0|_U$ are considered as divisors on $X_0 \times_{\mathbb{P}^1} U$. This sheaf homomorphism is injective, because different points on smooth fibres are never linearly equivalent, and in fact induces an isomorphism

$$H^1(\mathbb{P}^1, \mathcal{X}_0) \xrightarrow{\sim} H^1(\mathbb{P}^1, R^1\pi_*\mathbb{G}_m). \tag{5.5}$$

To prove this, note first that the subgroups of vertical divisors in $\text{Pic}(X_0 \times_{\mathbb{P}^1} U)$ form a subsheaf of $R^1\pi_*\mathbb{G}_m$ concentrated in the finitely many singular values $t \in \mathbb{P}^1$ of π. Hence, $H^1(\mathbb{P}^1, R^1\pi_*\mathbb{G}_m) \xrightarrow{\sim} H^1(\mathbb{P}^1, R^1\pi_*\mathbb{G}_m/\text{vert})$. Now the cokernel of the induced injection $\mathcal{X}_0 \hookrightarrow R^1\pi_*\mathbb{G}_m/\text{vert}$ is just $\mathbb{Z}$, measuring the fibre degree. Taking H^1 yields (5.5), as $H^1_{\text{ét}}(\mathbb{P}^1, \mathbb{Z}) = 0$ and $H^0(\mathbb{P}^1, R^1\pi_*\mathbb{G}_m/\text{vert}) \longrightarrow \mathbb{Z}$ is surjective, for $\mathcal{O}(C_0) \longmapsto 1$.

Composing (5.3) with (5.5) and (5.4) proves the first isomorphism of the proposition. For the second, see Section **18**.1.1. □

Corollary 5.7 *The Tate–Šafarevič group* $\text{III}(X_0)$ *is a torsion group.*

Proof This can, of course, be seen as a consequence of $\text{III}(X_0) \simeq H^2(X_0, \mathbb{G}_m)$, as the latter is known to be a torsion group; see Example **18**.1.4.

However, one can also argue geometrically as follows: let $(X, \varphi) \in \text{III}(X_0)$, i.e. $X \longrightarrow \mathbb{P}^1$ is an elliptic K3 surface with $\varphi : X_0 \xrightarrow{\sim} J(X)$. Then $d \cdot [(X, \varphi)]$ is represented by $J^d(X) \longrightarrow \mathbb{P}^1$ (see Remark 5.2) and $J^{d_0}(X) \simeq J(X)$ if d_0 is the index of the elliptic fibration $X \longrightarrow \mathbb{P}^1$; see Remark 4.4. Hence, $[(X, \varphi)] \in \text{III}(X_0)$ is of finite order dividing d_0. (In fact, it is of order exactly d_0; see Remark 5.9.) □

Despite this result, $\text{III}(X_0)$ is difficult to grasp. For complex K3 surfaces the analytic description of the Brauer group gives some insight, but over other fields, e.g. finite ones, the Tate–Šafarevič group remains elusive.

Remark 5.8 In the proof of the proposition we used the sheaf of étale local sections $\mathcal{X}_0$ of the elliptic K3 surface $\pi : X_0 \longrightarrow \mathbb{P}^1$ and the isomorphism $\text{III}(X_0) \simeq H^1(\mathbb{P}^1, \mathcal{X}_0)$; see (5.4). On the other hand, the Mordell–Weil group $\text{MW}(X_0)$ is by definition $H^0(\mathbb{P}^1, \mathcal{X}_0)$, so Mordell–Weil group and Tate–Šafarevič group are cohomology groups of the same sheaf on $\mathbb{P}^1$. Let us elaborate on this a little more.

Fibrewise multiplication by n yields a short exact sequence

$$0 \longrightarrow \mathcal{X}_0[n] \longrightarrow \mathcal{X}_0 \longrightarrow \mathcal{X}_0 \longrightarrow 0,$$

where $\mathcal{X}_0[n]$ is the sheaf of sections through n-torsion points in the fibres. Taking the long exact cohomology sequence gives the exact sequence

$$0 \longrightarrow \mathrm{MW}(X_0)/n \cdot \mathrm{MW}(X_0) \longrightarrow H^1(\mathbb{P}^1, \mathcal{X}_0[n]) \longrightarrow \mathrm{III}(X_0)[n] \longrightarrow 0.$$

Here, $\mathrm{III}(X_0)[n] \subset \mathrm{III}(X_0)$ is the subgroup of elements of order dividing n. The cohomology $H^1(\mathbb{P}^1, \mathcal{X}_0[n])$ linking $\mathrm{MW}(X_0)$ and $\mathrm{III}(X_0)$ is the analogue of the *Selmer group* of an elliptic curve for the elliptic K3 surface $\pi \colon X_0 \longrightarrow \mathbb{P}^1$. So one could introduce

$$S^n(X_0) := H^1(\mathbb{P}^1, \mathcal{X}_0[n])$$

and call it the Selmer group of the elliptic surface $\pi \colon X_0 \longrightarrow \mathbb{P}^1$.

Remark 5.9 For $\alpha \in \mathrm{III}(X_0)$, let $\pi \colon X_\alpha \longrightarrow \mathbb{P}^1$ be the associated elliptic K3 surface together with the natural isomorphism $\mathrm{J}(X_\alpha) \simeq X_0$. The techniques in the proof of Proposition 5.6 yield an exact sequence (see also [227, (4.35)] or [20, Prop. 1.6])

$$0 \longrightarrow \mathcal{X}_0 \longrightarrow R^1\pi_*\mathbb{G}_{m/\mathrm{vert}} \longrightarrow \mathbb{Z} \longrightarrow 0,$$

which splits for trivial α. For general α, taking cohomology one obtains an exact sequence

$$\mathrm{Pic}(X_\alpha) \longrightarrow \mathbb{Z} \longrightarrow \mathrm{III}(X_0) \longrightarrow H^1(\mathbb{P}^1, R^1\pi_*\mathbb{G}/\mathrm{vert}).$$

By definition, the image of $\mathrm{Pic}(X_\alpha) \longrightarrow \mathbb{Z}$ is generated by the index d_0 and the boundary of $\overline{1} \in \mathbb{Z}/d_0\mathbb{Z}$ yields $\alpha \in \mathrm{III}(X_0)$. In particular, d_0 is the order of $\alpha \in \mathrm{III}(X_0) \simeq \mathrm{Br}(X_0)$. Moreover, as in the proof of Proposition 5.6, one obtains the short exact sequence

$$0 \longrightarrow \langle \alpha \rangle \longrightarrow \mathrm{Br}(X_0) \longrightarrow \mathrm{Br}(X_\alpha) \longrightarrow 0. \tag{5.6}$$

This is a special case of (2.1) in Section **10.2.2**.

Viewing $\mathrm{J}(X_\alpha) \simeq X_0$ as the moduli space of sheaves on X_α, the class α can also be interpreted as the obstruction class to the existence of a universal sheaf on $X_0 \times X_\alpha$ and hence by (4.8)

$$\mathrm{D}^{\mathrm{b}}(X_0, \alpha) \simeq \mathrm{D}^{\mathrm{b}}(X_\alpha).$$

Donagi and Pantev in [156] generalized this equivalence to

$$D^b(X_\beta, \overline{\alpha}) \simeq D^b(X_\alpha, \overline{\beta}), \tag{5.7}$$

for arbitrary $\alpha, \beta \in \text{III}(X_0)$. Here, $\overline{\alpha}, \overline{\beta}$ denote their images under the natural maps $\text{Br}(X_0) \longrightarrow \text{Br}(X_\beta)$ and $\text{Br}(X_0) \longrightarrow \text{Br}(X_\alpha)$, respectively.

The following special case of (5.7) has been observed earlier. Consider the elliptic K3 surface $J^d(X_\alpha)$. As has been explained before, there is a natural isomorphism $J(J^d(X_\alpha)) \simeq J(X_\alpha) \simeq X_0$. Thus, $J^d(X_\alpha)$ corresponds to some class $\beta \in \text{III}(X_0)$ and in fact $\beta = d\alpha$. Therefore, in this case $\overline{\beta}$ is trivial and (5.7) becomes $D^b(J^d(X), \alpha_d) \simeq D^b(X)$ with $X = X_\alpha$ and $\alpha_d = \overline{\alpha}$, as in (4.8).

Remark 5.10 The famous conjecture of Birch and Swinnerton-Dyer (see the announcement as one of the Clay Millenium Problems by Wiles in [637]) predicts that for an elliptic curve over a number field K the rank of the Mordell–Weil group $E(K)$ equals the order of the L-series $L(E, s)$ at $s = 1$. Its generalization links the first non-trivial coefficient of the Taylor expansion of $L(E, s)$ to the order of the Tate–Šafarevič group $\text{III}(E)$. In particular, $\text{III}(E)$ is expected to be finite.

The function field analogue of it leads to the conjecture that for an elliptic curve E over $\mathbb{F}_q(t)$ the Tate–Šafarevič group $\text{III}(E)$ should be finite. Combined with Proposition 5.6 it therefore predicts that for an elliptic K3 surface $X_0 \longrightarrow \mathbb{P}^1$ over $\mathbb{F}_q$ the Brauer group $\text{Br}(X_0)$ is finite. This has been generalized by Artin and Tate to the conjecture that the Brauer group of any surface over a finite field should be finite; see Tate's [596, Sec. 1] and the discussion in Section **18**.2.2, especially Remark **18**.2.9.

5.3 We change the setting and consider complex elliptic K3 surfaces $\pi : X \longrightarrow \mathbb{P}^1$. Recall that X is projective if and only if its index (cf. Definition 4.3) d_0 is finite. In particular, an elliptic K3 surface $X_0 \longrightarrow \mathbb{P}^1$ with a section C_0 is always algebraic. Analogously to the definition of $\text{III}(X_0)$ in the algebraic setting one has the following definition.

Definition 5.11 The *analytic Tate–Šafarevič group* $\text{III}^{\text{an}}(X_0)$ of a complex elliptic K3 surface $X_0 \longrightarrow \mathbb{P}^1$ with a section is the set of elliptic K3 surfaces $\pi : X \longrightarrow \mathbb{P}^1$ such that the π-smooth part $X' \longrightarrow \mathbb{P}^1$ is endowed with the structure of an $X_0' \longrightarrow \mathbb{P}^1$ torsor.

We stress that, although X_0 is algebraic, an elliptic K3 surface X representing an element in $\text{III}^{\text{an}}(X_0)$ may very well be non-algebraic. However, as in Proposition 5.4, X is automatically a K3 surface. Arguing via moduli spaces of twisted sheaves is tricky in the non-algebraic setting, but the fact that X and $J(X)$ are locally (this time in the analytic topology) isomorphic fibrations still holds. See also Remark 5.15.

Most of what has been said above in the algebraic setting holds true in the analytic one, by replacing étale topology, cohomology, etc., by their analytic versions. However, there are also striking differences, as becomes clear immediately. First, the proof of Proposition 5.6 goes through in the analytic version and the asserted isomorphism then reads (see also [517, Ch. VII.8])

$$\text{III}^{\text{an}}(X_0) \simeq H^2(X_0, \mathcal{O}_{X_0}^*). \tag{5.8}$$

In fact, the intermediate isomorphisms (5.4), (5.4), and (5.5) also hold:

$$\text{III}^{\text{an}}(X_0) \simeq H^1(\mathbb{P}^1, \mathcal{X}_0^{\text{an}}) \simeq H^1(\mathbb{P}^1, R^1\pi_*\mathcal{O}_{X_0}^*) \simeq H^2(X_0, \mathcal{O}_{X_0}^*).$$

Here, $\mathcal{X}_0^{\text{an}}$ denotes the sheaf of analytic sections of $X_0 \longrightarrow \mathbb{P}^1$.

Corollary 5.12 *For a complex elliptic K3 surface $X_0 \longrightarrow \mathbb{P}^1$ with a section there exists a short exact sequence*

$$0 \longrightarrow \text{NS}(X_0) \longrightarrow H^2(X_0, \mathbb{Z}) \longrightarrow H^2(X_0, \mathcal{O}_{X_0}) \longrightarrow \text{III}^{\text{an}}(X_0) \longrightarrow 0.$$

In particular,

$$\text{III}^{\text{an}}(X_0) \simeq \mathbb{C}/\mathbb{Z}^{22-\rho(X_0)}.$$

Proof This follows from the exponential sequence, $H^3(X_0, \mathbb{Z}) = 0$, and (5.8).   □

Remark 5.13 The standard comparison of the analytic cohomology of $\mathcal{O}^*$ with the étale cohomology of $\mathbb{G}_m$ relates the analytic with the algebraic Tate–Šafarevič group. As a motivation, start with the well-known

$$H^1(X_0, \mathbb{G}_m) \simeq H^1(X_0, \mathcal{O}_{X_0}^*). \tag{5.9}$$

The two sides are naturally isomorphic to $\text{Pic}(X_0)$, which is the same for both topologies. However, in degree two this becomes

$$H^2(X_0, \mathbb{G}_m) \simeq H^2(X_0, \mathcal{O}_{X_0}^*)_{\text{tors}}$$

and hence

$$\text{III}(X_0) \simeq \text{III}^{\text{an}}(X_0)_{\text{tors}}. \tag{5.10}$$

To prove (5.10), we use the usual comparison morphism $\xi \colon H^2(X_0, \mathbb{G}_m) \longrightarrow H^2(X_0, \mathcal{O}_{X_0}^*)$, the Kummer sequence $0 \longrightarrow \mu_n \longrightarrow \mathbb{G}_m \longrightarrow \mathbb{G}_m \longrightarrow 0$, and the fact that étale and analytic cohomology coincide for finite abelian groups. This then yields immediately that ξ surjects onto the torsion of $H^2(X_0, \mathcal{O}_{X_0}^*)$. To prove injectivity, apply (5.9) and the fact that $H^2(X_0, \mathbb{G}_m)$ is a torsion group. (The latter follows from Corollary 5.7 for elliptic X_0, but of course holds in general for the Brauer group of a smooth surface; see Section **18**.1.1.)

In particular, one finds (see Section **18**.1.2)

$$\text{III}(X_0) \simeq (\mathbb{Q}/\mathbb{Z})^{22-\rho(X_0)}.$$

The identification

$$\text{Ш}(X_0) \simeq \text{Ш}^{\text{an}}(X_0)_{\text{tors}} \subset \text{Ш}^{\text{an}}(X_0)$$

can also be explained geometrically. Clearly, one has a natural inclusion $\text{Ш}(X_0) \subset \text{Ш}^{\text{an}}(X_0)$ and we have remarked already that $\text{Ш}(X_0)$ is a torsion group. On the other hand, if $X \longrightarrow \mathbb{P}^1$ defines a torsion class in $\text{Ш}^{\text{an}}(X_0)$, then there exists a finite $d > 0$ such that $J^d(X)$ admits a section; see proof of Corollary 5.7. This section gives rise to a line bundle of degree d on each fibre which then can be shown to glue to a line bundle L on X. Moreover, $\pi_* L \otimes \mathcal{O}_{\mathbb{P}^1}(n)$ for $n \gg 0$ admits non-trivial global sections. Interpreted as sections of $L \otimes \pi^* \mathcal{O}_{\mathbb{P}^1}(n)$, their zero sets are divisors on X of positive fibre degree. Hence X is algebraic and, therefore, is contained in $\text{Ш}(X_0)$.

Remark 5.14 Note that for a non-algebraic elliptic K3 surface $X \longrightarrow \mathbb{P}^1$ its generic fibre E is ill-defined. However, as $J(X) \longrightarrow \mathbb{P}^1$ is always algebraic, $\text{Jac}(E)$ makes perfect sense nevertheless.

Remark 5.15 The surjection

$$H^2(X_0, \mathcal{O}_{X_0}) \longrightarrow\!\!\!\!\!\longrightarrow H^2(X_0, \mathcal{O}_{X_0}^*) \simeq \text{Ш}^{\text{an}}(X_0)$$

allows one to write (however, not effectively) a family of elliptic surfaces over the line $\mathbb{C} \simeq H^2(X_0, \mathcal{O}_{X_0})$ parametrizing all elliptic surfaces $X \longrightarrow \mathbb{P}^1$ with $J(X) \simeq X_0$.

A sketch of the argument can be found in [186, Ch. 1.5]; it roughly goes as follows: pick a fine enough open cover $\mathbb{P}^1 = \bigcup U_i$ such that classes in

$$H^2(X_0, \mathcal{O}_{X_0}) \simeq H^1(X_0, R^1 \pi_* \mathcal{O}_{X_0})$$

can be represented by sections of $R^1 \pi_* \mathcal{O}_{X_0}$ over $U_i \cap U_j$ and such that for every singular fibre X_t there exists a unique U_i containing t. Now use

$$R^1 \pi_* \mathcal{O}_{X_0}|_{U_i \cap U_j} \longrightarrow \mathcal{X}_0^{\text{an}}|_{U_i \cap U_j}$$

to translate the glueing maps over $U_i \cap U_j$ defining X_0 by the section of $\mathcal{X}_0^{\text{an}}$ obtained as images of classes in $H^2(X_0, \mathcal{O}_{X_0})$. This yields new elliptic surfaces and one checks that their classes in $\text{Ш}^{\text{an}}(X_0) \simeq H^2(X_0, \mathcal{O}_{X_0}^*)$ are given by the image under the exponential map $H^2(X_0, \mathcal{O}_{X_0}) \longrightarrow H^2(X_0, \mathcal{O}_{X_0}^*)$.

It is worth noting that the family constructed in this way really is a family of elliptic surfaces; i.e. it comes with compatible projections to $\mathbb{P}^1$. Also note that this approach to $\text{Ш}^{\text{an}}(X_0)$ shows that all elliptic surfaces $X \longrightarrow \mathbb{P}^1$ in $\text{Ш}^{\text{an}}(X_0)$ (and so in particular all in $\text{Ш}(X_0)$) are deformation equivalent to $X_0 \longrightarrow \mathbb{P}^1$ and, therefore, are K3 surfaces as well. This is an alternative argument for Proposition 5.4 when $k = \mathbb{C}$. In this family, the algebraic surfaces are dense, because

$$\text{III}(X_0) \simeq (\mathbb{Q}/\mathbb{Z})^{22-\rho(X_0)} \subset \text{III}^{\text{an}}(X_0) \simeq \mathbb{C}/\mathbb{Z}^{22-\rho(X_0)}$$

induced by $H^2(X_0, \mathbb{Q}) \hookrightarrow H^2(X_0, \mathbb{R}) \twoheadrightarrow H^2(X_0, \mathcal{O}_{X_0}) \simeq \mathbb{C}$ is dense.

References and Further Reading

In practice, it can be very difficult to determine or describe all elliptic fibrations of a given K3 surface. This is only partially due to the automorphism group. For Kummer surfaces associated with the Jacobian Jac(C) of a generic genus two curve C this was recently studied in detail by Kumar in [339]. For elliptic fibrations of Kummer surfaces associated with a product of elliptic curves, see [344, 460, 467] and [297] for elliptic fibrations of a generic double plane with ramification over six lines.

Is a semistable (i.e. only I_n-fibres occur) extremal elliptic K3 surface determined by its configuration of singular fibres? This question has been treated by Miranda and Persson [416] and Artal Bartolo, Tokunaga, and Zhang [12]; in the latter article one finds more on the possible Mordell–Weil groups. In [341] Kuwata exhibits examples of elliptic quartic surfaces with Mordell–Weil groups of rank at least 12.

The description of NS(J(X)) by Keum [291] was motivated by Belcastro's thesis [57]. However, in the latter J(X) was linked to a moduli space of bundles with Mukai vector $(d_0, [X_t], 0)$. The relation between the two approaches can be explained in terms of elementary transformations as in Section **9.2.2** or, more abstractly, by the spherical twist $T_{\mathcal{O}}$; see Section **16.2.3**.

Questions and Open Problems

To the best of my knowledge, not all of the statements for complex elliptic K3 surfaces that should hold in positive characteristic too have actually been worked out in full detail; see e.g. Remarks 3.11 and 4.10.

It would be interesting to compute periods of non-projective $(X \longrightarrow \mathbb{P}^1) \in \text{III}^{\text{an}}(X_0)$.

As mentioned in Section 4.3, there are things left to check to view $J^d(X)$ as a moduli space of sheaves in the non-algebraic setting.

12

Chow Ring and Grothendieck Group

This chapter starts with a quick review of the basic facts on Chow and Grothendieck groups. In particular, we mention Roitman's result about torsion freeness, which we formulate only for K3 surfaces, and prove divisibility of the homologically trivial part. Section 2 outlines Mumford's result about $CH^2(X)$ being big for complex K3 surfaces and contrasts it with the Bloch–Beĭlinson conjecture for K3 surfaces over number fields. This section also contains two approaches, due to Bloch and Green–Griffiths–Paranjape, to prove that $CH^2(X)$ grows under transcendental base field extension. The last section discusses more recent results of Beauville and Voisin on a natural subring of $CH^*(X)$ that naturally splits the cycle map.

1 General Facts on $CH^*(X)$ and $K(X)$

We consider an algebraic K3 surface X over an arbitrary field k and study its Chow ring $CH^*(X)$ and its Grothendieck group $K(X)$. In this first section we recall standard definitions and results and explain what they say for K3 surfaces.

1.1 The ultimate reference for intersection theory and Chow groups is Fulton's book [192]. A brief outline summarizing the basic functorial properties of the Chow ring can be found in [236, App. A].

For an arbitrary variety Y over a field k, a *cycle of codimension n* is a finite linear combination $Z = \sum n_i[Z_i]$ with $n_i \in \mathbb{Z}$ and $Z_i \subset Y$ closed integral subvarieties of codimension n. The group of all such cycles shall be denoted $Z^n(Y)$.

Let $\nu \colon \widetilde{V} \longrightarrow V \subset Y$ be the normalization of a subvariety $V \subset Y$. Recall that two divisors D, D' on $\widetilde{V}$, i.e. cycles of codimension one on the normal variety $\widetilde{V}$, are linearly equivalent if $D - D'$ is a principal divisor (which for the Cartier divisor is equivalent to $\mathcal{O}(D) \simeq \mathcal{O}(D')$). In this case, the image cycles $Z := \nu_*D$ and $Z' := \nu_*D'$ on Y are called *rationally equivalent.* The equivalence relation generated by this is *rational equivalence* and is denoted $Z \sim Z'$.

The *Chow group* of cycles of codimension n on a variety Y is by definition the group of all cycles of codimension n modulo rational equivalence:

$$\mathrm{CH}^n(Y) := Z^n(Y)/\sim.$$

For a smooth variety Y the map $D \longmapsto \mathcal{O}(D)$ yields an injection $\mathrm{CH}^1(Y) \hookrightarrow \mathrm{Pic}(Y)$ (by definition rational equivalence equals linear equivalence for codimension one cycles), which is in fact an isomorphism for integral Y; see [236, II.Prop. 6.15]. One can define the first Chern class $c_1 : \mathrm{Pic}(Y) \longrightarrow \mathrm{CH}^1(Y)$ as its inverse.

For a smooth quasi-projective variety Y it is possible to define the intersection of cycles modulo rational equivalence which endows

$$\mathrm{CH}^*(Y) := \bigoplus \mathrm{CH}^n(Y)$$

with the structure of a graded commutative ring. For two subvarieties $Z, Z' \subset Y$ meeting transversally this is given by the naive intersection $Z \cap Z'$. If the two subvarieties do not intersect transversally or even in the wrong codimension, one needs to deform them first according to Chow's moving lemma (for algebraically closed k) which requires working modulo rational equivalence; cf. [192, Ch. 11] or [621, 21.2]. Another approach to the intersection product uses deformation to the normal cone.

If Y is of dimension d, then any $Z \in \mathrm{CH}^d(Y)$ can be written as a finite sum $Z = \sum n_i[y_i]$ with closed points $y_i \in Y$. The *degree* of Z is then defined as

$$\deg\left(\sum n_i[y_i]\right) := \sum n_i[k(y_i) : k],$$

which does not depend on the chosen representative. It defines a group homomorphism

$$\deg : \mathrm{CH}^d(Y) \longrightarrow \mathbb{Z},$$

the kernel of which is denoted

$$\mathrm{CH}^d(Y)_0 := \mathrm{Ker}\left(\deg : \mathrm{CH}^d(Y) \longrightarrow \mathbb{Z}\right).$$

Let us now specialize to the case that Y is a K3 surface X. For dimension reasons one has

$$\mathrm{CH}^*(X) = \mathrm{CH}^0(X) \oplus \mathrm{CH}^1(X) \oplus \mathrm{CH}^2(X).$$

Clearly, $\mathrm{CH}^0(X) \simeq \mathbb{Z}$, which is naturally generated by $[X]$, and $\mathrm{CH}^1(X) \simeq \mathrm{Pic}(X)$ via the first Chern class.[1]

---

[1] For a complex non-projective K3 surface X it might happen that X does not contain any curve, and in this sense $Z^1(X) = 0$ and $\mathrm{CH}^1(X) = 0$, but nevertheless one could have $\mathrm{Pic}(X) \neq 0$. For example, consider a K3 surface with $\mathrm{Pic}(X) = \mathrm{NS}(X)$ generated by a line bundle L with $(L)^2 = -4$; cf. Example **3.3.2**.

Remark 1.1 For $k = \bar{k}$ rational equivalence of 0-cycles can be understood more explicitly as follows. A cycle Z of codimension zero is rationally equivalent to 0 if there exists a morphism $f \colon \mathbb{P}^1 \longrightarrow S^n(X)$ such that $f(0) - f(\infty) = Z$. The equivalence relation generated by this condition really is rational equivalence. Here, $S^n(X)$ denotes the symmetric product of the surface X (cf. Section **10**.3.3), and the cycle $f(t)$ is $\sum[x_i]$ if the image of t under f is the point $(x_1, \ldots, x_n) \in S^n(X)$; see [192, Ex. 1.6.3] or [439].

The intersection product with $\mathrm{CH}^0(X) = \mathbb{Z}$ is obvious and for dimension reasons $\mathrm{CH}^2(X)$ intersects trivially with $\mathrm{CH}^1(X) \oplus \mathrm{CH}^2(X)$. Thus, the only interesting intersection product on a surface X is

$$\mathrm{CH}^1(X) \times \mathrm{CH}^1(X) \longrightarrow \mathrm{CH}^2(X).$$

If $C_1, C_2 \subset X$ are two curves, then $[C_1] \cdot [C_2] \in \mathrm{CH}^2(X)$ can be described as the image $\nu_*[D]$ under the normalization $\nu \colon \tilde{C}_1 \longrightarrow C_1 \subset X$ of any divisor D with $\mathcal{O}(D) \simeq \nu^* \mathcal{O}(C_2)$. Note that in this case, $\deg([C_1] \cdot [C_2]) = \deg \mathcal{O}(D) = (C_1.C_2)$; see Section **1**.2.1. In Section 3, we shortly return to the intersection product of codimension one cycles and describe its image in $\mathrm{CH}^2(X)$.

1.2 The really mysterious part of the Chow ring of a K3 surface is $\mathrm{CH}^2(X)$. (See Section **17**.2 for a discussion of the group $\mathrm{CH}^1(X) \simeq \mathrm{Pic}(X)$.) If X contains a k-rational point, then $\deg \colon \mathrm{CH}^2(X) \longrightarrow \mathbb{Z}$ is surjective. Otherwise its image is a finite index subgroup. In any case, the essential part of $\mathrm{CH}^2(X)$ is the kernel $\mathrm{CH}^2(X)_0$.

The following observation, although stated here only for K3 surfaces, holds in full generality; see e.g. [192, Ex. 1.6.6].

Proposition 1.2 *If k is algebraically closed, then the group $\mathrm{CH}^2(X)_0$ is divisible.*

Proof Clearly, $\mathrm{CH}^2(X)_0$ is generated by cycles of the form $[x] - [y]$ with $x, y \in X$. Choose a smooth irreducible curve $x, y \in C \subset X$. Then $\mathcal{O}(x - y) \in \mathrm{Pic}^0(C)$. The abelian variety $\mathrm{Pic}^0(C)$ is divisible, for multiplication by n defines a finite and hence surjective morphism $\mathrm{Pic}^0(C) \longrightarrow \mathrm{Pic}^0(C)$. The push-forward of a divisor corresponding to the nth root of $\mathcal{O}(x - y)$ yields $(1/n)([x] - [y]) \in \mathrm{CH}^2(X)$. $\qquad\square$

The next theorem, originally due to Roitman [512], is much harder. It is again only a special case of a completely general statement that involves the Albanese variety (which is trivial for K3 surfaces).

Theorem 1.3 *If k is separably closed, then $\mathrm{CH}^2(X)$ is torsion free.*

Proof See Roitman's original article [512] and Bloch's version [66, 67] showing that there is no torsion prime to the characteristic. The general statement was established by Milne [406]. A brief account was given by Colliot-Thélène in [122], but see also Voisin's [621, Sec. 22.1.2] in the complex setting. $\qquad\square$

Summarizing, for a K3 surface over an algebraically closed field the Chow groups

$$\text{CH}^0(X) \simeq \mathbb{Z}, \quad \text{CH}^1(X) \simeq \text{Pic}(X) = \text{NS}(X) \simeq \mathbb{Z}^{\rho(X)}, \quad \text{and} \quad \text{CH}^2(X)$$

are torsion free. Moreover, the degree map yields an exact sequence

$$0 \longrightarrow \text{CH}^2(X)_0 \longrightarrow \text{CH}^2(X) \longrightarrow \mathbb{Z} \longrightarrow 0$$

with $\text{CH}^2(X)_0$ a divisible group.

Remark 1.4 The torsion of $\text{CH}^2(Y)$ for arbitrary surfaces has been studied intensively. For a survey, see [122]. We only briefly mention the following results applicable to K3 surfaces. So, we shall assume that X is a K3 surface over an arbitrary field k, although the following results hold under more general assumptions.

(i) If ℓ is prime to the characteristic of k, then $\text{CH}^2(X)[\ell^\infty]$ (the part of $\text{CH}^2(X)$ annihilated by some power of ℓ) is a subquotient of $H^3_{et}(X, \mathbb{Q}_\ell/\mathbb{Z}_\ell(2))$; see [122, Thm. 3.3.2].

(ii) If k is a finite field, then the torsion subgroup of $\text{CH}^2(X)$ is finite; cf. [122, Thm. 5.2] and [126]. See also Proposition 2.16, asserting that in fact $\text{CH}^2(X)$ is torsion free in this situation.

(iii) I am not aware of any finiteness results or instructive examples for the torsion of $\text{CH}^2(X)$ for a K3 surface X over a number field or over $\mathbb{F}_q(t)$. See the comments at the end of this chapter.

1.3 The *Grothendieck group* $K(Y)$ of a variety (or a Noetherian scheme) Y is the free abelian group generated by coherent sheaves F on Y divided by the subgroup generated by elements of the form $[F_2] - [F_1] - [F_3]$ whenever there exists a short exact sequence $0 \longrightarrow F_1 \longrightarrow F_2 \longrightarrow F_3 \longrightarrow 0$. Elements of $K(Y)$ are represented by finite linear combinations $\sum n_i[F_i]$ with $n_i \in \mathbb{Z}$ and $F_i \in \text{Coh}(Y)$.

By the very construction, $K(Y)$ is in fact a group that is naturally associated with the abelian category $\text{Coh}(Y)$. Indeed, for an arbitrary abelian category $\mathcal{A}$ one defines its Grothendieck group $K(\mathcal{A})$ as the quotient of the free abelian group generated by objects of $\mathcal{A}$ by the subgroup generated by elements of the form $[A_2] - [A_1] - [A_3]$ for all short exact sequences $0 \longrightarrow A_1 \longrightarrow A_2 \longrightarrow A_3 \longrightarrow 0$. Note that in particular $[A] = [A']$ in $K(\mathcal{A})$ if $A \simeq A'$ and $[A \oplus B] = [A] + [B]$. Thus, clearly

$$K(Y) = K(\text{Coh}(Y)).$$

There is yet another categorical interpretation of $K(Y)$ which relies on the bounded derived category $\text{D}^b(Y) := \text{D}^b(\text{Coh}(Y))$ viewed as a triangulated category. For an arbitrary (small) triangulated category $\mathcal{D}$ one defines $K(\mathcal{D})$ as the quotient of the free abelian group generated by the objects of $\mathcal{D}$ modulo the subgroup generated by elements of the form $[A_2] - [A_1] - [A_3]$ for all exact triangles $A_1 \longrightarrow A_2 \longrightarrow A_3 \longrightarrow A_1[1]$.

For the notion of a triangulated category and, in particular, of exact triangles, see [208, 613] and Section **16**.1.1. Since the identity $A = A$ gives rise to an exact triangle $A \longrightarrow 0 \longrightarrow A[1] \longrightarrow A[1]$, one has $[A[1]] = -[A]$ for all objects A. As any object in the bounded derived category $D^b(\mathcal{A})$ of an abelian category $\mathcal{A}$ admits a finite filtration with 'quotients' isomorphic to shifts of objects in $\mathcal{A}$, there is a natural isomorphism $K(D^b(\mathcal{A})) \simeq K(\mathcal{A})$. Applied to our case, one finds

$$K(Y) = K(\mathrm{Coh}(Y)) \simeq K(D^b(Y)).$$

For a smooth and quasi-projective variety Y, the Grothendieck group can equivalently be defined as the free abelian group generated by locally free sheaves modulo short exact sequences as before. Indeed, any coherent sheaf on Y admits a finite locally free resolution $0 \longrightarrow F_n \longrightarrow \cdots \longrightarrow F_0 \longrightarrow F \longrightarrow 0$ and thus $[F] = \sum (-1)^i [F_i]$. The advantage of working with locally free sheaves only is that the tensor product induces on $K(Y)$ the structure of a commutative ring by

$$[F] \cdot [F'] := [F \otimes F'].$$

The Grothendieck group and the Chow group can be compared via the *Chern character*. The Chern character defines a ring homomorphism

$$\mathrm{ch} \colon K(Y) \longrightarrow \mathrm{CH}^*(Y)_\mathbb{Q}.$$

(Recall that for abelian groups G we use the shorthand $G_\mathbb{Q} := G \otimes_\mathbb{Z} \mathbb{Q}$.) The Chern classes $c_i(F)$ of a coherent sheaf F itself are elements in $\mathrm{CH}^i(Y)$, but the Chern character $\mathrm{ch}(F)$ has non-trivial denominators in general. However, it induces a ring isomorphism

$$\mathrm{ch} \colon K(Y)_\mathbb{Q} \xrightarrow{\sim} \mathrm{CH}^*(Y)_\mathbb{Q}.$$

Observe that $\mathrm{ch}(\mathcal{O}_Z) = [Z]$ mod $\mathrm{CH}^{*>n}(Y)$ for any subvariety $Z \subset Y$ of codimension n.

1.4   Let us come back to the case of a K3 surface X. Then the Chern character of a sheaf F on X is given as

$$\mathrm{ch}(F) = \mathrm{rk}(F) + c_1(F) + \frac{(c_1^2 - 2c_2)(F)}{2}.$$

Here, $\mathrm{rk}(F)$ is the dimension of the fibre of F at the generic point $\eta \in X$, i.e. $\mathrm{rk}(F) = \dim_{K(X)}(F_\eta)$, and $c_1(F) = c_1(\det(F))$. If F is globally generated and locally of rank two, then $c_2(F)$ can be represented by $[Z(s)]$, where $Z(s)$ is the zero locus of a regular section $s \in H^0(X, F)$. Proposition 1.2 can be used to show that Chern characters of sheaves on K3 surfaces are in fact integral, at least for $k = \bar{k}$.

Corollary 1.5 *Let X be a K3 surface over an algebraically closed field k. Then the Chern character naturally defines an isomorphism of rings*

$$\text{ch}\colon K(X) \xrightarrow{\sim} \text{CH}^*(X).$$

Proof By the Riemann–Roch formula $\deg(c_1(L)^2) = (L)^2$ is even. Thus, for algebraically closed k it is divisible by two in the image of the surjection $\deg\colon \text{CH}^2(X) \twoheadrightarrow \mathbb{Z}$. On the other hand, by Proposition 1.2 the kernel of deg, i.e. $\text{CH}^2(X)_0$, is divisible for $k = \bar{k}$ and hence $(1/2)c_1(L)^2$ exists uniquely, due to the absence of torsion in $\text{CH}^2(X)$; see Theorem 1.3.

Next we prove that $\text{ch}\colon K(X) \longrightarrow \text{CH}^*(X)$ is surjective. Indeed, the generator $1 = [X]$ of $\text{CH}^0(X)$ equals $\text{ch}(\mathcal{O}_X)$ and $[x] = \text{ch}(k(x))$ for all closed points $x \in X$. Thus $\text{CH}^0(X) \oplus \text{CH}^2(X)$ is contained in the image. As $(1/2)c_1(L)^2 \in \text{CH}^2(X)$ for all $L \in \text{Pic}(X)$ and $\text{ch}(L) = 1 + c_1(L) + (1/2)c_1(L)^2$, all first Chern classes $c_1(L)$ are in the image of the Chern character, i.e. $\text{CH}^1(X) \subset \text{Im}(\text{ch})$.

To prove injectivity, one shows that for any smooth surface Y there are natural isomorphisms

$$\text{rk}\colon F^0 K(Y)/F^1 K(Y) \xrightarrow{\sim} \mathbb{Z},$$

$$c_1\colon F^1 K(Y)/F^2 K(Y) \xrightarrow{\sim} \text{Pic}(Y), \quad \text{and} \quad c_2\colon F^2 K(Y) \xrightarrow{\sim} \text{CH}^2(Y).$$

Here, $F^i K(Y)$ is the subgroup generated by sheaves with support of codimension $\geq i$. In particular, $\text{ch}\colon K(Y) \longrightarrow \text{CH}^*(Y)_{\mathbb{Q}}$ is always injective. See [192, Ex. 15.3.6]. □

It would be interesting to find a direct proof for the torsion freeness of $K(X)$.

2 Chow Groups: Mumford and Bloch–Beĭlinson

After these general results, we now pass to things that are more specific to K3 surfaces. In fact, although we shall state the results for K3 surfaces only, often the condition $p_g(X) := h^0(X, \omega_X) > 0$ suffices.

2.1 We start with a celebrated result of Mumford for K3 surfaces over $\mathbb{C}$ (or over any uncountable algebraically closed field of characteristic zero). In [439] he disproves an old claim of Severi that the group of 0-cycles modulo rational equivalence is always finite-dimensional by showing that the dimension of the image of the natural map

$$\sigma_n\colon X^n \times X^n \longrightarrow \text{CH}^2(X)_0, \quad ((x_1, \ldots, x_n), (y_1, \ldots, y_n)) \longmapsto \sum ([x_i] - [y_i])$$

cannot be bounded. To make this precise, we need a few preparations. See [621] for details and more general results.

Clearly, the map $\sigma_n \colon X^n \times X^n \longrightarrow CH^2(X)_0$ factorizes over the symmetric product

$$\sigma_n \colon S^n(X) \times S^n(X) \longrightarrow CH^2(X)_0$$

and we shall rather work with the latter.

Proposition 2.1   *The fibres of the map $\sigma_n \colon S^n(X) \times S^n(X) \longrightarrow CH^2(X)_0$ are countable unions of closed subvarieties. Moreover, there exists a countable union $Y \subset S^n(X) \times S^n(X)$ of proper subvarieties such that for all points (Z_1, Z_2) in the complement of Y the maximal dimension of $\sigma_n^{-1}\sigma_n(Z_1, Z_2)$ is constant.*

Proof   The very rough idea goes as follows. Cycles $Z_1, Z_2 \in S^n(X)$ that define the same class $\alpha \in CH^2(X)_0$ are obtained by adding cycles of the form $\mathrm{div}_0(f) + D$ and $\mathrm{div}_\infty(f) + D$ to Z_1 and Z_2, respectively. Here, f is a rational function on some curve C in X, $\mathrm{div}_0(f)$ and $\mathrm{div}_\infty(f)$ are its zero and pole divisor, and D is just some divisor on C.

These data are parametrized by certain Hilbert schemes and thus form a countable set of varieties. For more details, see [621, Lem. 22.7]. $\square$

Let now f_n be the dimension of the generic fibre $\sigma_n^{-1}\sigma_n(Z_1, Z_2)$ in the sense of the proposition. Although the image of σ_n does not have the structure of a variety, one can talk about its dimension.

Definition 2.2   The image dimension of σ is defined as

$$\dim(\mathrm{Im}(\sigma_n)) := \dim(S^n(X) \times S^n(X)) - f_n = 4n - f_n.$$

The following result then says that the 'dimension' of $CH^2(X)_0$ is infinite.

Theorem 2.3 (Mumford)   *For a complex K3 surface X one has*

$$\lim \dim(\mathrm{Im}(\sigma_n)) = \infty.$$

Proof   The key idea is the following. The fibres of σ_n are countable unions of subvarieties. The generator of $H^0(X, \Omega_X^2)$ induces a non-degenerate regular two-form on $X^n \times X^n$ which is symmetric and hence descends to a generically non-degenerate two-form on $S^n(X) \times S^n(X)$. (Restrict to the smooth part to avoid the singularities.) Morally (but not literally!), the components of the fibres of σ_n tend to be rationally connected, for they parametrize rationally equivalent cycles. Since a rationally connected variety does not admit any non-trivial two-form (see [312, IV Cor. 3.8]), the components of the fibres should be of dimension at most $(1/2) \dim(S^n(X) \times S^n(X))$. $\square$

Remark 2.4   It turns out that for general surfaces $CH^2(X)_0$ is finite-dimensional if and only if σ_n is surjective for large n; see [621, Prop. 22.10]. Furthermore, this condition for a K3 surface is equivalent to $CH^2(X)_0 = 0$ or, still equivalent, to $\mathrm{Jac}(C) \longrightarrow CH^2(X)_0$ for some ample curve $C \subset X$. The latter is the notion of finite-dimensionality used by Bloch in [67]. So K3 surfaces over $\mathbb{C}$ have infinite-dimensional $CH^2(X)$.

2.2 In contrast to Mumford's result for K3 surfaces over $\mathbb{C}$ (or, more generally, uncountable algebraically closed fields of characteristic zero) the situation is expected to be completely different for K3 surfaces over global fields, e.g. over number fields.

Conjecture 2.5 (Bloch–Beĭlinson) *If X is a K3 surface over a number field k (i.e. a finite field extension of $\mathbb{Q}$), then the degree map defines an isomorphism*

$$\mathrm{CH}^2(X)_{\mathbb{Q}} \simeq \mathbb{Q}.$$

If X is a K3 surface over $\overline{\mathbb{Q}}$, then $\mathrm{CH}^2(X) \simeq \mathbb{Z}$.

This is only a special case of much deeper conjectures generalizing the conjecture of Birch and Swinnerton-Dyer for elliptic curves; see [56, 68, 504]. However, there is essentially no evidence for this conjecture. There is not a single K3 surface X known that is defined over a number field and has $\mathrm{CH}^2(X)_{\mathbb{Q}} \simeq \mathbb{Q}$. In fact, it seems we do not even have examples where any kind of finiteness result for $\mathrm{CH}^2(X)_0$ has been established. As shall be briefly mentioned below, $\mathrm{CH}^2(X)_0$ often contains torsion classes which after base change to $\overline{\mathbb{Q}}$ become trivial.

Remark 2.6 It is expected that the conjecture fails when one replaces $\overline{\mathbb{Q}}$ by the minimal algebraically closed field of definition. But to the best of my knowledge, there has never been given an explicit example for this; i.e. there does not seem to be known an example of a K3 surface X defined over an algebraically closed field k with $\mathrm{trdeg}_{\mathbb{Q}}(k) > 0$ and not over any field of smaller transcendence degree with $\mathrm{CH}^2(X) \neq \mathbb{Z}$.[2]

Remark 2.7 There is a different set of finiteness conjectures due to Bass.[3] For a smooth projective variety X over a field k which is finitely generated over its prime field, the Grothendieck group $K(X)$ is conjectured to be finitely generated; see [35, Chap. XIII]. Note that this in particular predicts that for a K3 surface over a number field $\mathrm{CH}^2(X)$ should be finitely generated, but (up to torsion) the conjecture of Bloch–Beĭlinson is more precise. A priori Bass's conjectures do not explain why passing from a number field to $\overline{\mathbb{Q}}$ the rank of $\mathrm{CH}^2(X)$ does not increase. On the other hand, Bass's conjectures also predict that $\mathrm{CH}^2(X)$ is finitely generated for fields which are finite extensions of $\mathbb{Q}(t)$ or $\mathbb{F}_p(t_1, t_2)$ (or other purely transcendental extensions of the prime field of finite transcendence degree). Compare this with the results in Sections 2.3 and 2.4.

---

[2] In [279, App. B] one finds an example due to Schoen of a K3 surface X over some finite extension of $\mathbb{Q}(t)$ that cannot be defined over $\overline{\mathbb{Q}}$ and for which $\mathrm{CH}^2(X)_0$ is of infinite rank. However, this example becomes isotrivial after passing to the algebraic closure of $\mathbb{Q}(t)$. Thanks to Stefan Schreieder for pointing this out.

[3] I wish to thank Jean-Louis Colliot-Thélène for the reference and explanations.

Example 2.8 Note that while we do not have a single example confirming the Bloch–Beĭlinson conjecture, we have plenty of examples confirming the conjecture of Bass. For example, it is not difficult to show that $CH^2(X) \simeq \mathbb{Z}$ for the generic fibre $X := \mathcal{X}_\eta$ of the universal quartic $\mathcal{X} \subset |\mathcal{O}(4)| \times \mathbb{P}^3_\mathbb{Q}$. So, here the base field is the finitely generated field $k(\eta) = \mathbb{Q}(t_1, \ldots, t_{34})$.[4] Note also that considering the same situation over $\mathbb{C}$ yields an example of a K3 surface with $CH^2(X) \simeq \mathbb{Z}$ over the field $\mathbb{C}(t_1, \ldots, t_{34})$, which certainly is not finitely generated.

2.3 The Chow group can change under base field extension. Suppose a K3 surface X is defined over a field k and $k \subset K$ is a field extension. The pull-back defines a natural homomorphism

$$CH^*(X) \longrightarrow CH^*(X_K), \; Z \longmapsto Z_K.$$

Clearly, $CH^0(X) \xrightarrow{\sim} CH^0(X_K)$ and $CH^1(X) \hookrightarrow CH^1(X_K)$; see Section **17**.2.1. In degree two the map is in general neither injective nor surjective. However, its kernel is purely torsion, due to the following easy lemma.

Lemma 2.9 *For any field extension $k \subset K$ the pull-back map*

$$CH^2(X)_\mathbb{Q} \hookrightarrow CH^2(X_K)_\mathbb{Q}$$

is injective.

Proof Consider first a finite extension $k \subset K$. Then the natural projection $\pi : X_K \longrightarrow X$ is a finite morphism of degree $[K : k]$ and thus satisfies

$$\pi_* \pi^* \alpha = [K : k] \cdot \alpha$$

for all $\alpha \in CH^*(X)$. This is a special case of the projection formula; see e.g. [236, p. 426]. Hence, if $\pi^* \alpha \in CH^2(X_K)$ is zero, then $\alpha \in CH^2(X)$ was at least torsion. This proves the result for any finite (and then also for any algebraic) field extension. Below we reduce the general result to this case.

Let now $k \subset K$ be an arbitrary field extension $k \subset K$. If $Z \in CH^2(X)$ is in the kernel of the pull-back $CH^2(X) \longrightarrow CH^2(X_K)$, then Z becomes trivial after a finitely generated field extension $k \subset L \subset K$. Indeed, the rational equivalence making Z_K trivial over K involves only finitely many curves C_i and rational functions on them. The finitely many coefficients needed to define these curves with the rational functions generate a field L. In fact, we may assume that L is the quotient field of a finitely generated k-algebra A and the curves C_i are defined over A. Now think of X_L as the generic fibre of the 'spread' $X \times_k \operatorname{Spec}(A) \longrightarrow \operatorname{Spec}(A)$. In particular, for any closed point $a \in \operatorname{Spec}(A)$

---

[4] We emphasize again that no explicit examples of K3 surfaces over a number field seem to be known for which $CH^2(X)$ is finitely generated.

the restriction of $Z_{\mathrm{Spec}(A)}$ to the fibre $X \times \mathrm{Spec}(k(a))$ is rationally equivalent to zero by means of the restriction of the curves C_i. As $k(a)$ is a finite field extension of k and the restriction of $Z_{\mathrm{Spec}(A)}$ to the fibre over a is nothing but $Z_{k(a)}$, this shows by step one that Z was torsion. □

Note that for finite Galois extensions K/k with Galois group G, the cokernel of the base-change map $\mathrm{CH}^2(X) \longrightarrow \mathrm{CH}^2(X_K)^G$ is torsion (see [122, §2] and compare this with the discussion in Section 17.2.2), i.e.

$$\mathrm{CH}^2(X)_{\mathbb{Q}} \xrightarrow{\ \sim\ } \mathrm{CH}^2(X_K)^G_{\mathbb{Q}}.$$

The following result due to Bloch; see [67].

Proposition 2.10 *Let X be a K3 surface over an arbitrary field k such that $\rho(X \times_k \bar{k}) < 22$. If $K = k(X)$ denotes the function field of X, then*

$$\mathrm{CH}^2(X)_{\mathbb{Q}} \longrightarrow \mathrm{CH}^2(X_K)_{\mathbb{Q}}$$

is not surjective.

Proof The construction of an extra cycle is very explicit. Consider the diagonal $\Delta \subset X \times X$ and its restriction Δ_K to the generic fibre $X_K = X \times \mathrm{Spec}(K) \subset X \times X$ of the second projection. One now proves that $[\Delta_K] \in \mathrm{CH}^2(X_K)_{\mathbb{Q}}$ is not contained in $\mathrm{CH}^2(X)_{\mathbb{Q}}$. For this, one can certainly pass to the algebraic closure of k, and, therefore, we may simply assume $k = \bar{k}$.

Suppose $[\Delta_K]$ was contained in $\mathrm{CH}^2(X)_{\mathbb{Q}}$, i.e. $[\Delta_K] = \sum n_i[x_i]$ in $\mathrm{CH}^2(X_K)_{\mathbb{Q}}$ for certain $n_i \in \mathbb{Q}$ and closed points $x_i \in X$ with their associated classes $[x_i] \in \mathrm{CH}^2(X)_{\mathbb{Q}} \subset \mathrm{CH}^2(X_K)_{\mathbb{Q}}$. In other words, there exist curves $C_i \subset X_K$ and rational functions $f_i \in K(C_i)$ such that $\Delta_K = \sum n_i x_i + \sum \mathrm{div}_{C_i}(f_i)$ as cycles on $X_K \subset X \times X$. Taking the closure in $X \times X$ yields

$$\Delta = \sum n_i(\{x_i\} \times X) + \sum D_i + V \tag{2.1}$$

as cycles on $X \times X$. Here, $V \subset X \times X$ does not meet the generic fibre (and therefore does not dominate the second factor) and $D_i := \mathrm{div}_{\overline{C}_i}(f_i)$ with $\overline{C}_i$ the closure of C_i.

Both sides of the equation can be viewed as cohomological correspondences. In characteristic zero one could pass to the associated complex surfaces and use singular cohomology. Otherwise use ℓ-adic étale cohomology, $\ell \neq \mathrm{char}(k)$.

Clearly, $[\Delta]_*$ is the identity on $H^2_{\text{ét}}(X, \mathbb{Q}_\ell(1))$. On the other hand, $[\{x_i\} \times X]_*$ acts trivially on $H^2_{\text{ét}}(X, \mathbb{Q}_\ell(1))$ for degree reasons and the D_i are rationally and hence homologically trivial.[5] Thus, $\mathrm{id} = [\Delta]_* = [V]_*$.

---

[5] At this point one uses that the cycle map factors through the Chow ring and, in this case more precisely, that $\mathrm{CH}^2(X \times X) \longrightarrow H^4_{\text{ét}}(X \times X, \mathbb{Q}_\ell(2))$ is well defined.

Under the assumption $\rho(X) < 22$, the first Chern class induces a proper inclusion $\mathrm{NS}(X)_{\mathbb{Q}_\ell} \subset H^2_{\acute{e}t}(X, \mathbb{Q}_\ell(1))$; cf. Section **17**.2.2. Since the image of V under the second projection is supported in dimension ≤ 1, the image of $[V]_*$ is contained in $\mathrm{NS}(X)_{\mathbb{Q}_\ell}$. Contradiction. $\qquad\qquad\qquad\qquad\qquad\qquad\qquad\qquad\qquad\qquad\qquad\qquad\qquad\square$

Remark 2.11   Clearly, in characteristic zero the assumption on the Picard group is superfluous. One always has $\rho(X \times_k \bar{k}) \leq 20$, which can be proved by Hodge theory over $\mathbb{C}$; see Sections **1**.3.3 or **17**.1.1.

However, for K3 surfaces in positive characteristic the case $\rho(X \times_k \bar{k}) = 22$ may occur and the above argument breaks down. In fact, it was conjectured and has now been proved by Liedtke in [373] that for a K3 surface X the condition $\rho(X \times_k \bar{k}) = 22$ is equivalent to X being unirational; see Proposition **17**.2.7 and Section **18**.3.5. Any unirational surface satisfies $\mathrm{CH}^2(X)_{\mathbb{Q}} \simeq \mathbb{Q}$ and, since a unirational variety remains unirational after base change, the Chow group does indeed not grow after passage to $X_{k(X)}$ or any other field extension.

As an immediate consequence, we obtain the following weak form of Mumford's result; cf. Theorem 2.3.

Corollary 2.12   *Let X be a K3 surface over $\mathbb{C}$. Then $\dim_{\mathbb{Q}} \mathrm{CH}^2(X)_{\mathbb{Q}} = \infty$.*

Proof   Indeed, X is defined over the algebraic closure k_0 of a finitely generated extension of $\mathbb{Q}$, i.e. $X = X_0 \times_{k_0} \mathbb{C}$, and by choosing inductively k_i to be the algebraic closure of $K(X_0 \times_{k_0} k_{i-1})$ and embeddings $k_0 \subset k_1 \subset \cdots \subset \mathbb{C}$ one obtains a strictly ascending chain of vector spaces

$$\mathrm{CH}^2(X_0)_{\mathbb{Q}} \subsetneqq \mathrm{CH}^2(X_0 \times_{k_0} k_1)_{\mathbb{Q}} \subsetneqq \cdots \subsetneqq \mathrm{CH}^2(X_0 \times_{k_0} \mathbb{C})_{\mathbb{Q}}. \qquad\qquad \square$$

Remark 2.13   (i) The same arguments show that for every K3 surface X over an algebraically closed field k of infinite transcendence degree over its prime field, one has $\dim_{\mathbb{Q}} \mathrm{CH}^2(X)_{\mathbb{Q}} = \infty$ provided that $\rho(X) < 22$.

(ii) In fact, Bloch uses similar methods to prove the full result of Mumford, i.e. that there is no curve $C \subset X$ (possibly disconnected) such that $\mathrm{Pic}^0(C) \longrightarrow \mathrm{CH}^2(X)_0$ is surjective; cf. Remark 2.4. For details, see [67, App. Lect. 1].

In characteristic zero, an analogous construction can be used to show that the Chow group increases already after base change to an algebraically closed field of transcendence degree one. The following is based on the paper by Green, Griffiths, and Paranjape [217] and works more generally for surfaces with $p_g \neq 0$.

Proposition 2.14   *Let X be a K3 surface over a field k of characteristic zero. If K is an algebraically closed extension of k with $\mathrm{trdeg}_k(K) \geq 1$, then*

$$\mathrm{CH}^2(X)_{\mathbb{Q}} \longrightarrow \mathrm{CH}^2(X_K)_{\mathbb{Q}}$$

is not surjective.

Proof Similar to the proof of Proposition 2.10, one constructs a certain cycle $Z \subset X \times C$, whose generic fibre over C defines a class that is not contained in the image of the pull-back $CH^2(X)_\mathbb{Q} \longrightarrow CH^2(X_{k(C)})_\mathbb{Q}$. Here, C is a smooth curve with function field $k(C)$. For the following we can assume that k is algebraically closed of finite transcendence degree with an embedding $k \subset \mathbb{C}$, which allows us to use Hodge theory for the complex manifolds $X_\mathbb{C}$ and $C_\mathbb{C}$.

The cycle Z is constructed as follows. First, consider the diagonal $\Delta \subset X \times X$ and its action $[\Delta]_*$ on $H^*(X_\mathbb{C}, \mathbb{Q})$. In degree two it respects the decomposition $H^2(X_\mathbb{C}, \mathbb{Q}) = \mathrm{Pic}(X_\mathbb{C})_\mathbb{Q} \oplus T(X_\mathbb{C})_\mathbb{Q}$. Here, $T(X_\mathbb{C})$ is the transcendental lattice; cf. Section **3**.2.2. On $\mathrm{Pic}(X_\mathbb{C})_\mathbb{Q}$ one can describe $[\Delta]_*$ as the action of a cycle of the form $\sum m_i(C_i \times D_i)$ with curves $C_i, D_i \subset X$ and $m_i \in \mathbb{Q}$. Since $\mathrm{Pic}(X) \simeq \mathrm{Pic}(X_\mathbb{C})$ (cf. Lemma **17**.2.2), we may assume that indeed these curves exist over k. Now, fix a closed point $x \in X$ and let

$$Y := \Delta - \sum m_i(C_i \times D_i) - \{x\} \times X.$$

Viewed as a correspondence from the second to the first factor, it acts trivially on $H^4(X_\mathbb{C}, \mathbb{Q})$ and $\mathrm{Pic}(X)$ and as the identity on the transcendental part $T(X_\mathbb{C})$.[6] Now pick a smooth curve $C \subset X$ and let Z be the pull-back of Y under the natural inclusion $X \times C \subset X \times X$.

One checks that Z is homologically trivial, i.e. $0 = [Z]_* : H^*(C_\mathbb{C}, \mathbb{Q}) \longrightarrow H^*(X_\mathbb{C}, \mathbb{Q})$. For example, for the generator $[C] \in H^0(C_\mathbb{C}, \mathbb{Z})$ one computes, using the projection formula, that $[Z]_*[C] = [Y]_*[C]$, where on the right-hand side $[C] \in H^2(X_\mathbb{C}, \mathbb{Z})$ is contained in the Picard group and hence $[Y]_*[C] = 0$. A similar argument works for the generator of $H^2(C_\mathbb{C}, \mathbb{Z})$. Furthermore, the image of $H^1(C_\mathbb{C}, \mathbb{Z})$ is contained in $H^3(X_\mathbb{C}, \mathbb{Z})$ and hence is trivial.

As a homologically trivial cycle, Z on the complex threefold $X_\mathbb{C} \times C_\mathbb{C}$ is the boundary $\partial\Gamma$ of a real three-dimensional cycle $\Gamma \subset X_\mathbb{C} \times C_\mathbb{C}$. This yields a map

$$H^2(X_\mathbb{C}) \times H^1(C_\mathbb{C}) \longrightarrow \mathbb{C}, \quad (\alpha, \beta) \longmapsto \int_\Gamma p_X^*\alpha \wedge p_C^*\beta, \qquad (2.2)$$

which is well defined up to classes in $H_3(X_\mathbb{C} \times C_\mathbb{C}, \mathbb{Z})$. In other words, we are considering the Abel–Jacobi class of Z in the intermediate Jacobian $J^3(X_\mathbb{C} \times C_\mathbb{C})$; see [621, Ch. 12]. At this point one has to check that the pairing (2.2) is non-trivial on $T(X_\mathbb{C}) \times H^1(C_\mathbb{C})$ (up to integral classes) for sufficiently generic C. In fact, it suffices to choose a generic member of a pencil on X. For the details of this part of the argument, see [217].

The rest is similar to the arguments in the proof of Proposition 2.10. Suppose $Z_{k(C)} \in CH^2(X_{k(C)})_\mathbb{Q}$ is of the form $\sum n_i x_i$ for certain closed points $x_i \in X$. Since

---

[6] This construction is inspired by Murre's decomposition of the diagonal for surfaces; see [445].

Z is homologically trivial, one automatically has $\sum n_i = 0$. Then the closure of Z in $X \times C$ is of the form

$$\sum n_i(\{x_i\} \times C) + \sum D_i + V$$

with $[D_i] = 0$ in $\mathrm{CH}^2(X \times C)$ and such that the image of V under $X \times C \longrightarrow C$ consists of a finite number of points. This is the analogue of (2.1).

The Abel–Jacobi map is defined on the homologically trivial part of $\mathrm{CH}^2(X \times C)$ and in particular trivial on the rationally trivial cycles D_i. One now shows that also $\sum n_i(\{x_i\} \times C)$ and V are trivial under the Abel–Jacobi map. More precisely, they define trivial pairings on $T(X_\mathbb{C}) \times H^1(C_\mathbb{C})$. Indeed, a cycle $\Gamma_0 \subset X_\mathbb{C} \times C_\mathbb{C}$ with $\partial \Gamma_0 = \sum n_i(\{x_i\} \times C_\mathbb{C})$ can be obtained, by connecting the points $x_i \in X$ by real paths γ and then taking the product with $C_\mathbb{C}$. Clearly, the integral $\int_{\Gamma_0}$ is then trivial on classes of the form $\alpha \wedge \beta$, as the two-form α vanishes when restricted to the paths γ.

As the vertical cycle V lives over finitely many points $y_i \in C$, it is of the form $\sum m_i(C_i \times \{y_i\})$. Using paths $\gamma \subset C_\mathbb{C}$, one constructs a cycle Γ_1 with $\partial \Gamma_1 = V_\mathbb{C}$ with components of the form $C_i \times \gamma$. Now use that $T(X_\mathbb{C})$ is contained in the kernel of the restriction map $H^2(X_\mathbb{C}, \mathbb{Q}) \longrightarrow H^2(C_i, \mathbb{Q})$ to deduce that $\int_{C_i \times \gamma} \alpha \wedge \beta = 0$ for all $\alpha \in T(X_\mathbb{C})$.   □

As we shall briefly mention below, the assumption on the characteristic is essential, e.g. for $k = \overline{\mathbb{F}}_p$ the result does not hold; cf. Remark 2.18.

Remark 2.15   Passing to an algebraically closed extension of positive transcendence degree not only does one make the Chow group bigger, but one even expects $\mathrm{CH}^2(X)_\mathbb{Q}$ to become infinite-dimensional right away. An explicit example has been worked out by Schoen [530]: for the Fermat elliptic curve E over $\mathbb{Q}$ it is shown that $\mathrm{CH}^2((E \times E)_{\overline{\mathbb{Q}(E)}})_0$ is infinite-dimensional, i.e. not concentrated on a curve; see Remark 2.4.

Summarizing, one can say that cohomological methods can be used to prove non-triviality of classes, but there are no techniques known, cohomological or other, that would prove triviality of cycles in an effective way, i.e. that potentially could lead to a proof of the Bloch–Beĭlinson conjecture 2.5.

2.4   Let us add a few comments on the situation over finite fields. The following is a folklore result.

Proposition 2.16   *Let X be a smooth projective variety of dimension n over a finite field k. Then $\mathrm{CH}^n(X)_0$ is torsion (and in fact finite). In particular, if X is a K3 surface, then $\mathrm{CH}^2(X)_0$ is torsion (and in fact trivial).*

Proof   For any cycle $Z = \sum n_i[x_i] \in \mathrm{CH}^n(X)$, there is a curve $C \subset X$ defined over some finite extension k' of k such that all points x_i are contained in $C(k')$. For simplicity we shall assume that C is smooth, otherwise work with its normalization. If Z is of degree zero, i.e. $Z \in \mathrm{CH}^n(X)_0$, then Z defines a k'-rational point of $\mathrm{Pic}^0(C')$, where $C' := C \times_k k'$. However, the group of k'-rational points of $\mathrm{Pic}^0(C')$ is finite for a finite field k'. Hence, Z as an element in $\mathrm{Pic}^0(C')$ must be torsion. Since the push-forward $\mathrm{CH}^*(C') \longrightarrow \mathrm{CH}^*(X_{k'})$ is additive, this shows that $Z \in \mathrm{CH}^n(X_{k'})_0$ is torsion. Since the kernel of $\mathrm{CH}^n(X) \longrightarrow \mathrm{CH}^n(X_{k'})$ is torsion by Lemma 2.9, this proves the assertion.

To prove finiteness of $\mathrm{CH}^n(X)_0$ and triviality of $\mathrm{CH}^2(X)_0$ for K3 surfaces, one uses a result of Kato and Saito (cf. [122, Thm. 5.3]), which describes $\mathrm{CH}^n(X)_0$ as the kernel of the natural map $\pi_1^{ab}(X) \longrightarrow \hat{\mathbb{Z}}$, which is trivial for K3 surfaces. Alternatively, at least for the ℓ-torsion, one can use that $\mathrm{CH}^2(X)[\ell] \hookrightarrow H^4_{\acute{e}t}(X, \mathbb{Z}_\ell(2))$ due to a result of Colliot-Thélène, Sanscu, and Soulé [126, Cor. 3]. □

Corollary 2.17 *Let X be a K3 surface over $\overline{\mathbb{F}}_p$. Then $\mathrm{CH}^2(X) \simeq \mathbb{Z}$.*

Proof Let $x, y \in X$ be closed points. Then there exists a finite field extension $\mathbb{F}_q$ of $\mathbb{F}_p$ such that X and x, y are defined over $\mathbb{F}_q$; i.e. there exists a K3 surface X_0 over $\mathbb{F}_q$ with $X = X_0 \times_{\mathbb{F}_q} \overline{\mathbb{F}}_p$ and such that x, y are obtained by base changing some $\mathbb{F}_q$-rational points $x_0, y_0 \in X_0$. Thus, the class $[x] - [y] \in \mathrm{CH}^2(X)_0$ is contained in the image of $\mathrm{CH}^2(X_0)_0 \longrightarrow \mathrm{CH}^2(X)_0$.

However, by Proposition 2.16, $\mathrm{CH}^2(X_0)_0$ is torsion and hence $[x] - [y] \in \mathrm{CH}^2(X)_0$ is torsion. Theorem 1.3 then shows $\mathrm{CH}^2(X)_0 = 0$ and thus $\mathrm{CH}^2(X) \simeq \mathbb{Z}$. □

Remark 2.18 The Bloch–Beĭlinson conjecture predicts properties of cycle groups over global fields. In particular, it would say that for a K3 surface over finite extensions of $\mathbb{F}_p(t)$, the group $\mathrm{CH}^2(X)_0$ is torsion. Equivalently, one expects that for a K3 surface over the algebraic closure of $\mathbb{F}_p(t)$ the group $\mathrm{CH}^2(X)_0$ is trivial.

Note that in particular Proposition 2.14 is not expected to generalize to positive characteristic even for K3 surfaces with $\rho(X) < 22$; cf. Remark 2.11. Schoen shows in [530, Prop. 3.2] that for a K3 surface X over $\mathbb{F}_q$ that is dominated by a product of curves (e.g. a Kummer surface) the Chow group is just $\mathbb{Z}$ after base change to the algebraic closure of $\mathbb{F}_p(t)$. This holds true for all K3 surfaces which are finite-dimensional in the sense of Kimura–O'Sullivan; see [262].

3 Beauville–Voisin Ring

Due to Mumford's result, the Chow group $\mathrm{CH}^2(X)$ of a complex K3 surface X is infinite-dimensional; see Theorem 2.3. Besides this fact, very little is known about $\mathrm{CH}^2(X)$. In this section we discuss a result of Beauville and Voisin showing that the cycle map $\mathrm{CH}^*(X) \longrightarrow H^*(X, \mathbb{Z})$ can be split multiplicatively by a natural subring

$R(X) \subset \mathrm{CH}^*(X)$. Moreover, the ring $R(X)$ contains many interesting characteristic classes of bundles that we have encountered earlier.

3.1 Let X be a complex algebraic K3 surface.

Definition 3.1 The *Beauville–Voisin ring*

$$R(X) \subset \mathrm{CH}^*(X)$$

is the subring generated by the Chow–Mukai vectors

$$v^{\mathrm{CH}}(L) := \mathrm{ch}(L)\sqrt{\mathrm{td}(X)} \in \mathrm{CH}^*(X)$$

of all line bundles $L \in \mathrm{Pic}(X)$.

Theorem 3.2 (Beauville–Voisin) *Let X be a complex algebraic K3 surface. Then the cycle map $\mathrm{CH}^*(X) \longrightarrow H^*(X, \mathbb{Z})$ induces an isomorphism of rings*

$$R(X) \xrightarrow{\sim} H^0(X, \mathbb{Z}) \oplus \mathrm{NS}(X) \oplus H^4(X, \mathbb{Z}).$$

The theorem, or rather its proof sketched below, is spelled out by the following corollary.

Corollary 3.3 *There exists a distinguished class (the Beauville–Voisin class)*

$$c_X \in \mathrm{CH}^2(X)$$

of degree one with the following properties:

  (i) *If $x \in X$ is contained in a (possibly singular) rational curve $C \subset X$, then $[x] = c_X$.*
  (ii) *For any $L \in \mathrm{Pic}(X)$, one has $c_1^2(L) \in \mathbb{Z} \cdot c_X \subset \mathrm{CH}^2(X)$.*
  (iii) *$c_2(X) = 24c_X$.* □

Proof Since $\mathrm{Pic}(X) \simeq \mathrm{NS}(X)$, the cycle map $R(X) \longrightarrow H^*(X, \mathbb{Z})$ is injective in degree at most one. Thus, only the injectivity of $R^2(X) \longrightarrow H^4(X, \mathbb{Z})$ needs to be proved. We present only an outline of the main arguments and refer to [53] for the details.

The first step of the proof consists in showing that the classes $c_1(L)^2 \in \mathrm{CH}^2(X)$ for line bundles $L \in \mathrm{Pic}(X)$ are all contained in a subgroup of $\mathrm{CH}^2(X)$ of rank one. This is equivalent to showing that for two line bundles L_1, L_2 the classes $c_1(L_1)^2$ and $c_1(L_2)^2$ are linearly dependent. Since $\mathrm{Pic}(X)$ is spanned by ample line bundles, it is enough to prove this for L_1 and L_2 ample.

Now use the theorem of Bogomolov and Mumford (see Theorem **13**.1.1 and Corollary **13**.1.5) which implies that any ample divisor is linearly equivalent to a sum of rational curves. Since any ample curve is 1-connected by Remark **2**.1.7, it suffices to show that for irreducible rational curves $C_1, C_2, C_3 \subset X$, the products

$$C_i.C_j = c_1(\mathcal{O}(C_i)).c_1(\mathcal{O}(C_j)) \in \mathrm{CH}^2(X)$$

are linearly dependent. As all points on an irreducible rational curve C are rationally equivalent, one has $c_1(\mathcal{O}(C)).c_1(\mathcal{O}(C_i)) = (C.C_i) \cdot [x]$ for any point $x \in C$.

This first part of the proof in particular shows that all points $x \in X$ contained in some rational curve are rationally equivalent; i.e. they all define the same class $[x] \in CH^2(X)$. This class is taken as the Beauville–Voisin class c_X.

The second part of the proof, more involved and using elliptic curves, shows that $c_2(X) = 24c_X$. Since $v^{CH}(L) = \exp(c_1(L)).(1 + c_2(X)/24)$, this clearly would prove the assertion of the theorem.

The key to this part is the following property of the class c_X; cf. [53, Cor. 2.3]: for a point x_0 contained in some rational curve (and thus $[x_0] = c_X$) let i and j be the embeddings $X \longrightarrow X \times X$, $x \longmapsto (x, x_0)$ and $x \longmapsto (x_0, x)$, respectively. Then for all $\xi \in CH^2(X \times X)$,

$$\Delta^*\xi = i^*\xi + j^*\xi + n \cdot c_X, \tag{3.1}$$

where $n = \deg(\Delta^*\xi - i^*\xi - j^*\xi)$. As for $\xi = [\Delta]$ one has $\Delta^*\xi = c_2(X)$; this proves the claim. We do not attempt to prove (3.1) here, but see below for examples where $c_2(X) \in \mathbb{Z} \cdot c_X$ can be checked easily. $\square$

Example 3.4 In the first version of [53] Beauville listed a number of specific examples of K3 surfaces for which $c_2(X) \in \mathbb{Z} \cdot c_X$ is easy to prove. Those include elliptic K3 surfaces, Kummer surfaces, and quartic hypersurfaces. For example, for a quartic $X \subset \mathbb{P}^3$ the normal bundle sequence $0 \longrightarrow \mathcal{T}_X \longrightarrow \mathcal{T}_{\mathbb{P}^3}|_X \longrightarrow \mathcal{O}(4)|_X \longrightarrow 0$ and $c_2(\mathbb{P}^3) = c_2(\mathcal{O}(1)^{\oplus 4})$ (use the Euler sequence) immediately yield $c_2(X) \in \mathbb{Z} \cdot c_1^2(\mathcal{O}(1)|_X)$. For a Kummer surface one simply uses the fact that away from the 16 rational curves $C_1, \ldots, C_{16}$ corresponding to the 16 two-torsion points, the tangent bundle of X is, after pull-back to the blow-up of the abelian surface in the two-torsion points, isomorphic to the tangent bundle of the abelian surface. The latter is trivial and thus only points on the C_i, $i = 1, \ldots, 16$, contribute to $c_2(X)$. However, as explained in the proof, $[x] = c_X$ for any point contained in a rational curve.

Note that if X satisfies $\rho(X) \geq 3$ and contains at least one smooth rational curve, then the existence result of ample rational curves can be avoided altogether, as then $NS(X)$ is in fact spanned by classes of smooth rational curves; see Corollary **8.3.12**.

Remark 3.5 As all points on rational curves $x \in C \subset X$ represent the same distinguished class $[x] = c_X$, one might ask for a possible converse. Curves with this property have been introduced as *constant cycle curves* in [262]. Although they share many properties of rational curves, they need not always be rational.

3.2 In [257] the result of Beauville and Voisin has been generalized to a statement on spherical objects in $D^b(X) = D^b(\mathrm{Coh}(X))$.

Definition 3.6   An object $E \in D^b(X)$ is called *spherical* if $\operatorname{Ext}^i(E,E) \simeq k$ for $i = 0, 2$ and zero otherwise; see Section **16.2.3**.

Example 3.7   Since $H^1(X, \mathcal{O}) = 0$, any line bundle L on a K3 surface is spherical. Also, if $C \subset X$ is a smooth(!) rational curve, then all $\mathcal{O}_C(i)$ are spherical. Every rigid bundle E, i.e. a bundle with no non-trivial deformation or, equivalently, with $\operatorname{Ext}^1(E,E) = 0$, is spherical provided it is simple. In particular, stable rigid bundles are spherical. It is known that for any class $\delta = (r, \ell, s) \in H^0(X, \mathbb{Z}) \oplus \operatorname{NS}(X) \oplus H^4(X, \mathbb{Z})$ with $\langle \delta, \delta \rangle = (\ell)^2 - 2rs = -2$, there exists a spherical complex $E \in D^b(X)$ with Mukai vector $v(E) = \delta$. See Section **10.3.1** for comments on the existence of spherical bundles.

We state the following theorem without proof. It was proved in [257] for $\rho(X) \geq 2$ and using Lazarsfeld's result that curves in primitive linear systems on K3 surfaces are Brill–Noether general; see Section **9.2**. Voisin in [626] gives a more direct argument for spherical vector bundles not relying on Brill–Noether theory and also covering the case $\rho(X) = 1$.

Theorem 3.8   *The Chow–Mukai vector of any spherical object E is contained in the Beauville–Voisin ring, i.e.*

$$v^{\mathrm{CH}}(E) = \operatorname{ch}(E) \cdot (1, 0, c_X) \in R(X) \subset \mathrm{CH}^*(X).$$

The interest in this generalization stems from the fact that the set of spherical objects in $D^b(X)$ is preserved under linear exact autoequivalences of $D^b(X)$, which is not the case for the set of line bundles. In [257] it was seen as evidence for the Bloch–Beĭlinson conjecture for K3 surfaces over number fields, because a spherical object on $X_{\mathbb{C}}$ for a K3 surface X over $\overline{\mathbb{Q}}$ is always defined over $\overline{\mathbb{Q}}$.

For a K3 surface X over $\overline{\mathbb{Q}}$, base change $\mathrm{CH}^*(X) \longrightarrow \mathrm{CH}^*(X_{\mathbb{C}})$ should identify $\mathrm{CH}^*(X)$ with the Beauville–Voisin ring $R(X)$. This would clearly prove Conjecture 2.5.

References and Further Reading

For a K3 surface X over a field $k \neq \overline{k}$ the Chow group $\mathrm{CH}^2(X)$ might have torsion. It would be interesting to have some explicit examples. We recommend [122] for a survey on the general question concerning torsion in Chow groups. It is generally believed that for a number field k the torsion should be finite. This has been shown for surfaces with $p_g = 0$ by Colliot-Thélène and Raskind [125] and Salberger [521] and for certain surfaces of the form $E \times E$ with E an elliptic curve by Langer and Saito [350]. It is not clear what to expect for other fields. In particular, in [21] Asakura and Saito give examples of hypersurfaces of degree ≥ 5 in $\mathbb{P}^3$ over p-adic fields such that for every $\ell \neq p$ the ℓ-torsion in $\mathrm{CH}^2(X)$ is infinite. It is not clear whether this also happens for $d = 4$, i.e. for quartic K3 surfaces. Note that it is conjectured that over p-adic fields $\mathrm{CH}^2(X)_0$ is the direct product of a finite group and a divisible group. The ℓ-divisibility for almost all ℓ was shown by Saito and Sato in [519]; see also [124]. Already in [505] Raskind finds examples of K3 surfaces over p-adic fields such that the torsion of $CH^2(X)_0$ prime to p is finite; see also [123]. For questions on the torsion of $CH^2(X)$ for varieties over finite fields, see the article [126] by Colliot-Thélène, Sansuc, and Soulé.

The results of Section 3 are part of a bigger picture. It seems that the conjectured Bloch filtration of $CH^*(X)$ for arbitrary varieties admits a natural splitting in the case of hyperkähler or irreducible symplectic manifolds. This has been put forward by Beauville in [50] and strengthened and verified in a series of examples by Voisin in [624].

It is an interesting problem to decide which points on a K3 surface X are rationally equivalent to a given point x_0. Maclean in her thesis [385] shows that for a generic complex projective K3 surface X and generic $x_0 \in X$ the set $\{x \mid x \sim x_0\}$ is always dense in the classical topology independent of whether $[x_0] = c_X$ or not; cf. Theorem **13**.5.2. But note that the set is not expected to have any reasonable topology. For example, when X is defined over a finitely generated field $K \subset \mathbb{C}$, then $\mathrm{Aut}_K(\mathbb{C})$ acts on $X(\mathbb{C})$ in a highly non-continuous way but leaves the set $\{x \mid [x] = c_X\}$ invariant.

We have not touched upon the general finiteness conjecture of Kimura and its applications in the case of K3 surfaces; see e.g. [489]. Also, we have left untouched the results confirming the Bloch–Beĭlinson conjectures for symplectic automorphisms acting on Chow groups; see e.g. [260, 265, 625].

We have also omitted the cohomological approach, which describes $CH^2(X)$ as the Zariski cohomology $H^2(X, \mathcal{K}_2)$ with $\mathcal{K}_2$ the sheaf associated with $U \longmapsto K_2(U)$ (Milnor K-theory). Passing to the formal version of it (cf. Section **18**.1.3) leads to the definition of the 'tangent space' of $CH^2(X)$ as $H^2(X, \mathcal{O}_X) \otimes \Omega_{k/\mathbb{Q}}$ for a K3 surface X defined over a field k of characteristic zero. See Bloch's lecture notes [67] for details.

Questions and Open Problems

It would be interesting to find a different approach to the torsion of $CH^2(X)$ via the torsion of $K(X)$ in the case of non-separably closed fields k. Nothing seems to be known in this direction. More generally, it would be interesting to see whether viewing $CH^*(X)$ as $K(X)$ or $K(D^b(X))$ sheds new light on certain aspects of cycles on K3 surfaces. For example, in [258] it is shown that the Bloch–Beĭlinson conjecture is equivalent to the existence of a bounded t-structure on the dense subcategory spanned by spherical objects.

In [620] Voisin conjectures that for a complex projective K3 surface X and any two closed points $x, y \in X$ the two points $(x, y), (y, x) \in X \times X$ satisfy $[(x, y)] = [(y, x)]$ in $CH^4(X \times X)$. This has been proved in [620] for interesting special cases, but the general assertion remains open.

13

Rational Curves on K3 Surfaces

For this chapter we highly recommend the paper [69] by Bogomolov, Hassett, and Tschinkel and Hassett's survey [239]. All K3 surfaces in this chapter are projective. As an introduction, we shall discuss in detail the two main conjectures concerning rational curves on K3 surfaces: there exist infinitely many rational curves on an arbitrary K3 surface, and all rational curves on the general K3 surface are nodal.

0.1 Let us begin with the following observation. Suppose there is a family

$$\mathcal{C} \subset B \times X, \ \mathcal{C}_b \subset X \ (b \in B)$$

of rational curves on a surface X parametrized by some variety B such that $\mathcal{C} \longrightarrow X$ is dominant. Then there exists a dominant rational map $D \times \mathbb{P}^1 \ -\ -\twoheadrightarrow X$ with D a curve. In characteristic zero this would imply $\mathrm{kod}(X) \leq \mathrm{kod}(D \times \mathbb{P}^1) = -\infty$. This is absurd if X is a K3 surface. Thus, rational curves on K3 surfaces in characteristic zero do not come in families.

The assumption on the characteristic cannot be dropped, but even in positive characteristic K3 surfaces that admit families of rational curves are rare and should be regarded as very special.

Example 0.1 Consider the Fermat quartic $x_0^4 + x_1^4 + x_2^4 + x_3^4 = 0$ over an algebraically closed field k of characteristic $p \equiv 3(4)$. It is unirational, i.e. there exists a dominant rational map

$$\mathbb{P}^2 \ -\ -\twoheadrightarrow X$$

or, equivalently, the function field $K(X)$ of X admits a (non-separable) extension $K(X) \subset k(T_1, T_2) \simeq K(\mathbb{P}^2)$. This example was first studied by Tate in [591, 592], who showed

---

* Parts of this chapter are based on a seminar during the winter term 2011–2012. I wish to thank the participants, in particular Stefanie Anschlag and Michael Kemeny, for stimulating discussions and interesting talks on the subject.

that in this case $\rho(X) = 22$. A detailed computation of the function field can be found in Shioda's article [560] or Section **17.2.3**; see also [239] for further examples. Shioda in fact observed that any unirational (or, a priori weaker, uniruled) K3 surface over an algebraically closed field has maximal Picard number 22; see Proposition **17.2.7**. The converse has recently been proved by Liedtke in [373], answering a question by Artin in [16, p. 552]; see Section **18.3.5**.[1]

Therefore, naively one would rather expect a K3 surface X to contain only finitely many rational curves, unless X is very special. However, the above discussion excludes only the existence of families of rational curves, not the existence of discrete and possibly infinite sets of them. And indeed, as we shall see, this is what seems to happen.

0.2   The following conjecture has proved to be a strong motivation for a number of interesting developments over the last years. It is trivial for unirational, or equivalently supersingular, K3 surfaces, but it is otherwise as interesting in positive characteristic as it is in characteristic zero.

Conjecture 0.2   *Every polarized K3 surface (X, H) over an algebraically closed field contains infinitely many integral rational curves C linearly equivalent to some multiple of H.*

Note that it is not even known whether every polarized K3 surface (X, H) admits an integral rational curve linearly equivalent to some multiple nH at all.

Remark 0.3   A weaker question would be to ask for infinitely many integral rational curves without requiring the curves to be linearly equivalent to a multiple of the fixed polarization. This provides more flexibility when $\rho(X) \geq 2$. As we shall see, Conjecture 0.2 has been verified for many K3 surfaces, but even the weaker version is still open in general.

Example 0.4   Here are two concrete examples of K3 surfaces containing infinitely many integral rational curves. Both, however, are not typical, as the curves are smooth and thus not ample.

(i) For the following, see [69]. Let C be a smooth curve of genus two over an algebraically closed field of characteristic $\neq 2$. Consider a hyperelliptic involution $\eta \colon C \xrightarrow{\sim} C$, so $\pi \colon C \longrightarrow C/\langle \eta \rangle \simeq \mathbb{P}^1$. Pick a ramification point $x_0 \in C$, i.e. $\eta(x_0) = x_0$, and let

$$i \colon C \hookrightarrow \operatorname{Pic}^0(C), \quad x \longmapsto \mathcal{O}(x - x_0)$$

---

[1] As was pointed out to me by Christian Liedtke, it is a priori not clear that a uniruled K3 surface is in fact unirational, but see the arguments in the proof of Proposition **9.4.6**.

be the induced closed embedding. For $n \in \mathbb{N}$ we denote by C_n the image of C under the morphism

$$\text{Pic}^0(C) \longrightarrow \text{Pic}^0(C), \ L \longmapsto L^n.$$

The standard involution $L \longmapsto L^*$ on $\text{Pic}^0(C)$ acts on C_n via η. Indeed, $\mathcal{O}(x - x_0)^* \simeq \mathcal{O}(\eta(x) - x_0)$, as $\mathcal{O}(x + \eta(x)) \simeq \pi^*\mathcal{O}(1) \simeq \mathcal{O}(2x_0)$. Hence, the image of C_n under the quotient

$$\text{Pic}^0(C) \longrightarrow \text{Pic}^0(C)/\pm$$

is a smooth rational curve and so is its strict transform in the Kummer surface X associated with the abelian surface $\text{Pic}^0(C)$. Note that this indeed yields infinitely many rational curves in X, for $L \longmapsto L^n$ does not respect C, i.e. $C_m \neq C_n$ for $m \neq n$. (For example, one could use that $C \subset \text{Pic}^0(C)$ is ample and thus its pull-back under the nth power cannot split off a component isomorphic to C.)

In fact, any Kummer surface contains infinitely many rational curves; see [69, Ex. 5]. The example is also interesting from the point of view of rational points. Bogomolov and Tschinkel show in [71, Thm. 4.2] that for $k = \overline{\mathbb{F}}_p$, $p \neq 2$, every point on X is contained in a rational curve.

(ii) Every elliptic surface $X \longrightarrow \mathbb{P}^1$ with a zero section $C_0 \subset X$ and a section $C \subset X$ that is of infinite order (e.g. as point in the generic fibre or as an element in the Mordell–Weil group $\text{MW}(X)$; see Section **11**.3.2) contains infinitely many rational curves. Indeed, the multiples $C_n := nC$ (with respect to the group structure of the fibres) yield infinitely many smooth rational curves.

Equivalently, C can be used to define an automorphism $f_C \colon X \stackrel{\sim}{\longrightarrow} X$ of infinite order by translation in the fibres (see Section **15**.4.2), and the infinitely many rational curves can be obtained as the image of C under the iterations of f_C.

More generally, K3 surfaces X with infinite $\text{Aut}(X)$ often provide examples of K3 surfaces with infinitely many rational curves; see Remark 1.6.

0.3 To the best of my knowledge, there is no general philosophy supporting Conjecture 0.2. However, it does fit well with other results and conjectures. The following two circles of considerations should be mentioned in this context; see also [98].

(i) Smooth complex projective varieties X with trivial canonical bundle are conjectured to be *non-hyperbolic*. Even stronger, one expects that through closed points $x \in X$ in a dense set there exists a non-constant holomorphic map $f \colon \mathbb{C} \longrightarrow X$. Any rational or elliptic curve yields such a holomorphic map. Thus, if the union of all rational curves is dense in X, then X is indeed non-hyperbolic in the stronger sense.

It is known that K3 surfaces are indeed non-hyperbolic in this strong sense, but this is proved via families of elliptic curves (see Corollary 2.2), although rational curves also play a role. In fact, it has been conjectured that complex K3 surfaces are *dominable*, i.e. that there exists a holomorphic map

$$\mathbb{C}^2 \longrightarrow X$$

such that the determinant of the Jacobian is not identically trivial; see [95]. This is true for some K3 surfaces, e.g. Kummer surfaces (which involves showing the non-trivial fact that the complement of any finite subset of a torus $\mathbb{C}^2 / \Gamma$ is dominable), but remains an open question in general.

(ii) Lang conjectured that the set

$$X(k) \subset X$$

of k-rational points of a variety X of general type defined over a number field k should not be Zariski dense. A stronger version predicts the existence of a proper closed subset $Z \subset X$ such that for any finite extension K/k, the set of K-rational points $X(K)$ is up to a finite number of points contained in Z.

So it is natural to wonder what happens for X not of general type, e.g. for a K3 surface. The general expectation, known as *potential density*, is that for a K3 surface X defined over a number field k, there always exists a finite extension K/k such that $X(K)$ is dense in X.

The relation between rational points and rational curves on K3 surfaces is not completely understood. It is not excluded and sometimes even conjectured (*Bogomolov's logical possibility*) that through any point $x \in X(\overline{\mathbb{Q}})$ there exists a rational curve. This would of course imply the existence of infinitely many rational curves and prove the Bloch–Beĭlinson conjecture for K3 surfaces; see Section **12.2.2**. It is not clear whether one should expect that rational curves defined over some finite extension K/k are already dense, but it would of course imply potential density.

0.4 The second question that shall be discussed in this chapter is concerned with 'good' rational curves and asks, more specifically, whether every rational curve can be deformed to a nodal one on some deformation of the underlying K3 surface. Since a smooth rational curve has negative self-intersection, nodal curves are the least singular rational curves that can be hoped for in an ample linear system.

Conjecture 0.5 *For the general polarized K3 surface* $(X, H) \in M_d(\mathbb{C})$ *all rational curves in the linear systems* $|nH|$ *are nodal.*

Since the Picard group of the general (X, H) is generated by H, the conjecture simply predicts that all rational curves on X are nodal. One could be more optimistic and relax 'general' to 'generic'; see Section 0.5 for these notions.

Conjecture 0.2 was triggered by the Yau–Zaslow conjecture, which gives a formula, invariant under deformations of (X, H), for the number of rational curves in $|H|$

counted with the right multiplicities dictated by Gromov–Witten theory. The Yau–Zaslow conjecture was verified under the assumption that all rational curves are nodal. See Section 4 for more details and references.

Remark 0.6   Let $C \subset X$ be a nodal integral rational curve with δ nodes and let

$$\mathbb{P}^1 \simeq \tilde{C} \longrightarrow C$$

be its normalization. Then (cf. Section **2**.1.3)

$$1 = \chi(\mathcal{O}_{\tilde{C}}) - \chi(\mathcal{O}_C) \mid \delta.$$

For $C \in |nH|$ with $(H)^2 = 2d$ this shows

$$\delta = n^2 d + 1.$$

For reducible curves the number of nodes increases. For example if $C = C_1 + C_2 \in |nH|$ is the sum of two smooth rational curves $C_i \simeq \mathbb{P}^1$ intersecting transversally, then C is a nodal rational curve with $\delta = n^2 d + 2$ nodes.

0.5   To conclude the introduction we elaborate on the difference between *generic* and *general*. A certain property for polarized K3 surfaces holds for the generic (X, H) if it holds for all (X, H) in a non-empty Zariski open subset of the moduli space M_d. It holds only for the general (X, H) if it holds for all (X, H) in the complement of a countable union of proper Zariski closed subsets.

So, $\rho(X) = 1$ for the general complex polarized K3 surface $X \in M_d$, but not for the generic one. Also, if for one polarized K3 surface (X_0, H_0) the linear system $|H_0|$ contains a nodal rational curve, then the same is true for the generic $(X, H) \in M_d$; cf. Section 2.3 However, if all (infinitely many) rational curves on X_0 are nodal, this a priori implies only that the same is true for the general complex polarized K3 surface $(X, H) \in M_d$.

Special care is needed if the ground field k is countable. Then in principle a countable union of proper closed subsets could contain all k-rational points. Also, if the polarized K3 surface corresponding to the scheme-theoretic (geometric) generic point $\eta \in M_d$ has a certain property, it usually does not imply that also the generic fibres (X, H) in the above sense have the same property.

1 Existence Results

We shall outline the standard argument to produce rational curves on arbitrary complex projective K3 surfaces by first constructing special nodal, but reducible, ones on particular Kummer surfaces. The resulting curves are linearly equivalent to the primitive polarization, but a more lattice theoretic and less explicit construction allows one to also prove the existence of integral rational curves linearly equivalent to multiples of the polarization, at least on the generic K3 surface.

We shall as well discuss the situation over arbitrary fields and state, but not prove, Chen's result on nodal rational curves.

1.1   The first result we have to mention is generally attributed to Bogomolov and Mumford and was worked out by Mori and Mukai in [423]. Note that (iii) below cannot be found in [423], but it was apparently known to the experts that essentially the same arguments proving (i) and (ii) would yield (iii) as well; see [69].[2]

Theorem 1.1   (i) *Every polarized K3 surface* $(X, H) \in M_d(\mathbb{C})$ *contains at least one rational curve* $C \in |H|$.

(ii) *The generic polarized K3 surface* $(X, H) \in M_d(\mathbb{C})$ *contains a nodal integral rational curve* $C \in |H|$.

(iii) *For fixed* $n > 0$, *the generic polarized K3 surface* $(X, H) \in M_d(\mathbb{C})$ *contains an integral rational curve* $C \in |nH|$.

Proof   To prove (i), one starts with a Kummer surface X_0 associated with an abelian surface of the form $E_1 \times E_2$. We choose elliptic curves E_1 and E_2 such that there exists an isogeny $\varphi \colon E_1 \longrightarrow E_2$ of degree $2d + 5$. To be completely explicit, one could take $E_1 = \mathbb{C}/((2d + 5)\mathbb{Z} + i\mathbb{Z})$ and $E_2 = \mathbb{C}/(\mathbb{Z} + i\mathbb{Z})$ with φ the natural projection.

Let now $\Gamma := \Gamma_\varphi \subset E_1 \times E_2$ be the graph of φ and let

$$C_1 \subset X_0$$

be the strict transform of its quotient $\Gamma/\pm$ by the standard involution on $E_1 \times E_2$. Note that Γ contains (exactly) four of the 16 fixed points and that, therefore, $C_1 \simeq \mathbb{P}^1$. Next consider the strict transform

$$C_2 \subset X_0$$

of the quotient $(E_1 \times \{0\})/\pm$. The four two-torsion points of E_1 yield the fixed points of the involution and, therefore, $C_2 \simeq \mathbb{P}^1$ as well.

Thus, X_0 contains two smooth rational curves C_1, C_2, for which, of course, $(C_1)^2 = (C_2)^2 = -2$. To compute $(C_1 + C_2)^2$, one observes that the transversal intersection $\Gamma \cap (E_1 \times \{0\})$ consists of the $2d + 5$ points $\varphi^{-1}(0)$. The order of the subgroup $E_1[2]$ of two-torsion points of E_1 contained in $\varphi^{-1}(0)$ must divide $2d + 5$ and 4. Thus, $E_1[2] \cap \varphi^{-1}(0) = \{0\}$. The point 0 does not contribute to $(C_1.C_2)$, as under the

---

[2] Maybe it was actually stated for the first time by Chen in [114]. The proof there uses more complicated arguments which in fact prove the existence of one nodal(!) integral rational curve in $|nH|$ for the generic (X, H). This does not come out of the proof reproduced here.

blow-up $X_0 \longrightarrow (E_1 \times E_2)/\pm$ the two curves $(E_1 \times \{0\})/\pm$ and $\Gamma/\pm$ get separated over the corresponding point. However, all others do and, hence, $(C_1.C_2) = d + 2$. Thus,

$$C_1 + C_2 \subset X$$

is a reducible rational nodal curve with $(C_1 + C_2)^2 = 2d$.

A further analysis reveals that $C_1 + C_2$ is big and nef (but not ample). Indeed, $(C_1 + C_2.D) \geq 0$ for all curves $D \neq C_i$ and $(C_1 + C_2.C_i) = d > 0$ (but some of the 16 exceptional curves of $X_0 \longrightarrow (E_1 \times E_2)/\pm$ are not met).

Also, the class of $C_1 + C_2$ is primitive, as the intersection numbers of $C_1 + C_2$ with the fibres of the projections

$$X_0 \longrightarrow E_1/\pm \text{ and } X_0 \longrightarrow E_2/\pm$$

are 2 and $2d + 5$, respectively.

Now use deformation theory for $C_1 + C_2 \subset X_0$, to be explained in Section 2.2, to conclude that $C_1 + C_2$ deforms sideways to a curve in $|H|$ on the generic $(X, H) \in M_d$. In fact, as rational curves can only specialize to rational curves, this yields that every complex polarized K3 surface $(X, H) \in M_d$ contains a (possibly non-nodal) rational curve in $|H|$.

For (ii) just note that being nodal and irreducible is an open property and thus it suffices to find one (X, H) with this property. Going back to (i), one observes that (X, H) is provided by a small deformation $C \subset X$ of $C_1 + C_2 \subset X_0$ for which $\rho(X) = 1$.

(iii) We follow [69]. Ideally, one would like to argue as above and construct a specific K3 surface X_0 containing two smooth integral rational curves C_1, C_2 intersecting transversally and such that $\mathcal{O}(C_1), \mathcal{O}(C_2)$ are linearly independent in $NS(X) \otimes \mathbb{Q}$, with $C_1 + C_2 = nH_0$, where H_0 is primitive and $(H_0)^2 = 2d$. Then, $C_1 + C_2$ would again be a nodal rational curve (with $n^2d + 2$ nodes) that deforms sideways to a curve $C \in |nH|$ on the generic deformation (X, H) of (X_0, H_0).

Since $\rho(X) = 1$ for general X and thus only multiples of $\mathcal{O}(H_0)$ deform, the components C_1, C_2 cannot be specializations of curves on X. Therefore, C would automatically be an integral nodal rational curve.

Unfortunately, an explicit construction of $C_1 + C_2 \subset X_0$ with these properties is not available. Instead, one has to argue using abstract existence results, which has the consequence that the curves C_1, C_2 are not known to intersect transversally. This makes the deformation theory more complicated and the resulting deformation $C \in |nH|$ on X is not known to be nodal. Details concerning the deformation problem for the non-nodal curve are contained in [69], but see Remark 2.6.

Let us work out the details of how to find $C_1 + C_2 \subset X_0$. First realize the lattice with intersection matrix

$$\begin{pmatrix} -2 & nd \\ nd & 2d \end{pmatrix}$$

(with respect to a basis x, y) as a primitive sublattice $\Gamma \subset \Lambda$ of the K3 lattice $\Lambda = E_8(-1)^{\oplus 2} \oplus U^{\oplus 3}$. For example, take

$$x = e_1 - f_1 \quad \text{and} \quad y = ndf_1 + (e_2 + df_2),$$

where e_1, f_1 and e_2, f_2 are the standard bases of the first two hyperbolic planes $U_1 \oplus U_2$ in $U^{\oplus 3}$. It is easy to see that $\Gamma \subset \Lambda$ is indeed primitive. Then consider the moduli space of marked lattice polarized K3 surfaces

$$N_\Gamma = \{(X, \varphi) \mid \varphi \colon H^2(X, \mathbb{Z}) \xrightarrow{\sim} \Lambda, \ \varphi^{-1}(\Gamma) \subset \mathrm{Pic}(X)\}/\sim,$$

which (modulo certain non-Hausdorff phenomena) can be identified with the intersection of the period domain $D \subset \mathbb{P}(\Lambda_{\mathbb{C}})$ with $\mathbb{P}(\Gamma_{\mathbb{C}}^{\perp})$; cf. Sections **6.3.3**, **7.2.1**, and the papers by Dolgachev [150] and Beauville [48].

Next apply the surjectivity of the period map to produce a marked K3 surface (X_0, φ) with $\varphi^{-1}(\Gamma) = \mathrm{Pic}(X_0)$. Then $\varphi^{-1}(y) \in \mathrm{Pic}(X_0)$ is a class of square $2d$. Using Corollary **8.2.9**, we may assume that $\varphi^{-1}(y) = [H_0]$ with H_0 nef. Since $(x)^2 = -2$ and $(x.y) > 0$, the class $[C_1] := \varphi^{-1}(x) \in \mathrm{Pic}(X_0)$ is effective and can thus be written as $[C_1] = \sum[D_i]$ for integral curves $D_i \subset X_0$. In particular, $(D_i.H_0) \geq 0$. Suppose $[C_1]$ is not represented by an integral curve. Then there is one component, say D_1, which is contained in the shaded region representing the set $\{a[C_1] + b[H_0] \mid b < 0, \ a \geq -2b/n\}$:

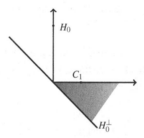

The second inequality is expressing $(D_1.H_0) \geq 0$. In particular, $(D_1)^2 < 0$ and, therefore, $(D_1)^2 = -2$ (see Section **2.1.3**), which for $D_1 \neq C_1$ leads to the contradiction $a^2 - 1 = abnd + b^2 d \leq -b^2 d$.

Hence, $[C_1]$ can indeed be represented by a smooth integral rational curve, i.e. $C_1 \simeq \mathbb{P}^1$. A similar computation shows that H_0 is not orthogonal to any (-2)-class and hence ample.

Finally, if X_0 has been found with $\mathrm{NS}(X_0) = \mathbb{Z}[C_1] + \mathbb{Z}[H_0]$, $C_1 \simeq \mathbb{P}^1$, and H_0 ample, then one shows that $n[H_0] - [C_1]$ can be represented by a smooth rational curve C_2. As its intersection with H_0 is positive and $(nH_0 - C_1)^2 = -2$, the class is effective, i.e. represented by a curve C_2 with irreducible components D_i (possibly occurring with positive multiplicities). Then $(D_i.H_0) \geq 0$ and, if $D_i \neq C_1$, also $(D_i.C_1) \geq 0$. At least one component C_2, say D_1, has a class in the shaded region representing $\{a[C_1] + b[H_0] \mid b \geq -(an)/2, \ a = -1, -2, \ b \in [0, n]\}$:

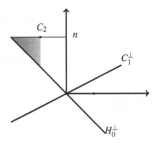

If $a = -2$, then $[D_1] = -2[C_1] + n[II_0]$ and hence $(D_1)^2 < -2$, which is absurd. Hence, $[D_1] = -[C_1] + b[H_0]$ with $0 \le b \le n$. Therefore,

$$[D_1]^2 = b^2 2d - b2nd - 2 = 2bd(b - n) - 2 \le -2$$

with equality only if $b = n$. Since $(D_1)^2 \ge -2$ (as for any integral curve), this shows $b = n$. Thus, $[D_1] = n[H_0] - [C_1]$; i.e. $C_2 = D_1$ is smooth and irreducible. □

Corollary 1.2 *The general complex polarized K3 surface $(X, H) \in M_d$ contains infinitely many integral rational curves linearly equivalent to some multiple nH.*

Proof For each $n > 0$, there exists a dense open dense set $U_n \subset M_d$ such that every $(X, H) \in U_n$ admits an integral rational curve $C \in |nH|$. Thus, the assertion holds for all $(X, H) \in \bigcap U_n$. □

The 'integral' is added to avoid counting a given rational curve C infinitely often by taking multiples nC. Of course, the corollary is equivalent to the assertion that there exist infinitely many reduced rational curves.

1.2 In Theorem 1.1 the K3 surfaces are assumed to be defined over $\mathbb{C}$. However, in characteristic zero, the (finitely generated) field of definition k of a given polarized K3 surface (X, H) can always be embedded into $\mathbb{C}$.

Remark 1.3 Since rational curves on K3 surfaces in characteristic zero are rigid, every rational curve in $|nH|$ on $X_{\mathbb{C}}$ is in fact defined over $\overline{k}$; cf. the arguments in Lemma **17**.2.2 and Section **16**.4.2. This shows that Theorem 1.1 holds for K3 surfaces defined over arbitrary algebraically closed fields of characteristic zero. Note however that Corollary 1.2 for countable fields, for example $k = \overline{\mathbb{Q}}$, is a void statement; cf. Remark 3.6.

Using the liftability of K3 surfaces from positive characteristic to characteristic zero (cf. Section **9**.5.3), one can obtain rational curves in positive characteristic as specializations of rational curves in characteristic zero. As specialization could turn irreducible curves into reducible or non-reduced ones, only the existence of finitely many rational curves can be obtained in this way; i.e. only (i) in Theorem 1.1 is a priori known in full generality. One might try to adapt the proof of (ii) to the case of positive characteristic at least under the assumption that the K3 surface can still be deformed to a Kummer surface of the required type. Presumably, $p|2d$ would need to be excluded for this.

Corollary 1.4 *Let (X, H) be a polarized K3 surface over an algebraically closed field k of arbitrary characteristic. Then $|H|$ contains a rational curve.* □

In [360] Li and Liedtke explain how the lifting to characteristic zero can be used to cover the case that H is only big and nef. This and the fact that for any effective line bundle L on a K3 surface there exist rational curves C_i such that $L = H + \sum n_i C_i$ with $n_i > 0$ and H nef (and hence either also big or satisfying $(H)^2 = 0$, in which case H is linearly equivalent to an effective sum of rational curves; cf. the proof of Proposition **2.3.10** and Remark **2.3.13**) can then be used to show the following statement; cf. [70, Prop. 2.5, Rem. 2.13] and [360, Thm. 1.1].

Corollary 1.5 *Let L be a non-trivial effective line bundle L on a K3 surface over an algebraically closed field. Then there exists a curve in $|L|$ which can be written as an effective sum of (possibly singular) rational curves.* □

One might also want to compare the corollary with Kovács's result in [327], (see Corollary **8.3.12**), saying that for $\rho(X) \geq 3$ there are either no smooth rational curves at all or the closure of the effective cone is spanned by them.

Remark 1.6 In [70] Bogomolov and Tschinkel prove the weak version of Conjecture 0.2 (cf. Remark 0.3), for K3 surfaces with infinite $\text{Aut}(X)$ (cf. Example 0.4, (ii)) and elliptic K3 surfaces.[3] In both cases, the rational curves are in general not linearly equivalent to some nH.

(i) If $\text{Aut}(X)$ is infinite (cf. [70, Thm. 4.10]), then one roughly proves that there is at least one integral rational curve C such that $\mathcal{O}(C)$ in $\text{Pic}(X)$ has an infinite orbit under the action of $\text{Aut}(X)$. Using Corollary 1.5, this eventually boils down to lattice theory. If X is known to contain one smooth rational curve, then it contains infinitely many by Corollary **8.4.7**.

(ii) For elliptic K3 surfaces, see [70, Sec. 3] and [239, Sec. 3.2]. To illustrate the idea, assume $X \longrightarrow \mathbb{P}^1$ is an elliptic fibration with a section $C_0 \subset X$ that serves as the zero section. Assume furthermore that there exists another rational curve $C \subset X$ for which the intersection $C \cap X_t$ with some smooth fibre X_t contains a non-torsion point. Consider its images C_n under the map $f^n \colon X \longrightarrow X$ fibrewise given by multiplication with n. Clearly, all curves C_n are rational and, as C is a non-torsion section, they are pairwise different. The main work in [70] consists of proving that any elliptic K3 surface

---

[3] Originally, $\rho \leq 19$ was assumed, but see [69, Rem. 6].

$X \longrightarrow \mathbb{P}^1$ is dominated by an elliptic K3 surface $X' \longrightarrow \mathbb{P}^1$ satisfying these additional assumptions.

Note that for a K3 surface X which is elliptic or has infinite $\mathrm{Aut}(X)$, the Picard number $\rho(X)$ is at least two. However, just assuming $\rho(X) \geq 2$ seems not quite enough to confirm the weak version of Conjecture 0.2. Although in this case there exists an infinite number of primitive ample classes H_i for which Theorem 1.1 proves the existence of rational curves $C_i \in |H_i|$, they could a priori all be linear combinations of just a finite number of rational curves. However, this does not happen for the general (X, H) of Picard number $\rho(X) \geq 2$ as remarked in [71, Rem. 4.7].

1.3   One would like to have to deal only with rational curves that are nodal. In Theorem 1.1, (ii), the existence of at least one nodal rational curve in $|H|$ was deduced for generic (X, H). But one could ask whether maybe every rational curve in $|H|$ for generic (X, H) is nodal. Also, what about nodal rational curves which are not primitive, i.e. which are contained in $|nH|$ with $n > 1$? These questions have been dealt with in Chen's papers [114] and [115]. The results, which partially answer Conjecture 0.5, are summarized as follows.

Theorem 1.7   *For K3 surfaces over $\mathbb{C}$ one has:*

(i) *For generic $(X, H) \in M_d$, every rational curve in $|H|$ is nodal.*
(ii) *For given $n > 0$, the generic $(X, H) \in M_d$ contains an integral nodal rational curve in $|nH|$.*

Chen's arguments in [114] are based on the degeneration of K3 surfaces to the union of two rational surfaces. Its main result is (ii), but also (i) is discussed. The deformation theory is more involved than in the approach outlined above and in Section 2 below.

In his sequel [115] degenerations of rational curves to curves on particular elliptic K3 surfaces are studied. This eventually allows Chen to prove (i), which should be seen as a strengthening of Theorem 1.1, (ii).

Similarly to Corollary 1.2 one obtains the following corollary.

Corollary 1.8   *The general complex polarized K3 surface $(X, H) \in M_d$ contains infinitely many integral nodal rational curves linearly equivalent to some multiple nH.*   □

2 Deformation Theory and Families of Elliptic Curves

The deformation theory of curves on K3 surfaces comes in various flavors. A curve $C \subset X$ can be deformed as a subvariety of X and the deformation theory is then completely described by the linear system $|\mathcal{O}(C)|$. More insight is gained by viewing C as the image of a morphism $f: \widetilde{C} \longrightarrow C \subset X$, e.g. from the normalization of C. Eventually, also X can be allowed to deform. We briefly sketch the basic principles and mention the results that are used in the context of this chapter.

2.1 If $C \subset X$ is a possibly singular curve contained in a K3 surface X, then the first-order deformations of C in X are parametrized by $H^0(C, \mathcal{N}_{C/X})$, where

$$\mathcal{N}_{C/X} := \mathcal{O}_C(C) \simeq \mathcal{O}(C)|_C$$

is the usual normal bundle $\mathcal{T}_X|_C / \mathcal{T}_C$ if C is smooth. The obstructions would a priori live in $H^1(C, \mathcal{N}_{C/X})$, but this space is often trivial and in any case all obstructions are trivial, for deformations of $C \subset X$ are described by the linear system $|\mathcal{O}(C)|$ which is of dimension $h^0(C, \mathcal{N}_{C/X}) = h^0(X, \mathcal{O}(C)) - 1$, as $H^1(X, \mathcal{O}_X) = 0$.

Proposition 2.1 *Let X be a K3 surface over an algebraically closed field of characteristic zero. Suppose there exists an integral nodal rational curve $C \subset X$ of arithmetic genus $g > 0$. Then there is a one-dimensional family of nodal elliptic curves in $|\mathcal{O}(C)|$.*

Proof The curve C is a stable curve of arithmetic genus $g = (C)^2/2 + 1$ and, as C is rational, it has g nodes; see Remark 0.6. Let $[C] \in \overline{\mathcal{M}}_g$ be the associated point in the moduli space of stable curves; see [10, 233]. As being stable is an open property, there exists an open set $C \in U \subset |\mathcal{O}(C)|$ parametrizing integral stable curves. Let

$$\varphi: U \longrightarrow \overline{\mathcal{M}}_g$$

be the induced classifying morphism. Then $\varphi[C]$ is contained in the locus of stable curves with at least $g - 1$ nodes, which is of codimension $g - 1$; see [10, Ch. XI] or [233, p. 50]. Hence, as $|\mathcal{O}(C)|$ is of dimension at least g, there exists a one-dimensional subvariety $C \in B \subset U$ parametrizing only nodal curves with at least $g - 1$ nodes. However, if C_b, $b \in B$, has more than $g - 1$ nodes, then C_b is rational. As X cannot be dominated by a family of rational curves, the generic C_b, $b \in B$, has exactly $g - 1$ nodes and hence is elliptic. $\qquad\square$

Corollary 2.2 *Let X be a K3 surface over an algebraically closed field of characteristic zero. Then there exist morphisms*

$$\widetilde{X} \overset{p}{\longrightarrow} X$$
$$q \downarrow $$
$$B $$

with $\widetilde{X}$ a smooth surface, $p: \widetilde{X} \longrightarrow X$ a surjective morphism, and $q: \widetilde{X} \longrightarrow B$ a fibration for which the generic fibre $\widetilde{X}_b$ is a smooth elliptic curve and such that $p: \widetilde{X}_b \longrightarrow X$ is generically injective.

In other words, any K3 surface in characteristic zero is dominated by a family of smooth elliptic curves.

Proof Due to Theorem 1.1, (ii), the generic K3 surface $(X, H) \in M_d(\mathbb{C})$ admits a nodal integral rational curve $C \in |H|$. Proposition 2.1 thus applies and yields a family $C \longrightarrow B \subset |H|$ with generic fibre C_b a nodal elliptic curve. Let $\widetilde{X} \longrightarrow C$ be the resolution of the normalization of the surface C. Then the generic closed fibre $\widetilde{X}_b$ over B is the normalization of C_b and hence a smooth elliptic curve.[4]

Inspection of the proof of Proposition 2.1 reveals that the argument works very well in families, i.e. over the open subset $U \subset M_d$ (or an appropriate finite cover of it) of (X, H) containing a nodal integral rational curve in $|H|$. Specialization to surfaces in $M_d \setminus U$ yields the assertion for all K3 surfaces. Indeed, the elliptic curves may specialize to curves with worse singularities and possibly to reducible ones. However, as in characteristic zero no K3 surface is covered by a family of rational curves, this still yields a dominating family of elliptic curves. If the generic fibre $\widetilde{X}_b$ is reducible, one has to pass to a finite cover of B to get irreducible fibres. Observe that in this last step one might end up with elliptic curves not linearly equivalent to any multiple nH. $\square$

Note that reduction to positive characteristic also essentially proves the assertion for K3 surfaces over algebraically closed fields k with $\mathrm{char}(k) > 0$. However, specializing might turn a dominating family of elliptic curves into a dominating family of rational curves. Of course, a family of rational curves can always be dominated by a family of elliptic curves, which, however, would not map generically injectively into X anymore.

2.2 If instead of a single K3 surface X one considers a family of K3 surfaces

$$\pi : \mathcal{X} \longrightarrow S,$$

say over a smooth connected base S, and a curve $C \subset X = \mathcal{X}_0$ in one of the fibres, then the situation becomes more interesting. We consider only the case that $H^1(X, \mathcal{O}(C)) = 0$, which holds if e.g. C is ample. Moreover, we assume that there exists a line bundle L on $\mathcal{X}$ with $L|_X \simeq \mathcal{O}(C)$. Then the deformations of C in the family $\pi : \mathcal{X} \longrightarrow S$ are parametrized by the projective bundle

$$P := \mathbb{P}(\pi_* L) \longrightarrow S$$

which, due to $H^1(X, \mathcal{O}(C)) = 0$, is fibrewise the linear system $|L|_{\mathcal{X}_t}|$ on $\mathcal{X}_t$ and hence of dimension $\dim(P) = \dim(S) + g$, where $g = (C)^2/2 + 1$.

It is straightforward to adopt the arguments in the proof of Proposition 2.1 to show that any nodal integral rational curve $C \in |H|$ on $X = \mathcal{X}_0$ deforms sideways to a nodal

---

[4] Usually, one mentions simultaneous resolutions at this point, which would involve passing to a finite cover of the base B. However, in the case of families of curves this is not necessary.

integral rational curve on the generic fibre $\mathcal{X}_t$. In order to complete the proof of Theorem 1.1, (i), we shall however discuss the slightly more complicated case of a nodal rational curve

$$C = C_1 + C_2 \subset X = \mathcal{X}_0$$

given as the union of two smooth rational curves C_1, C_2 intersecting transversally in $g + 1$ points, e.g. the curve on the Kummer surface associated with $E_1 \times E_2$ constructed in the proof of Theorem 1.1. Then C is a stable curve and thus corresponds to a point in the moduli space $\overline{\mathcal{M}}_g$ of stable curves of genus $g = (C)^2/2 + 1$. We now repeat the proof of Proposition 2.1.

Being stable is an open property and thus there exists an open neighbourhood $[C] \in U \subset P$ parametrizing only stable curves in nearby fibres $\mathcal{X}_t$. The universality of $\overline{\mathcal{M}}_g$ yields a classifying morphism $\varphi \colon U \longrightarrow \overline{\mathcal{M}}_g$. Then $\varphi[C]$ is contained in the locus of stable curves with at least g nodes, which is of codimension g; see [10, Ch. XI] or [233, p. 50]. Hence, there exists a subvariety $[C] \in T \subset U$ parametrizing curves with at least g nodes and such that $\dim(T) \geq \dim(S)$.

Since C as a curve with at least g nodes does not deform within $X = \mathcal{X}_0$ (they would all be rational), it must deform sideways to curves $C_t \subset \mathcal{X}_t$ with at least g nodes; i.e. $T \longrightarrow S$ is dominant.

If $\mathcal{X} \longrightarrow S$ and $C \subset X = \mathcal{X}_0$ are now given such that $\rho(\mathcal{X}_t) = 1$ for general $t \in S$ and C primitive, then C_t must be integral. This yields integral nodal rational curves in the general (and hence in the generic) fibre $\mathcal{X}_t$. The argument is taken from [69]. For a more local argument, which in fact underlies the bound for the dimension of the nodal locus, see [33, VIII.Ch. 23].

2.3 Let us now turn to stable maps. For the purpose of this chapter it suffices to consider *stable maps of arithmetic genus zero*. By definition this is a morphism

$$f \colon C \longrightarrow X$$

with only finitely many automorphisms and such that C is a connected projective curve with at most nodal singularities and $p_a(C) = 0$. Thus, the irreducible components $C_1, \ldots, C_n$ of C are all isomorphic to $\mathbb{P}^1$, the intersection of two components is always transversal, and there are no loops in C. Some of the components might be contracted by f, although stable maps of this type are of no importance for our discussion.

Clearly, the image of such a stable map $f \colon C \longrightarrow X$ is a rational, possibly singular, reducible, or even non-reduced curve in X. In characteristic zero or if X is not uniruled, the image $f(C)$ cannot be deformed without becoming non-rational. However, the map f itself can nevertheless have non-trivial deformations. But, although the isomorphism type of C may change, the components keep being isomorphic to $\mathbb{P}^1$.

There exists a moduli space $\mathcal{M}_0(X, \beta)$ of stable maps $f\colon C \longrightarrow X$ of arithmetic genus zero and such that $f_*[C] \in \mathrm{NS}(X)$ equals the given class β. We need a relative version of it. To simplify the discussion, we shall henceforth work in the complex setting. Then one can alternatively fix β as a cohomology class in $H^2(X, \mathbb{Z})$.

For a family of complex K3 surfaces $\mathcal{X} \longrightarrow S$ and $\beta \in H^2(\mathcal{X}, \mathbb{Z})$, let $\mathcal{M}_0(\mathcal{X}/S, \beta)$ denote the relative moduli space of stable maps. Thus, the fibre of

$$\mathcal{M}_0(\mathcal{X}/S, \beta) \longrightarrow S$$

over t is just $\mathcal{M}_0(\mathcal{X}_t, \beta_t)$. In particular, if β is a non-algebraic class on $\mathcal{X}_t$, then the fibre is empty.

It is known that $\mathcal{M}_0(\mathcal{X}/S, \beta)$ admits a coarse moduli space which is projective over S; cf. [193]. The construction as a proper algebraic space over S is easier and can be deduced from a standard Hilbert scheme construction and general existence results for quotients, e.g. the Keel–Mori result; cf. Theorem 5.2.6. In the discussion here we completely ignore the necessity of introducing markings in order to really obtain stable curves. But in any case, the global structure of $\mathcal{M}_0(\mathcal{X}/S, \beta)$ is of no importance for our purpose. The only thing that is needed is the following dimension count, which can be obtained from deformation theory for embedded rational curves or for maps between varieties; see [69, 312, 503] or [265, Prop. 2.1] for a short outline also valid for higher genus.

Theorem 2.3 *The dimension of each component of the moduli space $\mathcal{M}_0(\mathcal{X}/S, \beta)$ of stable maps to a family of K3 surfaces $\mathcal{X} \longrightarrow S$ is bounded from below:*

$$\dim(\mathcal{M}_0(\mathcal{X}/S, \beta)) \geq \dim(S) - 1.$$

Remark 2.4 The theorem works as well for families of K3 surfaces $\mathcal{X} \longrightarrow S$ in mixed characteristic. In fact in [69, 360] it is applied to the spread of a K3 surface X over a number field K, which is a family of K3 surfaces over an open set S of $\mathrm{Spec}(\mathcal{O}_K)$.

Geometrically the most instructive case is that of a universal deformation of a complex K3 surface X given by

$$\mathcal{X} \longrightarrow S = \mathrm{Def}(X);$$

see Section 6.2. As in the definition of the period map, we may assume that the cohomology of the fibres is trivialized, so that isomorphisms $H^2(\mathcal{X}_t, \mathbb{Z}) \simeq H^2(X, \mathbb{Z})$ are naturally given. A class $\beta \in H^2(X, \mathbb{Z})$ can thus be considered as a cohomology class on all the fibres. The bound on the dimension given above does a priori not allow one to conclude that $\mathcal{M}_0(\mathcal{X}/S, \beta) \longrightarrow S$ is surjective even when the fibres are zero-dimensional. And indeed, if β is an algebraic class on X, i.e. of type $(1,1)$, then it stays so only in a hypersurface $S_\beta \subset S$ which via the period map is obtained as the hyperplane section with $\beta^{\perp}$; see Section **6**.2.4. Thus, the image of $\mathcal{M}_0(\mathcal{X}/S, \beta) \longrightarrow S$ is contained in S_β and now the dimension bound reads

$$\dim(\mathcal{M}_0(\mathcal{X}/S, \beta)) \geq \dim(S) - 1 = \dim(S_\beta);$$

which under the condition that the fibre $\mathcal{M}_0(X, \beta)$ is zero-dimensional yields surjectivity of

$$\mathcal{M}_0(\mathcal{X}/S, \beta) \longrightarrow S_\beta.$$

Following Li and Liedtke [360] we say that a stable map $f\colon C \longrightarrow X$ is *rigid* if $\mathcal{M}_0(X, \beta)$ is of dimension zero in the point $[f\colon C \longrightarrow X] \in \mathcal{M}_0(X, \beta)$, but possibly non-reduced. As a special case of the above we state the following corollary.

Corollary 2.5 *Let $D \subset X$ be an integral rational curve. Then the normalization*

$$f\colon \mathbb{P}^1 \simeq \tilde{D} \longrightarrow D \subset X$$

is a rigid stable map. Therefore, if $[D]$ stays in $\mathrm{NS}(\mathcal{X}_t)$ for a family $\mathcal{X} \longrightarrow S$, then $f\colon \mathbb{P}^1 \longrightarrow X$ deforms sideways to $f_t\colon \mathbb{P}^1 \longrightarrow \mathcal{X}_t$ to the generic fibre. □

Note that in order to really obtain a deformation $\mathbb{P}^1 \times S \longrightarrow \mathcal{X}$ (over S) one might have to pass to an open subset of a finite covering of S.

Remark 2.6 Similarly, one can prove that every unramified stable map $f\colon C \longrightarrow X$ of arithmetic genus zero is rigid (in the stronger sense that it does not even admit first-order deformations); see [69]. This can be used to complete the proof of Theorem 1.1, (iii), by considering a stable map

$$f\colon C' = C_1' \cup C_2' \longrightarrow C = C_1 + C_2,$$

where C' is the nodal curve constructed as the union of two copies $C_i' \simeq \mathbb{P}^1$ intersecting transversally in one point and f induces $C_i' \xrightarrow{\sim} C_i$ and maps the node to one of the singularities of C.

A curve $D \subset X$ has a *rigid representative* if there exists a rigid stable map $f\colon C \longrightarrow X$ with image D (with multiplicities). Thus, an integral rational curve has a rigid representative and so has a reduced connected union of smooth rational curves $D_1 + \cdots + D_m$. In fact, as it turns out, any rational curve can be rigidified by an ample nodal rational curve according to the following result proved in [360].

Proposition 2.7 *Let $D_1, \ldots, D_m \subset X$ be integral rational curves (not necessarily distinct) and let $D \subset X$ be an integral ample nodal rational curve. Then for some $\ell \leq m$ the curve $\ell D + D_1 + \cdots + D_m$ has a rigid representative.*

For the details of the proof, we refer to [360, Thm. 2.9]. The idea is to view ℓD as the image of a stable map like this:

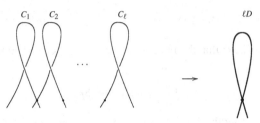

Here, the C_i are ℓ copies of the normalization of D. In a first step, they are glued in the pre-images of one node such that the resulting morphism is unramified. In the next step, the normalizations $\tilde{D}_i$ of the components D_i are glued to the C_i depending on the intersection $D \cap D_i$ and according to certain rules, of which the following drawings should convey the basic idea.

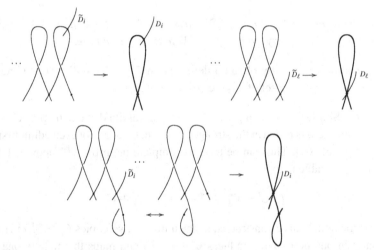

If one of the components D_i coincides with D, then it is dropped altogether, which might lead to $\ell < m$. Note that since D is ample, it really intersects all D_i non-trivially. The final outcome is a stable map that does not contract any component and for which the intersection of two components is *proper*. According to [360, Lem. 2.6], such a stable map is rigid.

3 Arithmetic Aspects

This section is devoted to an arithmetic approach towards the existence of infinitely many rational curves due to Bogomolov–Hassett–Tschinkel and developed further by

Li–Liedtke. First, the geometric part of the argument is explained and then we turn to the arithmetic aspects that involve reduction to positive characteristic and the Tate conjecture.

3.1 We begin this part with a result which is implicitly proved but not stated in [360]. Presenting it first allows one to concentrate on the geometric ideas of the argument in [69, 360] before combining them with more arithmetic considerations.

It is worth pointing out that the proof of the following theorem relies only on the existence of nodal rational curves in $|H|$ for generic (X, H) (see Theorem 1.1) and not on the more complicated result by Chen; cf. Theorem 1.7, evoked in [360].

Theorem 3.1 *Let (X, H) be a complex polarized K3 surface which cannot be defined over $\overline{\mathbb{Q}}$. Then X contains infinitely many integral rational curves (not necessarily ample or linearly equivalent to multiples of H).*

Proof The first step in the proof is to 'spread' the given K3 surface and thus view it as the (geometric) generic fibre of a non-isotrivial family. The second step consists of showing that the jump of the Picard number in this family gives rise to infinitely many rational curves in the generic fibre.

(i) To fix notation, write $(H)^2 = 2d$ and consider (X, H) as a $\mathbb{C}$-rational point of M_d, the moduli space of polarized K3 surfaces of degree $2d$; cf. Section **5**.1. For the purpose of this proof, M_d is considered over $\overline{\mathbb{Q}}$. Base change yields the moduli space of complex polarized K3 surfaces $M_d \times_{\overline{\mathbb{Q}}} \mathbb{C}$.

The polarized complex K3 surface (X, H) thus corresponds to a closed point of $M_d \times_{\overline{\mathbb{Q}}} \mathbb{C}$, but its image under

$$M_d \times_{\overline{\mathbb{Q}}} \mathbb{C} \longrightarrow M_d$$

is not closed, as X is not defined over $\overline{\mathbb{Q}}$. Let us denote by $T := \overline{\{(X, H)\}} \subset M_d$ the $\overline{\mathbb{Q}}$-variety (or a non-empty open subset of it) obtained as the closure of the image of $(X, H) \in M_d(\mathbb{C})$. Base changing T yields a subvariety $S := T_{\mathbb{C}} \subset M_d \times_{\overline{\mathbb{Q}}} \mathbb{C}$ and we denote its generic point by $\eta \in S$. Pass to an irreducible component of S if necessary.

Without using moduli spaces, this can be phrased as saying that a K3 surface X that cannot be defined over $\overline{\mathbb{Q}}$ can be 'spread out' over a positive-dimensional variety over $\overline{\mathbb{Q}}$ to yield a non-isotrivial family of K3 surfaces with generic fibre X. This can be base changed to yield a family of complex K3 surfaces; cf. the proof of Proposition 3.2.

The K3 surface corresponding to $\eta \in S$, i.e. a K3 surface defined over a certain finite extension of $k(\eta)$, can also be obtained directly by base changing X to $\mathbb{C} \subset k(\eta)$. Note that there is a bijection between rational curves in X and rational curves in $\mathcal{X}_\eta = X \times_{\mathbb{C}} k(\eta)$ (or in $\mathcal{X}_{\overline{\eta}}$), due to the rigidity of rational curves on K3 surfaces in characteristic zero; see Remark 1.3.

Now, by construction, the Picard number $\rho(\mathcal{X}_\eta)$ of the generic fibre is just $\rho(X)$. Moreover, using specialization or parallel transport we can view NS(X) as a sublattice of the Néron–Severi lattice NS$(\mathcal{X}_t)$ of any closed fibre $\mathcal{X}_t$; see Proposition **17.2.10**. In this sense, we shall view the polarization H on X also as a polarization on all the fibres $\mathcal{X}_t$.

For any such a family $\mathcal{X} \longrightarrow S$ the Noether–Lefschetz locus

$$S_0 := \{t \in S(\mathbb{C}) \mid \rho(\mathcal{X}_t) > \rho(X)\}$$

is a dense subset of S with $S \setminus S_0 \neq \emptyset$; see Sections **6.2.5** and **17.1.3**.

(ii) We are now going to produce rational curves on $\mathcal{X}_\eta$ (and hence on X) by detecting more and more curves on the special fibres $\mathcal{X}_t$. By Corollary 1.5, one knows that for a dense set of points $t \in S(\mathbb{C})$, the fibre $\mathcal{X}_t$ contains an integral rational curve $D_t \subset \mathcal{X}_t$ such that its class $D_t \in$ NS$(\mathcal{X}_t)$ is not contained in NS$(X) \subset$ NS$(\mathcal{X}_t)$.[5]

Moreover, the degree $(D_t.H)$ is unbounded and, more precisely, for any N the set $t \in S_0$ with $(D_t.H) \geq N$ is dense in S. To prove the latter, use that the moduli space

$$\mathrm{Mor}_{<N}(\mathbb{P}^1, \mathcal{X}/S) \longrightarrow S$$

of morphisms $f : \mathbb{P}^1 \longrightarrow \mathcal{X}_t$ of bounded degree $(f_*[\mathbb{P}^1].H) < N$ is a scheme of finite type over S, and that if the normalizations $f : \mathbb{P}^1 \longrightarrow D_t \subset \mathcal{X}_t$ of all the D_t were contained in it, then one component of $\mathrm{Mor}_{<N}(\mathbb{P}^1, \mathcal{X}/S)$ would dominate S, for $S_0 \subset S$ is dense. But whether $f_*[\mathbb{P}^1]$ is contained in NS$(X) \subset$ NS$(\mathcal{X}_t)$ is invariant under deformations and as NS$(X) =$ NS$(\mathcal{X}_t)$ for all $t \in S \setminus S_0$, this yields a contradiction.

For each rational curve $D_t \subset \mathcal{X}_t$ and $n \gg 0$ (depending on t) the divisor $nH - D_t$ is effective and hence, by Corollary 1.5, linearly equivalent to a sum $D_1 + \cdots + D_k$ of integral rational curves. The D_i are not necessarily distinct and k may depend on t.

Let now $U \subset M_d \times_{\overline{\mathbb{Q}}} \mathbb{C}$ be the dense open substack of polarized K3 surfaces (X', H') containing an integral nodal rational curve in $|H'|$; see Theorem 1.1, (ii). We shall first prove the assertion under the simplifying assumption that $U \cap S \neq \emptyset$. Then for any $N > 0$ there exists a $t \in U \cap S_0$ with $(D_t.H) \geq N$, integral rational curves $D_1, \ldots, D_k \subset \mathcal{X}_t$ such that $D_t + D_1 + \cdots + D_k \in |nH|$, and an integral nodal rational curve $D \subset \mathcal{X}_t$ in $|H|$. By [360, Thm. 2.9] (cf. Proposition 2.7), for some $\ell \leq k + 1$ the curve

$$\ell D + D_t + D_1 + \cdots + D_k \in |(n + \ell)H|$$

is the image of a rigid stable map $C_t \longrightarrow \mathcal{X}_t$ which therefore deforms to a stable map $f : C \longrightarrow \mathcal{X}_{\overline{\eta}}$ in the geometric generic fibre. Thus, in particular, there exists an integral component D_η of the image $f(C)$ which specializes to a curve in $\mathcal{X}_t$ that contains D_t

---

[5] According to Remark **17.2.13**, the cokernel of the specialization NS$(X) =$ NS$(\mathcal{X}_{\overline{\eta}}) \hookrightarrow$ NS$(\mathcal{X}_t)$ is torsion free. Hence, $\rho(X) = \rho(\mathcal{X}_t)$ if and only if NS$(X) \xrightarrow{\sim}$ NS$(\mathcal{X}_t)$.

as an irreducible component. This concludes the proof, because then D_η is an integral rational curve on $\mathcal{X}_{\bar\eta}$ of degree

$$(D_\eta.H) \geq (D_t.H) \geq N.$$

Note that we have no control over the linear equivalence class of D_η as a component of the curve $f(C) \subset \mathcal{X}_{\bar\eta}$ which is contained in the linear system $|(n + \ell)H|$.

If the open set U of (X', H') with nodal rational curves in $|H'|$ does not intersect S, then the argument has to be modified as follows. Through each $t \in S_0$ and for $D_t \subset \mathcal{X}_t$ chosen as above, there exists a codimension one subspace S_t in $M_d \times_{\bar{\mathbb{Q}}} \mathbb{C}$ such that the normalization $\mathbb{P}^1 \longrightarrow D_t \subset \mathcal{X}_t$ deforms to $\mathbb{P}^1 \longrightarrow \mathcal{X}_s$, for generic $s \in S_t$. Since the closed complement of U can accommodate only at most finitely many divisors S_t, we may assume that for almost all $t \in S_0$ the divisor S_t through t intersects U. Then one argues as before to obtain integral rational curves $D_\xi \subset \mathcal{X}_{\bar\xi}$ of arbitrary high degree in the fibre over the generic point $\xi \in U$ that specialize to a curve in $\mathcal{X}_t$ containing D_t. Specializing first to a curve in $\mathcal{X}_{\bar\eta}$ may decompose D_ξ, but one component D_η of it still specializes to a curve in $\mathcal{X}_t$ that contains D_t. This allows one to conclude as before that there exist integral rational curves in $\mathcal{X}_{\bar\eta}$ of unbounded degree. $\square$

3.2 The following is proved in [69]. As observed in [360], the proof also works if in addition the Picard number is fixed.

Proposition 3.2 *Conjecture 0.2 holds for any algebraically closed field k of characteristic zero if it holds for $k = \bar{\mathbb{Q}}$.*

Proof Any polarized K3 surface in characteristic zero is defined over a finitely generated field extension of $\bar{\mathbb{Q}}$. So we may assume that (X, H) is defined over the function field of an affine variety $B = \mathrm{Spec}(A)$ over $\bar{\mathbb{Q}}$. After shrinking B further if necessary, one obtains a smooth polarized family

$$(\mathcal{X}, \mathcal{H}) \longrightarrow B$$

with geometric generic fibre $X = \mathcal{X}_{\bar\eta}$.

If Conjecture 0.2 holds for K3 surfaces over $\bar{\mathbb{Q}}$, it holds for all closed fibres $\mathcal{X}_b$ of $(\mathcal{X}, \mathcal{H}) \longrightarrow B$. Thus, there exist integral rational curves $\mathbb{P}^1 \simeq \tilde{C} \longrightarrow C \subset \mathcal{X}_b$ in $|n\mathcal{H}_b|$ for arbitrary large n. By Corollary 2.5, these curves deform sideways as rational curves and, therefore, yield an integral rational curve $\mathbb{P}^1 \longrightarrow \mathcal{X}_{\bar\eta} = X$. $\square$

Remark 3.3 Also the weak version of Conjecture 0.2, not fixing the linear equivalence class of the curves C (see Remark 0.3), can be reduced to K3 surfaces over $\bar{\mathbb{Q}}$. For this one needs the existence of closed points $b \in B$ such that the specialization $\mathrm{NS}(X) \longrightarrow \mathrm{NS}(\mathcal{X}_b)$ is an isomorphism; see Remark 17.2.16.

The following results prove the conjecture in Remark 0.3 in many cases. It does not, however, address the stronger Conjecture 0.2, except when the Picard number is one. The first part is due to Bogomolov, Hassett, and Tschinkel [69] and the second to Li and Liedtke [360].

Theorem 3.4 *Let X be a K3 surface defined over an algebraically closed field k of characteristic zero. Then X contains infinitely many integral rational curves if one of the following conditions holds:*

(i) $\text{Pic}(X) = \mathbb{Z} \cdot H$ *with* $(H)^2 = 2$, *or*
(ii) $\rho(X)$ *is odd.*

Remark 3.5   As proved in [360], the assertion of Theorem 3.4 also holds in the following situation: $\text{char}(k) = p \geq 5$, $\rho(X)$ odd, and X is not supersingular. As since then supersingularity has been proved in [373] to be equivalent to unirationality and plenty of rational curves can be found on unirational K3 surfaces, the last condition is superfluous.

SKETCH OF PROOF OF THEOREM 3.4. First, one reduces the assertion to the case of K3 surfaces defined over a number field K; see Proposition 3.2. Then one considers the 'spread' $\mathcal{X} \longrightarrow S$ of X over an open set $S \subset \text{Spec}(\mathcal{O}_K)$ with $\mathcal{O}_K \subset K$ the ring of integers. Eventually one mimics the argument in the proof of Theorem 3.1. The Hodge theoretic argument to ensure that the Picard number jumps in a dense subset is in the arithmetic situation replaced by the Tate conjecture. Indeed, for any $\mathfrak{p} \in S$ the fibre $\mathcal{X}_{\mathfrak{p}}$ is a K3 surface over a finite field $k(\mathfrak{p})$ (after shrinking S to an open subset, if necessary) and the Tate conjecture implies that $\rho(\mathcal{X}_{\overline{k(\mathfrak{p})}})$ is even; see Corollary **17.2.9**. Thus, if $\rho(X)$ is odd, then the Picard number of the geometric closed fibres $\mathcal{X}_{\overline{k(\mathfrak{p})}}$ is always bigger.[6]

The rest of the proof is then almost identical to the one of Theorem 3.1. Note that in order to deform the additional rational curves in the special fibres, one still has to make sure that there are infinitely many fibres that are not supersingular.   □

Remark 3.6   Summarizing the above results, one can conclude that the only cases for which we do not yet know the existence of infinitely many integral rational curves (not necessarily linearly equivalent to some nH) are:

• K3 surfaces over $\overline{\mathbb{Q}}$ with Picard number $\rho(X) = 2$ or 4 and
• K3 surfaces over an algebraically closed field k with $\text{char}(k) = 2, 3$ or $\rho(X) \equiv 0\,(2)$.[7]

---

[6] When [360] appeared the Tate conjecture was still open, but the results of Bogomolov–Zarhin and Joshi–Rajan were enough to arrange the situation such that it holds for most fibres; cf. [360, Thm. 3.1].

[7] It seems reasonable to expect that the available techniques could prove the existence of infinitely many rational curves for elliptic K3 surfaces also in positive characteristic, so that also the case $\text{char}(k) \neq 2, 3$ and $\rho(X) \geq 5$ would be settled. But details would need to be worked out.

Indeed, for $\rho(X) = 1, 3$ use Theorem 3.4, (ii), and K3 surfaces with $\rho(X) \geq 5$ are elliptic (cf. Proposition **11**.1.3) for which the weak version (cf. Remark 0.3), is known; see Remark 1.6.

In fact, for $\rho(X) = 2$ only the case **(iv)** in Section **8**.3.2 remains open, as in **(i)**-**(iii)** either the K3 surface X has infinite $\mathrm{Aut}(X)$ (see Corollary **8**.4.8) or is elliptic. See also Example **8**.4.9.

In contrast, the stronger Conjecture 0.2 seems to be known only when $\rho(X) = 1$.

4 Counting of Rational Curves

As proved by Chen (cf. Theorem 1.7), for the generic polarized complex K3 surface $(X, H) \in M_d$ all rational curves in the linear system $|H|$ are nodal and, moreover, there always exists at least one. Since there are only finitely many of them, one would like to know how many there are exactly. The same question makes sense for arbitrary, i.e. non-generic, $(X, H) \in M_d$ possibly containing non-nodal rational curves in $|H|$, which would need to be counted with appropriate multiplicities.

4.1 If $C \subset X$ is a nodal rational curve in $|H|$ (or in some multiple), then C deforms with X to all nearby fibres $\mathcal{X}_t$ parametrized by M_d; see the discussion in Section 2.2. In other words, the number n_g of nodal rational curves in $|H|$ should be constant for generic (X, H).

It is customary to use the genus g instead of the degree d. Recall that $2d = 2g - 2$. Then the following formula was conjectured by Yau and Zaslow in [642]:

$$\sum_{g \geq 0} n_g q^g = \frac{q}{\Delta(q)} = \prod_{n \geq 1} (1 - q^n)^{-24}. \tag{4.1}$$

Here, $\Delta(q)$ is the modular form $q \prod_{n \geq 1} (1 - q^n)^{24}$, which can also be written in terms of the first two Eisenstein series

$$G_2 = (1/60)g_2 \quad \text{and} \quad G_3 = (1/140)g_3$$

as (see [547, Ch. VIII])

$$\Delta(q) = (1/2\pi)^{12}(g_2^3 - 27g_3^2).$$

Since $g = 0, 1$ correspond to $d = -2$ and $d = 0$, which are certainly not allowed for a polarized K3 surface, one simply sets $n_0 = 1$ and $n_1 = 24$. The latter, however, can be motivated geometrically as follows. The generic elliptic K3 surface $\pi : X \longrightarrow \mathbb{P}^1$ has exactly 24 singular fibres, each of which is a nodal rational curve with one node; see Remark **11**.1.12. So, if in this case one replaces the polarization H by the nef line bundle $\pi^* \mathcal{O}(1)$, the number 24 fits well.

Beauville in [46] succeeded to prove (4.1) under the assumption that all rational curves in $|H|$ are nodal, which later was proved by Chen; see Theorem 1.7. In [46] the number n_g is replaced by the Euler number $e(\mathrm{Jac}^g(\mathcal{C}/|H|))$ of the relative compactified Jacobian of the universal curve $\mathcal{C} \longrightarrow |H|$. Moreover, it is shown that for a non-rational curve $C \in |H|$ the Euler number $e(\mathrm{Jac}^g(C))$ is zero and, therefore, does not contribute to $n(g)$. If C is nodal rational, then indeed $e(\mathrm{Jac}^g(C)) = 1$, which proves that both definitions of n_g coincide if all rational curves in $|H|$ are nodal. The contribution for rational but more singular C was further discussed by Fantechi, Göttsche, and van Straten in [177].

Remark 4.1 The relative compactified Jacobian $\mathrm{Jac}^g(\mathcal{C}/|H|)$ can be viewed as a moduli space of simple or in fact stable sheaves on X with support on curves in $|H|$; cf. Section 10.1.5. It is known that these moduli spaces are deformation equivalent to the Hilbert scheme $\mathrm{Hilb}^n(X)$ of the same dimension; see Theorem **10.3.10** for the precise assumptions. In fact, $\mathrm{Jac}^g(\mathcal{C}/|H|)$ is even birational to $\mathrm{Hilb}^n(X)$. Now, since the Euler number of the Hilbert scheme was known (under the name of *Göttsche's formula*), Beauville could deduce (4.1) from [36] and [250].

4.2 Let us consider the first values of the numbers n_g which can be read off from the Yau–Zaslow formula (4.1)

$$\sum_{g \geq 0} n_g q^g = 1 + 24q + 324q^2 + 3200q^3 + 25650q^4 + \cdots .$$

As mentioned already, $n_1 = 24$ should be interpreted as the number of singular fibres, all of type I_1, in a generic elliptic K3 surface.

Next, $n_2 = 324$ should be the number of nodal rational curves in $|H|$ for the generic polarized (X, H) with $(H)^2 = 2$, i.e. in a double plane

$$\pi : X \longrightarrow \mathbb{P}^2$$

ramified over a generic sextic curve $C \subset \mathbb{P}^2$ with $H = \pi^* \mathcal{O}(1)$. As $H^0(X, \pi^*\mathcal{O}(1)) \simeq H^0(\mathbb{P}^2, \mathcal{O}(1))$, all curves in $|H|$ are double covers

$$\pi : C_\ell \longrightarrow \ell$$

of some line $\ell \subset \mathbb{P}^2$. Consider a point $x \in C \cap \ell$ and let $\tilde{x} \in C_\ell$ be its unique pre-image. If the intersection $C \cap \ell$ is transversal, then C_ℓ is smooth in $\tilde{x}$. So, in order to obtain a rational and thus in particular a singular curve $C_\ell \in |H|$, the line ℓ should be tangent to at least one point $x \in C$.

Suppose now that C_ℓ is rational. Then the composition with its normalization $\mathbb{P}^1 \simeq \tilde{C}_\ell \longrightarrow C_\ell \longrightarrow \ell$ has two ramification points. As the map ramifies over each transversal point of intersection $x \in C \cap \ell$, there are at most two of those. Note that, in particular, the 72 flexes do not lead to rational curves.[8]

So, we are looking for lines $\ell \subset \mathbb{P}^2$ with at most two transversal points of intersection with C and at least one that is not. Under the genericity assumption on C, these are exactly the 324 bitangents to a generic sextic, i.e. lines $\ell \subset \mathbb{P}^2$ that are bitangent to two points $x_1, x_2 \in C$ and have two further points of intersection $x_3, x_4 \in C$ (which are automatically transversal); see [236, Exer. IV.2.3]. As the pre-images $\tilde{x}_1, \tilde{x}_2 \in C_\ell$ are nodes of C_ℓ, this proves that for generic $C \subset \mathbb{P}^2$ all rational curves in $H = \pi^* \mathcal{O}(1)$ on the associated double plane $\pi : X \longrightarrow \mathbb{P}^2$ are nodal and there are 324 of them.[9]

The next case is that of a quartic $X \subset \mathbb{P}^3$ and rational curves in $|\mathcal{O}_X(1)|$. By Theorem 1.7, one knows that for a generic quartic all rational curves in this linear system are nodal (with exactly three nodes) and by (4.1) that there are precisely 3200 of them. All this seems to have been known classically, going back to Salmon, and can be verified by means of the Gauss map

$$\gamma : X \longrightarrow X^*, \ x \longmapsto T_x X.$$

Generically, $C_x := T_x X \cap X$ is a curve with a node at $x \in C_x$ and which is smooth elsewhere. Along a curve $D \subset X$, the curves $C_x, x \in D$, have generically two nodes and along D the Gauss map γ is of degree two. In exactly $3 \cdot 3200$ points $x \in X$, for generic choice of $X \subset \mathbb{P}^3$, the curve C_x has three nodes. Forgetting the points yields the 3200 nodal rational curves in $|\mathcal{O}_X(1)|$.

Remark 4.2 The irreducibility of the *Severi variety*, i.e. of the moduli space parametrizing $(X, H, C \in |H|)$ consisting of a polarized K3 surface $(X, H) \in M_d$ and a (nodal) rational curve C, is conjectured in general. It has been proved for $1 \leq g := d + 1 \leq 9$ and $g = 11$; see the paper [119] by Ciliberto and Dedieu. The case $g = 2$ is classical. For the computation of the monodromy group in this case, see Harris [232].

5 Density Results

We conclude by mentioning two density results: one for rational curves, building upon ideas of Bogomolov and Tschinkel, and another for points realizing the same rational equivalence class.

---

[8] According to [236, IV.Exer. 2.3], the generic sextic $C \subset \mathbb{P}^2$ contains 72 flexes (or inflection points), i.e. points $x \in C$ such that the intersection multiplicity of $T_x C$ with C at x is at least three. Generically only flexes of multiplicity three occur and a line $\ell = T_x C \subset \mathbb{P}^2$ of a flex is not tangent to C at any other point.

[9] Recall that the degree of the discriminant divisor in $|\mathcal{O}_{\mathbb{P}^3}(4)|$ is 108; see Section **17**.1.4. Is there a geometric interpretation for the curious equality $324 = 3 \cdot 108$?

5.1   The existence of infinitely many rational curves on K3 surfaces of Picard number one in particular proves that for all complex polarized K3 surfaces $(X, H) \in M_d$ in the complement of a countable union of hypersurfaces, the union of all rational curves (which can even be assumed to be linearly equivalent to nH for some n) is Zariski dense. This is strengthened by the following theorem of Chen and Lewis [117].

Theorem 5.1   *For $(X, H) \in M_d(\mathbb{C})$ in the complement of a countable union of nowhere dense subsets, the union of all rational curves in $\bigcup |nH|$ is analytically dense.*

The first step in the argument is an improvement of a technique of Bogomolov and Tschinkel showing the existence of infinitely many rational curves on an elliptic K3 surface with a non-torsion section; see Remark 1.6. Translating by a non-torsion section certainly gives an infinite number of rational curves, which automatically form a Zariski dense subset, but only if the non-torsion section hits the generic closed fibre in a sufficiently generic point is this set also dense in the classical topology.

Instead of simply translating in the fibres, Chen and Lewis map a point p in a generic fibre X_t to the point q for which $\mathcal{O}_{X_t}(q) \simeq H|_{X_t}(-((H.X_t) - 1)p)$, where H is a given ample line bundle. The existence of an elliptic K3 surface that is generic in this sense is proved by a dimension argument on a degeneration. In order to get density on the generic K3 surfaces, one rigidifies the rational curves on these elliptic K3 surfaces and uses again the theory of stable maps.

5.2   Theorem 5.1 compares nicely with the following result of Mclean [385]. See Chapter 12 for more on $CH^2(X)$.

Theorem 5.2   *For a generic complex projective K3 surface X and an arbitrary point $x \in X$, the set*

$$\{y \mid [x] = [y] \in CH^2(X)\}$$

is analytically (and hence Zariski) dense in X.

Proof   For a general K3 surface X the Picard group $\mathrm{Pic}(X)$ is generated by an ample divisor H. By Theorem 1.1, (ii), there exists a nodal rational curve in $|H|$ and hence by Proposition 2.1 a covering family of elliptic curves $C \longrightarrow B$. Since H generates $\mathrm{Pic}(X)$, all curves C_t are in fact integral.

The construction works in families and one can therefore assume the existence of such a family not only for a general polarized K3 surface (X, H) but in fact for a generic one. Now use Theorem 1.7 to deduce the existence of an integral nodal rational curve in $|2H|$. By applying Proposition 2.1 again, one finds another one-dimensional covering family $\mathcal{D} \longrightarrow B'$ of curves of geometric genus ≤ 1.

Let now $x \in X$ be an arbitrary point on such a generic polarized K3 surface (X, H) and choose one of the fibres, say $\mathcal{D}_0$, that contains x. A priori, $\mathcal{D}_0$ might be reducible, so choose an irreducible component $D \subset \mathcal{D}_0$ that contains x. Since the normalization $\widetilde{D}$ is either $\mathbb{P}^1$ or smooth elliptic, the set $S_{D,x}$ of points $y \in D$ with $[y] = [x]$ in $CH^2(X)$ is analytically and hence Zariski dense in D. Here one uses that $CH^2(X)$ is torsion free by Roitman's theorem **12**.1.3.

Now choose for every $y \in S_{D,x}$ a curve $y \in \mathcal{C}_y$ and use as above that the set of points $z \in \mathcal{C}_y$ with $[y] = [z]$ in $CH^2(X)$ is analytically dense in $\mathcal{C}_y$. A priori it might happen that $D = \mathcal{C}_y$. However, as the curves $\mathcal{C}_y$ are ample, we may choose them such that this is not the case for most y. Then, since $[y] = [x]$, this proves the assertion. Note that at this point one uses the irreducibility of all curves $\mathcal{C}_y$, since otherwise it could happen that D is a component of one fibre $\mathcal{C}_0$ and thus $\mathcal{C}_y$ could be $\mathcal{C}_0$ independently of y.[10] □

References and Further Reading

Benoist's talk in the Bourbaki seminar [58] contains a survey of parts of the material covered in this chapter.

In [216, Part B] Green and Griffiths prove that K3 surfaces of low genus are not hyperbolic. They outlined a completely geometric approach towards the existence of rational and elliptic curves on K3 surfaces that relies on a detailed understanding of the dual surface. To my knowledge, this has not been pursued any further. In [229, 230] Halic proposes a more geometric approach to Chen's result; see Theorem 1.7, (ii), on nodal (rational) curves on generic K3 surfaces.

Alternative proofs of the Yau–Zaslow formula (4.1) have been given by Bryan and Leung in [89] and by Chen in [115]. In fact, in [89] the authors prove a generalization of (4.1) to higher genus (but still curves in the primitive linear system $|H|$) conjectured by Göttsche [213]. It is expected that for general $(X, H) \in M_d$ all rational curves $C \in |nH|$ are nodal. Although this has not been proved yet, an appropriate generalization of the Yau–Zaslow formula (4.1) for Gromov–Witten invariants for non-primitive classes, i.e. $n > 1$, has been proved by Klemm, Maulik, Pandharipande, and Scheidegger [296]; cf. the survey [485] and references therein.

There are related classical numerical problems. For example, Segre in [542] asks, how many lines can a quartic $X \subset \mathbb{P}^3$ contain? Of course, the very general quartic in characteristic zero does not contain any line. It turns out that the maximal number of lines on an arbitrary smooth quartic is 64, which is achieved for the Fermat quartic. See [73, 501] for more details and a discussion of the possible configuration of these lines, and Sections 3.2.6 and 17.1.4.

Questions and Open Problems

It does not seem completely impossible to produce nodal rational curves in $|nH|$ with $n > 1$ by an explicit construction à la Bogomolov and Mumford as in the proof of Theorem 1.1 and in this way to provide a more direct argument for Theorem 1.7, (ii).

---

[10] The result should hold without any condition on X and can be proved if $[x] = c_X$ by modifying the above arguments; see [626, Lem. 2.3]. In fact, in [385] also x is assumed generic, but this can be avoided by interchanging the role of the two families $\mathcal{C}$ and $\mathcal{D}$ as done here.

It seems likely that density of rational curves in elliptic surfaces as proved by Bogomolov and Tschinkel in [70] continues to hold in positive characteristic, but details would need to be checked.

As mentioned in the introduction, Bogomolov apparently had asked back in 1981 whether every $\overline{\mathbb{Q}}$-rational point of a K3 surface defined over a number field lies on a rational curve. There does not seem to be much evidence for this, but attempts to disprove it have also failed; see e.g. the first version of [31]. Of course, this would immediately imply Conjecture **12.2.5**.

As a consequence of Lemma **17.2.6** one knows that all smooth rational curves on $X \times \overline{k}$, of which there might be infinitely many, live over a finite extension K/k. In many cases, the degree of K/k can be universally bounded. Is this still true for singular rational curves?

14

Lattices

Apart from Section 3, this chapter should be seen mainly as a reference. We collect all results from lattice theory that are used at some point in these notes. We try to give a readable survey of the relevant facts and a general feeling for the techniques, but we have often have to refer for details to the literature and for many things to Nikulin's influential paper [451]. Only the parts of lattice theory that are strictly relevant to the theory of K3 surfaces are touched upon; in particular most of the lattices are even and all are over $\mathbb{Z}$. In Section 3 one finds, among other things, a characterization of Picard lattices of small rank and a lattice theoretic description of Kummer surfaces.

Besides Nikulin's articles and earlier ones by Wall and Kneser, one could consult Dolgachev's survey article [149]; Serre's classic [547], covering the classification of even, unimodular lattices; and the textbooks [132, 161, 163, 302, 411], for example for the Leech lattice and its relatives.

0.1 A *lattice* Λ is by definition a free $\mathbb{Z}$-module of finite rank together with a symmetric bilinear form

$$(.): \Lambda \times \Lambda \longrightarrow \mathbb{Z},$$

which we always assume to be non-degenerate. A lattice Λ is called *even* if

$$(x)^2 := (x.x) \in 2\mathbb{Z}$$

for all $x \in \Lambda$, otherwise Λ is called *odd*. The determinant of the intersection matrix with respect to an arbitrary basis (over $\mathbb{Z}$) is called the *discriminant*, disc Λ.

A lattice Λ and the $\mathbb{R}$-linear extension of its bilinear form (.) give rise to the real vector space $\Lambda_{\mathbb{R}} := \Lambda \otimes_{\mathbb{Z}} \mathbb{R}$ endowed with a symmetric bilinear form. The latter can be diagonalized with only 1 and -1 on the diagonal, as we assumed that (.) is non-degenerate. The *signature* of Λ is (n_+, n_-), where $n_\pm$ is the number of ± 1 on the diagonal, and its *index* is $\tau(\Lambda) := n_+ - n_-$. The lattice Λ is called *definite* if either $n_+ = 0$ or $n_- = 0$ or, equivalently, if $\tau(\Lambda) = \pm \text{rk } \Lambda$. Otherwise, Λ is *indefinite*.

One defines an injection of finite index

$$i_\Lambda : \Lambda \hookrightarrow \Lambda^* := \mathrm{Hom}_{\mathbb{Z}}(\Lambda, \mathbb{Z}), \; x \mapsto (x. \;).$$

Alternatively, if Λ^* is viewed as the subset of all $x \in \Lambda_{\mathbb{Q}}$ of the $\mathbb{Q}$-vector space $\Lambda_{\mathbb{Q}}$ such that $(x.\Lambda) \subset \mathbb{Z}$, then i_Λ is just the natural inclusion $\Lambda \hookrightarrow \Lambda^* \hookrightarrow \Lambda_{\mathbb{Q}}$. The cokernel of $i_\Lambda : \Lambda \hookrightarrow \Lambda^*$ is called the *discriminant group*

$$A_\Lambda := \Lambda^* / \Lambda$$

of Λ, which is a finite group of order $|\mathrm{disc}\, \Lambda|$. A lattice is called *unimodular* if i_Λ defines an isomorphism $\Lambda \xrightarrow{\sim} \Lambda^*$ or, equivalently, if A_Λ is trivial or, still equivalently, if disc $\Lambda = \pm 1$. The minimal number of generators of the finite group A_Λ also plays a role and is denoted

$$\ell(A_\Lambda) = \ell(\Lambda).$$

Remark 0.1 It is an elementary but very useful observation that whenever disc Λ is square free, then $\ell(\Lambda) = 0$ or 1. Indeed, A_Λ is a finite abelian group and hence of the form $\bigoplus \mathbb{Z}/m_i\mathbb{Z}$. If disc Λ and hence $|A_\Lambda| = \prod m_i$ is square free, then the m_i are pairwise prime and, therefore, $A_\Lambda \simeq \mathbb{Z}/|A_\Lambda|\mathbb{Z}$.

The pairing $(\;.\;)$ on Λ induces a $\mathbb{Q}$-valued pairing on Λ^* and hence a pairing $A_\Lambda \times A_\Lambda \longrightarrow \mathbb{Q}/\mathbb{Z}$. If the lattice Λ is even, then the $\mathbb{Q}$-valued quadratic form on Λ^* yields

$$q_\Lambda : A_\Lambda \longrightarrow \mathbb{Q}/2\mathbb{Z},$$

which gives back the pairing on A_Λ. The finite group A_Λ together with q_Λ is called the *discriminant form* of Λ

$$(A_\Lambda, q_\Lambda : A_\Lambda \longrightarrow \mathbb{Q}/2\mathbb{Z}).$$

A finite abelian group A with a quadratic form $q \colon A \longrightarrow \mathbb{Q}/2\mathbb{Z}$ is called a *finite quadratic form*. The index $\tau(A, q) \in \mathbb{Z}/8\mathbb{Z}$ of it is well defined as the index $\tau(\Lambda)$ modulo 8 of any even lattice Λ with $(A_\Lambda, q_\Lambda) \simeq (A, q)$.[1]

Two lattices Λ and Λ' are said to have the same *genus*, $\Lambda \sim \Lambda'$, if $\Lambda \otimes \mathbb{Z}_p \simeq \Lambda' \otimes \mathbb{Z}_p$ for all prime p and $\Lambda \otimes \mathbb{R} \simeq \Lambda' \otimes \mathbb{R}$ (all isomorphisms are assumed to be compatible with the natural quadratic forms). The latter of course just means that Λ and Λ' have the same signature.

It is a classical result that there are at most finitely many isomorphism types of lattices with the same genus; cf. [105, Ch. 9.4] or [302, Satz 21.3]. Moreover, due to [451, Cor. 1.9.4], one knows that two even lattices Λ and Λ' have the same genus if their signatures

---

[1] See e.g. [631, Cor. 1 and 2]. This has two parts. First, any (A, q) can be realized as the discriminant form of an even lattice [629, Thm. 6]; cf. Theorem 1.5. Second, two lattices Λ_1 and Λ_2 with isomorphic discriminant forms are stably equivalent; i.e. there exist unimodular lattices Λ'_1, Λ'_2 such that $\Lambda_1 \oplus \Lambda'_1 \simeq \Lambda_2 \oplus \Lambda'_2$; cf. Corollary 1.7. This result in its various forms is due to Kneser, Durfee, and Wall.

coincide and $(A_\Lambda, q_\Lambda) \simeq (A_{\Lambda'}, q_{\Lambda'})$. In particular, for bounded $|\text{disc } \Lambda|$ and rk Λ there exist only finitely many isomorphism types of even lattices; see [105, Ch. 9].

0.2 For two lattices Λ_1 and Λ_2 the direct sum $\Lambda_1 \oplus \Lambda_2$ shall always denote the *orthogonal direct sum*, i.e. $(x_1 + x_2.y_1 + y_2)_{\Lambda_1 \oplus \Lambda_2} = (x_1.y_1)_{\Lambda_1} + (x_2.y_2)_{\Lambda_2}$. Clearly, there exists an isomorphism $A_{\Lambda_1 \oplus \Lambda_2} \simeq A_{\Lambda_1} \oplus A_{\Lambda_2}$ which for even lattices is compatible with the discriminant forms, i.e.

$$(A_{\Lambda_1 \oplus \Lambda_2}, q_{\Lambda_1 \oplus \Lambda_2}) \simeq (A_{\Lambda_1}, q_{\Lambda_1}) \oplus (A_{\Lambda_2}, q_{\Lambda_2}).$$

A morphism between two lattices $\Lambda_1 \longrightarrow \Lambda$ is by definition a linear map that respects the quadratic forms. If $\Lambda_1 \hookrightarrow \Lambda$ has finite index, then one proves

$$\text{disc } \Lambda_1 = \text{disc } \Lambda \cdot (\Lambda : \Lambda_1)^2, \tag{0.1}$$

e.g. by using the inclusions $\Lambda_1 \hookrightarrow \Lambda \hookrightarrow \Lambda^* \hookrightarrow \Lambda_1^*$.

An injective morphism $\Lambda_1 \hookrightarrow \Lambda$ is called a *primitive embedding* if its cokernel is torsion free. For example, the orthogonal complement $\Lambda_2 := \Lambda_1^\perp \subset \Lambda$ of any sublattice $\Lambda_1 \hookrightarrow \Lambda$ is a primitive sublattice intersecting Λ_1 trivially. Moreover, the induced embedding

$$\Lambda_1 \oplus \Lambda_2 \hookrightarrow \Lambda$$

is of finite index and in general not primitive, not even for primitive Λ_1. More precisely, as a consequence of (0.1) one finds

$$\text{disc } \Lambda_1 \cdot \text{disc } \Lambda_2 = \text{disc } \Lambda \cdot (\Lambda : \Lambda_1 \oplus \Lambda_2)^2. \tag{0.2}$$

Two even lattices Λ_1, Λ_2 are called *orthogonal* if there exists a primitive embedding $\Lambda_1 \hookrightarrow \Lambda$ into an even unimodular lattice with $\Lambda_1^\perp \simeq \Lambda_2$.

To give a taste of the kind of arguments that are often used, we prove the following standard result (cf. [451, Prop. 1.5.1, 1.6.1] or [161, Ch. 3.3]), which in particular shows that every even lattice Λ is orthogonal to $\Lambda(-1)$; see Section 0.3, **(iv)**.

Proposition 0.2 *Let Λ_1, Λ_2 be two even lattices.*

(i) *Then Λ_1 and Λ_2 are orthogonal if and only if $(A_{\Lambda_1}, q_{\Lambda_1}) \simeq (A_{\Lambda_2}, -q_{\Lambda_2})$.*
(ii) *More precisely, a primitive embedding $\Lambda_1 \hookrightarrow \Lambda$ into an even, unimodular lattice Λ with $\Lambda_1^\perp \simeq \Lambda_2$ is determined by an isomorphism $(A_{\Lambda_1}, q_{\Lambda_1}) \simeq (A_{\Lambda_2}, -q_{\Lambda_2})$.*

Proof As a first step one establishes for a fixed even lattice Λ' a bijective correspondence between finite index even overlattices $\Lambda' \subset \Lambda$ and isotropic subgroups $H \subset A_{\Lambda'}$

$$\{\Lambda' \subset \Lambda \mid (\Lambda : \Lambda') < \infty\} \longleftrightarrow \{H \subset A_{\Lambda'} \mid \text{isotropic}\}$$

given by $(\Lambda' \subset \Lambda) \mapsto (\Lambda/\Lambda' \subset \Lambda^*/\Lambda' \subset \Lambda'^*/\Lambda' = A_{\Lambda'})$. The inverse map is given by sending $H \subset A_{\Lambda'}$ to its pre-image under $\Lambda'^* \longrightarrow A_{\Lambda'}$.

Check that H is isotropic if and only if the quadratic form on Λ^* restricts to an integral even form on Λ. Moreover, by (0.1) Λ is unimodular if and only if $|H|^2 = |A_{\Lambda'}|$.

Next, with any primitive embedding $\Lambda_1 \subset \Lambda$ into an even unimodular lattice Λ such that $\Lambda_1^\perp \simeq \Lambda_2$ one associates the finite index sublattice $\Lambda' := \Lambda_1 \oplus \Lambda_2 \subset \Lambda$. The associated isotropic subgroup is $\Lambda/(\Lambda_1 \oplus \Lambda_2) \subset A_{\Lambda_1 \oplus \Lambda_2} \simeq A_{\Lambda_1} \oplus A_{\Lambda_2}$. The two projections

$$\Lambda/(\Lambda_1 \oplus \Lambda_2) \longrightarrow A_{\Lambda_i},$$

$i = 1, 2$, are injective, as both inclusions $\Lambda_i \subset \Lambda$ are primitive. They are surjective if and only if Λ is unimodular, as the canonical maps $\Lambda^* \longrightarrow \Lambda_i^*$ are surjective. Thus,

$$A_{\Lambda_1} \overset{\sim}{\longrightarrow} \Lambda/(\Lambda_1 \oplus \Lambda_2) \overset{\sim}{\longrightarrow} A_{\Lambda_2},$$

changing the sign of the quadratic form. Conversely, the graph of any isomorphism $(A_{\Lambda_1}, q_{\Lambda_1}) \simeq (A_{\Lambda_2}, -q_{\Lambda_2})$ is an isotropic subgroup $H \subset A_{\Lambda_1 \oplus \Lambda_2}$ which, therefore, gives rise to a finite index overlattice $\Lambda_1 \oplus \Lambda_2 \subset \Lambda$. Now, H being a graph of an isomorphism immediately translates into primitivity of the embeddings $\Lambda_i \subset \Lambda$ and unimodularity of Λ. $\qquad\square$

For the discriminants of two orthogonal lattices one thus finds

$$\text{disc } \Lambda_1 = \pm \text{disc } \Lambda_2, \tag{0.3}$$

where the sign is the discriminant of Λ, and, using (0.2), $|\text{disc } \Lambda_i| = (\Lambda : \Lambda_1 \oplus \Lambda_2)$.

0.3 To set up notations and to recall some basic facts, we include a list of standard examples of lattices that come up frequently in this book.

(i) By $\langle 1 \rangle$, $\mathbb{Z}$, or I_1 one denotes the lattice of rank one with intersection matrix 1. The direct sum $\langle 1 \rangle^{\oplus n}$ is often denoted I_n. See Corollary 1.3 for the related notation I_{n_+, n_-}.

(ii) The *hyperbolic plane* is the lattice

$$U := \begin{pmatrix} 0 & 1 \\ 1 & 0 \end{pmatrix},$$

i.e. $U \simeq \mathbb{Z}^2 = \mathbb{Z} \cdot e \oplus \mathbb{Z} \cdot f$ with the quadratic form given by $(e)^2 = (f)^2 = 0$ and $(e.f) = 1$. Clearly, disc $U = -1$. Other common notations for U are $\mathrm{II}_{1,1}$ or simply H.

Example 0.3 The hyperbolic plane is special in many ways. For example, if $U \hookrightarrow \Lambda$ is an arbitrary (not necessarily primitive) embedding, then

$$\Lambda = U \oplus U^\perp.$$

Indeed, if for $\alpha \in \Lambda$ one defines α' by $\alpha = (e.\alpha)f + (f.\alpha)e + \alpha'$, then $\alpha' \in U^\perp$. The same assertion holds for embeddings $U^{\oplus k} \hookrightarrow \Lambda$. See also Proposition 1.8.

(iii) The E_8-*lattice* is given by the intersection matrix

$$E_8 := \begin{pmatrix} 2 & -1 & & & & & & \\ -1 & 2 & -1 & & & & & \\ & -1 & 2 & -1 & -1 & & & \\ & & -1 & 2 & 0 & & & \\ & & -1 & 0 & 2 & -1 & & \\ & & & & -1 & 2 & -1 & \\ & & & & & -1 & 2 & -1 \\ & & & & & & -1 & 2 \end{pmatrix}$$

and is, therefore, even, unimodular, positive definite (i.e. $n_- = 0$) of rank eight with disc $E_8 = 1$. See also **(v)** below.

(iv) For any given lattice Λ the *twist* $\Lambda(m)$ is obtained by changing the intersection form $(\,.\,)$ of Λ by the integer m, i.e. $\Lambda = \Lambda(m)$ as $\mathbb{Z}$-modules but

$$(\,.\,)_{\Lambda(m)} := m \cdot (\,.\,)_\Lambda.$$

The discriminant of the twist is given by

$$\text{disc }\Lambda(m) = \text{disc }\Lambda \cdot m^{\text{rk }\Lambda},$$

which can be deduced from the exact sequence $0 \longrightarrow \Lambda/m\Lambda \longrightarrow A_{\Lambda(m)} \longrightarrow A_\Lambda \longrightarrow 0$ of groups. The latter also shows that, for example,

$$A_{U(m)} \simeq (\mathbb{Z}/m\mathbb{Z})^2.$$

We frequently use $\langle -1 \rangle := \langle 1 \rangle(-1)$, which is the rank one lattice with intersection matrix -1, $\mathbb{Z}(m) := \langle 1 \rangle(m)$, and $E_8(-1)$, which is negative definite, unimodular, and even. Note that $U(-1) \simeq U$.

(v) Lattices that are not unimodular play a role as well, for example, the lattices associated to the Dynkin diagrams A_n, D_n, E_6, E_7, and E_8. Only the last one gives rise to a unimodular lattice, which has been described above.

To any graph Γ with simple edges the lattice $\Lambda(\Gamma)$ associated with Γ has a basis e_i corresponding to the vertices with the intersection matrix given by $(e_i.e_j) = 2$ if $i = j$, $(e_i.e_j) = -1$ if e_i and e_j are connected by an edge, and $(e_i.e_j) = 0$ otherwise. So, for example, $A_1 \simeq \langle 2 \rangle$.

In fact, the graphs of ADE type as drawn below are the only connected graphs Γ for which the following holds: two vertices e_i, e_j of Γ are connected by at most one edge and the lattice $\Lambda(\Gamma)$ naturally associated with Γ is positive definite.

Geometrically, lattices of ADE type occur as configurations of exceptional divisors of minimal resolutions of rational double points. Recall that rational double points (or simple surface singularities, Kleinian singularities, etc.) are described explicitly by the following equations:

$A_{n \geq 1}$	$xy + z^{n+1}$	
$D_{n \geq 4}$	$x^2 + y(z^2 + y^{n-2})$	
E_6	$x^2 + y^3 + z^4$	
E_7	$x^2 + y(y^2 + z^3)$	
E_8	$x^2 + y^3 + z^5$	

The exceptional divisor of the minimal resolution of each of these singularities is a curve $\sum C_i$ with $C_i \simeq \mathbb{P}^1$, self-intersection $(C_i)^2 = -2$, and for $C_i \neq C_j$ one has $(C_i.C_j) = 0$ or 1. The vertices of the dual graph correspond to the irreducible components C_i and vertices are connected by an edge if the corresponding curves C_i and C_j intersect. The dual graph is depicted in each of the cases in the last column.

Alternatively, rational double points can be described as quotient singularities $\mathbb{C}^2/G$ by finite groups $G \subset \mathrm{SL}(2, \mathbb{C})$. For example, the A_n-singularity is isomorphic to the singularity of the quotient by the cyclic group of order n generated by

$$\begin{pmatrix} \xi_n & 0 \\ 0 & \xi_n^{-1} \end{pmatrix}$$

with ξ_n a primitive nth root of unity; see [393, Ch. 4.6] for more details and references. The lattice A_n can be realized explicitly as $(1, \ldots, 1)^\perp \subset I_{n+1} = \langle 1 \rangle^{\oplus n+1}$.

We also record the discriminant groups of lattices of ADE type (see [161]):

Λ	A_n	D_{2n}	D_{2n+1}	E_6	E_7	E_8
A_Λ	$\mathbb{Z}/(n+1)\mathbb{Z}$	$\mathbb{Z}/2\mathbb{Z} \oplus \mathbb{Z}/2\mathbb{Z}$	$\mathbb{Z}/4\mathbb{Z}$	$\mathbb{Z}/3\mathbb{Z}$	$\mathbb{Z}/2\mathbb{Z}$	$\{0\}$

(vi) The *K3 lattice*

$$\Lambda := E_8(-1)^{\oplus 2} \oplus U^{\oplus 3}$$

is an even, unimodular lattice of signature $(3, 19)$ and discriminant -1, which contains

$$\Lambda_d := E_8(-1)^{\oplus 2} \oplus U^{\oplus 2} \oplus \mathbb{Z}(-2d);$$

see Example 1.11. The *extended K3 lattice* or *Mukai lattice* is

$$\widetilde{\Lambda} := E_8(-1)^{\oplus 2} \oplus U^{\oplus 4},$$

which is even, unimodular of signature $(4, 20)$ and discriminant 1. Also the notations $\mathrm{II}_{3,19}$ for Λ and $\mathrm{II}_{4,20}$ for $\tilde{\Lambda}$ are common; cf. Corollary 1.3.

(vii) The *Enriques lattice* is the torsion-free part $H^2(Y, \mathbb{Z})_{\mathrm{tf}}$ of $H^2(Y, \mathbb{Z})$ of an Enriques surface Y. The universal cover $X \longrightarrow\!\!\!\!\!\rightarrow Y$ of any Enriques surface describes a K3 surface X and there exists an isomorphism $H^2(X, \mathbb{Z}) \overset{\sim}{\longrightarrow} \Lambda$ such that the covering involution $\iota \colon X \overset{\sim}{\longrightarrow} X$ acts on $\Lambda = E_8(-1) \oplus E_8(-1) \oplus U \oplus U \oplus U$ by

$$\iota^* \colon (x_1, x_2, x_3, x_4, x_5) \longmapsto (x_2, x_1, -x_3, x_5, x_4).$$

Thus, the invariant part is $\Lambda^{\iota} \simeq E_8(-2) \oplus U(2)$ and hence

$$H^2(Y, \mathbb{Z})_{\mathrm{tf}} \simeq E_8(-1) \oplus U \simeq \mathrm{II}_{1,9}.$$

See below for the notation II_{n_+, n_-}. Note that the Lefschetz fixed point formula shows that the invariant part has to be of rank 10. The easiest way to show that the action ι^* really is of the above form is by studying one particular Enriques surface and using that there is only one deformation class.

1 Existence, Uniqueness, and Embeddings of Lattices

In this section we recall some of the classical facts on unimodular lattices and their generalizations. We cannot give complete proofs, but sometimes sketch ad hoc arguments that may convey at least an idea of some of the techniques. The standard reference is Nikulin [451], where one finds many more and stronger results. We restrict to those parts that are used somewhere in this book. In particular, the local theory is only occasionally touched upon.

1.1 We begin with the classical result of Milnor concerning even, unimodular lattices; cf. [411, Ch. II] or [547, Ch. V].

Theorem 1.1 *Let (n_+, n_-) be given. Then there exists an even, unimodular lattice of signature (n_+, n_-) if and only if $n_+ - n_- \equiv 0 \, (8)$. If $n_\pm > 0$, then the lattice is unique.*

The key to this theorem is the description of the Grothendieck group of stable isomorphism classes of unimodular lattices. It turns out to be freely generated by $\langle \pm 1 \rangle$.

Corollary 1.2 *Suppose Λ_1, Λ_2 are two positive definite, even, unimodular lattices of the same rank. Then*

$$\Lambda_1 \oplus U \simeq \Lambda_2 \oplus U.$$

Proof Indeed, $\Lambda_i \oplus U$ are both even, unimodular lattices of signature $(\mathrm{rk}\, \Lambda_i + 1, 1)$ and by the theorem the isomorphism type of such lattices is unique. $\qquad \square$

Corollary 1.3 *Let Λ be an indefinite, unimodular lattice of signature (n_+, n_-).*

(i) *If Λ is even of index $\tau = n_+ - n_-$, then $\tau \equiv 0\,(8)$ and*

$$\Lambda \simeq E_8^{\oplus \frac{\tau}{8}} \oplus U^{\oplus n_-} \quad and \quad \Lambda \simeq E_8(-1)^{\oplus \frac{-\tau}{8}} \oplus U^{\oplus n_+},$$

according to the sign of τ.

(ii) *If Λ is odd, then*

$$\Lambda \simeq \langle 1 \rangle^{\oplus n_+} \oplus \langle -1 \rangle^{\oplus n_-}.$$

In the literature, also the notations I_{n_+, n_-} and II_{n_+, n_-} are used for an odd resp. even, indefinite, unimodular lattice of signature (n_+, n_-). So, for example (see Section 0.3)

$$I_{n_+, n_-} \simeq \langle 1 \rangle^{\oplus n_+} \oplus \langle -1 \rangle^{\oplus n_-} \quad and \quad II_{n_+, n_-} \simeq E_8^{\oplus \frac{n_+ - n_-}{8}} \oplus U^{\oplus n_-}$$

for $n_+ - n_- \geq 0$.

Example 1.4 (i) For complex K3 surfaces this can be used to describe the singular cohomology $H^2(X, \mathbb{Z})$ endowed with the intersection pairing as an abstract lattice (cf. Proposition 1.3.5):

$$H^2(X, \mathbb{Z}) \simeq E_8(-1)^{\oplus 2} \oplus U^{\oplus 3}.$$

(ii) The middle cohomology of a smooth cubic $Y \subset \mathbb{P}^5$ is (see [240, Prop. 2.12])

$$H^4(Y, \mathbb{Z}) \simeq I_{21,2} = \langle 1 \rangle^{\oplus 21} \oplus \langle -1 \rangle^{\oplus 2} \simeq E_8^{\oplus 2} \oplus U^{\oplus 2} \oplus \langle 1 \rangle^{\oplus 3}.$$

If $h \in H^4(Y, \mathbb{Z})$ is the square of the class of a hyperplane section, then $(h)^2 = 3$ and

$$h^\perp \simeq E_8^{\oplus 2} \oplus U^{\oplus 2} \oplus A_2.$$

One way to show this is to realize h as the vector $(1, 1, 1) \in \langle 1 \rangle^{\oplus 3} \simeq \mathbb{Z}^{\oplus 3}$, the orthogonal complement of which in $\langle 1 \rangle^{\oplus 3}$ is by definition the lattice A_2. Alternatively, one can argue that the lattice on the right-hand side has discriminant form $\mathbb{Z}/3\mathbb{Z}$ and, as $h^\perp$ is indeed even, Proposition 0.2 (or rather a version of it that also covers the odd lattice $\langle h \rangle$) shows that it is orthogonal to $\langle h \rangle \simeq \mathbb{Z}(3)$. However, any unimodular lattice of signature $(21, 2)$ is necessarily odd by Theorem 1.1 and hence must be isomorphic to $I_{21,2}$.

These classical results are generalized to the non-unimodular case by the following theorem.

Theorem 1.5 *Let (n_+, n_-) and a finite quadratic form (A, q) be given. Then there exists an even lattice Λ of signature (n_+, n_-) and discriminant form $(A_\Lambda, q_\Lambda) \simeq (A, q)$ if*

$$\ell(A) + 1 \leq n_+ + n_- \quad and \quad n_+ - n_- \equiv \tau(A, q)\,(8).$$

If $\ell(A) + 2 \leq n_+ + n_-$ and $n_\pm > 0$, then Λ is unique.

That every finite quadratic form is realized is a result due to Wall [629]. For the general case, see [451, Cor. 1.10.2 and 1.13.3]. Nikulin attributes parts of these results to Kneser, but [300] is difficult to read for non-specialists.

Remark 1.6 The uniqueness statement can be read as saying that any even, indefinite lattice Λ with $\ell(A_\Lambda) + 2 \leq \mathrm{rk}\,\Lambda$ is unique in its genus. In [451, Thm. 1.14.2] the condition $\ell(A_\Lambda) + 2 \leq \mathrm{rk}\,\Lambda$ is required only at all primes $p \neq 2$, and if at $p = 2$ the equality $\ell((A_\Lambda)_2) = \mathrm{rk}\,\Lambda$ holds, one requires that (A_Λ, q_Λ) splits off (as a direct summand) the discriminant form of $U(2)$ or of $A_2(2)$. A variant relaxing the condition at $p \neq 2$ is provided by [451, Thm. 1.13.2], which is of importance for the classification of Néron–Severi lattices of supersingular K3 surfaces; see Section **17.2.7**.

The non-unimodular analogue of Corollary 1.2 is the following.

Corollary 1.7 *Suppose Λ_1 and Λ_2 are two even lattices of the same signature (n_+, n_-) and with isomorphic discriminant forms $(A_{\Lambda_1}, q_{\Lambda_1}) \simeq (A_{\Lambda_2}, q_{\Lambda_2})$. Then*

$$\Lambda_1 \oplus U^{\oplus r} \simeq \Lambda_2 \oplus U^{\oplus r}$$

for $r \geq \max\{(1/2)(\ell(A_{\Lambda_i}) + 2 - n_+ - n_-), 1 - n_\pm\}$. □

1.2 Concerning embeddings of lattices, we start again with the classical unimodular case before turning to Nikulin's stronger versions.

Proposition 1.8 *For every even lattice Λ of rank r and every $r \leq r'$ there exists a primitive embedding*

$$\Lambda \hookrightarrow U^{\oplus r'}.$$

For $r < r'$ the embedding is unique up to automorphisms of $U^{\oplus r'}$.

Proof The following direct proof for the existence is taken from the article by Looijenga and Peters [380, Sec. 2]. Denote by e_i, f_i, $i = 1, \ldots, r$, the standard bases of the r copies of U. Then define the embedding on a basis $a_1, \ldots, a_r$ of Λ by

$$a_i \longmapsto e_i + \frac{1}{2}(a_i.a_i)f_i + \sum_{j<i}(a_j.a_i)f_j.$$

An explicit and elementary (but lengthy) proof for the uniqueness can also be found in [380] or [493, App. to Sec. 6.]. □

The result has direct geometric applications to the surjectivity of the period map for Kummer surfaces; cf. Corollary 3.20. It is also crucial for the proof of the following consequence, in which $U^{\oplus r'}$ is replaced by an arbitrary unimodular lattice.

Corollary 1.9 *Let Λ be an even, unimodular lattice of signature (n_+, n_-) and let Λ_1 be an even (not necessarily unimodular) lattice of signature (m_+, m_-). If $m_+ + m_- \leq \min\{n_+, n_-\}$, then there exists a primitive embedding*

$$\Lambda_1 \hookrightarrow \Lambda.$$

If the inequality is strict, then the embedding is unique up to automorphisms of Λ.

Proof Only the case $0 < n_\pm$ needs a proof. Also, we may assume that $\tau := n_+ - n_- \geq 0$, i.e. $n_- = \min\{n_+, n_-\}$. Thus, $\Lambda \simeq E_8^{\oplus \frac{\tau}{8}} \oplus U^{\oplus n_-}$ by Corollary 1.3. Then apply Proposition 1.8 to Λ_1, which yields a primitive embedding $\Lambda_1 \hookrightarrow U^{\oplus n_-} \hookrightarrow \Lambda$. For the uniqueness, see [278].   □

As a consequence of the existence part one obtains a result that is used in the description of the moduli space of polarized K3 surfaces; see Section 6.1.1. In Section 4.1 we show that the assumption $1 < n_\pm$ in the next corollary cannot be weakened; e.g. the assertion does not hold for $\Lambda = E_8(-1)^{\oplus 3} \oplus U$.

Corollary 1.10 *Let Λ be an even, unimodular lattice of signature (n_+, n_-) with $1 \leq n_\pm$. Then for any d there exists a primitive $\ell \in \Lambda$ with $(\ell)^2 = 2d$. If $1 < n_\pm$, then $\ell \in \Lambda$ is unique up to automorphisms of Λ.*[2]

Proof Consider $\Lambda_1 := \mathbb{Z}(2d)$, which is a lattice of signature $(1, 0)$ resp. $(0, 1)$, depending on the sign of d. Then giving a primitive $\ell \in \Lambda$ with $(\ell)^2 = 2d$ is equivalent to giving a primitive embedding $\Lambda_1 \hookrightarrow \Lambda$. The assertion thus follows directly from Corollary 1.9.

So strictly speaking, only the case $d \neq 0$ follows from the proposition, but see [628, Thm. 3], which also covers $d = 0$.   □

Example 1.11 The following two examples explain the lattice theory behind a curious relation between rationality of cubic fourfolds $Y \subset \mathbb{P}^5$ and K3 surfaces as suggested by Hassett in [240]. In particular, (1.1) below allows one to compare periods of special cubic fourfolds with periods of polarized K3 surfaces.

(i) Let $\Lambda \simeq H^2(X, \mathbb{Z})$ be the K3 lattice and let $\ell \in \Lambda$ be the primitive class of a polarization of degree $(\ell)^2 = 2d$. Then

$$\Lambda_d := \ell^\perp \simeq E_8(-1)^{\oplus 2} \oplus U^{\oplus 2} \oplus \mathbb{Z}(-2d).$$

Indeed, due to the corollary we may assume that $\ell = e + df \in U$ in one of the three copies of U and then use $(e + df)^\perp = \mathbb{Z} \cdot (e - df) \subset U$.

---

[2] The assumption that Λ is unimodular cannot be dropped, but for lattices of the form $\Lambda = U^{\oplus 2} \oplus \Lambda'$ the orbit $O(\Lambda) \cdot \ell$ of a primitive $\ell \in \Lambda$ is determined by $(\ell)^2$ and the class $(1/n)\ell \in A_\Lambda$, where $(\ell.\Lambda) = n\mathbb{Z}$. This is apparently due to Eichler [163, Ch. 10], but see [224, Prop. 3.3] for a proof in modern language.

(ii) Using the notation of Example 1.4, we consider a primitive lattice

$$\langle h, \beta \rangle \subset H^4(Y, \mathbb{Z}) \simeq I_{21,2}$$

of rank two and discriminant $2d$. Then it was shown in [240, Prop. 5.2.2] that there exists an $\ell \in \Lambda$ as in (i) with

$$\langle h, \beta \rangle^\perp \simeq \ell^\perp(-1) \tag{1.1}$$

if and only if $2d$ is not divisible by 4, 9, or any odd prime $p \equiv 2 \ (3)$.

Nikulin proves the following much stronger version of Corollary 1.9; see [451, Thm. 1.14.4].

Theorem 1.12 *Let Λ be an even, unimodular lattice of signature (n_+, n_-) and Λ_1 be an even lattice of signature (m_+, m_-). If $m_\pm < n_\pm$ and*

$$\ell(\Lambda_1) + 2 \leq \mathrm{rk}\, \Lambda - \mathrm{rk}\, \Lambda_1, \tag{1.2}$$

then there exists a primitive embedding $\Lambda_1 \hookrightarrow \Lambda$, which is unique up to automorphisms of Λ.

Remark 1.13 (i) In [451] the condition (1.2) is replaced by the corresponding local condition for all $p \neq 2$, and if for $p = 2$ the equality $\ell((\Lambda_1)_2) = \mathrm{rk}\, \Lambda - \mathrm{rk}\, \Lambda_1$ holds, one requires q_{Λ_1} to split off (as a direct summand) the discriminant form of $U(2)$ or of $A_2(2)$.

(ii) The assumption (1.2) clearly follows from the stronger

$$\mathrm{rk}\, \Lambda_1 + 2 \leq \mathrm{rk}\, \Lambda - \mathrm{rk}\, \Lambda_1. \tag{1.3}$$

Note that the right-hand side is just $\mathrm{rk}\, \Lambda_1^\perp$ once the embedding has been found.

(iii) In order to prove that a given primitive embedding $\Lambda_1 \hookrightarrow \Lambda$ is unique, it is enough to verify that $\Lambda_1^\perp$ contains a hyperbolic plane

$$U \hookrightarrow \Lambda_1^\perp.$$

Indeed, if so, then by Example 0.3 $\Lambda_1^\perp = U \oplus U^\perp$ and hence $\ell(\Lambda_1^\perp) = \ell(U^\perp) \leq \mathrm{rk}\, U^\perp = \mathrm{rk}\, \Lambda_1^\perp - 2$. However, $\ell(\Lambda_1^\perp) = \ell(\Lambda_1)$ by Proposition 0.2 and, therefore, (1.2) holds.

(iv) Note that the assumption $m_\pm < n_\pm$ cannot be weakened. For example, the embedding $U \hookrightarrow E_8(-1)^{\oplus 3} \oplus U$ is not unique; see Section 4.1.

Corollary 1.14 *Let Λ be an even, unimodular lattice of signature (n_+, n_-) and assume $0 \neq x \in \Lambda$ is primitive with $(x)^2 = 0$. Then there exists an isomorphism $\Lambda \simeq \Lambda' \oplus U$ that sends x to the first vector of the standard basis e, f of U.*

Proof Let us first prove this as a consequence of the general theory above assuming $n_\pm > 1$. Then the existence of a (unique) primitive embedding $U \hookrightarrow \Lambda$ follows from Theorem 1.12 or Corollary 1.9. For the direct sum decomposition, use Example 0.3. To conclude, apply Corollary 1.10 to find an automorphism of Λ that maps x to $e \in U$.

Now, assuming only $n_\pm \geq 1$ one argues as follows (and in fact it is more elementary): since Λ is unimodular, there exists a $y \in \Lambda$ with $(x.y) = 1$. Then $y' := y - ((y)^2/2)x$ still satisfies $(x.y') = 1$, but also $(y')^2 = 0$. Thus, $U \xrightarrow{\sim} \langle x, y' \rangle$ via $e \longmapsto x$ and $f \longmapsto y'$. For the direct sum decomposition, use again Example 0.3.   □

It is also useful to know when an even lattice can be embedded at all in some unimodular lattice. As a prototype, we state the following theorem [451, Cor. 1.12.3].

Theorem 1.15 *Let Λ_1 be an even lattice of signature (m_+, m_-). Then there exists a primitive embedding $\Lambda_1 \hookrightarrow \Lambda$ into an even, unimodular lattice Λ of signature (n_+, n_-) if*

(i) $n_+ - n_- \equiv 0\,(8)$.

(ii) $m_\pm \leq n_\pm$.

(iii) $\ell(\Lambda_1) < \operatorname{rk} \Lambda - \operatorname{rk} \Lambda_1$.

By Proposition 0.2, the existence of an embedding is in fact equivalent to the existence of an even lattice Λ_2 with signature $(n_+ - m_+, n_- - m_-)$ and $(A_{\Lambda_1}, q_{\Lambda_1}) \simeq (A_{\Lambda_2}, -q_{\Lambda_2})$. Of course, if $0 < n_\pm$, then Λ is unique; cf. Corollary 1.3.

Remark 1.16 We briefly explain the relation between the two Theorems 1.12 and 1.15. Assume (i), (ii), and (iii) in Theorem 1.15. If $m_\pm < n_\pm$ (and hence $0 < n_\pm$) and the stronger (1.2) instead of (iii) holds, then Theorem 1.12 can be used directly. If not, then the assumptions of Theorem 1.12 hold for Λ_1 and $\Lambda \oplus U$ for any even, unimodular lattice Λ of signature (n_+, n_-), the existence of which is due to Theorem 1.1. Moreover, for two such lattices Λ and Λ' one has $\Lambda \oplus U \simeq \Lambda' \oplus U$ by Corollary 1.2. Now Theorem 1.12 implies the existence of a (unique) primitive embedding $\varphi \colon \Lambda_1 \hookrightarrow \Lambda \oplus U$, and to conclude the proof of Theorem 1.15 one has to show that the projection into one of the Λ' with $\Lambda \oplus U \simeq \Lambda' \oplus U$ yields an embedding $\Lambda_1 \hookrightarrow \Lambda'$.

Remark 1.17 The more precise version [451, Thm. 1.12.2] of the above theorem replaces (iii) by the weaker $\ell(\Lambda_1) \leq \operatorname{rk} \Lambda - \operatorname{rk} \Lambda_1$ and adds conditions on the discriminant at every p for which equality is attained; e.g. if equality holds at $p = 2$, i.e. $\ell((\Lambda_1)_2) = \operatorname{rk} \Lambda - \operatorname{rk} \Lambda_1$, then it suffices to assume that $(A_{\Lambda_1}, q_{\Lambda_1})$ splits off the discriminant group of $\langle 1 \rangle \langle 2 \rangle$. For most applications, but not quite for all, the above version suffices.

The following is a very useful consequence of Theorem 1.15; see [451, Thm. 1.12.4].

Corollary 1.18 *Let Λ_1 be an even lattice of signature (m_+, m_-). Assume (n_+, n_-)
satisfies*

(i) $n_+ - n_- \equiv 0 \, (8)$.

(ii) $m_\pm \leq n_\pm$.

(iii) $\text{rk} \, \Lambda_1 \leq \frac{1}{2}(n_+ + n_-)$.

*Then there exists a primitive embedding $\Lambda_1 \hookrightarrow \Lambda$ into an even, unimodular lattice Λ
of signature (n_+, n_-).*

Proof Use $\ell(\Lambda_1) \leq \text{rk} \, \Lambda_1 \leq (1/2)(n_+ + n_-) \leq (n_+ + n_-) - \text{rk} \, \Lambda_1$. If at least one
of these inequalities is strict, Theorem 1.15 applies directly. For the remaining case,
see [451]. □

2 Orthogonal Group

Next, we collect some standard facts concerning the group of automorphisms $\text{O}(\Lambda)$ of
a lattice Λ. So by definition $\text{O}(\Lambda)$ is the group of all $g \colon \Lambda \xrightarrow{\sim} \Lambda$ with $(g(x).g(y)) = (x.y)$ for all $x, y \in \Lambda$. Clearly, $\text{O}(\Lambda)$ is a discrete subgroup of the real Lie group
$\text{O}(\Lambda_{\mathbb{R}}) \simeq \text{O}(n_+, n_-)$. In particular, if Λ is definite, then $\text{O}(\Lambda_{\mathbb{R}})$ is compact and,
therefore, $\text{O}(\Lambda)$ is finite.

The theory of automorphs of binary quadratic forms may serve as a motivation. In
modern terms, one considers a lattice Λ of rank two, which can also be thought of as a
quadratic equation $ax^2 + 2bxy + cy^2$ with $a, b, c \in \mathbb{Z}$, and an *automorph* is nothing but an
element $g \in \text{O}(\Lambda)$ (sometimes assumed to have $\det(g) = 1$). Interestingly, $\text{O}(\Lambda)$ can be
finite even for indefinite Λ. In fact, it is finite if and only if $d := -\text{disc} \, \Lambda = b^2 - ac > 0$
is a square, which is equivalent to the existence of $0 \neq x \in \Lambda$ with $(x)^2 = 0$. The
idea behind this assertion is to link elements $g \in \text{O}(\Lambda)$ to solutions of *Pell's equation*
$x^2 - dy^2 = 1$, which has a unique (up to sign) solution if and only if d is a square and
otherwise has infinitely many. See e.g. [105, Ch. 13.3].

Example 2.1 It is easy to check that

$$\text{O}(U) \simeq (\mathbb{Z}/2\mathbb{Z})^2.$$

Indeed, $g \in \text{O}(U)$ is uniquely determined by the image $g(e)$ of the first standard basis
vector, which has to be contained in $\{\pm e, \pm f\}$.

2.1 Let Λ be a lattice. A *root* of Λ, also called (-2)-*class*, is an element $\delta \in \Lambda$ with
$(\delta)^2 = -2$. The set of roots is denoted by Δ, so

$$\Delta := \{\delta \in \Lambda \mid (\delta)^2 = -2\}.$$

The *root lattice* of Λ is the sublattice $R \subset \Lambda$ (not necessarily primitive) spanned by Δ.

For any $\delta \in \Delta$ one defines the *reflection*

$$s_\delta : x \longmapsto x + (x.\delta)\delta,$$

which is an orthogonal transformation, i.e. $s_\delta \in O(\Lambda)$. Clearly, $s_\delta = s_{-\delta}$. The subgroup

$$W := \langle s_\delta \mid \delta \in \Delta \rangle \subset O(\Lambda)$$

is called the *Weyl group* of Λ. See also the discussion in Section **8.2.3**.

For $\Lambda = U$, the Weyl group is the proper subgroup $\mathbb{Z}/2\mathbb{Z} \subset O(U) \simeq (\mathbb{Z}/2\mathbb{Z})^2$ generated by the reflection $s_{e-f} \colon e \mapsto f$.

Theorem 2.2 Let Λ be the K3 lattice $E_8(-1)^{\oplus 2} \oplus U^{\oplus 3}$. Then any $g \in O(\Lambda)$ with *trivial spinor norm can be written as a product $\prod s_{\delta_i}$ of reflections associated with (-2)-classes $\delta_i \in \Delta \subset \Lambda$.*

For the notion of the spinor norm, see Section **7.5.4**. This is essentially a special case of results applicable to a large class of unimodular lattices due to Wall [630, 4.7]. More precisely, Wall proves that $O(E_8(-1)^{\oplus m} \oplus U^{\oplus n})$ is generated by reflections s_δ with $(\delta)^2 = \pm 2$ for $m, n \geq 2$. The result was later generalized to certain non-unimodular lattices by Ebeling [160] and Kneser [301, Satz 4], which also contains the above stronger form of Wall's result using only (-2)-classes. Sometimes the condition $\det(g) = 1$ is added, which, however, can always be achieved by passing from g to $s_\delta \circ g$ for some (-2)-class δ.

Example 2.3 For a definite lattice Λ, e.g. with $\Lambda(-1)$ of ADE type, the orthogonal group $O(\Lambda)$, and hence the Weyl group, is finite. For instance, the Weyl group of $A_n(-1)$ is the symmetric group $\mathfrak{S}_{n+1}$, and for the lattice $E_8(-1)$, which contains 240 roots, the Weyl group is of order $2^{14} \cdot 3^5 \cdot 5^2 \cdot 7 = 4! \cdot 6! \cdot 8!$ and equals $O(E_8(-1))$. The quotient by its center, which is of order four, is a simple group. Similarly, $E_6(-1)$ and $E_7(-1)$ contain 72 and 126, roots, respectively. See [132] for more details, and for more examples of root lattices and Weyl groups of definite lattices not of ADE type, see also Section 4.3.

Note that frequently the inclusion $W \subset O(\Lambda)$ is proper and even of infinite index. In Section **15.2** one finds examples where Λ is NS(X) of a K3 surface X and $W \subset O(\mathrm{NS}(X))$ is of finite index if and only if Aut(X) is finite; see Theorem **15.2.6**. Even from a purely lattice theoretic point of view it is an interesting question for which lattices the Weyl group is essentially (i.e. up to finite index) the orthogonal group. For lattices of signature $(1, \rho - 1)$, see Theorem **15.2.10**.

2.2 Any $g \in O(\Lambda)$ naturally induces $g^* \in O(\Lambda^*)$ by $g^* \varphi \colon x \longmapsto \varphi(g^{-1}x)$. With this definition $g^*|_\Lambda = g$ for the natural embedding $\Lambda \hookrightarrow \Lambda^*$. Hence, g induces an automorphism $\overline{g}$ of A_Λ. If Λ is even and, therefore, A_Λ endowed with the discriminant form q_Λ, then $\overline{g}$ respects q_Λ. This yields a natural homomorphism

$$O(\Lambda) \longrightarrow O(A_\Lambda),$$

which is often surjective due to the next result.

Theorem 2.4 *Let Λ be an even indefinite lattice with $\ell(A_\Lambda) + 2 \leq \text{rk}\,\Lambda$. Then*

$$O(\Lambda) \longrightarrow\!\!\!\!\rightarrow O(A_\Lambda)$$

is surjective. See [451, Thm. 1.14.2].

Observe that the assumption $\ell(A_\Lambda) + 2 \leq \text{rk}\,\Lambda$ is the same as in Theorem 1.5 so that Λ is determined by its signature and (A_Λ, q_Λ). The result is proved by first lifting to the p-adic lattices [451, Cor. 1.9.6]. Those then glue due to a result by Nikulin [450, Thm. 1.2′].

Let us consider the situation of Proposition 0.2; i.e. let $\Lambda_1 \subset \Lambda$ be a primitive sublattice of an even, unimodular lattice Λ with orthogonal complement $\Lambda_2 := \Lambda_1^\perp$. Using $(A_{\Lambda_1}, q_{\Lambda_1}) \simeq (A_{\Lambda_2}, -q_{\Lambda_2})$, one can identify $O(A_{\Lambda_1}) \simeq O(A_{\Lambda_2})$. This yields

$$O(\Lambda_1) \xrightarrow{\;r_1\;} O(A_{\Lambda_1}) \simeq O(A_{\Lambda_2}) \xleftarrow{\;r_2\;} O(\Lambda_2) \qquad (2.1)$$

with $r_i(g_i) := \overline{g}_i$.

For future reference we state the following obvious lemma.

Lemma 2.5 *If $g \in O(\Lambda)$ preserves Λ_1 and hence Λ_2, then the two automorphisms $g_i := g|_{\Lambda_i}$, $i = 1, 2$, satisfy $\overline{g}_1 = \overline{g}_2$ in (2.1).* □

The converse is also true, due to the next result; see [451, Thm. 1.6.1, Cor. 1.5.2.].

Proposition 2.6 *An automorphism $g_1 \in O(\Lambda_1)$ can be extended to an automorphism $g \in O(\Lambda)$ if and only if $\overline{g}_1 \in \text{Im}(r_2)$. If $\overline{g}_1 = \text{id}$, then g_1 can be lifted to $g \in O(\Lambda)$ with $g|_{\Lambda_2} = \text{id}$.*

Proof Let us prove the second assertion. For this observe first that an element $y \in \Lambda_\mathbb{Q} = \Lambda_\mathbb{Q}^* = \Lambda_{1\mathbb{Q}} \oplus \Lambda_{2\mathbb{Q}} = \Lambda_{1\mathbb{Q}}^* \oplus \Lambda_{2\mathbb{Q}}^*$ that is contained in $\Lambda_1^* \oplus \Lambda_2^*$ is in fact contained in Λ if and only if its class $\overline{y} \in A_{\Lambda_1} \oplus A_{\Lambda_2}$ is contained in the isotropic subgroup $\Lambda/(\Lambda_1 \oplus \Lambda_2)$ (cf. proof of Proposition 0.2). The given $g_1 \in O(\Lambda_1)$ can be extended to g as asserted if and only if for any $x = x_1 + x_2 \in \Lambda$ with $x_i \in \Lambda_i^* \subset \Lambda_{i\mathbb{Q}}$ and also $y := g_1(x_1) + x_2$ is contained in Λ. But $\overline{y} = \overline{x} \in A_{\Lambda_1} \oplus A_{\Lambda_2}$ if $\overline{g}_1 = \text{id}$. □

This has the following immediate consequence.

Corollary 2.7 *Let Λ be an even, unimodular lattice and $\ell \in \Lambda$ with $(\ell)^2 \neq 0$. Then*

$$\{g \in O(\ell^\perp) \mid \text{id} = \overline{g} \in O(A_{\ell^\perp})\} = \{g|_{\ell^\perp} \mid g \in O(\Lambda),\, g(\ell) = \ell\}. \qquad \Box$$

The corollary is relevant for moduli space considerations; see e.g. Section **6**.3.2 where it is applied to $\ell = e + df \in U \subset E_8(-1)^{\oplus 2} \oplus U^{\oplus 3}$. The group on the right-hand side is $\widetilde{O}(\Lambda_d)$ with $\Lambda_d = \ell^\perp$ in the notation there.

2.3   There is another kind of orthogonal transformation of lattices of the form $\Lambda \oplus U$ with Λ even. Those play an important role in mirror symmetry of K3 surfaces and so we briefly mention them here. The example one should have in mind is the K3 lattice $\Lambda = E_8(-1)^{\oplus 2} \oplus U^{\oplus 3}$. Then $\Lambda \oplus U$ can be thought of as the Mukai lattice $\widetilde{H}(X, \mathbb{Z})$ or the usual cohomology $H^*(X, \mathbb{Z})$ of a K3 surface X.

With e and f denoting the standard basis of the extra copy U (which is $(H^0 \oplus H^4)(X, \mathbb{Z})$ in the geometric example), one defines a ring structure on $\Lambda \oplus U$ by

$$(\lambda e + x + \mu f) \cdot (\lambda' e + y + \mu' f) := (\lambda \lambda') e + (\lambda y + \lambda' x) - (\lambda \mu' + \lambda' \mu + (x.y)) f.$$

Of course, in the example this gives back the usual ring structure on $H^*(X, \mathbb{Z})$.

Next, for any $B \in \Lambda$ one defines $\exp(B) := e + B + ((B.B)/2) f \in \Lambda \oplus U$ and denotes multiplication with it by the same symbol:

$$\exp(B) \colon \Lambda \oplus U \longrightarrow \Lambda \oplus U.$$

A direct computation reveals the following lemma.

Lemma 2.8   *The B-field shift $\exp(B)$ is an orthogonal transformation of $\Lambda \oplus U$, i.e.* $\exp(B) \in O(\Lambda \oplus U)$. $\square$

According to Wall [630], one furthermore has the following proposition.

Proposition 2.9   *Let Λ be an even, unimodular lattice of signature (n_+, n_-) with $n_\pm \geq 2$. Then $O(\Lambda \oplus U)$ is generated by the subgroups*

$$O(\Lambda), \quad O(U), \quad \text{and} \quad \{\exp(B) \mid B \in \Lambda\}.$$

The result applies to the extended Mukai lattice and was used by Aspinwall and Morrison in [25] to describe the symmetries of conformal field theories associated with K3 surfaces, see [252] for further references.

3 Embeddings of Picard, Transcendental, and Kummer Lattices

The above discussion shall now be applied to Picard and transcendental lattices of K3 surfaces. The first question here is which lattices can be realized at all. Later we discuss the case of Kummer surfaces and in particular the Kummer lattice containing all exceptional curves.

3.1 We start with results for small Picard numbers. As an immediate consequence of Theorem 1.12 Morrison in [424] proves the following corollary.

Corollary 3.1 *Let N be an even lattice of signature $(1, \rho - 1)$ with $\rho \leq 10$. Then there exists a complex projective K3 surface X with $\mathrm{NS}(X) \simeq N$. Moreover, the primitive embedding $N \hookrightarrow H^2(X, \mathbb{Z})$ is unique up to the action of $\mathrm{O}(H^2(X, \mathbb{Z}))$.*

Proof Let Λ be the K3 lattice. Then by Theorem 1.12, or rather Remark 1.13, (ii), there exists a unique primitive embedding $N \hookrightarrow \Lambda$. Next, choose a Hodge structure of K3 type on $T := N^\perp$ and view it as a Hodge structure on Λ with N purely of type $(1, 1)$. Then by the surjectivity of the period map, Theorem 7.4.1, there exists a K3 surface X together with a Hodge isometry $H^2(X, \mathbb{Z}) \simeq \Lambda$. Under this isomorphism $N \hookrightarrow \mathrm{NS}(X)$. If now the Hodge structure on T is chosen as sufficiently general, i.e. in the complement of the countable union $\bigcup_{\alpha \in T} \alpha^\perp \cap D \subset D \subset \mathbb{P}(T_\mathbb{C})$ of proper closed subsets, then $T^{1,1} \cap T = 0$ and, therefore, $N \simeq \mathrm{NS}(X)$. Clearly, X is projective, as by assumption $\mathrm{NS}(X) \simeq N$ contains a class of positive square; see Remark 8.1.3. $\qquad\square$

The arguments also apply to negative definite lattices N of rank $\mathrm{rk}\, N \leq 10$, only that then the K3 surface X is of course not projective. See Remark 3.7 for the case $\mathrm{rk}\, N = 11$.

Remark 3.2 (i) The realization problem is more difficult over other fields, even algebraically closed ones of characteristic zero. The rank one case for K3 surfaces over number fields was settled by Ellenberg [168](cf. Proposition 17.2.15) but can also be deduced from general existence results applicable to higher Picard rank; cf. Remark 17.2.16.

(ii) This result definitely does not hold (in this form) in positive characteristic. For example, the Picard number of a K3 surface over $\overline{\mathbb{F}}_p$ is always even; see Corollary 17.2.9. However, if X is a K3 surface over $\overline{\mathbb{Q}}$ such that a lattice N as above embeds into $\mathrm{NS}(X)$, then, as reduction modulo p is injective on the Picard group (see Proposition 17.2.10), it also embeds into $\mathrm{NS}(X_p)$ of any smooth reduction X_p modulo p.

Corollary 3.3 *For a complex projective K3 surface X of Picard number $\rho(X) \leq 10$ the isomorphism type of its transcendental lattice $T := T(X)$ (without its Hodge structure) is uniquely determined by $\rho(X)$ and its discriminant form $(A_T, q_T) \simeq (A_{\mathrm{NS}(X)}, -q_{\mathrm{NS}(X)})$.*

Proof Theorem 1.5 can be applied, as $\ell(T) + 2 = \ell(\mathrm{NS}(X)) + 2 \leq \rho(X) + 2 \leq 12 \leq \mathrm{rk}\, T(X)$. $\qquad\square$

Remark 3.4 Transcendental lattices of *conjugate K3 surfaces* have the same genus. Recall that for a complex projective K3 surface X and any automorphism $\sigma \in \mathrm{Aut}(\mathbb{C})$ the base change $X^\sigma := X \times_{\mathbb{C}, \sigma} \mathbb{C}$ is again a K3 surface. Then $\mathrm{Pic}(X) \simeq \mathrm{Pic}(X^\sigma)$, as $X \simeq X^\sigma$ as schemes and, as the Picard group of a K3 surface determines the genus of its transcendental lattice, $T(X)$ and $T(X^\sigma)$ have indeed the same genus.

The converse holds for K3 surfaces with maximal Picard number, i.e. if $T(X)$ and $T(Y)$ of two K3 surfaces have the same genus (or, equivalently, if $\mathrm{Pic}(X)$ and $\mathrm{Pic}(Y)$ have the same genus) and $\rho(X) = 20$, then X and Y are conjugate to each other (and consequently $\mathrm{Pic}(X) \simeq \mathrm{Pic}(Y)$). This follows from Corollary 3.21 and results by Schütt and Shimada in [537, 554] showing that the transcendental lattices of conjugate K3 surfaces of maximal Picard number account for all lattices in a given genus; cf. Remark 3.23.

Interchanging the role of $\mathrm{NS}(X)$ and $T(X)$ in the two previous corollaries yields the analogous statements for large Picard number. Again, as a consequence of Theorem 1.12 one obtains the following corollary.

Corollary 3.5 *Let T be an even lattice of signature $(2, 20-\rho)$ with $12 \le \rho \le 20$. Then there exists a complex projective K3 surface X with $T(X) \simeq T$. Moreover, the primitive embedding $T(X) \hookrightarrow H^2(X, \mathbb{Z})$ is unique up to the action of $\mathrm{O}(H^2(X, \mathbb{Z}))$.* □

And as an analogue of Corollary 3.3, one finds the following.

Corollary 3.6 *For a complex projective K3 surface X of Picard number $12 \le \rho(X)$ the isomorphism type of $N := \mathrm{NS}(X)$ is uniquely determined by $\rho(X)$ and its discriminant form (A_N, q_N).* □

Remark 3.7 Clearly, similar results can be stated for other lattices. For example, when N in Corollary 3.1 is of rank $\rho = 11$, but one knows in addition $\ell(N) < 10$, then the assertion still holds. In fact, Morrison notes in [424, Rem. 2.11] that also every even lattice of signature $(1, 10)$ can be realized as $\mathrm{NS}(X)$. However, the embedding into $H^2(X, \mathbb{Z})$ may not be unique. Similarly, if in Corollary 3.6 the Picard number $\rho(X) < 12$ but X admits an elliptic fibration with a section, then the uniqueness is still valid. Indeed, by Example 0.3 one has $\mathrm{NS}(X) \simeq U \oplus N'$ and hence $\ell(\mathrm{NS}(X)) = \ell(N') \le \mathrm{rk} N' = \rho(X) - 2$.

Corollary 3.8 *If a complex projective K3 surface X satisfies $12 \le \rho(X)$, then there exists an embedding $U \hookrightarrow \mathrm{NS}(X)$ and in fact*

$$\mathrm{NS}(X) \simeq U \oplus N'.$$

In particular, there exists a (-2)-class $\delta \in \mathrm{NS}(X)$ and, moreover and more precisely, X admits an elliptic fibration with a section.

Proof We follow Kovács in [327, Lem. 4.1] and apply Corollary 1.18 to $\Lambda_1 = T(X)$ and $\Lambda = E_8(-1)^{\oplus 2} \oplus U^{\oplus 2}$. As indeed $\mathrm{rk} T(X) \le 10 = (1/2)\mathrm{rk} \Lambda$, there exists a primitive embedding $T(X)$ into some even, unimodular lattice of signature $(2, 18)$.

However, by Corollary 1.3 one knows that Λ is the only such lattice. Hence, there exists a primitive embedding

$$T(X) \hookrightarrow \Lambda \hookrightarrow \Lambda \oplus U \simeq H^2(X, \mathbb{Z})$$

with $U \subset T(X)^\perp \subset H^2(X, \mathbb{Z})$. But $\ell(T(X)) + 2 \leq \operatorname{rk} T(X) + 2 \leq 12 \leq \rho(X) = \operatorname{rk} H^2(X, \mathbb{Z}) - \operatorname{rk} T(X)$ and hence by Theorem 1.12 the embedding $T(X) \hookrightarrow H^2(X, \mathbb{Z})$ is unique up to automorphisms of the lattice $H^2(X, \mathbb{Z})$. Therefore, also the natural embedding $T(X) \subset H^2(X, \mathbb{Z})$ has the property that there exists a hyperbolic plane $U \subset T(X)^\perp = \operatorname{NS}(X)$. The direct sum decomposition follows from Example 0.3.

Eventually, let $\delta = e - f \in U \subset \operatorname{NS}(X)$, which is a (-2)-class, and use Remark **8**.2.13 for the existence of the elliptic fibration. Up to the action of the Weyl group and up to sign, the class e is realized as the fibre class of an elliptic fibration and, using $(\delta.e) = 1$, $\pm\delta$ is the class of a section.

If one is only after the (-2)-classes, one could use the existence of an embedding $T(X) \oplus A_1(-1) \hookrightarrow H^2(X, \mathbb{Z})$, which exists according to Theorem 1.15 (or rather the stronger version alluded to in Remark 1.17). $\qquad \square$

Remark 3.9 As a consequence of Corollary 3.8 and of the density of the Noether–Lefschetz locus (see Proposition **6**.2.9 and Section **17**.1.3), one concludes that elliptic K3 surfaces with a section are dense in the moduli space of (marked) K3 surfaces as well as in the moduli space M_d of polarized K3 surfaces.

3.2 The following is a result due to Mukai [429, Prop. 6.2] (up to a missing sign).

Corollary 3.10 *Let X and X' be complex projective K3 surfaces with $12 \leq \rho(X) = \rho(X')$. Then any Hodge isometry $\varphi \colon T(X) \xrightarrow{\sim} T(X')$ (see Section 3.2.2) can be extended to a Hodge isometry*

$$\widetilde{\varphi} \colon H^2(X, \mathbb{Z}) \xrightarrow{\sim} H^2(X', \mathbb{Z}).$$

Moreover, one can choose $\widetilde{\varphi}$ such that there exists an isomorphism $f \colon X' \xrightarrow{\sim} X$ with $\widetilde{\varphi} = \pm f^*$.

Proof The existence of $\widetilde{\varphi}$ is a purely lattice theoretic question, for $\operatorname{NS}(X) = T(X)^\perp$ has trivial Hodge structure. Now, (1.3) in Remark 1.13 holds for $T(X)$, and, hence, the induced embedding $T(X) \xrightarrow{\sim} T(X') \subset H^2(X', \mathbb{Z})$ can be extended to $H^2(X, \mathbb{Z})$ (as abstract lattices $H^2(X, \mathbb{Z})$ and $H^2(X', \mathbb{Z})$ are of course isomorphic).

For the second one we need the global Torelli theorem. Due to Proposition **8**.2.6 we can modify any given extension $\widetilde{\varphi}$ by reflections s_δ for appropriate (-2)-classes $\delta \in \operatorname{NS}(X)$, such that $\widetilde{\varphi}$ maps an ample class to an ample class (possibly after a further sign change) and then Theorem 7.5.3 applies. Note that the reflections s_δ do not alter φ. $\qquad \square$

Remark 3.11 The conclusion also holds for smaller rank. For example, if X is an elliptic K3 surface with a section, then the existence of a Hodge isometry $T(X) \simeq T(X')$ implies $X \simeq X'$. Indeed, then $\mathrm{NS}(X)$ contains a hyperbolic plane spanned by the classes of the fibre and of the section, and, therefore, Remark 1.13, (iii), applies.

In Chapter 16 we are interested in derived equivalences between K3 surfaces. This requires the following modification of the above, which turns out to work without any restriction on the Picard number.

Recall that the Mukai lattice $\widetilde{H}(X, \mathbb{Z})$ is the lattice given by the Mukai pairing $\langle \, , \, \rangle$ on $H^*(X, \mathbb{Z})$ (see Definition 9.1.4) together with the Hodge structure of weight two defined by

$$\widetilde{H}^{1,1}(X) = H^{1,1}(X) \oplus H^0(X) \oplus H^4(X) \quad \text{and} \quad \widetilde{H}^{2,0}(X) = H^{2,0}(X).$$

In particular, the transcendental lattice of $\widetilde{H}(X, \mathbb{Z})$ is just $T(X)$ and as abstract lattices

$$\widetilde{H}(X, \mathbb{Z}) \simeq E_8(-1)^{\oplus 2} \oplus U^{\oplus 4}.$$

Corollary 3.12 *Let X and X' be arbitrary complex projective K3 surfaces. Then any Hodge isometry $\varphi \colon T(X) \xrightarrow{\sim} T(X')$ can be extended to a Hodge isometry*

$$\widetilde{\varphi} \colon \widetilde{H}(X, \mathbb{Z}) \xrightarrow{\sim} \widetilde{H}(X', \mathbb{Z}).$$

Proof Remark 1.13, (iii), applies, as $(H^0 \oplus H^4)(X, \mathbb{Z})$ is a hyperbolic plane and the transcendental lattices of complex projective K3 surfaces are non-degenerate. □

3.3 We shall briefly discuss the Kummer lattice K, which can be approached in two ways. It was first studied in [493, App. to Sec. 5] and [449].

The geometric description of K goes as follows: take an abelian surface A (or a two-dimensional complex torus) and let $X \longrightarrow A/\iota$ be the minimal resolution of the quotient by the standard involution $\iota \colon x \longmapsto -x$; i.e. X is the Kummer surface associated with A (see Example 1.1.3). The 16 exceptional curves $\mathbb{P}^1 \simeq \overline{E}_i \subset X$ and their classes

$$e_i := [\overline{E}_i] \in H^2(X, \mathbb{Z})$$

span a lattice of rank 16 which is abstractly isomorphic to $A_1(-1)^{\oplus 16} \simeq \langle -2 \rangle^{\oplus 16}$.

Definition 3.13 The *Kummer lattice* K is the saturation of $\langle e_i \rangle \subset H^2(X, \mathbb{Z})$, i.e. the smallest primitive sublattice of $H^2(X, \mathbb{Z})$ that contains all classes e_i:

$$\langle -2 \rangle^{\oplus 16} \simeq \bigoplus \mathbb{Z} \cdot e_i \subset K \subset H^2(X, \mathbb{Z}).$$

Equivalently, K is the double orthogonal $(\bigoplus \mathbb{Z} \cdot e_i)^{\perp\perp}$. Recall from Section 3.2.5 that $(\bigoplus \mathbb{Z} \cdot e_i)^{\perp} = \pi_* H^2(A, \mathbb{Z})$. Here, $\pi \colon \widetilde{A} \longrightarrow X$ is the projection from the blow-up in the two-torsion points of A, which induces a natural inclusion $H^2(A, \mathbb{Z}) \hookrightarrow H^2(\widetilde{A}, \mathbb{Z})$.

An alternative and more algebraic description of the Kummer lattice is available. Indeed, one can define K as the sublattice $K \subset \bigoplus \mathbb{Q} \cdot e_i$ spanned by the basis e_i and all elements of the form $\frac{1}{2} \sum_{i \in W} e_i$ with $W \subset \mathbb{F}_2^{\oplus 4}$ a hyperplane. Here, the set $\{e_i\}$ is identified with the set of two-torsion points of A which in turn is viewed as the $\mathbb{F}_2$-vector (or rather affine) space $(\mathbb{Z}/2\mathbb{Z})^{\oplus 4}$.

In order to see that both definitions amount to the same lattice, one first observes that

$$\bigoplus \mathbb{Z} \cdot e_i \subset K \subset K^* \subset \bigoplus \mathbb{Z} \cdot (e_i/2) \subset H^2(X, \mathbb{Q}),$$

as $(e_i)^2 = -2$. For the detailed argument, we refer to [33, VIII. 4 and 5], [54, Exp. VIII], or [493, Sec. 5].

Note that in particular the geometric description defines a lattice that is independent of the abelian surface and that $\bigoplus \mathbb{Z} \cdot e_i$ is the root lattice of K.

Proposition 3.14 *The Kummer lattice satsfies the following conditions:*

(i) *The orthogonal complement $K^\perp$ of K in $H^2(X, \mathbb{Z})$ is isomorphic to $U(2)^{\oplus 3}$.*
(ii) *The inclusion $\bigoplus \mathbb{Z} \cdot e_i \subset K$ has index 2^5.*
(iii) *The lattice K is negative definite with disc $K = 2^6$.*
(iv) *The discriminant form satisfies $q_K \simeq q_{U(2)}^{\oplus 3}$. In particular,*

$$A_K \simeq (\mathbb{Z}/2\mathbb{Z})^{\oplus 6} \quad and \quad \ell(K) = 6.$$

Proof The geometric description of K yields (i), for

$$\pi_* H^2(A, \mathbb{Z}) \simeq \left(\bigoplus \mathbb{Z} \cdot e_i \right)^\perp \subset H^2(X, \mathbb{Z}),$$

$\pi_* H^2(A, \mathbb{Z}) \simeq H^2(A, \mathbb{Z})(2)$, and $H^2(A, \mathbb{Z}) \simeq U^{\oplus 3}$; cf. Section 3.2.5.

Due to (0.3), (i) implies (iii), for clearly disc $U(2)^{\oplus 3} = -2^6$ and disc $H^2(X, \mathbb{Z}) = -1$. Also (i) implies (iv), for $q_K \simeq -q_{K^\perp} \simeq -q_{U(2)}^{\oplus 3} = q_{U(2)}^{\oplus 3}$. Use (0.1) and disc $\bigoplus \mathbb{Z} \cdot e_i = 2^{16}$ to deduce (ii) from (i). $\qquad \square$

As another application of the general lattice theory outlined in Section 1.2 we state the following corollary.

Corollary 3.15 *The primitive embeddings*

$$K \hookrightarrow H^2(X, \mathbb{Z}) \text{ and } K \hookrightarrow H^*(X, \mathbb{Z})$$

are unique up to isometries of $H^2(X, \mathbb{Z}) \simeq E_8(-1)^{\oplus 2} \oplus U^{\oplus 3}$ and $H^*(X, \mathbb{Z}) \simeq E_8(-1)^{\oplus 2} \oplus U^{\oplus 4}$, respectively.

Proof The assertion for the embedding into the bigger lattice $H^*(X, \mathbb{Z})$ follows directly from Theorem 1.12. For the embedding into $H^2(X, \mathbb{Z})$ the finer version of it alluded to in Remark 1.13, (i), and the existence of $U \subset K^\perp \subset H^2(X, \mathbb{Z})$ have to be used. $\qquad\square$

See Example 4.8 for an embedding of the Kummer lattice into a distinguished Niemeier lattice.

Remark 3.16 To motivate the next result, note that a K3 surface X that contains 16 disjoint smooth rational curves $C_1, \ldots, C_{16} \subset X$ with $(1/2) \sum [C_i] \in \mathrm{NS}(X)$ is in fact a Kummer surface. (The existence of the square root is automatic, as we shall explain below; see Remark 3.19.) Indeed, the existence of the root of $\mathcal{O}(\sum C_i)$ can be used to prove the existence of a double cover $\widetilde{X} \longrightarrow X$ ramified over the $\bigcup C_i \subset X$. As the branch locus is smooth, $\widetilde{X}$ is smooth. The inverse images $\widetilde{C}_i \subset \widetilde{X}$ of the curves C_i are 16 disjoint (-1)-curves and so can be blown down $\widetilde{X} \longrightarrow A$ to a smooth surface A. Using the Kodaira–Enriques classification, one shows that A has to be a torus. See the arguments in Example 1.1.3 and [33, VIII.Prop. 6.1] or [54, Exp. VIII].

The following result was first stated by Nikulin in [449].

Theorem 3.17 *Let X be a complex K3 surface. Consider the following conditions:*

  (i) *The surface X is isomorphic to a Kummer surface.*
 (ii) *There exists a primitive embedding $K \hookrightarrow \mathrm{NS}(X)$.*
(iii) *There exists a primitive embedding $T(X) \hookrightarrow U(2)^{\oplus 3}$.*

Then (i) and (ii) are equivalent and imply (iii). If X is also projective, then the converse holds as well.[3]

Proof By the above discussion, (i) implies (ii). Since by Corollary 3.15 the embedding $K \hookrightarrow H^2(X, \mathbb{Z})$ is unique up to automorphism, its orthogonal complement is always isomorphic to $U(2)^{\oplus 3}$. Taking orthogonal complements, one finds that (ii) implies (iii).

Suppose a primitive embedding $T(X) \hookrightarrow U(2)^{\oplus 3}$ is given. Choose an embedding $U(2)^{\oplus 2} \hookrightarrow H^2(X, \mathbb{Z})$ with orthogonal complement K. By Corollary 3.5, for which we need X to be projective, the standard embedding $T(X) \subset H^2(X, \mathbb{Z})$ differs from the composition

$$T(X) \hookrightarrow U(2)^{\oplus 3} \hookrightarrow H^2(X, \mathbb{Z})$$

by an isometry of $H^2(X, \mathbb{Z})$ and, therefore, also the former contains K in its orthogonal complement; i.e. there exists a primitive embedding $K \hookrightarrow T(X)^\perp = \mathrm{NS}(X)$. So, (iii) implies (ii).

---

[3] As Matthias Schütt points out, (i) and (ii) are also equivalent for non-supersingular K3 surfaces in char $\neq 2$. Indeed, the inclusion $K \hookrightarrow \mathrm{NS}(X)$ can be lifted to characteristic zero, which is enough to conclude.

The difficult part is to deduce (i) from the purely lattice theoretic statements (ii) or (iii). We take a short cut by assuming the global Torelli theorem **7.5.3** and the surjectivity of the period map for two-dimensional complex tori. Historically, of course, the global Torelli theorem was first proved for Kummer surfaces; cf. [449] and the comments in Section **7.6**. So, suppose $K \subset \mathrm{NS}(X) \subset H^2(X,\mathbb{Z})$. Again using the uniqueness of the embedding, one finds an isomorphism $K^\perp \simeq U(2)^{\oplus 3}$, which comes with the natural Hodge structure of weight two on $K^\perp$. Then there exists a complex torus A and a Hodge isometry $H^2(A,\mathbb{Z})(2) \simeq U(2)^{\oplus 3} \simeq K^\perp$ (see Section **3.2.4**) which can be extended to a Hodge isometry $H^2(Y,\mathbb{Z}) \simeq H^2(X,\mathbb{Z})$ by Corollary 3.10. Here Y denotes the Kummer surface associated with A. Therefore, $X \simeq Y$, and, in particular, X is a Kummer surface. □

Example 3.18 The transcendental lattice of the Fermat quartic $X \subset \mathbb{P}^3$ defined by $x_0^4 + \cdots + x_3^4 = 0$ has been described as $T(X) \simeq \mathbb{Z}(8) \oplus \mathbb{Z}(8)$ (see Section **3.2.6**) which evidently embeds into $U(2)^{\oplus 3}$. Hence, the corollary can be used to show that the complex Fermat quartic X is indeed a Kummer surface, which suffices to deduce the result over arbitrary algebraically closed fields. An explicit isomorphism, following Mizukami, has been described in [273].

Remark 3.19 In [449] Nikulin also shows that a complex K3 surface is Kummer if and only if there exist 16 disjoint smooth rational curves $C_1, \ldots, C_{16} \subset X$. Here is an outline of the argument.[4] In order to prove that $(1/2) \sum C_i \in \mathrm{NS}(X)$, which is enough by Remark 3.16, consider the lattice $\Lambda' := \bigoplus \mathbb{Z} \cdot [C_i] \subset H^2(X,\mathbb{Z})$ and its saturation $\Lambda' \subset \Lambda \subset H^2(X,\mathbb{Z})$, which, due to the proof of Proposition 0.2, is determined by the isotropic subgroup $\mathbb{F}_2^\ell \simeq \Lambda/\Lambda' \subset A_{\Lambda'} \simeq \mathbb{F}_2^{16}$. Using the exact sequence

$$0 \longrightarrow A_\Lambda \longrightarrow A_{\Lambda'}/(\Lambda/\Lambda') \longrightarrow \Lambda'^*/\Lambda^* \simeq \mathbb{F}_2^\ell \longrightarrow 0$$

and the inequality

$$\ell(A_\Lambda) = \ell(A_{\Lambda^\perp}) \leq \mathrm{rk}\,\Lambda^\perp = 6,$$

one finds $\ell \geq 5$.

Similarly to the argument based on the Kodaira–Enriques classification in Remark 3.16, one shows that $|I| = 0, 8$, or 16 for any subset $I \subset \{1, \ldots, 16\}$ with $(1/2) \sum_{i \in I} [C_i] \in \mathrm{NS}(X)$; see [449, Lem. 3]. An elementary result in coding theory then allows one to identify $\Lambda/\Lambda' \subset A_{\Lambda'} \simeq \mathbb{F}_2^{16}$ as the code $\mathcal{D}_5$ (for the notation and the result, see [44]) which by construction contains $(1, \ldots, 1) \in \mathbb{F}_2^{16} \simeq A_{\Lambda'}$, i.e. $(1/2) \sum [C_i] \in \Lambda \subset \mathrm{NS}(X)$.

For the following consequences, see Morrison [424] and for part (i) also the paper by Looijenga and Peters [380, Prop. 6.1]. Recall that a Kummer surface X associated with a torus A satisfies $\rho(X) = \rho(A) + 16$, and if it is algebraic even $\rho(X) \geq 17$.

---

[4] Thanks to Jonathan Wahl for explaining this to me.

Corollary 3.20   *Let X be a complex projective K3 surface.*

(i) *Assume $\rho(X) = 19$ or 20. Then X is a Kummer surface if and only if*

$$T(X) \simeq T(2)$$

for some even lattice T.

(ii) *Assume $\rho(X) = 18$. Then X is a Kummer surface if and only if*

$$T(X) \simeq T(2) \oplus U(2)$$

for some even lattice T of rank two.

(iii) *Assume $\rho(X) = 17$. Then X is a Kummer surface if and only if*

$$T(X) \simeq T(2) \oplus U(2)^{\oplus 2}$$

for some even lattice T of rank one, i.e. $T \simeq \mathbb{Z}(2k)$.

Proof   The theorem shows that X is a K3 surface if and only if there exists a primitive sublattice $T' \subset U^{\oplus 3}$ with $T(X) \simeq T'(2)$. The three cases correspond to $\mathrm{rk}\, T(X) = 3$ or $2, = 4$, and $= 5$, respectively.

If $\mathrm{rk}\, T(X) = 3$ or 2 and $T(X) \simeq T(2)$, then use Proposition 1.8 to embed T into $U^{\oplus 3}$. Conversely, if X is the Kummer surface associated with the abelian surface A, then $T(X) \simeq T(A)(2)$.

If $\mathrm{rk}\, T(X) = 4$ and $T(X) \simeq T(2) \oplus U(2)$, then by Proposition 1.8 there exists a primitive embedding $T \hookrightarrow U^{\oplus 2}$ and, therefore, an embedding $T(X) \hookrightarrow U(2)^{\oplus 3}$. Thus, by the theorem, X is isomorphic to a Kummer surface. Conversely, if X is a Kummer surface associated with an abelian surface A and $\rho(X) = 18$, then $T(X) \simeq T(A)(2)$ and there exists a primitive embedding $T(A) \hookrightarrow U^{\oplus 3}$. Now deduce from Theorem 1.5 applied to $T(A)$ and $T \oplus U$, where $T := T(A)^{\perp}(-1) \hookrightarrow U^{\oplus 3}$, that $T(A) \simeq T \oplus U$.

The argument for $\mathrm{rk}\, T(X) = 5$ is similar. $\square$

3.4   The classification of complex K3 surfaces of maximal Picard number $\rho(X) = 20$ in terms of their transcendental lattices is particularly simple. This is a result due to Shioda and Inose [568].

Corollary 3.21   *The map that associates with a complex K3 surface X with $\rho(X) = 20$ its transcendental lattice $T(X)$ describes a bijection*

$$\{X \mid \rho(X) = 20\} \longleftrightarrow \{T \mid \text{positive definite, even, oriented lattice, } \mathrm{rk}\, T = 2\}, \quad (3.1)$$

where both sides are up to isomorphisms.

Proof   Clearly, by the Hodge index theorem, $T(X)$ of a K3 surface X with $\rho(X) = 20$ is a positive definite, even lattice of rank two. It comes with a natural orientation by declaring real and imaginary parts of a generator of $T(X)^{2,0}$ to be positively oriented in $T(X)_{\mathbb{R}}$; cf. Section 6.1.2.[5]

The map $X \longmapsto T(X)$ is surjective by Corollary 3.5. To prove injectivity, suppose there exists an isometry $T(X) \simeq T(X')$ that respects the orientation. But then, due to $\mathrm{rk}\, T(X) = \mathrm{rk}\, T(X') = 2$, the isometry is automatically compatible with the Hodge structures. (The period domain $D \subset \mathbb{P}(T_{\mathbb{C}})$ for each given lattice T consists of precisely two points.) Due to Corollary 3.10, this Hodge isometry extends to a Hodge isometry $H^2(X, \mathbb{Z}) \simeq H^2(X', \mathbb{Z})$ and, therefore, $X \simeq X'$ by the global Torelli theorem 7.5.3. (Note that due to Corollary 3.6 we knew already that $\mathrm{NS}(X)$ is determined by $T(X)$.)   □

Of course, the set on the right-hand side of (3.1) can be identified with the set of integral matrices of the form

$$\begin{pmatrix} 2a & b \\ b & 2c \end{pmatrix} \quad \text{with } \Delta := b^2 - 4ac < 0 \quad \text{and} \quad a, c > 0 \tag{3.2}$$

up to conjugation by matrices in $\mathrm{SL}(2, \mathbb{Z})$. The largest values of Δ are -4 and -3, which have been studied by Vinberg in [618].

Remark 3.22   The proof of the surjectivity in [568] is more explicit. For any T given by a matrix as in (3.2) a K3 surface X with $T(X) \simeq T$ is constructed as a double cover of a Kummer surface associated with the abelian surface $A = E_1 \times E_2$, where $E_i := \mathbb{C}/(\mathbb{Z} + \tau_i \mathbb{Z})$ with $\tau_1 = (-b + \sqrt{\Delta})/2a$ and $\tau_2 = (b + \sqrt{\Delta})/2$.

Furthermore, in [275] Inose showed that X is birational to a quartic surface given by an explicit equation $f(x_0, \ldots, x_3)$ with coefficients algebraic over $\mathbb{Q}(j(E_1), j(E_2))$.

Remark 3.23   If instead of the isomorphism type of the lattice T in (3.1) one considers only its genus, then, as was shown independently by Schütt [537, Thm. 15] and Shimada [554], on the right-hand side one distinguishes the K3 surfaces S only up to conjugation.

Remark 3.24   Due to Corollary 3.20, those K3 surfaces X with $\rho(X) = 20$ for which $T(X)$ is of the form $T(2)$ are actually Kummer surfaces. In other words, if $T(X)$ is of rank two, then X is a Kummer surface if and only if $(\alpha)^2 \equiv 0\,(4)$ for all $\alpha \in T(X)$.

Using this characterization of these so-called *extremal Kummer surfaces* enables one to prove density of complex projective Kummer surfaces, which was important in early proofs of the global Torelli theorem; see Section 7.6.1. So, using the notation in Chapters 6 and 7, this can be phrased by saying that

$$\mathcal{P} \colon \{(X, \varphi) \mid X \text{ Kummer}\} \longrightarrow D$$

has a dense image.

---

[5] Observe that the lattice with the reversed orientation is realized by the complex conjugate K3 surface $\overline{X}$.

The proof proceeds in two steps.

(i) The set of periods $x \in D \subset \mathbb{P}(\Lambda_{\mathbb{C}})$ for which the corresponding positive plane $P \subset \Lambda_{\mathbb{R}}$ (see Proposition 6.1.5) is defined over $\mathbb{Q}$ and which has the property that $(\alpha)^2 \equiv 0\,(4)$ for all $\alpha \in P \cap \Lambda$ is dense in D. For the intriguing but elementary proof, we refer to [54, Exp. IX], [33, Ch. VIII], or [380, Sec. 6]. Compare this with similar density results, e.g. Propositions 6.2.9 and 7.1.3.

(ii) From this one immediately concludes: the set of marked K3 surfaces $(X, \varphi) \in N$ in the moduli space of marked K3 surfaces (see Section 7.2.1) for which X is a projective Kummer surface (of maximal Picard number $\rho(X) = 20$) is dense in N.

Remark 3.25 A similar statement can be proved for the polarized case. For any $d > 0$ the image of the set of marked polarized Kummer surfaces under

$$\mathcal{P}_d \colon N_d \lhook\joinrel\longrightarrow D_d$$

is dense in D_d and hence Kummer surfaces are dense in N_d.

In [380, Rem. 6.5] the authors point out a gap in [493, Sec. 6]. However, if [380, Thm. 2.4] is replaced by the stronger Theorem 1.12 or Remark 1.13, then this gap can be filled and the arguments in [380] go through unchanged. Indeed, the above arguments can be adapted to prove the density of rational periods in $D_d \subset \mathbb{P}(\Lambda_{d\mathbb{C}})$ with the same divisibility property. One needs the stronger lattice theory result to ensure that the orthogonal complement $\alpha^\perp \subset \Lambda_d$ of an arbitrary class $\alpha \in \Lambda_d$ still contains a hyperbolic plane; cf. the proof of [380, Prop. 6.2]. Alternatively, one could use the density in Remark 3.24 combined with the density of abelian surfaces of fixed degree in the space of two-dimensional tori; cf. [5].

4 Niemeier Lattices

Books have been written on Niemeier lattices and the Leech lattice in particular. We highlight only a few aspects that seem relevant for our purposes. For more information, the reader should consult e.g. [132].

As it better fits the applications to K3 surfaces, we adopt the convention that Niemeier lattices are negative definite.

4.1 By definition a *Niemeier lattice* is an even, unimodular, negative definite lattice N of rank 24. The easiest example is $E_8(-1)^{\oplus 3}$.

Corollary 4.1 *Let N be a Niemeier lattice. Then*

$$N \oplus U \simeq E_8(-1)^{\oplus 3} \oplus U =: \mathrm{II}_{1,25}.$$

Conversely, if $0 \neq w \in \mathrm{II}_{1,25}$ *is a primitive vector with* $(w)^2 = 0$, *then there exists an isomorphism* $\mathrm{II}_{1,25} \simeq N \oplus U$ *with* N *a Niemeier lattice and such that* w *corresponds to* $e \in U$.

Proof The isomorphism follows from the classification of indefinite lattices; see Corollary 1.2 or 1.3. The second statement is a consequence of Corollary 1.14. □

This gives, at least in principle, a way to construct all Niemeier lattices. It turns out that there are exactly 24 primitive $0 \neq w \in \mathrm{II}_{1,25}$ with $(w)^2 = 0$ up to the action of $O(\mathrm{II}_{1,25})$, which eventually leads to a complete classification. Note that for w as above, the corresponding Niemeier lattice is isomorphic to $w^{\perp}/\langle w \rangle$.

A list of all Niemeier lattices was first given by Niemeier [448]. A more conceptual approach, which in particular clarifies the role of the root lattices R, is due to Venkov [610]. See also [132, Ch. 16] or [161, Ch. 3.4] for detailed proofs.

Theorem 4.2 (Niemeier, Venkov) *There exist precisely* 24 *isomorphism classes of Niemeier lattices* N, *each of which is uniquely determined by its root lattice* $R \subset N$ *which is either trivial or of rank* 24.

As in Section 2.1, the root lattice $R \subset N$ is the sublattice spanned by all (-2) classes, so

$$R := \langle \delta \mid \delta \in N \text{ with } (\delta)^2 = -2 \rangle \subset N.$$

Note that the root lattice can be trivial, i.e. $R = 0$. This is the case if and only if N is the *Leech lattice*; see Section 4.4 for more on the Leech lattice. In fact, for all other Niemeier lattices $R \subset N$ is of finite index. Conversely, each Niemeier lattice, except for the Leech lattice, can be seen as the minimal unimodular overlattice of the corresponding root lattice.

The existence of the Leech lattice, i.e. of a Niemeier lattice without roots, can be shown by a procedure that changes a Weyl vector in a way that forces the number of roots to decrease.

Remark 4.3 The following list of the 24 root lattices R, or rather of $R(-1)$, can be found in [132, 322, 450]:[6]

$$0, A_1^{\oplus 24}, A_2^{\oplus 12}, A_3^{\oplus 8}, A_4^{\oplus 6}, A_6^{\oplus 4}, A_8^{\oplus 3}, A_{12}^{\oplus 2}, A_{24}, D_4^{\oplus 6}, D_6^{\oplus 4}, D_8^{\oplus 3}, D_{12}^{\oplus 2}, D_{24}, E_6^{\oplus 4}, E_8^{\oplus 3},$$
$$A_5^{\oplus 4} \oplus D_4, A_7^{\oplus 2} \oplus D_5^{\oplus 2}, A_9^{\oplus 2} \oplus D_6, A_{15} \oplus D_9, A_{17} \oplus E_7, D_{10} \oplus E_7^{\oplus 2}, D_{16} \oplus D_8,$$
$$A_{11} \oplus D_7 \oplus E_6.$$

Due to the general classification, R is a direct sum of lattices of ADE type (see Section 0.3), but obviously not all of those that are of rank 24 occur in the above list.

---

[6] The lattices are usually listed according to their Coxeter number, so that 0, the root lattice of the Leech lattice, would be the last one in the list and $A_1(-1)^{\oplus 24}$ the penultimate.

Later, for example, in Section **15**.3.2, we often explain arguments for the case of the Niemeier lattice with root lattice $A_1(-1)^{\oplus 24}$.

Remark 4.4 The lattice $\mathrm{II}_{1,25} \simeq E_8(-1)^{\oplus 3} \oplus U$ has signature $(1, 25)$ and the general theory of Section **8**.2 applies. In particular, one can consider the positive cone $\mathcal{C} \subset \mathrm{II}_{1,25} \otimes \mathbb{R}$ (one of the two connected components of the set of all x with $(x)^2 > 0$) and its chamber decomposition.

A fundamental domain for the action of the Weyl group W of $\mathrm{II}_{1,25}$, i.e. one chamber $\mathcal{C}_0 \subset \mathcal{C}$, can be described in terms of Leech roots. A *Leech root* is a (-2)-class $\delta \in \mathrm{II}_{1,25}$ such that $(\delta . e) = 1$ for $e \in U$ the first basis vector of the standard basis of U. Then one chamber $\mathcal{C}_0$ consists of all $x \in \mathcal{C} \subset \mathrm{II}_{1,25} \otimes \mathbb{R}$ with $(x.\delta) > 0$ for all Leech roots δ. The closure of every chamber, in particular of $\mathcal{C}_0$, contains precisely 24 primitive classes $0 \neq w_i \in \mathrm{II}_{1,25} \cap \partial \cap \partial \mathrm{C}$, i.e. 24 primitive square zero lattice elements are contained in the boundary of $\mathcal{C}_0$. They give rise to 24 decompositions $\mathrm{II}_{1,25} \simeq N_i \oplus U$ with the 24 Niemeier lattices $N_0, \ldots, N_{23}$.

A variant of the above allows one to write a bijection between the set of all primitive $w \in \mathrm{II}_{1,25} \cap \partial \mathcal{C}$ with $w^\perp / \langle w \rangle$ isomorphic to the Leech lattice N_0 (see below) and the set of chambers. It is given by

$$w \longmapsto \mathcal{C}_w := \{x \mid (x.\delta) > 0 \text{ if } (\delta . w) = 1\},$$

where on the right-hand side all roots δ are considered; see [131]. Note that the chamber decomposition of $\mathcal{C}$ is not rational polyhedral. For details and proofs, see [77, 132].

4.2 The importance of Niemeier lattices in the theory of K3 surfaces becomes clear by the following consequence of Nikulin's Theorem 1.15. Note that in the application in Section **15**.3.2 in fact all Niemeier lattices occur, with the exception of the Leech lattice.

Corollary 4.5 *Let Λ_1 be a negative definite, even lattice with $\ell(\Lambda_1) < 24 - \mathrm{rk}\,\Lambda_1$. Then there exists a primitive embedding into a Niemeier lattice $\Lambda_1 \hookrightarrow N$.*   □

Sometimes, this version of the corollary is not quite sufficient, but the finer version alluded to in Remark 1.17 usually is.

Example 4.6 Consider the Néron–Severi lattice $\mathrm{NS}(X)$ of a complex projective K3 surface. Then for every Niemeier lattice N there exists a primitive embedding (see [555, Prop. 8.1])

$$\mathrm{NS}(X) \hookrightarrow N \oplus U \simeq E_8(-1)^{\oplus 3} \oplus U \simeq \mathrm{II}_{1,25}.$$

This follows from Theorem 1.15, as $\mathrm{NS}(X)$ is even of signature $(1, \rho(X) - 1)$ and

$$\ell(\mathrm{NS}(X)) = \ell(T(X)) \le 22 - \rho(X) < 26 - \rho(X).$$

The possibility to embed NS(X) into $\mathrm{II}_{1,25}$ prompts the question of how the chamber decompositions of the positive cones of the two lattices are related. Very roughly, although the chamber decomposition of $\mathrm{II}_{1,25}$ is not rational polyhedral, it sometimes (under certain conditions on the orthogonal complement NS(X)$^\perp \subset \mathrm{II}_{1,25}$) induces a rational polyhedral decomposition of the ample cone which turns out to be a chamber decomposition for the action of Aut(X); compare this to the cone conjecture Theorem **8**.4.2. See the papers by Borcherds [77] and Shimada [555] for details.

Example 4.7 For a recent discussion of the following applications to embeddings of Néron–Severi lattices, see Nikulin's [456].

(i) Let $\Lambda_1 := $ NS(X) of a (non-projective) complex K3 surface X with negative definite NS(X). In this case $\ell(\mathrm{NS}(X)) = \ell(T(X)) \leq 22 - \rho(X) < 24 - \rho(X)$, and, therefore, there exists a primitive embedding

$$\mathrm{NS}(X) \hookrightarrow N$$

into some Niemeier lattice N.

(ii) For a complex projective K3 surface X the result can be applied to $\Lambda_1 := \ell^\perp \subset$ NS(X) for any $\ell \in$ NS(X) with $(\ell)^2 > 0$. Again one checks $\ell(\ell^\perp) \leq \ell(T(X)) + 1 \leq 22 - \mathrm{rk}\,\ell^\perp < 24 - \mathrm{rk}\,\ell^\perp$. So, there exists a primitive embedding

$$\ell^\perp \hookrightarrow N$$

into some Niemeier lattice N. In fact, by applying the corollary instead to $\Lambda_1 = \ell^\perp \oplus A_1(-1)$ one can exclude the Leech lattice (Kondō's trick).

Note that in both cases the lattice spanned by (-2)-classes (in NS(X) or in $\ell^\perp$, respectively) becomes a sublattice of the root lattice of a Niemeier lattice.

Example 4.8 The Kummer lattice K (see Section 3.3) can be embedded

$$K \hookrightarrow N$$

into the Niemeier lattice N with root lattice $A_1(-1)^{\oplus 24}$. See [456, 589]. The existence of a primitive embedding into one of the Niemeier lattices follows from Theorem 1.15.

4.3 The Niemeier lattice that is of special interest in the context of K3 surface is the Niemeier lattice with root lattice $R \simeq A_1(-1)^{\oplus 24} \simeq \langle -2 \rangle^{\oplus 24}$. It can be constructed as the set $N \subset R_\mathbb{Q}$ of all vectors of the form $\frac{1}{2} \sum n_i e_i$ with $n_i \in \mathbb{Z}$ and such that $(\bar{n}_1, \ldots, \bar{n}_{24}) \in \mathbb{F}_2^{24}$ is an element of the (extended binary) *Golay code* $W \subset \mathbb{F}_2^{24}$, i.e.

$$W = N/R \subset R^*/R \simeq \mathbb{F}_2^{24}.$$

By definition, the Golay code W is a 12-dimensional linear subspace with the property that for all $0 \neq w = (w_i) \in W$ one has $|\{i \mid w_i \neq 0\}| \geq 8$. The subspace W can be written explicitly and it is unique up to linear coordinate change; see e.g. [132].

Remark 4.9   The only roots in this Niemeier lattice are the elements $\pm e_i$. Hence the Weyl group of N is $(\mathbb{Z}/2\mathbb{Z})^{\oplus 24}$.

Viewing the symmetric group $\mathfrak{S}_{24}$, as usual, as a subgroup of $GL(\mathbb{F}_2^{24})$, one defines the *Mathieu group* as (see also Section 15.3.1)

$$M_{24} := \{\sigma \in \mathfrak{S}_{24} \mid \sigma(W) = W\}.$$

It is a simple sporadic group of order

$$|M_{24}| = 244.823.040 = 2^{10} \cdot 3^3 \cdot 5 \cdot 7 \cdot 11 \cdot 23.$$

Proposition 4.10   *The orthogonal group $O(N)$ of the Niemeier lattices N with root lattice $R = A_1(-1)^{\oplus 24}$ is naturally isomorphic to*

$$O(N) \simeq M_{24} \ltimes (\mathbb{Z}/2\mathbb{Z})^{\oplus 24}.$$

Proof   The group $\mathfrak{S}_{24} \ltimes (\mathbb{Z}/2\mathbb{Z})^{\oplus 24}$ acts naturally on the root lattice $R = A_1(-1)^{\oplus 24}$ by permutation of the basis vectors e_i and sign change $e_i \mapsto -e_i$. Also, as $\operatorname{rk} R = 24$, any $g \in O(N)$ is determined by its action on the roots and thus $O(N) \hookrightarrow \mathfrak{S}_{24} \ltimes (\mathbb{Z}/2\mathbb{Z})^{\oplus 24}$. By definition of N in terms of R and the Golay code, the image is contained in $M_{24} \ltimes (\mathbb{Z}/2\mathbb{Z})^{\oplus 24}$, and, conversely, every element in $M_{24} \ltimes (\mathbb{Z}/2\mathbb{Z})^{\oplus 24}$ defines an orthogonal transformation of N.   □

For more comments and references concerning $O(N)/W(N)$, for the other Niemeier lattices N, see Section 15.3.1.

4.4   There are various ways of constructing the Leech lattice N_0. The easiest is maybe the one due to Conway (see [132, Ch. 27]) that describes

$$N_0(-1) \subset II_{25,1} := E_8^{\oplus 3} \oplus U \subset \mathbb{R}^{25,1}$$

as $w^\perp/\mathbb{Z}w$ with $w := (0, 1, \ldots, 24, 70)$.

The Leech lattice does not contain any roots and so its Weyl group is trivial. The orthogonal group $O(N_0)$ of the Leech lattice is called the *Conway group* Co_0. The quotient by its center $Co_1 := Co_0/\{\pm 1\}$ is a simple sporadic group of order

$$|Co_1| = 4.157.776.806.543.360.000 = 2^{21} \cdot 3^9 \cdot 5^4 \cdot 7^2 \cdot 11 \cdot 13 \cdot 23.$$

The Conway group contains a subgroup isomorphic to the Mathieu group M_{24}; cf. [130].

References and Further Reading

In papers by Sarti [522, 523] one finds explicit computations of Picard and transcendental lattices of K3 surfaces with $\rho = 19$.

A K3 surface is a *generalized Kummer surface* if it is isomorphic to the minimal resolution of the quotient A/G of an abelian surface (or a two-dimensional complex torus) A by a finite group G. The possible groups G, all finite subgroups of $SL(2, \mathbb{C})$, can be classified (cf. [190]) as: (i) cyclic groups of order 2, 3, 4, or 6; (ii) binary dihedral groups $(2, 2, n = 2, 3)$ (leading to a D_4 and D_5 singularity); and (iii) binary tetrahedral group (leading to an E_6-singularity). In positive characteristic the list is slightly longer; see [284] or [71, Prop. 4.4]. For a classification of general finite subgroups $G \subset \mathrm{Aut}(A)$, see Fujiki's article [190]. In [60] Bertin studies the analogue K_n of the Kummer lattice for $G = \mathbb{Z}/n\mathbb{Z}$, $n = 3, 4, 6$. Note that rk $K_n = 18$ in all three cases. It is proved that a K3 surface is the minimal resolution of A/G if and only if there exists a primitive embedding $K_n \hookrightarrow NS(X)$ (which is the analogue of Theorem 3.17). A description of the generalized Kummer lattice in the remaining cases has been given by Wendland [635] and Garbagnati [200].

Persson in [491] shows that a K3 surface is a 'maximizing' double plane if $NS(X)_\mathbb{Q}$ is spanned by effective curves $H, E_1, \ldots, E_{19}$ with $(H)^2 = 2$, $(H.E_i) = 0$, and H irreducible.

Coming back to Example 4.6, one finds bits on the relation between the positive cones in $NS(X)$ and $II_{1,25}$ (see [77, 555, 556]), but I am not aware of a concise treatment of it in the literature.

15

Automorphisms

Let X be a complex K3 surface or an algebraic K3 surface over a field k. By $\text{Aut}(X)$ we denote the group of all automorphisms $X \xrightarrow{\sim} X$. An automorphism of a complex K3 surface is simply a biholomorphic map, and an automorphism of an algebraic K3 surface over k is an isomorphism of k-schemes.

Then $\text{Aut}(X)$ has the structure of a complex Lie group or of an algebraic group, respectively. However, as $H^0(X, \mathcal{T}_X) = 0$, it is simply a discrete, reduced group. The same argument also shows that for a K3 surface X over an algebraically closed field k the automorphism group does not change under base change; i.e. $\text{Aut}(X) \simeq \text{Aut}(X \times_k K)$ for any field extension $k \subset K$. For the general theory of transformation groups of complex manifolds, see [306, Ch. III] and in the algebraic context see [228, Sec. C.2].

There are two kinds of automorphisms, symplectic and non-symplectic ones. Section 1 treats the group $\text{Aut}_s(X)$ of symplectic automorphisms, mostly for complex K3 surfaces, and explains that at least for projective K3 surfaces the subgroup $\text{Aut}_s(X) \subset \text{Aut}(X)$ is of finite index. Section 2 describes $\text{Aut}(X)$ of a complex K3 surface in terms of isometries of the Hodge structure $H^2(X, \mathbb{Z})$. This allows one to classify K3 surfaces with finite $\text{Aut}(X)$ for which a classification of the possible $\text{NS}(X)$ is also known. Mukai's result on the classification of all finite groups occurring as subgroups $G \subset \text{Aut}_s(X)$ can be found in Section 3. Concrete examples are constructed in the final Section 4.

1 Symplectic Automorphisms

This first section collects a number of elementary observations on symplectic automorphisms, crucial for any further investigation.

* Thanks to Giovanni Mongardi and Matthias Schütt for detailed comments on this chapter.

Definition 1.1 An automorphism $f \colon X \xrightarrow{\sim} X$ of a K3 surface is called *symplectic* if the induced action on $H^0(X, \Omega_X^2)$ is the identity; i.e. for a generator $\sigma \in H^0(X, \Omega_X^2)$ one has $f^* \sigma = \sigma$.

Note that this definition makes sense for complex K3 surfaces as well as for algebraic ones. One thinks of the two-form σ as a holomorphic or algebraic symplectic structure, hence the name.

The subgroup of all symplectic automorphisms shall be denoted

$$\mathrm{Aut}_s(X) \subset \mathrm{Aut}(X).$$

Remark 1.2 For complex K3 surfaces, one can equivalently use the transcendental lattice $T(X)$, which by definition is the minimal sub-Hodge structure of $H^2(X, \mathbb{Z})$ containing $H^{2,0}(X)$; see Section 3.2.2. As the natural inclusion $H^0(X, \Omega_X^2) \simeq H^{2,0}(X) \subset T(X)_{\mathbb{C}}$ is compatible with the action of f, one finds that f is symplectic if and only if $f^* = \mathrm{id}$ on $T(X)$. Indeed, one direction is obvious, and for the other use that for f symplectic $\mathrm{Ker}((f^* - \mathrm{id})|_{T(X)}) \subset T(X)$ is a sub-Hodge structure containing $H^{2,0}(X)$; cf. Lemma 3.3.3.

Remark 1.3 The action of an automorphism $f \colon X \xrightarrow{\sim} X$ on the one-dimensional space $H^0(X, \Omega_X^2)$ is linear. Therefore, if f is of finite order n, its action on $H^0(X, \Omega_X^2)$ is given by multiplication by an nth root of unity $\zeta \in k$. Thus, as there are no non-trivial pth roots of unity in a field of characteristic $\mathrm{char}(k) = p > 0$, any automorphism of order p is in this case automatically symplectic.

1.1 Concerning the local structure of symplectic automorphisms of finite order of a complex K3 surface, the following elementary fact is useful (see also the proof of Proposition 3.11 for a slightly stronger statement).

Lemma 1.4 *Let X be a complex K3 surface. Assume $f \in \mathrm{Aut}_s(X)$ is of finite order $n := |f|$ and $x \in X$ is a fixed point of f. Then there exists a local holomorphic coordinate system z_1, z_2 around x such that $f(z_1, z_2) = (\lambda_x z_1, \lambda_x^{-1} z_2)$ with λ_x a primitive nth root of unity.*

Proof The following argument has been taken from Cartan [104]. Consider a small open ball around $x = 0 \in U \subset \mathbb{C}^2$. We first show that there are coordinates (z_1, z_2) with $f(z_1, z_2) = d_0 f \cdot (z_1, z_2)$. Define $g(y) := (1/n) \sum_{i=1}^n (d_0 f)^{-i} \cdot f^i(y)$ and check $d_0 g = (1/n) \sum_{i=1}^n (d_0 f)^{-i} d_0(f^i) = \mathrm{id}$. Hence, $(z_1, z_2) := g(y)$ can be used as local coordinate functions around 0. Then note

$$g(f(y)) = (1/n) \sum_{i=1}^{n} (d_0 f)^{-i} \cdot f^{i+1}(y)$$

$$= (d_0 f) \cdot \left((1/n) \sum_{i=1}^{n} (d_0 f)^{-i-1} \cdot f^{i+1}(y) \right) = d_0 f \cdot g(y).$$

Thus, with respect to the new coordinate system $(z_1, z_2) = g(y)$, we can think of f as the linear map $d_0 f$ and, after a further linear coordinate change, $f(z_1, z_2) = (\lambda_1 z_1, \lambda_2 z_2)$.

Now deduce from $f \in \mathrm{Aut}_s(X)$ that $\det(d_0 f) = 1$, i.e. $\lambda_x := \lambda_1 = \lambda_2^{-1}$, and from $|f| = n$ that λ_x is an nth root of unity. If $\lambda_x^k = 1$ for some $k < n$, then $f^k = \mathrm{id}$ in a neighbourhood of x and hence $f^k = \mathrm{id}$ globally. Contradiction. Hence, every λ_x is a primitive nth root of unity. $\square$

In particular, the fixed point set of a non-trivial symplectic automorphism of a complex K3 surface consists of a finite set of reduced points. This does not hold any longer in positive characteristic; see Remark 1.9.

Mukai in [430, Sec. 1] deduces from this the following result; see also [450, Sec. 5].

Corollary 1.5 *Let* $\mathrm{id} \neq f \in \mathrm{Aut}_s(X)$ *be of finite order* $n := |f|$. *Then the fixed point set* $\mathrm{Fix}(f)$ *is finite and non-empty. More precisely,* $1 \leq |\mathrm{Fix}(f)| \leq 8$ *and in fact*

$$|\mathrm{Fix}(f)| = \frac{24}{n} \prod_{p|n} \left(1 + \frac{1}{p} \right)^{-1}, \tag{1.1}$$

which in particular depends only on the order n.

Proof The local description provided by the lemma shows that all fixed points are isolated; i.e. $\mathrm{Fix}(f)$ is finite, and non-degenerate, as $\lambda_x \neq 1$ for all $x \in \mathrm{Fix}(f)$.

Thus, the *Lefschetz fixed point formula* for biholomorphic automorphisms (see [221, p. 426]) applies and reads in the present case

$$\sum_i (-1)^i \mathrm{tr}\left(f^* |_{H^i(X, \mathcal{O})} \right) = \sum_{x \in \mathrm{Fix}(f)} \det\left(\mathrm{id} - d_x f \right)^{-1}$$

$$= \sum_{x \in \mathrm{Fix}(f)} \frac{1}{(1 - \lambda_x)(1 - \lambda_x^{-1})}.$$

Here, the sum on the right-hand side runs over the (finite) set $\mathrm{Fix}(f)$ of fixed points of f. The sum on the left-hand side equals 2, for $H^i(X, \mathcal{O})$ is one-dimensional for $i = 0, 2$ (and trivial otherwise), and the symplectic automorphism f acts as id on $H^2(X, \mathcal{O})$. Hence $\mathrm{Fix}(f)$ is non-empty. Since $|\lambda_x| = 1$, one has $|1 - \lambda_x^{\pm 1}| \leq 2$, which immediately proves $2 \geq (1/4)|\mathrm{Fix}(f)|$, i.e. $|\mathrm{Fix}(f)| \leq 8$.

If k is prime to n, then $\mathrm{Fix}(f^k) = \mathrm{Fix}(f)$. Therefore, also

$$\sum_{x \in \mathrm{Fix}(f)} \frac{1}{(1 - \lambda_x^k)(1 - \lambda_x^{-k})} = 2$$

and hence

$$\sum_{x \in \mathrm{Fix}(f)} \frac{1}{\varphi(n)} \sum_{(k,n)=1} \frac{1}{(1 - \lambda_x^k)(1 - \lambda_x^{-k})} = 2,$$

where $\varphi(n)$ denotes the Euler function. For any primitive nth root of unity λ one has[1]

$$\sum_{(k,n)=1} \frac{1}{(1 - \lambda^k)(1 - \lambda^{-k})} = \frac{n^2}{12} \prod_{p|n} \left(1 - \frac{1}{p^2}\right)$$

(see [430, Lem. 1.3]), which combined with $\varphi(n) = n \prod_{p|n} \left(1 - \frac{1}{p}\right)$ yields (1.1). □

The above techniques already yield a weak version of Proposition 2.1.

Corollary 1.6 *Let $f \colon X \xrightarrow{\sim} X$ be an automorphism of finite order of a complex K3 surface and assume that $f^* = \mathrm{id}$ on $H^2(X, \mathbb{Z})$. Then f is the identity.*

Proof By assumption, f is in particular symplectic, and, hence, the lemma and the previous corollary apply. If $f \neq \mathrm{id}$, then the topological Lefschetz fixed point formula

$$\sum (-1)^i \mathrm{tr}\left(f^*|_{H^i(X,\mathbb{Q})}\right) = |\mathrm{Fix}(f)| \tag{1.2}$$

(see [221, p. 421]) would yield the contradiction $24 = |\mathrm{Fix}(f)| \leq 8$. □

Remark 1.7 Corollary 1.6 holds for automorphisms of K3 surfaces over an arbitrary algebraically closed field k with singular cohomology replaced by étale cohomology; see [510, Prop. 3.4.2]. See also Remark 2.2.

1.2 For the following see Nikulin [450, Sec. 5], where one also finds a discussion of a number of special cases of finite, cyclic and non-cyclic, symplectic group actions.

Corollary 1.8 *If $f \in \mathrm{Aut}_s(X)$ is of finite order n, then $n \leq 8$.*

Proof Since f is symplectic, $(H^0 \oplus H^{2,0} \oplus H^{0,2} \oplus H^4)(X)$ is contained in the invariant part $H^*(X, \mathbb{C})^{\langle f \rangle} \subset H^*(X, \mathbb{C})$. Moreover, for any ample or Kähler class $\alpha \in H^{1,1}(X)$

---

[1] This looks like a standard formula in number theory, which in [430] is deduced by Möbius inversion from $\sum_{i=1}^{n-1}((1-\lambda^i)(1-\lambda^{-i}))^{-1} = (n^2 - 1)/12$. Up to the factor $(1/12)$ the right-hand side is the Jordan totient function J_2. Curiously, the formula does not seem to appear in any of the standard number theory books.

the sum $\sum_{i=1}^{n}(f^i)^*\alpha$ is f-invariant and ample (resp. Kähler) and hence non-trivial.[2] Together this shows dim $H^*(X, \mathbb{C})^{\langle f \rangle} \geq 5$. Next, viewing $(1/n)\sum_{i=1}^{n}(f^i)^*$ as a projector onto $H^*(X, \mathbb{C})^{\langle f \rangle}$ one finds

$$\sum_{i=1}^{n} \text{tr}((f^i)^*|_{H^*(X,\mathbb{C})}) = n \cdot \dim H^*(X, \mathbb{C})^{\langle f \rangle},$$

which due to (1.2) can also be written as $24 + \sum_{i=1}^{n-1}|\text{Fix}(f^i)| = n \cdot \dim H^*(X, \mathbb{C})^{\langle f \rangle}$. Hence, $24 + \sum_{i=1}^{n-1}|\text{Fix}(f^i)| \geq 5 \cdot n$, which together with (1.1) yields the result. Indeed, for example, for $n = p$ the inequality becomes $24 + (p-1)24/(p+1) \geq 5 \cdot p$, i.e. $p \leq 43/5$. One reduces the general case to $n = p$ by passing to powers of f, but this part of the argument is not particularly elegant and is left to the reader.   □

Combining Corollaries 1.5 and 1.8 one finds that only the following tuples $(n, |\text{Fix}(f)|)$ for symplectic automorphisms of finite order of a complex projective K3 surface can occur:

n	2	3	4	5	6	7	8
$|\text{Fix}(f)|$	8	6	4	4	2	3	2
$\rho(X) \geq$	9	13	15	17	17	19	19

The table has been completed by a lower bound for the Picard number $\rho(X)$, which follows from $T(X)_{\mathbb{C}} \subset H^2(X, \mathbb{C})^{\langle f \rangle}$ and the resulting inequality $24 + \sum_{i=1}^{n-1}|\text{Fix}(f^i)| = n \cdot \dim H^*(X, \mathbb{C})^{\langle f \rangle} \geq n \cdot (2 + \text{rk}\,T(X) + 1) = n \cdot (25 - \rho(X))$. So, complex K3 surfaces with symplectic automorphisms tend to have rather high Picard number. In fact, in Corollary 2.12 below we shall see that K3 surfaces of Picard rank $\rho(X) = 1$ essentially have no automorphisms at all. See Section 4 for concrete examples of symplectic automorphisms of finite order and more results on NS(X) in these cases.

Remark 1.9 Dolgachev and Keum in [152, Thm. 3.3] show that the above discussion carries over to *tame* symplectic automorphisms. More precisely, for a K3 surface X over an algebraically closed field k of characteristic $p = \text{char}(k) > 0$ and a symplectic automorphism $f \in \text{Aut}_s(X)$ of finite order $n := |f|$ prime to p exactly the same values for $(n, |\text{Fix}(f)|)$ as recorded in the above table can (and do) occur.

For wild automorphisms, i.e. those $f \in \text{Aut}(X)$ with p dividing the order of f, the situation is more difficult. In [152, Thm. 2.1], however, it is proved that if there exists an $f \in \text{Aut}(X)$ of order p, then $p \leq 11$ or, equivalently, for $p > 11$ no $f \in \text{Aut}(X)$ exists whose order is divisible by p.

---

[2] This is an observation of independent interest: for any symplectic automorphism $f: X \dashrightarrow X$ of finite order there exists a Kähler class α and hence a hyperkähler structure determined by a hyperkähler metric g (see Section 7.3.2) that is invariant under f. In particular, f then acts as a biholomorphic automorphism on the twistor space $\mathcal{X}(\alpha) \longrightarrow T(\alpha) \simeq \mathbb{P}^1$.

If X admits a wild automorphism of order $p = 11$, then $\rho(X) = 2, 12$, or 22. According to Schütt [539] the generic Picard number really is $\rho(X) = 2$ in this case. In [153] Dolgachev and Keum give an explicit example of an automorphisms of order 11 in characteristic 11, which then is automatically symplectic, provided by a hypersurface of degree 12 in $\mathbb{P}(1, 1, 4, 6)$ with the automorphism given by translation $(t_0 : t_1 : x : y) \mapsto (t_0 : t_0 + t_1 : x : y)$; see also Better Oguiso [468].

It is maybe worth pointing out that in the wild case the fixed point set of a symplectic automorphism is not necessarily discrete. See also the more recent article [292] by Keum where all possible orders of automorphisms of K3 surfaces in characteristic > 3 are determined.

1.3   We now attempt to explain the difference between $\mathrm{Aut}_s(X)$ and the full $\mathrm{Aut}(X)$.

Corollary 1.10   *Let X be a complex K3 surface and $f \in \mathrm{Aut}(X)$.*

(i) *If X is projective, then there exists an integer $n > 0$ such that $f^{*n} = \mathrm{id}$ on $T(X)$.*
(ii) *If X is not projective, then $f^* = \mathrm{id}$ on $T(X)$ or f^* has infinite order on $T(X)$.*

Proof   The first assertion is an immediate consequence of Corollary 3.3.4. For the second assertion choose a class $\alpha \in H^2(X, \mathbb{Q})$ such that its $(1, 1)$-part is a Kähler class. If f^* has finite order n on $T(X)$, then the $(1, 1)$-part of the finite sum $h := \sum_{i=1}^{n} (f^i)^*(\alpha)$ is still a Kähler class and its $(2, 0)$-part is f^*-invariant. However, if $f^* \neq \mathrm{id}$ on $T(X)$, then there are no f^*-invariant $(2, 0)$-classes. Indeed, $\mathrm{Ker}((f^* - \mathrm{id})|_{T(X)}) \subset T(X)$ would contradict the minimality of $T(X)$. Hence, h must be a rational Kähler class and, therefore, X is projective. Alternatively, one can divide X by the action of the non-symplectic f which gives either an Enriques surface or a rational surface; see Section 4.3 and [450]. In either case, the quotient and hence X itself would be projective.   □

Example 1.11   It can indeed happen that a non-projective K3 surface X admits automorphisms $f : X \xrightarrow{\sim} X$ such that f^* does not act by a root of unity on $H^{2,0}(X)$ or, equivalently, is not of finite order on $T(X)$. One can use [643] to produce examples on complex tori and then pass to the associate Kummer surface; cf. [321, Rem. 4.8].

Remark 1.12   Essentially the same argument has been applied by Esnault and Srinivas in [173] to prove the following result, which we state only for K3 surfaces: let (X, L) be a polarized K3 surface over an algebraically closed field k and let $f \in \mathrm{Aut}(X)$. Then the induced action f^* on the largest f-stable subspace $V \subset c_1(L)^{\perp} \subset H^2(X, \mathbb{Q}_\ell(2))$, $\ell \neq \mathrm{char}(k)$, has finite order. However, this does not seem to imply that on a projective K3 surface in positive characteristic Corollary 1.10 still holds, i.e. that any automorphism becomes symplectic after passing to some finite power.

Corollary 3.3.4 in fact describes the group of Hodge isometries of $T(X)$ as a finite cyclic group. Hence, $\mathrm{Aut}(X)$ of a complex projective K3 surface acts on $T(X)$ via a finite cyclic group; i.e. there exists a short exact sequence

$$1 \longrightarrow \mathrm{Aut}_s(X) \longrightarrow \mathrm{Aut}(X) \longrightarrow \mu_m \longrightarrow 1, \tag{1.3}$$

where $\mathrm{Aut}_s(X)$ acts trivially and μ_m faithfully on $T(X)$.

Remark 1.13 It is also interesting to consider the kernel of $\mathrm{Aut}(X) \longrightarrow \mathrm{O}(\mathrm{NS}(X))$ which by Proposition 5.3.3 is finite for projective X. Using Proposition 2.1 below, it can be identified, via its action on $T(X)$, with a subgroup $\mu_n \subset \mu_m$. Then $n = m$ if and only if (1.3) splits. This is the case if $\mathrm{NS}(X)$ or, equivalently, $T(X)$ is unimodular. It would be interesting to write an explicit example in the non-unimodular case where $n \ne m$.

If X is not projective, a similar exact sequence can be written for any finite subgroup $G \subset \mathrm{Aut}(X)$, but then automatically $m = 1$ due to Corollary 1.10, (ii).

But also for projective X, the possibilities for m are quite limited; see [450, Thm. 3.1, Cor. 3.2].

Corollary 1.14 *Let X be a complex projective K3 surface. The order of the cyclic group μ_m in (1.3) satisfies*

$$\varphi(m) \le \mathrm{rk}\, T(X) = 22 - \rho(X)$$

and in fact $\varphi(m) \mid \mathrm{rk}\, T(X)$. In particular, $m \le 66$.

Proof Let $f \in \mathrm{Aut}(X)$ act on $T(X)$ by a primitive mth root of unity ζ_m. Then the minimal polynomial Φ_m of ζ_m divides the characteristic polynomial of f^* on $T(X)$. Hence, $\varphi(m) = \deg \Phi_m \le \mathrm{rk}\, T(X)$. In order to prove $\varphi(m) | \mathrm{rk}\, T(X)$, one simply remarks that all irreducible subrepresentations of μ_m on $T(X)_{\mathbb{Q}}$ are of rank $\varphi(m)$. Indeed, otherwise for some $n < m$ there exists an $\alpha \in T(X)$ with $f^{n*}(\alpha) = \alpha$. Then pair α with $H^{2,0}(X)$, on which f^n acts non-trivially, to deduce the contradiction $\alpha \in H^{1,1}(X, \mathbb{Z})$. In other words, as a representation of μ_m one has $T(X) \simeq \mathbb{Z}[\zeta_m]^{\oplus r}$ with $r = \mathrm{rk}\, T(X)/\varphi(m)$. The last assertion follows from $\varphi(m) > 21$ for $m > 66$. $\qquad\qquad\square$

Remark 1.15 In addition, it has been shown by Machida, Oguiso, Xiao, and Zhang in [384, 639, 653] that for a given m a complex K3 surface X together with an automorphism $f \in \mathrm{Aut}(X)$ of order m and such that $f^* = \zeta_m \cdot \mathrm{id}$ on $H^{2,0}(X)$ with ζ_m an m-primitive root of unity exists if and only if $\varphi(m) \le 21$ and $m \ne 60$.

Example 1.16 In [319] Kondō proves more precise results for the case that $\mathrm{NS}(X)$ (or, equivalently, $T(X)$) is unimodular. Then m divides 66, 44, 42, 36, 28, or 12 and actually equals one of them if $\varphi(m) = \mathrm{rk}\, T(X)$. In the latter case, X is uniquely determined. In [627] Vorontsov announced restrictions on m in the case that $T(X)$ is not unimodular. This was worked out by Oguiso and Zhang in [477], who furthermore showed that again X is uniquely determined if $\varphi(m) = \operatorname{rk} T(X)$. So, a complete classification of all complex projective K3 surfaces with $\varphi(m) = \operatorname{rk} T(X)$ exists. Note that in [477] and elsewhere the authors work with $\mu_m := \operatorname{Ker}(\operatorname{Aut}(X) \longrightarrow O(\operatorname{NS}(X)))$.

For an explicit example of an automorphism with $m = 66$, see papers by Kondō and Keum [319], [321, Ex. 4.9], or [292, Ex. 3.1]. The K3 surface is elliptic of Picard number two, more precisely $\operatorname{NS}(X) \simeq U$, and can be described by the Weierstrass equation $y^2 = x^3 + t^{12} - t$.

2 Automorphisms via Periods

Describing $\operatorname{Aut}(X)$ of a complex K3 surface in terms of the Hodge structure $H^2(X, \mathbb{Z})$ is done in two steps. One first shows that the natural representation is faithful, i.e. $\operatorname{Aut}(X) \hookrightarrow O(H^2(X, \mathbb{Z}))$, and then describes the image in terms of the Hodge structure (plus some additional data). As it turns out, up to finite index the group of all Hodge isometries of $H^2(X, \mathbb{Z})$ is the semi-direct product of $\operatorname{Aut}(X)$ and the Weyl group W; cf. Section **8**.2.3. As an application, we review results of Nikulin and Kondō on the classification of complex projective K3 surfaces with finite automorphism group.

2.1 We begin with the faithfulness of the natural representation. The next result is a strengthening of Corollary 1.6; see [54] or [493].

Proposition 2.1 *Let f be an automorphism of a complex K3 surface X. If $f^* = \operatorname{id}$ on $H^2(X, \mathbb{Z})$, then $f = \operatorname{id}$. In other words, the natural action yields an injective map*

$$\operatorname{Aut}(X) \hookrightarrow O(H^2(X, \mathbb{Z})).$$

Proof Let us first give an argument for the case that X is projective. By assumption f fixes one (and in fact every) ample line bundle L, i.e. $f^*L \simeq L$. However, as we have seen, automorphisms of polarized K3 surfaces have finite order (see Proposition **5**.3.3) and hence Corollary 1.6 can be applied.

As mentioned in [54, Exp. IX], any $f \in \operatorname{Aut}(X)$ with $f^* = \operatorname{id}$ deforms sideways, which allows one to reduce to the projective case. If one wants to avoid deforming X, one shows instead that f is of finite order and then applies again Corollary 1.6. To prove $|f| < \infty$, one either uses a general result due to Fujiki [189] and Lieberman [362] saying that the group of components of $\operatorname{Aut}(X)$ of an arbitrary compact Kähler manifold X that fix a Kähler class $\alpha \in H^{1,1}(X, \mathbb{R})$ is finite or the following arguments more specific to K3 surfaces: every Kähler class $\alpha \in \mathcal{K}_X \subset H^{1,1}(X, \mathbb{R})$ is uniquely represented by a Ricci-flat Kähler class ω; see Theorem **9**.4.11. The uniqueness of ω (which is the easy part of the Calabi conjecture) yields $f^*\omega = \omega$. Writing $\omega = g(I, \,)$

with I the complex structure on the underlying differentiable manifold M and viewing f as a diffeomorphism of M with $f^*I = I$ yields $f^*g = g$; i.e. f can be seen as an isometry of the Riemannian manifold (M, g). However, $O(M, g)$ is a compact group; see [61] for references. Thus, the complex Lie group of all $f \in \mathrm{Aut}(X)$ with $f^* = \mathrm{id}$ (or, weaker, $f^*\alpha = \alpha$) is a subgroup of the compact $O(M, g)$. At the same time, it is discrete due to $H^0(X, \mathcal{T}_X) = 0$ and hence finite. The finiteness shows that any f with $f^* = \mathrm{id}$ is of finite order.   □

Remark 2.2   (i) Of course, using the usual compatibility between singular and étale cohomology one can show that for any K3 surface over a field k of char$(k) = 0$ also the natural map

$$\mathrm{Aut}(X) \hookrightarrow O(H^2_{\acute{e}t}(X, \mathbb{Z}_\ell)) \tag{2.1}$$

is injective for any prime ℓ; see [510, Lem. 3.4.1].

(ii) Note that Proposition **5.3.3** is valid in positive characteristic; i.e. any $f \in \mathrm{Aut}(X)$ with $f^*L \simeq L$ for some ample line bundle is of finite order. More precisely, the kernel of

$$\mathrm{Aut}(X) \longrightarrow O(\mathrm{NS}(X)) \tag{2.2}$$

is finite; cf. Remark **5.3.4**. In [479] Ogus shows that (2.2) is in fact injective for supersingular K3 surfaces; cf. Remark **2.5**.

(iii) By lifting to characteristic zero, Ogus also shows that for K3 surfaces over an algebraically closed field k of char$(k) = p > 2$ the natural map

$$\mathrm{Aut}(X) \longrightarrow \mathrm{Aut}(H^2_{\mathrm{cr}}(X/W)) \tag{2.3}$$

injective: see [478, 2. Cor. 2.5] and Section **18.3.2** for the notation. This was later used by Rizov to prove injectivity of (2.1) in characteristic p and for $\ell \neq p$. Of course, it is enough to prove this for the finite subgroup of $f \in \mathrm{Aut}(X)$ with $f^*L \simeq L$ and this is precisely [510, Prop. 3.4.2]. See also Keum's account of it [292, Thm. 0.4].

(iv) Suppose X is a K3 surface over an algebraically closed field k. Then for any field extension K/k base change yields an isomorphism

$$\mathrm{Aut}(X) \simeq \mathrm{Aut}(X \times_k K).$$

This follows from $H^0(X, \mathcal{T}_X) = 0$. A similar statement holds for line bundles, and the argument is spelled out in this case in the proof of Lemma **17.2.2**.

2.2   The next step to understand $\mathrm{Aut}(X)$ completely is to characterize it as a subgroup of $H^2(X, \mathbb{Z})$ purely in terms of the intersection pairing and the Hodge structure on $H^2(X, \mathbb{Z})$. The following is the automorphism part of the global Torelli theorem, so a special case of Theorem **7.5.3**. The injectivity is just Proposition 2.1.

Corollary 2.3 *For a complex K3 surface X the map $f \longmapsto f^*$ induces an isomorphism*

$$\mathrm{Aut}(X) \xrightarrow{\sim} \{g \in \mathrm{O}(H^2(X, \mathbb{Z})) \mid \textit{Hodge isometry with } g(\mathcal{K}_X) \cap \mathcal{K}_X \neq \emptyset\}$$

of $\mathrm{Aut}(X)$ *with the group of all Hodge isometries* $g \colon H^2(X, \mathbb{Z}) \xrightarrow{\sim} H^2(X, \mathbb{Z})$ *for which there exists an ample (or Kähler) class* $\alpha \in H^2(X, \mathbb{Z})$ *with* $g(\alpha)$ *again ample (resp. Kähler).* □

Note that a Hodge isometry g of $H^2(X, \mathbb{Z})$ maps one ample (or Kähler) class to an ample (resp. Kähler) class if and only if it does so for every class. Indeed, a Hodge isometry that preserves the positive cone $\mathcal{C}_X$ also preserves its chamber decomposition. Hence, either a chamber and its image under g are disjoint or they coincide. Also, for projective X this condition is equivalent to saying that g preserves the set of effective classes. See Section **8**.1.2 for a discussion of the ample cone.

Combined with the description of the ample cone in Proposition **8**.5.5, this implies the following result (cf. [577, Prop. 2.2]).

Corollary 2.4 *The group* $\mathrm{Aut}(X)$ *of a complex projective K3 surface X is finitely generated.*

Proof Consider the subgroup $G \subset \mathrm{O}(\mathrm{NS}(X))$ of all $g \in \mathrm{O}(\mathrm{NS}(X))$ preserving the set $\Delta_+ := \{[C] \mid C \simeq \mathbb{P}^1\}$ and such that $g = \mathrm{id}$ on $A_{\mathrm{NS}(X)}$. This is an arithmetic group and thus finitely generated [81]. We show that $\mathrm{Aut}_s(X) \simeq G$, and, as $\mathrm{Aut}_s(X) \subset \mathrm{Aut}(X)$ is of finite index (see Section 1.3), this is enough to conclude that also $\mathrm{Aut}(X)$ is finitely generated.

Any $f \in \mathrm{Aut}_s(X)$ acts as identity on the discriminant $A_{\mathrm{NS}(X)} = \mathrm{NS}(X)^*/\mathrm{NS}(X)$, because this action coincides with the one on $A_{T(X)}$ under the natural isomorphism $A_{\mathrm{NS}(X)} \simeq A_{T(X)}$; see Lemma **14**.2.5. Furthermore, f^* preserves the set Δ_+, because $f^*[C] = [f^{-1}(C)]$.

Conversely, every $g \in G$ can be extended to an isometry of $\tilde{g}$ of $H^2(X, \mathbb{Z})$ that acts as id on $T(X)$; see Proposition **14**.2.6. But then $\tilde{g}$ is a Hodge isometry which, due to Corollary **8**.1.7, preserves the ample cone. Therefore, $\tilde{g} = f^*$ for some $f \in \mathrm{Aut}(X)$ by Corollary 2.3 and in fact $f \in \mathrm{Aut}_s(X)$. In other words, $\mathrm{Aut}_s(X)$ is isomorphic to G and thus finitely generated. □

Remark 2.5 (i) In [369, Thm. 6.1] Lieblich and Maulik verified that the arguments carry over to the case of positive characteristic. So, also for a K3 surface X over an algebraically closed field of positive characteristic $\mathrm{Aut}(X)$ is finitely generated.

However, as there is no direct analogue of the Hodge structure on $H^2(X, \mathbb{Z})$ and as automorphisms can in general not be lifted to characteristic zero, there is no explicit description for $\mathrm{Aut}(X)$. However, it is known that the kernel of $\mathrm{Aut}(X) \longrightarrow \mathrm{O}(\mathrm{NS}(X))$ is finite (see Remark 2.2) and its image has finite index in the subgroup preserving the ample cone (cf. Theorem 2.6 and Remark 2.8). For supersingular K3 surfaces

Ogus proved that the image equals the subgroup of orthogonal transformations that respect not only the ample cone, but also the two-dimensional kernel of $c_1 \colon \mathrm{NS}(X) \otimes k \longrightarrow H^1(X, \Omega_X)$ (which in this sense plays the role of the $(2,0)$-part of the Hodge structure for complex K3 surfaces).

(ii) The result also holds true for non-projective complex K3 surfaces, as has been observed by Oguiso in [473, Thm. 1.5].

2.3 Consider a complex K3 surface X and its Weyl group $W \subset O(\mathrm{NS}(X))$, i.e. the subgroup generated by all reflections $s_{[C]}$ with $\mathbb{P}^1 \simeq C \subset X$; cf. Sections **14**.2.1 and **8**.2.4. The following result was first stated by Pjateckiĭ-Šapiro and Šafarevič in [493, Sec. 7]. A sketch of the argument can be found in the proof of Theorem **8**.4.2. It relies heavily on Corollary 2.3.

Theorem 2.6 *Let X be a complex projective K3 surface. Then the natural map sending f to f^* induces a homomorphism*

$$\mathrm{Aut}(X) \longrightarrow O(\mathrm{NS}(X))/W$$

with finite kernel and finite cokernel.

Another way to phrase this is to say that

$$\mathrm{Aut}_s(X) \ltimes W \subset O(\mathrm{NS}(X))$$

is a finite index subgroup. As an immediate consequence one finds the following (see Corollary **8**.4.7).

Corollary 2.7 *The group of automorphisms $\mathrm{Aut}(X)$ of a complex projective K3 surface is finite if and only if $O(\mathrm{NS}(X))/W$ is finite.* □

Remark 2.8 Both Theorem 2.6 and Corollary 2.7 hold for projective K3 surfaces over algebraically closed fields of positive characteristic; see [369].

For non-projective K3 surfaces, however, the situation is different. As was mentioned before, the quotient of $\mathrm{Aut}_s(X) \subset \mathrm{Aut}(X)$ may contain elements of infinite order. So, a priori, it could happen that the kernel of $\mathrm{Aut}(X) \longrightarrow O(\mathrm{NS}(X))/W$ contains elements of infinite order, just because $\mathrm{NS}(X)$ is too small, e.g. $\mathrm{NS}(X) = 0$. Explicit examples can presumably be found among those mentioned in Example 1.11.

2.4 It turns out that complex projective K3 surfaces with finite $\mathrm{Aut}(X)$ can be classified. More precisely, their Picard lattices and the groups occurring as $\mathrm{Aut}(X)$ can be (more or less) explicitly described. Due to Corollary 2.7, the question of whether $\mathrm{Aut}(X)$ is finite becomes a question about the lattice $\mathrm{NS}(X)$ and its Weyl group $W \subset O(\mathrm{NS}(X))$.

In the following, we let $W \subset O(N)$ be the Weyl group of a lattice N of signature $(1, \rho - 1)$; see Section **8**.2.3.

Definition 2.9 By $\mathcal{F}^\rho$ one denotes the set of isomorphism classes of even lattices N of signature $(1, \rho - 1)$ such that $O(N)/W$ is finite.

The main result is the following theorem due to Nikulin. For the statement and its proof, see the relevant articles [452] and [453] by Nikulin. The second part of the result is [453, Thm. 10.1.1].

Theorem 2.10 *The set $\mathcal{F}^\rho$ is empty for $\rho \geq 20$ and non-empty but finite for $3 \leq \rho \leq 19$. Every $N \in \mathcal{F}^\rho$ can be realized as $N \simeq \mathrm{NS}(X)$ of some K3 surface X.*

Example 2.11 The cases $\rho(X) = 1, 2$ are rather easy, and, in particular, the list of possible lattices $\mathrm{NS}(X)$ is infinite in both cases.

(i) If $\rho(X) = 1$, then the Weyl group is trivial and $O(\mathrm{NS}(X)) = \{\pm 1\}$. Hence, $\mathrm{Aut}(X)$ is finite by Corollary 2.7; see also the more precise Corollary 2.12. Note that in particular $\mathbb{Z}(2d) \in \mathcal{F}^1$ for all $d > 0$.

(ii) If $\rho(X) = 2$, then according to [493, Sec. 7] (see also [198, Cor. 1]), $\mathrm{Aut}(X)$ is finite if and only if there exists $0 \neq \alpha \in \mathrm{NS}(X)$ with $(\alpha)^2 = 0$ or $(\alpha)^2 = -2$; see the discussion in Sections **14**.2, **8**.3.2, and Example **8**.4.9.

In [454] Nikulin carried out the classification for $\rho(X) = 3$, in particular $|\mathcal{F}^3| = 26$; see also [457]. The case $\rho(X) = 4$ is due to Vinberg and was published only many years later in [619], where it is also shown that $|\mathcal{F}^4| = 14$.

The case $\rho(X) \geq 5$ is treated by Nikulin in [453]; see also the announcement in his earlier paper [452]. The most explicit form of this result can be found in [457, Thm. 1, 2];[3] see also [321, Thm. 6.2]. From the explicit list given there one sees immediately that $\mathcal{F}^\rho$ is non-empty for all $1 \leq \rho \leq 19$.

Note that the theorem in particular says that complex K3 surfaces with $\rho(X) = 20$ have infinite $\mathrm{Aut}(X)$, which was first shown by Shioda and Inose in [568, Thm. 5]. In fact, a K3 surface X with $\rho(X) = 20$ is a rational double cover of a Kummer surface Y associated with the product of two elliptic curves $E_1 \times E_2$ (see Example **11**.1.2 and Remark **14**.3.22) and often X itself is of this form. For Y, translation by a non-torsion section of $Y \longrightarrow \mathbb{P}^1 = E_1/\pm$ is an automorphism of infinite order. The existence of a non-torsion section follows from the Shioda–Tate formula; see Example **11**.3.5. For the description of $\mathrm{Aut}(X)$ for X with $\rho(X) = 20$ and small discriminant, see the papers by Borcherds and Vinberg [77, 618].

Similarly, for (Shioda) supersingular (or, equivalently, unirational) K3 surfaces, i.e. K3 surfaces with $\rho(X) = 22$ (see Section **18**.3.5), the group $\mathrm{Aut}(X)$ is as well infinite; see [276].

---

[3] Thanks to Jürgen Hausen for this reference.

The following was first observed in [453, Cor. 10.1.3]; see also [386, Lem. 3.7]. Note that there is no analogue for this in positive characteristic or, at least, not over $\overline{\mathbb{F}}_p$ as there the Picard number is always even (see Corollary **17**.2.9).

Corollary 2.12   *Let X be a complex projective K3 surface with* $\mathrm{Pic}(X) \simeq \mathbb{Z} \cdot H$. *Then*

$$\mathrm{Aut}(X) = \begin{cases} \{\mathrm{id}\} & \text{if } (H)^2 > 2, \\ \mathbb{Z}/2\mathbb{Z} & \text{if } (H)^2 = 2. \end{cases} \tag{2.4}$$

Proof   By Corollary **3**.3.5, any $f \in \mathrm{Aut}(X)$ acts as $\pm\mathrm{id}$ on $T(X)$. On the other hand, $f^* = \mathrm{id}$ on $\mathrm{NS}(X)$, for the pull-back of the ample generator H has to be ample. But the action of f^* on the discriminant groups $A_{T(X)}$ and on $A_{\mathrm{NS}(X)} \simeq \mathbb{Z}/(H)^2\mathbb{Z}$ coincide under the natural isomorphism; cf. Lemma **14**.2.5. For $(H)^2 \neq 2$ this excludes $f^* \neq \mathrm{id}$ on $T(X)$. Hence, $f^* = \mathrm{id}$ on $H^2(X, \mathbb{Z})$ and Proposition 2.1 yields the result.

If $(H)^2 = 2$, then X is a double plane (see Remark **2**.2.4) and the covering involution $i \colon X \xrightarrow{\sim} X$ indeed acts as $-\mathrm{id}$ on $T(X)$. For any other automorphism with $f^* = -\mathrm{id}$, the above argument can be applied to the composition $i \circ f$, which shows $f = i$.   □

An explicit description of all the possible finite $\mathrm{Aut}(X)$ was eventually given by Kondō in [320, Sec. 4]; see also [318]. The final result should be read as saying that K3 surfaces with finite $\mathrm{Aut}(X)$ are rather special. In particular, all the intriguing finite groups $G \subset \mathrm{Aut}_s(X)$ classified by Mukai (see Section 3) can be realized only on K3 surfaces X with infinite $\mathrm{Aut}(X)$.

Theorem 2.13   *Suppose $\mathrm{Aut}(X)$ of a complex projective K3 surface is finite. Then the symplectic automorphism group $\mathrm{Aut}_s(X) \subset \mathrm{Aut}(X)$ is isomorphic to one of the following groups:*

$$\{1\}, \quad \mathbb{Z}/2\mathbb{Z}, \quad or \quad \mathfrak{S}_3.$$

The two cases $15 \leq \rho(X)$ and $9 \leq \rho(X) \leq 14$ are treated separately, where the lower bound for $\rho(X)$ follows from the table in Section 1.2. For the first case one has $\mathrm{Aut}_s(X) \simeq \{1\}, \mathbb{Z}/2\mathbb{Z}$, or $\mathfrak{S}_3$, whereas for the second $\mathrm{Aut}_s(X) \simeq \{1\}$ or $\mathbb{Z}/2\mathbb{Z}$. The result as phrased by Kondō in [320] distinguishes instead between the two cases that the index m of $\mathrm{Aut}_s(X) \subset \mathrm{Aut}(X)$ satisfies $m \leq 2$, in which case $\mathrm{Aut}(X) \simeq \mathrm{Aut}_s(X) \times \mathbb{Z}/m\mathbb{Z}$, or $m > 2$. In the course of the proof Kondō in particular notices that for K3 surfaces X with finite $\mathrm{Aut}(X)$ the lattice $\mathrm{NS}(X)$ determines the group $\mathrm{Aut}(X)$.

Example 2.14   There are many explicit examples of K3 surfaces with finite automorphism groups in the literature. Galluzzi and Lombardo show in [197] that $\mathrm{Aut}(X)$ for $\mathrm{NS}(X)$ with intersection matrix

$$\begin{pmatrix} 2 & d \\ d & -2 \end{pmatrix}$$

with $d \equiv 1 \ (2)$ is isomorphic to $\mathbb{Z}/2\mathbb{Z}$.

2.5 We conclude this section by reviewing some examples of K3 surfaces with infinite automorphism groups.

(i) In [632] Wehler considers a K3 surface X given as a complete intersection of the Fano variety of lines $\mathbb{F} \subset \mathbb{P}^2 \times \mathbb{P}^{2*}$ on $\mathbb{P}^2$ with a hypersurface of type $(2, 2)$. K3 surfaces of this form come in an 18-dimensional family and for the general member NS(X) is of rank two with intersection matrix

$$\begin{pmatrix} 2 & 4 \\ 4 & 2 \end{pmatrix}$$

and Aut$(X) = (\mathbb{Z}/2\mathbb{Z}) * (\mathbb{Z}/2\mathbb{Z})$.

The two generators correspond to the covering involutions σ_1, σ_2 of the projections $X \longrightarrow \mathbb{P}^2$ to the two factors. In particular, σ_1, σ_2 are not symplectic, but $\sigma_1 \circ \sigma_2$ is symplectic (and of infinite order). Compare this example with Corollary 2.12.

(ii) In [197, Thm. 4] the result of Wehler is complemented by showing that Aut$(X) \simeq (\mathbb{Z}/2\mathbb{Z}) * (\mathbb{Z}/2\mathbb{Z})$ for any K3 surface with NS(X) of rank two and intersection matrix

$$\begin{pmatrix} 2 & d \\ d & 2 \end{pmatrix}$$

with $d > 1$ odd. For a K3 surface with intersection form

$$\begin{pmatrix} 2 & d \\ d & -2 \end{pmatrix}$$

on NS(X) and d odd it is shown that Aut$(X) \simeq \mathbb{Z}/2\mathbb{Z}$; see [197, Thm. 3].

(iii) According to Bini [62], any K3 surface with NS$(X) \simeq \mathbb{Z}(2nd) \oplus \mathbb{Z}(-2n)$ with $n \geq 2$ and d not a square satisfies Aut$(X) \simeq \mathbb{Z}$.

(iv) A systematic investigation of the case $\rho(X) = 2$ was undertaken by Galluzzi, Lombardo, and Peters in [198, Cor. 1]. In particular it is proved that the only infinite Aut(X) that can occur are $\mathbb{Z}$ and $(\mathbb{Z}/2\mathbb{Z}) * (\mathbb{Z}/2\mathbb{Z})$.

(v) In papers by Festi et al. and Oguiso [178, 474] one finds an automorphism f of a certain quartic $X \subset \mathbb{P}^3$ with $\rho(X) = 2$ which is of infinite order and without fixed points. By Corollary 1.10, some finite power of it is also symplectic. Is f itself symplectic?

(vi) As mentioned before, Aut(X) is infinite if $\rho(X) = 20$; see [568, Thm. 5]. In this paper, Shioda and Inose also noted that there exist K3 surfaces with $\rho(X) = 18$ and finite Aut(X). An earlier example can be found in [493, Sec. 7].

In view of Corollary **3.3.5** it seems reasonable to expect that in general automorphism groups of K3 surfaces with odd Picard number should be easier to study, at least the kernel of Aut$(X) \longrightarrow O(NS(X))$ is at most $\{\pm 1\}$. See Shimada [555] for examples with $\rho(X) = 3$.

3 Finite Groups of Symplectic Automorphisms

The goal of this section is to convey an idea of the celebrated result by Mukai [430] concerning finite groups realized by symplectic automorphisms of K3 surfaces which generalizes earlier results of Nikulin [450] for finite abelian groups. We first state the result and discuss some of its consequences, and later provide the background for it and present the main ingredients of its proof.

As was noted before, the list of finite groups occurring as $\mathrm{Aut}(X)$ or $\mathrm{Aut}_s(X)$ is rather short; cf. Theorem 2.13. In particular, most of the interesting finite groups occurring as subgroups of $\mathrm{Aut}_s(X)$ are subgroups of an infinite $\mathrm{Aut}_s(X)$.

Theorem 3.1 (Mukai) *For a finite group G the following conditions are equivalent:*

  (i) *There exists a complex (projective) K3 surface X such that G is isomorphic to a subgroup of* $\mathrm{Aut}_s(X)$.
 (ii) *There exists an injection* $G \hookrightarrow M_{23}$ *into the Mathieu group* M_{23} *such that the induced action of G on* $\Omega := \{1, \ldots, 24\}$ *has at least five orbits.*

There are exactly 11 maximal subgroups of finite groups acting faithfully and symplectically on a complex (projective) K3 surface, i.e. that satisfy (i) or, equivalently, (ii). An explicit list can be found in [430, Ex. 0.4] or [391, Sec. 4]. The orders of these maximal groups are

$$|G| = 48, 72, 120, 168, 192, 288, 360, 384, 960 \tag{3.1}$$

(some appearing twice). They can all be realized on explicitly described K3 surfaces (namely on quartics, complete intersections, and double covers) with the group action given on an ambient projective space. The existence can also be proved via the global Torelli theorem and the surjectivity of the period map; see Section 3.3 for comments. An explicit list of all 79 non-trivial possible finite groups (without making the link to M_{23}) was given by Xiao in [640, Sec. 2]; see also Hashimoto [238, Sec. 10.2].[4]

Example 3.2   The complex Fermat quartic has Picard number $\rho(X) = 20$ (cf. Section 3.2.6) and thus $\mathrm{Aut}(X)$ and $\mathrm{Aut}_s(X)$ are both infinite due to Theorem 2.10 or using the Shioda–Inose structure [568, Thm. 5]. However, $\mathrm{Aut}_s(X)$ contains the maximal finite subgroup $(\mathbb{Z}/4\mathbb{Z})^2 \rtimes \mathfrak{S}_4$ of order 384. Here, $(a, b, c, d) \in (\mathbb{Z}/4\mathbb{Z})^4$ acts by $[x_0 : x_1 : x_2 : x_3] \mapsto [\zeta^a x_0 : \zeta^b x_1 : \zeta^c x_2 : \zeta^d x_3]$, where ζ is a primitive fourth root of unity. Of course, this is effectively only an action of $(\mathbb{Z}/4\mathbb{Z})^3$ and imposing further $a + b + c + d = 0$ yields a symplectic action of $(\mathbb{Z}/4\mathbb{Z})^2$. The factor $\mathfrak{S}_4$ acts by permutation of the coordinates. See [430, Ex. 0.4] for more details.

---

[4] The list has in fact 81 entries as it also records the discriminant of the invariant part of its action which is not unique in exactly two cases; cf. [238, Prop. 3.8] and Remark 3.14.

In [472] Oguiso shows that this large finite group essentially characterizes the Fermat quartic. For a discussion of the group of automorphisms of the polarized $(X, \mathcal{O}(1))$, see [325] and the references therein.

Remark 3.3   Using the analogue $0 \longrightarrow G_0 \longrightarrow G \longrightarrow \mu_m \longrightarrow 0$ of (1.3) for an arbitrary finite subgroup $G \subset \text{Aut}(X)$ of a complex projective K3 surface X one finds $|G| = |G_0| \cdot m$. Corollary 1.14 together with $|G_0| \leq 960$ from (3.1) yields the a priori bound $|G| \leq 960 \cdot 66$. However, in [640] Xiao shows the stronger inequality $|G| \leq 5760$ by combining $\varphi(m)|(22 - \rho(Y))$ for the minimal resolution $Y \longrightarrow X/G_0$ (see Proposition 3.11) with $\rho(Y) \geq \text{rk}\, L_{G_0} + 1$ (see (3.6) below).[5] In [323] Kondō improved this to

$$|G| \leq 3840$$

by excluding the case $|G| = 5760$. Moreover, he showed that a certain extension G of $\mathbb{Z}/4\mathbb{Z}$ by the group M_{20}, which satisfies $|G| = 3840$, acts on the Kummer surfaces associated with $(\mathbb{C}/(\mathbb{Z} + i\mathbb{Z}))^2$.

Remark 3.4   In positive characteristic the situation is completely different. For example, Kondō in [324] shows that any subgroup $G \subset M_{23}$ acting on Ω with three orbits can be realized as a finite group of symplectic automorphisms of a (supersingular) K3 surface (with Artin invariant one and over an appropriate prime p). Examples include M_{11} and M_{22}. Moreover, not every finite group of symplectic automorphisms can be realized as a subgroup of M_{23}.

For the proof of Theorem 3.1, we shall follow Kondō's approach in [322], which is more lattice theoretic than Mukai's original proof. See also Mason [391] for a detailed and somewhat simplified account of the latter. Only the main steps are sketched and in the final argument we restrict to the discussion of only one out of the possible 23 Niemeier lattices; see Section **14**.4.

3.1   Let us begin with the necessary background on Mathieu groups and Niemeier lattices; see also Section **14**.4.

The *Mathieu groups* M_{11}, M_{12}, M_{22}, M_{23}, and M_{24} are the 'first generation of the happy family' of finite simple sporadic groups. They were discovered by Mathieu in 1861 in [392], who did most of his work in mathematical physics.[6] Only M_{23} and M_{24} are (so far) relevant for K3 surfaces.

---

[5] I am sure it must be a purely numerical coincidence that 5760 also comes up in the second denominator of $\sqrt{\text{td}} = 1 + \frac{1}{24}c_2 + \frac{1}{5760}(7c_2^2 - 4c_4) + \cdots$ of a hyperkähler manifold.

[6] His obituary [159] contains timeless remarks on mathematics: 'He was the champion of a science that was out of fashion.'

One way to define M_{24}, which is a simple group of order

$$|M_{24}| = 244.823.040 = 2^{10} \cdot 3^3 \cdot 5 \cdot 7 \cdot 11 \cdot 23,$$

is to start with the (extended binary) *Golay code* $W \subset \mathbb{F}_2^{24}$. By definition, the Golay code is a 12-dimensional linear subspace with the property that $|\{i \mid w_i \neq 0\}| \geq 8$ for all $0 \neq w = (w_i) \in W$. The subspace W can be written explicitly. It it is unique up to linear coordinate change; see e.g. [132].

The symmetric group $\mathfrak{S}_{24}$ with its action on $\Omega := \{1, \ldots, 24\}$ can be viewed as subgroup of $\mathrm{GL}(\mathbb{F}_2^{24})$ by permuting the vectors $e_1, \ldots, e_{24}$ of the standard basis. Then one defines

$$M_{24} := \{\sigma \in \mathfrak{S}_{24} \mid \sigma(W) = W\}.$$

It is known that M_{24} still acts transitively on Ω. In fact, it acts 5-transitively on Ω, i.e. for two ordered tuples $(i_1, \ldots, i_5)$ and $(j_1, \ldots, j_5)$ of distinct numbers $i_k, j_k \in \Omega$ there exists an $\sigma \in M_{24}$ with $\sigma(i_k) = j_k, k = 1, \ldots, 5$.

Remark 3.5 Alternatively, M_{24} can be introduced as the automorphism group of the *Steiner system* $S(5, 8, 24)$. More precisely, a Steiner system $S(5, 8, 24)$ is a subset of $\mathcal{P}(\Omega)$ consisting of subsets $M \subset \Omega$ (the blocks) with $|M| = 8$ and such that any $N \subset \Omega$ with $|N| = 5$ is contained in exactly one $M \in S(5, 8, 24)$. Up to the action of $\mathfrak{S}_{24}$ on $\mathcal{P}(\Omega)$, the Steiner system $S(5, 8, 24)$ is unique and $|S(5, 8, 24)| = 759$. Then

$$M_{24} = \{\sigma \in \mathfrak{S}_{24} \mid \sigma(S(5, 8, 24)) = S(5, 8, 24)\}.$$

All Mathieu groups can be described in terms of Steiner systems. See [132, Chs. 3 and 10].

The Mathieu group M_{23} is now defined as the stabilizer of one element in Ω, say e_{24}:

$$M_{23} := \mathrm{Stab}(e_{24}) \subset M_{24}.$$

Then M_{23} is a simple group of index 24 in M_{24} and thus of order

$$|M_{23}| := 10.200.960 = 2^7 \cdot 3^2 \cdot 5 \cdot 7 \cdot 11 \cdot 23.$$

Clearly, M_{23} acts 4-transitively on the remaining $\{1, \ldots, 23\}$. Observe that apart from the prime factors 11 and 23 only prime factors $p < 8$ occur which are the only prime orders of symplectic automorphisms of a complex K3 surface; see Section 1.1.[7]

The following observation, which may have triggered Mukai's results in [430], is a first sign of the intriguing relation between symplectic automorphisms of K3 surfaces and groups contained in the sporadic group M_{23}. Consider the permutation action of

---

[7] Curiously, 11 comes up as the order of an automorphism of a K3 surface in characteristic 11, but 23 does not, according to Keum [292].

$M_{23} \subset M_{24} \subset \mathfrak{S}_{24}$ on $\mathbb{Q}^{24}$ (instead of $\mathbb{F}_2^{24}$) and its character. Then one can compute that for elements $\sigma \in M_{23}$ of order up to eight the trace $\chi(\sigma) := \mathrm{tr}(\sigma) = |\Omega^\sigma|$ is given by (1.1).

The relation between the Mathieu group M_{23} and Niemeier lattices can be best exemplified by a result already stated in Section **14**.4.3: The orthogonal group $O(N)$ of the Niemeier lattices N with root lattice $R = A_1(-1)^{\oplus 24}$ is naturally isomorphic to

$$O(N) \simeq M_{24} \ltimes (\mathbb{Z}/2\mathbb{Z})^{\oplus 24}. \tag{3.2}$$

For all Niemeier lattices N the quotients $O(N)/W(N)$ by the corresponding Weyl group is known. For example, if the twist $R(-1)$ of the root lattice R of a Niemeier lattice N is not one of the following $A_1^{\oplus 24}, A_2^{\oplus 12}, A_3^{\oplus 8}, A_8^{\oplus 4}, D_4^{\oplus 6}, A_5^{\oplus 4} \oplus D_4, E_6^{\oplus 4}, A_6^{\oplus 4}, A_8^{\oplus 3}$, then $O(N)/W(N)$ is of order at most eight or isomorphic to $\mathfrak{S}_4$ and hence contained in M_{23}. See [322, Sec. 3] for details and references.

3.2 Let now X be a complex K3 surface. Then the same arguments used to prove Corollary 1.8 also show the following lemma.

Lemma 3.6 *For the invariant part $H^*(X, \mathbb{C})^G$ of a finite group $G \subset \mathrm{Aut}_s(X)$ one has* $\dim H^*(X, \mathbb{C})^G \geq 5$ *and*

$$\frac{1}{|G|} \sum_{f \in G} |\mathrm{Fix}(f)| = \dim H^*(X, \mathbb{C})^G,$$

where by convention $|\mathrm{Fix}(\mathrm{id})| = 24$.

Proof Clearly, $(H^0 \oplus H^4)(X)$ is invariant under G and, since G is symplectic, also $(H^{2,0} \oplus H^{0,2})(X)$ is. As G is finite, the sum $\tilde{\alpha} = \sum_{f \in G} f^* \alpha$ is well defined and non-trivial for any Kähler (or ample) class α. This shows the first assertion. For the second consider the linear projector $\alpha \mapsto (1/|G|) \sum f^* \alpha$ onto $H^*(X)^G$ to show

$$|G| \cdot \dim H^*(X, \mathbb{C})^G = \sum_{f \in G} \mathrm{tr}(f^*|_{H^*(X, \mathbb{C})}) = \sum_{f \in G} |\mathrm{Fix}(f)|$$

by taking traces and applying the Lefschetz fixed point formula. $\qquad\square$

The result is again not optimal, e.g. for X projective $\dim H^*(X, \mathbb{C})^G \geq 25 - \rho(X)$, as G acts trivially on $T(X)$. The following lemma collects further elementary observations; cf. [450].

Lemma 3.7 *For a complex K3 surface X and a finite subgroup $G \subset \mathrm{Aut}_s(X)$, consider the orthogonal complement*

$$L_G := (H^2(X, \mathbb{Z})^G)^\perp.$$

(i) *Then L_G is negative definite and without (-2)-classes. Moreover, $\mathrm{rk}\, L_G \leq 19$.*

(ii) *The group G acts trivially on the discriminant group A_{L_G} of L_G.*

(iii) *The minimal number of generators of L_G satisfies $\ell(L_G) \leq 22 - \mathrm{rk}L_G$.*

Proof As $G \subset \mathrm{Aut}_s(X)$, one has $T(X) \subset H^2(X, \mathbb{Z})^G$ and hence $L_G \subset T(X)^\perp \subset$ $\mathrm{NS}(X)$; cf. Section 3.3.1. If X is projective, one finds an invariant ample class $\alpha \in$ $\mathrm{NS}(X)^G$ and, by Hodge index and using $L_G \subset \alpha^\perp \subset \mathrm{NS}(X)$, the lattice L_G is negative definite of rank ≤ 19. If X is not projective, one still finds an invariant Kähler class $\alpha \in \mathcal{K}_X$ and, as $(H^{2,0} \oplus H^{0,2})(X)_\mathbb{R} \oplus \mathbb{R}\alpha$ is positive definite, L_G is contained in the negative definite orthogonal complement of it. Once more, L_G is negative definite of rank ≤ 19.

The lattice L_G does not contain any effective classes, because those would intersect positively with the invariant ample or Kähler class α. Since for a (-2)-class $\delta \in \mathrm{NS}(X)$ either δ or $-\delta$ is effective, this proves that L_G does not contain any (-2)-classes.

As now L_G is negative definite, $H^2(X, \mathbb{Z})^G$ is non-degenerate. Indeed, if $0 \neq$ $x \in H^2(X, \mathbb{Z})^G$ with $(x.y) = 0$ for all $y \in H^2(X, \mathbb{Z})^G$, then $x \in L_G$ and hence $(x)^2 < 0$. Thus, Proposition 14.0.2 can be applied and shows that there exists a natural isomorphism

$$A_{L_G} \simeq A_{H^2(X,\mathbb{Z})^G}, \tag{3.3}$$

which right away shows

$$\ell(A_{L_G}) = \ell(A_{H^2(X,\mathbb{Z})^G}) \leq \mathrm{rk}\, H^2(X, \mathbb{Z})^G = 22 - \mathrm{rk}\, L_G.$$

Moreover, by Lemma 14.2.5, (3.3) is compatible with the action of $O(H^*(X, \mathbb{Z}))$. As G acts trivially on $H^*(X, \mathbb{Z})^G$, it also acts trivially on A_{L_G}. □

Corollary 3.8 *There exists a primitive embedding*

$$L_G \oplus A_1(-1) \hookrightarrow N \tag{3.4}$$

into a Niemeier lattice N. Moreover, the action of G on L_G extends by id on the orthogonal complement of (3.4) to an action on N.

Proof Apply Corollary 14.4.5 to $\Lambda_1 := L_G \oplus A_1(-1)$. The last lemma shows

$$\ell(\Lambda_1) = \ell(L_G) + 1 \leq 24 - \mathrm{rk}(L_G \oplus A_1(-1)). \tag{3.5}$$

So literally Corollary 14.4.5 applies only if for some reason $\ell(H^2(X, \mathbb{Z})^G) <$ $\mathrm{rk}\, H^2(X, \mathbb{Z})^G$, but the finer version mentioned in Remark 14.1.17 always does, as due to the factor $A_1(-1)$ the local conditions at odd primes p as well as at $p = 2$ are trivially satisfied. This yields a primitive embedding $L_G \oplus A_1(-1) \hookrightarrow N$.

As the action of G on L_G is trivial on A_{L_G}, it can be extended as desired due to Proposition 14.2.6. □

SKETCH OF PROOF OF THEOREM 3.1. We prove that (i) implies (ii). For the other direction, see Section 3.3. The Niemeier lattice N in Corollary 3.8 is not unique, and a priori all Niemeier lattices with the exception of the Leech lattice can occur. To give an idea of the proof of Theorem 3.1 we pretend that N is the Niemeier lattice with root lattice $R = A_1(-1)^{\oplus 24}$; see Section **14.4.3**.

Then $O(N) \simeq M_{24} \ltimes (\mathbb{Z}/2\mathbb{Z})^{\oplus 24}$ (see (3.2)), which in particular yields an inclusion $G \hookrightarrow M_{24} \ltimes (\mathbb{Z}/2\mathbb{Z})^{\oplus 24}$. Suppose there exists an element $\sigma \in G$ with $\sigma(e_i) = -e_i$ for some i, then $e_i \in (N^G)^{\perp} = L_G$. But L_G does not contain any (-2)-class by Lemma 3.7. Hence, $G \hookrightarrow M_{24}$. Clearly, the root that corresponds to $A_1(-1)$ in the direct sum $L_G \oplus A_1(-1)$, which we shall call e_{24}, is fixed by the action of G. Thus, $G \hookrightarrow M_{23} = \mathrm{Stab}(e_{24})$. In order to conclude it remains to show that G has at least five orbits. For this use $\mathrm{rk}\, N^G \geq 5$, so that one can choose (the beginning of) a basis of $N_{\mathbb{Q}}^G$ of the form $e_1 + \sum_{i \geq 5} a_{1,i} e_i, \ldots, e_5 + \sum_{i \geq 5} a_{5,i} e_i$ (after possibly permuting the e_i). Then for $i \neq j \in \{1, \ldots, 5\}$ the orbits of e_i and e_j are disjoint.

In the case that the Niemeier lattice in Corollary 3.8 is not the one with root lattice $A_1(-1)^{\oplus 24}$ Kondō finds similar arguments. The important input is an explicit description of $O(N)$ in all 23 cases. The Leech lattice is excluded by the root in $A_1(-1)$ and, in fact, one quickly reduces to nine of the Niemeier lattices, as the other ones have very small automorphism group, so that the assertion becomes trivial. □

Remark 3.9 For generalizations of the result to finite groups of autoequivalences of $D^b(X)$ (see [261]), it is worth pointing out that embedding $L_G \oplus A_1(-1)$, and not merely L_G, into some Niemeier lattice N is crucial for three reasons.

First and on a purely technical level, the factor $A_1(-1)$ ensured that the local conditions in Nikulin's criterion hold when equality holds in (3.5), so that also in this case an embedding (3.4) can be found.

Second, since G acts trivially on $A_1(-1)$ it ensures that not only $G \hookrightarrow M_{24}$ but indeed $G \hookrightarrow M_{23}$ exists.

Third, and maybe most important, from the extra $A_1(-1)$ one deduces that the Niemeier lattice cannot be the Leech lattice N_0 which does not contain any (-2)-class. If one allowed the Leech lattice at this point, one would get an embedding $G \hookrightarrow Co_0 = O(N_0)$ into the Conway group Co_0, which is a group of order $|Co_0| = 2^{22} \cdot 3^9 \cdot 5^4 \cdot 7^2 \cdot 11 \cdot 13 \cdot 23$ and thus much larger than M_{24}.

For the generalization to finite groups of symplectic derived equivalences (see the comments at the end of Chapter 16), it is useful to phrase the above discussion and in particular Theorem 3.1 in the more general situation of a finite subgroup $G \subset O(\widetilde{H}(X, \mathbb{Z}))$ with invariant part $\widetilde{H}(X, \mathbb{Z})^G$ containing four positive directions (or, equivalently, with negative definite orthogonal complement $\widetilde{L}_G$) and without (-2)-classes in $\widetilde{L}_G$. In the case that $G \subset \mathrm{Aut}_s(X)$, the invariant part furthermore contains a hyperbolic plane U (namely $(H^0 \oplus H^4)(X, \mathbb{Z})$), which eventually ensures the embedding into M_{23}.

3.3 As mentioned before and as claimed by Theorem 3.1, all subgroups $G \subset M_{23}$ acting with at least five orbits on Ω indeed occur as groups of symplectic automorphisms. In his original paper [430], Mukai gives an explicit construction for each of the maximal groups and in the appendix to Kondō's paper [322] he describes a more abstract argument that relies on the surjectivity of the period map. Here is a sketch of the latter.

In a first step, one writes down the 11 maximal groups $G \subset M_{23}$. For each of them the action on Ω has exactly five orbits and those can be described explicitly. Then one considers the natural action of G on the Niemeier lattice N with root lattice $A_1(-1)^{\oplus 24}$. The invariant part N^G is of rank five, and, therefore, its orthogonal complement N_G is a negative definite lattice with rk $N_G = 19$.

Next, and this is where most of the work is, one has to analyze the discriminant form $(A, q) := (A_{N_G}, -q_{N_G})$ to show that Theorem **14**.1.5 can be applied to $(n_+, n_-) = (3, 0)$ and (A, q), which thus yields an even positive definite lattice Λ_1 with rk $\Lambda_1 = 3$ and $(A_{\Lambda_1}, q_{\Lambda_1}) \simeq (A, q)$.

Due to Proposition **14**.0.2, the two lattices N_G and Λ_1 are orthogonal to each other inside an even, unimodular lattice Λ, which then has signature $(3, 19)$ and is, therefore, isomorphic to the K3 lattice. Hence, there exists a finite index embedding

$$N_G \oplus \Lambda_1 \subset E_8(-1)^{\oplus 2} \oplus U^{\oplus 3}$$

with N_G and Λ_1 primitive.

The surjectivity of the period map (see Theorem 7.4.1) can be used to show that every Hodge structure of weight two on Λ_1 can be realized as the Hodge structure of a K3 surface. More precisely, for any $0 \neq \alpha \in \Lambda_{1\mathbb{R}}$ there exists a K3 surface X with an isometry $H^2(X, \mathbb{Z}) \simeq E_8(-1)^{\oplus 2} \oplus U^{\oplus 3}$ such that $(H^{2,0} \oplus H^{0,2})(X)_{\mathbb{R}} \simeq \alpha^\perp \subset \Lambda_{1\mathbb{R}}$. If the line $\mathbb{R} \cdot \alpha$ is not rational, then $NS(X) \simeq N_G$ and hence X is a non-projective K3 surface of Picard number $\rho(X) = 19$. In this case, α (up to sign) is a Kähler class. Otherwise, X is projective with $\rho(X) = 20$, but one might have to apply elements of the Weyl group to make sure that α is Kähler, i.e. ample.

The action of G on N_G induces the trivial action on A_{N_G} and can, therefore, be extended by id_{Λ_1} to an action of G on $H^2(X, \mathbb{Z})$; cf. Proposition **14**.2.6. Thus, G is a group of Hodge isometries. For $\rho(X) = 19$ the action of G on $H^2(X, \mathbb{Z})$ leaves invariant $(H^{2,0} \oplus H^{0,2})(X)$ and the Kähler class α. So the global Torelli theorem 7.5.3 or rather Corollary 2.3 applies and G can be interpreted as a subgroup of $\text{Aut}_s(X)$.

In order to get an action of G on a projective K3 surface, one argues that any automorphism of a non-projective X that leaves invariant the Kähler class and acts as the identity on $(H^{2,0} \oplus H^{0,2})(X)$ is an automorphism of each of the fibres of the twistor family; see Section 7.3.2. This takes care of the Weyl group action mentioned before for the case that $\mathbb{R} \cdot \alpha$ is not rational.

Remark 3.10 (i) It is curious to observe that for the existence result only the Niemeier lattice N with root lattice $A_1(-1)^{\oplus 24}$ is involved, whereas in Corollary 3.8 a priori every Niemeier lattice apart from the Leech lattice can occur.

So, every L_G can in fact be embedded into the particular Niemeier lattice N_1 with root lattice $A_1(-1)^{\oplus 24}$, but in order to embed $L_G \oplus A_1(-1)$ others are needed. In the derived setting one rather uses embeddings into the Leech lattice; see Remark 3.9.

(ii) The proof also shows that any of the finite groups occurring in Theorem 3.1 can in fact be realized as a group of symplectic automorphisms acting on a K3 surface X of maximal Picard number $\rho(X) = 20$. This can also be seen as a consequence of the fact that automorphisms of polarized K3 surfaces specialize; see Section **5**.2.3.

3.4 The starting point for Nikulin's approach to the classification of finite abelian groups of symplectic automorphisms, which was later extended by Xiao in [640] to the non-abelian case, is the following proposition. It in particular shows that the set of finite groups acting faithfully and symplectically on K3 surfaces is closed under quotients.

Proposition 3.11 *Let X be a complex K3 surface and $G \subset \mathrm{Aut}_s(X)$ be a finite subgroup. Then the quotient X/G has only rational double point singularities and its minimal desingularization*

$$Y \longrightarrow X/G$$

is again a K3 surface. Moreover, if $G \subset G'$ is a normal subgroup of a finite $G' \subset \mathrm{Aut}_s(X)$, then G'/G acts symplectically on Y.

Proof First, it is an easy exercise to generalize the proof of Lemma 1.4 to see that for a fixed point $x \in X$ of a finite $G \subset \mathrm{Aut}(X)$ there exists a local holomorphic coordinate system (z_1, z_2) in which G acts linearly. If G is symplectic, then $G \subset \mathrm{SL}(2, \mathbb{C})$. The local structure of $\mathbb{C}^2/G$ for finite subgroups $G \subset \mathrm{SL}(2, \mathbb{C})$ is of course well known; see e.g. [33, Ch. III]. In particular, the canonical bundle of the minimal resolution $\widetilde{\mathbb{C}^2/G} \longrightarrow \mathbb{C}^2/G$ is trivial; cf. Section **14**.0.3, **(v)**. Globally, as $\omega_X \simeq \mathcal{O}_X$ is G-invariant, this proves that Y has trivial canonical bundle. In fact, Y has to be a K3 surface, as any holomorphic one-form on Y would induce a holomorphic one-form on X. The second assertion is clear. $\qquad\square$

Remark 3.12 In the situation of the proposition, let $E_i \subset Y$ be the exceptional curves, i.e. the irreducible curves contracted under $Y \longrightarrow X/G$. The lattice spanned by their classes $[E_i] \in H^2(Y, \mathbb{Z})$ is a direct sum of lattices of ADE type; see Section **14**.0.3. Its saturation shall be called $M \subset H^2(Y, \mathbb{Z})$. Then for $L_G := (H^2(X, \mathbb{Z})^G)^{\perp} \subset \mathrm{NS}(X)$ one finds

$$\mathrm{rk}\, L_G = \mathrm{rk}\, M, \tag{3.6}$$

which is of course just the number of components E_i. See Whitcher's account of it [636, Prop. 2.4].

Note, however, that L_G and M are very different lattices. Indeed, by Lemma 3.7 the former does not contain any (-2)-classes, whereas the latter has a root lattice of the same rank. According to [636, Thm. 2.1] and [199, Prop. 2.4] there is an exact sequence

$$M \longrightarrow H^2(Y, \mathbb{Z}) \longrightarrow H^2(X, \mathbb{Z})^G \longrightarrow H^3(G, \mathbb{Z}) \longrightarrow 0.$$

The idea of [450] and [640] is then to study the configuration of the singular points of X/G to eventually get a classification of all possible finite $G \subset \mathrm{Aut}_s(X)$.

The following is the main result of Nikulin [450].

Theorem 3.13 *There are exactly* 14 *non-trivial finite abelian(!) groups G that can be realized as subgroups of* $\mathrm{Aut}_s(X)$ *of a complex K3 surface X. Moreover, the induced action on the abstract lattice* $H^2(X, \mathbb{Z})$ *is unique up to orthogonal transformations.*

Apart from the cyclic groups $\mathbb{Z}/n\mathbb{Z}$, $2 \leq n \leq 8$, the list comprises the following groups:

$$(\mathbb{Z}/2\mathbb{Z})^2, \quad (\mathbb{Z}/2\mathbb{Z})^3, \quad (\mathbb{Z}/2\mathbb{Z})^4, \quad (\mathbb{Z}/3\mathbb{Z})^2, \quad (\mathbb{Z}/4\mathbb{Z})^4,$$
$$\mathbb{Z}/2\mathbb{Z} \times \mathbb{Z}/4\mathbb{Z}, \quad \text{and} \quad \mathbb{Z}/2\mathbb{Z} \times \mathbb{Z}/6\mathbb{Z}.$$

Remark 3.14 In principle at least, it is possible to describe abstractly the action of all these 14 groups on the K3 lattice $E_8(-1)^{\oplus 2} \oplus U^{\oplus 3}$. For the cyclic groups, see the article by Garbagnati and Sarti [201] and Section 4.1.

For non-abelian groups $G \subset \mathrm{Aut}_s(X)$ the uniqueness of the induced action on the lattice $H^2(X, \mathbb{Z})$ was addressed by Hashimoto in [238]. It turns out that with the exception of five groups (of which three are among the 11 maximal symplectic groups) the uniqueness continues to hold also for non-abelian finite groups $G \subset \mathrm{Aut}_s(X)$. That the action for non-abelian group actions might not be unique had been observed also in [636, 640].

4 Nikulin Involutions, Shioda–Inose Structures, and More

In what follows we describe some concrete and geometrically interesting examples of automorphisms of K3 surface and highlight further results in special situations. For proofs and details we often refer to the original sources.

4.1 Let $f \colon X \xrightarrow{\sim} X$ be a symplectic automorphism of prime order $p := |f|$. Then the invariant part $H^2(X, \mathbb{Z})^{(f)}$ and its orthogonal complement

$$L := (H^2(X, \mathbb{Z})^{(f)})^\perp$$

can be completely classified as abstract lattices. In fact, the action of f on the lattice $H^2(X, \mathbb{Z})$ is independent of X itself (up to orthogonal transformation); cf. Theorem 3.13.

The explicit descriptions of $H^2(X, \mathbb{Z})^{(f)}$ and L can be found in papers by Garbagnati, Sarti, and Nikulin [201, 450]. Let us look a bit closer at the case $p = 2$.

A *Nikulin involution* on a K3 surface is a symplectic automorphism $\iota \colon X \xrightarrow{\sim} X$ of order two. According to Corollary 1.5 a Nikulin involution of a complex K3 surface has eight fixed points $x_1, \ldots, x_8 \in X$ and the quotient $X/\langle \iota \rangle$ has therefore eight A_1-singularities. Thus, the minimal resolution $Y \longrightarrow X/\langle \iota \rangle$ has an exceptional divisor consisting of eight (-2)-curves $E_i \simeq \mathbb{P}^1$.

By the table in Section 1.2, a K3 surface admitting a Nikulin involution has Picard number $\rho(X) \geq 9$. Moreover, the induced action $\iota^* \colon H^2(X, \mathbb{Z}) \xrightarrow{\sim} H^2(X, \mathbb{Z})$ (which, as an abstract isometry, is independent of X) satisfies

$$H^2(X, \mathbb{Z})^{\langle \iota \rangle} \simeq E_8(-2) \oplus U^{\oplus 3} \quad \text{and} \quad L \simeq E_8(-2)$$

with $L \subset \mathrm{NS}(X)$.

In [205] van Geemen and Sarti show that $\mathrm{NS}(X)$ contains $E_8(-2) \oplus \mathbb{Z}(2d)$ as a sublattice (with both factors primitive but not necessarily the sum) and that for general X, i.e. $\rho(X) = 9$, one has

$$\mathrm{NS}(X) \simeq E_8(-2) \oplus \mathbb{Z}(2d) \quad \text{or} \quad (\mathrm{NS}(X) : E_8(-2) \oplus \mathbb{Z}(2d)) = 2. \tag{4.1}$$

The second case can occur only for d even. The summand $\mathbb{Z}(2d)$ corresponds to an ι-invariant ample line bundle L. Although invariant, L might not descend to a line bundle on the quotient $X/\langle \iota \rangle$ (due to the possibly non-trivial action of ι on the fibres of L over the fixed points x_i) and this is when $\mathrm{NS}(X)$ is only an overlattice of $E_8(-2) \oplus \mathbb{Z}(2d)$ of index two. The article [205] also contains a detailed discussion of the moduli spaces of K3 surfaces with Nikulin involution. This was generalized by Garbagnati and Sarti in [201] to symplectic automorphisms of prime order p.

Remark 4.1 The following sufficient criterion for the existence of a Nikulin involution on a complex algebraic K3 surface is due to Morrison (see [424, Thm. 5.7]): if there exists a primitive embedding

$$E_8(-1)^{\oplus 2} \hookrightarrow \mathrm{NS}(X), \tag{4.2}$$

then X admits a Nikulin involution. Furthermore, by [424, Thm. 6.3], (4.2) is equivalent to the existence of a primitive embedding

$$T(X) \hookrightarrow U^{\oplus 3}. \tag{4.3}$$

Indeed, for example, any embedding (4.3) induces $T(X) \hookrightarrow U^{\oplus 3} \hookrightarrow \Lambda = E_8(-1)^{\oplus 2} \oplus U^{\oplus 3}$, which by Corollary 14.3.5 is unique and hence (4.2) exists. Note that the existence of either of the two embeddings is equivalent to the existence of a

Shioda–Inose structure, i.e. a Nikulin involution with a quotient birationally equivalent to a Kummer surface. Hence, by Proposition **14**.1.8 a complex algebraic K3 surface admits a Shioda–Inose structure if $\rho(X) = 19$ or 20.

As for an abelian surface $H^2(A, \mathbb{Z}) \simeq U^{\oplus 3}$ (see Section **3**.2.3), the existence of a Shioda–Inose structure on X is also equivalent to the existence of a Hodge isometry $T(X) \simeq T(A)$ for some abelian surface A; see [424, Thm. 6.3].

4.2 Consider an elliptic K3 surface $\pi : X \longrightarrow \mathbb{P}^1$ with a section $C_0 \subset X$. As usual, for a smooth fibre X_t we consider the point of intersection of C_0 with X_t as the origin of the elliptic curve X_t. Assume now that there exists another section $C \subset X$, i.e. a non-trivial element in $MW(X)$; see Section **11**.3.2. The intersection of C with X_t provides another point $x_t \in X_t$ which may be torsion or not. We say that C is a torsion section of order n if $x_t \in X_t \cap C$ is a torsion point of order n for most geometric fibres X_t or, equivalently, if $C \in MW(X)$ is an element of order n.

Definition 4.2 To any section $C \in MW(X)$ one associates

$$f_C : X \overset{\sim}{\dashrightarrow} X$$

by translating a point $y \in X_t$ to $x_t + y \in X_t$, where $X_t \cap C = \{x_t\}$.

A priori, f_C is only a rational (or meromorphic) map, but as K3 surfaces have trivial canonical class, it extends to an automorphism. Of course, the order of f_C equals the order of $C \in MW(X)$, and, in particular, if $x_t \in X_t$ is of infinite order for one fibre X_t then $|f_C| = \infty$.

Example 4.3 This provides probably the easiest way to produce examples of K3 surfaces with infinite $\mathrm{Aut}(X)$. Indeed, take the Kummer surface associated with $E_1 \times E_2$ and assume $\rho(X) > 18$, e.g. $E_1 \simeq E_2$. Its Mordell–Weil rank is positive due to the Shioda–Tate formula **11**.3.4 (cf. Example **11**.3.5), and, therefore, $X \longrightarrow \mathbb{P}^1 = E_1/\pm$ admits a section C of infinite order which yields an automorphism f_C with $|f_C| = \infty$. (A section C like this can be described explicitly as the quotient of the diagonal $\Delta \subset E \times E$.)

Lemma 4.4 *The automorphism $f_C : X \overset{\sim}{\dashrightarrow} X$ associated with a section $C \in MW(X)$ is symplectic.*

Proof We sketch the argument in the complex case. It is enough to prove $f_C^* \sigma = \sigma$, for some $0 \neq \sigma \in H^0(X, \Omega_X^2)$, in the dense open set of points $x \in X$ contained in a smooth fibre X_t. For the restriction $\sigma|_{X_t} \in H^0(X_t, \Omega_X^2|_{X_t})$ this amounts to showing $(f_C^* \sigma)|_{X_t} = \sigma|_{X_t}$ in $H^0(X_t, \Omega_X^2|_{X_t})$.

Now, let $X_t \subset U$ be an open neighbourhood of the form $U \simeq R^1 \pi_* \mathbb{C}_U / R^1 \pi_* \mathbb{Z}_U$ such that $C_0 \cap U$ is the image of the zero section of $R^1 \pi_* \mathbb{C}_U$. Then any point $x \in X_t$ can be extended to a flat section $C_x \subset U$ which then by translation induces an isomorphism $f_{C_x}: U \xrightarrow{\sim} U$. Pulling back $\sigma|_U$ via f_{C_x} and then restricting back to X_t yields a holomorphic map

$$X_t \longrightarrow H^0(X_t, \Omega^2_X|_{X_t}), \quad x \longmapsto (f^*_{C_x}(\sigma|_U))|_{X_t},$$

which, as X_t is compact, has to be constant. Hence, $f^*_{C_t}(\sigma|_U)|_{X_t} \equiv f^*_{C_0}(\sigma|_U)|_{X_t} = \sigma|_{X_t}$.

As C was not assumed to be flat, the section C_x associated with the intersection point $x \in C \cap X_t$ might differ from C. Nevertheless, $f^*_{C_x}(\sigma|_U)|_{X_t} = f^*_C(\sigma|_U)|_{X_t}$. This is perhaps best seen in local coordinates z_1, z_2 with C and C_x given by holomorphic maps $z_1 \longmapsto (z_1, g(z_1))$ and $z_1 \longmapsto (z_1, g_x(z_1))$, respectively. If $\sigma = F(z_1, z_2) \cdot dz_1 \wedge dz_2$, then $f^*_C \sigma = F(z_1, z_2 + g(z_1)) \cdot dz_1 \wedge dz_2$ and $f^*_{C_x} \sigma = F(z_1, z_2 + g_x(z_1)) \cdot dz_1 \wedge dz_2$. But of course for $x \in C \cap X_t$ we have $F(t, z_2 + g(t)) = F(t, z_2 + g_x(t))$. Hence, $(f^*_C \sigma)|_{X_t} = (f^*_{C_x}(\sigma|_U))|_{X_t} = \sigma|_{X_t}$. $\qquad\square$

Remark 4.5 The construction yields an injection

$$\mathrm{MW}(X) \hookrightarrow \mathrm{Aut}_s(X),$$

with its image clearly contained in the abelian part of $\mathrm{Aut}_s(X)$. The inclusion also allows one to tie Cox's computation of the order of elements in $\mathrm{MW}(X)_{\mathrm{tors}}$ (see Remark **11**.3.11) to Corollary 1.8. This shows that in characteristic zero $\mathrm{MW}(X)_{\mathrm{tors}} \simeq \mathbb{Z}/n\mathbb{Z} \times \mathbb{Z}/m\mathbb{Z}$ with $m, n \leq 8$. In positive characteristic the upper bound has to be modified according to Remark 1.9.

Remark 4.6 Combining the lemma with Lemma 1.4, one gets an alternative proof for the fact that distinct sections C_0, C_1 of an elliptic fibration $X \longrightarrow \mathbb{P}^1$ that on the generic fibre differ by torsion do not intersect; see Remark **11**.3.8. Indeed, if C_1 is a torsion section, then f_{C_1} is a symplectic automorphism of finite order which has only isolated fixed points. However, if C_0 and C_1 meet a closed fibre X_t in the same (automatically smooth) point, then translation on this fibre is constant, and, therefore, X_t would be contained in $\mathrm{Fix}(f_{C_1})$, which is absurd.

Example 4.7 To have at least one concrete example, consider an elliptic K3 surface $X \longrightarrow \mathbb{P}^1$ described by an equation of the form $y^2 = x(x^2 + a(t)x + b(t))$. A zero section C_0 can be given by $x = z = 0$ and a two-torsion section C by $x = y = 0$. The K3 surface X with the associated involution $f_C: X \xrightarrow{\sim} X$ has been studied by van Geemen and Sarti in [205, Sec. 4], where f_C is shown to be symplectic, because the quotient $X/\langle f_C \rangle$ turns out to be a (singular) K3 surface. One also finds that in this case $E_8(-2) \oplus \mathbb{Z}(2d) \subset \mathrm{NS}(X)$ in (4.1) is of index two.

4.3 Here are the most basic examples of non-symplectic automorphisms of finite order of complex K3 surfaces. Recall that due to Corollary 1.10 those can exist only on projective K3 surfaces.

(i) Let $X \longrightarrow \mathbb{P}^2$ be a double plane, i.e. a K3 surface given as the double cover of $\mathbb{P}^2$ ramified over a (say smooth) sextic; see Example **1.1.3**. The covering involution $\iota \colon X \xrightarrow{\sim} X$ is of order two, and, since the generator of $H^0(X, \Omega_X^2)$ does not descend to $\mathbb{P}^2$, ι cannot be symplectic.

(ii) Let $\iota \colon X \xrightarrow{\sim} X$ be a fixed point free involution of a K3 surface X over a field of characteristic $\neq 2$. Then the quotient $\overline{X} := X/\langle \iota \rangle$ is an Enriques surface and every Enriques surface can be constructed in this way. Due to Corollary 1.5, ι cannot be symplectic and, therefore, $H^{2,0}(\overline{X}) = 0$. See Section **14.0.3** for a description of the Enriques lattice $H^2(\overline{X}, \mathbb{Z})$.

Of course, there exist non-symplectic automorphisms of higher order, but at least birationally their quotients are always of the above form. More precisely, using the classification of surfaces, one proves the following lemma.

Lemma 4.8 *Let $f \colon X \xrightarrow{\sim} X$ be a non-symplectic automorphism of finite order. Then $X/\langle f \rangle$ is rational or birational to an Enriques surface.* $\quad\square$

Remark 4.9 In [317] Kondō proves that any complex K3 surface cover X of an Enriques surface Y, i.e. a K3 surface with a fixed point free involution, has infinite $\mathrm{Aut}(X)$. However, $\mathrm{Aut}(Y)$ might be finite.

As a special case of the results proved by Machida and Oguiso resp. Zhang in [384, 653], based on similar arguments as in the symplectic case (see Section 1.1 and Remark 1.15), we mention the following lemma.

Lemma 4.10 *If $f \colon X \xrightarrow{\sim} X$ is a non-symplectic automorphism of prime order p, then $p = 2, 3, 5, 7, 11, 13, 17,$ or 19.*

The invariant part $\mathrm{NS}(X)^{\langle f \rangle} = H^2(X, \mathbb{Z})^{\langle f \rangle}$ of non-symplectic automorphisms of prime order p has been completely determined. For $p = 2$ this is due to Nikulin and the classification was completed by Artebani, Sarti, and Taki in [14], which also contains a detailed analysis of the fixed-point sets. As it turns out, K3 surfaces with non-symplectic automorphisms of finite order often also admit symplectic involutions; cf. [148, 202].

References and Further Reading

Instead of automorphisms one could look at endomorphisms and, more precisely, at rational dominant maps $f \colon X \longrightarrow X$. Recently, Chen [116] has shown that a very general complex projective K3 surface does not admit any rational endomorphism of degree > 1. In [140] Dedieu studies an interesting link to the irreducibility of the Severi variety.

The behavior of $\mathrm{Aut}(X)$ under deformations was addressed by Oguiso in [471]. In particular it is shown that in any non-trivial deformation $X \longrightarrow S$ of projective K3 surfaces the set $\{t \in S \mid |\mathrm{Aut}(X_t)| = \infty\}$ is dense.

In [180] Frantzen classifies all finite $G \subset \mathrm{Aut}_s(X)$ all elements of which commute with a non-symplectic involution with fixed points.

Symplectic and non-symplectic automorphisms of higher-dimensional generalizations of K3 surfaces provided by irreducible symplectic manifolds have recently attracted a lot of attention; see e.g. [51, 72, 419].

The global structure of an infinite $\text{Aut}(X)$ is not completely clear. Borcherds found an example of a K3 surface for which $\text{Aut}(X)$ is not isomorphic to an arithmetic group; see [77, Ex. 5.8] and [604, Ex. 6.3].

Automorphisms act not only on cohomology, but also on Chow groups. Standard results in Hodge theory can be used to show that any non-symplectic $f \in \text{Aut}(X)$ acts non-trivially on $\text{CH}^2(X)$. The converse is more difficult, but for $|f| < \infty$ it has been verified in [260, 265, 625].

For highly non-projective K3 surfaces X, namely those with $\rho(X) = 0$, one knows that $\text{Aut}(X)$ is either trivial or isomorphic to $\mathbb{Z}$; see the survey by Macrì and Stellari [386]. More generally, Oguiso showed in [473] that for any non-projective K3 surface X with $\text{NS}(X)$ negative definite $\text{Aut}(X)$ is either finite or a finite extension of $\mathbb{Z}$. If $\text{NS}(X)$ is allowed to have an isotropic direction, then $\text{Aut}(X)$ is isomorphic to $\mathbb{Z}^n$, $n \leq \rho(X) - 1$, up to finite index (almost abelian).

Liftability of groups of automorphisms from positive characteristic to characteristic zero has been addressed in a paper by Esnault and Oguiso [172]. In particular, it is shown that there exist special lifts of any K3 surface that essentially exclude all non-trivial automorphism from lifting to characteristic zero.

It would be worth another chapter to talk about the dynamical aspects of automorphisms (and more generally endomorphisms) of K3 surfaces. This started with the two articles by Cantat and McMullen [102, 403], but see also [475] for recent progress and references.

Questions and Open Problems

As far as I can see, the relation between $\text{Aut}_s(X)$ and $\text{Aut}(X)$ as discussed in Section 1.3 has not yet been addressed in positive characteristic. In general, as mentioned repeatedly, there are still a few open questions in positive characteristic and over non-algebraically closed fields.

16

Derived Categories

According to a classical result due to Gabriel [194] the abelian category $\mathrm{Coh}(X)$ determines X. More precisely, if X and Y are two varieties over a field k and $\mathrm{Coh}(X) \xrightarrow{\sim} \mathrm{Coh}(Y)$ is a k-linear equivalence, then X and Y are isomorphic varieties. The situation becomes more interesting when instead of the abelian category $\mathrm{Coh}(X)$ one considers its bounded derived category $\mathrm{D}^b(X)$. Then Gabriel's theorem is no longer valid in general, and, in fact, there exist non-isomorphic K3 surfaces X and Y with equivalent bounded derived categories. In this chapter we outline the main results concerning derived categories of coherent sheaves on K3 surfaces. As the general theory of Fourier–Mukai transforms has been presented in detail in various surveys and in particular in the two monographs [34, 254], we look for ad hoc arguments highlighting the special features of K3 surfaces.

1 Derived Categories and Fourier–Mukai Transforms

We start with a brief recap of the main concepts of the theory of bounded derived categories of coherent sheaves, but for a serious introduction the reader is advised to consult one of the standard sources, e.g. [208, 613]. For more details on Fourier–Mukai transforms, see [254].

1.1 Let X be a smooth projective variety of dimension n over a field k. By $\mathrm{Coh}(X)$ we denote the category of coherent sheaves on X, which is viewed as a k-linear abelian category. Note that all Hom-spaces $\mathrm{Hom}(E, F)$ for $E, F \in \mathrm{Coh}(X)$ are k-vector spaces of finite dimension. The *bounded derived category* of X is by definition the bounded derived category of the abelian category $\mathrm{Coh}(X)$:

$$\mathrm{D}^b(X) := \mathrm{D}^b(\mathrm{Coh}(X)),$$

which is viewed as a k-linear triangulated category.

To be a little more precise, one first introduces the category $\mathrm{Kom}^b(X)$ of bounded complexes $E^\bullet = \cdots \longrightarrow E^{i-1} \longrightarrow E^i \longrightarrow E^{i+1} \longrightarrow \cdots$, where $E^i \in \mathrm{Coh}(X)$ and $E^i = 0$ for $|i| \gg 0$. Morphisms in $\mathrm{Kom}^b(X)$ are given by commutative diagrams

$$
\begin{array}{ccccccccc}
E^\bullet & \cdots \longrightarrow & E^{i-1} & \longrightarrow & E^i & \longrightarrow & E^{i+1} & \longrightarrow & \cdots \\
\downarrow{\scriptstyle\varphi} & & \downarrow{\scriptstyle\varphi^{i-1}} & & \downarrow{\scriptstyle\varphi^i} & & \downarrow{\scriptstyle\varphi^{i+1}} & & \\
F^\bullet & \cdots \longrightarrow & F^{i-1} & \longrightarrow & F^i & \longrightarrow & F^{i+1} & \longrightarrow & \cdots .
\end{array}
$$

There exists a natural functor $\mathrm{Kom}^b(X) \longrightarrow \mathrm{D}^b(X)$ which identifies the objects of both categories, so

$$\mathrm{Ob}(\mathrm{D}^b(X)) = \mathrm{Ob}(\mathrm{Kom}^b(X)),$$

and, roughly, inverts quasi-isomorphisms.

Recall that a morphism of complexes $\varphi \colon E^\bullet \longrightarrow F^\bullet$ is called a *quasi-isomorphism* (qis) if the induced morphisms $H^i(\varphi) \colon H^i(E^\bullet) \longrightarrow H^i(F^\bullet)$ between the cohomology sheaves are isomorphisms in all degrees i.

However, as an intermediate step in the passage from $\mathrm{Kom}^b(X)$ to $\mathrm{D}^b(X)$ one constructs the homotopy category $\mathrm{K}^b(X)$. It has again the same objects as $\mathrm{Kom}^b(X)$, but

$$\mathrm{Hom}_{\mathrm{K}^b(X)}(E^\bullet, F^\bullet) = \mathrm{Hom}_{\mathrm{Kom}^b(X)}(E^\bullet, F^\bullet)/\!\sim,$$

where a homotopy $\sim$ between two morphisms of complexes $\varphi, \psi \colon E^\bullet \longrightarrow F^\bullet$ is given by morphisms $h^i \colon E^i \longrightarrow F^{i-1}$ with $\varphi^i - \psi^i = h^{i+1} \circ d_E^i + d_F^{i-1} \circ h^i$. It can be shown that a morphism $E^\bullet \longrightarrow F^\bullet$ in $\mathrm{D}^b(X)$ is an equivalence class of roofs $E^\bullet \xleftarrow{\ \psi\ } G^\bullet \xrightarrow{\ \varphi\ } F^\bullet$ in $\mathrm{K}^b(X)$ with ψ a quasi-isomorphism. Two roofs are equivalent if they can be dominated by a third making all diagrams commutative in $\mathrm{K}^b(X)$ (so, up to homotopy only). Of course, at this point many details need to be checked. In particular, one has to define the composition of roofs and show that it behaves well with respect to the equivalence of roofs. In any case, the composition

$$\mathrm{Kom}^b(X) \longrightarrow \mathrm{K}^b(X) \longrightarrow \mathrm{D}^b(X)$$

identifies the objects of all three categories, and, on the level of homomorphisms, one first divides out by homotopy and then localizes quasi-isomorphisms.

What makes $\mathrm{D}^b(X)$ a *triangulated category* is the existence of the *shift*

$$E^\bullet \longmapsto E^\bullet[1],$$

defined by $E^\bullet[1]^i = E^{i+1}$ and $d_{E[1]}^i = -d_E^{i+1}$, and of *exact* (or distinguished) *triangles*. A triangle in $\mathrm{D}^b(X)$ is given by morphisms $E^\bullet \longrightarrow F^\bullet \longrightarrow G^\bullet \longrightarrow E^\bullet[1]$. A triangle is exact if it is isomorphic, in $\mathrm{D}^b(X)$, to a triangle of the form

$$A^\bullet \xrightarrow{\ \varphi\ } B^\bullet \xrightarrow{\ \tau\ } C(\varphi) \xrightarrow{\ \pi\ } A^\bullet[1],$$

where $C(\varphi)$ with $C(\varphi)^i := A^{i+1} \oplus B^i$ is the mapping cone of a morphism φ in $\mathrm{Kom}^b(X)$ and τ and π are the natural morphisms. Again, a number of things need to be checked to make this a useful notion, e.g. that rotating an exact triangle $E^\bullet \longrightarrow F^\bullet \longrightarrow G^\bullet \longrightarrow E^\bullet[1]$ yields again an exact triangle $F^\bullet \longrightarrow G^\bullet \longrightarrow E^\bullet[1] \longrightarrow F^\bullet[1]$ (with appropriate signs). The properties of the shift functor and the collection of exact triangles in $\mathrm{D}^b(X)$ can be turned into the notion of a triangulated category satisfying axioms TR1–TR4.

To conclude this brief reminder of the construction of $\mathrm{D}^b(X)$, recall that there exists a fully faithful functor

$$\mathrm{Coh}(X) \lhook\joinrel\longrightarrow \mathrm{D}^b(X)$$

satisfying $\mathrm{Ext}^i(E, F) \simeq \mathrm{Hom}_{\mathrm{D}^b(X)}(E, F[i])$. For this reason we also use the notation $\mathrm{Ext}^i(E^\bullet, F^\bullet) := \mathrm{Hom}_{\mathrm{D}^b(X)}(E^\bullet, F^\bullet[i])$ for complexes $E^\bullet$ and $F^\bullet$.

1.2   For the following, see also the discussion in Section **12.1.3**.

The *Grothendieck group* $K(\mathrm{Coh}(X))$ of the abelian category $\mathrm{Coh}(X)$ of, say, a smooth projective variety X is defined as the quotient of the free abelian group generated by all $[E]$, with $E \in \mathrm{Coh}(X)$, divided by the subgroup generated by expressions of the form $[F] - [E] - [G]$ for short exact sequences $0 \longrightarrow E \longrightarrow F \longrightarrow G \longrightarrow 0$.

Similarly, the Grothendieck group $K(\mathrm{D}^b(X))$ of the triangulated category $\mathrm{D}^b(X)$ is the quotient of the free abelian group generated by all $[E^\bullet]$, with $E^\bullet \in \mathrm{D}^b(X)$, divided by the subgroup generated by expressions of the form $[F^\bullet] - [E^\bullet] - [G^\bullet]$ for exact triangles $E^\bullet \longrightarrow F^\bullet \longrightarrow G^\bullet \longrightarrow E^\bullet[1]$. Note that $[E^\bullet[1]] = -[E^\bullet]$ in $K(\mathrm{D}^b(X))$. Using the full embedding $\mathrm{Coh}(X) \lhook\joinrel\longrightarrow \mathrm{D}^b(X)$ one obtains a natural isomorphism

$$K(X) := K(\mathrm{Coh}(X)) \xrightarrow{\sim} K(\mathrm{D}^b(X)),$$

the inverse of which is given by $[E^\bullet] \longmapsto \sum (-1)^i [E^i]$.

The *Euler pairing*

$$\chi(E^\bullet, F^\bullet) := \sum (-1)^i \dim \mathrm{Ext}^i(E^\bullet, F^\bullet)$$

is well defined for bounded complexes and, by using additivity for exact sequences, can be viewed as a bilinear form $\chi(\ ,\ )$ on $K(X)$. Note that Serre duality implies $\chi(E^\bullet, F^\bullet) = (-1)^n \chi(F^\bullet, E^\bullet \otimes \omega_X)$, where $n = \dim(X)$.

The *numerical Grothendieck group* (cf. Section **10.2**)

$$N(X) := K(X)/\!\sim$$

is defined as the quotient by the radical of $\chi(\ ,\ )$. This is well defined, for if $\chi(E^\bullet, F^\bullet) = 0$ for fixed $E^\bullet$ and all $F^\bullet$, then also $\chi(F^\bullet, E^\bullet) = (-1)^n \chi(E^\bullet, F^\bullet \otimes \omega_X) = 0$.

From now on our notation does not distinguish between sheaves F and complexes of sheaves $F^\bullet$ – both are usually denoted by just F.

1.3   For a smooth projective variety X of dimension n the composition

$$S \colon D^b(X) \xrightarrow{\sim} D^b(X), \quad E \longmapsto E \otimes \omega_X[n]$$

is a *Serre functor*; i.e. for all complexes E and F there exist functorial isomorphisms

$$\operatorname{Hom}_{D^b(X)}(E, F) \xrightarrow{\sim} \operatorname{Hom}_{D^b(X)}(F, E \otimes \omega_X[n])^*.$$

This, in particular, yields the more traditional form of Serre duality (cf. the discussion in Section **9**.1.2)

$$\operatorname{Ext}^i(E, F) \xrightarrow{\sim} \operatorname{Ext}^{n-i}(F, E \otimes \omega_X)^*.$$

For a K3 surface X the Serre functor is isomorphic to the double shift:

$$S \colon E \longmapsto E^\bullet[2].$$

Grothendieck–Verdier duality, a natural generalization of Serre duality, is crucial for the following discussion of Fourier–Mukai functors; see [254, Sec. 3.4] for the formulation and references.

Definition 1.1   Let X and Y be two smooth projective varieties over k and let $\mathcal{P} \in D^b(X \times Y)$. Then the associated *Fourier–Mukai transform*

$$\Phi := \Phi_{\mathcal{P}} \colon D^b(X) \longrightarrow D^b(Y)$$

is the exact functor given as the composition of derived functors

$$E \longmapsto Lq^*E \longmapsto Lq^*E \otimes^L \mathcal{P} \longmapsto Rp_*(Lq^*E \otimes^L \mathcal{P}),$$

where q and p denote the two projections to X and Y.

Under our assumptions on X and Y, all functors are well defined and indeed map bounded complexes to bounded complexes. As all functors on the level of derived categories have to be considered as derived functors anyway, one often simply writes

$$\Phi_{\mathcal{P}}(E) := p_*(q^*E \otimes \mathcal{P}).$$

The kernel $\mathcal{P}$ can also be used to define a Fourier–Mukai transform in the other direction $D^b(Y) \longrightarrow D^b(X)$, which, by abuse of notation, is also denoted $\Phi_{\mathcal{P}}$.

Remark 1.2   For proofs and details of the following facts, we refer to [254].

(i) A Fourier–Mukai functor $\Phi_{\mathcal{P}} \colon D^b(X) \longrightarrow D^b(Y)$ admits left and right adjoints which can be described as Fourier–Mukai transforms

$$\Phi_{\mathcal{P}_L}, \Phi_{\mathcal{P}_R} \colon D^b(Y) \longrightarrow D^b(X)$$

with

$$\mathcal{P}_L := \mathcal{P}^* \otimes p^*\omega_Y[\dim(Y)] \quad \text{and} \quad \mathcal{P}_R := \mathcal{P}^* \otimes q^*\omega_X[\dim(X)].$$

Here, $\mathcal{P}^*$ denotes the derived dual $R\mathcal{H}om(\mathcal{P}, \mathcal{O})$.

(ii) The composition of two Fourier–Mukai transforms

$$\Phi_\mathcal{P}: D^b(X) \longrightarrow D^b(Y) \quad \text{and} \quad \Phi_\mathcal{Q}: D^b(Y) \longrightarrow D^b(Z)$$

with $\mathcal{P} \in D^b(X \times Y)$ and $\mathcal{Q} \in D^b(Y \times Z)$ is again a Fourier–Mukai transform

$$\Phi_\mathcal{R} := \Phi_\mathcal{Q} \circ \Phi_\mathcal{P}: D^b(X) \longrightarrow D^b(Z).$$

The Fourier–Mukai kernel $\mathcal{R}$ can be described as the convolution of $\mathcal{P}$ and $\mathcal{Q}$:

$$\mathcal{R} \simeq \pi_{XZ*}(\pi_{XY}^*\mathcal{P} \otimes \pi_{YZ}^*\mathcal{Q}),$$

where, for example, π_{XY} denotes the projection $X \times Y \times Z \longrightarrow X \times Y$.

In the following, X and Y are always smooth projective varieties over a field k. They are called *derived equivalent* if there exists a k-linear exact equivalence

$$D^b(X) \xrightarrow{\sim} D^b(Y).$$

Recall that a functor between triangulated categories is exact if it commutes with shift functors and maps exact triangles to exact triangles.

Due to a result of Orlov, any k-linear exact equivalence is isomorphic to a Fourier–Mukai transform $\Phi_\mathcal{P}$ (even for non-algebraically closed fields). This theorem has been improved and generalized; see [101] for a recent survey and references. So, no information is lost when restricting to the seemingly more manageable class of Fourier–Mukai transforms.

Remark 1.3 It seems that there is not a single example of an exact equivalence $D^b(X) \xrightarrow{\sim} D^b(Y)$ known that has been described without using the Fourier–Mukai formalism.

1.4 To test whether a given exact functor is fully faithful it is often enough to control the images of objects in a spanning class. A collection of objects $\Omega \subset D^b(X)$ on a K3 surface is called a *spanning class* if for all $F \in D^b(X)$ the following condition is satisfied: if $\text{Hom}(E, F[i]) = 0$ for all $E \in \Omega$ and all i, then $F \simeq 0$.[1]

The following criterion due to Orlov (see [254, Prop. 1.49] for the proof and references) is often the only method that allows one to decide whether a given functor is fully faithful.

---

[1] Due to Serre duality, the condition is equivalent to: if $\text{Hom}(F, E[i]) = 0$ for all $E \in \Omega$ and all i, then $F \simeq 0$. This needs to be added if ω_X is not trivial.

Lemma 1.4 *Let* $\Phi\colon \mathrm{D}^b(X) \longrightarrow \mathrm{D}^b(Y)$ *be a Fourier–Mukai transform and let* $\Omega \subset \mathrm{D}^b(X)$ *be a spanning class. Then* Φ *is fully faithful if and only if* Φ *induces isomorphisms*

$$\mathrm{Hom}(E, F[i]) \xrightarrow{\sim} \mathrm{Hom}(\Phi(E), \Phi(F)[i])$$

for all $E, F \in \Omega$ *and all* i.

Example 1.5 Here are the three most frequent examples of spanning classes in $\mathrm{D}^b(X)$.

(i) The set $\Omega := \{k(x) \mid x \in X \text{ closed}\}$ is a spanning class. Indeed, for any non-trivial coherent sheaf F and a closed point $x \in X$ in its support $\mathrm{Hom}(F, k(x)) \neq 0$. For complexes one argues similarly using a non-trivial homomorphism from the maximal non-vanishing cohomology sheaf of F to some $k(x)$.

(ii) For any ample line bundle L on X, the set $\Omega := \{L^i \mid i \in \mathbb{Z}\}$ is a spanning class. Indeed, $\mathrm{Hom}(L^i, F) \neq 0$ for any non-trivial coherent sheaf F and $i \ll 0$. For complexes one argues via the minimal non-vanishing cohomology sheaf of F.

(iii) Let $E \in \mathrm{D}^b(X)$ be any object and

$$E^\perp := \{F \mid \mathrm{Hom}(E, F[i]) = 0 \text{ for all } i\}.$$

It is easy to see that $\Omega = \{E\} \cup E^\perp$ is a spanning class.

For the first example of a spanning class, the following result due to Bondal and Orlov is a surprising strengthening of Lemma 1.4. However, for K3 surfaces one can often get away without it.

Proposition 1.6 *Let* $\Phi\colon \mathrm{D}^b(X) \longrightarrow \mathrm{D}^b(Y)$ *be a Fourier–Mukai transform of smooth projective varieties over an algebraically closed field k.*

Then, Φ *is fully faithful if and only if* $\mathrm{Hom}(\Phi(k(x)), \Phi(k(x))) \simeq k$ *for arbitrary closed points* $x, y \in X$ *and* $\mathrm{Ext}^i(\Phi(k(x)), \Phi(k(y))) = 0$ *for* $x \neq y$, $i < 0$, *or* $i > \dim(X)$.

In order to apply Lemma 1.4 directly, one would also need to ensure that the maps

$$T_x X \simeq \mathrm{Ext}^1(k(x), k(x)) \longrightarrow \mathrm{Ext}^1(\Phi(k(x)), \Phi(k(x))) \tag{1.1}$$

are isomorphisms for all $x \in X$. The map (1.1) compares first-order deformations of $k(x)$ and of $\Phi(k(x))$ via Φ. This point of view emphasizes that a Fourier–Mukai transform $\Phi_{\mathcal{P}}$ defines an equivalence if $\mathcal{P}$ can be seen as a universal family of complexes on Y parametrized by X and vice versa.

For later use note that for an equivalence Φ the isomorphisms (1.1) glue to an isomorphism between the tangent bundle and the relative Ext-sheaf:

$$\mathcal{T}_X \xrightarrow{\sim} \mathcal{E}xt^1_q(\mathcal{P}, \mathcal{P}). \tag{1.2}$$

Similarly, one constructs an isomorphism $\mathcal{O}_X \xrightarrow{\sim} \mathcal{E}xt^0_q(\mathcal{P}, \mathcal{P})$. See [254, Ch. 11.1] for details.

Typically, the hardest part in proving that a given functor $\Phi_{\mathcal{P}} \colon D^b(X) \longrightarrow D^b(Y)$ is an equivalence is in proving that it is fully faithful. That the functor is then an equivalence is often deduced from the following result; cf. [254, Cor. 1.56].

Lemma 1.7 *Let* $\Phi \colon D^b(X) \longrightarrow D^b(Y)$ *be a fully faithful Fourier–Mukai transform which commutes with Serre functors, i.e.* $\Phi \circ S_X \simeq S_Y \circ \Phi$. *Then* Φ *is an equivalence.*

Proof We sketch the main steps of the proof. To simplify notations, write $G := \Phi_{\mathcal{P}_L}$ and $H := \Phi_{\mathcal{P}_R}$ for the left and right adjoint of Φ. First, one shows that $H(F) = 0$ implies $G(F) = 0$. Indeed, if $H(F) = 0$, then $\operatorname{Hom}(E, H(F)) = 0$ for all $E \in D^b(X)$. Using adjunction twice, Serre duality, and the compatibility of Φ with S_X and S_Y, one gets $\operatorname{Hom}(G(F), S_X(E)) = 0$ for all E and, therefore, by the Yoneda lemma $G(F) = 0$.

Next, define full triangulated subcategories $D_1, D_2 \subset D^b(Y)$ as follows. Let

$$D_1 := \operatorname{Im}(\Phi) := \{\Phi(E) \mid E \in D^b(X)\} \quad \text{and} \quad D_2 := \operatorname{Ker}(H) := \{F \mid H(F) = 0\}.$$

Using the adjunction morphism $\Phi \circ H \longrightarrow \operatorname{id}$, every object $F \in D^b(Y)$ can be put in an exact triangle $\Phi(H(F)) \longrightarrow F \longrightarrow F'$ with $H(F') = 0$ (use $\operatorname{id} \simeq H \circ \Phi$ for fully faithful Φ). However, then by the first step and adjunction $\operatorname{Hom}(F', \Phi(H(F))[1]) = \operatorname{Hom}(G(F'), H(F)[1]) = 0$ and thus $F \simeq \Phi(H(F)) \oplus F'$. This eventually yields a direct sum decomposition $D^b(Y) \simeq D_1 \oplus D_2$. Studying the induced decomposition of $\mathcal{O}_Y$ and of all point sheaves $k(y)$, $y \in Y$, one proves $D_2 = 0$, i.e. $\Phi \colon D^b(X) \xrightarrow{\sim} D^b(Y)$. □

2 Examples of (Auto)equivalences

Before attempting a classification of all Fourier–Mukai partners of a fixed K3 surface X and a description of the group $\operatorname{Aut}(D^b(X))$ of autoequivalences of its derived category, we prove one basic fact and describe a few important examples which form the building blocks for both problems.

For the rest of the section, all K3 surfaces are assumed to be projective over a fixed field k.

2.1 Let us first show that derived categories of K3 surfaces cannot be realized by any other type of varieties.

Proposition 2.1 *Suppose X is a K3 surface and Y is a smooth projective variety derived equivalent to X. Then Y is a K3 surface.*

Proof Any Fourier–Mukai equivalence $\Phi = \Phi_{\mathcal{P}} \colon D^b(X) \xrightarrow{\sim} D^b(Y)$ commutes with Serre functors, i.e. $\Phi \circ S_X \simeq S_Y \circ \Phi$. Hence, $S_Y \simeq \Phi \circ S_X \circ \Phi^{-1}$. However, as S_X is the shift $E \longmapsto E[2]$ and Φ commutes with shifts, also the Serre functor

$S_Y: F \longmapsto F \otimes \omega_Y[\dim(Y)]$ is just $F \longmapsto F[2]$. Hence, $\omega_Y \simeq \mathcal{O}_Y$, $\dim(Y) = 2$, and, by Enriques classification, Y is either a K3 or an abelian surface.

To exclude abelian surfaces, one uses the two spectral sequences, where as before p and q denote the two projections:

$$E_2^{ij} = H^i(X, \mathcal{E}xt_q^j(\mathcal{P}, \mathcal{P})) \Rightarrow \mathrm{Ext}_{X \times Y}^{i+j}(\mathcal{P}, \mathcal{P}) \qquad (2.1)$$

and

$$E_2^{ij} = H^i(Y, \mathcal{E}xt_p^j(\mathcal{P}, \mathcal{P})) \Rightarrow \mathrm{Ext}_{X \times Y}^{i+j}(\mathcal{P}, \mathcal{P}). \qquad (2.2)$$

Writing $\mathcal{E}xt_p^j(\mathcal{P}, \mathcal{P}) = R^j p_*(\mathcal{P}^* \otimes \mathcal{P})$, etc., they can be viewed as Leray spectral sequences for the two projections.

From (2.1) one then deduces the exact sequence

$$0 \longrightarrow H^1(X, \mathcal{E}xt_q^0(\mathcal{P}, \mathcal{P})) \longrightarrow \mathrm{Ext}_{X \times Y}^1(\mathcal{P}, \mathcal{P}) \longrightarrow H^0(X, \mathcal{E}xt_q^1(\mathcal{P}, \mathcal{P})) \longrightarrow \cdots.$$

Using (1.2) and the assumption that X is a K3 surface, one finds $H^1(X, \mathcal{E}xt_q^0(\mathcal{P}, \mathcal{P})) \simeq H^1(X, \mathcal{O}_X) = 0$ and $H^0(X, \mathcal{E}xt_q^1(\mathcal{P}, \mathcal{P})) \simeq H^0(X, \mathcal{T}_X) = 0$. Hence, $\mathrm{Ext}_{X \times Y}^1(\mathcal{P}, \mathcal{P}) = 0$. Now using the analogous exact sequence obtained from (2.2) one finds $H^1(Y, \mathcal{O}_Y) = H^1(Y, \mathcal{E}xt_p^0(\mathcal{P}, \mathcal{P})) \hookrightarrow \mathrm{Ext}_{X \times Y}^1(\mathcal{P}, \mathcal{P}) = 0$ and hence $H^1(Y, \mathcal{O}_Y) = 0$. Therefore, Y is indeed a K3 surface.

Alternatively, one can use the induced isomorphism between singular or étale cohomology (see Sections 3.1 and 4.3) to exclude Y from being an abelian surface. □

Definition 2.2 Let X be a K3 surface. Any K3 surface Y for which there exists a k-linear exact equivalence $\mathrm{D}^b(X) \simeq \mathrm{D}^b(Y)$ is called a *Fourier–Mukai partner* of X. The set of all such Y up to isomorphisms is denoted

$$\mathrm{FM}(X) := \{Y \mid \mathrm{D}^b(X) \simeq \mathrm{D}^b(Y)\}/_{\simeq}.$$

Note that of course $X \in \mathrm{FM}(X)$ and so this set is never empty.

2.2 Recall the notion of moduli spaces of stable sheaves; see Section **10**.2. Suppose the moduli space $M_H(v) = M_H(v)^s$ of H-stable sheaves E with Mukai vector $v(E) = v$ is projective and two-dimensional. Then $\langle v, v \rangle = 0$ and $M_H(v)$ is in fact a K3 surface; see Corollaries **10**.2.1 and **10**.3.5. Assume furthermore that there exists a universal sheaf $\mathcal{E}$ on $M_H(v) \times X$, i.e. that $M_H(v)$ is a fine moduli space.[2]

Proposition 2.3 *The universal family $\mathcal{E}$ on $M_H(v) \times X$ induces an exact equivalence*

$$\Phi_{\mathcal{E}}: \mathrm{D}^b(M_H(v)) \xrightarrow{\sim} \mathrm{D}^b(X).$$

[2] For k algebraically closed, the moduli space is fine if there exists a $v' \in N(X)$ with $\langle v, v' \rangle = 1$ and H is generic; see Section 10.2.2.

Proof We may assume that k is algebraically closed; cf. the discussion in Section 4.2. As both X and $M_H(v)$ are smooth surfaces with trivial canonical bundle, it suffices to show that $\Phi_\mathcal{E}$ is fully faithful; cf. Lemma 1.7. We want to apply Lemma 1.4 using the spanning class $\Omega := \{k(t) \mid t \in M_H(v)\}$.

For closed points $t_1 \neq t_2$, the corresponding sheaves $E_1 := \mathcal{E}|_{\{t_1\} \times X}$ and $E_2 := \mathcal{E}|_{\{t_2\} \times X}$ are non-isomorphic stable sheaves of the same Mukai vector v. Hence, $\mathrm{Hom}(E_1, E_2) = 0 = \mathrm{Hom}(E_2, E_1)$ and by Serre duality $\mathrm{Ext}^2(E_1, E_2) = 0$. As $\chi(E_1, E_2) = -\langle v, v \rangle = 0$, also $\mathrm{Ext}^1(E_1, E_2) = 0$.

If $t_1 = t_2 =: t$ and so $E_1 \simeq E_2 =: F$, one has $\mathrm{Hom}(E, E) = k$ and by Serre duality $\mathrm{Ext}^2(E, E) = k$. Moreover, the induced map $T_t M_H(v) \xrightarrow{\sim} \mathrm{Ext}^1_X(E, E)$ is an isomorphism; see Proposition **10.1.11**. $\square$

One could also reverse the order of arguments by first proving $\mathrm{D}^b(X) \simeq \mathrm{D}^b(M_H(v))$ under the assumption that $M_H(v) = M_H(v)^s$ is a projective surface and that a universal family exists. Proposition 2.1 then would imply that $M_H(v)$ is a K3 surface; cf. Corollary **10.3.5**.

Example 2.4 Consider an elliptic K3 surface $X \longrightarrow \mathbb{P}^1$ and let $\mathrm{J}^d(X) \longrightarrow \mathbb{P}^1$ be its Jacobian fibration of degree d; see Section **11.4.2**. As was explained there, $\mathrm{J}^d(X) \simeq M_H(v_d)$ for $v_d = (0, [X_t], d)$ and H generic. The moduli space $M_H(v_d) = M_H(v_d)^s$ is fine if there exists a vector v' with $\langle v, v' \rangle = 1$. The latter is equivalent to d and the index d_0 (see Definition 11.4.3) is coprime. Hence,

$$\mathrm{D}^b(\mathrm{J}^d(X)) \simeq \mathrm{D}^b(X)$$

for g.c.d.$(d, d_0) = 1$.

2.3 Let us exhibit some standard Fourier–Mukai transforms before introducing spherical twists, responsible for the rich structure of the group of autoequivalences of derived categories of K3 surfaces.

(i) For any morphism $f : X \longrightarrow Y$ the direct image functor $f_* : \mathrm{D}^b(X) \longrightarrow \mathrm{D}^b(Y)$ and the pull-back $f^* : \mathrm{D}^b(Y) \longrightarrow \mathrm{D}^b(X)$ (both derived) are Fourier–Mukai transforms with kernel $\mathcal{O}_{\Gamma_f} \in \mathrm{Coh}(X \times Y)$, the structure sheaf of the graph $\Gamma_f \subset X \times Y$.

(ii) If $L \in \mathrm{Pic}(X)$, then $\mathrm{D}^b(X) \xrightarrow{\sim} \mathrm{D}^b(X)$, $E \longmapsto L \otimes E$ defines a Fourier–Mukai auto-equivalence with Fourier–Mukai kernel $\Delta_* L$. Here, $\Delta : X \hookrightarrow X \times X$ denotes the diagonal embedding.

(iii) The shift functor $\mathrm{D}^b(X) \xrightarrow{\sim} \mathrm{D}^b(X)$, $E \longmapsto E[1]$ is the Fourier–Mukai transform with kernel $\mathcal{O}_\Delta[1]$. Similarly, the Serre functor for a K3 surface X is the Fourier–Mukai transform with kernel $\mathcal{O}_\Delta[2]$.

Definition 2.5 An object $E \in D^b(X)$ on a K3 surface X is called *spherical* if

$$\text{Ext}^i(E, E) \simeq \begin{cases} k & \text{if } i = 0, 2, \\ 0 & \text{else.} \end{cases}$$

Consider the cone of the composition of the restriction to the diagonal with the trace

$$\mathcal{P}_E := C\left(E^* \boxtimes E \longrightarrow (E^* \boxtimes E)|_\Delta \xrightarrow{\text{tr}} \mathcal{O}_\Delta \right) \in D^b(X \times X).$$

Definition 2.6 The *spherical twist*

$$T_E : D^b(X) \xrightarrow{\sim} D^b(X)$$

associated with a spherical object $E \in D^b(X)$ is the Fourier–Mukai equivalence with kernel $\mathcal{P}_E$, i.e. $T_E := \Phi_{\mathcal{P}_E}$.

The easiest argument to show that T_E is indeed an equivalence uses the spanning class $\{E\} \cup E^\perp$. It is straightforward to check that

$$T_E(E) \simeq E[-1] \quad \text{and} \quad T_E(F) \simeq F \quad \text{for } F \in E^\perp,$$

from which the assumptions of Lemma 1.4 can be easily verified. This argument is due to Ploog and simplifies the original one of Seidel and Thomas; see [254, Prop. 8.6] for details and references. Note that due to Lemma 1.7 full faithfulness of T_E immediately implies that it is an equivalence.

Example 2.7 (i) Any line bundle L on a K3 surface X can be considered as a spherical object in $D^b(X)$, for $\text{Ext}^1(L, L) \simeq H^1(X, \mathcal{O}) = 0$. Note that the two autoequivalences, T_L and $L \otimes (\ )$, associated with a line bundle L are different for all L.

(ii) If $\mathbb{P}^1 \simeq C \subset X$ is a smooth rational curve, then $\mathcal{O}_C(\ell)$ considered as a sheaf on X, or rather as an object in $D^b(X)$, is spherical for all ℓ. Obviously, $\text{Ext}^i(\mathcal{O}_C(\ell), \mathcal{O}_C(\ell))$ is one-dimensional for $\ell = 0, 2$. The vanishing of Ext^1 can be deduced from $\langle v, v \rangle = -2$ for $v = v(\mathcal{O}_C(\ell)) = (0, [C], \ell + 1)$ (see below) or, more geometrically, by observing that $\mathcal{O}_C(\ell)$ really has no first-order deformations.

The group of all k-linear exact autoequivalences of $D^b(X)$ up to isomorphisms shall be denoted

$$\text{Aut}(D^b(X)) := \{\Phi : D^b(X) \xrightarrow{\sim} D^b(X) \mid k\text{-linear, exact}\}/_\simeq.$$

2.4 Any Fourier–Mukai transform $\Phi : D^b(X) \longrightarrow D^b(Y)$ induces a natural map

$$\Phi^K : K(X) \longrightarrow K(Y), \quad [F] \longmapsto [\Phi(F)].$$

For $\Phi = f_*$ or $\Phi = L\otimes(\ )$ this is of course given by the push-forward $f_*\colon K(X) \longrightarrow K(Y)$ and the tensor product with L, respectively. For the spherical twist $\Phi = T_E$ one has

$$T_E^K[F] = [F] - \chi(F,E) \cdot [E];$$

i.e. T_E^K is the reflection associated with the (-2)-class $[E] \in K(X)$.

For an equivalence Φ, one has $\chi(E,F) = \chi(\Phi(E),\Phi(F))$. Thus, the induced Φ^K descends to a homomorphism between the numerical Grothendieck groups

$$\Phi^N \colon N(X) \longrightarrow N(Y).$$

To make this more explicit let us first show that $N(X)$ with the induced pairing $\chi(\ ,\ )$ is isomorphic to the *extended Néron–Severi group* $\mathbb{Z} \oplus \mathrm{NS}(X) \oplus \mathbb{Z} = \mathrm{NS}(X) \oplus U$ with U the hyperbolic plane. We use the Mukai vector to define a map

$$K(X) \longrightarrow \mathbb{Z} \oplus \mathrm{NS}(X) \oplus \mathbb{Z},$$

$$[E] \longmapsto v(E),$$

where $v(E)$ is the Mukai vector of E:

$$
\begin{aligned}
v(E) &= (\mathrm{rk}(E), c_1(E), \chi(E) - \mathrm{rk}(E)) \\
&= (\mathrm{rk}(E), c_1(E), c_1(E)^2/2 - c_2(E) + \mathrm{rk}(E)).
\end{aligned}
$$

Recall, $\langle v_1, v_2 \rangle$ for $v_i := (r_i, \ell_i, s_i)$ is the Mukai pairing $(\ell_1.\ell_2) - r_1 s_2 - s_1 r_2$; cf. Section **9.1.2**. Thus, indeed

$$N(X) = K(X)/_\sim \ \xrightarrow{\ \sim\ } \ \mathbb{Z} \oplus \mathrm{NS}(X) \oplus \mathbb{Z} \simeq \mathrm{NS}(X) \oplus U$$

and we henceforth think of $N(X)$ rather as $N(X) = \mathbb{Z} \oplus \mathrm{NS}(X) \oplus \mathbb{Z}$.

For a Fourier–Mukai equivalence $\Phi\colon \mathrm{D}^b(X) \longrightarrow \mathrm{D}^b(Y)$ the induced homomorphism

$$\Phi^N \colon N(X) \longrightarrow N(Y)$$

sends $v(E)$ to $v(\Phi_{\mathcal{P}}(E))$. In this description the spherical twist T_E acts again as the reflection associated with the (-2)-class $v(E) \in N(X)$:

$$T_E^N \colon N(X) \ \xrightarrow{\ \sim\ } \ N(X), \quad v \longmapsto v + \langle v, v(E) \rangle \cdot v(E). \tag{2.3}$$

Using the multiplicative structure of $N(X)$, one finds that the tensor product $\Phi = L\otimes(\ )$ acts by $\Phi^N \colon v \longmapsto \exp(\ell) \cdot v$, where $\exp(\ell) := \mathrm{ch}(L) = (1, \ell, \ell^2/2)$ for $\ell := c_1(L)$. This is an example of a B-field shift; cf. Section **14.2.3**.

Corollary 2.8 *Any equivalence* $\Phi\colon \mathrm{D}^b(X) \xrightarrow{\ \sim\ } \mathrm{D}^b(Y)$ *between K3 surfaces induces an isometry of the extended Néron–Severi lattices*

$$\Phi^N \colon N(X) \xrightarrow{\ \sim\ } N(Y).$$

In particular, $\rho(X) = \rho(Y)$. $\qquad\qquad\qquad\qquad\qquad\qquad\qquad\qquad\qquad\qquad\quad\square$

Using the Mukai vector of the Fourier–Mukai kernel of an arbitrary Fourier–Mukai transform $\Phi \colon D^b(X) \longrightarrow D^b(Y)$, one can define a homomorphism $\Phi^N \colon N(X) \longrightarrow N(Y)$ (cf. Proposition 3.2) without assuming Φ to be an equivalence. However, Φ^N is in general neither injective nor compatible with the Mukai pairing.

Corollary 2.9 *Any (-2)-class in the numerical Grothendieck group $N(X)$ of a K3 surface over an algebraically closed field can be realized (non-uniquely) as the Mukai vector $v(E)$ of a spherical object $E \in D^b(X)$.*

Proof According to Remark 10.3.3, any (-2)-class $\delta = (r, \ell, s) \in N(X)$ with $r > 0$ can be realized as $v(E)$ of a rigid simple bundle E. The case $r < 0$ follows by taking $v(E[1])$ for $v(E) = -\delta$. If $r = 0$, then after tensoring with a high power of an ample line bundle one can assume $s \neq 0$. If $v(E) = -(s, \ell, 0)$ (which is still a (-2)-class), then $v(T_{\mathcal{O}}(E)) = \delta$. Compare this with the arguments in the proof of Proposition 3.5. □

Remark 2.10 However, in general the usual Néron–Severi lattices $NS(X)$ and $NS(Y)$ are not isomorphic. In fact, in [470] Oguiso shows that for any $n > 0$ there exist n derived equivalent complex projective K3 surfaces $X_1, \ldots, X_n$ with pairwise non-isometric Néron–Severi lattices $NS(X_1), \ldots, NS(X_n)$. See also Stellari's article [575].

Example 2.11 We come back to the Jacobian fibration $J^d(X) \longrightarrow \mathbb{P}^1$ of degree d of an elliptic K3 surface $X \longrightarrow \mathbb{P}^1$. For g.c.d.$(d, d_0) = 1$ we have noted in Example 2.4 that $D^b(J^d(X)) \simeq D^b(X)$ and, therefore, $N(J^d(X)) \simeq N(X)$. In particular,

$$\operatorname{disc} NS(X) = \operatorname{disc} NS(J^d(X)),$$

confirming (4.6) in Section **11.4.2**.

3 Action on Cohomology

It is not surprising that for complex K3 surfaces more detailed information about derived (auto)equivalences can be obtained from Hodge theory. As it turns out, by a beautiful theorem combining work of Mukai and Orlov, whether two complex projective K3 surfaces have equivalent derived categories is determined by their Hodge structures. This is a derived version of the global Torelli theorem **7.5.3**.

In this section we are mostly concerned with complex projective K3 surfaces. For results over other fields, see Section 4.

3.1 When the integral cohomology of a complex K3 surface X

$$H^*(X, \mathbb{Z}) = H^0(X, \mathbb{Z}) \oplus H^2(X, \mathbb{Z}) \oplus H^4(X, \mathbb{Z}) \simeq H^2(X, \mathbb{Z}) \oplus U$$

is viewed with the *Mukai pairing* (cf. Section **9**.1.2)

$$\langle \alpha, \beta \rangle := (\alpha_2.\beta_2) - (\alpha_0.\beta_4) - (\alpha_4.\beta_0)$$

and the cohomological grading is suppressed, it is denoted $\widetilde{H}(X,\mathbb{Z})$. It comes with a weight-two Hodge structure defined by

$$\widetilde{H}^{1,1}(X) := H^{1,1}(X) \oplus (H^0 \oplus H^4)(X) \quad \text{and} \quad \widetilde{H}^{2,0}(X) := H^{2,0}(X),$$

which is in fact determined by $\widetilde{H}^{2,0}(X)$ and the condition that $\widetilde{H}^{2,0}(X)$ and $\widetilde{H}^{1,1}(X)$ are orthogonal with respect to the Mukai pairing.

Moreover, the numerical Grothendieck (or extended Néron–Severi) group can then be identified as

$$N(X) \simeq \widetilde{H}^{1,1}(X) \cap \widetilde{H}(X,\mathbb{Z}) = (H^{1,1}(X) \cap H^2(X,\mathbb{Z})) \oplus (H^0 \oplus H^4)(X,\mathbb{Z})$$

and the Mukai vector $v(E)$ of any complex $E \in D^b(X)$ can be written as $v(E) = (\mathrm{rk}(E), c_1(E), c_1(E)^2/2 - c_2(E) + \mathrm{rk}(E)) \in N(X) \subset \widetilde{H}(X,\mathbb{Z})$. That $v(E)$ is indeed an integral class can be deduced from the fact that the intersection pairing on $H^2(X,\mathbb{Z})$ is even or by using $c_1(E)^2/2 - c_2(E) + \mathrm{rk}(E) = \chi(E) - \mathrm{rk}(E)$. As it turns out, this still holds for objects on the product of two K3 surfaces, because of the following technical lemma due Mukai; see [254, Lem. 10.6].

Lemma 3.1 *For any complex* $\mathcal{P} \in D^b(X \times Y)$ *on the product of two K3 surfaces* X *and* Y, *the Mukai vector* $v(\mathcal{P}) := \mathrm{ch}(\mathcal{P})\sqrt{\mathrm{td}(X \times Y)} \in H^*(X \times Y, \mathbb{Q})$ *is integral, i.e. contained in* $H^*(X \times Y, \mathbb{Z})$.

Proposition 3.2 *Let* $\Phi_{\mathcal{P}} \colon D^b(X) \xrightarrow{\sim} D^b(Y)$ *be a derived equivalence between K3 surfaces* X *and* Y. *Then the cohomological Fourier–Mukai transform* $\alpha \longmapsto p_*(q^*\alpha.v(\mathcal{P}))$ *defines an isomorphism of Hodge structures*

$$\Phi_{\mathcal{P}}^H \colon \widetilde{H}(X,\mathbb{Z}) \xrightarrow{\sim} \widetilde{H}(Y,\mathbb{Z})$$

which is compatible with the Mukai pairing, i.e. $\Phi_{\mathcal{P}}^H$ *is a Hodge isometry.*

Proof Since by the lemma $v(\mathcal{P})$ is integral, $\Phi_{\mathcal{P}}^H$ maps $H^*(X,\mathbb{Z})$ to $H^*(Y,\mathbb{Z})$. Note however that it usually does not respect the grading. Applying the same argument to its inverse $\Phi_{\mathcal{P}_L} = \Phi_{\mathcal{P}_R}$ and using that $\Phi_{\mathcal{O}_\Delta}^H = \mathrm{id}$, one finds that $\Phi_{\mathcal{P}}^H$ is an isomorphism of $\mathbb{Z}$-modules.

We have to prove that it preserves the Mukai pairing and the Hodge structures. As $v(\mathcal{P})$ is an algebraic class and thus $v(\mathcal{P}) \in \bigoplus H^{i,i}(X \times Y)$, clearly $\Phi_{\mathcal{P}}^H(H^{2,0}(X)) = H^{2,0}(Y)$. As $\widetilde{H}^{1,1} \perp \widetilde{H}^{2,0}$, it remains to verify the compatibility with the Mukai pairing, which for classes in $N(X) \subset \widetilde{H}(X,\mathbb{Z})$ follows from $\chi(E,F) = \chi(\Phi(E), \Phi(F))$. To prove it for arbitrary classes in $\widetilde{H}(X,\mathbb{Z})$, it suffices to check that $\langle \Phi_{\mathcal{P}}^H(\alpha), \beta \rangle = \langle \alpha, \Phi_{\mathcal{P}}^{H^{-1}}(\beta) \rangle$, which can be proved by using $\Phi_{\mathcal{P}}^{-1} = \Phi_{\mathcal{P}_L}$ and applying the projection formula for the two projections of $X \times Y$ on cohomology. See [254, Prop. 5.44] for details. □

Remark 3.3 The cohomological Fourier–Mukai transform $\Phi_{\mathcal{P}}^H \colon \widetilde{H}(X,\mathbb{Z}) \xrightarrow{\sim} \widetilde{H}(Y,\mathbb{Z})$, $\alpha \longmapsto p_*(q^*\alpha.v(\mathcal{P}))$ does indeed extend $\Phi_{\mathcal{P}}^N \colon N(X) \longrightarrow N(Y)$, $v(E) \longmapsto v(\Phi_{\mathcal{P}}(E))$. This is a consequence of the Grothendieck–Riemann–Roch formula applied to the projection $p \colon X \times Y \longrightarrow Y$, which shows $\mathrm{ch}(p_*(q^*E \otimes \mathcal{P}))\mathrm{td}(Y) = p_*(\mathrm{ch}(q^*E \otimes \mathcal{P})\mathrm{td}(X \times Y))$. Note that the Grothendieck–Riemann–Roch formula is also used for the diagonal embedding $\Delta \colon X \hookrightarrow X \times X$ to show that $\Phi_{\mathcal{O}_\Delta}^H = \mathrm{id}$, which was used in the above proof.

Example 3.4 (i) For a spherical twist $T_E \colon \mathrm{D}^b(X) \xrightarrow{\sim} \mathrm{D}^b(X)$ the induced action on cohomology is again just the reflection

$$T_E^H = s_{v(E)} \colon \alpha \longmapsto \alpha + \langle \alpha, v(E) \rangle \cdot v(E)$$

in the hyperplane orthogonal to the (-2)-class $v(E) \in N(X) = \widetilde{H}^{1,1}(X,\mathbb{Z})$; cf. (2.3). In particular, $T_{\mathcal{O}}^H$ is the identity on $H^2(X,\mathbb{Z})$ and acts by $(r,0,s) \longmapsto (-s,0,-r)$ on $(H^0 \oplus H^4)(X,\mathbb{Z})$. Another important example is the case $E = \mathcal{O}_C(-1)$ with $C \simeq \mathbb{P}^1$. Then T_E^H is the reflection $s_{[C]}$; cf. Remark 8.2.10.

(ii) The autoequivalence of $\mathrm{D}^b(X)$ given by $E \longmapsto L \otimes E$ for some line bundle L acts on cohomology via multiplication with $\exp(\ell) = (1, \ell, \ell^2/2) \in \widetilde{H}^{1,1}(X,\mathbb{Z})$, where $\ell = c_1(L)$.

(iii) Suppose a K3 surface X is isomorphic to a fine moduli space $M_H(v) = M_H(v)^s$ of stable sheaves on Y and $\mathcal{P} = \mathcal{E}$ is a universal family. For the induced equivalence $\Phi_{\mathcal{E}} \colon \mathrm{D}^b(X) \xrightarrow{\sim} \mathrm{D}^b(Y)$ (see Proposition 2.3) one has $\Phi_{\mathcal{E}}^H(0,0,1) = v$, because $(0,0,1) = v(k(t))$ for any $t \in X = M_H(v)^s$.

3.2 Using the global Torelli theorem 7.5.3, the above proposition has been completed by Orlov in [483] to yield the following proposition.

Proposition 3.5 *Two complex projective K3 surfaces X and Y are derived equivalent if and only if there exists a Hodge isometry $\widetilde{H}(X,\mathbb{Z}) \simeq \widetilde{H}(Y,\mathbb{Z})$.*

Proof We copy the proof from [254, Sec. 10.2]. Only the 'if' remains to be verified. So suppose $\varphi \colon \widetilde{H}(X,\mathbb{Z}) \xrightarrow{\sim} \widetilde{H}(Y,\mathbb{Z})$ is a Hodge isometry. It will be changed by Hodge isometries induced by Fourier–Mukai equivalences until the classical global Torelli theorem applies. We let $v := (r, \ell, s) := \varphi(0,0,1)$.

(i) Assume first that $v = \pm(0,0,1)$. As $(0,0,1)^\perp = H^2 \oplus H^4$, the Hodge isometry φ then induces a Hodge isometry $H^2(X,\mathbb{Z}) \xrightarrow{\sim} H^2(Y,\mathbb{Z})$. Hence, $X \simeq Y$ by the global Torelli theorem 7.5.3 and, in particular, $\mathrm{D}^b(X) \simeq \mathrm{D}^b(Y)$.

(ii) Next suppose $r \neq 0$. Changing φ by a sign if necessary, one can assume $r > 0$. Then consider the moduli space $M := M_H(v)^s$ of stable sheaves on Y with Mukai vector v. As $\langle v, v \rangle = \langle(0,0,1),(0,0,1)\rangle = 0$, M is two-dimensional. Using that v is primitive and choosing H generic, M is seen to be projective and hence a K3 surface;

see Corollary **10**.3.5. It follows from the existence of $v' := \varphi(-1,0,0)$ with $\langle v, v' \rangle = 1$ that there exists a universal sheaf $\mathcal{E}$ on $Y \times M$; see Section **10**.2.2. By Proposition 2.3 the Fourier–Mukai transform $\Phi_{\mathcal{E}} : D^b(M) \xrightarrow{\sim} D^b(Y)$ is an equivalence with $\Phi_{\mathcal{E}}^H(0,0,1) = v$. Hence, for the composition one finds $\Phi_{\mathcal{E}}^{H^{-1}}(\varphi(0,0,1)) = (0,0,1)$. Therefore, $X \simeq M$ by step (i) and in particular $D^b(X) \simeq D^b(M) \simeq D^b(Y)$.

(iii) If $v = (0, \ell, s)$ with $\ell \neq 0$, then compose φ first with the Hodge isometry $\exp(c_1(L))$ for some $L \in \mathrm{NS}(Y)$ such that $s + (c_1(L).\ell) \neq 0$ and then with the Hodge isometry $T_{\mathcal{O}_Y}^H$; cf. Example 3.4. The new Hodge isometry satisfies the assumption of step (ii). As $\exp(c_1(L))$ is induced by the equivalence $L \otimes (\ )$ and $T_{\mathcal{O}_Y}^H$ by the spherical twist $T_{\mathcal{O}_Y}$, one concludes also in this case that $D^b(X) \simeq D^b(Y)$. (Alternatively, one could try to use a moduli space $M_H(v)$ for $v = (0, \ell, s)$ with $s \neq 0$ and then argue as in (ii); see Proposition **10**.2.5. However, for the non-emptiness one would need to add hypotheses on ℓ as in Theorem **10**.2.7.)   □

Remark 3.6   The proof actually reveals that for derived equivalent K3 surfaces X and Y either $X \simeq Y$ or X is isomorphic to a moduli space of stable sheaves of positive rank on Y, i.e. $X \simeq M_{H_Y}(v)^s$. In [255] it has been shown that in the latter case X is in fact isomorphic to a moduli space of μ-stable(!) vector bundles(!) on Y.

Using more lattice theory, the above result can also be stated as follows.

Corollary 3.7   *Two complex projective K3 surfaces X and Y are derived equivalent if and only if there exists a Hodge isometry $T(X) \simeq T(Y)$ between their transcendental lattices.*

Proof   This has been stated already as Corollary **14**.3.12. Here is the proof again.

Let $\varphi : T(X) \xrightarrow{\sim} T(Y)$ be a Hodge isometry. We use Nikulin's theorem **14**.1.12 and Remark **14**.1.13, (iii), to conclude that due to the existence of the hyperbolic plane $(H^0 \oplus H^4)(X, \mathbb{Z})$ in $T(X)^\perp \subset \widetilde{H}(X, \mathbb{Z})$, the two embeddings

$$T(X) \hookrightarrow \widetilde{H}(X, \mathbb{Z}) \quad \text{and} \quad T(X) \xrightarrow{\sim} T(Y) \hookrightarrow \widetilde{H}(Y, \mathbb{Z})$$

into the two lattices $\widetilde{H}(X, \mathbb{Z})$ and $\widetilde{H}(Y, \mathbb{Z})$, which are abstractly isomorphic, differ by an isometry; i.e. the isometry φ extends to an isometry $\widetilde{\varphi} : \widetilde{H}(X, \mathbb{Z}) \xrightarrow{\sim} \widetilde{H}(Y, \mathbb{Z})$. Automatically, $\widetilde{\varphi}$ is also compatible with Hodge structures.   □

Similar lattice theoretic tricks, e.g. using the existence of hyperbolic planes in the orthogonal complement of the transcendental lattice, can be used to show that particular K3 surfaces do not admit any non-trivial Fourier–Mukai partners.

Corollary 3.8   *In the following cases, a K3 surface X does not admit any non-isomorphic Fourier–Mukai partners, i.e.*

$$\mathrm{FM}(X) = \{X\}.$$

(i) X admits an elliptic fibration with a section.

(ii) $\rho(X) \geq 12$.

(iii) $\rho(X) \geq 3$ and disc $NS(X)$ is square free.

Proof (i) The Picard lattice $NS(X)$ contains a hyperbolic plane $U \hookrightarrow NS(X)$ spanned by X_t and a section C_0. Therefore, $FM(X) = \{X\}$; see Remark **14**.3.11. Note, however, that for elliptic K3 surfaces without a section the situation is of course different; see Example 2.4.

(ii) Any Hodge isometry $T(X) \xrightarrow{\sim} T(Y)$ extends to a Hodge isometry $H^2(X, \mathbb{Z}) \simeq H^2(Y, \mathbb{Z})$ and hence by the global Torelli theorem $X \simeq Y$. Compare this with Corollary **14**.3.10.

(iii) By Remark **14**.0.1, $A(T(X))$ is cyclic, i.e. $\ell(T(X)) \leq 1$, and, therefore, by Theorem **14**.1.12 the embedding $T(X) \hookrightarrow H^2(X, \mathbb{Z})$ is unique. $\quad\quad\square$

Example 3.9 For a Kummer surface X associated with a complex abelian surface one therefore has $FM(X) = \{X\}$. This observation can be used to prove that for two complex abelian surfaces A and B and their associated Kummer surfaces X and Y one has $D^b(A) \simeq D^b(B)$ if and only if $X \simeq Y$. See the articles by Hosono et al. and Stellari [245, 576] and the discussion in Section **3**.2.5.

Proposition 3.10 *Let X be a projective K3 surface over an algebraically closed field k. Then X has only finitely many Fourier–Mukai partners, i.e.*

$$|FM(X)| < \infty.$$

Proof We sketch the argument due to Bridgeland and Maciocia [87] for complex projective K3 surfaces. For arbitrary fields, see the article [371] by Lieblich and Olsson.

First recall that an equivalence $D^b(X) \simeq D^b(Y)$ induces (Hodge) isometries $T(X) \simeq T(Y)$ and $N(X) \simeq N(Y)$. From the latter one deduces that $NS(X)$ and $NS(Y)$ have the same genus, so that there are only finitely many lattices realized as $NS(Y)$ with $Y \in FM(X)$; cf. Section **14**.0.1. Hence, it is enough to show that for any K3 surface X there are only finitely many isomorphism classes of K3 surfaces Y with $D^b(X) \simeq D^b(Y)$ and $NS(X) \simeq NS(Y)$. Now, $H^2(Y, \mathbb{Z})$ of such a Y sits in

$$T(X) \oplus NS(X) \simeq T(Y) \oplus NS(Y) \subset H^2(Y, \mathbb{Z}) \subset (T(X) \oplus NS(X))^*.$$

The isomorphism is a Hodge isometry and all inclusions are of finite index. Hence, since the Hodge structure of $H^2(Y, \mathbb{Z})$ is determined by the one on $T(Y) \oplus NS(Y)$, there are only finitely many Hodge structures that can be realized as $H^2(Y, \mathbb{Z})$. But the isomorphism type of Y is determined by the Hodge structure $H^2(Y, \mathbb{Z})$ due to the global Torelli theorem **7**.5.3. $\quad\quad\square$

Remark 3.11 Mukai applies the same techniques in [429] to prove special cases of the Hodge conjecture for the product $X \times Y$ of two K3 surfaces X and Y. More precisely he proves that any class in $H^{2,2}(X \times Y, \mathbb{Q})$ that induces an isometry $T(X)_\mathbb{Q} \xrightarrow{\sim} T(Y)_\mathbb{Q}$ is algebraic provided $\rho(X) \geq 11$. In [455] Nikulin was able to weaken the hypothesis to $\rho(X) \geq 5$.

3.3 Analogously to the finer version of the global Torelli theorem **7.5.3**, saying that a Hodge isometry $H^2(X, \mathbb{Z}) \xrightarrow{\sim} H^2(Y, \mathbb{Z})$ can be lifted to a (unique) isomorphism if it maps a Kähler class to a Kähler class, one can refine the above technique to determine which Hodge isometries $\varphi \colon \widetilde{H}(X, \mathbb{Z}) \xrightarrow{\sim} \widetilde{H}(Y, \mathbb{Z})$ are induced by derived equivalences. The first step, due to Hosono et al. [246] and Ploog, is to show that for any φ there exists a Fourier–Mukai equivalence $\Phi_\mathcal{P} \colon D^b(X) \xrightarrow{\sim} D^b(Y)$ with $\Phi_\mathcal{P}^H = \varphi \circ (\pm \mathrm{id}_{H^2})$; cf. [254, Cor. 10.2]. Here, $-\mathrm{id}_{H^2}$ denotes the Hodge isometry of $\widetilde{H}(X, \mathbb{Z})$ that acts as id on $(H^0 \oplus H^4)(X, \mathbb{Z})$ and as $-\mathrm{id}$ on $H^2(X, \mathbb{Z})$.

For the next step, one needs to introduce the orientation of the positive directions of $\widetilde{H}(X, \mathbb{Z})$. Recall that the Mukai pairing on $\widetilde{H}(X, \mathbb{Z})$ has signature $(4, 20)$; in particular, there exist four-dimensional real subspaces $H_X \subset \widetilde{H}(X, \mathbb{R})$ with $\langle \, , \, \rangle|_H$ positive definite. Although the space H_X is not unique, orientations of, say, $H_X \subset \widetilde{H}(X, \mathbb{R})$ and of $H_Y \subset \widetilde{H}(Y, \mathbb{R})$ can be compared via isometries $\varphi \colon \widetilde{H}(X, \mathbb{Z}) \xrightarrow{\sim} \widetilde{H}(Y, \mathbb{Z})$ using the orthogonal projection $\varphi(H_X) \xrightarrow{\sim} H_Y$.

In fact, associated with an ample class $\ell \in H^{1,1}(X, \mathbb{Z})$ there is a natural $H_X(\ell) \subset \widetilde{H}(X, \mathbb{R})$ spanned by $\mathrm{Re}(\sigma), \mathrm{Im}(\sigma), \mathrm{Re}(\exp(i\ell))$, and $\mathrm{Im}(\exp(i\ell))$, where $0 \neq \sigma \in H^{2,0}(X)$ and $\exp(i\ell) = (1, i\ell, -\ell^2/2)$. Moreover, $H_X(\ell)$ comes with a natural orientation fixed by the given ordering of the generators. It is straightforward to see that under orthogonal projections these orientations of $H_X(\ell)$ and $H_X(\ell')$ for two ample classes (or, more generally, classes ℓ, ℓ' in the positive cone $\mathcal{C}_X \subset H^{1,1}(X, \mathbb{R})$) coincide. We call this the *natural orientation* of the four positive directions of $\widetilde{H}(X, \mathbb{Z})$.

Combining the above results with a deformation theoretic argument developed in [268] one eventually obtains the following theorem.

Theorem 3.12 *Let X and Y be complex projective K3 surfaces. For a Hodge isometry $\varphi \colon \widetilde{H}(X, \mathbb{Z}) \xrightarrow{\sim} \widetilde{H}(Y, \mathbb{Z})$ the following conditions are equivalent:*

(i) *There exists a Fourier–Mukai equivalence $\Phi_\mathcal{P} \colon D^b(X) \xrightarrow{\sim} D^b(Y)$ with $\varphi = \Phi_\mathcal{P}^H$.*
(ii) *The natural orientations of the four positive directions of $\widetilde{H}(X, \mathbb{Z})$ and $\widetilde{H}(Y, \mathbb{Z})$ coincide under φ.*

For $X = Y$ this has the following immediate consequence.

Corollary 3.13 *The image of the map*

$$\mathrm{Aut}(D^b(X)) \longrightarrow \mathrm{Aut}(\widetilde{H}(X, \mathbb{Z})), \quad \Phi \longmapsto \Phi^H$$

is the subgroup $\mathrm{Aut}^+(\widetilde{H}(X,\mathbb{Z})) \subset \mathrm{Aut}(\widetilde{H}(X,\mathbb{Z}))$ *of all orientation preserving Hodge isometries of* $\widetilde{H}(X,\mathbb{Z})$. $\qquad\qquad\square$

3.4 Viewing $\Phi \longmapsto \Phi^H$ as a representation

$$\rho \colon \mathrm{Aut}(\mathrm{D}^{\mathrm{b}}(X)) \longtwoheadrightarrow \mathrm{Aut}^+(\widetilde{H}(X,\mathbb{Z})),$$

the study of $\mathrm{Aut}(\mathrm{D}^{\mathrm{b}}(X))$ reduces to a description of $\mathrm{Ker}(\rho)$. First note that this kernel is non-trivial. Apart from the double shift $E \longmapsto E[2]$, also squares T_E^2 of all spherical twists T_E are contained in $\mathrm{Ker}(\rho)$ (and usually not, and probably never, contained in $\mathbb{Z}[2]$). Due to the abundance of spherical objects (recall that all line bundles are spherical) $\mathrm{Ker}(\rho)$ is a very rich and complicated group. A conjecture of Bridgeland [86] describes it as a fundamental group. We state this conjecture in a slightly different form following [263, Sec. 5].

First consider the finite index subgroup

$$\mathrm{Aut}_s(\mathrm{D}^{\mathrm{b}}(X)) \subset \mathrm{Aut}(\mathrm{D}^{\mathrm{b}}(X))$$

of all Φ with $\Phi^H = \mathrm{id}$ on $T(X)$. The fact that its quotient is a finite cyclic group is based on the same argument that proves that $\mathrm{Aut}_s(X) \subset \mathrm{Aut}(X)$ is of finite index; see Section **15**.1.3. Observe that all spherical twists T_E as well as the equivalences $L \otimes (\ )$ are contained in $\mathrm{Aut}_s(\mathrm{D}^{\mathrm{b}}(X))$.

Next let $D \subset \mathbb{P}(N(X)_\mathbb{C})$ be the period domain as defined in Section **6**.1.1. Note that, as $N(X)$ has signature $(2, \rho(X))$, this period domain has two connected components $D = D^+ \sqcup D^-$ which are interchanged by complex conjugation. Then let

$$D_0 := D \setminus \bigcup_{\delta \in \Delta} \delta^\perp,$$

where $\Delta := \{\delta \in N(X) \mid \delta^2 = -2\}$. Compare this with Remark **6**.3.7. The discrete group

$$\widetilde{\mathrm{O}}(N(X)) := \{g \in \mathrm{O}(N(X)) \mid \overline{g} = \mathrm{id} \text{ on } N(X)^*/N(X)\}$$

acts on the period domain D and preserves the open subset $D_0 \subset D$. The quotient of this action is considered as an orbifold or a stack $[\widetilde{\mathrm{O}}(N(X)) \setminus D_0]$. The analogy with the moduli space of polarized K3 surfaces as described in Section **6**.4 (see in particular Corollary **6**.4.3 and Remark **6**.3.7) is intentional and motivated by mirror symmetry.

Conjecture 3.14 (Bridgeland)   *For a complex projective K3 surface X there exists a natural isomorphism*

$$\mathrm{Aut}_s(\mathrm{D}^{\mathrm{b}}(X))/\mathbb{Z}[2] \simeq \pi_1^{\mathrm{st}}[\widetilde{\mathrm{O}}(N(X)) \setminus D_0].$$

The advantage of this form of Bridgeland's conjecture is that it describes the finite index subgroup $\mathrm{Aut}_s(\mathrm{D}^b(X)) \subset \mathrm{Aut}(\mathrm{D}^b(X))$ as the fundamental group of a smooth Deligne–Mumford stack, which is very close to the fundamental group of a quasi-projective variety and in particular finitely generated. In contrast, the original form of the conjecture in [86] describes $\mathrm{Ker}(\rho)$ as a fundamental group of a complex manifold that is not quasi-projective, and, indeed, $\mathrm{Ker}(\rho) \subset \mathrm{Aut}_s(\mathrm{D}^b(X))$ is a subgroup that is not finitely generated anymore.

Remark 3.15 For $\rho(X) = 1$ the conjecture has recently been proved by Bayer and Bridgeland [40]. In this case Kawatani [288] had shown that the conjecture is equivalent to the statement that $\mathrm{Ker}(\rho)$ is the product of $\mathbb{Z}$ (the even shifts) and the free group generated by all T_E^2 with locally free spherical sheaves E.

4 Twisted, Non-projective, and in Positive Characteristic

In this final section we consider K3 surfaces that are twisted or defined over fields other than $\mathbb{C}$ or that are not projective. Most of what has been explained for complex projective K3 surfaces still holds, or at least is expected to hold, for twisted projective K3 surfaces and for projective K3 surfaces over arbitrary algebraically closed fields, but non-projective complex K3 surfaces behave slightly differently.

4.1 Let X be a projective K3 surface and let $\mathrm{Br}(X)$ be its Brauer group. As explained in the Appendix, to any class $\alpha \in \mathrm{Br}(X)$ one can associate the abelian category $\mathrm{Coh}(X, \alpha)$ of α-twisted sheaves. Its bounded derived category shall be denoted by $\mathrm{D}^b(X, \alpha)$. Two twisted K3 surfaces (X, α) and (Y, β) are called *derived equivalent* if there exists an exact linear equivalence $\mathrm{D}^b(X, \alpha) \simeq \mathrm{D}^b(Y, \beta)$. The Fourier–Mukai formalism can be developed in the twisted context and, as proved by Canonaco and Stellari in [100], any exact linear equivalence between twisted derived categories is of Fourier–Mukai type.

It is quite natural to consider twisted K3 surfaces even if one is a priori interested only in untwisted ones. This was first advocated by Căldăraru in [96]. Suppose X is a K3 surface and $v \in N(X)$ is a primitive Mukai vector with $\langle v, v \rangle = 0$. For a generic ample line bundle H the moduli space of stable sheaves $M_H(v)$ is a smooth projective surface and in fact a K3 surface; see Corollary **10**.3.5. However, $M_H(v)$ is in general not a fine moduli space and, therefore, a priori not derived equivalent to X. The obstruction to the existence of a universal family is a Brauer class $\alpha \in \mathrm{Br}(M_H(v))$ (see Section **10**.2.2), and an $\alpha \boxtimes 1$-twisted universal sheaf on $M_H(v) \times X$ exists. The analogous Fourier–Mukai formalism then yields an equivalence

$$\mathrm{D}^b(M_H(v), \alpha^{-1}) \xrightarrow{\sim} \mathrm{D}^b(X);$$

see [96]. As non-fine moduli spaces do exist, one then is naturally led to also consider twisted K3 surfaces. In fact, an untwisted K3 surface X usually has more truly twisted Fourier–Mukai partners than untwisted ones. For example, in [383] Ma proved that a K3 surface X (untwisted) with $\mathrm{Pic}(X) \simeq \mathbb{Z} \cdot H$ with $(H)^2 = 2d$ is derived equivalent to a twisted K3 surface (Y, β) with $\mathrm{ord}(\beta) = n$ if and only if $n^2 | d$; cf. Corollary 3.8.

In order to approach derived equivalences of twisted complex K3 surfaces via Hodge structures, as done before in the untwisted case, let us first identify the Brauer group $\mathrm{Br}(X)$ of a complex K3 surface X with the subgroup

$$\mathrm{Br}(X) \simeq H^2(X, \mathcal{O}_X^*)_{\mathrm{tor}} \subset H^2(X, \mathcal{O}_X^*) = H^2(X, \mathcal{O}_X)/H^2(X, \mathbb{Z})$$

of all torsion classes; cf. Section **18**.1.2.

Via the exponential sequence, any class $\alpha \in \mathrm{Br}(X)$ can thus be realized as the $(0, 2)$-part of a class $B \in H^2(X, \mathbb{Q})$, which is unique up to $\mathrm{NS}(X)_\mathbb{Q}$ and $H^2(X, \mathbb{Z})$:

$$H^2(X, \mathbb{Q}) \longrightarrow \mathrm{Br}(X), \quad B \longmapsto \alpha_B.$$

Definition 4.1 The natural Hodge structure of weight two $\widetilde{H}(X, \alpha_B, \mathbb{Z})$ associated with a twisted K3 surface (X, α_B) is the Mukai lattice $\widetilde{H}(X, \mathbb{Z})$ with (p, q)-part given by

$$\widetilde{H}^{p,q}(X, \alpha_B) := \exp(B) \cdot \widetilde{H}^{p,q}(X).$$

Here, $\exp(B) := 1 + B + B^2/2 \in H^*(X, \mathbb{Q})$ acts by multiplication on $\widetilde{H}(X, \mathbb{C})$ which, as is straightforward to check, preserves the Mukai pairing. For the abstract version, see Section 14.2.3 and also [256] for a survey of various aspects of this construction and further references.

More concretely, $\widetilde{H}^{2,0}(X, \alpha_B)$ is spanned by

$$\sigma + B \wedge \sigma \in H^{2,0}(X) \oplus H^4(X, \mathbb{C}),$$

where $0 \neq \sigma \in H^{2,0}(X)$, and its orthogonal complement is

$$\widetilde{H}^{1,1}(X, \alpha_B) := (\widetilde{H}^{2,0} \oplus \widetilde{H}^{0,2})(X, \alpha_B)^\perp \subset \widetilde{H}(X, \mathbb{C}).$$

Clearly, $\widetilde{H}^{2,0}(X, \alpha_B) = H^{2,0}(X)$ for $B \in \mathrm{NS}(X)_\mathbb{Q}$. Also, multiplication with $\exp(B)$ for $B \in H^2(X, \mathbb{Z})$ defines an isometry of Hodge structures $\widetilde{H}(X, \mathbb{Z}) \simeq \widetilde{H}(X, \alpha_B, \mathbb{Z})$. Similarly, one finds that the isomorphism type of the Hodge structure $\widetilde{H}(X, \alpha_B, \mathbb{Z})$ depends only on $\alpha_B \in \mathrm{Br}(X) \subset H^2(X, \mathcal{O}_X^*)$ and not on the class B. So we shall simply write $\widetilde{H}(X, \alpha, \mathbb{Z})$ for the Hodge structure of weight two of a twisted K3 surface. Note that the natural grading of $H^*(X, \mathbb{Z})$ has been lost, so that in general $\widetilde{H}^{1,1}(X, \alpha, \mathbb{Z})$ does not contain the (or in fact any) hyperbolic plane $(H^0 \oplus H^4)(X, \mathbb{Z})$.

The following analogue of Proposition 3.5 was proved in [272]. Unlike the untwisted version, the natural orientation of the four positive directions (cf. Section 3.3) $\widetilde{H}(X, \mathbb{R})$, which via $\exp(B)$ is isometric to $\widetilde{H}(X, \alpha_B, \mathbb{R})$, does matter here.

Proposition 4.2 *Two twisted complex projective K3 surfaces* (X, α) *and* (Y, β) *are derived equivalent if and only if there exists a Hodge isometry* $\tilde{H}(X, \alpha, \mathbb{Z}) \simeq \tilde{H}(Y, \beta, \mathbb{Z})$ *respecting the natural orientation of the four positive directions.*

Remark 4.3   (i) Originally it was conjectured that also the analogue of Corollary 3.7 would hold for twisted K3 surfaces, but see [271, Rem. 4.10].

(ii) The orientation of the four positive direction matters, as there is in general no Hodge isometry of $\tilde{H}(X, \alpha, \mathbb{Z})$ similar to $-\mathrm{id}_{H^2}$ in the untwisted case that would reverse it.

(iii) The analogue of Proposition 3.10 holds also in the twisted case. So, for a twisted complex projective K3 surface $(X, \alpha \in \mathrm{Br}(X))$ there exist only finitely many isomorphism classes of twisted complex projective K3 surfaces $(Y, \beta \in \mathrm{Br}(Y))$ with $\mathrm{D}^b(X, \alpha) \simeq \mathrm{D}^b(Y, \beta)$.

Remark 4.4   An untwisted K3 surface always admits spherical object; e.g. every line bundle, even the trivial one, is an example. However, many twisted K3 surfaces do not. In particular, on any twisted K3 surface (X, α) locally free α-twisted sheaves are all of rank at least the order of $\alpha \in \mathrm{Br}(X)$; see Remark **18**.1.3. In this case, one expects the group $\mathrm{Aut}(\mathrm{D}^b(X, \alpha))$ to be rather simple, and in particular the kernel of

$$\rho \colon \mathrm{Aut}(\mathrm{D}^b(X, \alpha)) \longrightarrow \mathrm{Aut}(\tilde{H}(X, \alpha, \mathbb{Z}))$$

should be spanned by [2]. That this is indeed the case has been proved in [267], which in particular proves Conjecture 3.14 for all twisted K3 surfaces (X, α) with $\Delta = \emptyset$ or, equivalently, $D_0 = D$.

4.2   It was only in Section 3 that the K3 surfaces were assumed to be defined over $\mathbb{C}$. We shall now explain how one can reduce to this case if $\mathrm{char}(k) = 0$. First of all, every K3 surface X over a field k is defined over a finitely generated field k_0; i.e. there exists a K3 surface X_0 over k_0 such that $X \simeq X_0 \times_{k_0} k$. Similarly, if $\Phi_{\mathcal{P}} \colon \mathrm{D}^b(X) \xrightarrow{\sim} \mathrm{D}^b(Y)$ is a Fourier–Mukai equivalence, then there exists a finitely generated field k_0 such that X, Y, and $\mathcal{P}$ are defined over k_0. It is not difficult to show that then the k_0-linear Fourier–Mukai transform $\Phi_{\mathcal{P}_0} \colon \mathrm{D}^b(X_0) \xrightarrow{\sim} \mathrm{D}^b(Y_0)$ is an exact equivalence as well.

Suppose k_0 is algebraically closed and $E \in \mathrm{D}^b(X_0 \times_{k_0} k)$ is spherical. Then, as for line bundles (see Lemma 17.2.2), there exists a spherical object $E_0 \in \mathrm{D}^b(X_0)$ which yields E after base change to k. Similarly, every automorphism f of $X_0 \times_{k_0} k$ is obtained by base change from an isomorphism $f_0 \colon X_0 \xrightarrow{\sim} X_0$ over k_0; see Remark 15.2.2. Since analogous statements hold for smooth rational curves and for universal families of stable sheaves, all autoequivalences described explicitly before are defined over the smaller algebraically closed(!) field k_0. In fact, as the kernel $\mathcal{P}$ of any Fourier–Mukai equivalence $\Phi_{\mathcal{P}} \colon \mathrm{D}^b(X_0 \times_{k_0} k) \xrightarrow{\sim} \mathrm{D}^b(Y_0 \times_{k_0} k)$ is rigid, i.e. $\mathrm{Ext}^1(\mathcal{P}, \mathcal{P}) = 0$ (see the proof of Proposition 2.1), any Fourier–Mukai equivalence descends to k_0. The details of the argument are spelled out (for line bundles) in the proof of Lemma **17.2.2**.

Hence, for a K3 surface X_0 over the algebraic closure k_0 of a finitely generated field extension of $\mathbb{Q}$ and for any choice of an embedding $k_0 \hookrightarrow \mathbb{C}$, which always exists, one has

$$\mathrm{Aut}(\mathrm{D}^b(X_0 \times_{k_0} k)) \simeq \mathrm{Aut}(\mathrm{D}^b(X_0)) \simeq \mathrm{Aut}(\mathrm{D}^b(X_0 \times_{k_0} \mathbb{C})).$$

In this sense, for K3 surfaces over algebraically closed fields k with $\mathrm{char}(k) = 0$ the situation is identical to the case of complex K3 surfaces. Understanding the difference between autoequivalences of $\mathrm{D}^b(X_0)$ and $\mathrm{D}^b(X_0 \times_{k_0} k)$ for a non-algebraically closed field k_0 is more subtle. It is related to questions about the field of definitions of line bundles and smooth rational curves; cf. Lemma **17.2.2**.

In [242] Hassett and Tschinkel describe an example with $\mathrm{Pic}(X) \simeq \mathrm{Pic}(X \times_{k_0} \overline{k_0})$ admitting a spherical sheaf on $X \times_{k_0} \overline{k_0}$ not descending to X. The paper is the first one that studies this type of question systematically (for K3 surfaces). Special emphasis is put on the question of (density of) rational points. For example, in the article one finds an example of derived equivalent K3 surfaces $\mathrm{D}^b(X) \simeq \mathrm{D}^b(Y)$ for which $|\mathrm{Aut}(X)| < \infty$ but $|\mathrm{Aut}(Y)| = \infty$. The recent paper [22] contains examples of derived equivalent twisted K3 surfaces (X, α) and (Y, β) over $\mathbb{Q}$ with $(X, \alpha)(\mathbb{Q}) = \emptyset$; i.e. the Brauer class is non-trivial in all $\mathbb{Q}$-rational points, and $(Y, \beta)(\mathbb{Q}) \neq \emptyset$, contrary to an expectation expressed in [242].

4.3 Let us complement the above by a brief discussion of the case of positive characteristic. For the time being, there is only one paper that deals with this, namely Lieblich and Olsson [371], to which we refer for details.

The following is [371, Thm. 5.1], which can be seen as a characteristic-free version of the first step of the proof of Proposition 3.5. Recall that $N(X) \simeq \mathbb{Z} \oplus \mathrm{NS}(X) \oplus \mathbb{Z}$ with $v(k(x)) = (0, 0, 1)$.

Proposition 4.5 *Let* $\Phi \colon \mathrm{D}^b(X) \xrightarrow{\sim} \mathrm{D}^b(Y)$ *be a derived equivalence between K3 surfaces over an algebraically closed field k such that the induced map $\Phi^N \colon N(X) \xrightarrow{\sim} N(Y)$ satisfies $\Phi^N(0, 0, 1) = \pm(0, 0, 1)$. Then $X \simeq Y$.*

Proof As a first step we claim that Φ can be modified by autoequivalences of $\mathrm{D}^b(Y)$ such that the new Φ^N respects the direct sum decomposition $N = \mathbb{Z} \oplus \mathrm{NS} \oplus \mathbb{Z}$. For this, one rearranges the arguments in the proof of Proposition 3.5: if $\Phi^N(1, 0, 0) = (r, L, s)$, then necessarily $r = \pm 1$, and after composing Φ with $L^* \otimes (\ )$ or $L \otimes (\ )$ one can assume that $\Phi^N(1, 0, 0) = \pm(1, 0, 0)$. Then use that $\mathrm{NS} = (\mathbb{Z} \oplus \mathbb{Z})^\perp$.

If $k = \mathbb{C}$, then the Hodge isometry $\Phi^H \colon \widetilde{H}(X, \mathbb{Z}) \xrightarrow{\sim} \widetilde{H}(Y, \mathbb{Z})$ of any Φ that respects the decomposition $N = \mathbb{Z} \oplus \mathrm{NS} \oplus \mathbb{Z}$ restricts to a Hodge isometry $H^2(X, \mathbb{Z}) \xrightarrow{\sim} H^2(Y, \mathbb{Z})$. Hence, $X \simeq Y$ by the global Torelli theorem **7.5.3**.

Let now k be an arbitrary algebraically closed field k of characteristic zero. As X, Y, and the Fourier–Mukai kernel of Φ are described by finitely many equations, there exist K3 surfaces X', Y' over the algebraic closure k' of some finitely generated field k_0 and an equivalence $\Phi'\colon \mathrm{D}^b(X') \xrightarrow{\sim} \mathrm{D}^b(Y')$ such that its base change with respect to $k' \subset k$ yields $\Phi\colon \mathrm{D}^b(X) \xrightarrow{\sim} \mathrm{D}^b(Y)$. Clearly, Φ' still respects the decomposition $N = \mathbb{Z} \oplus \mathrm{NS} \oplus \mathbb{Z}$.

Next choose an embedding $k' \hookrightarrow \mathbb{C}$. The base change $\Phi'_{\mathbb{C}}\colon \mathrm{D}^b(X'_{\mathbb{C}}) \xrightarrow{\sim} \mathrm{D}^b(Y'_{\mathbb{C}})$ of Φ' still respects $N = \mathbb{Z} \oplus \mathrm{NS} \oplus \mathbb{Z}$ and, therefore, $X'_{\mathbb{C}} \simeq Y'_{\mathbb{C}}$. But as both surfaces X' and Y' arc defined over k', also $X' \simeq Y'$ and, therefore, $X \simeq Y$.

In the case of positive characteristic, one first modifies Φ further. By composing it with spherical twists of the form $T_{\mathcal{O}_C(-1)}$ for (-2)-curves $\mathbb{P}^1 \simeq C \subset X$, one can assume that Φ^N maps an ample line bundle L on X to M with M or M^* ample on Y; see Corollary 8.2.9. For this one needs that $T^N_{\mathcal{O}_C(-1)}$ is the reflection in the hyperplane $v(\mathcal{O}_C(-1))^\perp = (0, [C], 0)^\perp$; cf. Example 2.7. In characteristic zero, this step corresponds to changing a Hodge isometry $H^2(X, \mathbb{Z}) \xrightarrow{\sim} H^2(Y, \mathbb{Z})$ to one that maps the ample cone to the ample cone up to sign.

The technically difficult part, which we do not explain in detail, is a lifting argument, which allows one to lift X and Y together with the Fourier–Mukai kernel $\mathcal{P} \in \mathrm{D}^b(X \times Y)$ to characteristic zero $\mathcal{P}_K \in \mathrm{D}^b(X_K \times Y_K)$. As being an equivalence is an open condition, $\Phi_{\mathcal{P}_K}\colon \mathrm{D}^b(X_K) \xrightarrow{\sim} \mathrm{D}^b(Y_K)$ is still an equivalence and it satisfies $\Phi(0, 0, 1) = \pm(0, 0, 1)$. Hence, $X_K \simeq Y_K$ and reducing back to k, while using [396], yields $X \simeq Y$.   □

The following corresponds to [371, Thm. 1.2], which there was proved using crystalline cohomology. For the sake of variety, we present a proof using étale cohomology (which also takes care of the technical assumption char $\neq 2$).[3]

Proposition 4.6 *Assume X and Y are K3 surfaces over a finite field $\mathbb{F}_q$ with equivalent derived categories $\mathrm{D}^b(X) \simeq \mathrm{D}^b(Y)$. Then their Zeta functions (see Section 4.4.1) coincide,*

$$Z(X, t) = Z(Y, t),$$

and, in particular,

$$|X(\mathbb{F}_q)| = |X(\mathbb{F}_q)|.$$

Proof By Orlov's result, the assumed equivalence $\Phi\colon \mathrm{D}^b(X) \xrightarrow{\sim} \mathrm{D}^b(Y)$ is a Fourier–Mukai transform. We let $\mathcal{P} \in \mathrm{D}^b(X \times Y)$ be its kernel. Its Mukai vector

$$v(\mathcal{P}) \in \bigoplus H^{2k}_{\acute{e}t}(\overline{X} \times_{\mathbb{F}_q} \overline{Y}, \mathbb{Q}_\ell(k))$$

---

[3] This proof was prompted by a question of Mircea Mustaţă.

is invariant under the Frobenius action, as $\mathcal{P}$ is defined over $\mathbb{F}_q$. Here, $\overline{X} = X \times_{\mathbb{F}_q} \overline{\mathbb{F}}_q$ and $\overline{Y} = Y \times_{\mathbb{F}_q} \overline{\mathbb{F}}_q$.

Reasoning as for complex K3 surfaces with their singular cohomology, one finds that $\Phi_{\mathcal{P}}$ induces an (ungraded) isomorphism

$$H^0_{\text{ét}}(\overline{X}, \mathbb{Q}_\ell) \oplus H^2_{\text{ét}}(\overline{X}, \mathbb{Q}_\ell(1)) \oplus H^4_{\text{ét}}(\overline{X}, \mathbb{Q}_\ell(2))$$
$$\simeq H^0_{\text{ét}}(\overline{Y}, \mathbb{Q}_\ell) \oplus H^2_{\text{ét}}(\overline{Y}, \mathbb{Q}_\ell(1)) \oplus H^4_{\text{ét}}(\overline{Y}, \mathbb{Q}_\ell(2)),$$

which is compatible with the Frobenius action.

If $\alpha_{i,j}$ and $\beta_{i,j}$ denote the eigenvalues of the Frobenius action on $H^i_{\text{ét}}(\overline{X}, \mathbb{Q}_\ell)$ and $H^i_{\text{ét}}(\overline{Y}, \mathbb{Q}_\ell)$, respectively, then the isomorphism shows

$$\left\{ \alpha_{0,1}, \frac{\alpha_{2,1}}{q}, \dots, \frac{\alpha_{2,22}}{q}, \alpha_{4,1} \right\} = \left\{ \beta_{0,1}, \frac{\beta_{2,1}}{q}, \dots, \frac{\beta_{2,22}}{q}, \beta_{4,1} \right\}.$$

But, clearly, $\alpha_{0,1} = \beta_{0,1}$ and $\alpha_{4,1} = \beta_{4,1}$ and hence $\{\alpha_{2,j}/q\} = \{\beta_{2,j}/q\}$.

Conclude by using the Weil conjectures; see Theorem 4.4.1. $\qquad\square$

Remark 4.7 Note that the same principle applies to derived equivalences for higher-dimensional varieties X and Y over a finite field $\mathbb{F}_q$. However, the induced identification between various sets of eigenvalues of the Frobenius is not enough to conclude equality of their Zeta functions. This corresponds to the problem in characteristic zero that for derived equivalent varieties X and Y one knows $\sum_{p-q=i} h^{p,q}(X) = \sum_{p-q=i} h^{p,q}(Y)$ for all i but not whether the individual Hodge numbers satisfy $h^{p,q}(X) = h^{p,q}(Y)$.

4.4 A complex K3 surface X is projective if and only if there exists a line bundle $L \in \text{Pic}(X)$ with $(L)^2 > 0$; see Remark 1.3.4. A very general complex projective K3 surface has $\text{Pic}(X) \simeq \mathbb{Z}$, whereas the very general complex (not assuming projectivity) K3 surface has $\text{Pic}(X) = 0$. The category $\text{Coh}(X)$ of a complex K3 surface $\text{Pic}(X) = 0$ seems to be much smaller than for a complex projective K3 surface; e.g. its numerical Grothendieck group is

$$N(X) \simeq \mathbb{Z} \oplus \mathbb{Z} \simeq (H^0 \oplus H^4)(X, \mathbb{Z}).$$

However, its Grothendieck group $K(X)$ is in fact much bigger, due to the absence of curves on X. Another way to look at this is to say that, although $\text{Pic}(X) = 0$, moduli spaces of μ-stable vector bundles E with $v(E) = v := (r, 0, s)$, in particular $\det(E) \simeq \mathcal{O}$, still exist and are of dimension $\langle v, v \rangle + 2 = -2rs + 2$ as in the projective situation. They contribute non-trivially to $K(X)$.

However, as it turns out, $\text{Pic}(X) = 0$ has the bigger impact on the situation. So, it is not hard to show that under this assumption $\mathcal{O}_X$ is the only spherical object in $\text{Coh}(X)$ and in fact, up to shift, in $D^b(X)$. Thus, Bridgeland's conjecture in particular suggests that $\text{Aut}(D^b(X))$ should be essentially trivial, which was indeed proved in [267].

Theorem 4.8 *Let X be a complex K3 surface with* $\mathrm{Pic}(X) = 0$. *Then*

$$\mathrm{Aut}(\mathrm{D}^{\mathrm{b}}(X)) \simeq \mathbb{Z} \oplus \mathbb{Z} \oplus \mathrm{Aut}(X).$$

The first two factors of $\mathrm{Aut}(\mathrm{D}^{\mathrm{b}}(X))$ are generated by the shift functor and the spherical shift $T_{\mathcal{O}}$. The group $\mathrm{Aut}(X)$ is either trivial or $\mathbb{Z}$; see [386]. Note that the same techniques were used in [267] to determine $\mathrm{Aut}(\mathrm{D}^{\mathrm{b}}(X))$ for the generic fibre of the generic formal deformation of a projective K3 surface, which is the key to Corollary 3.13 and the main result of [268].

Remark 4.9 It is worth pointing out that in the non-algebraic setting Gabriel's theorem fails; i.e. the abelian category $\mathrm{Coh}(X)$ of a non-projective complex K3 surface does not necessarily determine the complex manifold X. In fact, in [611] Verbitsky shows that two very general non-projective complex K3 surfaces X and Y have equivalent abelian categories $\mathrm{Coh}(X) \simeq \mathrm{Coh}(Y)$.

5 Appendix: Twisted K3 Surfaces

It has turned out to be interesting to introduce the notion of twisted K3 surfaces. This works in both the algebraic and the analytic setting.

Definition 5.1 A *twisted K3 surface* (X, α) consists of a K3 surface X and a Brauer class $\alpha \in \mathrm{Br}(X)$.

The notion makes sense for arbitrary X, and a general theory of sheaves on (X, α), so-called twisted sheaves, can be set up. This is of particular importance for K3 surfaces.

There are various ways of defining twisted sheaves and the abelian category $\mathrm{Coh}(X, \alpha)$ of those, but they all require an additional choice, of either an Azumaya algebra $\mathcal{A}$, a Čech cycle representing α, or a $\mathbb{G}_m$-gerbe. The following is copied from [256].

(i) Choose a Čech cocycle $\{\alpha_{ijk} \in \mathcal{O}^*(U_{ijk})\}$ representing $\alpha \in \mathrm{Br}(X)$ with respect to an étale or analytic cover $\{U_i\}$ of X. Then an $\{\alpha_{ijk}\}$-*twisted coherent sheaf* $(\{E_i\}, \{\varphi_{ij}\})$ consists of coherent sheaves E_i on U_i and $\varphi_{ij} \colon E_j|_{U_{ij}} \xrightarrow{\sim} E_i|_{U_{ij}}$ such that

$$\varphi_{ii} = \mathrm{id}, \quad \varphi_{ji} = \varphi_{ij}^{-1}, \quad \text{and} \quad \varphi_{ij} \circ \varphi_{jk} \circ \varphi_{ki} = \alpha_{ijk} \cdot \mathrm{id}.$$

Defining morphisms in the obvious way, $\{\alpha_{ijk}\}$-twisted sheaves form an abelian category, and, as different choices of $\{\alpha_{ijk}\}$ representing α lead to equivalent categories, this is taken as $\mathrm{Coh}(X, \alpha)$.

(ii) For any Azumaya algebra $\mathcal{A}$ representing α one can consider the abelian category $\mathrm{Coh}(X, \mathcal{A})$ of $\mathcal{A}$-modules which are coherent as $\mathcal{O}_X$-modules. To see that this category is equivalent to the above, pick a locally free coherent α-twisted sheaf G and let $\mathcal{A}_G :=$

$G \otimes G^*$, which is an Azumaya algebra representing α. For an α-twisted sheaf E (with respect to the same choice of the cycle representing α), $E \otimes G^*$ is an untwisted sheaf with the structure of an $\mathcal{A}_G$-module. This eventually leads to an equivalence

$$\mathrm{Coh}(X, \mathcal{A}_G) \simeq \mathrm{Coh}(X, \alpha).$$

(iii) To Azumaya algebras $\mathcal{A}$ but also to Čech cocycles $\{\alpha_{ijk}\}$ representing a Brauer class α one can associate $\mathbb{G}_m$-gerbes over X, denoted $\mathcal{M}_{\mathcal{A}}$ and $\mathcal{M}_{\{\alpha_{ijk}\}}$, respectively.

The gerbe $\mathcal{M}_{\mathcal{A}}$ associates with $T \longrightarrow X$ the category $\mathcal{M}_{\mathcal{A}}(T)$ whose objects are pairs (E, ψ) with E a locally free coherent sheaf on T and $\psi : \mathcal{E}nd(E) \simeq \mathcal{A}_T$ an isomorphism of $\mathcal{O}_T$-algebras; see [210, 405]. A morphism $(E, \psi) \longrightarrow (E', \psi')$ is given by an isomorphism $E \overset{\sim}{\longrightarrow} E'$ that commutes with the $\mathcal{A}_T$-actions induced by ψ and ψ', respectively. It is easy to see that the group of automorphisms of an object (E, ψ) is $\mathcal{O}^*(T)$.

The gerbe $\mathcal{M}_{\{\alpha_{ijk}\}}$ associates with $T \longrightarrow X$ the category $\mathcal{M}_{\{\alpha_{ijk}\}}(T)$ whose objects are collections $\{\mathcal{L}_i, \varphi_{ij}\}$, where $\mathcal{L}_i \in \mathrm{Pic}(T_{U_i})$ and $\varphi_{ij} : \mathcal{L}_j|_{T_{U_{ij}}} \overset{\sim}{\longrightarrow} \mathcal{L}_i|_{T_{U_{ij}}}$ with $\varphi_{ij} \cdot \varphi_{jk} \cdot \varphi_{ki} = \alpha_{ijk}$; see [365]. A morphism $\{\mathcal{L}_i, \varphi_{ij}\} \longrightarrow \{\mathcal{L}'_i, \varphi'_{ij}\}$ is given by isomorphisms $\mathcal{L}_i \overset{\sim}{\longrightarrow} \mathcal{L}'_i$ compatible with φ_{ij} and φ'_{ij}. For another construction of a gerbe associated to α, see [136].

Any sheaf $\mathcal{F}$ on a $\mathbb{G}_m$-gerbe $\mathcal{M} \longrightarrow X$ comes with a natural $\mathbb{G}_m$-action and thus decomposes as $F = \bigoplus F^k$, where the $\mathbb{G}_m$-action on F^k is given by the character $\lambda \longmapsto \lambda^k$. The category of coherent sheaves of weight k, i.e. with $F = F^k$, on a $\mathbb{G}_m$-gerbe $\mathcal{M}$ is denoted $\mathrm{Coh}(\mathcal{M})_k$. There are natural equivalences

$$\mathrm{Coh}(X, \alpha) \simeq \mathrm{Coh}(\mathcal{M}_{\mathcal{A}})_1 \simeq \mathrm{Coh}(\mathcal{M}_{\{\alpha_{ijk}\}})_1.$$

Moreover, $\mathrm{Coh}(X, \alpha^\ell) \simeq \mathrm{Coh}(\mathcal{M}_{\mathcal{A}})_\ell \simeq \mathrm{Coh}(\mathcal{M}_{\{\alpha_{ijk}\}})_\ell$; see [136, 156, 365] for more details.

(iv) A realization in terms of Brauer–Severi varieties has been explained in [646]. Suppose $E = (\{E_i\}, \{\varphi_{ij}\})$ is a locally free $\{\alpha_{ijk}\}$-twisted sheaf. The projective bundles $\pi_i : \mathbb{P}(E_i) \longrightarrow U_i$ glue to the Brauer–Severi variety $\pi : \mathbb{P}(E) \longrightarrow X$ and the relative $\mathcal{O}_{\pi_i}(1)$ glue to a $\{\pi^*\alpha_{ijk}^{-1}\}$-twisted line bundle $\mathcal{O}_\pi(1)$ on $\mathbb{P}(E)$. As for an $\{\alpha_{ijk}\}$-twisted sheaf F the product $\pi^* F \otimes \mathcal{O}_\pi(1)$ is naturally an untwisted sheaf, one obtains an equivalence

$$\mathrm{Coh}(X, \alpha) \simeq \mathrm{Coh}(\mathbb{P}(E)/X)$$

with the full subcategory of $\mathrm{Coh}(\mathbb{P}(E))$ of all coherent sheaves F' on $\mathbb{P}(E)$ for which the natural morphism $\pi^* \pi_* (F' \otimes (\pi^* E \otimes \mathcal{O}_\pi(1))^*) \longrightarrow F' \otimes (\pi^* E \otimes \mathcal{O}_\pi(1))^*$ is an isomorphism. Note that the bundle $\pi^* E \otimes \mathcal{O}_\pi(1)$ can be described as the unique nontrivial extension $0 \longrightarrow \mathcal{O}_{\mathbb{P}(E)} \longrightarrow \pi^* E \otimes \mathcal{O}_\pi(1) \longrightarrow \mathcal{T}_{\mathbb{P}(E)} \longrightarrow 0$ and thus depends only on the Brauer–Severi variety $\pi : \mathbb{P}(E) \longrightarrow X$.

References and Further Reading

We have noted in Remark 2.10 that $|\mathrm{FM}(X)|$ can be arbitrarily large. In fact even the number of non-isometric Néron–Severi lattices realized by surfaces in $\mathrm{FM}(X)$ cannot be bounded. We have also noted that under certain lattice theoretic conditions $|\mathrm{FM}(X)| = 1$; cf. Corollary 3.8. There are also results that give precise numbers; e.g. in [247, 470] Oguiso et al. prove that $|\mathrm{FM}(X)| = 2^{\tau(d)-1}$ for a K3 surface with $\mathrm{NS}(X) \simeq \mathbb{Z} \cdot H$ such that $(H)^2 = 2d$. Here, $\tau(d)$ is the number of distinct primes dividing d. See also Stellari [575] for similar computations in the case of $\rho(X) = 2$ and Ma [382] for an identification of $\mathrm{FM}(X)$ with the set of cusps of the Kähler moduli space. For a polarized K3 surface (X, H) one can ask for the subset of $\mathrm{FM}(X)$ of all Y that admit a polarization of the same degree $(H)^2$. The analogous counting problem was addressed by Hulek and Ploog in [249].

The discussion of $\mathrm{Aut}(\mathrm{D}^b(X))$ in Section 3.4 is best viewed from the perspective of stability conditions on $\mathrm{D}^b(X)$, a notion that can be seen as a refinement of bounded t-structures on $\mathrm{D}^b(X)$. See the original paper by Bridgeland [86] or the survey [263].

In analogy to Mukai's description of finite groups of symplectic automorphisms of a K3 surface reviewed in Section **15**.3 a complete description of all finite subgroups $G \subset \mathrm{Aut}_s(\mathrm{D}^b(X))$ fixing a stability condition has been given in [261].

Triangulated categories that are quite similar to the bounded derived category $\mathrm{D}^b(X)$ of a K3 surface also occur in other situations. Most prominently, for a smooth cubic fourfold $Z \subset \mathbb{P}^5$ the orthogonal complement $\mathcal{A}_Z \subset \mathrm{D}^b(Z)$ of $\mathcal{O}_Z$, $\mathcal{O}_Z(1)$, and $\mathcal{O}_Z(2)$ is such a category. It has been introduced by Kuznetsov in [345] and studied quite a lot, as it seems to be related to the rationality question for cubic fourfolds. Although the Hodge theory of $\mathcal{A}_Z$ has been introduced (see [2, 264]) and $\mathcal{A}_Z$ is a 'deformation of $\mathrm{D}^b(X)$', basic facts like the existence of stability conditions on $\mathcal{A}_Z$ have not yet been established.

Questions and Open Problems

It is not known whether for a spherical object $E \in \mathrm{D}^b(X)$, or even for a spherical sheaf $E \in \mathrm{Coh}(X)$, the orthogonal $E^\perp$ contains non-trivial objects. If yes, which is expected, this would be a quick way to show that T_E^{2k} is never a simple shift for any $k \neq 0$.

If $\rho(X) = 1$, then Mukai shows in [429] that any spherical sheaf E with $\mathrm{rk}(E) \neq 0$ is in fact a μ-stable vector bundle. One could wonder if for $\rho(X) > 1$ any spherical sheaf is μ-stable with respect to some polarization. Is there a way to 'count' spherical vector bundles with given Mukai vector v?

It would be interesting to have an explicit example of two non-isomorphic K3 surfaces X and Y over a field K with $\mathrm{D}^b(X) \simeq \mathrm{D}^b(Y)$ (over K) with $X_{\overline{K}} \simeq Y_{\overline{K}}$. A related question is whether Proposition 3.10 remains true over non-algebraically closed fields (of characteristic zero).

Of course, the main open problem in the area is Conjecture 3.14 for $\rho(X) > 1$.

17

Picard Group

As remarked by Zariski in [652], 'The evaluation of ρ [the Picard number] for a given surface presents in general grave difficulties'. This is still valid and to a lesser extent also for K3 surfaces. The Picard number or the finer invariant provided by the Néron–Severi lattice NS(X) is the most basic invariant of a K3 surface, from which one can often read off basic properties of X, e.g. whether X admits an elliptic fibration or is projective. Line bundles also play a distinguished role in the derived category $D^b(X)$, as the easiest kind of spherical objects, and for the description of many other aspects of the geometry of X.

In this chapter we collect the most important results on the Picard group of a K3 surface. A number of results are sensitive to the ground field, whether it is algebraically closed or of characteristic zero. Accordingly, we first deal in Section 1 with the case of complex K3 surfaces, where the description of the Picard group reduces to Hodge theory, which nevertheless may be complicated to fully understand even for explicitly given K3 surfaces. Later, in Section 2, we switch to more algebraic aspects and finally to the Tate conjecture, the analogue of the Lefschetz theorem on (1, 1)-classes for finitely generated fields. In the latter two parts we often refer to Chapter 18 on Brauer groups. These two chapters are best read together.

1 Picard Groups of Complex K3 Surfaces

We start out with a few recollections concerning the Picard group of complex K3 surfaces.

1.1 For any K3 surface X, complex or algebraic over an arbitrary field, the Picard group Pic(X) is isomorphic to the Néron–Severi group NS(X). In other words, any line

* Thanks to François Charles and Matthias Schütt for detailed comments on this chapter.

bundle L on a K3 surface X that is algebraically equivalent to the trivial line bundle $\mathcal{O}_X$ is itself trivial; see Section 1.2. For projective K3 surfaces, a stronger statement holds: any numerically trivial line bundle is trivial. So, in this case

$$\text{Pic}(X) \simeq \text{NS}(X) \simeq \text{Num}(X);$$

see Proposition 1.2.4. For complex non-algebraic K3 surfaces the last isomorphism does not hold in general; see Remark 1.3.4.

Let us now consider arbitrary complex K3 surfaces, projective or not. Then the Lefschetz theorem on $(1, 1)$-classes yields an isomorphism

$$\text{Pic}(X) \simeq \text{NS}(X) \simeq H^{1,1}(X, \mathbb{Z}) := H^{1,1}(X) \cap H^2(X, \mathbb{Z}).$$

As $h^{1,1}(X) = 20$, this in particular shows that

$$\text{Pic}(X) \simeq \mathbb{Z}^{\oplus \rho(X)} \quad \text{with} \quad 0 \leq \rho(X) \leq 20. \tag{1.1}$$

Moreover, as we shall see, every possible Picard number is attained by some complex K3 surface.

Complex K3 surfaces with maximal Picard number $\rho(X) = 20$ are sometimes called *singular K3 surfaces*, although of course they are smooth as are all K3 surfaces. In the physics literature, they are also called *attractive K3 surfaces*; see e.g. [24]. As outlined in Section 14.3.4, complex K3 surfaces with $\rho(X) = 20$ can be classified in terms of their transcendental lattice.

Remark 1.1 If X is projective, then $1 \leq \rho(X)$ and the Hodge index theorem asserts that the usual intersection pairing on $H^2(X, \mathbb{Z})$ restricted to $\text{NS}(X)$ is even and non-degenerate of signature $(1, \rho(X) - 1)$; cf. Proposition 1.2.4.

In general, the number of positive eigenvalues can never exceed one but can very well be zero. Even worse, there exist K3 surfaces such that $\text{NS}(X) \simeq \mathbb{Z} \cdot L$ with $(L)^2 = 0$, e.g. elliptic K3 surfaces without any multisection (see Example 3.3.2), and so $\text{NS}(X)$ is degenerate with neither positive nor negative eigenvalues.

Remark 1.2 Also recall that every K3 surface X is Kähler (cf. Section 7.3.2) and that X is projective if and only if there exists a line bundle L such that $c_1(L)$ is contained in the Kähler cone $\mathcal{K}_X$ or, a priori weaker but equivalent, that there exists a line bundle L with $c_1(L)$ contained in the positive cone $\mathcal{C}_X$ or simply satisfying $(L)^2 > 0$.

Another sufficient condition for the projectivity of X is the existence of a line bundle L with $c_1(L) \in \partial \mathcal{C}_X \setminus \overline{\mathcal{K}}_X$. Indeed, if L is not nef, then by Theorem 8.5.2 there exists a (-2)-curve C with $(C.L) < 0$ and thus $(L^n(-C))^2 > 0$ for $n \gg 0$.

1.2 It is rather difficult to decide which lattices of rank ≤ 20 can be realized as $\text{NS}(X)$. For any complex K3 surface X, the lattice $\text{NS}(X)$ is even with at most one positive eigenvalue and with at most one isotropic direction, which immediately leads to the following rough classification.

Proposition 1.3 *Let X be a complex K3 surface. Then one of the following cases occurs:*

(i) $\operatorname{sign} \mathrm{NS}(X) = (1, \rho(X) - 1)$, *which is the case if and only if X is projective or, equivalently,* $\operatorname{trdeg} K(X) = 2$.

(ii) *The kernel of* $\mathrm{NS}(X) \longrightarrow \mathrm{Num}(X)$ *is of rank one and* $\mathrm{Num}(X)$ *is negative definite. This is the case if and only if* $\operatorname{trdeg} K(X) = 1$.

(iii) $\mathrm{NS}(X)$ *is negative definite, which is the case if and only if* $K(X) \simeq \mathbb{C}$. $\qquad \square$

The last assertion has been first observed by Nikulin in [450, Sec. 3.2]. For higher-dimensional generalizations to hyperkähler manifolds; see [99].

For the reader's convenience we list the following results that have been mentioned in other chapters already.

(i) Any even lattice N of signature $(1, \rho - 1)$ with $\rho \leq 11$ can be realized as $\mathrm{NS}(X)$; see Corollary **14**.3.1 and Remark **14**.3.7.

(ii) For a complex projective K3 surface X of Picard number $\rho(X) \geq 12$ the isomorphism type of $N := \mathrm{NS}(X)$ is uniquely determined by its rank $\rho(X)$ and the discriminant group (A_N, q_N); see Corollary **14**.3.6.

(iii) The last fact in particular applies to the case of K3 surfaces with $\rho(X) = 20$. Then the transcendental lattice $T := T(X)$ is a positive definite, even lattice of rank two which uniquely determines X up to conjugation; see Corollary **14**.3.21 and Remark **14**.3.4. The discriminant group (A_T, q_T) uniquely determines $\mathrm{NS}(X)$.

(iv) If X is of algebraic dimension zero, i.e. $K(X) \simeq \mathbb{C}$, then there exists a primitive embedding $\mathrm{NS}(X) \hookrightarrow N$ into some Niemeier lattice; see Example **14**.4.7.

(v) If N is an even lattice of signature $(1, \rho - 1)$ such that its Weyl group $W(N) \subset O(N)$ is of finite index, then $N \simeq \mathrm{NS}(X)$ for some K3 surface X; see Theorem **15**.2.10.

1.3 Let $f : X \longrightarrow S$ be a smooth proper family of complex K3 surfaces over a connected base S, e.g. an analytic disk. Recall from Proposition **6**.2.9 that the Noether–Lefschetz locus

$$\mathrm{NL}(X/S) \subset S$$

of all points $t \in S$ with $\rho(X_t) > \rho_0$ is dense if the family is not isotrivial (i.e. locally the period map is non-constant). Here, ρ_0 is the minimum of all Picard numbers $\rho(X_t)$.

Let now S be simply connected and fix an isomorphism $R^2 f_* \mathbb{Z} \simeq \Lambda$ of local systems. Then for $0 \neq \alpha \in \Lambda$ the locus $\{t \mid \alpha \in H^{1,1}(X_t)\}$ is either S, empty, or of codimension one. This process can be iterated. For any $\rho \geq \rho_0$ the set

$$S(\rho) := \{t \mid \rho(X_t) \geq \rho\} \subset S$$

is a countable union of closed subsets of codimension $\leq \rho - \rho_0$.

For the universal family of marked K3 surfaces $X \longrightarrow N$ (see Section **6.3.3**), the picture becomes very clean. In this case

$$N(20) \subset N(19) \subset \cdots \subset N(1) = \mathrm{NL}(X/N) \subset N$$

defines a stratification by countable unions of closed subsets with $\dim N(\rho) = 20 - \rho$. Moreover, $N(\rho+1) \setminus N(\rho) = \{t \mid \rho(X_t) = \rho+1\}$. Applying Proposition **6**.2.9 repeatedly shows that $N(20) \subset N(i)$ is dense for all $0 \leq i \leq 20$.

Let $\alpha \in \mathcal{K}_X$ be a Kähler class and $\mathcal{X}(\alpha) \longrightarrow T(\alpha) \simeq \mathbb{P}^1$ the associated twistor space; see Section **7.3.2**. A fixed isometry $H^2(X, \mathbb{Z}) \simeq \Lambda$ induces a marking of the family and hence a morphism $T(\alpha) \hookrightarrow N$. For very general $\alpha \in \mathcal{K}_X$, i.e. not contained in any hyperplane orthogonal to a class $0 \neq \ell \in H^{1,1}(X, \mathbb{Z})$, the twistor line is not contained in the Noether–Lefschetz locus $N(1) = \mathrm{NL}(X/N)$ and so $\rho(\mathcal{X}(\alpha)_t) = 0$ for all except countably many $t \in T(\alpha)$.

The polarized case can be dealt with similarly. If $X \longrightarrow N_d$ is the universal family of marked polarized K3 surfaces (see Section **6.3.4**), then $N_d = N_d(1)$ and

$$N_d(20) \subset N_d(19) \subset \cdots \subset N_d(1) = N_d$$

with $\dim N_d(\rho) = 20 - \rho$. Again, $N_d(2)$ is dense in $N_d(1) = N_d$ and more generally $N_d(\rho + 1)$ in $N(\rho)$, $\rho = 1, \ldots, 19$.

Remark 1.4 However, there exist examples of families of K3 surfaces for which the Picard rank does not jump as possibly suggested by the above. For example, there exist one-dimensional families of K3 surfaces $f : X \longrightarrow S$ for which $\rho(X_t) \geq \rho_0 + 2$ for all $t \in \mathrm{NL}(X/S)$. For an explicit example, see [469, Ex. 5], for which $\rho_0 = 18$. This phenomenon can be explained[1] in terms of the Mumford–Tate group. Roughly, for any family $X \longrightarrow S$ that comes with an action of a fixed field K on the Hodge structures given by the transcendental lattices $T(X_t)_{\mathbb{Q}}$, the Picard number $\rho(X_t)$ can jump only by multiples of $[K : \mathbb{Q}]$. Indeed, $T(X_t)_{\mathbb{Q}}$ is a vector space over K and hence $\dim T(X_t)_{\mathbb{Q}} = d_t \cdot [K : \mathbb{Q}]$. Therefore $\rho(X_t) = 22 - d_t \cdot [K : \mathbb{Q}]$.

1.4 The Néron–Severi lattice of a complex K3 surface X can be read off from its period, but it is usually difficult to determine $\mathrm{NS}(X)$ or the period of X when X is given by equations (even very explicit ones). There is no general recipe for doing this, but the computations have been carried out in a number of non-trivial examples.

Kummer surfaces. For the Kummer surface X associated with a torus A, the existence of the Kummer lattice $K \subset \mathrm{NS}(X)$ (see Section **14.3.3**) shows that $\rho(X) \geq 16$. If A is

---

[1] As I learned from François Charles.

an abelian surface, then $\rho(X) \geq 17$, and, in fact, for general A this is an equality. For arbitrary A one has

$$\rho(X) = 16 + \rho(A).$$

For the Kummer surface X associated with a product $E_1 \times E_2$ of elliptic curves, this becomes

$$\rho(X) = \begin{cases} 18 & \text{if } E_1 \not\sim E_2, \\ 19 & \text{if } E_1 \sim E_2 \text{ without CM}, \\ 20 & \text{if } E_1 \sim E_2 \text{ with CM}. \end{cases} \tag{1.2}$$

Also recall from the discussion in Section **14.3.3** that a complex K3 surface is a Kummer surface if and only if there exists a primitive embedding of the Kummer lattice $K \hookrightarrow \mathrm{NS}(X)$ or, slightly suprisingly, if and only if there exists 16 disjoint smooth rational curves $C_1, \ldots, C_{16} \subset X$.

Quartics. Due to the discussion above, the very general complex quartic $X \subset \mathbb{P}^3$ satisfies $\rho(X) = 1$ and in fact $\mathrm{Pic}(X) = \mathbb{Z} \cdot \mathcal{O}(1)|_X$.

On the other hand, recall from Section **3.2.6** that the Fermat quartic $X \subset \mathbb{P}^3$, $x_0^4 + \cdots + x_3^4 = 0$, has $\rho(X) = 20$. In fact, $\mathrm{NS}(X)$ is generated by lines $\ell \subset X$ and

$$\mathrm{NS}(X) \simeq E_8(-1)^{\oplus 2} \oplus U \oplus \mathbb{Z}(-8) \oplus \mathbb{Z}(-8).$$

The detailed computation has been carried out by Schütt, Shioda, and van Luijk in [541].

The Fermat quartic is a member of the *Dwork (or Fermat) pencil*:

$$X_t \subset \mathbb{P}^3, \quad x_0^4 + \cdots + x_3^4 - 4t \prod x_i = 0.$$

Note that X_t is smooth except for $t^4 = 1$ and $t = \infty$ (which at first glance is surprising as the discriminant divisor in $|\mathcal{O}_{\mathbb{P}^3}(4)|$ has degree 108 by [207, 13.Thm. 2.5]). Generically one has $\rho(X_t) = 19$, but $\rho(X_t) = 20$ for an analytically dense set of countably many points $t \in \mathbb{P}^1$. In fact, if $\rho(X_t) = 20$, then $t \in \mathbb{P}^1(\bar{\mathbb{Q}})$; see Proposition 2.14. The Néron–Severi lattice of the very general X_t has been described by Bini and Garbagnati in [63]; e.g. its discriminant group is isomorphic to $(\mathbb{Z}/8\mathbb{Z})^2 \times \mathbb{Z}/4\mathbb{Z}$. Using Section **14.3.3**, one can show that all X_t are Kummer surfaces; see [63, Cor. 4.3].

Kuwata in [341] studied quartics of the form

$$X = X_{\lambda_1, \lambda_2} \subset \mathbb{P}^3, \quad \phi_1(x_0, x_1) = \phi_2(x_2, x_3).$$

After coordinate changes one can assume that $\phi_i(x, y) = yx(y - x)(y - \lambda_i x)$ with $\lambda_i \in \mathbb{C} \setminus \{0, 1\}$. If $E_i, i = 1, 2$, are the elliptic curves defined by $t^2 = \phi_i(1, x)$, then the Kummer surface associated with $E_1 \times E_2$ is birational to the quotient of X by the involution $\iota \colon (x_0 : x_1 : x_2 : x_3) \longmapsto (x_0 : x_1 : -x_2 : -x_3)$:

$$X/\iota \sim (E_1 \times E_2)/\pm.$$

(So, X has a Shioda–Inose structure.) This allows one to use (1.2) to compute $\rho(X_{\lambda_1,\lambda_2})$. In particular, $\rho(X_{\lambda_1,\lambda_2}) \geq 18$. In [341, Prop. 1.4] it is proved that the number of lines $\ell \subset X$ is 16, 32, 48, or 64 depending on $j(E_i)$.

In [74] Boissière and Sarti later showed that the vector space $\mathrm{NS}(X_\lambda) \otimes \mathbb{Q}$ of $X_\lambda :=$ $X_{\lambda,\lambda}$ is spanned by lines if and only if (i) $\lambda \notin \overline{\mathbb{Q}}$, (ii) $\lambda \in \{-1, 2, 1/2, e^{2\pi i/3}, e^{-2\pi i/3}\}$, or (iii) $\lambda \in \overline{\mathbb{Q}} \setminus \{-1, 2, 1/2, e^{2\pi i/3}, e^{-2\pi i/3}\}$ and $\rho(X_\lambda) = 19$. Furthermore, the lattice $\mathrm{NS}(X_\lambda)$ is generated by lines only in case (ii).

Explicit equations tend to have coefficients in $\mathbb{Q}$ or $\overline{\mathbb{Q}}$, but it was an open problem for a long time (apparently going back to Mumford) whether there exists a quartic with $\rho(X) = 1$ with algebraic coefficients. We come back to this in Section 2.6. Over larger fields, quartics with $\rho(X) = 1$ can be found more easily. For example, the generic fibre of the universal quartic $\mathcal{X} \longrightarrow |\mathcal{O}(4)|$ is a K3 surface over $\mathbb{Q}(t_1, \ldots, t_{34})$ of geometric Picard number one.

Double planes. The very general double plane $X \longrightarrow \mathbb{P}^2$ (see Example 1.1.3) satisfies $\rho(X) = 1$. This can be deduced either from a general Noether–Lefschetz principle for cyclic coverings as e.g. in [92, 174] or from counting dimensions and realizing that the family of double planes modulo isomorphisms is of dimension 19, as is the moduli space M_2 of polarized K3 surfaces of degree two.

If the branching curve of $X \longrightarrow \mathbb{P}^2$ consists of six general lines, the minimal resolution of the double cover is a K3 surface of Picard number $\rho(X) = 16$. The exceptional curves over the 15 double points of the double cover (over the 15 intersection points of pairs of lines) span a lattice inside $\mathrm{NS}(X)$ isomorphic to $\mathbb{Z}(-2)^{\oplus 15}$. For some observations on Picard groups of K3 surfaces appearing as minimal resolutions of singular double covers of $\mathbb{P}^2$ branched over a sextic, especially those with maximal Picard number 20, see Persson's article [491]. Using the density of K3 surfaces of maximal Picard number $\rho(X) = 20$ in the moduli space M_2 of polarized K3 surfaces of degree two, one concludes that there do exist smooth double planes ramified over a smooth sextic with maximal Picard number.

Complete intersections. Recall from Example 1.1.3 that besides quartics the only other non-degenerate complete intersection K3 surfaces are complete intersections either of a quadric and a cubic in $\mathbb{P}^4$ or of three quadrics in $\mathbb{P}^5$. As for quartics and double planes, the very general complete intersection K3 surface X satisfies $\rho(X) = 1$.

In [85] Bremner studies rational points on the hypersurface defined by $u_0^6 + u_1^6 + u_2^6 = v_0^6 + v_1^6 + v_2^6$ by relating it to a particular intersection $X = Q_1 \cap Q_2 \cap Q_3 \subset \mathbb{P}^5$ of three quadrics. The Picard rank of this particular X (over $\mathbb{C}$) is 19; see also [343].

Elliptic K3 surfaces. The description of the Néron–Severi lattice of an elliptic K3 surface $X \longrightarrow \mathbb{P}^1$ with a section can essentially be reduced to the description of its Mordell–Weil group $\mathrm{MW}(X)$ via the short exact sequence

$$0 \longrightarrow A \longrightarrow \mathrm{NS}(X) \longrightarrow \mathrm{MW}(X) \longrightarrow 0,$$

where A is the subgroup generated by vertical divisors and the section; see Proposition **11.3.2**.

For the general elliptic K3 surface with a section C_0 there exists an isometry $\mathrm{NS}(X) \simeq U$ that maps X_t and $C_0 + X_t$ to the standard generators of the hyperbolic plane U. The Picard group of a general, in particular not projective, complex elliptic K3 surface is generated by the fibre class, and so $\rho(X) = 1$ with trivial intersection form, i.e. $\mathrm{NS}(X) \simeq \mathbb{Z}(0)$.

2 Algebraic Aspects

Let now X be an algebraic K3 surface over an arbitrary field k. Then

$$\mathrm{Pic}(X) \simeq \mathrm{NS}(X) \simeq \mathrm{Num}(X) \simeq \mathbb{Z}^{\oplus \rho(X)} \quad \text{with} \quad 1 \le \rho(X) \le 22$$

is endowed with an even, non-degenerate pairing of signature $(1, \rho(X) - 1)$; see Proposition **1.2.4**, Remark **1.3.7**, and Section 2.2 below. Which lattices can be realized depends very much on the field k. Moreover, the Néron–Severi lattice can grow under base change to a larger field. We pay particular attention to finite fields and number fields.

Before starting, we mention in passing a characteristic p version of the injectivity $c_1 \colon \mathrm{NS}(X) \hookrightarrow H^{1,1}(X)$ for complex K3 surfaces; see Section **1.3.3**. As $H^{1,1}(X)$ and also the de Rham cohomology $H^2_{\mathrm{dR}}(X)$ are vector spaces over the base field, the first Chern class must be trivial on $p \cdot \mathrm{NS}(X)$. For a proof of the following statement, see Ogus's original in [478, Cor. 1.4] or the more concrete in [514, Sec. 7] or [607, Sec. 10].

Proposition 2.1 *For any K3 surface over a field of characteristic $p > 0$ the first Chern class induces an injection*

$$c_1 \colon \mathrm{NS}(X)/p \cdot \mathrm{NS}(X) \hookrightarrow H^2_{\mathrm{dR}}(X).$$

Moreover, its image is contained in $F^1 H^2_{\mathrm{dR}}(X)$ and if X is not supersingular, it induces an injection $\mathrm{NS}(X)/p \cdot \mathrm{NS}(X) \hookrightarrow H^{1,1}(X) = H^1(X, \Omega_X)$.

2.1 For a field extension K/k base change yields the K3 surface $X_K := X \times_k K$ over K, and for every line bundle L on X a line bundle L_K on X_K, the pull-back of L under the projection $X_K \longrightarrow X$. By flat base change, $H^i(X_K, L_K) = H^i(X, L) \otimes_k K$. Using that L is trivial if and only if $H^0(X, L) \ne 0 \ne H^0(X, L^*)$ and similarly for L_K, one finds that base change defines an injective homomorphism

$$\mathrm{Pic}(X) \hookrightarrow \mathrm{Pic}(X_K), \quad L \longmapsto L_K, \tag{2.1}$$

which is compatible with the intersection pairing.

Lemma 2.2 *If k is algebraically closed, then the base change map (2.1) is bijective.*

Proof Defining a line bundle M on X_K involves only a finite number of equations. Hence, we may assume that K is finitely generated over k and, therefore, can be viewed as the quotient field of a finitely generated k-algebra A. Also, as k is algebraically closed, any closed point $t \in \mathrm{Spec}(A)$ has residue field $k(t) \simeq k$. Localizing A with respect to finitely many denominators if necessary, we may in fact assume that M is a line bundle on $X_A := X \times_k \mathrm{Spec}(A)$ and thus can be viewed as a family of line bundles on X parametrized by $\mathrm{Spec}(A)$.

Consider the classifying morphism $f \colon \mathrm{Spec}(A) \longrightarrow \mathrm{Pic}_X$ for this family; see Section **10**.1.1. The Picard scheme Pic_X of a K3 surface X is reduced and zero-dimensional, as its tangent space at a point $[L] \in \mathrm{Pic}_X(k) = \mathrm{Pic}(X)$ is $\mathrm{Ext}^1(L, L) \simeq H^1(X, \mathcal{O}) = 0$; see Section **10**.1.6. Thus, f is a constant morphism with image a k-rational point of Pic_X. Therefore, M is a constant family and, in particular, $M \simeq L_K$ for some $L \in \mathrm{Pic}(X)$. □

In Section **16**.4.2 the assertion of the lemma has been applied to the more general class of spherical objects in the derived category $\mathrm{D}^{\mathrm{b}}(X_K)$. The proof is valid in this broader generality, replacing Pic_X with the stack of simple complexes.

Definition 2.3 The *geometric Picard number* of a K3 surface X over a field k is $\rho(X_{\bar{k}})$, where $\bar{k}$ is the algebraic closure of k or, equivalently, $\rho(X_K)$ for any algebraically closed field K containing k.

Clearly, any K3 surface X over k can be obtained by base change from a K3 surface X_0 over some finitely generated field k_0. By the lemma, $\mathrm{Pic}(X) \hookrightarrow \mathrm{Pic}(X_{\bar{k}}) \simeq \mathrm{Pic}(X_0 \times \bar{k}_0)$. If $\mathrm{char}(k) = 0$, one can choose an embedding $k_0 \hookrightarrow \mathbb{C}$, which yields an injection

$$\mathrm{Pic}(X) \hookrightarrow \mathrm{Pic}(X_0 \times \bar{k}_0) \hookrightarrow \mathrm{Pic}(X_0 \times \mathbb{C}).$$

In particular, in characteristic zero every K3 surface X satisfies $\rho(X) \leq 20$ by (1.1).

Remark 2.4 For a purely inseparable extension K/k of degree $[K : k] = q = p^n$ the cokernel of the base change map $\mathrm{Pic}(X) \hookrightarrow \mathrm{Pic}(X_K)$ is annihilated by p^n and so $\rho(X) = \rho(X_K)$. Indeed, if L is described by a cocycle $\{\psi_{ij}\}$, then L^q is described by $\{\psi_{ij}^q\}$ which is defined on X, and hence L^q is base changed from X.

2.2 To study $\mathrm{Pic}(X)$ by cohomological methods one uses the usual isomorphism $\mathrm{Pic}(X) \simeq H^1(X, \mathbb{G}_m)$. Then, for n prime to $\mathrm{char}(k)$ the Kummer sequence

$$0 \longrightarrow \mu_n \longrightarrow \mathbb{G}_m \overset{(\ )^n}{\longrightarrow} \mathbb{G}_m \longrightarrow 0$$

induces an injection $\text{Pic}(X) \otimes \mathbb{Z}/n\mathbb{Z} \hookrightarrow H^2_{et}(X, \mu_n)$. Applied to $n = \ell^m$, for a prime $\ell \neq \text{char}(k)$, and taking limits yields injections

$$\text{Pic}(X) \hookrightarrow \text{Pic}(X) \otimes \mathbb{Z}_\ell \hookrightarrow H^2_{et}(X, \mathbb{Z}_\ell(1)).$$

This proves $\rho(X) \leq 22$, which is a special case of a classical result due to Igusa [274].

As remarked earlier (see Remark 1.3.7), the Kummer sequence and the fact that $\text{Pic}(X)$ is torsion free also show that $H^1_{et}(X, \mu_n) \simeq k^*/(k^*)^n$. For separably closed k, this shows $H^1_{et}(X, \mu_n) = 0$ and by duality also $H^3_{et}(X, \mu_n) = 0$.

For a finite Galois extension K/k with Galois group G the Hochschild–Serre spectral sequence

$$E^{p,q}_2 = H^p(G, H^q(X_K, \mathbb{G}_m)) \Rightarrow H^{p+q}(X, \mathbb{G}_m),$$

and Hilbert 90, i.e. $H^1(G, \mathbb{G}_m) = 0$, can be used to construct an exact sequence

$$0 \longrightarrow \text{Pic}(X) \longrightarrow \text{Pic}(X_K)^G \longrightarrow H^2(G, K^*),$$

which in particular shows again the injectivity of $\text{Pic}(X) \longrightarrow \text{Pic}(X_K)$ in this situation. It also shows

$$\text{Pic}(X) \otimes \mathbb{Q} \xrightarrow{\sim} \text{Pic}(X_K)^G \otimes \mathbb{Q}, \qquad (2.2)$$

for $H^2(G, K^*)$ is torsion; cf. [549]. In other words, $\text{Pic}(X) \subset \text{Pic}(X_K)^G$ is a subgroup of finite index.[2]

By Wedderburn's theorem (see Remark **18**.2.1), the Brauer group of a finite field is trivial, i.e. $H^2(G, K^*) = 0$. Therefore,

$$\text{Pic}(X) = \text{Pic}(X_K)^G$$

for extensions K/k of finite fields. Also, $\text{Pic}(X) = \text{Pic}(X_K)^G$ whenever $X(K) \neq \emptyset$. See Section **18**.1.1 for further results relying on the Hochschild–Serre spectral sequence.

Remark 2.5 Suppose X is a K3 surface over a field k of characteristic zero, sufficiently small to admit an embedding $\sigma : k \hookrightarrow \mathbb{C}$. Then each such embedding yields a complex K3 surface X_σ. If k is algebraically closed, then $\text{Pic}(X_\sigma) \simeq \text{Pic}(X)$, and, in particular, the isomorphism type of the lattice $\text{Pic}(X_\sigma)$ is independent of σ. But what about the transcendental lattice $T(X_\sigma)$? Clearly, the genus of $T(X_\sigma)$ is also independent of σ, as its orthogonal complement in $H^2(X_\sigma, \mathbb{Z})$ is $\text{Pic}(X_\sigma) \simeq \text{Pic}(X)$. However, its isomorphism type can change. See Remark **14**.3.23.

Lemma 2.6 *Let X be a K3 surface over an arbitrary field k. Then there exists a finite extension $k \subset K$ such that $\text{Pic}(X_K) \simeq \text{Pic}(X_{\bar{k}})$ and, in particular, $\rho(X_K) = \rho(X_{\bar{k}})$.*

---

[2] To have a concrete example for which the map is indeed not surjective, consider the quadric $x_0^2 + x_1^2 + x_2^2$ over $\mathbb{R}$ and its base change $X_\mathbb{C} \simeq \mathbb{P}^1_\mathbb{C}$. Then $\mathcal{O}(1)$ is Galois invariant, because $\mathcal{O}(2)$ descends, but it is not linearizable.

Moreover, equality $\rho(X_K) = \rho(X_{\bar{k}})$ *can be achieved with a universal bound on* $[K : k]$. *If* char$(k) = 0$ *and* $X(k) \neq \emptyset$ *or if* k *is finite, the same holds for* Pic$(X_K) \simeq$ Pic$(X_{\bar{k}})$.

Proof Indeed, Pic$(X \times \bar{k})$ is finitely generated and the defining equations for any finite set of generators $L_1, \ldots, L_\rho \in$ Pic$(X \times \bar{k})$ involve only finitely many coefficients which then generate a finite extension K/k.

If k is a field of characteristic zero or a finite field, then there exists a finite Galois extension $k \subset K$ such that Pic$(X_K) \simeq$ Pic$(X_{\bar{k}})$. Now, the image of the action

$$\rho: G := \mathrm{Gal}(K/k) \longrightarrow O(\mathrm{Pic}(X_K)) \subset \mathrm{GL}(\rho(X_K), \mathbb{Z})$$

is a finite subgroup and it is known classically that every finite subgroup of GL$(n, \mathbb{Z})$ injects into GL$(n, \mathbb{F}_3)$. Therefore, $|\mathrm{Im}(\rho)|$ is universally bounded. Hence, for $H := \mathrm{Ker}(\rho)$ also $[K^H : k]$ is universally bounded. If one assumes in addition that $X(K^H) \neq \emptyset$ for char$(k) = 0$, then Pic$(X_{K^H}) \simeq$ Pic$(X_K)^H \simeq$ Pic(X_K). Thus $k \subset K' := K^H$ is a finite extension of universally bounded degree with Pic$(X_{K'}) \simeq$ Pic$(X_{\bar{k}})$.

For arbitrary k, the same argument shows Pic$(X_{K'}) \otimes \mathbb{Q} \simeq$ Pic$(X_{\bar{k}}) \otimes \mathbb{Q}$ (cf. Remark 2.4), and hence $\rho(X_{K'}) = \rho(X_{\bar{k}})$. $\qquad\square$

2.3 We follow Shioda [560] to describe the easiest example of a unirational K3 surface. Consider the Fermat quartic $X \subset \mathbb{P}^3$ over an algebraically closed field of characteristic $p = 3$ defined by $x_0^4 + \cdots + x_3^4 = 0$ or, after coordinate change, by $x_0^4 - x_1^4 = x_2^4 - x_3^4$. A further coordinate change $y_0 = x_0 - x_1, y_1 = x_0 + x_1, y_2 = x_2 - x_3$, $y_3 = x_2 + x_3$ turns this into $y_0 y_1 (y_0^2 + y_1^2) = y_2 y_3 (y_2^2 + y_3^2)$. Setting $y_3 = 1, y_1 = y_0 u$, and $y_2 = uv$, one sees that the function field of X is isomorphic to the function field of the affine variety given by $y_0^4 (1 + u^2) = v((uv)^2 + 1)$, which in turn can be embedded into the function field of the variety defined by $u^2(t^4 - v)^3 - v + t^{12}$ with $t^3 = y_0$ (where one uses $p = 3$). Now switch to coordinates $s = u(t^4 - v)$, t, and v in which the equation becomes $s^2(t^4 - v) - v + t^{12}$. Thus, v can be written as a rational function in s and t. All together this yields an injection $K(X) \hookrightarrow k(s, t)$ which geometrically corresponds to a dominant rational map $\mathbb{P}^2 -\!-\!\to X$.

A similar computation proves the unirationality of the Fermat hypersurface $Y \subset \mathbb{P}^3$ of degree $p + 1$ over a field of characteristic p. For $p \equiv 3$ (4) the endomorphism $[x_0 : \cdots : x_3] \longmapsto [x_0^{(p+1)/4} : \cdots : x_3^{(p+1)/4}]$ of $\mathbb{P}^3$ defines a dominant map $Y \longrightarrow X$ and hence X is unirational for all $p \equiv 3$ (4).

Contrary to the case of complex K3 surfaces, there exist K3 surfaces over fields of positive characteristic with $\rho(X) = 22$. The Fermat quartic is such an example, which was first observed by Tate in [591, 592]. It can also be seen as a consequence of the following more general result due to Shioda. For more on these surfaces, see Section 2.7 below.

Proposition 2.7 *Let X be a unirational K3 surface over an algebraically closed field. Then $\rho(X) = 22$.*

Proof We follow Shioda [560] and show first more generally that for any dominant rational map $Y - \dashrightarrow X$ between smooth surfaces one has $b_2(Y) - \rho(Y) \geq b_2(X) - \rho(X)$. First, by resolving the indeterminacies of the rational map by blowing up and using that by the standard blow-up formulae for Pic and H^2 the expression $b_2(Y) - \rho(Y)$ does not change in the process, one reduces to the case of a dominant morphism $\pi: Y \longrightarrow X$. Then use that $\pi^*: H^2_{\text{ét}}(X, \mathbb{Q}_\ell(1)) \hookrightarrow H^2_{\text{ét}}(Y, \mathbb{Q}_\ell(1))$ is injective sending $\text{NS}(X) \otimes \mathbb{Q}_\ell$ into $\text{NS}(Y) \otimes \mathbb{Q}_\ell$. Moreover, the induced map (of $\mathbb{Q}_\ell$-vector spaces)

$$H^2_{\text{ét}}(X, \mathbb{Q}_\ell(1))/\text{NS}(X) \otimes \mathbb{Q}_\ell \hookrightarrow H^2_{\text{ét}}(Y, \mathbb{Q}_\ell(1))/\text{NS}(Y) \otimes \mathbb{Q}_\ell$$

is still injective, for $\pi_* \circ \pi^* = \deg(\pi) \cdot \text{id}$.

Apply this general observation to $Y = \mathbb{P}^2$, for which $b_2(Y) = \rho(Y) = 1$, and a unirational K3 surface X. Then $\rho(X) = b_2(X) = 22$. □

We leave it to the reader to check that the arguments also apply to uniruled K3 surfaces and so the proposition and the following corollary hold true more generally. However, a posteriori, one knows that uniruled K3 surfaces are in fact unirational.

Corollary 2.8 *Unirational K3 surfaces satisfy the Tate conjecture 3.1.* □

Already in [591, 592] Tate mentions that the Fermat quartic over a field of characteristic $p \equiv 3\,(4)$ has Picard number $\rho = 22$. In fact, both Shioda and Tate deal with Fermat quartics over fields of characteristic p such that $p^n \equiv 3\,(4)$ for some n (and Fermat hypersurfaces of higher degree as well).

2.4 For K3 surfaces over algebraically closed fields, the Picard number satisfies $\rho(X) \neq 21$. This has been observed by Artin in [16] and follows from the inequality $\rho(X) \leq 22 - 2h(X)$ for K3 surfaces of finite height; see Lemma **18**.3.6 and also Remark **18**.3.12. Thus, algebraic K3 surfaces over algebraically closed fields have Picard number

$$\rho = 1, 2, \ldots, 19, 20, 22,$$

and all values are realized over suitable algebraically closed fields. For non-algebraically closed fields and even for finite fields also $\rho = 21$ is possible; see Remark 2.23.

For a K3 surface X over a finite field $\mathbb{F}_q$ the Galois group $\text{Gal}(\overline{\mathbb{F}}_q/\mathbb{F}_q)$ is a cyclic group generated by the Frobenius. By the Weil conjectures (cf. Section **4**.4), the eigenvalues of the induced action f^* on $H^2_{\text{ét}}(\overline{X}, \mathbb{Q}_\ell(1))$ are of absolute value $|\alpha_i| = 1$. Moreover, as the second Betti number of a K3 surface is even, $\alpha_1 = \cdots = \alpha_{2k} = \pm 1$ for an even number of the eigenvalues; see Theorem **4**.4.1. After base change to a finite extension $\mathbb{F}_{q'}/\mathbb{F}_q$, we may assume $\alpha_1 = \cdots = \alpha_{2k} = 1$ (and no roots of unity among the $\alpha_{i>2k}$). Then the Tate conjecture 3.4 implies the following corollary.

Corollary 2.9   *The Picard number of a K3 surface X over* $\overline{\mathbb{F}}_p$, $p \neq 2$, *is always even, i.e.* $\rho(X) \equiv 0\,(2)$. □

Compare this with the statement that for complex K3 surfaces of odd Picard number, automorphisms act as $\pm$id on the transcendental lattice $T(X)$; see Corollary **3**.3.5.

This simple corollary has turned out to have powerful consequences, e.g. for the existence of rational curves on complex K3 surfaces of odd Picard number; cf. Section **13**.3.2.

2.5   Consider a flat proper morphism $X \longrightarrow \mathrm{Spec}(A)$ with A an integral domain. Denote the generic point by $\eta \in \mathrm{Spec}(A)$, so its residue field $k(\eta)$ is the quotient field of A, and pick a closed point $t \in \mathrm{Spec}(A)$. Assume that the two fibres X_η and X_t are K3 surfaces (over the fields $k(\eta)$ and $k(t)$, respectively). Taking the closure of Weil divisors on X_η yields a homomorphism $\mathrm{Pic}(X_\eta) \longrightarrow \mathrm{Pic}(X)$ which, when composed with the restriction to the closed fibre X_t, yields the *specialization homomorphism*

$$\mathrm{sp}\colon \mathrm{Pic}(X_\eta) \longrightarrow \mathrm{Pic}(X_t).$$

Passing to finite extensions of $k(\eta)$ or, geometrically, finite integral coverings of $\mathrm{Spec}(A)$, one obtains a specialization homomorphism between the geometric fibres $X_{\bar\eta} := X_\eta \times \overline{k(\eta)}$ and $X_{\bar t} := X_t \times \overline{k(t)}$

$$\overline{\mathrm{sp}}\colon \mathrm{Pic}(X_{\bar\eta}) \longrightarrow \mathrm{Pic}(X_{\bar t}).$$

Proposition 2.10   *The specialization homomorphisms*

$$\mathrm{sp}\colon \mathrm{Pic}(X_\eta) \hookrightarrow \mathrm{Pic}(X_t) \quad and \quad \overline{\mathrm{sp}}\colon \mathrm{Pic}(X_{\bar\eta}) \hookrightarrow \mathrm{Pic}(X_{\bar t})$$

are injective and compatible with the intersection product.

Proof   The intersection form can be expressed in terms of Euler–Poincaré characteristics; see (2.1) in Section **1**.2.1. As those stay constant in flat families, specialization indeed preserves the intersection form. But if sp and $\overline{\mathrm{sp}}$ are compatible with the intersection form, which is non-degenerate, then they are automatically injective.   □

Consider the case of a trivial family $X \longrightarrow \mathrm{Spec}(A)$; i.e. A is a k-algebra and $X \simeq X_0 \times_k \mathrm{Spec}(A)$ for some K3 surface X_0 over k. Then for any $t \in \mathrm{Spec}(A)$ the residue field $k(t)$ is an extension of k and $X_\eta \simeq X_0 \times_k K$, where $K = k(\eta)$ is the quotient field of A. Moreover, the composition of pull-back and specialization $\mathrm{Pic}(X_0) \longrightarrow \mathrm{Pic}(X_\eta) \longrightarrow \mathrm{Pic}(X_t)$ is nothing but the base change map for the extension

$k \subset k(t)$. In particular, if k is algebraically closed and t is a closed point (and so $k(t) \simeq k$), the injectivity of sp shows once more that $\text{Pic}(X_0) \longrightarrow \text{Pic}(X_0 \times_k K)$ is bijective; cf. Lemma 2.2.

Remark 2.11 Geometrically, the proposition is related to the fact that for complex K3 surfaces the Noether–Lefschetz locus is a countable union of closed subsets; see Section 1.3. More arithmetically, the proposition is often applied to proper flat families $X \longrightarrow \text{Spec}(\mathcal{O}_K)$ over the integers of a number field K such that the generic fibre X_η is a K3 surface over $K \subset \bar{\mathbb{Q}}$. Then for all but finitely many $\mathfrak{p} \in \text{Spec}(\mathcal{O}_K)$ the reduction $X_\mathfrak{p}$ is a K3 surface over the finite field $k(\mathfrak{p}) \subset \bar{\mathbb{F}}_p$ and specialization defines an injection

$$\overline{\text{sp}} \colon \text{Pic}(X_{\bar{\mathbb{Q}}}) \lhook\joinrel\longrightarrow \text{Pic}(X_{\bar{\mathbb{F}}_p}).$$

Due to Corollary 2.9 this can never be an isomorphism if $\rho(X_{\bar{\mathbb{Q}}})$ is odd and $p \neq 2$. However, even when $\rho(X_{\bar{\mathbb{Q}}})$ is even, the Picard number can jump for infinitely many primes. This has been studied by Charles in [112].

Remark 2.12 The obstructions to deform a line bundle L on a closed fibre X_t sideways to finite order are contained in the one-dimensional space $H^2(X_t, \mathcal{O}_{X_t})$. When these obstructions vanish the line bundle deforms to a line bundle on the formal neighbourhood of $X_t \subset X$; see [225] or [176, Ch. 8]. However, this does not necessarily imply the existence of a deformation of L to a Zariski open neighbourhood, which usually requires passing to a covering of $\text{Spec}(A)$ or, alternatively, passing to a finite extension of $k(\eta)$. In particular, even for a geometric closed point t, the images of sp and $\overline{\text{sp}}$ in $\text{Pic}(X_t) = \text{Pic}(X_{\bar{t}})$ might be different.

Remark 2.13 It is not difficult to see that for $X \longrightarrow \text{Spec}(A)$ over an algebraically closed field of characteristic zero the cokernel of $\overline{\text{sp}} \colon \text{Pic}(X_{\bar{\eta}}) \longrightarrow \text{Pic}(X_t)$ is torsion free. Indeed, in this case the obstruction space $H^2(X_t, \mathcal{O}_{X_t})$ is divisible and therefore the obstructions to deform L to finite-order neighbourhoods vanish if and only if they do so for an arbitrary non-trivial power L^k.

One also knows that for a K3 surface X over $\mathbb{Q}$ with good reduction at $p \neq 2$ the cokernel of $\overline{\text{sp}} \colon \text{Pic}(X_{\bar{\mathbb{Q}}}) \longrightarrow \text{Pic}(X_{\bar{\mathbb{F}}_p})$ is torsion free. In [170, Thm. 1.4] Elsenhans and Jahnel deduce this from a result of Raynaud [508], which applies to K3 surfaces over discrete valuation rings of unequal characteristic and ramification index $< p - 1$.

2.6 We next shall discuss K3 surfaces defined over number fields. Clearly, any K3 surface X over $\bar{\mathbb{Q}}$ is isomorphic to the base change of a K3 surface defined over some number field. It is however a non-trivial task to determine the number field or even its degree.

The following is an observation going back to Shioda and Inose in [568]. For more information on the field of definition k_0, see Schütt [537] and Shimada [554].

Proposition 2.14 *Let X be a K3 surface over an algebraically closed field k of characteristic zero. If $\rho(X) = 20$, then X is defined over a number field; i.e. there exists a K3 surface X_0 over a number field k_0, an embedding $k_0 \hookrightarrow k$, and an isomorphism*

$$X \simeq X_0 \times_{k_0} k.$$

Moreover, we may assume that base change yields $\mathrm{Pic}(X_0) \xrightarrow{\sim} \mathrm{Pic}(X)$.

Proof In [568] the assertion is reduced to Kummer surfaces via Remark **15**.4.1, but this can be avoided. It is enough to show that X can be defined over $\overline{\mathbb{Q}}$ for some embedding $\overline{\mathbb{Q}} \hookrightarrow k$. Consider a base $L_1, \ldots, L_{20}$ of $\mathrm{Pic}(X)$. We may assume that X and all L_i are defined over a finitely generated integral $\overline{\mathbb{Q}}$-algebra A. In particular, X is defined over the quotient field $Q(A)$ of A. After spreading and localizing A if necessary, one obtains a smooth family $\mathcal{X} \longrightarrow \mathrm{Spec}(A)$ of K3 surfaces with line bundles $\mathcal{L}_1, \ldots, \mathcal{L}_{20}$.

Specialization yields injections

$$\mathrm{Pic}(X_{\overline{k}}) \simeq \mathrm{Pic}(X_{\overline{\eta}}) \hookrightarrow \mathrm{Pic}(\mathcal{X}_t)$$

for all closed points $t \in \mathrm{Spec}(A)$. Thus, $\mathcal{X} \longrightarrow \mathrm{Spec}(A)$ is a smooth family of K3 surfaces of maximal Picard number $\rho = 20$. However, the locus of polarized K3 surfaces of maximal Picard number in characteristic zero is zero-dimensional. This can be seen by abstract deformation theory for $(\mathcal{X}; \mathcal{L}_1, \ldots, \mathcal{L}_{20})_t$ or by first base changing to $\mathbb{C}$ and then applying the period description; cf. Section 1.3.

Hence, there exists a K3 surface X_0 over $\overline{\mathbb{Q}}$ such that all fibres $\mathcal{X}_t$ over closed points $t \in \mathrm{Spec}(A)$ are isomorphic to X_0. Hence, after localizing A further and finite étale base changing $\mathrm{Spec}(A') \longrightarrow \mathrm{Spec}(A)$ the two families $\mathcal{X} \times_{\mathrm{Spec}(A)} \mathrm{Spec}(A')$ and $X_0 \times \mathrm{Spec}(A')$ are isomorphic.[3] Therefore, the generic fibre $\mathcal{X}_{\eta'}$ of the first one is isomorphic to $X_0 \times Q(A')$. Choosing an embedding $Q(A') \hookrightarrow k$ (which exists as k is algebraically closed) and base changing to k eventually yields $X \simeq \mathcal{X}_\eta \times k \simeq \mathcal{X}_{\eta'} \times k \simeq X_0 \times k$. $\square$

In the moduli space M_d of polarized K3 surfaces of degree $2d$ the set of points corresponding to K3 surfaces with geometric Picard number at least two, i.e. the Noether–Lefschetz locus, is a countable union of hypersurfaces; see Section 1.3. As the set $M_d(\overline{\mathbb{Q}})$ parametrizing K3 surfaces defined over $\overline{\mathbb{Q}}$ is countable it could a priori be contained in the Noether–Lefschetz locus. That this is not the case was shown by Ellenberg [168] for any d and for $d = 2, 3, 4$ by Terasoma in [598] (who proves existence over $\mathbb{Q}$ and not only over some number field).

---

[3] Compare this with the proof of Lemma 2.2. The role of Pic_X is here played by the moduli space M_d. The étale base change $\mathrm{Spec}(A') \longrightarrow \mathrm{Spec}(A)$ is necessary as M_d only corepresents the moduli functor and so M_d^{lev} has to be used; cf. Section **6**.4.2 and below.

Proposition 2.15 *For any $d > 0$ there exists a number field k and a polarized K3 surface (X, L) of degree $2d$ of geometric Picard number one over k.*

Proof Consider the moduli space M_d^{lev} of polarized K3 surfaces (X, L) with a level structure $H^2(X, \mathbb{Z}/\ell^N\mathbb{Z})_{\text{p}} \simeq \Lambda_d \otimes \mathbb{Z}/\ell^N\mathbb{Z}$ and its natural projection

$$\pi : M_d^{\text{lev}} \longrightarrow M_d,$$

which is a Galois covering with Galois group $O(\Lambda_d \otimes \mathbb{Z}/\ell^N\mathbb{Z})$. In Section **6.4.2** this was constructed via period domains in the complex setting, but it exists over $\overline{\mathbb{Q}}$, as needed here, and in fact over a number field, say $k_0/\mathbb{Q}$, over which it is still a Galois covering.

Next, pick a generically finite morphism $M_d \longrightarrow \mathbb{P}^{19}$ and assume for simplicity that the composition

$$p : M_d^{\text{lev}} \longrightarrow M_d \longrightarrow \mathbb{P}^{19}$$

is a Galois covering with Galois group, say G. By Hilbert's irreducibility theorem there exists a Zariski dense subset of points $t \in \mathbb{P}^{19}(k_0)$ with $p^{-1}(t) = \{y\}$ such that $k(y)/k_0$ is a Galois extension with Galois group G. Then for $x := \pi(y) \in M_d$ and $k := k(x)$, one has $\text{Gal}(k(y)/k) \simeq O(\Lambda_d \otimes \mathbb{Z}/\ell^N\mathbb{Z})$. We simplify the discussion by assuming that M_d is a fine moduli space, otherwise pass to some finite covering. Then x corresponds to a polarized K3 surface (X, L) defined over k.

Consider the Galois representation $\rho : \text{Gal}(\overline{\mathbb{Q}}/k) \longrightarrow O(H_{\text{ét}}^2(X_{\overline{\mathbb{Q}}}, \mathbb{Z}_\ell(1))_{\text{p}})$ and its image $\text{Im}(\rho)_N$ in $O(H_{\text{ét}}^2(X_{\overline{\mathbb{Q}}}, \mu_{\ell^N})_{\text{p}}) \simeq O(\Lambda_d \otimes \mathbb{Z}/\ell^N\mathbb{Z})$. As it contains $\text{Gal}(k(y)/k)$, in fact $\text{Im}(\rho)_N \simeq O(\Lambda_d \otimes \mathbb{Z}/\ell^N\mathbb{Z})$.

However, by an argument from p-adic Lie group theory (see [168, Lem. 3]), one can show that for $N \gg 0$ any closed subgroup of $O(\Lambda_d \otimes \mathbb{Z}_\ell)$ that surjects onto $O(\Lambda_d \otimes \mathbb{Z}/\ell^N\mathbb{Z})$ is in fact $O(\Lambda_d \otimes \mathbb{Z}_\ell)$. Hence, $\text{Im}(\rho)$ equals $O(H_{\text{ét}}^2(X_{\overline{\mathbb{Q}}}, \mathbb{Z}_\ell(1))_{\text{p}})$[4] which is enough to conclude that $\rho(X_{\overline{\mathbb{Q}}}) = 1$. Indeed, otherwise $\rho(X_{k'}) > 1$ for some finite extension k'/k for which $\rho(\text{Gal}(\overline{\mathbb{Q}}/k'))$ is, on the one hand, a finite index subgroup of $O(H_{\text{ét}}^2(X_{\overline{\mathbb{Q}}}, \mathbb{Z}_\ell(1))_{\text{p}})$ and, on the other hand, contained in the subgroup fixing $c_1(M)$ of a line bundle linearly independent of L, which is not of finite index. $\qquad\square$

Remark 2.16 It is possible to adapt the above arguments to prove that any lattice that occurs as $\text{NS}(X)$ of a complex algebraic K3 surface X can also be realized by a K3 surface over a number field or, equivalently, over $\overline{\mathbb{Q}}$.

---

[4] More informally, the argument could be summarized as follows. As M_d (over $\mathbb{C}$) is constructed as an open subset of the quotient of D_d by the orthogonal $\widetilde{O}(\Lambda_d)$ the monodromy on $H^2(X, \mathbb{Z})_{\text{p}}$ is the orthogonal group. Then use the relation between the monodromy group and the Galois group of the function field of M_d (see e.g. [232]) to show that the latter acts as $\widetilde{O}(\Lambda_d \otimes \mathbb{Z}_\ell)$ on the cohomology of the geometric generic fibre. Hilbert's irreducibility theorem then ensures the existence of a K3 surface over a number field with this property; cf. [598, Thm. 2].

The proposition and this generalization can also be deduced from a more general result due to André in [8], where he studies the specialization for arbitrary smooth and proper morphisms $X \longrightarrow T$ of varieties over an algebraically closed field of characteristic zero. In particular his results imply that there always exists a closed point $t \in T$ with $\rho(X_{\overline{\eta}}) = \rho(X_t)$ or, equivalently, for which $\overline{\mathrm{sp}} \colon \mathrm{NS}(X_{\overline{\eta}}) \xrightarrow{\sim} \mathrm{NS}(X_t)$ is an isomorphism. See also the article [400] by Maulik and Poonen which contains a p-adic proof of this consequence.

Remark 2.17   The proposition is a sheer existence result that does not give any control over the number field k nor tell one how to explicitly construct examples. But K3 surfaces over $\mathbb{Q}$ of geometric Picard number one of low degree have been constructed. The first explicit example ever is due to van Luijk [609]. We briefly explain the main idea of his construction.

A K3 surface X over $\mathbb{Q}$ can be described by equations with coefficients in $\mathbb{Q}$. By clearing denominators, these equations define a scheme over $\mathbb{Z}$ with generic fibre X. The closed fibres are the reductions $X_{\mathbb{F}_p}$ of X modulo p. Specialization yields injective maps

$$\overline{\mathrm{sp}} \colon \mathrm{Pic}(X_{\overline{\mathbb{Q}}}) \hookrightarrow \mathrm{Pic}(X_{\overline{\mathbb{F}}_p})$$

for all primes p with $X_{\mathbb{F}_p}$ smooth; see Proposition 2.10. Of course, $\rho(X_{\overline{\mathbb{Q}}}) = 1$ holds if $\rho(X_{\overline{\mathbb{F}}_p}) = 1$, which however for $p \neq 2$ is excluded by Corollary 2.9. If $\rho(X_{\overline{\mathbb{F}}_p}) = 2$, then either $\rho(X_{\overline{\mathbb{Q}}}) = 1$ or $\mathrm{Pic}(X_{\overline{\mathbb{Q}}}) \hookrightarrow \mathrm{Pic}(X_{\overline{\mathbb{F}}_p})$ is a sublattice of finite index d_p with $d_p^2 = \mathrm{disc}\,\mathrm{Pic}(X_{\overline{\mathbb{Q}}})/\mathrm{disc}\,\mathrm{Pic}(X_{\overline{\mathbb{F}}_p})$; see Section **14**.0.2. For any two primes $p \neq p'$ with good reduction $X_{\overline{\mathbb{F}}_p}$ and $X_{\overline{\mathbb{F}}_{p'}}$ of Picard number two, this shows that

$$d_{p,p'} := \frac{\mathrm{disc}\,\mathrm{Pic}(X_{\overline{\mathbb{F}}_p})}{\mathrm{disc}\,\mathrm{Pic}(X_{\overline{\mathbb{F}}_{p'}})} = \left(\frac{d_{p'}}{d_p}\right)^2$$

is a square. Thus, in order to prove $\rho(X_{\overline{\mathbb{Q}}}) = 1$, it suffices to find two such primes for which $d_{p,p'}$ is not a square.

As by the Chinese remainder theorem any two smooth quartics over $\mathbb{F}_p$ and $\mathbb{F}_{p'}$ with $p \neq p'$ are reductions of a smooth quartic over $\mathbb{Q}$; quartics seem particularly accessible. There are two steps to carry this out. First, by the Weil conjectures (see Theorem **4**.4.1), computing $\rho(X_{\overline{\mathbb{F}}_p})$ (or rather the rank of the Frobenius invariant part of $H^2_{\text{ét}}(X_{\overline{\mathbb{F}}_p}, \mathbb{Z}_\ell(1))$) is in principle possible by an explicit count of points $|X(\mathbb{F}_{p^n})|$. Note that the Tate conjecture is not used here, because if the Frobenius invariant part of $H^2_{\text{ét}}(X_{\overline{\mathbb{F}}_p}, \mathbb{Q}_\ell(1))$ is of dimension two, then either $\rho(X_{\overline{\mathbb{F}}_p}) = 1$, in which case we immediately have $\rho(X_{\overline{\mathbb{Q}}}) = 1$, or $\rho(X_{\overline{\mathbb{F}}_p}) = 2$. Second, one has to decide whether $d_{p,p'} \in \mathbb{Q}^{*2}$. Fortunately, for this one does not need a complete description of $\mathrm{NS}(X_{\overline{\mathbb{F}}_p})$, which might be tricky. Indeed, due to (0.1) in Section **14**.0.2 it suffices to compute the discriminant of a finite index sublattice in each of the two Néron–Severi lattices. This can often be achieved by exhibiting explicit curves on the surfaces. Alternatively, following Kloosterman [298] one can use the Artin–Tate conjecture **18**.2.4 to conclude (using that $|\mathrm{Br}(X_{\mathbb{F}_q})|$ is a square; see Remark **18**.2.8).

For explicit equations of quartics, see van Luijk's original article [609]. For an explicit equation of a double plane, see the articles by Elsenhans and Jahnel [169, 170]. In the latter, reduction modulo one prime only is used based on the authors' result that the cokernel of the specialization map is torsion free; see Remark 2.13. None of the available equations describing K3 surfaces over $\mathbb{Q}$ of geometric Picard number one is particularly simple.

2.7 We next discuss K3 surfaces X over an algebraically closed field k with maximal Picard number $\rho(X) = 22$. They are sometimes called *Shioda supersingular* K3 surfaces. It is comparatively easy to show that a K3 surface of maximal Picard number $\rho(X) = 22$ is (Artin) supersingular; see Corollary **18**.3.9. (The much harder converse had been open for a long time as the last step in the proof of the Tate conjecture.)

As an immediate consequence of Proposition **11**.1.3 one has the following corollary.

Corollary 2.18 *Let X be a K3 surface over an algebraically closed field k. If $\rho(X) = 22$, then X admits an elliptic fibration.* □

Note, however, that not every such surface admits an elliptic fibration with a section, as was observed by Kondō and Shimada in [326].

The following result is due to Artin [16]; see also the article of Rudakov and Šafarevič [514]. It predates the Tate conjecture, which can be used to cover supersingular K3 surfaces.

Proposition 2.19 *Let X be a K3 surface over an algebraically closed field k of characteristic p with $\rho(X) = 22$. Then there exists an integer $1 \leq \sigma(X) \leq 10$, the Artin invariant, such that*

$$\mathrm{disc}\,\mathrm{NS}(X) = -p^{2\sigma(X)}.$$

Moreover, $A_{\mathrm{NS}(X)} = \mathrm{NS}(X)^*/\mathrm{NS}(X) \simeq (\mathbb{Z}/p\mathbb{Z})^{2\sigma(X)}$.

Proof As $\rho(X) = 22$, the natural map $\mathrm{NS}(X) \otimes \mathbb{Q}_\ell \longrightarrow H^2_{\acute{e}t}(X, \mathbb{Q}_\ell(1))$, $\ell \neq p$, is an isomorphism and by Proposition 3.5 in fact

$$\mathrm{NS}(X) \otimes \mathbb{Z}_\ell \xrightarrow{\sim} H^2_{\acute{e}t}(X, \mathbb{Z}_\ell(1)).$$

Hence, $\mathrm{disc}\,\mathrm{NS}(X) = \pm p^r$ and, by the Hodge index theorem, the sign has to be negative.

The hardest part is to show that $p\mathrm{NS}(X)^* \subset \mathrm{NS}(X)$ implying $A_{\mathrm{NS}(X)} \simeq (\mathbb{Z}/p\mathbb{Z})^r$, which we skip here. In [16] this is deduced from a then still conjectural duality statement for flat cohomology. In [478, 514] crystalline cohomology is used instead; cf. Section **18**.3.2. Note that in particular $0 \leq r \leq 22$.

Next one proves that r is even, so $r = 2\sigma(X)$. In [514] the natural inclusion $NS(X) \otimes W(k) \hookrightarrow H^2_{cr}(X/W(k))$ is considered as a finite index inclusion of lattices over the Witt ring $W(k)$. Combining it with the appropriate version of the elementary fact (0.1) in Section **14**.0.2 and the fact that the natural pairing on $H^2_{cr}(X/W(k))$ is unimodular yields the assertion. Artin's proof in [16, Sec. 6] instead relies on the pairing on the Brauer group $Br(X_0)$ (see Remark **18**.2.8) for the specialization X_0 of X to a K3 surface over a finite field.

Note that $\sigma(X) = 0$ if and only if $NS(X)$ is unimodular. As there is no unimodular even lattice of signature $(1, 21)$ (cf. Theorem 14.1.1), one has $1 \leq \sigma(X)$. Similarly, $\sigma(X) = 11$ cannot occur, as then $NS(X)(p^{-1})$ would be unimodular, even, and of signature $(1, 21)$. Hence, $1 \leq \sigma(X) \leq 10$.   □

The proposition can be combined with a purely lattice theoretic result; cf. Corollary **14**.3.6. For $\sigma < 10$ the following is a direct consequence of Nikulin's theorem **14**.1.5. In [478, Sec. 3] a more direct proof was given by Ogus; see also the survey [514, Sec. 1] by Rudakov and Šafarevič for explicit descriptions.

Proposition 2.20   *For any prime number $p > 2$ and any integer $1 \leq \sigma \leq 10$, there exists a unique lattice $N_{p,\sigma}$ with the following properties:*

(i) *The lattice $N_{p,\sigma}$ is even and non-degenerate.*
(ii) *The signature of $N_{p,\sigma}$ is $(1, 21)$.*
(iii) *The discriminant group of $N_{p,\sigma}$ is isomorphic to $(\mathbb{Z}/p\mathbb{Z})^{2\sigma}$.*   □

The lattice $N_{p,\sigma}$ is often called the *Rudakov–Šafarevič lattice*.

Corollary 2.21   *Let X be a supersingular K3 surface over an algebraically closed field k of characteristic $p > 2$. Then*

$$NS(X) \simeq N_{p,\sigma},$$

where $\sigma = \sigma(X)$ is the Artin invariant of X.   □

We refer to the original sources [514] for results in the case $p = 2$. The following result is due to Kondō and Shimada [326].

Corollary 2.22   *If $\sigma + \sigma' = 11$, then $N_{p,\sigma}$ is isomorphic to $N^*_{p,\sigma'}(p)$.*

Proof   One has to show that $N^*_{p,\sigma'}(p)$ satisfies the conditions (i)–(iii) above. First, $N^*_{p,\sigma'}(p)$ is a lattice, as $pN^*_{p,\sigma'} \subset N_{p,\sigma'}$, and it clearly is non-degenerate of signature $(1, 21)$. Moreover, if A is the intersection matrix of $N_{p,\sigma'}$, then the one of $N^*_{p,\sigma'}(p)$ is pA^{-1} and, hence, $\text{disc } N^*_{p,\sigma'}(p) = -p^{2\sigma}$. Then, as $p(pA^{-1})^{-1} = A$ is an integral matrix, the discriminant group of $N^*_{p,\sigma'}(p)$ is isomorphic to $(\mathbb{Z}/p\mathbb{Z})^{2\sigma}$. For $p \neq 2$ the lattice is obviously even. See [326] for the case $p = 2$.   □

Remark 2.23 Ogus in [478, 479] proved that a supersingular K3 surface with Artin invariant $\sigma(X) = 1$ is unique up to isomorphisms.[5] In [538] Schütt shows that this surface has a model over $\mathbb{F}_p$ with Picard number 21.

3 Tate Conjecture

Together with the Hodge conjecture and the Grothendieck standard conjectures, the general Tate conjecture is one of the central open questions in algebraic geometry. If true, it would allow one to read off the space of algebraic cycles modulo homological equivalence of a variety over a finitely generated field from the Galois action on its cohomology. In this sense, it is an arithmetic analogue of the Hodge conjecture. We shall state only the case of degree two. Although it is the arithmetic analogue of the well-known Lefschetz theorem on $(1, 1)$-classes, it is wide open for general smooth projective varieties.

Conjecture 3.1 (Tate conjecture in degree two) *Let X be a smooth projective variety over a finitely generated field k. Denote by k_s its separable closure and let $G :=$ $\mathrm{Gal}(k_s/k)$. Then for all prime numbers $\ell \neq \mathrm{char}(k)$ the natural cycle class map induces an isomorphism*

$$\mathrm{NS}(X) \otimes \mathbb{Q}_\ell \xrightarrow{\sim} H^2_{\acute{e}t}(X \times k_s, \mathbb{Q}_\ell(1))^G. \tag{3.1}$$

It has also been conjectured by Tate (attributed to Grothendieck and Serre) (see [591, 595]) that the action of G is semi-simple, which is sometimes formulated as part of the Tate conjecture.

Remark 3.2 Let k'/k be a finite Galois extension in k_s. If (3.1) holds for $X \times k'$, then it holds for X as well. Indeed, for $G' := \mathrm{Gal}(k_s/k')$ and using (2.2) one has

$$
\begin{array}{ccc}
\mathrm{NS}(X) \otimes \mathbb{Q}_\ell & \longrightarrow & H^2_{\acute{e}t}(X \times k_s, \mathbb{Q}_\ell(1))^G \\
\parallel & & \parallel \\
\left(\mathrm{NS}(X \times k') \otimes \mathbb{Q}_\ell\right)^{\mathrm{Gal}(k'/k)} & \xrightarrow{\sim} & \left(H^2_{\acute{e}t}(X \times k_s, \mathbb{Q}_\ell(1))^{G'}\right)^{\mathrm{Gal}(k'/k)}.
\end{array}
$$

A similar argument works when k'/k is just separable. Eventually, this proves that the Tate conjecture is equivalent to

$$\mathrm{NS}(X \times k_s) \otimes \mathbb{Q}_\ell \xrightarrow{\sim} \bigcup H^2_{\acute{e}t}(X \times k_s, \mathbb{Q}_\ell(1))^H,$$

where the union is over all open subgroups $H \subset G$.

---

[5] Ogus also proved a Torelli-type theorem; cf. Section **18**.3.6. See also [514].

Remark 3.3 It is known for a product $X = Y_1 \times Y_2$ that the Tate conjecture (in degree two) for X is equivalent to the Tate conjecture for the two factors. Moreover, the Tate conjecture holds for a variety X if it can be rationally dominated by a variety for which it holds; see [595, Thm. 5.2] or [605, Sec. 12]. Thus, for example, the Tate conjecture holds for all varieties that are dominated by a product of curves (DPC). As mentioned already at the end of Chapter 4, there are no K3 surfaces that are known not to be DPC.

For elliptic K3 surfaces $X \longrightarrow \mathbb{P}^1$ over finite fields, the Tate conjecture for X is equivalent to the function field analogue of the Birch–Swinnerton-Dyer conjecture for the generic fibre $E = X_\eta$; see [596, 605] or Remark **18**.2.9 for more details and a more general version.

3.1 K3 surfaces, as abelian varieties, have always served as a testing ground for fundamental conjectures. This was the case for the Weil conjectures and is certainly also true for the Tate conjecture. For many K3 surfaces the Tate conjecture had been verified in the early 1980s, but the remaining cases have been settled only recently. Due to the effort of many people (see below for precise references), one now has the following theorem.

Theorem 3.4 *The Tate conjecture holds true for K3 surfaces in characteristic $p \neq 2$.*

3.2 In characteristic zero, the Tate conjecture for K3 surfaces follows from the Tate conjecture for abelian varieties proved by Faltings and the Lefschetz theorem on $(1, 1)$-classes. This is a folklore argument; see [7, Thm. 1.6.1] or [595, Thm. 5.6], which we reproduce here. For number fields the proof is due to Tankeev [584], who proves a Lie algebra version of Tate's conjecture asserting that $NS(X \times k_s) \otimes \mathbb{Q}_\ell$ is the part of $H^2_{et}(X \times k_s, \mathbb{Q}_\ell(1))$ invariant under the Lie algebra of the Galois group.

First, choose an embedding $k \hookrightarrow \mathbb{C}$. Then any K3 surface X over k induces a complex K3 surface $X_{\mathbb{C}}$. The Kuga–Satake construction induces an embedding of Hodge structures

$$H^2(X_{\mathbb{C}}, \mathbb{Q}(1)) \hookrightarrow \operatorname{End}(H^1(KS(X_{\mathbb{C}}), \mathbb{Q})). \tag{3.2}$$

Here, $KS(X_{\mathbb{C}})$ is the Kuga–Satake variety associated with the Hodge structure $H^2(X_{\mathbb{C}}, \mathbb{Z})$; cf. Section **4**.2.6. Recall that it is not known whether this correspondence really is always algebraic (cf. Conjecture **4**.2.11), but this is not needed for the argument here. As the Hodge structures are polarized, (3.2) can be split by a morphism of Hodge structures $\pi : \operatorname{End}(H^1(KS(X_{\mathbb{C}}), \mathbb{Q})) \twoheadrightarrow H^2(X_{\mathbb{C}}, \mathbb{Q}(1))$.

In fact, the Kuga–Satake variety $KS(X_{\mathbb{C}})$ is obtained by base change $KS(X) = A \times_k \mathbb{C}$ from an abelian variety A over k (up to some finite extension); see Proposition **4**.4.3 and

Remark **4.4.4.** Then, similar to (3.2), there exists a Galois invariant inclusion

$$H^2_{\text{ét}}(X \times k_s, \mathbb{Q}_\ell(1)) \hookrightarrow \text{End}(H^1_{\text{ét}}(A \times k_s, \mathbb{Q}_\ell)); \qquad (3.3)$$

see Remark **4.4.5.** Moreover, (3.2) and (3.3) are compatible via the natural comparison morphisms

$$H^2(X_{\mathbb{C}}, \mathbb{Q}(1)) \hookrightarrow H^2_{\text{ét}}(X \times k_s, \mathbb{Q}_\ell(1))$$

and

$$\text{End}(H^1(A_{\mathbb{C}}, \mathbb{Q})) \hookrightarrow \text{End}(H^1_{\text{ét}}(A \times k_s, \mathbb{Q}_\ell)).$$

Now take $\alpha \in H^2_{\text{ét}}(X \times k_s, \mathbb{Q}_\ell(1))^G$ and consider its image $f \in \text{End}(H^1_{\text{ét}}(A \times k_s, \mathbb{Q}_\ell))$ under (3.3), which due to Faltings's result [175, Ch. VI] can be written as a $\mathbb{Q}_\ell$-linear combination $f = \sum \lambda_i f_i$ of endomorphisms of $A \times k_s$. In particular, the f_i are actually Hodge classes in $\text{End}(H^1(A_{\mathbb{C}}, \mathbb{Q}))$. Their projections $\pi(f_i) \in H^{1,1}(X_{\mathbb{C}}, \mathbb{Q})$ are algebraic by the Lefschetz theorem on $(1,1)$-classes and they remain of course algebraic when considered as classes in $H^2_{\text{ét}}(X \times k_s, \mathbb{Q}_\ell(1))$. But then $\alpha = \sum \lambda_i \pi(f_i)$ is a $\mathbb{Q}_\ell$-linear combination of algebraic classes and hence contained in $\text{NS}(X \times k') \otimes \mathbb{Q}_\ell \subset H^2_{\text{ét}}(X \times k_s, \mathbb{Q}_\ell(1))$ for some finite extension k'/k; see Lemma 2.6. However, α as a cohomology class is G-invariant and hence also as a class in $\text{NS}(X \times k') \otimes \mathbb{Q}_\ell$ contained in the invariant part $\text{NS}(X) \otimes \mathbb{Q}_\ell = (\text{NS}(X \times k') \otimes \mathbb{Q}_\ell)^{\text{Gal}(k'/k)}$.

The Tate conjecture in characteristic zero, together with the Lefschetz theorem on $(1,1)$-classes, at least morally implies the Mumford–Tate conjecture; see Theorem 3.3.11.

3.3 So, most of the attention was focussed on K3 surfaces over finite fields, in which the Galois invariant part of $H^2_{\text{ét}}(\overline{X}, \mathbb{Q}_\ell(1))$ is just the eigenvalue one eigenspace

$$H^2_{\text{ét}}(\overline{X}, \mathbb{Q}_\ell(1))^{f^* - \text{id}} \subset H^2_{\text{ét}}(\overline{X}, \mathbb{Q}_\ell(1))$$

of the Frobenius. For the notation and the definition of the Frobenius action, see Section 4.4.1.

Proposition 3.5 *For a smooth projective surface X over a finite field $\mathbb{F}_q$ the following conditions are equivalent:*

(i) $\text{NS}(X) \otimes \mathbb{Q}_\ell \xrightarrow{\sim} H^2_{\text{ét}}(X \times k_s, \mathbb{Q}_\ell(1))^{f^* - \text{id}}$ *(Tate conjecture).*

(ii) $\text{NS}(X) \otimes \mathbb{Z}_\ell \xrightarrow{\sim} H^2_{\text{ét}}(X \times k_s, \mathbb{Z}_\ell(1))^{f^* - \text{id}}$ *(integral Tate conjecture).*

(iii) $\text{rk NS}(X) = -\text{ord}_{s=1} Z(X, q^{-s})$. *(See Section 4.4.1.)*

(iv) *The Brauer group $\text{Br}(X)$ is finite (Artin conjecture).*

In particular, the Tate conjecture for surfaces is independent of $\ell \neq p$.

Proof   Clearly, (ii) implies (i) and by the Weil conjectures $-\text{ord}_{s=1}Z(X, q^{-s})$ equals the dimension of the (generalized) eigenspace to the eigenvalue q of the action of the Frobenius on $H^2_{\acute{e}t}(\overline{X}, \mathbb{Q}_\ell)$; see Section **4**.4.1. Hence, (iii) implies (i). Moreover, one also knows that the action of the Frobenius is semisimple on the generalized eigenspace for the eigenvalue q; cf. Remark **4**.4.2.

Let us prove that (i) implies (iii) for which we follow Tate's Bourbaki article [596, p. 437]; cf. [595, Sec. 2] and [605, Lect. 2, Prop. 9.2]. To shorten the notation, set $H := H^2_{\acute{e}t}(X \times k_s, \mathbb{Q}_\ell(1))$, $N := \text{NS}(X) \otimes \mathbb{Q}_\ell$, and let

$$H^G := \{\alpha \mid f^*\alpha = \alpha\} \quad \text{and} \quad H_G := H/\{f^*\alpha - \alpha \mid \alpha \in H\}$$

be the invariant resp. coinvariant part of the Galois action. First observe that the non-degenerate pairing $H \times H \longrightarrow \mathbb{Q}_\ell$ given by Poincaré duality naturally leads to an isomorphism $H_G \xrightarrow{\sim} \text{Hom}(H^G, \mathbb{Q}_\ell)$. Next, check that the composition

$$N \xrightarrow{\ c\ } H^G \longrightarrow H_G \xrightarrow{\ \sim\ } \text{Hom}(H^G, \mathbb{Q}_\ell) \xhookrightarrow{\ c^*\ } \text{Hom}(N, \mathbb{Q}_\ell)$$

is the map induced by the non-degenerate intersection pairing on N. Here, c is the cycle class map, which is in fact bijective by assumption. Eventually use that the natural map $H^G \longrightarrow H_G$ is injective if and only if the generalized eigenspace is just H^G.

That (i) also implies the a priori stronger condition (ii) follows from the fact that the quotient of the inclusion

$$\text{Pic}(X) \otimes \mathbb{Z}_\ell \hookrightarrow H^2_{\acute{e}t}(X, \mathbb{Z}_\ell(1)) \simeq H^2_{\acute{e}t}(\overline{X}, \mathbb{Z}_\ell(1))^G$$

is the Tate module $T_\ell\text{Br}(X)$, which is free. See (1.8) in Section **18**.1.1 and also Remark **18**.2.3 and the proof of Lemma **18**.2.5. The equivalence of (i) (or (ii) or (iii)) with (iv) is proved in Section **18**.2.2. Eventually note that (iii) (and (iv)) is independent of ℓ and so are all other statements.   □

Remark 3.6   (i) For varieties over arbitrary finitely generated fields, Tate in [591] points out that if the conjecture holds true for one $\ell \neq p$ and the action of the Frobenius is semi-simple, then the Tate conjecture holds true for all $\ell \neq p$. Over finite fields this is due to the fact that the Zeta function does not depend on ℓ, as seen above. For varieties over arbitrary fields the proof is more involved.

(ii) For K3 surfaces over finite fields the Frobenius action can be shown to be semi-simple directly; see Remark **4**.4.2. In fact, for K3 surfaces over arbitrary finitely generated field k the action of the Galois group $\text{Gal}(k_s/k)$ on $H^2_{\acute{e}t}(X \times k_s, \mathbb{Q}_\ell(1))$ is semi-simple. This follows from the analogous statement for abelian varieties (see [175]) and the Galois invariant embedding (3.3).

First attempts to prove the Tate conjecture for K3 surfaces go back to Artin and Swinnerton-Dyer in [20], where it is proved for K3 surfaces over finite fields admitting an elliptic fibration with a section. In [16] Artin proved the conjecture for supersingular elliptic K3 surfaces. In [515] Rudakov, Zink, and Šafarevič treated the case of K3 surfaces with a polarization of degree two in characteristic ≥ 3. Nygaard in [463] proved the Tate conjecture for ordinary K3 surfaces over finite fields, i.e. for those of height $h(X) = 1$; see Section **18**.3.1. This was in [464] extended by Nygaard and Ogus to all K3 surfaces of finite height, i.e. $h(X) < 11$, over finite fields of characteristic ≥ 5. So it 'only' remained to verify the conjecture for (non-elliptic) supersingular K3 surfaces.

This was almost 30 years later addressed by Maulik in [398], who eventually proved the Tate conjecture for supersingular K3 surfaces over finite fields k (or rather Artin's conjecture) with a polarization of degree d satisfying $2d + 4 < \mathrm{char}(k)$. In [110] Charles built upon Maulik's approach and removed the dependence of the characteristic on the degree (and also avoiding the reduction of the supersingular case to the case of elliptic K3 surfaces dealt with in [20]). Eventually only $p \geq 5$ had to be assumed. An independent approach was pursued by Madapusi Pera [387], who proved Theorem 3.4 for all K3 surfaces in characteristic $\neq 2$.

We do not go into details of any of these proofs, but recommend Benoist's Bourbaki survey [58] for a first introduction. Note that all recent proofs of the Tate conjecture rely on the Kuga–Satake construction to some extent. The proofs in [398, 110] deal only with the remaining supersingular K3 surfaces, whereas in [387] there is no need to distinguish between supersingular and non-supersingular K3 surfaces. The more recent paper of Charles [111] contains another approach, in spirit closer to the paper by Artin and Swinnerton-Dyer. See Section **18**.2.3 for more information. Note that in order to prove the Tate conjecture over arbitrarily finitely generated fields in positive characteristic, it is indeed enough to prove it for finite fields; see [58, Prop. 2.6].

3.4 An immediate consequence of the Tate conjecture is that the Picard number of a K3 surface over a finite field is always even; see Corollary 2.9. The following is another consequence.

Corollary 3.7 *Let X be a K3 surface over an algebraically closed field of characteristic ≥ 3. Then X is supersingular if and only if $\rho(X) = 22$.*

Proof Since $\rho(X) \leq 22 - 2h(X)$ for K3 surfaces of finite height (see Lemma **18**.3.6), any K3 surface of maximal Picard number $\rho(X) = 22$ has to be supersingular. For the converse assume that X is defined over a finite field $\mathbb{F}_q$. Using the arguments in the proof of Theorem **18**.3.10, one finds that the action of the Frobenius on $H^2_{\text{ét}}(X \times \overline{\mathbb{F}}_q, \mathbb{Q}_\ell(1))$ is of finite order and after passing to a certain finite extension $\mathbb{F}_{q^r}$ we can even assume it is trivial. Hence by the Tate conjecture $\mathrm{NS}(X) \otimes \mathbb{Q}_\ell \simeq H^2_{\text{ét}}(X \times \overline{\mathbb{F}}_q, \mathbb{Q}_\ell(1))$ and, therefore, $\rho(X) = 22$.

In order to reduce to the case of finite fields one has to show that the specialization (see Proposition 2.10) for a family of K3 surfaces over a finite field with supersingular geometric generic fibre is not only injective but that the Picard number stays in fact constant. For this one has to use that the Brauer group of the generic fibre is annihilated by a power of p, which again relies on the Tate conjecture. See [16, Thm. 1.1] for details. □

In [370] Lieblich, Maulik, and Snowden observed that the Tate conjecture for K3 surfaces over finite fields is equivalent to the finiteness of these surfaces. Due to the usual boundedness results, the set of isomorphism classes of polarized K3 surfaces (X, H) over a fixed finite field $\mathbb{F}_q$ and with fixed degree $(H)^2$ is finite. Using the Tate conjecture, the main result of [370] becomes the following.

Proposition 3.8 *There exist only finitely many isomorphism types of K3 surfaces over any fixed finite field of characteristic $p \geq 5$.*

In [111] Charles reversed the argument and proved the finiteness of K3 surfaces over a finite field. According to [370] this then implies the Tate conjecture. See Section 18.2.3 for a rough outline of these approaches.

References and Further Reading

André in [7, Lem. 2.3.1] provides some information; which power of a Galois invariant line bundle on X_K descends to X itself; cf. Section 2.2.

In [609] van Luijk also shows that the set of quartic K3 surfaces over $\mathbb{Q}$ of geometric Picard number one and with infinitely many rational points is in fact dense in the moduli space of quartics.

In [583, Thm. 3.3] Tankeev proves an analogue of the Tate conjecture for the generic fibre of one-dimensional families $X \longrightarrow C$ of complex projective K3 surfaces under certain conditions on the stable reduction at some point of C.

Zarhin studies in [649] the action of the Frobenius on the 'transcendental part' of $H^2_{et}(X \times k_s, \mathbb{Q}_\ell(1))$, i.e. the orthogonal complement of the invariant part, and proves that for ordinary K3 surfaces the characteristic polynomial is irreducible. In [650] he builds upon Nygaard's work and proves that for ordinary K3 surfaces X the Tate conjecture holds true for all self-products $X \times \cdots \times X$ in all degrees(!). This is somewhat surprising as the Hodge conjecture for self-products of complex K3 surfaces is not known in such generality. For the square $X \times X$ the results actually show that the Galois invariant part of $H^4_{et}((X \times X) \times k_s, \mathbb{Q}_\ell(2))$ is spanned by products of divisor classes on the two factors, graphs of powers of the Frobenius, and the trivial classes $X \times x$, $x \times X$.

Questions and Open Problems

The results of Ellenberg and van Luijk leave one question open: are there K3 surfaces of geometric Picard number one defined over $\mathbb{Q}$ of arbitrarily high degree? If the condition on the Picard number is dropped, the existence becomes easy. More generally, one could ask which lattices $NS(X_{\overline{\mathbb{Q}}})$ can be realized by K3 surfaces X defined over $\mathbb{Q}$. A related conjecture by Šafarevič [516] asks whether for any d there exist only finitely many lattices N realized as $NS(X \times \overline{k})$ of a K3 surface X defined over a number field k of degree $\leq d$. It is relatively easy to prove that there are only finitely many such lattices of maximal rank 20.

18

Brauer Group

The Brauer group of a K3 surface X, complex or algebraic, is an important invariant of the geometry and the arithmetic of X. Quite generally, for an arbitrary variety (or scheme, or complex manifold, etc.) the Brauer group can be seen as a higher degree version of the group $\mathrm{Pic}(X)$ of isomorphism classes of invertible sheaves L on X, which can be cohomologically is described as $\mathrm{Pic}(X) \simeq H^1(X, \mathbb{G}_m)$ or $\mathrm{Pic}(X) \simeq H^1(X, \mathcal{O}_X^*)$.

Similarly, the Brauer group $\mathrm{Br}(X)$, for example of a K3 surface, is geometrically defined as the set of equivalence classes of sheaves of Azumaya algebras over X and cohomologically identified as $\mathrm{Br}(X) \simeq H^2(X, \mathbb{G}_m)$ in the algebraic setting and as $\mathrm{Br}(X) \simeq H^2(X, \mathcal{O}_X^*)_{\mathrm{tors}}$ in the analytic. However, contrary to $\mathrm{Pic}(X)$, the Brauer group is a torsion abelian group. Its formal version leads to the notion of the height.

In the following, cohomology with coefficients in $\mathbb{G}_m$ or μ_n always means étale cohomology. For a variety X over a field k we denote by $\overline{X}$ the base change $X \times_k k_s$ to a separable closure of k_s/k

1 General Theory: Arithmetic, Geometric, Formal

For general information on Brauer groups of schemes, see Grothendieck's original [227], Milne's account [405, Ch. IV], or the more recent notes of Poonen [495]. Here, we shall first briefly sketch the main facts and constructions. The analytic theory is less well documented. We concentrate on those aspects that are strictly necessary for the purpose of this book.

1.1 Algebraic and arithmetic. To define the Brauer group of a scheme X let us first recall the notion of an *Azumaya algebra* over X, which by definition is an

---

* Thanks to François Charles and Christian Liedtke for comments and discussions.

$\mathcal{O}_X$-algebra $\mathcal{A}$ that is coherent as an $\mathcal{O}_X$-module and étale locally isomorphic to the matrix algebra $M_n(\mathcal{O}_X)$. Note that by definition an Azumaya algebra is associative but rarely commutative. The fibre $\mathcal{A}(x) := \mathcal{A} \otimes k(x)$ at every point $x \in X$ is a central simple algebra over $k(x)$.[1]

By the Skolem–Noether theorem, any automorphism of the k-algebra $M_n(k)$ is of the form $a \mapsto g \cdot a \cdot g^{-1}$ for some $g \in GL_n(k)$, i.e. $\mathrm{Aut}(M_n(k)) \simeq PGL_n(k)$. Hence, the usual Čech cocycle description yields a bijection between the set of isomorphism classes of Azumaya algebras and the first étale cohomology of $PGL_n := GL_n/\mathbb{G}_m$:

$$\{\mathcal{A} \mid \text{Azumaya algebra of rank } n^2\} \simeq H^1_{\acute{e}t}(X, PGL_n).\text{[2]} \qquad (1.1)$$

Unlike GL_n, étale cohomology of PGL_n differs from its cohomology with respect to the Zariski topology; i.e. an étale PGL_n-bundle is usually not Zariski locally trivial.

An Azumaya algebra is called *trivial* if it is isomorphic to $\mathcal{E}nd(E)$ for some locally free sheaf E, and two Azumaya algebras $\mathcal{A}_1$ and $\mathcal{A}_2$ are called *equivalent*, $\mathcal{A}_1 \sim \mathcal{A}_2$, if there exist locally free sheaves E_1 and E_2 such that

$$\mathcal{A}_1 \otimes \mathcal{E}nd(E_1) \simeq \mathcal{A}_2 \otimes \mathcal{E}nd(E_2)$$

as Azumaya algebras.

Definition 1.1 The *Brauer group* of X is the set of equivalence classes of Azumaya algebras

$$\mathrm{Br}(X) := \{\mathcal{A} \mid \text{Azumaya algebra}\}/_\sim$$

with the group structure on $\mathrm{Br}(X)$ given by the tensor product $\mathcal{A}_1 \otimes \mathcal{A}_2$.

Note that for the opposite algebra $\mathcal{A}^\circ$ there exists a natural isomorphism

$$\mathcal{A} \otimes_{\mathcal{O}_X} \mathcal{A}^\circ \xrightarrow{\sim} \mathcal{E}nd_{\mathcal{O}_X}(\mathcal{A}), \ a_1 \otimes a_2 \mapsto (a \mapsto a_1 \cdot a \cdot a_2)$$

which makes $\mathcal{A}^\circ$ the inverse of $\mathcal{A}$ in $\mathrm{Br}(X)$.

Remark 1.2 Due to Wedderburn's theorem, any central simple k-algebra is a matrix algebra $M_n(D)$ over a uniquely defined division k-algebra D. As $M_n(D) \simeq M_n(k) \otimes_k D$, one has $M_n(D) \sim D$ and so $[D] = [M_n(D)] \in \mathrm{Br}(k)$. Note that $M_n(D) \sim M_n(D')$ if and only if $D \simeq D'$. See [299, Ch. II] or [1, Tag 074J].

Remark 1.3 There are two numerical invariants attached to a class $\alpha \in \mathrm{Br}(X)$, its *period* and its *index*. The period (or exponent) $\mathrm{per}(\alpha)$ is by definition the order of α as

---

[1] Recall that a central simple k-algebra (always finite-dimensional in our context) is an associative k-algebra with centre k and without any proper non-trivial two-sided ideal. It is known that a k-algebra A is a central simple algebra if and only if there exists a Galois extension k'/k with $A \otimes_k k' \simeq M_n(k')$ for some $n > 0$.

[2] See [405, IV Prop. 1.4] for the generalization of the Skolem–Noether theorem to Azumaya algebras over arbitrary rings, which is needed here.

an element in the group $\mathrm{Br}(X)$, whereas the index $\mathrm{ind}(\alpha)$ of α is the minimal $\sqrt{\mathrm{rk}(\mathcal{A})}$ of all Azumaya algebras $\mathcal{A}$ representing α. For $X = \mathrm{Spec}(k)$ the index equals the minimal degree of a Galois extension k'/k such that $\alpha_{k'} = 0$; see [545, Thm. 10].

Due to (1.5) below, the period always divides the index

$$\mathrm{per}(\alpha) \mid \mathrm{ind}(\alpha).$$

Moreover, it is known that their prime factors coincide. Classically it is also known that in general $\mathrm{per}(\alpha) \ne \mathrm{ind}(\alpha)$ (see e.g. Kresch's example of a three-dimensional variety in characteristic zero in [329]), and the notorious period-index problem asks under which conditions $\mathrm{per}(\alpha) = \mathrm{ind}(\alpha)$. For function fields of surfaces and for surfaces over finite fields this has been addressed by de Jong [137] and Lieblich [366]. In particular, $\mathrm{per}(\alpha) = \mathrm{ind}(\alpha)$ for Brauer classes $\alpha \in \mathrm{Br}(X)$ with $\mathrm{per}(\alpha)$ prime to q on K3 surfaces over $\mathbb{F}_q$. Also note that for complex K3 surfaces one always has $\mathrm{per}(\alpha) = \mathrm{ind}(\alpha)$; see [270].

The *cohomological Brauer group* of an arbitrary scheme X is the (torsion part of the) étale cohomology

$$\mathrm{Br}'(X) := H^2(X, \mathbb{G}_m)_{\mathrm{tors}}.$$

The two Brauer groups can be compared via a natural group homomorphism

$$\mathrm{Br}'(X) \longrightarrow \mathrm{Br}'(X), \tag{1.2}$$

which is constructed by means of the short exact sequence

$$0 \longrightarrow \mathbb{G}_m \longrightarrow \mathrm{GL}_n \longrightarrow \mathrm{PGL}_n \longrightarrow 0, \tag{1.3}$$

the bijection (1.1), and the induced boundary operator[3]

$$H^1(X, \mathrm{GL}_n) \longrightarrow H^1_{\acute{e}t}(X, \mathrm{PGL}_n) \xrightarrow{\ \delta_n\ } H^2(X, \mathbb{G}_m). \tag{1.4}$$

Indeed, using that $\mathrm{PGL}_n = \mathrm{GL}_n/\mathbb{G}_m \simeq \mathrm{SL}_n/\mu_n$, one finds a factorization

$$\delta_n \colon H^1_{\acute{e}t}(X, \mathrm{PGL}_n) \longrightarrow H^2(X, \mu_n) \longrightarrow H^2(X, \mathbb{G}_m)$$

which in particular shows that

$$\mathrm{Im}(\delta_n) \subset H^2(X, \mathbb{G}_m)[n]. \tag{1.5}$$

The first arrow in (1.4) can be interpreted as the map that sends a locally free sheaf E to $\mathcal{E}nd(E)$, which implies that (1.2) is injective; see [405, IV.Thm. 2.5] for details.

---

[3] The standard reference for non-abelian cohomology of sheaves like PGL_n is [210].

Grothendieck in [227] proved the surjectivity of (1.2) for curves and regular surfaces, but (1.2) is in fact an isomorphism, so

$$\mathrm{Br}(X) \xrightarrow{\sim} \mathrm{Br}'(X)$$

for any quasi-compact and separated X with an ample line bundle. The result is usually attributed to Gabber but the only available proof is de Jong's [136].

Example 1.4 (i) It is not hard to show that for an arbitrary field k the natural map (1.2) yields an isomorphism

$$\mathrm{Br}(k) := \mathrm{Br}(\mathrm{Spec}(k)) \xrightarrow{\sim} H^2(\mathrm{Spec}(k), \mathbb{G}_m) \simeq H^2(\mathrm{Gal}(k_s/k), k_s^*).$$

In particular, all groups involved are torsion. See [549, Ch. X.5].

(ii) More generally, $H^2(X, \mathbb{G}_m)$ is torsion if X is regular and integral. Indeed, in this case the restriction to the generic point of X defines an injection

$$\mathrm{Br}(X) \hookrightarrow \mathrm{Br}(K(X))$$

(see [405, IV.Cor. 2.6]), and the latter group is torsion. So, in all cases relevant to us

$$\mathrm{Br}(X) \simeq H^2(X, \mathbb{G}_m).$$

Let X be a variety over an arbitrary field k. Then for any n prime to $\mathrm{char}(k)$ the exact *Kummer sequence* $0 \longrightarrow \mu_n \longrightarrow \mathbb{G}_m \longrightarrow \mathbb{G}_m \longrightarrow 0$ yields a short exact sequence

$$0 \longrightarrow H^1(X, \mathbb{G}_m) \otimes \mathbb{Z}/n\mathbb{Z} \longrightarrow H^2(X, \mu_n) \longrightarrow \mathrm{Br}(X)[n] \longrightarrow 0. \tag{1.6}$$

If X is proper, then this in particular shows that

$$|\mathrm{Br}(X)[n]| < \infty.$$

Example 1.5 For a K3 surface X and n prime to the characteristic, the vanishing of $H^1(\overline{X}, \mu_n)$ (see Remark 1.3.7), the Kummer sequence, and Poincaré duality imply $\mathrm{Br}(\overline{X}) \otimes \mathbb{Z}/n\mathbb{Z} \subset H^3_{\acute{e}t}(\overline{X}, \mu_n) = 0$, so the torsion part of $\mathrm{Br}(\overline{X})$ prime to p is a divisible group.

For finite fields one even expects $\mathrm{Br}(X)$ to be finite altogether, due to the following very general (and widely open in this generality) corollary.

Conjecture 1.6 (Artin) *For any proper scheme X over* $\mathrm{Spec}(\mathbb{Z})$ *the Brauer group* $\mathrm{Br}(X)$ *is finite.*

A more precise form in the case of smooth projective surfaces over finite fields is given by the Artin–Tate conjecture 2.4. As the conjecture assumes properness over $\mathrm{Spec}(\mathbb{Z})$, it does not apply to varieties over number fields, and, indeed, the Brauer group of a number field itself is large; see Remark 2.1 and Section 2.4.

For a prime $\ell \neq \operatorname{char}(k)$ one defines the Tate module as the inverse limit

$$T_\ell \operatorname{Br}(X) := \varprojlim \operatorname{Br}(X)[\ell^n], \tag{1.7}$$

which is a free $\mathbb{Z}_\ell$-module. Taking limits and using that the inverse system $\operatorname{NS}(X) \otimes (\mathbb{Z}/\ell^n\mathbb{Z})$ satisfies the Mittag-Leffler condition, one deduces from (1.6) the short exact sequence

$$0 \longrightarrow \operatorname{Pic}(X) \otimes \mathbb{Z}_\ell \longrightarrow H^2_{\text{ét}}(X, \mathbb{Z}_\ell(1)) \longrightarrow T_\ell \operatorname{Br}(X) \longrightarrow 0, \tag{1.8}$$

which bears a certain resemblance to the finite index inclusion $\operatorname{NS}(X) \oplus T(X) \subset H^2(X, \mathbb{Z})$ for a complex projective K3 surface; cf. Remark 1.10 and Section **3**.3.

Let now X be proper and geometrically integral over an arbitrary field k with separable closure k_s/k and let $\overline{X} := X \times_k k_s$. Note that $\operatorname{Pic}(\overline{X}) \otimes \mathbb{Z}_\ell \simeq \operatorname{NS}(\overline{X}) \otimes \mathbb{Z}_\ell$, as the kernel of $\operatorname{Pic}(\overline{X}) \longrightarrow \operatorname{NS}(\overline{X})$ is an ℓ-divisible group. The short exact sequence (1.8) for $\overline{X}$ is a sequence of $G := \operatorname{Gal}(k_s/k)$-modules and the Tate conjecture predicts that $\operatorname{NS}(X) \otimes \mathbb{Q}_\ell \simeq H^2_{\text{ét}}(\overline{X}, \mathbb{Q}_\ell(1))^G$ if k is finitely generated; cf. Section **17**.3 and Section 2.2 below for the relation to the finiteness of $\operatorname{Br}(X)$.

The Brauer groups of k, X, and $\overline{X}$ are compared via the Hochschild–Serre spectral sequence[4]

$$E_2^{p,q} = H^p(k, H^q(\overline{X}, \mathbb{G}_m)) \Rightarrow H^{p+q}(X, \mathbb{G}_m). \tag{1.9}$$

Using Hilbert 90, i.e. $H^1(k, \mathbb{G}_m) = 0$, it yields an exact sequence:

$$\operatorname{Pic}(X) \hookrightarrow \operatorname{Pic}(\overline{X})^G \longrightarrow \operatorname{Br}(k) \longrightarrow \operatorname{Br}_1(X) \longrightarrow H^1(k, \operatorname{Pic}(\overline{X})) \longrightarrow H^3(k, \mathbb{G}_m). \tag{1.10}$$

Here, by definition $\operatorname{Br}_1(X) := \operatorname{Ker}\big(\operatorname{Br}(X) \longrightarrow \operatorname{Br}(\overline{X})^G\big)$, which is part of a natural filtration

$$\operatorname{Br}_0(X) \subset \operatorname{Br}_1(X) \subset \operatorname{Br}(X)$$

with $\operatorname{Br}_0(X) := \operatorname{Im}\big(\operatorname{Br}(k) \longrightarrow \operatorname{Br}(X)\big)$. Then there exist inclusions

$$\operatorname{Br}_1(X)/\operatorname{Br}_0(X) \hookrightarrow H^1(k, \operatorname{Pic}(\overline{X})) \tag{1.11}$$

(which is an isomorphism if $H^3(k, \mathbb{G}_m) = 0$, e.g. for all local and global fields) and

$$\operatorname{Coker}\big(\operatorname{Br}(X) \longrightarrow \operatorname{Br}(\overline{X})^G\big) \hookrightarrow H^2(k, \operatorname{Pic}(\overline{X})). \tag{1.12}$$

Classes in $\operatorname{Br}_1(X)$ are called algebraic, and all others, i.e. those giving non-trivial classes in $\operatorname{Br}(X)/\operatorname{Br}_1(X) \simeq \operatorname{Im}(\operatorname{Br}(X) \longrightarrow \operatorname{Br}(\overline{X}))$, transcendental.

Clearly, if $X(k) \neq \emptyset$, then (see also [443, App. I] for a direct proof)

$$\operatorname{Br}(k) \hookrightarrow \operatorname{Br}(X) \quad \text{and} \quad \operatorname{Pic}(X) \xrightarrow{\sim} \operatorname{Pic}(\overline{X})^G. \tag{1.13}$$

---

[4] Which is the usual spectral sequence associated with the composition of two functors. In the present case use that for a sheaf F on X the composition of $F \longmapsto F(\overline{X})$ with $M \longmapsto M^G$ equals $F(X)$.

1.2 Analytic. For a complex possibly non-algebraic K3 surface or more generally a compact complex manifold X, the definition of $\mathrm{Br}(X)$ as the group of equivalence classes of Azumaya algebras on X translates literally, replacing étale topology by the classical topology. However, the cohomological Brauer group, defined as

$$\mathrm{Br}'(X) := H^2(X, \mathcal{O}_X^*)_{\mathrm{tors}},$$

is strictly smaller than $H^2(X, \mathcal{O}_X^*)$ (unless completely trivial), in contrast to the étale cohomology group $H^2(X, \mathbb{G}_m)_{\mathrm{tors}} = H^2(X, \mathbb{G}_m)$.

As in the algebraic setting, the Brauer group and the cohomological Brauer group can be compared by means of a long exact sequence. The relevant short exact sequence, the analytic analogue of (1.3), is

$$0 \longrightarrow \mathcal{O}_X^* \longrightarrow \mathrm{GL}_n \longrightarrow \mathrm{PGL}_n \longrightarrow 0,$$

which yields

$$H^1(X, \mathrm{GL}_n) \longrightarrow H^1(X, \mathrm{PGL}_n) \longrightarrow H^2(X, \mathcal{O}_X^*)$$

and consequently a natural injective homomorphism $\mathrm{Br}(X) \hookrightarrow \mathrm{Br}'(X)$. As in the algebraic setting, this morphism is expected to be an isomorphism in general, i.e.

$$\mathrm{Br}(X) \xrightarrow{\sim} \mathrm{Br}'(X),$$

which has been proved in [270] for arbitrary complex K3 surfaces.

Example 1.7 To get a feeling for certain torsion parts of $\mathrm{Br}(X)$, we mention a result of van Geemen [608]. For the generic double plane $X \longrightarrow \mathbb{P}^2$ branched over a smooth sextic $C \subset \mathbb{P}^2$ there exists a short exact sequence

$$0 \longrightarrow \mathrm{Jac}(C)[2] \longrightarrow \mathrm{Br}(X)[2] \longrightarrow \mathbb{Z}/2\mathbb{Z} \longrightarrow 0.$$

For a complex K3 surface X, the exponential sequence $0 \longrightarrow \mathbb{Z} \longrightarrow \mathcal{O}_X \longrightarrow \mathcal{O}_X^* \longrightarrow 0$ induces a long exact sequence (cf. Section **1**.3.2)

$$0 \longrightarrow H^1(X, \mathcal{O}_X^*) \longrightarrow H^2(X, \mathbb{Z}) \longrightarrow H^2(X, \mathcal{O}_X) \longrightarrow H^2(X, \mathcal{O}_X^*) \longrightarrow 0.$$

As $H^1(X, \mathcal{O}_X^*) \simeq \mathrm{Pic}(X) \simeq \mathbb{Z}^{\oplus \rho(X)}$ and $H^2(X, \mathbb{Z}) \simeq \mathbb{Z}^{\oplus 22}$, one finds that $H^2(X, \mathcal{O}_X^*) \simeq \mathbb{C}/\mathbb{Z}^{\oplus 22 - \rho(X)}$ and so

$$\mathrm{Br}(X) \simeq \mathrm{Br}'(X) = H^2(X, \mathcal{O}_X^*)_{\mathrm{tors}} \simeq (\mathbb{Q}/\mathbb{Z})^{\oplus 22 - \rho(X)}. \tag{1.14}$$

This is a divisible group, which could also be deduced from the exact Kummer sequence $0 \longrightarrow \mu_n \longrightarrow \mathcal{O}_X^* \longrightarrow \mathcal{O}_X^* \longrightarrow 0$ as in the algebraic setting.

Note that the composition $\mathbb{Q} \longrightarrow \mathcal{O}_X \longrightarrow \mathcal{O}_X^*$ yields a surjection

$$H^2(X, \mathbb{Q}) \longrightarrow\!\!\!\rightarrow \mathrm{Br}'(X) \simeq \mathrm{Br}(X), \quad B \longmapsto \alpha_B$$

and it can indeed be useful to represent a Brauer class by a lift in $H^2(X, \mathbb{Q})$; cf. Section **16**.4.1.

Thinking of $\text{Br}(X)$ as a geometric replacement for the transcendental part $T(X)$ of the Hodge structure associated with X can be made more precise as follows: lifting a class $\alpha \in \text{Br}(X)$ to a class $B \in H^2(X, \mathbb{Q})$, i.e. $\alpha = \alpha_B$, and using the intersection product on $T(X) \subset H^2(X, \mathbb{Z})$ yields an isomorphism

$$\text{Br}(X) \xrightarrow{\sim} \text{Hom}(T(X), \mathbb{Q}/\mathbb{Z}).$$

This holds more generally for all X with $H^3(X, \mathbb{Z}) = 0$ and $\text{Br}(X) \simeq \text{Br}'(X)$.

Remark 1.8   Note that only for $\rho(X) = 20$ the group $H^2(X, \mathcal{O}_X^*)$ has a reasonable geometric structure, namely that of a complex elliptic curve. In fact, in this case X is a double cover of a Kummer surface associated with a product $E_1 \times E_2$ of two CM elliptic curves E_1, E_2 isogenous to $H^2(X, \mathcal{O}_X^*)$; see Remark **14**.3.22.

Remark 1.9   In case X is a projective complex K3 surface, there is the algebraic Brauer group $H^2(X, \mathbb{G}_m)$ and the analytic one $H^2(X, \mathcal{O}_X^*)_{\text{tors}}$. They are isomorphic,

$$H^2(X, \mathbb{G}_m) \simeq H^2(X, \mathcal{O}_X^*)_{\text{tors}},$$

which can either be seen by comparing Azumaya algebras in the étale and analytic topology or by comparing the two cohomology groups directly; see Remark **11**.5.13.

Remark 1.10   For a K3 surface X over an arbitrary algebraically closed field k, e.g. $k = \overline{\mathbb{F}}_p$, one has by (1.8)

$$T_\ell \text{Br}(X) \simeq \mathbb{Z}_\ell^{\oplus 22 - \rho(X)},$$

as in this case $H^2_{\text{ét}}(X, \mathbb{Z}_\ell(1))$ is of rank 22. In the complex case, this can be also deduced from (1.14) above.

1.3 Formal.

1.3 Formal.   Contrary to the case of the Picard group, the Brauer group (of a K3 surface) cannot be given the structure of an algebraic group, as over a separably closed field it is torsion and divisible. However, its 'formal completion' can be constructed as a formal group scheme. This was made rigorous by Artin and Mazur in [19]. The result relies on Schlessinger's theory of pro-representable functors of which we briefly recall the basic features. See the original articles [17, 527] or [237], but the most suitable account for our purpose is [176, Ch. 6]. For K3 surfaces, the formal Brauer group is a smooth one-dimensional formal group which in positive characteristic allows one to introduce the height as an auxiliary invariant.

Let us briefly review the classical theory of the Picard functor in the easiest case of a smooth projective variety X over a field k. The Picard functor is the sheafification (which is needed only when X comes without a k-rational point) of the contravariant functor

$$\mathbf{Pic}_X \colon (Sch/k)^o \longrightarrow (Ab), \quad S \longmapsto \mathrm{Pic}(X_S)/\sim.$$

Here, $X_S := X \times S$ with the second projection $p \colon X_S \longrightarrow S$ and $L \sim L'$ if there exists a line bundle M on S with $L \simeq L' \otimes p^*M$. Alternatively, one could introduce directly $\mathbf{Pic}_X(S)$ as $H^0(S, R^1 p_* \mathbb{G}_m)$. Compare Sections **10**.1.1 and **11**.4.1.

The Picard functor is represented by a scheme Pic_X; cf. [82, 176]. The connected component containing the point that corresponds to $\mathcal{O}_X$ is a projective k-scheme Pic_X^0. The Zariski tangent space of Pic_X at a point corresponding to a line bundle L on $X_{k'}$ is naturally isomorphic to $H^1(X_{k'}, \mathcal{O}_{X_{k'}})$; cf. Proposition **10**.1.11. Although the obstruction space $H^2(X, \mathcal{O}_X)$ need not be zero in general, it is not for a K3 surface; the Picard scheme is smooth if $\mathrm{char}(k) = 0$. In this case, Pic_X^0 is an abelian variety of dimension $h^1(X, \mathcal{O}_X)$.

Example 1.11 For a K3 surface X over an arbitrary field k, Pic_X is zero-dimensional and reduced. In particular, Pic_X^0 consists of just one k-rational reduced point which corresponds to $\mathcal{O}_X$. Other points of Pic_X might not be k-rational, but they are all reduced. Compare Sections **10**.1.6 and **17**.2.1.

Let k be any field and denote by (Art/k) the category of local Artin k-algebras. A *deformation functor* is a covariant functor

$$F \colon (Art/k) \longrightarrow (Sets)$$

such that $F(k)$ is a single point.

A deformation functor is *pro-representable* if there exists a local k-algebra R with residue field $k \simeq R/\mathfrak{m}$ and finite-dimensional Zariski tangent space $(\mathfrak{m}/\mathfrak{m}^2)^*$ such that $F \simeq h_R$; i.e. there are functorial (in $A \in (Art/k)$) bijections

$$F(A) \simeq \mathrm{Mor}_{k\text{-}alg}(R, A).$$

Note that if $F \simeq h_R$, then also $F \simeq h_{\hat{R}}$, for the $\mathfrak{m}$-adic completion $\hat{R}$ of R. Hence, if F is pro-representable at all, it is pro-representable by a complete local k-algebra R.

To understand F, one needs to study whether objects defined over some Artinian ring A, i.e. elements in $F(A)$, can be lifted to bigger Artinian rings $A' \longrightarrow A$, and if at all, in how many ways. It usually suffices to consider *small extensions*, i.e. for which the kernel I of the quotient $A' \longrightarrow A$ in (Art/k) satisfies $\mathfrak{m}_{A'} \cdot I = 0$.

A *tangent-obstruction theory* for a deformation functor F consists of two finite-dimensional k-vector spaces T_1 and T_2 such that for any small extension $I \longrightarrow A' \longrightarrow A$ in (Art/k) there exists an exact sequence of sets

$$T_1 \otimes_k I \longrightarrow F(A') \longrightarrow F(A) \longrightarrow T_2 \otimes_k I \tag{1.15}$$

which is assumed to be left exact for $A = k$ and which implicitly assumes that $T_1 \otimes_k I$ acts transitively on the fibres of $F(A') \longrightarrow F(A)$. Note that a tangent-obstruction theory need not exist, and when it does, T_2 is not unique. Recall that the functor F is formally smooth if $F(A') \longrightarrow F(A)$ is surjective for all $A' \longrightarrow A$. Therefore, if F admits a tangent-obstruction theory with $T_2 = 0$, then F is formally smooth. Also, if $F \simeq h_R$, then $T_1 \simeq F(k[x]/x^2)$ is isomorphic to the Zariski tangent space $(\mathfrak{m}/\mathfrak{m}^2)^*$ and R is smooth if $T_2 = 0$. One of the main results of [527] is the following one, which we phrase in the language of [176] that replaces Schlessinger's conditions (H1)–(H4) by the condition on the tangent-obstruction theory.

Theorem 1.12 (Schlessinger)   *A deformation functor F is pro-representable if and only if F admits a tangent-obstruction theory for which the sequence (1.15) is left exact for all small extensions $I \longrightarrow A' \longrightarrow A$.*

If the deformation functor takes values in the category of abelian groups (Ab), then under the same assumptions Schlessinger's theory yields a complete local k-algebra R such that the formal spectrum $\mathrm{Spf}(R)$ has a group structure.

(i) Let us test this for the formal completion of the Picard functor (using that $\mathrm{Pic}(A)$ is trivial)

$$\widehat{\mathbf{Pic}}_X \colon (Art/k) \longrightarrow (Ab), \ A \longmapsto \mathrm{Ker} \left(\mathrm{Pic}(X_A) \longrightarrow \mathrm{Pic}(X) \right).$$

The restriction $\mathrm{Pic}(X_A) \longrightarrow \mathrm{Pic}(X)$ is part of an exact sequence

$$\longrightarrow H^1(X, 1 + \mathcal{O}_X \otimes_k \mathfrak{m}_A) \longrightarrow \mathrm{Pic}(X_A) \longrightarrow \mathrm{Pic}(X) \longrightarrow H^2(X, 1 + \mathcal{O}_X \otimes_k \mathfrak{m}_A) \longrightarrow$$

induced by the short exact sequence $0 \longrightarrow 1 + \mathcal{O}_X \otimes_k \mathfrak{m}_A \longrightarrow \mathcal{O}^*_{X_A} \longrightarrow \mathcal{O}^*_X \longrightarrow 1$. So, in fact, for complete X

$$\widehat{\mathbf{Pic}}_X(A) = H^1(X, 1 + \mathcal{O}_X \otimes_k \mathfrak{m}_A) \simeq H^1(X, \mathcal{O}_X) \otimes_k \mathfrak{m}_A.$$

Since (the sheafification of) the global Picard functor $\mathbf{Pic}_X$ on $(Sch/k)^o$ is representable by a scheme Pic_X, the formal completion $\mathrm{Spf}(\widehat{\mathcal{O}}_{\mathrm{Pic}_X,0})$ of Pic_X at the origin (or rather the complete k-algebra $\widehat{\mathcal{O}}_{\mathrm{Pic}_X,\mathcal{O}_X}$) pro-represents $\widehat{\mathbf{Pic}}_X$.

But Schlessinger's theory can in fact be applied to $\widehat{\mathbf{Pic}}_X$ directly. A tangent-obstruction theory in this case is given by $T_1 := H^1(X, \mathcal{O}_X)$ and $T_2 := H^2(X, \mathcal{O}_X)$. Indeed, for a complete variety X the short exact sequence

$$0 \longrightarrow 1 + \mathcal{O}_X \otimes_k I \longrightarrow \mathcal{O}^*_{X_{A'}} \longrightarrow \mathcal{O}^*_{X_A} \longrightarrow 1 \tag{1.16}$$

associated with a small extension $I \longrightarrow A' \longrightarrow A$ yields (1.15).

For a K3 surface, $H^1(X, \mathcal{O}_X) = 0$ and hence Pic_X consists of isolated reduced points. In particular, $\widehat{\mathbf{Pic}}_X$ is formally smooth and, in fact, pro-representable by $\mathrm{Spf}(k)$.

(ii) The local approach works equally well for the functor described by the Brauer group of a variety X. Consider

$$\widehat{\mathbf{Br}}_X \colon (Art/k) \longrightarrow (Ab), \quad A \longmapsto \mathrm{Ker}\,(\mathrm{Br}(X_A) \longrightarrow \mathrm{Br}(X))\,.$$

Note that in this case one cannot hope to represent the global version on (Sch/k), but Schlessinger's local theory applies to $\widehat{\mathbf{Br}}_X$. A tangent-obstruction theory is this time provided by $T_1 := H^2(X, \mathcal{O}_X)$ and $T_2 := H^3(X, \mathcal{O}_X)$. Indeed, by using the long cohomology sequence associated with (1.16) one finds

$$\longrightarrow H^2(X, \mathcal{O}_X) \otimes_k I \longrightarrow \widehat{\mathbf{Br}}_X(A') \longrightarrow \widehat{\mathbf{Br}}_X(A) \longrightarrow H^3(X, \mathcal{O}_X) \otimes_k I \longrightarrow .$$

Moreover, $H^2(X, \mathcal{O}_X) \otimes_k I \longrightarrow \widehat{\mathbf{Br}}_X(A')$ is injective if and only if $\widehat{\mathbf{Pic}}_X(A') \longrightarrow \widehat{\mathbf{Pic}}_X(A)$ is surjective. This immediately yields the following consequence of Theorem 1.12.

Corollary 1.13 *Let X be a complete variety over an arbitrary field k.*

 (i) *If $\widehat{\mathbf{Pic}}_X$ is formally smooth, then $\widehat{\mathbf{Br}}_X$ is pro-representable.*
 (ii) *If $\widehat{\mathbf{Br}}_X$ is pro-representable, then its Zariski tangent space is naturally isomorphic to $H^2(X, \mathcal{O}_X)$.*
 (iii) *If $H^3(X, \mathcal{O}_X) = 0$, then $\widehat{\mathbf{Br}}_X$ is formally smooth.* □

If $\widehat{\mathbf{Br}}_X$ is pro-representable by a complete k-algebra R, one writes $\widehat{\mathrm{Br}}_X$ for the formal group $\mathrm{Spf}(R)$ and calls it the *formal Brauer group* of X.

Remark 1.14 Assume $\mathrm{char}(k) = 0$. Then the exponential map $\exp \colon \mathcal{O}_X \otimes_k \mathfrak{m}_A \xrightarrow{\sim} 1 + \mathcal{O}_X \otimes_k \mathfrak{m}_A$ yields an isomorphism of group functors $H^2(X, \mathcal{O}_X) \otimes_k \mathfrak{m}_A \xrightarrow{\sim} \widehat{\mathbf{Br}}_X(A)$. As then

$$\widehat{\mathbf{Br}}_X(A) \simeq \mathrm{Hom}_k(H^2(X, \mathcal{O}_X)^*, \mathfrak{m}_A) \simeq \mathrm{Hom}_{k\text{-}alg}(S^* H^2(X, \mathcal{O}_X)^*, A),$$

this shows directly

$$\widehat{\mathrm{Br}}_X \simeq \mathrm{Spf}\left(\widehat{S}^* H^2(X, \mathcal{O}_X)^*\right).$$

However, in positive characteristic the situation is different.

As $H^3(X, \mathcal{O}_X) = 0$ for a K3 surface X and the discrete and reduced Pic_X is obviously smooth, the general theory applied to K3 surfaces becomes the following corollary.

Corollary 1.15 *Let X be a K3 surface over an arbitrary field k. Then $\widehat{\mathbf{Br}}_X$ is pro-representable by a smooth, one-dimensional formal group $\widehat{\mathrm{Br}}_X \simeq \mathrm{Spf}(R)$.* □

2 Finiteness of Brauer Group

This section is devoted to finiteness results and conjectures for Brauer groups of K3 surfaces over finitely generated fields k. Most of the results hold in broader generality, but, due to the vanishing of various cohomology groups of odd degree, the picture is often simpler for K3 surfaces. The case $\mathrm{char}(k) = 0$, for which number fields provide the most interesting examples, is discussed in Section 2.4. For the case $\mathrm{char}(k) > 0$ (in fact, mostly finite fields and their algebraic closure), see Section 2.2.

2.1 We shall begin by recalling basic facts on the Brauer group of the relevant base fields.

Remark 2.1 (o) Separably closed fields have trivial Brauer groups.

(i) Brauer groups of local fields are known to be

$$\mathrm{Br}(\mathbb{R}) \simeq \mathbb{Z}/2\mathbb{Z}, \ \ \mathrm{Br}(\mathbb{C}) = 0, \ \text{and} \ \ \mathrm{Br}(k) \simeq \mathbb{Q}/\mathbb{Z}$$

for non-archimedean local fields k, i.e. finite extensions of $\mathbb{Q}_p$ or $\mathbb{F}_q((T))$, where the isomorphism is given by the Hasse invariant; see e.g. [410, Ch. IV].

(ii) For any global field k, i.e. a finite extension of $\mathbb{Q}$ or $\mathbb{F}_q(T)$, there exists a short exact sequence

$$0 \longrightarrow \mathrm{Br}(k) \longrightarrow \bigoplus \mathrm{Br}(k_v) \longrightarrow \mathbb{Q}/\mathbb{Z} \longrightarrow 0. \tag{2.1}$$

It shows in particular that the Brauer group of a number field is big. That (2.1) is exact follows from the cohomological part of local and global class field theory; see e.g. [106].

(iii) By Wedderburn's theorem, $\mathrm{Br}(k) = 0$ for any finite field $k = \mathbb{F}_q$; see e.g. [410, 549]. In fact, in this case $H^q(k, \mathbb{G}_m) = 0$ for all $q \geq 1$; see [1, Tag 0A2M].

(iv) If k_0 is algebraically closed and $\mathrm{trdeg}(k/k_0) = 1$, then $\mathrm{Br}(k) = 0$ (Tsen's theorem). See [549, Ch. X.7], [410, Ch. IV], or [1, Tag 03RD].

Of course, the arithmetic information of k encoded by $\mathrm{Br}(k)$ disappears when passing to $\overline{X} = X \times_k k_s$, i.e. the composition $\mathrm{Br}(k) \longrightarrow \mathrm{Br}(X) \longrightarrow \mathrm{Br}(\overline{X})$ induced by the Hochschild–Serre spectral sequence (1.9) is trivial, simply because it factors through $\mathrm{Br}(k) \longrightarrow \mathrm{Br}(k_s)$.

The following is the first general finiteness result.

Lemma 2.2 *For any K3 surface X over an arbitrary field the groups*

$$H^1(k, \mathrm{Pic}(\overline{X})) \quad \textit{and} \quad \mathrm{Br}_1(X)/\mathrm{Br}_0(X) \tag{2.2}$$

are finite.

Proof There exists a finite Galois extension k'/k such that $\mathrm{NS}(\overline{X}) \simeq \mathrm{NS}(X \times_k k')$; see Lemma **17.2.6**. Hence, the action of $G = \mathrm{Gal}(k_s/k)$ on $\mathrm{NS}(\overline{X})$ factors over a finite quotient G/H. The Hochschild–Serre spectral sequence yields an exact sequence

$$0 \longrightarrow H^1(G/H, \mathrm{NS}(\overline{X})) \longrightarrow H^1(G, \mathrm{NS}(\overline{X})) \longrightarrow H^0(G/H, H^1(H, \mathrm{NS}(\overline{X}))).$$

However, $H^1(H, \mathrm{NS}(\overline{X})) = \mathrm{Hom}(H, \mathrm{NS}(\overline{X})) = 0$, as $\mathrm{NS}(\overline{X}) \simeq \mathbb{Z}^{\oplus \rho(\overline{X})}$ is a torsion-free trivial H-module and H is profinite. Then use that $H^q(G', A)$ is finite for any finitely generated G'-module A over a finite group G' and $q > 0$; see [106, Ch. IV].

For the second assertion, use the inclusion (1.11). □

The two groups in (2.2) coincide if $H^3(k, \mathbb{G}_m) = 0$.[5]

Remark 2.3 Coming back to the exact sequence (1.8), note that the finitely generated free $\mathbb{Z}_\ell$-module $T_\ell \mathrm{Br}(X)$ is trivial if and only if the ℓ-primary part $\mathrm{Br}(X)[\ell^\infty] := \bigcup \mathrm{Br}(X)[\ell^n]$ of $\mathrm{Br}(X)$, is finite, which is equivalent to

$$\mathrm{Pic}(X) \otimes \mathbb{Z}_\ell \xrightarrow{\ \sim\ } H^2_{\text{ét}}(X, \mathbb{Z}_\ell(1))$$

and also to

$$\mathrm{Pic}(X) \otimes \mathbb{Q}_\ell \xrightarrow{\ \sim\ } H^2_{\text{ét}}(X, \mathbb{Q}_\ell(1)).$$

2.2 Let us now consider the case of finite fields. The following conjecture is motivated by the conjecture of Birch and Swinnerton-Dyer for elliptic curves; see [596] and Remark 2.9 below.

Conjecture 2.4 (Artin–Tate conjecture) *Let X be a smooth geometrically connected projective surface over a finite field $k = \mathbb{F}_q$. Then $\mathrm{Br}(X)$ is finite and*

$$\frac{|\mathrm{Br}(X)| \cdot |\mathrm{disc}\, \mathrm{NS}(X)|}{q^{\alpha(X)} \cdot |\mathrm{NS}(X)_{\text{tors}}|^2} \cdot (1 - q^{1-s})^{\rho(X)} \sim P_2(q^{-s}) \tag{2.3}$$

for $s \longrightarrow 1$.

Here, $\alpha(X) := \chi(X, \mathcal{O}_X) - 1 + \dim \mathrm{Pic}^0(X)$ and

$$P_2(t) = \det(1 - f^* t \mid H^2_{\text{ét}}(\overline{X}, \mathbb{Q}_\ell)) = \prod_{i=1}^{b_2(X)} (1 - \alpha_i t),$$

where $\overline{X} := X \times \overline{\mathbb{F}}_q$ and f^* denotes the action of the Frobenius. By the Weil conjecture (see Section 4.4.1), one knows that $\alpha_i \in \overline{\mathbb{Q}}$ with $|\alpha_i| = q$ (for all embeddings $\overline{\mathbb{Q}} \hookrightarrow \mathbb{C}$)

---

[5] One could try to prove finiteness of $\mathrm{Coker}(\mathrm{Br}(X) \longrightarrow \mathrm{Br}(\overline{X})^G)$ and $H^2(k, \mathrm{Pic}(\overline{X}))$ in a similar way, but $H^2(H, \mathrm{NS}(\overline{X}))^G$ coming up in the spectral sequence is a priori not finite. However, see Theorem 2.10 and Remark 2.11.

and $NS(X) \otimes \mathbb{Q}_\ell$ is contained in the Galois invariant part of $H^2_{\text{ét}}(\overline{X}, \mathbb{Q}_\ell(1))$. So, we may assume that $\alpha_1 = \cdots = \alpha_{\rho(X)} = q$ and (2.3) becomes

$$\frac{|\text{Br}(X)| \cdot |\text{disc NS}(X)|}{|NS(X)_{\text{tors}}|^2} = q^{\alpha(X)} \cdot \prod_{i=\rho(X)+1}^{b_2(X)} (1 - \alpha_i q^{-1}). \tag{2.4}$$

Note that for a K3 surface $NS(X)_{\text{tors}} = 0$ and $\alpha(X) = 1$ and so in this case (2.4) reads

$$|\text{Br}(X)| \cdot |\text{disc NS}(X)| = q \cdot \prod_{i=\rho(X)+1}^{22} (1 - \alpha_i q^{-1}). \tag{2.5}$$

In any case, the left-hand side of (2.4) is clearly non-zero, which shows that the Artin–Tate conjecture implies the (degree two) Tate conjecture **17**.3.1 saying $\text{rk NS}(X) = -\text{ord}_{s=1} Z(X, q^{-s})$ or, equivalently,

$$NS(X) \otimes \mathbb{Q}_\ell \simeq H^2_{\text{ét}}(\overline{X}, \mathbb{Q}_\ell(1))^G, \tag{2.6}$$

where $G = \text{Gal}(\overline{k}/k) \simeq \hat{\mathbb{Z}}$. See Proposition **17**.3.5.

To prove the converse, namely that the Tate conjecture (2.6) implies the Artin–Tate conjecture (2.3), one needs the following lemma.

Lemma 2.5 *Let X be a smooth geometrically connected projective surface over a finite field $k = \mathbb{F}_q$ and $\ell \neq p$. Then $\text{Br}(X)[\ell^\infty]$ is finite if and only if $NS(X) \otimes \mathbb{Q}_\ell \simeq H^2_{\text{ét}}(\overline{X}, \mathbb{Q}_\ell(1))^G$. Moreover, if the finiteness holds for one $\ell \neq p$, then it holds for all $\ell \neq p$.*

Proof For the proof compare Tate's survey [595]. To simplify the exposition we shall assume that X is a K3 surface.

As $k = \mathbb{F}_q$, one knows that $H^p(G, \Lambda) = 0$ for any finite $G(\simeq \hat{\mathbb{Z}})$-module Λ and $p \neq 0, 1$; see [549, Ch. XIII.1]. Thus, by the Hochschild–Serre spectral sequence (cf. (1.9)) for μ_{ℓ^n}, there exists a short exact sequence

$$0 \longrightarrow H^1(k, H^1(\overline{X}, \mu_{\ell^n})) \longrightarrow H^2(X, \mu_{\ell^n}) \longrightarrow H^2(\overline{X}, \mu_{\ell^n})^G \longrightarrow 0.$$

As $H^1(\overline{X}, \mu_{\ell^n})$ is the ℓ^n-torsion part of $NS(\overline{X})$, it is trivial for K3 surfaces. Therefore,

$$NS(X) \otimes \mathbb{Z}_\ell \hookrightarrow H^2_{\text{ét}}(X, \mathbb{Z}_\ell(1)) \simeq H^2_{\text{ét}}(\overline{X}, \mathbb{Z}_\ell(1))^G.$$

Hence, by Remark 2.3,

$$NS(X) \otimes \mathbb{Z}_\ell \simeq H^2_{\text{ét}}(\overline{X}, \mathbb{Z}_\ell(1))^G \tag{2.7}$$

if and only if the Tate module $T_\ell \mathrm{Br}(X)$ is trivial, which is equivalent to $|\mathrm{Br}(X)[\ell^\infty]| < \infty$. As observed earlier, (2.7) is equivalent to $\mathrm{NS}(X) \otimes \mathbb{Q}_\ell \simeq H^2_{\text{ét}}(X, \mathbb{Q}_\ell(1)) \simeq H^2_{\text{ét}}(\overline{X}, \mathbb{Q}_\ell(1))^G$, which due to Proposition **17**.3.5 is independent of ℓ.[6]   □

Next combine $0 \longrightarrow \mathrm{NS}(X) \otimes \mathbb{Z}/\ell\mathbb{Z} \longrightarrow H^2(X, \mu_\ell) \longrightarrow \mathrm{Br}(X)[\ell] \longrightarrow 0$ (see (1.6)) with the isomorphism $H^2(X, \mu_\ell) \simeq H^2(\overline{X}, \mu_\ell)^G$. If one could now show that the Tate conjecture $\mathrm{NS}(X) \otimes \mathbb{Z}_\ell \xrightarrow{\sim} H^2_{\text{ét}}(\overline{X}, \mathbb{Z}_\ell(1))^G$ automatically yields isomorphisms $\mathrm{NS}(X) \otimes \mathbb{Z}/\ell\mathbb{Z} \simeq H^2(\overline{X}, \mu_\ell)^G$ for most ℓ, the finiteness of all (or, equivalently, of one) $\mathrm{Br}(X)[\ell^\infty]$ as in the lemma would imply the finiteness of the whole $\mathrm{Br}(X)$. This part is quite delicate and Tate in [596] uses the compatibility of the intersection pairing on $\mathrm{NS}(X)$ and the cup-product $H^2(X, \mu_\ell) \times H^3(X, \mu_\ell) \longrightarrow \mathbb{Z}/\ell\mathbb{Z}$ (see Remark 2.8) to show that this indeed holds, proving simultaneously the equation (2.3) up to powers of p. The p-torsion part was later dealt with by Milne in [404], using crystalline, flat, and Witt vector cohomology.[7]

Theorem 2.6 (Tate)   *The Tate conjecture* **17**.*3.1 implies the Artin–Tate conjecture 2.4. More precisely, if* $\mathrm{NS}(X) \otimes \mathbb{Q}_\ell \simeq H^2_{\text{ét}}(\overline{X}, \mathbb{Q}_\ell(1))^G$ *for one* $\ell \neq p$, *then* $\mathrm{Br}(X)$ *is finite and (2.3) holds.*

This allows one to confirm the Artin–Tate conjecture in the easiest case, namely for $\rho(\overline{X}) = 22$, e.g. for unirational K3 surfaces (see Proposition **17**.2.7). Indeed, then automatically $\mathrm{Br}(X)$ is finite. In any case, as the Tate conjecture has been proved for K3 surfaces (see Section **17**.3.3), one has

Corollary 2.7   *The Artin–Tate conjecture 2.4 holds for all K3 surfaces over finite fields.*

□

Remark 2.8   The proof of the Theorem 2.6 involves a certain alternating pairing $\mathrm{Br}(X) \times \mathrm{Br}(X) \longrightarrow \mathbb{Q}/\mathbb{Z}$, which is defined by lifting $\alpha, \beta \in \mathrm{Br}(X)[n]$ to classes in $H^2(X, \mu_n)$, projecting one to $H^3(X, \mu_n)$ via the natural boundary operator, and then using the cup-product to $H^5(X, \mu_n^{\otimes 2}) \simeq \mathbb{Z}/n\mathbb{Z}$. The kernel consists of the divisible elements; see [596, Thm. 5.1] and [404, Thm. 2.4]. So, once the Artin–Tate conjecture has been confirmed, the pairing is non-degenerate. Moreover, $|\mathrm{Br}(X)|$ is then a square or twice a square. The latter can be excluded due to the work of Liu, Lorenzini, and Raynaud [376].

Remark 2.9   The famous Birch–Swinnerton-Dyer conjecture (see [637]), has a function field analogue. Consider for example an elliptic K3 surface $X \longrightarrow \mathbb{P}^1$ over a

---

[6] Note that historically Tate's Bourbaki article [596] precedes Deligne's proof of the Weil conjectures. However, for K3 surfaces the independence follows at once from $Z(X, t) = P_2(X, t)^{-1}$, and, for arbitrary surfaces, Weil's conjecture for abelian varieties suffices to conclude.

[7] Tate in [596] writes: 'The problem ... for $\ell = p$ should furnish a good test for any p-adic cohomology theory...'

finite field $\mathbb{F}_q$ with generic fibre E over the function field $K = \mathbb{F}_q(T)$. The Weierstrass model $\overline{X} \longrightarrow \mathbb{P}^1$ (see Section **11.2.2**), has only integral closed fibres $\overline{X}_t$, which are either smooth elliptic or rational with one ordinary double point or one cusp. The function $L(E, s)$ of the elliptic curve E over K counts the number of rational points on the closed fibres:

$$L(E, s) := \prod_{\substack{X_t \text{ smooth}}} (1 - a_t q_t^{-s} + q_t^{1-2s})^{-1} \cdot \prod_{\substack{X_t \text{ singular}}} (1 - a_t q_t^{-s})^{-1}.$$

Here, q_t is the cardinality of the residue field of $t \in \mathbb{P}^1$, i.e. $k(t) \simeq \mathbb{F}_{q_t}$; $a_t := q_t + 1 - |X_t(k(t))|$ if X_t is smooth; $a_t = \pm 1$ if $\overline{X}_t$ has an ordinary double point with rational or irrational tangents, respectively; and $a_t = 0$ if $\overline{X}_t$ has a cusp. The Birch–Swinnerton-Dyer conjecture then asserts (for an arbitrary non-constant elliptic curve over a function field) that

$$\text{rk } E(K) = \text{ord}_{s=1} L(E, s).$$

Recall from Section **11.2.3** that the group of K-rational points $E(K)$ of the generic fibre can also be interpreted as the Mordell–Weil group $\text{MW}(X)$ of the elliptic fibration $X \longrightarrow \mathbb{P}^1$. The Shioda–Tate formula, Corollary **11.3.4**, expresses the rank of $\text{MW}(X)$ as

$$\text{rk } \text{MW}(X) = \rho(X) - 2 - \sum_t (r_t - 1).$$

On the other hand, the Weil conjectures for X (see Section **4.4**) and for the various fibres of $X \longrightarrow \mathbb{P}^1$ lead to a comparison of $L(E, s)$ and the Zeta function of X:

$$\text{ord}_{s=1} L(E, s) = -\text{ord}_{s=1} Z(X, q^{-s}) - 2 - \sum_t (r_t - 1);$$

see for example [605, Lect. 3.6] for details. Therefore,

$$\text{rk } E(K) - \text{ord}_{s=1} L(E, s) = \rho(X) + \text{ord}_{s=1} Z(X, q^{-s}),$$

which shows that the Birch–Swinnerton-Dyer conjecture for the elliptic curve E over the function field K is equivalent to the Tate conjecture (and hence to the Artin–Tate conjecture) for the K3 surface X (use Proposition **17.3.5**). Compare this with Remark **17.3.3**.

This can be a pushed a bit further to show that the Birch–Swinnerton-Dyer conjecture for elliptic curves over function fields $K(B)$ of arbitrary curves B over $\mathbb{F}_q$ implies the Tate conjecture for arbitrary K3 surfaces. Indeed, by Corollary **13.2.2** any K3 surface is covered by an elliptic surface $\widetilde{X} \longrightarrow X$. As above, the Birch–Swinnerton-Dyer conjecture for the generic fibre of the elliptic fibration $\widetilde{X} \longrightarrow B$ implies the Tate conjecture for $\widetilde{X}$. However, it is known that the Tate conjecture for any surface rationally dominating X implies the Tate conjecture for X itself; see [595, Thm. 5.2] or [605, Sec. 12].

2.3 We now sketch the main ideas of the paper [20] by Artin and Swinnerton-Dyer, proving the Tate conjecture for elliptic K3 surfaces with a section,[8] explain how similar ideas have been used by Lieblich, Maulik, and Snowden in [370] to relate the Artin–Tate conjecture to finiteness results for K3 surfaces over finite fields, and, at the end, briefly touch upon Charles's more recent approach [111] to the Tate conjecture relying on similar ideas. We suppress many technical subtleties but hope to convey the main ideas.

(i) Let $X_0 \longrightarrow \mathbb{P}^1$ be an elliptic K3 surface and assume that $T_\ell \mathrm{Br}(X_0)$, $\ell \neq p$, is not trivial. Due to the exact sequence (1.8), one can then choose a class $\alpha \in H^2_{\acute{e}t}(X_0, \mathbb{Z}_\ell(1))$ that is orthogonal to $\mathrm{NS}(X_0) \otimes \mathbb{Z}_\ell \hookrightarrow H^2_{\acute{e}t}(X_0, \mathbb{Z}_\ell(1))$ with respect to the intersection pairing and projects onto a non-trivial class $(\alpha_n) \in T_\ell \mathrm{Br}(X_0)$. We may assume that $\alpha_n \in \mathrm{Br}(X_0)$ has order $d_n := \ell^n$ and $\ell \alpha_n = \alpha_{n-1}$.

Now use $\mathrm{III}(X_0) \simeq \mathrm{Br}(X_0)$; see Section **11**.5.2.[9] Thus, every $\alpha_n \in \mathrm{Br}(X_0)$ gives rise to an elliptic K3 surface

$$X_n := X_{\alpha_n} \longrightarrow \mathbb{P}^1$$

over k with Jacobian fibration $\mathrm{J}(X_n) \simeq X_0$. As hinted at in Remark **11**.5.9, the index of $X_n \longrightarrow \mathbb{P}^1$, i.e. the minimal positive fibre degree of a line bundle on X_n, equals d_n.

On each of the X_n one constructs a distinguished multisection D_n of fibre degree d_n, by using $\mathrm{J}^{d_n}(X_n) \simeq \mathrm{J}(X_n) \simeq X_0$ and the given zero-section of $X_0 \longrightarrow \mathbb{P}^1$. A crucial observation in [20] then says that $(D_n)^2 \equiv (\alpha)^2 (d_n)$. This is a central point in the argument and it is proved in [20] by lifting to characteristic zero and studying elliptic fibrations from a differentiable point of view. As a consequence, changing D_n by a fibre class, one finds a divisor L_n on X_n with positive $(L_n)^2$ bounded independently of n. Using the action of the Weyl group, we may assume that L_n is big and nef; cf. Corollary **8**.2.9 (observe that we may assume that $\mathrm{NS}(X_n) \simeq \mathrm{NS}(\overline{X}_n)$). However, due to the boundedness results for (quasi-)polarized K3 surfaces (see Section **5**.2.1 and Theorem **2**.2.7), up to isomorphisms there exist only finitely many (quasi-) polarized K3 surfaces (X, L) of bounded degree $(L)^2$ over any fixed finite field. Hence, infinitely many of the elliptic fibrations $X_n \longrightarrow \mathbb{P}^1$ are actually defined on the same K3 surface X.

In [20] the argument concludes by observing that up to the action of $\mathrm{Aut}(X)$ there exist only finitely many elliptic fibrations on any K3 surface; see Proposition **11**.1.3. Alternatively, one could use (4.5) in Section **11**.4.2 showing that

$$|\mathrm{disc}\, \mathrm{NS}(X_n)| = d_n^2 \cdot |\mathrm{disc}\, \mathrm{NS}(X_0)|$$

---

[8] In [16] Artin explains how to use this to cover also supersingular K3 surfaces. Then [20] applies and shows $\rho(\mathrm{J}(\overline{X})) = 22$. But $\rho(\overline{X}) = \rho(\mathrm{J}(\overline{X}))$ and, therefore, also $\rho(\overline{X}) = 22$, which proves the Tate and hence the Artin–Tate conjecture.

[9] As noted there already, the simplifying assumption that the ground field is algebraically closed is not needed and everything works in our situation of K3 surfaces over a finite field $k = \mathbb{F}_q$.

(reflecting the equivalence $D^b(X_n) \simeq D^b(X_0, \alpha_n)$ (see Remarks **11**.4.9 and **11**.5.9)), which of course excludes X_n and X_m from being isomorphic to each other for all $n \neq m$.
(ii) In [370] Lieblich, Maulik, and Snowden follow a similar strategy to relate finiteness of the Brauer group of a K3 surface over a finite field and finiteness of the set of isomorphism classes of K3 surfaces over a given finite field (not fixing the degree). Suppose X_0 is an arbitrary K3 surface defined over a finite field $k = \mathbb{F}_q$ with $T_\ell \mathrm{Br}(X_0) \neq 0$. As above, there exists a non-trivial class $(\alpha_n) \in T_\ell \mathrm{Br}(X_0)$ with α_n of order $d_n := \ell^n$. If X_0 comes without an elliptic fibration, it is a priori not clear how to associate with the (infinitely many) classes $\alpha_n \in \mathrm{Br}(X_0)$ K3 surfaces defined over the given field. However, the isomorphism $J(X_n) \simeq X_0$ from above can also be expressed by saying that X_0 is the non-fine moduli space of stable sheaves with Mukai vector $(0, [X_{nt}], d_n)$ on X_n; see Section **11**.4.2. Furthermore, α_n can be viewed as the obstruction to the existence of a universal family on $X_n \times X_0$; see Section **16**.4.1.

Reversing the role of the two factors, X_n can be considered as a fine moduli space of stable α_n-twisted sheaves on X_0. And this now works also in the non-elliptic case as well. More specifically, in [370] for any α_n a Mukai vector v_n is found for which the moduli space $X_n := M(v_n)$ of stable α_n-twisted sheaves on X_0 is fine, projective, and of dimension two (and hence a K3 surface). The general theory then yields equivalences

$$D^b(X_n) \simeq D^b(M(v_n)) \simeq D^b(X_0, \alpha_n).$$

If now one assumes that there exist only finitely many K3 surfaces over the fixed finite field k, then infinitely many of the X_n have to be isomorphic to a fixed K3 surface, say, X. However, then $D^b(X) \simeq D^b(X_0, \alpha_n)$ for infinitely many Brauer class, which in [370] is excluded by lifting to characteristic zero and then using a finiteness result in [271]. Alternatively, one could use the isometry $N(X_n) \simeq N(X) \simeq N(X_0, \alpha_n)$, deduced from the derived equivalence, to conclude that $|\mathrm{disc}\, \mathrm{NS}(X_n)| = d_n^2 \cdot |\mathrm{disc}\, \mathrm{NS}(X_0)|$ and, thus, to exclude isomorphisms between the X_ns for different n.

So, in order to prove the Artin–Tate conjecture this way, it remains to prove that there exist only finitely many K3 surfaces (of a priori unbounded degree) over any given finite field. The authors of [370] also show that, conversely, this finiteness is implied by the Artin–Tate conjecture.
(iii) The approach was more recently refined by Charles in [111]. In order to obtain finiteness without a priori bounding the degree of the K3 surfaces, four-dimensional moduli spaces $M(v)$ of stable sheaves are used. As $\rho(M(v)) = \rho(X) + 1$, this provides more freedom to choose an appropriate polarization of bounded degree. So, once again, starting with a non-trivial $(\alpha_n) \in T_\ell \mathrm{Br}(X_0)$ one constructs infinitely many K3 surfaces X_n for which the four-dimensional moduli spaces have bounded degree and of which, therefore, one cannot have infinitely many over a fixed finite field if Matsusaka's theorem were known in positive characteristic. The numerical considerations are quite intricate and Charles has again to resort to the Kuga–Satake construction, as the

birational geometry of the four-dimensional irreducible symplectic varieties is not well enough understood in order to deduce from the existence of a class of bounded degree also the existence of a (quasi)-polarization of bounded degree for which one would need certain results from MMP that are not available in positive characteristic.

2.4 To conclude this section, let us briefly touch upon the case of K3 surfaces over number fields. The Brauer group of a K3 surface over a number field k is certainly not finite due to the exact sequence $\mathrm{NS}(\overline{X})^G \longrightarrow \mathrm{Br}(k) \longrightarrow \mathrm{Br}(X)$; see (1.10), and Remark 2.1. However, up to $\mathrm{Br}(k)$ certain general finiteness results for K3 surfaces in characteristic zero have been proved by Skorobogatov and Zarhin in [573].

Proposition 2.10 *Let X be a K3 surface over a finitely generated field k of characteristic zero. Then $\mathrm{Br}(\overline{X})^G$, $\mathrm{Br}(X)/\mathrm{Br}_1(X)$, and $\mathrm{Br}_1(X)/\mathrm{Br}_0(X)$ are finite groups.*

Proof As $\mathrm{Br}(X)/\mathrm{Br}_1(X)$ is a subgroup of $\mathrm{Br}(\overline{X})^G$ and $\mathrm{Br}_1(X)/\mathrm{Br}_0(X)$ is finite due to Lemma 2.2, only the finiteness of $\mathrm{Br}(\overline{X})^G$ needs to be checked.

Due to the Tate conjecture (in characteristic zero; see Section **17**.3.2) $\mathrm{NS}(X) \otimes \mathbb{Z}_\ell \simeq H^2_{\acute{e}t}(\overline{X}, \mathbb{Z}_\ell(1))^G$ and hence $\mathrm{Br}(\overline{X})^G[\ell^\infty]$ is finite. As for K3 surfaces over finite fields, one concludes that then $\mathrm{Br}(\overline{X})^G$ itself is finite. Indeed, one shows that $\mathrm{NS}(X) \otimes \mathbb{Z}/\ell\mathbb{Z} \xrightarrow{\sim} H^2(\overline{X}, \mu_\ell)^G$ for most ℓ, which is enough to conclude. This part is easier than in the case of positive characteristic, as one can use the comparison to singular cohomology and the transcendental lattice. $\square$

Remark 2.11 (i) In [574] the authors extend their result to finitely generated fields in positive characteristic $p \neq 2$ and prove that the non-p-torsion part of $\mathrm{Br}(X)/\mathrm{Br}_0(X)$ is finite.

(ii) Artin's conjecture 1.6 predicts finiteness of $\mathrm{Br}(X)$ for any proper scheme over $\mathrm{Spec}(\mathbb{Z})$. The Tate conjecture proves it for K3 surfaces over $\mathbb{F}_q$, and, as was explained to me by François Charles, this proves it for any projective family with generic fibre a K3 surface over an open subset of $\mathrm{Spec}(\mathcal{O}_K)$ for any number field K.

3 Height

For a K3 surface X the formal Brauer group $\widehat{\mathrm{Br}}_X$ is a smooth one-dimensional formal group which can be studied in terms of its formal group law. Basic facts concerning formal group laws are recalled and then used to define the height, which is a notion that is of interest only in positive characteristic and that can alternatively be defined in terms of crystalline cohomology. Roughly, K3 surfaces in positive characteristic behave as complex K3 surfaces as long as their height is finite. Those of infinite height, so-called supersingular K3 surfaces, are of particular interest. Some of their most fundamental properties are discussed. For example, for a long time supersingular K3 surfaces were the only K3 surfaces for which the Tate conjecture had not been known.

3.1 A formal group structure on $\mathrm{Spf}(R)$ is given by a morphism $R \longrightarrow R \hat{\otimes}_k R$ of k-algebras. If $\mathrm{Spf}(R)$ is smooth of dimension one and an isomorphism of k-algebras $R \simeq k[[T]]$ is chosen, then the morphism is given by the image of T, which can be thought of as a power series $F(X, Y) \in k[[X, Y]] \simeq R \hat{\otimes}_k R$ in two variables. In fact, $F(X, Y)$ is a *formal group law*. In particular,

$$F(X, F(Y, Z)) = F(F(X, Y), Z) \quad \text{and} \quad F(X, Y) = X + Y + \text{ higher order terms.}$$

Example 3.1 (i) The formal completion of the additive group $\mathbb{G}_a = \mathrm{Spec}(k[T])$ is described by $\widehat{\mathbb{G}}_a = \mathrm{Spf}(k[[T]])$ with the formal group law

$$F(X, Y) = X + Y.$$

(ii) The formal completion of the multiplicative group $\mathbb{G}_m = \mathrm{Spec}(k[t, t^{-1}])$ is described by $\widehat{\mathbb{G}}_m = \mathrm{Spf}(k[[T]])$, $t = 1 + T$, with the formal group law

$$F(X, Y) = X + Y + XY.$$

After choosing $G_i \simeq \mathrm{Spf}(k[[T]])$, a morphism between two smooth one-dimensional formal groups $G_1 \longrightarrow G_2$, with formal group laws F_1 and F_2, can be represented by a power series $f(T)$ satisfying

$$(f \otimes f)(F_2(X, Y)) = F_1(f(X), f(Y)).$$

This allows one to speak of isomorphisms of formal groups and to prove that in characteristic zero any smooth one-dimensional formal group is in fact isomorphic to $\widehat{\mathbb{G}}_a$; cf. Remark 1.14.

If $\mathrm{char}(k) = p > 0$, formal group laws can be classified according to their *height*. For this, consider multiplication by $[m] \colon G \longrightarrow G$ or its associated power series $[m](T)$, which can recursively be determined by $[m + 1](T) = F([m](T), T)$. One then shows that $[p](T)$ is either zero or of the form

$$[p](T) = aT^{p^h} + \text{ higher order terms}$$

with $a \neq 0$. The two cases correspond to $[p] \colon G \longrightarrow G$ being trivial or having as its kernel a finite group scheme of order p^h. The *height* of a smooth one-dimensional formal group G is then defined as

$$h(G) := \begin{cases} \infty & \text{if } [p](T) = 0, \\ h & \text{if } [p](T) = aT^{p^h} + \cdots, \ a \neq 0. \end{cases}$$

Over a separably closed field k of characteristic $p > 0$, the height $h(G)$ of a smooth one-dimensional formal group G determines G up to isomorphisms (see e.g. [353]):

$$G_1 \simeq G_2 \quad \text{if and only if} \quad h(G_1) = h(G_2).$$

Moreover, all positive integers $h = 1, 2, \ldots$ and ∞ can be realized.

Example 3.2 (i) In Example 3.1 one finds

$$h(\widehat{\mathbb{G}}_a) = \infty \quad \text{and} \quad h(\widehat{\mathbb{G}}_m) = 1.$$

(ii) For an elliptic curve E over a separably closed field and its formal completion $\widehat{E}$ at the origin, either

$$h(\widehat{E}) = 2 \quad \text{or} \quad h(\widehat{E}) = 1.$$

In the first case, which is equivalently described by $E[p^r] = 0$ for all $r \geq 1$, the curve E is called *supersingular*. In the second case, when $E[p^r] \simeq \mathbb{Z}/p^r\mathbb{Z}$, it is called *ordinary*. See e.g. [569] for details on why the height, which a priori is defined in terms of the action of the Frobenius, can be read off from the p-torsion points; cf. Section 3.4.

Definition 3.3 The *height $h(X)$ of a K3 surface X* defined over a separably closed field k of characteristic $p > 0$ is defined as the height

$$h(X) := h(\widehat{\mathrm{Br}}_X)$$

of its formal Brauer group. A K3 surface X is called *supersingular* (or *Artin supersingular*) if $h(X) = \infty$, i.e. $\widehat{\mathrm{Br}}_X \simeq \widehat{\mathbb{G}}_a$, and *ordinary* if $h(X) = 1$, i.e. $\widehat{\mathrm{Br}}_X \simeq \widehat{\mathbb{G}}_m$.

Example 3.4 Consider the Fermat quartic $X \subset \mathbb{P}^3$ defined by $\sum_i x_i^4 = 0$ in characteristic $p > 2$. Then

$$h(X) = \begin{cases} \infty & \text{if } p \equiv 3(4), \\ 1 & \text{if } p \equiv 1(4). \end{cases}$$

The discriminants of $\mathrm{NS}(X)$ in the two cases are -64 (as in characteristic zero; see Section 3.2.6) and $-p^2$, respectively. Shioda's argument in [564] uses that the Fermat quartic is a Kummer surfaces associated with a product of two elliptic curves; see Example 14.3.18.

3.2 For any perfect field k of characteristic $p > 0$, let $W = W(k)$ be its ring of Witt vectors, which is a complete DVR with uniformizing parameter $p \in W$, residue field k, and fraction field K of characteristic zero. So, for example, $W(\mathbb{F}_p) \simeq \mathbb{Z}_p = \varprojlim W_n(\mathbb{F}_p)$ with $W_n(\mathbb{F}_p) = \mathbb{Z}/p^n\mathbb{Z}$ and, in general, $W(k) = \varprojlim W_n(k)$. The Frobenius morphism $F: k \longrightarrow k$, $a \longmapsto a^p$ lifts to the Frobenius $F: W(k) \longrightarrow W(k)$ (by functoriality), which is a ring homomorphism and thus induces an automorphism of the fraction field.

Consider the ring $K\{T\}$, which is the usual polynomial ring $K[T]$ but with $T \cdot \lambda = F(\lambda) \cdot T$ for all $\lambda \in K$. An *F-isocrystal* consists of a finite-dimensional vector space V over K with a $K\{T\}$-module structure. In other words, V comes with a lift of the Frobenius $F \colon V \longrightarrow V$ such that $F(\lambda \cdot v) = F(\lambda) \cdot F(v)$ for $\lambda \in K$ and $v \in V$. The standard example of an F-isocrystal is provided by

$$V_{r,s} := K\{T\}/(T^s - p^r),$$

which is of dimension s and *slope* r/s. The Frobenius action on $V_{r,s}$ is given by left-multiplication by T.

If k is algebraically closed, then any irreducible F-isocrystal is isomorphic to $V_{r,s}$ with $(r, s) = 1$, for example $V_{mr,ms} \simeq V_{r,s}^{\oplus m}$. Moreover, due to a result of Dieudonné and Manin [388] any F-isocrystal V with bijective F is a direct sum of those. One writes

$$V \simeq \bigoplus V_{r_i, s_i} \tag{3.1}$$

with slopes $r_1/s_1 < \cdots < r_n/s_n$. The *Newton polygon* of an F-isocrystal V, which determines it uniquely, is the following convex polygon with slopes r_i/s_i:

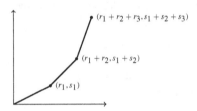

The *height* $h(V)$ of an F-isocrystal V is the dimension of the subspace of slope strictly less than one: $V_{[0,1)} := \bigoplus_{r_i/s_i < 1} V_{r_i, s_i}$. One defines $h(V) = \infty$ if this space is zero.

For the F-isocrystals we are interested in, Poincaré duality holds; i.e. they are endowed with a non-degenerate pairing $V \times V \longrightarrow K$ satisfying $(Fx.Fy) = p^d(x.y)$, where later $d = 1$ for elliptic curves and $d = 2$ for K3 surfaces. This leads to the condition $\sum r_i = (d/2) \sum s_i$. For $d = 2$ one checks that in the decomposition (3.1) the pairing can be non-trivial only between $V_{r,s}$ and $V_{r',s'}$ with $r/s + r'/s' = 2$.

Let X be a smooth projective variety over an algebraically closed field k of characteristic $p > 0$. Assume that there exists a lift to a smooth proper morphism $\pi \colon \mathcal{X} \longrightarrow \mathrm{Spf}(W)$. Then the crystalline cohomology $H_{\mathrm{cr}}^*(X/W)$ of X can be computed via the de Rham cohomology. More precisely, if $H_{\mathrm{dR}}^*(\mathcal{X}/W)$ denotes the relative de Rham cohomology of π, then there exists an isomorphism of (finite type) W-modules $H_{\mathrm{cr}}^*(X/W) \simeq H_{\mathrm{dR}}^*(\mathcal{X}/W)$. The Frobenius of k first lifted to W can then be further lifted to a homomorphism $F^* H_{\mathrm{dR}}^*(\mathcal{X}/W) \xrightarrow{\sim} H_{\mathrm{dR}}^*(\mathcal{X}/W)$. On the generic fibre, it yields a semi-linear endomorphism of

$$H^*_{cr}(X) := H^*_{dR}(\mathcal{X}/W) \otimes_W K,$$

which in each degree defines an F-isocrystal equipped with Poincaré duality. As above, the F-isocrystals $H^i_{cr}(X)$ are uniquely described by their Newton polygons and they are independent of the chosen lift $\pi : \mathcal{X} \longrightarrow \mathrm{Spf}(W)$ (if there is one at all). The Frobenius action on $H^*_{cr}(X)$ is not induced by any Frobenius action on $\mathcal{X}$, which usually does not exist, but by functoriality. See Mazur's introduction [402] for more details.[10]

Remark 3.5 For a K3 surface X, the cohomology $H^2_{cr}(X/W)$ is torsion free. Hence,

$$c_1 : \mathrm{NS}(X) \longhookrightarrow H^2_{cr}(X/W) \tag{3.2}$$

is injective. Indeed, $H^2_{cr}(X/W) \otimes k \simeq H^2_{dR}(X)$, and then use Proposition **17**.2.1.

The Newton polygon of a variety X, i.e. of $H^*_{cr}(X)$, can be compared with the Hodge polynomial of X encoding the Hodge numbers $h^{p,q}(X)$. For an elliptic curve E and a K3 surface X the Hodge polygons (in degree one and two, respectively) encode the Hodge numbers $h^{0,1}(E) = h^{1,0}(E) = 1$ and $h^{0,2}(X) = 1$, $h^{1,1}(X) = 20$, and $h^{2,0}(X) = 1$, respectively. They look like this:

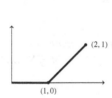 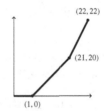

A famous conjecture of Katz, proved by Mazur [402], asserts that the Newton polygon of a variety X always lies above the Hodge polygon of X. For an elliptic curve E this leaves the following two possibilities:

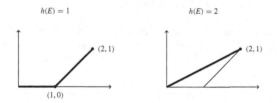

---

[10] For a K3 surface X, the W-modules $H^i_{dR}(\mathcal{X}/W)$ are free of rank $1, 0, 22, 0, 1$ for $i = 0, 1, 2, 3, 4$, respectively, and the Hodge spectral sequence $E^{p,q}_1 = H^q(\mathcal{X}, \Omega^p_{\mathcal{X}/W}) \Rightarrow H^{p+q}_{dR}(\mathcal{X}/W)$ degenerates. The latter is an immediate consequence of the vanishing $H^0(X, \Omega_X) = 0$ (see Section **9**.5.1), as was observed by Deligne [143].

For a K3 surface the following three are in principle possible:

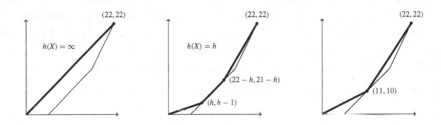

However, using Remark 3.5, the last one can be excluded. Indeed, as the Frobenius pull-back acts by multiplication by p on $NS(X)$, there exist non-trivial classes in $H^2_{cr}(X)$ on which the Frobenius acts by multiplication by p. This immediately proves

$$h(X) = \infty \quad \text{or} \quad h(X) = 1, \ldots, 10, \tag{3.3}$$

where $h(X)$ here is defined as the height of the F-isocrystal $H^2_{cr}(X)$ (which below will be shown to coincide with $h(\widehat{\mathrm{Br}}_X)$). More precisely, the argument proves the following lemma.

Lemma 3.6 *For any non-supersingular K3 surface X defined over an algebraically closed field of characteristic $p > 0$, Picard number and height of X can be compared via*

$$\rho(X) \leq b_2(X) - 2h(X) = 22 - 2h(X). \tag{3.4}$$

$\square$

This should be regarded as the analogue of

$$\rho(X) \leq b_2(X) - 2h^{0,2}(X)$$

for a smooth complex projective variety. The latter follows from Hodge decomposition $H^2(X, \mathbb{C}) = H^{0,2}(X) \oplus H^{1,1}(X) \oplus H^{2,0}(X)$, $h^{2,0}(X) = h^{0,2}(X)$, and $\rho(X) \leq h^{1,1}(X)$.

3.3 We have encountered two definitions of the height $h(X)$ of a K3 surface X in positive characteristic, as the height $h(\widehat{\mathrm{Br}}_X)$ of the formal Brauer group of X and as the height of the F-isocrystal $H^2_{cr}(X)$. These two notions coincide, but this is a non-trivial fact due to Artin and Mazur [19] and we can give only a rough sketch of the arguments that prove it.

First, Dieudonné theory establishes an equivalence

$$\mathbb{D}_K : \{ \text{ formal groups}/k \text{ with } h < \infty \} \xrightarrow{\sim} \{ F\text{-isocrystals with } V = V_{[0,1)} \}$$

by sending a formal group G first to the W-module $\mathbb{D}G := \text{Hom}(G, \widehat{CW})$ of all formal group scheme maps and then to $\mathbb{D}_K G := \mathbb{D}G \otimes_W K$. Here, $\widehat{CW}$ is the formal group scheme representing $A \longmapsto \varprojlim CW(A/\mathfrak{m}_A^n)$ with $CW(A)$ the group of A-valued Witt covectors; cf. [179]. The $K\{T\}$-module structure on $\mathbb{D}G$ is induced by the natural one on $\widehat{CW}$.

This equivalence is compatible with the notion of heights on the two sides, which on the right-hand side is just the dimension. For example, $\mathbb{D}\widehat{\mathbb{G}}_m \simeq W$ and so $\mathbb{D}_K \widehat{\mathbb{G}}_m$ is the trivial F-isocrystal $K = V_{0,1}$ and indeed $h(\widehat{\mathbb{G}}_m) = 1 = h(V_{0,1})$. Note that $\mathbb{D}_K$ applied to the formal group $\widehat{\mathbb{G}}_a$ (which is not p-divisible or, equivalently, has $h(\widehat{\mathbb{G}}_a) = \infty$) yields the infinite-dimensional $K[[T]]$.

To compare the two definitions of $h(X)$, Artin and Mazur in [19] find an isomorphism

$$\mathbb{D}_K \widehat{\text{Br}}_X \simeq H^2_{\text{cr}}(X)_{[0,1)}. \tag{3.5}$$

Using their more suggestive notation $\Phi^2(X, \mathbb{G}_m) := \widehat{\text{Br}}_X$, the first step towards (3.5) consists of proving a series of isomorphisms

$$\mathbb{D}\widehat{\text{Br}}_X \simeq \mathbb{D}\Phi^2(X, \mathbb{G}_m) \simeq \mathbb{D}\Phi^2(X, \widehat{\mathbb{G}}_m) \simeq H^2_{\text{ét}}(X, \mathbb{D}\widehat{\mathbb{G}}_m) \simeq H^2_{\text{ét}}(X, W\mathcal{O}_X).$$

The last isomorphism is a sheaf version for $\mathbb{D}\widehat{\mathbb{G}}_m \simeq W$, where $W\mathcal{O}_X$ denotes the sheaf of Witt vectors. Witt vector cohomology can be compared with crystalline cohomology by the Bloch–Illusie (or slope) spectral sequence

$$E_1^{p,q} = H^q(X, W\Omega_X^p) \Rightarrow H^{p+q}_{\text{cr}}(X/W).$$

The spectral sequence is compatible with the Frobenius action (appropriately defined on the left-hand side) and yields in particular an isomorphism

$$H^2(X, W\mathcal{O}_X) \otimes_W K \simeq H^2_{\text{cr}}(X)_{[0,1)}. \tag{3.6}$$

We recommend Chambert–Loir's survey [109] for more details and references; see also Liedtke's notes [372]. To conclude the discussion, one finds that indeed

$$h(\widehat{\text{Br}}_X) = h(H^2_{\text{cr}}(X)). \tag{3.7}$$

3.4 Katsura and van der Geer in [607] derive from (3.6) a rather concrete description of the height as

$$h(X) = \min\{n \mid 0 \neq F \colon H^2(X, W_n\mathcal{O}_X) \longrightarrow H^2(X, W_n\mathcal{O}_X)\}. \tag{3.8}$$

Here, F is the usual Frobenius action. This description covers the supersingular case, as in this case $F = 0$ on $H^2(X, W\mathcal{O}_X)$ can be shown to imply $F = 0$ on all $H^2(X, W_n\mathcal{O}_X)$.

This approach fits well with the comparison of ordinary and supersingular elliptic curves based on the *Hasse invariant*. For an elliptic curve E in characteristic $p > 0$, the absolute Frobenius $E \longrightarrow E$ induces an action $F \colon H^1(E, \mathcal{O}_E) \longrightarrow H^1(E, \mathcal{O}_E)$. By definition, the Hasse invariant of E is zero if this map is zero, and it is one if it is bijective. Then it is known that E is supersingular if and only if its Hasse invariant is zero, or equivalently

$$h(E) = 1 \quad \text{if and only if} \quad F \colon H^1(E, \mathcal{O}_E) \xrightarrow{\sim} H^1(E, \mathcal{O}_E).$$

See for example [236, IV.Exer. 4.15]. For a K3 surface X, a very similar statement describes ordinary K3 surfaces:

$$h(X) = 1 \quad \text{if and only if} \quad F \colon H^2(X, \mathcal{O}_X) \xrightarrow{\sim} H^2(X, \mathcal{O}_X).$$

However, in order to distinguish between the remaining cases $h(X) = 2, \ldots, 10, \infty$, one has to consider the action on $H^2(X, W_n \mathcal{O}_X)$, $n = 2, \ldots, 10$.

The height can be used to stratify the moduli space of polarized K3 surfaces. We continue to assume k algebraically closed of characteristic $p > 0$ and consider the moduli stack M_d, $p \nmid 2d$, of polarized K3 surfaces (X, L) of degree $(L)^2 = 2d$ over k; see Chapter 5. Using the formal group law for $\widehat{\mathrm{Br}}_X$, Artin showed in [16] that $h(X) \geq h$ is a closed condition of codimension $\leq h - 1$. This remains valid in the case $h = \infty$ if one sets $h = 11$. Applying (3.8), van der Geer and Katsura give an alternative proof of this fact which in addition endows the set of polarized K3 surfaces (X, L) of height $h(X) \geq h$ with a natural scheme structure. We summarize these results by the following theorem.

Theorem 3.7 *For $h = 1, \ldots, 10$,*

$$M_d^h := \{(X, L) \mid h(X) \geq h\} \subset M_d$$

is empty or a closed substack of dimension $\dim M_d^h = 20 - h$ *which is smooth outside the supersingular locus. The supersingular locus*

$$M_d^\infty := \{(X, L) \mid h(X) = \infty\} \subset M_d$$

is of dimension $\dim M_d^\infty = 9$.

Remark 3.8 Although Picard number and height are intimately related via the inequality (3.4), their behavior as a function on M_d is very different. Whereas $h(X) \geq h$ is a closed condition, the condition $\rho(X) \geq \rho$ is not; it rather defines a countable union of closed sets.

As $\rho(X) \leq 22 - 2h$ for $(X, H) \in M_d^h$, one in particular has $\rho(X) \leq 2$ on M_d^{10} (which for $k = \overline{\mathbb{F}}_p$ is equivalent to $\rho(X) = 2$; see Corollary 17.2.9). However, many more (X, H) not contained in M_d^{10} satisfy $\rho(X) \leq 2$; in fact, the general one should have this property. Also note the slightly counterintuitive behavior of $\rho(X)$ on M_d^{10} which satisfies $\rho(X) \leq 2$ on the open set $M_d^{10} \setminus M_d^\infty \subset M_d^{10}$ but jumps to $\rho(X) = 22$ on M_d^∞; see below. It is also a remarkable fact that over the nine-dimensional M_d^∞ the Picard number stays constant, which in characteristic zero is excluded by Proposition **6**.2.9; see also Section **17**.1.3.

Related to this, recall that for K3 surfaces in characteristic zero the maximal Picard number is $\rho(X) = 20$ and that these surfaces are rigid; see Section **17**.1.3. In positive characteristic, the maximal Picard number is $\rho(X) = 22$ and surfaces of this type are not rigid.

Artin [16] also explored the possibility to extend the stratification

$$M_d^\infty \subset M_d^{10} \subset \cdots \subset M_d^2 \subset M_d^1 = M_d \qquad (3.9)$$

obtained in this way by taking the Artin invariant of supersingular K3 surfaces into account. Recall from Section **17**.2.7 that disc $NS(X) = -p^{2\sigma(X)}$ for a supersingular K3 surface X with the Artin invariant $\sigma(X)$ satisfying $1 \le \sigma(X) \le 10$. As the discriminant goes down under specialization (use (0.1) in Section **14**.0.2), $\sigma(X) \le \sigma$ is a closed condition. Defining $M_d^{\infty,\sigma} := \{(X, H) \mid h(X) = \infty, \sigma(X) \le \sigma\}$ yields a stratification

$$M_d^{\infty,1} \subset \cdots \subset M_d^{\infty,10}. \qquad (3.10)$$

Combining both stratification (3.9) and (3.10), one obtains a stratification

$$M_d^{\infty,1} \subset \cdots \subset M_d^{\infty,10} = M_d^\infty \subset M_d^{10} \subset \cdots \subset M_d^2 \subset M_d^1 = M_d.$$

For a detailed analysis of this filtration, see the article [166] by Ekedahl and van der Geer. Ogus in [480] shows in addition that the singular locus of M_d^h is contained in $M_d^{\infty,h-1}$. Already in [564] Shioda proves that all values $h(X) = 1, \ldots, 10, \infty$, and $1 \le \sigma(X) \le 10$ are actually realized in every characteristic $p > 2$.

3.5 We conclude with a few remarks on supersingular K3 surfaces. First, using an argument from Section **17**.3.4, we prove the following corollary.

Corollary 3.9 *Assume X is a Shioda supersingular K3 surface, i.e. $\rho(X) = 22$, then X is (Artin) supersingular. In particular, any unirational K3 surface is supersingular.*

Proof As $h(X)$ is positive, (3.4) implies the first assertion. For the second, use Proposition **17**.2.7. □

The converse was conjectured by Artin [16] and finally proved by Maulik, Charles, and Madapusi Pera. As the Tate conjecture had previously been proved for K3 surfaces of finite height, this finished the proof of the Tate conjecture for K3 surfaces. See also the discussion in Section **17**.3.3 and Corollary **17**.3.7.

Theorem 3.10 *Let X be a K3 surface over an algebraically closed field of characteristic $p > 2$. Then*

$$h(X) = \infty \quad \text{if and only if} \quad \rho(X) = 22.$$

Proof We recommend Benoist's Bourbaki talk [58] for an overview. Let us explain how the 'only if' is implied by the Tate conjecture.[11] So, assume that X is a K3 surface over a finite field $\mathbb{F}_q$ with $h(\overline{X}) = \infty$, i.e. $\widehat{\mathrm{Br}}_{\overline{X}} \simeq \widehat{\mathbb{G}}_a$. It is difficult to extract from the description of the formal Brauer group $\widehat{\mathrm{Br}}_{\overline{X}}$ any information on $\mathrm{Br}(X)$ directly. Instead, one uses the alternative description of $h(\overline{X})$ as the height of the F-isocrystal $H^2_{\mathrm{cr}}(\overline{X})$, which in this case says that some power of the action of $(1/p)F$ on $H^2_{\mathrm{cr}}(\overline{X})$ is trivial. Finally, one uses that the eigenvalues of the Frobenius action on $H^2_{\mathrm{cr}}(\overline{X})$ and $H^2_{\acute{e}t}(\overline{X}, \mathbb{Q}_\ell)$ coincide, a general result due to Katz and Messing [286]. Therefore, some power of $f^*\colon H^2_{\acute{e}t}(\overline{X}, \mathbb{Q}_\ell(1)) \longrightarrow H^2_{\acute{e}t}(\overline{X}, \mathbb{Q}_\ell(1))$ is trivial, and, hence, after passing to a finite extension $\mathbb{F}_{q^r}$ the Galois action on $H^2_{\acute{e}t}(\overline{X}, \mathbb{Q}_\ell(1))$ is trivial. The Tate conjecture then implies $\mathrm{NS}(X \times \mathbb{F}_{q^r}) \otimes \mathbb{Q}_\ell \simeq H^2_{\acute{e}t}(\overline{X}, \mathbb{Q}_\ell(1))$, which yields $\rho(\overline{X}) = 22$.   □

The proof shows that in the context of the Tate conjecture the formal Brauer group plays the role of a supporting actor, morally but not factually explaining the role of the (geometric) Brauer group.

Remark 3.11 Artin also developed an approach to reduce the Tate conjecture for supersingular K3 surfaces to the case of supersingular elliptic K3 surfaces. His [16, Thm. 1.1] asserts that for a connected family of supersingular K3 surfaces the Picard number stays constant. Thus, if the locus M_d^∞ of supersingular K3 surfaces can be shown to be irreducible (or at least connected) or if every component parametrizes at least one elliptic K3 surface, then the Tate conjecture for supersingular K3 surfaces would be implied by the elliptic case. This idea, going back to [515], has been worked out by Maulik in [398], where he shows that every component contains a complete curve, proving the Tate conjecture for all K3 surfaces of degree $2d$ with $p > 2d + 4$.

Remark 3.12 The proof of Corollary 3.9 actually shows that already $\rho(X) \geq 21$ implies that X is supersingular. But according to the theorem, supersingularity implies $\rho(X) = 22$. Hence, for all K3 surfaces over an algebraically closed field k,

$$\rho(X) \neq 21.$$

Recall that for complex K3 surfaces this follows immediately from $\rho(X) \leq 20$. See also Section **17.2.4**.

Remark 3.13 It is not difficult to see that a K3 surface X with $\rho(X) = 22$ over a separably closed field k of characteristic p has no classes of ℓ^n torsion in $\mathrm{Br}(X)$ for $\ell \neq p$. Indeed, by the usual Kummer sequence there exists an exact sequence

$$\mathrm{NS}(X) \otimes \mathbb{Z}/\ell^n\mathbb{Z} \longrightarrow H^2_{\acute{e}t}(X, \mu_{\ell^n}) \longrightarrow \mathrm{Br}(X)[\ell^n] \longrightarrow 0.$$

---

[11] As mentioned before, Artin (with Swinnerton–Dyer) had proved the result for supersingular elliptic K3 surfaces. Using that any K3 surface with $\rho(X) \geq 5$ is actually elliptic (see Proposition **11**.1.3), it is enough to argue that $h(X) = \infty$ implies $\rho(X) \geq 5$.

As $H^3_{\text{ét}}(X, \mu_\ell) = 0$, the maps $H^2_{\text{ét}}(X, \mu_{\ell^n}) \longrightarrow H^2_{\text{ét}}(X, \mu_{\ell^{n-1}})$ are surjective and by Proposition **17**.3.5

$$\text{NS}(X) \otimes \mathbb{Z}_\ell \xrightarrow{\sim} H^2_{\text{ét}}(X, \mathbb{Z}_\ell(1)). \text{ Hence, } \text{NS}(X) \otimes \mathbb{Z}/\ell^n\mathbb{Z} \longrightarrow H^2_{\text{ét}}(X, \mu_{\ell^n})$$

is surjective and, therefore, $\text{Br}(X)[\ell^n] = 0$. Of course, if $k = \overline{\mathbb{F}}_p$ and if one is willing to accept the Artin–Tate conjecture, then (2.5) applied to a model of X over any finite field $\mathbb{F}_q$ immediately yields that $\text{Br}(X)$ is p-primary.

Remark 3.14 We have seen that a unirational K3 surface over an algebraically closed field always has maximal Picard number $\rho(X) = 22$. The converse of this assertion has been proved by Liedtke in [373]. See also Lieblich's articles [367, 368] for further information. This had been checked earlier for various special cases, for example by Shioda in [562] for Kummer surfaces.

The upshot is that for K3 surfaces over an algebraically closed field of positive characteristic all three concepts of supersingular are equivalent:

$$\boxed{X \text{ is unirational} \iff h(X) = \infty \iff \rho(X) = 22.}$$

3.6 Due to the lack of space, the beautiful work of Ogus [478, 479], proving a global Torelli theorem for supersingular K3 surfaces, will not be discussed here. We recommend Liedtke's notes [372] for an introduction. The final result is that two supersingular K3 surfaces X and X' are isomorphic if and only if there exists an isomorphism of W-modules

$$H^2_{\text{cr}}(X/W) \simeq H^2_{\text{cr}}(X'/W),$$

which is compatible with the Frobenius action and the intersection pairing. Compare Remark **17**.2.23.

References and Further Reading

Formal completion of Chow has been studied by Stienstra [579]. In the supersingular case the Artin invariant enters its description.

The group $\text{Br}(X)/\text{Br}_1(X) \simeq \text{Im}(\text{Br}(X) \longrightarrow \text{Br}(\overline{X})^G)$ can indeed be non-trivial, and classes in this group have been used by Hassett and Várilly-Alvarado to construct examples of K3 surfaces over number fields for which the Hasse principle fails. In [241] one finds examples of effective bounds on the order

$$|\text{Br}(X)/\text{Br}_0(X)| = |\text{Br}_1(X)/\text{Br}_0(X)| \cdot |\text{Br}(X)/\text{Br}_1(X)|.$$

The Brauer–Manin obstruction, which has not been discussed in this book, is based upon the sequence (2.1). See [241] for references.

In [607, Thm. 15.1] the class $[M_d^h] \in \text{CH}^{h-1}(M_d)$ is expressed as a multiple of $c_1^{h-1}(\pi_* \omega_{X/M_d})$. In [166] this was extended to the smaller strata $M_d^{\infty,\sigma}$.

Lieblich in [368] proves that the supersingular locus M_d^∞ is rationally connected.

References

[1] The Stacks Project. http://stacks.math.columbia.edu, 2014. (Cited on pages 94, 411, and 420.)

[2] N. Addington and R. Thomas. Hodge theory and derived categories of cubic fourfolds. *Duke Math. J.*, 163(10):1885–1927, 2014. (Cited on page 384.)

[3] A. Altman and S. Kleiman. Compactifying the Picard scheme. II. *Amer. J. Math.*, 101(1):10–41, 1979. (Cited on page 238.)

[4] K. Amerik and M. Verbitsky. Morrison–Kawamata cone conjecture for hyperkähler manifolds. 2014. arXiv:1408.3892. (Cited on page 174.)

[5] S. Anan'in and M. Verbitsky. Any component of moduli of polarized hyperkähler manifolds is dense in its deformation space. 2010. arXiv:1008.2480. (Cited on page 324.)

[6] Y. André. Mumford–Tate groups of mixed Hodge structures and the theorem of the fixed part. *Compositio Math.*, 82(1):1–24, 1992. (Cited on page 122.)

[7] Y. André. On the Shafarevich and Tate conjectures for hyper-Kähler varieties. *Math. Ann.*, 305(2):205–248, 1996. (Cited on pages 54, 58, 77, 79, 85, 92, 404, and 408.)

[8] Y. André. Pour une théorie inconditionnelle des motifs. *Inst. Hautes Études Sci. Publ. Math.*, 83:5–49, 1996. (Cited on pages 79 and 400.)

[9] M. Aprodu. Brill–Noether theory for curves on K3 surfaces. In *Contemporary geometry and topology and related topics*, pages 1–12. Cluj University Press, Cluj-Napoca, 2008. (Cited on page 183.)

[10] E. Arbarello, M. Cornalba, and P. Griffiths. *Geometry of algebraic curves.* Vol. II, vol. 268 of *Grundlehren der Mathematischen Wissenschaften*. Springer-Verlag, Heidelberg, 2011. With a contribution by J. Harris. (Cited on pages 283 and 285.)

[11] E. Arbarello, M. Cornalba, P. Griffiths, and J. Harris. *Geometry of algebraic curves. Vol. I*, vol. 267 of *Grundlehren der Mathematischen Wissenschaften*. Springer-Verlag, New York, 1985. (Cited on pages 17 and 183.)

[12] E. Artal Bartolo, H. Tokunaga, and D.-Q. Zhang. Miranda–Persson's problem on extremal elliptic K3 surfaces. *Pacific J. Math.*, 202(1):37–72, 2002. (Cited on page 253.)

[13] M. Artebani, J. Hausen, and A. Laface. On Cox rings of K3 surfaces. *Compositio Math.*, 146(4):964–998, 2010. (Cited on pages 170 and 174.)

[14] M. Artebani, A. Sarti, and S. Taki. K3 surfaces with non-symplectic automorphisms of prime order. *Math. Z.*, 268(1-2):507–533, 2011. With an appendix by S. Kondō. (Cited on page 356.)

[15] E. Artin. *Geometric algebra*. Interscience Publishers, New York and London, 1957. (Cited on pages 59, 61, and 142.)

[16] M. Artin. Supersingular K3 surfaces. *Ann. Sci. École Norm. Sup. (4)*, 7:543–567 (1975), 1974. (Cited on pages 11, 273, 395, 401, 402, 407, 408, 425, 434, 435, and 436.)

[17] M. Artin. Versal deformations and algebraic stacks. *Invent. Math.*, 27:165–189, 1974. (Cited on pages 90 and 416.)

[18] M. Artin. Néron models. In *Arithmetic geometry (Storrs, CT, 1984)*, pages 213–230. Springer-Verlag, New York, 1986. (Cited on page 232.)

[19] M. Artin and B. Mazur. Formal groups arising from algebraic varieties. *Ann. Sci. École Norm. Sup. (4)*, 10(1):87–131, 1977. (Cited on pages 416, 432, and 433.)

[20] M. Artin and P. Swinnerton-Dyer. The Shafarevich–Tate conjecture for pencils of elliptic curves on K3 surfaces. *Invent. Math.*, 20:249–266, 1973. (Cited on pages 249, 406, 407, and 425.)

[21] M. Asakura and S. Saito. Surfaces over a p-adic field with infinite torsion in the Chow group of 0-cycles. *Algebra Number Theory*, 1(2):163–181, 2007. (Cited on page 270.)

[22] K. Ascher, K. Dasaratha, A. Perry, and R. Zhou. Derived equivalences and rational points of twisted K3 surfaces. 2015. arXiv:1506.01374. (Cited on page 379.)

[23] A. Ash, D. Mumford, M. Rapoport, and Y.-S. Tai. *Smooth compactifications of locally symmetric varieties*. Cambridge Mathematical Library. Cambridge University Press, Cambridge, second edition, 2010. With the collaboration of Peter Scholze. (Cited on page 167.)

[24] P. Aspinwall and R. Kallosh. Fixing all moduli for M-theory on K3×K3. *J. High Energy Phys.*, (10):001, 20 pp. (electronic), 2005. (Cited on page 386.)

[25] P. Aspinwall and D. Morrison. String theory on K3 surfaces. In *Mirror symmetry, II*, vol. 1 of *AMS/IP Stud. Adv. Math.*, pages 703–716. Amer. Math. Soc., Providence, RI, 1997. (Cited on pages x and 314.)

[26] M. Atiyah. On analytic surfaces with double points. *Proc. Roy. Soc. London. Ser. A*, 247:237–244, 1958. (Cited on page xi.)

[27] M. Atiyah and F. Hirzebruch. Analytic cycles on complex manifolds. *Topology*, 1:25–45, 1962. (Cited on page 38.)

[28] L. Bădescu. *Algebraic surfaces*. Universitext. Springer-Verlag, New York, 2001. (Cited on pages 3, 8, 15, 29, 33, and 221.)

[29] W. Baily and A. Borel. Compactification of arithmetic quotients of bounded symmetric domains. *Ann. Math. (2)*, 84:442–528, 1966. (Cited on pages 82 and 105.)

[30] A. Baragar. The ample cone for a K3 surface. *Canad. J. Math.*, 63(3):481–499, 2011. (Cited on page 173.)

[31] A. Baragar and D. McKinnon. K3 surfaces, rational curves, and rational points. *J. Number Theory*, 130(7):1470–1479, 2010. (Cited on page 298.)

[32] P. Bardsley and R. Richardson. Étale slices for algebraic transformation groups in characteristic p. *Proc. Lond. Math. Soc. (3)*, 51(2):295–317, 1985. (Cited on page 94.)

[33] W. Barth, K. Hulek, C. Peters, and A. Van de Ven. *Compact complex surfaces*. Ergebnisse der Mathematik und ihrer Grenzgebiete. Springer-Verlag, Berlin, 2004. (Cited on pages x, 2, 3, 4, 7, 11, 15, 26, 142, 146, 149, 157, 158, 173, 219, 221, 224, 244, 285, 319, 320, 324, and 351.)

[34] C. Bartocci, U. Bruzzo, and D. Hernández Ruipérez. *Fourier–Mukai and Nahm transforms in geometry and mathematical physics*, vol. 276 of *Progr. Math.* Birkhäuser, Boston, 2009. (Cited on page 358.)

[35] H. Bass. *Algebraic K-theory.* W. A. Benjamin, New York and Amsterdam, 1968. (Cited on page 261.)

[36] V. Batyrev. Birational Calabi–Yau *n*-folds have equal Betti numbers. In *New trends in algebraic geometry (Warwick, 1996)*, vol. 264 of *Lond. Math. Soc. Lecture Note Ser.*, pages 1–11. Cambridge University Press, Cambridge, 1999. (Cited on page 294.)

[37] T. Bauer. Seshadri constants on algebraic surfaces. *Math. Ann.*, 313(3):547–583, 1999. (Cited on page 168.)

[38] T. Bauer, S. Di Rocco, and T. Szemberg. Generation of jets on K3 surfaces. *J. Pure Appl. Algebra*, 146(1):17–27, 2000. (Cited on page 33.)

[39] T. Bauer and M. Funke. Weyl and Zariski chambers on K3 surfaces. *Forum Math.*, 24(3):609–625, 2012. (Cited on page 174.)

[40] A. Bayer and T. Bridgeland. Derived automorphism groups of K3 surfaces of Picard rank 1. 2013. arXiv:1310.8266. (Cited on page 376.)

[41] A. Bayer and E. Macrì. MMP for moduli of sheaves on K3s via wall-crossing: nef and movable cones, Lagrangian fibrations. *Invent. Math.*, 198(3):505–590, 2014. (Cited on page 218.)

[42] A. Bayer and E. Macrì. Projectivity and birational geometry of Bridgeland moduli spaces. *J. Amer. Math. Soc.*, 27(3):707–752, 2014. (Cited on page 218.)

[43] A. Beauville. *Surfaces algébriques complexes*, vol. 54 of *Astérisque*. Soc. Math. France, Paris, 1978. (Cited on pages 3, 15, 18, 21, and 32.)

[44] A. Beauville. Sur le nombre maximum de points doubles d'une surface dans $\mathbb{P}^3$ ($\mu(5) = 31$). In *Journées de Géometrie Algébrique d'Angers, Juillet 1979*, pages 207–215. Sijthoff & Noordhoff, Alphen aan den Rijn, 1980. (Cited on page 321.)

[45] A. Beauville. Variétés Kähleriennes dont la première classe de Chern est nulle. *J. Differential Geom.*, 18(4):755–782 (1984), 1983. (Cited on page 216.)

[46] A. Beauville. Counting rational curves on K3 surfaces. *Duke Math. J.*, 97(1):99–108, 1999. (Cited on page 294.)

[47] A. Beauville. Some stable vector bundles with reducible theta divisor. *Manuscripta Math.*, 110(3):343–349, 2003. (Cited on page 187.)

[48] A. Beauville. Fano threefolds and K3 surfaces. In *The Fano Conference*, pages 175–184. University of Torino, Turin, 2004. (Cited on pages 99 and 279.)

[49] A. Beauville. La conjecture de Green générique [d'après C. Voisin], Séminaire Bourbaki, Exposé 924, 2003/2004. *Astérisque*, 299:1–14, 2005. (Cited on pages x and 184.)

[50] A. Beauville. On the splitting of the Bloch–Beilinson filtration. In *Algebraic cycles and motives. Vol. 2*, vol. 344 of *Lond. Math. Soc. Lecture Note Ser.*, pages 38–53. Cambridge University Press, Cambridge, 2007. (Cited on page 271.)

[51] A. Beauville. Antisymplectic involutions of holomorphic symplectic manifolds. *J. Topol.*, 4(2):300–304, 2011. (Cited on page 357.)

[52] A. Beauville and R. Donagi. La variété des droites d'une hypersurface cubique de dimension 4. *C. R. Acad. Sci. Paris Sér. I Math.*, 301(14):703–706, 1985. (Cited on page 14.)

[53] A. Beauville and C. Voisin. On the Chow ring of a K3 surface. *J. Algebraic Geom.*, 13(3):417–426, 2004. (Cited on pages 268 and 269.)

[54] A. Beauville et al. *Géométrie des surfaces K3: modules et périodes*, vol. 126 of *Astérisque*. Soc. Math. France, Paris, 1985. (Cited on pages x, 15, 28, 47, 106, 114, 130, 131, 134, 138, 146, 147, 172, 173, 319, 320, 324, and 337.)

[55] K. Behrend, B. Conrad, D. Edidin, W. Fulton, B. Fantechi, L. Göttsche, and A. Kresch. *Introduction to stacks*. In progress. (Cited on pages 95, 96, and 97.)

[56] A. Beĭlinson. Higher regulators and values of *L*-functions. In *Current problems in mathematics, Vol. 24*, Itogi Nauki i Tekhniki, pages 181–238. Akad. Nauk SSSR Vsesoyuz. Inst. Nauchn. i Tekhn. Inform., Moscow, 1984. (Cited on page 261.)

[57] S. Belcastro. *Picard lattices of families of K3 surfaces*. ProQuest LLC, Ann Arbor, MI, 1997. PhD thesis, University of Michigan. (Cited on page 253.)

[58] O. Benoist. Construction de courbes sur les surfaces K3 [d'après Bogomolov–Hassett–Tschinkel, Charles, Li–Liedtke, Madapusi Pera, Maulik...], Séminaire Bourbaki, Exposé 1081, 2013–2014. 2014. (Cited on pages 99, 297, 407, and 436.)

[59] N. Bergeron, Z. Li, J. Millson, and C. Moeglin. The Noether–Lefschetz conjecture and generalizations. 2014. arXiv:1412.3774. (Cited on page 127.)

[60] J. Bertin, J.-P. Demailly, L. Illusie, and C. Peters. *Introduction to Hodge theory*, vol. 8 of *SMF/AMS Texts and Monographs*. Amer. Math. Soc., Providence, RI, 2002. (Cited on pages 106, 108, and 329.)

[61] A. Besse. *Einstein manifolds*. Classics in Mathematics. Springer-Verlag, Berlin, 2008. (Cited on page 338.)

[62] G. Bini. On automorphisms of some K3 surfaces with Picard number two. *MCFA Annals*, 4:1–3, 2007. (Cited on page 343.)

[63] G. Bini and A. Garbagnati. Quotients of the Dwork pencil. *J. Geom. Phys.*, 75:173–198, 2014. (Cited on page 389.)

[64] C. Birkenhake and H. Lange. *Complex tori*, vol. 177 of *Progr. Math.* Birkhäuser, Boston 1999. (Cited on pages 10 and 58.)

[65] C. Birkenhake and H. Lange. *Complex abelian varieties*, vol. 302 of *Grundlehren der Mathematischen Wissenschaften*. Springer-Verlag, Berlin, second edition, 2004. (Cited on page 58.)

[66] S. Bloch. Torsion algebraic cycles and a theorem of Roitman. *Compositio Math.*, 39(1):107–127, 1979. (Cited on page 256.)

[67] S. Bloch. *Lectures on algebraic cycles*. Duke University Mathematics Series, IV. Duke University Mathematics Department, Durham, NC, 1980. (Cited on pages 256, 260, 263, 264, and 271.)

[68] S. Bloch. Algebraic cycles and values of *L*-functions. *J. Reine Angew. Math.*, 350:94–108, 1984. (Cited on page 261.)

[69] F. Bogomolov, B. Hassett, and Y. Tschinkel. Constructing rational curves on K3 surfaces. *Duke Math. J.*, 157(3):535–550, 2011. (Cited on pages 272, 273, 274, 277, 278, 281, 285, 286, 287, 289, 291, and 292.)

[70] F. Bogomolov and Y. Tschinkel. Density of rational points on elliptic K3 surfaces. *Asian J. Math.*, 4(2):351–368, 2000. (Cited on pages 281 and 298.)

[71] F. Bogomolov and Y. Tschinkel. Rational curves and points on K3 surfaces. *Amer. J. Math.*, 127(4):825–835, 2005. (Cited on pages 274, 282, and 329.)

[72] S. Boissière, M. Nieper-Wißkirchen, and A. Sarti. Higher dimensional Enriques varieties and automorphisms of generalized Kummer varieties. *J. Math. Pures Appl. (9)*, 95(5): 553–563, 2011. (Cited on page 357.)

[73] S. Boissière and A. Sarti. Counting lines on surfaces. *Ann. Sc. Norm. Super. Pisa Cl. Sci. (5)*, 6(1):39–52, 2007. (Cited on page 297.)

[74] S. Boissière and A. Sarti. On the Néron–Severi group of surfaces with many lines. *Proc. Amer. Math. Soc.*, 136(11):3861–3867, 2008. (Cited on page 390.)

[75] C. Borcea. Diffeomorphisms of a K3 surface. *Math. Ann.*, 275(1):1–4, 1986. (Cited on page 143.)

[76] C. Borcea. K3 surfaces and complex multiplication. *Rev. Roumaine Math. Pures Appl.*, 31(6):499–505, 1986. (Cited on page 52.)

[77] R. Borcherds. Coxeter groups, Lorentzian lattices, and K3 surfaces. *Internat. Math. Res. Notices*, 19:1011–1031, 1998. (Cited on pages 326, 327, 329, 341, and 357.)

[78] R. Borcherds, L. Katzarkov, T. Pantev, and N. Shepherd-Barron. Families of K3 surfaces. *J. Algebraic Geom.*, 7(1):183–193, 1998. (Cited on page 111.)

[79] A. Borel. Some metric properties of arithmetic quotients of symmetric spaces and an extension theorem. *J. Differential Geometry*, 6:543–560, 1972. Collection of articles dedicated to S. S. Chern and D. C. Spencer on their sixtieth birthdays. (Cited on page 118.)

[80] A. Borel. *Introduction aux groupes arithmétiques.* Publ. de l'Institut de Math. de l'Université de Strasbourg, XV. Actualités Sci. Ind., No. 1341. Hermann, Paris, 1969. (Cited on page 104.)

[81] A. Borel and Harish-Chandra. Arithmetic subgroups of algebraic groups. *Ann. of Math. (2)*, 75:485–535, 1962. (Cited on page 339.)

[82] S. Bosch, W. Lütkebohmert, and M. Raynaud. *Néron models*, vol. 21 of *Ergebnisse der Mathematik und ihrer Grenzgebiete (3)*. Springer-Verlag, Berlin, 1990. (Cited on pages 86, 199, 238, 239, and 417.)

[83] N. Bourbaki. *Lie groups and Lie algebras.* Chapters 4–6. Elements of Mathematics (Berlin). Springer-Verlag, Berlin, 2002. (Cited on page 152.)

[84] N. Bourbaki. *Éléments de mathématique. Algèbre.* Chapitre 9. Springer-Verlag, Berlin, 2007. (Cited on page 59.)

[85] A. Bremner. A geometric approach to equal sums of sixth powers. *Proc. Lond. Math. Soc. (3)*, 43(3):544–581, 1981. (Cited on page 390.)

[86] T. Bridgeland. Stability conditions on K3 surfaces. *Duke Math. J.*, 141(2):241–291, 2008. (Cited on pages 375, 376, and 384.)

[87] T. Bridgeland and A. Maciocia. Complex surfaces with equivalent derived categories. *Math. Z.*, 236(4):677–697, 2001. (Cited on page 373.)

[88] T. Bröcker and T. tom Dieck. *Representations of compact Lie groups*, vol. 98 of *Graduate Texts in Mathematics*. Springer-Verlag, New York, 1995. (Cited on page 59.)

[89] J. Bryan and N. C. Leung. The enumerative geometry of K3 surfaces and modular forms. *J. Amer. Math. Soc.*, 13(2):371–410 (electronic), 2000. (Cited on page 297.)

[90] N. Buchdahl. On compact Kähler surfaces. *Ann. Inst. Fourier (Grenoble)*, 49(1):287–302, 1999. (Cited on pages 135, 137, and 171.)

[91] N. Buchdahl. Compact Kähler surfaces with trivial canonical bundle. *Ann. Global Anal. Geom.*, 23(2):189–204, 2003. (Cited on pages 144 and 146.)

[92] A. Buium. Sur le nombre de Picard des revêtements doubles des surfaces algébriques. *C. R. Acad. Sci. Paris Sér. I Math.*, 296(8):361–364, 1983. (Cited on page 390.)

[93] D. Burns and M. Rapoport. On the Torelli problem for kählerian K3 surfaces. *Ann. Sci. École Norm. Sup. (4)*, 8(2):235–273, 1975. (Cited on pages 44, 116, 133, and 145.)

[94] D. Burns, Jr. and J. Wahl. Local contributions to global deformations of surfaces. *Invent. Math.*, 26:67–88, 1974. (Cited on page 142.)

[95] G. Buzzard and S. Lu. Algebraic surfaces holomorphically dominable by $\mathbb{C}^2$. *Invent. Math.*, 139(3):617–659, 2000. (Cited on page 275.)

[96] A. Căldăraru. *Derived categories of twisted sheaves on Calabi–Yau manifolds.* ProQuest LLC, Ann Arbor, MI, 2000. PhD thesis, Cornell University. (Cited on pages 208, 242, 376, and 377.)

[97] C. Camere. About the stability of the tangent bundle of $\mathbb{P}^n$ restricted to a surface. *Math. Z.*, 271(1–2):499–507, 2012. (Cited on pages 179, 186, and 187.)

[98] F. Campana. Orbifolds, special varieties and classification theory. *Ann. Inst. Fourier (Grenoble)*, 54(3):499–630, 2004. (Cited on page 274.)

[99] F. Campana, K. Oguiso, and T. Peternell. Non-algebraic hyperkähler manifolds. *J. Differential Geom.*, 85(3):397–424, 2010. (Cited on page 387.)

[100] A. Canonaco and P. Stellari. Twisted Fourier–Mukai functors. *Adv. Math.*, 212(2):484–503, 2007. (Cited on page 376.)

[101] A. Canonaco and P. Stellari. Fourier–Mukai functors: a survey. In *Derived categories in algebraic geometry*, EMS Ser. Congr. Rep., pages 27–60. Eur. Math. Soc., Zürich, 2012. (Cited on page 362.)

[102] S. Cantat. Sur la dynamique du groupe d'automorphismes des surfaces K3. *Transform. Groups*, 6(3):201–214, 2001. (Cited on page 357.)

[103] J. Carlson, S. Müller-Stach, and C. Peters. *Period mappings and period domains*, vol. 85 of *Cambridge Studies in Advanced Mathematics*. Cambridge University Press, 2003. (Cited on page 102.)

[104] H. Cartan. Quotient d'un espace analytique par un groupe d'automorphismes. In *Algebraic geometry and topology*, pages 90–102. Princeton University Press, 1957. A symposium in honor of S. Lefschetz. (Cited on page 331.)

[105] J. Cassels. *Rational quadratic forms*, vol. 13 of *Lond. Math. Soc. Monographs*. Academic Press (Harcourt Brace Jovanovich), London, 1978. (Cited on pages 300, 301, and 311.)

[106] J. Cassels and A. Fröhlich, editors. *Algebraic number theory*. Academic Press (Harcourt Brace Jovanovich), London, 1986. Reprint of the 1967 original. (Cited on pages 420 and 421.)

[107] F. Catanese and B. Wajnryb. Diffeomorphism of simply connected algebraic surfaces. *J. Differential Geom.*, 76(2):177–213, 2007. (Cited on page 129.)

[108] E. Cattani, P. Deligne, and A. Kaplan. On the locus of Hodge classes. *J. Amer. Math. Soc.*, 8(2):483–506, 1995. (Cited on page 122.)

[109] A. Chambert-Loir. Cohomologie cristalline: un survol. *Exposition. Math.*, 16(4):333–382, 1998. (Cited on page 433.)

[110] F. Charles. The Tate conjecture for K3 surfaces over finite fields. *Invent. Math.*, 194(1):119–145, 2013. Erratum 202(1):481–485, 2015. (Cited on page 407.)

[111] F. Charles. Birational boundedness for holomorphic symplectic varieties, Zarhin's trick for K3 surfaces, and the Tate conjecture. 2014. arXiv:1407.0592. (Cited on pages 27, 79, 197, 407, 408, 425, and 426.)

[112] F. Charles. On the Picard number of K3 surfaces over number fields. *Algebra Number Theory*, 8(1):1–17, 2014. (Cited on page 397.)

[113] F. Charles and C. Schnell. Notes on absolute Hodge classes. In *Hodge theory*, vol. 49 of *Math. Notes*, pages 469–530. Princeton University Press, Princeton, NJ, 2014. (Cited on pages 70, 78, and 79.)

[114] X. Chen. Rational curves on K3 surfaces. *J. Algebraic Geom.*, 8(2):245–278, 1999. (Cited on pages 277 and 282.)

[115] X. Chen. A simple proof that rational curves on K3 are nodal. *Math. Ann.*, 324(1):71–104, 2002. (Cited on pages 282 and 297.)

[116] X. Chen. Self rational maps of K3 surfaces. 2010. arXiv:1008.1619. (Cited on page 356.)

[117] X. Chen and J. Lewis. Density of rational curves on K3 surfaces. *Math. Ann.*, 356(1):331–354, 2013. (Cited on page 296.)

[118] J. Choy and Y.-H. Kiem. Nonexistence of a crepant resolution of some moduli spaces of sheaves on a K3 surface. *J. Korean Math. Soc.*, 44(1):35–54, 2007. (Cited on page 217.)

[119] C. Ciliberto and T. Dedieu. On universal Severi varieties of low genus K3 surfaces. *Math. Z.*, 271(3–4):953–960, 2012. (Cited on page 295.)

[120] C. Ciliberto and G. Pareschi. Pencils of minimal degree on curves on a K3 surface. *J. Reine Angew. Math.*, 460:15–36, 1995. (Cited on page 184.)

[121] H. Clemens, J. Kollár, and S. Mori. *Higher-dimensional complex geometry*, vol. 166 of *Astérisque*. Soc. Math. France, Paris, 1988. (Cited on page 25.)

[122] J.-L. Colliot-Thélène. Cycles algébriques de torsion et K-théorie algébrique. In *Arithmetic algebraic geometry (Trento, 1991)*, vol. 1553 of *Lecture Notes in Math.*, pages 1–49. Springer-Verlag, Berlin, 1993. (Cited on pages 256, 257, 263, 267, and 270.)

[123] J.-L. Colliot-Thélène. L'arithmétique du groupe de Chow des zéro-cycles. *J. Théor. Nombres Bordeaux*, 7(1):51–73, 1995. Les Dix-huitièmes Journées Arithmétiques (Bordeaux, 1993). (Cited on page 271.)

[124] J.-L. Colliot-Thélène. Groupe de Chow des zéro-cycles sur les variétés p-adiques [d'après S. Saito, K. Sato et al.], Séminaire Bourbaki, Exposé 1012, 2009/2010. *Astérisque*, 339: 1–30, 2011. (Cited on page 270.)

[125] J.-L. Colliot-Thélène and W. Raskind. Groupe de Chow de codimension deux des variétés définies sur un corps de nombres: un théorème de finitude pour la torsion. *Invent. Math.*, 105(2):221–245, 1991. (Cited on page 270.)

[126] J.-L. Colliot-Thélène, J.-J. Sansuc, and C. Soulé. Torsion dans le groupe de Chow de codimension deux. *Duke Math. J.*, 50(3):763–801, 1983. (Cited on pages 257, 267, and 271.)

[127] P. Colmez and J.-P. Serre, editors. *Correspondance Grothendieck–Serre*. Documents Mathématiques, 2. Soc. Math. France, Paris, 2001. (Cited on page 80.)

[128] D. Comenetz. Two algebraic deformations of a K3 surface. *Nagoya Math. J.*, 82:1–26, 1981. (Cited on page 84.)

[129] B. Conrad, M. Lieblich, and M. Olsson. Nagata compactification for algebraic spaces. *J. Inst. Math. Jussieu*, 11(4):747–814, 2012. (Cited on page 83.)

[130] J. Conway. A group of order $8,315,553,613,086,720,000$. *Bull. Lond. Math. Soc.*, 1:79–88, 1969. (Cited on page 328.)

[131] J. Conway. The automorphism group of the 26-dimensional even unimodular Lorentzian lattice. *J. Algebra*, 80(1):159–163, 1983. (Cited on page 326.)

[132] J. Conway and N. Sloane. *Sphere packings, lattices and groups*, vol. 290 of *Grundlehren der Mathematischen Wissenschaften*. Springer-Verlag, New York, third edition, 1999. With additional contributions by E. Bannai, R. E. Borcherds, J. Leech, S. P. Norton, A. M. Odlyzko, R. A. Parker, L. Queen, and B. B. Venkov. (Cited on pages 299, 312, 324, 325, 326, 327, 328, and 346.)

[133] F. Cossec and I. Dolgachev. *Enriques surfaces. I*, vol. 76 of *Progr. Math.* Birkhäuser Boston Inc., 1989. (Cited on pages 219, 242, and 246.)

[134] D. Cox. Mordell–Weil groups of elliptic curves over $\mathbb{C}(t)$ with $p_g = 0$ or 1. *Duke Math. J.*, 49(3):677–689, 1982. (Cited on pages 232 and 237.)

[135] D. Cox and S. Zucker. Intersection numbers of sections of elliptic surfaces. *Invent. Math.*, 53(1):1–44, 1979. (Cited on page 237.)

[136] J. de Jong. A result of Gabber. 2003. Preprint. (Cited on pages 383 and 413.)

[137] J. de Jong. The period-index problem for the Brauer group of an algebraic surface. *Duke Math. J.*, 123(1):71–94, 2004. (Cited on page 412.)

[138] O. Debarre. *Higher-dimensional algebraic geometry*. Universitext. Springer-Verlag, New York, 2001. (Cited on page 25.)

[139] O. Debarre. *Complex tori and abelian varieties*, vol. 11 of *SMF/AMS Texts and Monographs*. Amer. Math. Soc., Providence, RI, 2005. (Cited on pages 10 and 58.)

[140] T. Dedieu. Severi varieties and self-rational maps of K3 surfaces. *Internat. J. Math.*, 20(12):1455–1477, 2009. (Cited on page 356.)

[141] P. Deligne. La conjecture de Weil pour les surfaces K3. *Invent. Math.*, 15:206–226, 1972. (Cited on pages 68, 73, 74, 78, 79, and 122.)

[142] P. Deligne. *Cohomologie étale*, vol. 569 of *Lecture Notes in Math.* Springer-Verlag, Berlin, 1977. Séminaire de Géométrie Algébrique du Bois-Marie SGA $4\frac{1}{2}$, Avec la collaboration de J. F. Boutot, A. Grothendieck, L. Illusie et J.-L. Verdier. (Cited on page 247.)

[143] P. Deligne. Relèvement des surfaces K3 en caractéristique nulle. In *Algebraic surfaces (Orsay, 1976–78)*, vol. 868 of *Lecture Notes in Math.*, pages 58–79. Springer-Verlag, Berlin and New York, 1981. Prepared for publication by L. Illusie. (Cited on pages 75 and 431.)

[144] P. Deligne and L. Illusie. Relèvements modulo p^2 et décomposition du complexe de de Rham. *Invent. Math.*, 89(2):247–270, 1987. (Cited on pages 21, 195, and 196.)

[145] P. Deligne, J. Milne, A. Ogus, and K. Shih. *Hodge cycles, motives, and Shimura varieties*, vol. 900 of *Lecture Notes in Math.* Springer-Verlag, Berlin, 1982. (Cited on pages 53, 58, 78, and 79.)

[146] P. Deligne and D. Mumford. The irreducibility of the space of curves of given genus. *Inst. Hautes Études Sci. Publ. Math.*, 36:75–109, 1969. (Cited on pages 84, 94, 96, and 97.)

[147] J.-P. Demailly and M. Paun. Numerical characterization of the Kähler cone of a compact Kähler manifold. *Ann. of Math. (2)*, 159(3):1247–1274, 2004. (Cited on pages 136, 171, and 172.)

[148] J. Dillies. On some order 6 non-symplectic automorphisms of elliptic K3 surfaces. *Albanian J. Math.*, 6(2):103–114, 2012. (Cited on page 356.)

[149] I. Dolgachev. Integral quadratic forms: applications to algebraic geometry [after V. Nikulin], Séminaire Bourbaki, Exposé 611, 1982/1983. *Astérisque*, 105:251–278, 1983. (Cited on pages 170 and 299.)

[150] I. Dolgachev. Mirror symmetry for lattice polarized K3 surfaces. *J. Math. Sci.*, 81(3): 2599–2630, 1996. Algebraic geometry, 4. (Cited on pages 99, 127, and 279.)

[151] I. Dolgachev. *Classical algebraic geometry – a modern view*. Cambridge University Press, Cambridge, 2012. (Cited on page 15.)

[152] I. Dolgachev and J. Keum. Finite groups of symplectic automorphisms of K3 surfaces in positive characteristic. *Ann. of Math. (2)*, 169(1):269–313, 2009. (Cited on page 334.)

[153] I. Dolgachev and J. Keum. K3 surfaces with a symplectic automorphism of order 11. *J. Eur. Math. Soc. (JEMS)*, 11(4):799–818, 2009. (Cited on page 335.)

[154] I. Dolgachev and S. Kondō. Moduli of K3 surfaces and complex ball quotients. In *Arithmetic and geometry around hypergeometric functions*, vol. 260 of *Progr. Math.*, pages 43–100. Birkhäuser, Basel, 2007. (Cited on pages 99 and 127.)

[155] R. Donagi and D. Morrison. Linear systems on K3-sections. *J. Differential Geom.*, 29(1):49–64, 1989. (Cited on page 184.)

[156] R. Donagi and T. Pantev. Torus fibrations, gerbes, and duality. *Mem. Amer. Math. Soc.*, 193(901):vi+90, 2008. With an appendix by D. Arinkin. (Cited on pages 250 and 383.)

[157] S. Donaldson. Polynomial invariants for smooth four-manifolds. *Topology*, 29(3):257–315, 1990. (Cited on page 144.)

[158] S. Donaldson. Scalar curvature and projective embeddings. I. *J. Differential Geom.*, 59(3):479–522, 2001. (Cited on page 98.)

[159] P. Duhem. Émile Mathieu, his life and works. *Bull. Amer. Math. Soc.*, 1(7):156–168, 1892. (Cited on page 345.)

[160] W. Ebeling. *The monodromy groups of isolated singularities of complete intersections*, vol. 1293 of *Lecture Notes in Math.* Springer-Verlag, Berlin, 1987. (Cited on page 312.)

[161] W. Ebeling. *Lattices and codes.* Advanced Lectures in Mathematics. Friedr. Vieweg & Sohn Braunschweig, revised edition, 2002. A course partially based on lectures by F. Hirzebruch. (Cited on pages 299, 301, 304, and 325.)

[162] D. Edidin. Notes on the construction of the moduli space of curves. In *Recent progress in intersection theory (Bologna, 1997)*, Trends Math., pages 85–113. Birkhäuser, Boston, 2000. (Cited on page 97.)

[163] M. Eichler. *Quadratische Formen und orthogonale Gruppen*, vol. 63 of *Grundlehren der Mathematischen Wissenschaften.* Springer-Verlag, Berlin, second edition, 1974. (Cited on pages 299 and 308.)

[164] L. Ein and R. Lazarsfeld. Stability and restrictions of Picard bundles, with an application to the normal bundles of elliptic curves. In *Complex projective geometry (Trieste, 1989/Bergen, 1989)*, vol. 179 of *Lond. Math. Soc. Lecture Note Ser.*, pages 149–156. Cambridge University Press, 1992. (Cited on page 187.)

[165] T. Ekedahl. Foliations and inseparable morphisms. In *Algebraic geometry, Bowdoin, 1985 (Brunswick, ME, 1985)*, vol. 46 of *Proc. Sympos. Pure Math.*, pages 139–149. Amer. Math. Soc., Providence, RI, 1987. (Cited on page 190.)

[166] T. Ekedahl and G. van der Geer. Cycle classes on the moduli of K3 surfaces in positive characteristic. *Selecta Math. (N.S.)*, 21(1):245–291, 2015. (Cited on pages 435 and 437.)

[167] F. El Zein. *Introduction à la théorie de Hodge mixte.* Actualités Mathématiques. Hermann, Paris, 1991. (Cited on pages 58 and 102.)

[168] J. Ellenberg. K3 surfaces over number fields with geometric Picard number one. In *Arithmetic of higher-dimensional algebraic varieties (Palo Alto, CA, 2002)*, vol. 226 of *Progr. Math.*, pages 135–140. Birkhäuser, Boston, 2004. (Cited on pages 315, 398, and 399.)

[169] A. Elsenhans and J. Jahnel. On the computation of the Picard group for K3 surfaces. *Math. Proc. Cambridge Philos. Soc.*, 151(2):263–270, 2011. (Cited on page 401.)

[170] A. Elsenhans and J. Jahnel. The Picard group of a K3 surface and its reduction modulo p. *Algebra Number Theory*, 5(8):1027–1040, 2011. (Cited on pages 397 and 401.)

[171] A. Elsenhans and J. Jahnel. Examples of K3 surfaces with real multiplication. *LMS J. Comput. Math.*, 17(suppl. A):14–35, 2014. (Cited on page 58.)

[172] H. Esnault and K. Oguiso. Non-liftability of automorphism groups of a K3 surface in positive characteristic. *Math. Ann.*, 363(3–4):1187–1206, 2015. (Cited on page 357.)

[173] H. Esnault and V. Srinivas. Algebraic versus topological entropy for surfaces over finite fields. *Osaka J. Math.*, 50(3):827–846, 2013. (Cited on page 335.)

[174] J. Esser. *Noether–Lefschetz-Theoreme für zyklische Überlagerungen.* University Essen, Fachbereich Mathematik und Informatik, Essen, 1993. (Cited on page 390.)

[175] G. Faltings, G. Wüstholz, F. Grunewald, N. Schappacher, and U. Stuhler. *Rational points.* Aspects of Mathematics, E6. Friedr. Vieweg & Sohn, Braunschweig, third edition, 1992. (Cited on pages 233, 405, and 406.)

[176] B. Fantechi, L. Göttsche, L. Illusie, S. Kleiman, N. Nitsure, and A. Vistoli. *Fundamental algebraic geometry, Grothendieck's FGA explained*, vol. 123 of *Mathematical Surveys and Monographs*. Amer. Math. Soc., Providence, RI, 2005. (Cited on pages 86, 93, 195, 199, 204, 238, 239, 397, 416, 417, and 418.)

[177] B. Fantechi, L. Göttsche, and D. van Straten. Euler number of the compactified Jacobian and multiplicity of rational curves. *J. Algebraic Geom.*, 8(1):115–133, 1999. (Cited on page 294.)

[178] D. Festi, A. Garbagnati, B. van Geemen, and R. van Luijk. The Cayley–Oguiso automorphism of positive entropy on a K3 surface. *J. Mod. Dyn.*, 7(1):75–97, 2013. (Cited on page 343.)

[179] J.-M. Fontaine. *Groupes p-divisibles sur les corps locaux*. Société Mathématique de France, Paris, 1977. Astérisque, No. 47-48. (Cited on page 433.)

[180] K. Frantzen. Classification of K3 surfaces with involution and maximal symplectic symmetry. *Math. Ann.*, 350(4):757–791, 2011. (Cited on page 356.)

[181] R. Friedman. A degenerating family of quintic surfaces with trivial monodromy. *Duke Math. J.*, 50(1):203–214, 1983. (Cited on page 126.)

[182] R. Friedman. Global smoothings of varieties with normal crossings. *Ann. of Math. (2)*, 118(1):75–114, 1983. (Cited on page 126.)

[183] R. Friedman. A new proof of the global Torelli theorem for K3 surfaces. *Ann. of Math. (2)*, 120(2):237–269, 1984. (Cited on pages 99, 116, 127, and 146.)

[184] R. Friedman. On threefolds with trivial canonical bundle. In *Complex geometry and Lie theory (Sundance, UT, 1989)*, vol. 53 of *Proc. Sympos. Pure Math.*, pages 103–134. Amer. Math. Soc., Providence, RI, 1991. (Cited on page 142.)

[185] R. Friedman. *Algebraic surfaces and holomorphic vector bundles*. Universitext. Springer Verlag, New York, 1998. (Cited on pages 15, 175, 198, and 221.)

[186] R. Friedman and J. Morgan. *Smooth four-manifolds and complex surfaces*, vol. 27 of *Ergebnisse der Mathematik und ihrer Grenzgebiete (3)*. Springer-Verlag, Berlin, 1994. (Cited on pages 13, 15, 129, 219, 226, 244, 247, and 252.)

[187] R. Friedman and D. Morrison. The birational geometry of degenerations: an overview. In *The birational geometry of degenerations (Cambridge, MA, 1981)*, vol. 29 of *Progr. Math.*, pages 1–32. Birkhäuser, Boston, 1983. (Cited on pages 125 and 126.)

[188] R. Friedman and F. Scattone. Type III degenerations of K3 surfaces. *Invent. Math.*, 83(1): 1–39, 1986. (Cited on page 126.)

[189] A. Fujiki. On automorphism groups of compact Kähler manifolds. *Invent. Math.*, 44(3):225–258, 1978. (Cited on page 337.)

[190] A. Fujiki. Finite automorphism groups of complex tori of dimension two. *Publ. Res. Inst. Math. Sci.*, 24(1):1–97, 1988. (Cited on page 329.)

[191] T. Fujita. On polarized manifolds whose adjoint bundles are not semipositive. In *Algebraic geometry, Sendai, 1985*, vol. 10 of *Adv. Stud. Pure Math.*, pages 167–178. North-Holland, Amsterdam, 1987. (Cited on page 24.)

[192] W. Fulton. *Intersection theory*, vol. 2 of *Ergebnisse der Mathematik und ihrer Grenzgebiete (3)*. Springer-Verlag, Berlin, second edition, 1998. (Cited on pages 254, 255, 256, and 259.)

[193] W. Fulton and R. Pandharipande. Notes on stable maps and quantum cohomology. In *Algebraic geometry – Santa Cruz 1995*, vol. 62 of *Proc. Sympos. Pure Math.*, pages 45–96. Amer. Math. Soc., Providence, RI, 1997. (Cited on page 286.)

[194] P. Gabriel. Des catégories abéliennes. *Bull. Soc. Math. France*, 90:323–448, 1962. (Cited on page 358.)

[195] C. Galati and A. Knutsen. Seshadri constants of K3 surfaces of degrees 6 and 8. *Int. Math. Res. Not. IMRN*, 17:4072–4084, 2013. (Cited on page 34.)

[196] F. Galluzzi. Abelian fourfold of Mumford-type and Kuga–Satake varieties. *Indag. Math. (N.S.)*, 11(4):547–560, 2000. (Cited on page 79.)

[197] F. Galluzzi and G. Lombardo. On automorphisms group of some K3 surfaces. *Atti Accad. Sci. Torino Cl. Sci. Fis. Mat. Natur.*, 142:109–120 (2009), 2008. (Cited on pages 342 and 343.)

[198] F. Galluzzi, G. Lombardo, and C. Peters. Automorphs of indefinite binary quadratic forms and K3 surfaces with Picard number 2. *Rend. Semin. Mat. Univ. Politec. Torino*, 68(1): 57–77, 2010. (Cited on pages 341 and 343.)

[199] A. Garbagnati. Symplectic automorphisms on Kummer surfaces. *Geom. Dedicata*, 145:219–232, 2010. (Cited on page 352.)

[200] A. Garbagnati. On K3 surface quotients of K3 or abelian surfaces. 2015. arXiv:1507.03824. (Cited on page 329.)

[201] A. Garbagnati and A. Sarti. Symplectic automorphisms of prime order on K3 surfaces. *J. Algebra*, 318(1):323–350, 2007. (Cited on pages 99, 352, and 353.)

[202] A. Garbagnati and A. Sarti. On symplectic and non-symplectic automorphisms of K3 surfaces. *Rev. Mat. Iberoam.*, 29(1):135–162, 2013. (Cited on page 356.)

[203] B. van Geemen. Kuga–Satake varieties and the Hodge conjecture. In *The arithmetic and geometry of algebraic cycles (Banff, AB, 1998)*, vol. 548 of *NATO Sci. Ser. C Math. Phys. Sci.*, pages 51–82. Kluwer Academic, Dordrecht, 2000. (Cited on pages 39, 41, 53, 58, 65, 66, and 68.)

[204] B. van Geemen. Real multiplication on K3 surfaces and Kuga–Satake varieties. *Michigan Math. J.*, 56(2):375–399, 2008. (Cited on pages 52, 54, 58, and 79.)

[205] B. van Geemen and A. Sarti. Nikulin involutions on K3 surfaces. *Math. Z.*, 255(4): 731–753, 2007. (Cited on pages 99, 353, and 355.)

[206] G. van der Geer and T. Katsura. Note on tautological classes of moduli of K3 surfaces. *Mosc. Math. J.*, 5(4):775–779, 972, 2005. (Cited on page 99.)

[207] I. Gelfand, M. Kapranov, and A. Zelevinsky. *Discriminants, resultants and multidimensional determinants*. Modern Birkhäuser Classics. Birkhäuser, Boston 2008. (Cited on page 389.)

[208] S. Gelfand and Y. Manin. *Methods of homological algebra*. Springer Monographs in Mathematics. Springer-Verlag, Berlin, second edition, 2003. (Cited on pages 258 and 358.)

[209] J. Giansiracusa. The diffeomorphism group of a K3 surface and Nielsen realization. *J. Lond. Math. Soc. (2)*, 79(3):701–718, 2009. (Cited on pages 144 and 145.)

[210] J. Giraud. *Cohomologie non abélienne*, vol. 179 of *Grundlehren der Mathematischen Wissenschaften*. Springer-Verlag, Berlin and New York, 1971. (Cited on pages 383 and 412.)

[211] M. Gonzalez-Dorrego. (16, 6) configurations and geometry of Kummer surfaces in $\mathbb{P}^3$. *Mem. Amer. Math. Soc.*, 107(512):vi+101, 1994. (Cited on page 15.)

[212] L. Göttsche. The Betti numbers of the Hilbert scheme of points on a smooth projective surface. *Math. Ann.*, 286(1-3):193–207, 1990. (Cited on page 217.)

[213] L. Göttsche. A conjectural generating function for numbers of curves on surfaces. *Comm. Math. Phys.*, 196(3):523–533, 1998. (Cited on page 297.)

[214] L. Göttsche and D. Huybrechts. Hodge numbers of moduli spaces of stable bundles on K3 surfaces. *Internat. J. Math.*, 7(3):359–372, 1996. (Cited on page 216.)

[215] H. Grauert. On the number of moduli of complex structures. In *Contributions to function theory (Internat. Colloq. Function Theory, Bombay, 1960)*, pages 63–78. Tata Institute of Fundamental Research, Bombay, 1960. (Cited on pages xi and 109.)

[216] M. Green and P. Griffiths. Two applications of algebraic geometry to entire holomorphic mappings. In *The Chern Symposium 1979*, pages 41–74. Springer-Verlag, New York, 1980. (Cited on page 297.)

[217] M. Green, P. Griffiths, and K. Paranjape. Cycles over fields of transcendence degree 1. *Michigan Math. J.*, 52(1):181–187, 2004. (Cited on pages 264 and 265.)

[218] M. Green and R. Lazarsfeld. Special divisors on curves on a K3 surface. *Invent. Math.*, 89(2):357–370, 1987. (Cited on page 184.)

[219] P. Griffiths. Periods of integrals on algebraic manifolds: summary of main results and discussion of open problems. *Bull. Amer. Math. Soc.*, 76:228–296, 1970. (Cited on pages 126 and 147.)

[220] P. Griffiths, editor. *Topics in transcendental algebraic geometry*, vol. 106 of *Annals of Mathematics Studies*, Princeton University Press, Princeton, NJ, 1984. (Cited on page 102.)

[221] P. Griffiths and J. Harris. *Principles of algebraic geometry*. Wiley-Interscience, New York, 1978. (Cited on pages 43, 58, 171, 332, and 333.)

[222] P. Griffiths and J. Harris. On the variety of special linear systems on a general algebraic curve. *Duke Math. J.*, 47(1):233–272, 1980. (Cited on page 183.)

[223] V. Gritsenko, K. Hulek, and G. Sankaran. The Kodaira dimension of the moduli of K3 surfaces. *Invent. Math.*, 169(3):519–567, 2007. (Cited on page 99.)

[224] V. Gritsenko, K. Hulek, and G. Sankaran. Abelianisation of orthogonal groups and the fundamental group of modular varieties. *J. Algebra*, 322(2):463–478, 2009. (Cited on page 308.)

[225] A. Grothendieck. *Fondements de la géométrie algébrique. [Extraits du Séminaire Bourbaki, 1957–1962.]*. Secrétariat Mathématique, Paris, 1962. (Cited on pages 85, 93, 204, and 397.)

[226] A. Grothendieck. Éléments de géométrie algébrique. IV. Étude locale des schémas et des morphismes de schémas IV. *Inst. Hautes Études Sci. Publ. Math.*, 32:361, 1967. (Cited on page 98.)

[227] A. Grothendieck. Le groupe de Brauer. III. Exemples et compléments. In *Dix Exposés sur la Cohomologie des Schémas*, pages 88–188. North-Holland, Amsterdam, 1968. (Cited on pages 247, 249, 410, and 413.)

[228] A. Grothendieck. Technique de descente et théorèmes d'existence en géométrie algébrique. II. Le théorème d'existence en théorie formelle des modules. In *Séminaire Bourbaki, Vol. 5*, Exp. No. 195, pages 369–390. Soc. Math. France, Paris, 1995. (Cited on page 330.)

[229] M. Halic. A remark about the rigidity of curves on K3 surfaces. *Collect. Math.*, 61(3): 323–336, 2010. (Cited on page 297.)

[230] M. Halic. Some remarks about curves on K3 surfaces. In *Teichmüller theory and moduli problem*, vol. 10 of *Ramanujan Math. Soc. Lect. Notes Ser.*, pages 373–385. Ramanujan Math. Soc., Mysore, 2010. (Cited on page 297.)

[231] A. Harder and A. Thompson. The geometry and moduli of K3 surfaces. 2015. arXiv:1501.04049. (Cited on page 126.)

[232] J. Harris. Galois groups of enumerative problems. *Duke Math. J.*, 46(4):685–724, 1979. (Cited on pages 295 and 399.)

[233] J. Harris and I. Morrison. *Moduli of curves*, vol. 187 of *Graduate Texts in Mathematics*. Springer-Verlag, New York, 1998. (Cited on pages 283 and 285.)

[234] R. Hartshorne. *Residues and duality*, vol. 20 of *Lecture Notes in Math.* Springer-Verlag, Berlin, 1966. Lecture notes of a seminar on the work of A. Grothendieck, given at Harvard 1963/1964. With an appendix by P. Deligne. (Cited on page 6.)

[235] R. Hartshorne. *Ample subvarieties of algebraic varieties*, vol. 156 of *Lecture Notes in Math.* Springer-Verlag, Berlin, 1970. Notes written in collaboration with C. Musili. (Cited on pages 2, 4, 19, and 149.)

[236] R. Hartshorne. *Algebraic geometry*, vol. 52 of *Graduate Texts in Mathematics*. Springer-Verlag, New York, 1977. (Cited on pages 2, 3, 4, 5, 6, 8, 17, 19, 24, 27, 28, 29, 32, 73, 149, 189, 213, 219, 221, 222, 239, 254, 255, 262, 295, and 434.)

[237] R. Hartshorne. *Deformation theory*, vol. 257 of *Graduate Texts in Mathematics*. Springer-Verlag, New York, 2010. (Cited on pages 92, 110, and 416.)

[238] K. Hashimoto. Finite symplectic actions on the K3 lattice. *Nagoya Math. J.*, 206:99–153, 2012. (Cited on pages 344 and 352.)

[239] B. Hassett. Rational curves on K3 surfaces. *Lecture notes.* (Cited on pages 272, 273, and 281.)

[240] B. Hassett. Special cubic fourfolds. *Compositio Math.*, 120(1):1–23, 2000. (Cited on pages 306, 308, and 309.)

[241] B. Hassett, A. Kresch, and Y. Tschinkel. Effective computation of Picard groups and Brauer–Manin obstructions of degree two K3 surfaces over number fields. *Rend. Circ. Mat. Palermo (2)*, 62(1):137–151, 2013. (Cited on page 437.)

[242] B. Hassett and Y. Tschinkel. Rational points on K3 surfaces and derived equivalence. 2014. arXiv:1411.6259. (Cited on pages 126 and 379.)

[243] N. Hitchin, A. Karlhede, U. Lindström, and M. Roček. Hyper-Kähler metrics and supersymmetry. *Comm. Math. Phys.*, 108(4):535–589, 1987. (Cited on page 136.)

[244] E. Horikawa. Surjectivity of the period map of K3 surfaces of degree 2. *Math. Ann.*, 228(2):113–146, 1977. (Cited on page 116.)

[245] S. Hosono, B. Lian, K. Oguiso, and S.-T. Yau. Kummer structures on K3 surface: an old question of T. Shioda. *Duke Math. J.*, 120(3):635–647, 2003. (Cited on page 373.)

[246] S. Hosono, B. Lian, K. Oguiso, and S.-T. Yau. Autoequivalences of derived category of a K3 surface and monodromy transformations. *J. Algebraic Geom.*, 13(3):513–545, 2004. (Cited on page 374.)

[247] S. Hosono, B. H. Lian, K. Oguiso, and S.-T. Yau. Fourier–Mukai partners of a K3 surface of Picard number one. In *Vector bundles and representation theory (Columbia, MO, 2002)*, vol. 322 of *Contemp. Math.*, pages 43–55. Amer. Math. Soc., Providence, RI, 2003. (Cited on pages 218 and 384.)

[248] R. Hudson. *Kummer's quartic surface*. Cambridge Mathematical Library. Cambridge University Press, Cambridge, 1990. With a foreword by W. Barth, revised reprint of the 1905 original. (Cited on page 15.)

[249] K. Hulek and D. Ploog. Fourier–Mukai partners and polarised K3 surfaces. In *Arithmetic and geometry of K3 surfaces and Calabi–Yau threefolds*, vol. 67 of *Fields Inst. Commun.*, pages 333–365. Springer-Verlag, New York, 2013. (Cited on page 384.)

[250] D. Huybrechts. Compact hyperkähler manifolds: basic results. *Invent. Math.*, 135(1): 63–113, 1999. (Cited on pages 115, 133, 138, 217, and 294.)

[251] D. Huybrechts. Compact hyperkähler manifolds. In *Calabi–Yau manifolds and related geometries (Nordfjordeid, 2001)*, Universitext, pages 161–225. Springer-Verlag, Berlin, 2003. (Cited on pages 147 and 216.)

[252] D. Huybrechts. Moduli spaces of hyperkähler manifolds and mirror symmetry. In *Intersection theory and moduli*, ICTP Lect. Notes, XIX, pages 185–247 (electronic). Abdus Salam Int. Cent. Theoret. Phys., Trieste, 2004. (Cited on pages 140 and 314.)

[253] D. Huybrechts. *Complex geometry*. Universitext. Springer-Verlag, Berlin, 2005. (Cited on pages 39, 40, 41, 58, 171, 191, and 192.)

[254] D. Huybrechts. *Fourier–Mukai transforms in algebraic geometry*. Oxford Mathematical Monographs. The Clarendon Press, Oxford University Press, Oxford, 2006. (Cited on pages 214, 358, 361, 362, 363, 364, 367, 370, 371, and 374.)

[255] D. Huybrechts. Derived and abelian equivalence of K3 surfaces. *J. Algebraic Geom.*, 17(2):375–400, 2008. (Cited on page 372.)

[256] D. Huybrechts. The global Torelli theorem: classical, derived, twisted. In *Algebraic geometry – Seattle 2005. Part 1*, vol. 80 of *Proc. Sympos. Pure Math.*, pages 235–258. Amer. Math. Soc., Providence, RI, 2009. (Cited on pages 377 and 382.)

[257] D. Huybrechts. Chow groups of K3 surfaces and spherical objects. *J. Eur. Math. Soc. (JEMS)*, 12(6):1533–1551, 2010. (Cited on pages 269 and 270.)

[258] D. Huybrechts. A note on the Bloch–Beilinson conjecture for K3 surfaces and spherical objects. *Pure Appl. Math. Q.*, 7(4, Special Issue: In memory of Eckart Viehweg): 1395–1405, 2011. (Cited on page 271.)

[259] D. Huybrechts. A global Torelli theorem for hyperkähler manifolds [after M. Verbitsky], Séminaire Bourbaki, Exposé 1040, 2010/2011. *Astérisque*, 348:375–403, 2012. (Cited on pages 128, 132, 133, 135, and 139.)

[260] D. Huybrechts. Symplectic automorphisms of K3 surfaces of arbitrary finite order. *Math. Res. Lett.*, 19(4):947–951, 2012. (Cited on pages 271 and 357.)

[261] D. Huybrechts. On derived categories of K3 surfaces and Mathieu groups. 2013. arXiv:1309.6528. (Cited on pages 349 and 384.)

[262] D. Huybrechts. Curves and cycles on K3 surfaces. *Algebraic Geometry*, 1(1):69–106, 2014. With an appendix by C. Voisin. (Cited on pages 267 and 269.)

[263] D. Huybrechts. Introduction to stability conditions. In *Moduli spaces*, vol. 411 of *Lond. Math. Soc. Lecture Note Ser.*, pages 179–229. Cambridge University Press, 2014. (Cited on pages 375 and 384.)

[264] D. Huybrechts. The K3 category of a cubic fourfold. 2015. arXiv:1505.01775. (Cited on page 384.)

[265] D. Huybrechts and M. Kemeny. Stable maps and Chow groups. *Doc. Math.*, 18:507–517, 2012. (Cited on pages 271, 286, and 357.)

[266] D. Huybrechts and M. Lehn. *The geometry of moduli spaces of sheaves*. Cambridge Mathematical Library. Cambridge University Press, Cambridge, second edition, 2010. (Cited on pages 175, 176, 179, 180, 181, 182, 186, 199, 203, 204, 207, 208, 209, 210, 212, 213, 214, 215, 216, 218, and 243.)

[267] D. Huybrechts, E. Macrì, and P. Stellari. Stability conditions for generic K3 categories. *Compositio Math.*, 144(1):134–162, 2008. (Cited on pages 378, 381, and 382.)

[268] D. Huybrechts, E. Macrì, and P. Stellari. Derived equivalences of K3 surfaces and orientation. *Duke Math. J.*, 149(3):461–507, 2009. (Cited on pages 374 and 382.)

[269] D. Huybrechts and M. Nieper-Wißkirchen. Remarks on derived equivalences of Ricci-flat manifolds. *Math. Z.*, 267(3-4):939–963, 2011. (Cited on pages 13 and 216.)

[270] D. Huybrechts and S. Schröer. The Brauer group of analytic K3 surfaces. *Int. Math. Res. Not.*, 50:2687–2698, 2003. (Cited on pages 412 and 415.)

[271] D. Huybrechts and P. Stellari. Equivalences of twisted K3 surfaces. *Math. Ann.*, 332(4):901–936, 2005. (Cited on pages 378 and 426.)

[272] D. Huybrechts and P. Stellari. Proof of Căldăraru's conjecture. Appendix: "Moduli spaces of twisted sheaves on a projective variety" by K. Yoshioka. In *Moduli spaces and arithmetic geometry*, vol. 45 of *Adv. Stud. Pure Math.*, pages 31–42. Math. Soc. Japan, Tokyo, 2006. (Cited on page 377.)

[273] E. Ieronymou, A. Skorobogatov, and Y. Zarhin. On the Brauer group of diagonal quartic surfaces. *J. Lond. Math. Soc. (2)*, 83(3):659–672, 2011. With an appendix by P. Swinnerton-Dyer. (Cited on page 321.)

[274] J. Igusa. Betti and Picard numbers of abstract algebraic surfaces. *Proc. Nat. Acad. Sci. U.S.A.*, 46:724–726, 1960. (Cited on page 393.)

[275] H. Inose. Defining equations of singular K3 surfaces and a notion of isogeny. In *Proceedings of the International Symposium on Algebraic Geometry (Kyoto Univ., Kyoto, 1977)*, pages 495–502. Kinokuniya Book Store, Tokyo, 1978. (Cited on page 323.)

[276] H. Ito. On automorphisms of supersingular K3 surfaces. *Osaka J. Math.*, 34(3):713–724, 1997. (Cited on page 341.)

[277] B. Iversen. *Cohomology of sheaves*. Universitext. Springer-Verlag, Berlin, 1986. (Cited on page 12.)

[278] D. James. On Witt's theorem for unimodular quadratic forms. *Pacific J. Math.*, 26:303–316, 1968. (Cited on page 308.)

[279] U. Jannsen. *Mixed motives and algebraic K-theory*, vol. 1400 of *Lecture Notes in Math.* Springer-Verlag, Berlin, 1990. With appendices by S. Bloch and C. Schoen. (Cited on page 261.)

[280] J.-P. Jouanolou. *Théorèmes de Bertini et applications*, vol. 42 of *Progr. Math.* Birkhäuser, Boston, 1983. (Cited on page 27.)

[281] D. Kaledin, M. Lehn, and C. Sorger. Singular symplectic moduli spaces. *Invent. Math.*, 164(3):591–614, 2006. (Cited on pages 210 and 217.)

[282] A. Kas. Weierstrass normal forms and invariants of elliptic surfaces. *Trans. Amer. Math. Soc.*, 225:259–266, 1977. (Cited on pages 226 and 229.)

[283] T. Katsura. On Kummer surfaces in characteristic 2. In *Proc. Int. Sympos Algebraic Geom. (Kyoto Univ., 1977)*, pages 525–542. Kinokuniya Book Store, Tokyo, 1978. (Cited on page 3.)

[284] T. Katsura. Generalized Kummer surfaces and their unirationality in characteristic p. *J. Fac. Sci. University Tokyo Sect. IA Math.*, 34(1):1–41, 1987. (Cited on page 329.)

[285] N. Katz. Review of ℓ-adic cohomology. In *Motives (Seattle, WA, 1991)*, vol. 55 of *Proc. Sympos. Pure Math.*, pages 21–30. Amer. Math. Soc., Providence, RI, 1994. (Cited on page 74.)

[286] N. Katz and W. Messing. Some consequences of the Riemann hypothesis for varieties over finite fields. *Invent. Math.*, 23:73–77, 1974. (Cited on page 436.)

[287] Y. Kawamata. On the cone of divisors of Calabi–Yau fiber spaces. *Internat. J. Math.*, 8(5):665–687, 1997. (Cited on pages 160, 167, and 168.)

[288] K. Kawatani. A hyperbolic metric and stability conditions on K3 surfaces with $\rho = 1$. 2012. arXiv:1204.1128. (Cited on page 376.)

[289] S. Keel and S. Mori. Quotients by groupoids. *Ann. of Math. (2)*, 145(1):193–213, 1997. (Cited on pages 83, 84, 90, 98, and 204.)

[290] G. Kempf, F. Knudsen, D. Mumford, and B. Saint-Donat. *Toroidal embeddings. I.* Lecture Notes in Mathematics, vol. 339. Springer-Verlag, Berlin and New York, 1973. (Cited on page 125.)

[291] J. Keum. A note on elliptic K3 surfaces. *Trans. Amer. Math. Soc.*, 352(5):2077–2086, 2000. (Cited on pages 242 and 253.)

[292] J. Keum. Orders of automorphisms of K3 surfaces. 2012. arXiv:1203.5616v8. (Cited on pages 335, 337, 338, and 346.)

[293] V. Kharlamov. Topology, moduli and automorphisms of real algebraic surfaces. *Milan J. Math.*, 70:25–37, 2002. (Cited on pages x and 147.)

[294] F. Kirwan. Moduli spaces of degree d hypersurfaces in $\mathbb{P}_n$. *Duke Math. J.*, 58(1):39–78, 1989. (Cited on page 99.)

[295] S. Kleiman. The standard conjectures. In *Motives (Seattle, WA, 1991)*, vol. 55 of *Proc. Sympos. Pure Math.*, pages 3–20. Amer. Math. Soc., Providence, RI, 1994. (Cited on page 73.)

[296] A. Klemm, D. Maulik, R. Pandharipande, and E. Scheidegger. Noether–Lefschetz theory and the Yau–Zaslow conjecture. *J. Amer. Math. Soc.*, 23(4):1013–1040, 2010. (Cited on page 297.)

[297] R. Kloosterman. Classification of all Jacobian elliptic fibrations on certain K3 surfaces. *J. Math. Soc. Japan*, 58(3):665–680, 2006. (Cited on page 253.)

[298] R. Kloosterman. Elliptic K3 surfaces with geometric Mordell–Weil rank 15. *Canad. Math. Bull.*, 50(2):215–226, 2007. (Cited on pages 237 and 401.)

[299] A. Knapp. *Advanced algebra*. Cornerstones. Birkhäuser, Boston, 2007. (Cited on page 411.)

[300] M. Kneser. Klassenzahlen indefiniter quadratischer Formen in drei oder mehr Veränderlichen. *Arch. Math. (Basel)*, 7:323–332, 1956. (Cited on page 307.)

[301] M. Kneser. Erzeugung ganzzahliger orthogonaler Gruppen durch Spiegelungen. *Math. Ann.*, 255(4):453–462, 1981. (Cited on page 312.)

[302] M. Kneser. *Quadratische Formen*. Springer-Verlag, Berlin, 2002. Revised and edited in collaboration with Rudolf Scharlau. (Cited on pages 299 and 300.)

[303] A. Knutsen and A. Lopez. A sharp vanishing theorem for line bundles on K3 or Enriques surfaces. *Proc. Amer. Math. Soc.*, 135(11):3495–3498, 2007. (Cited on page 33.)

[304] S. Kobayashi. First Chern class and holomorphic tensor fields. *Nagoya Math. J.*, 77:5–11, 1980. (Cited on page 193.)

[305] S. Kobayashi. *Differential geometry of complex vector bundles*, vol. 15 of *Publications of the Math. Soc. of Japan*. Princeton University Press, Princeton, NJ, 1987. Kanô Memorial Lectures, 5. (Cited on pages 191 and 192.)

[306] S. Kobayashi. *Transformation groups in differential geometry*. Classics in Mathematics. Springer-Verlag, Berlin, 1995. Reprint of the 1972 edition. (Cited on page 330.)

[307] K. Kodaira. On compact analytic surfaces. II, III. *Ann. of Math. (2) 77 (1963), 563–626;* ibid., 78:1–40, 1963. (Cited on pages 225 and 244.)

[308] K. Kodaira. On the structure of compact complex analytic surfaces. I. *Amer. J. Math.*, 86:751–798, 1964. (Cited on pages xi, 109, and 129.)

[309] K. Kodaira. On homotopy K3 surfaces. In *Essays on topology and related topics (Mémoires dédiés à Georges de Rham)*, pages 58–69. Springer-Verlag, New York, 1970. (Cited on page 13.)

[310] K. Kodaira. *Complex manifolds and deformation of complex structures*, vol. 283 of *Grundlehren der Mathematischen Wissenschaften*. Springer-Verlag, New York, 1986. With an appendix by D. Fujiwara. (Cited on page 109.)

[311] K. Koike, H. Shiga, N. Takayama, and T. Tsutsui. Study on the family of K3 surfaces induced from the lattice $(D_4)^3 \oplus \langle -2 \rangle \oplus \langle 2 \rangle$. *Internat. J. Math.*, 12(9):1049–1085, 2001. (Cited on page 79.)

[312] J. Kollár. *Rational curves on algebraic varieties*, vol. 32 of *Ergebnisse der Mathematik und ihrer Grenzgebiete (3)*. Springer-Verlag, Berlin, 1996. (Cited on pages 92, 189, 260, and 286.)

[313] J. Kollár. Quotient spaces modulo algebraic groups. *Ann. of Math. (2)*, 145(1):33–79, 1997. (Cited on page 83.)

[314] J. Kollár. Non-quasi-projective moduli spaces. *Ann. of Math. (2)*, 164(3):1077–1096, 2006. (Cited on page 91.)

[315] J. Kollár and N. Shepherd-Barron. Threefolds and deformations of surface singularities. *Invent. Math.*, 91(2):299–338, 1988. (Cited on page 127.)

[316] J. Kollár et al. *Flips and abundance for algebraic threefolds*, vol. 211 of *Astérisque*. Soc. Math. France, Paris, 1992. (Cited on page 188.)

[317] S. Kondō. Enriques surfaces with finite automorphism groups. *Japan. J. Math. (N.S.)*, 12(2):191–282, 1986. (Cited on page 356.)

[318] S. Kondō. On algebraic K3 surfaces with finite automorphism groups. *Proc. Japan Acad. Ser. A Math. Sci.*, 62(9):353–355, 1986. (Cited on page 342.)

[319] S. Kondō. On automorphisms of algebraic K3 surfaces which act trivially on Picard groups. *Proc. Japan Acad. Ser. A Math. Sci.*, 62(9):356–359, 1986. (Cited on pages 336 and 337.)

[320] S. Kondō. Algebraic K3 surfaces with finite automorphism groups. *Nagoya Math. J.*, 116:1–15, 1989. (Cited on page 342.)

[321] S. Kondō. Quadratic forms and K3. Enriques surfaces [translation of Sûgaku **42** (1990), no. 4, 346–360; MR1083944 (92b:14018)]. *Sugaku Expositions*, 6(1):53–72, 1993. Sugaku Expositions. (Cited on pages 335, 337, and 341.)

[322] S. Kondō. Niemeier lattices, Mathieu groups, and finite groups of symplectic automorphisms of K3 surfaces. *Duke Math. J.*, 92(3):593–603, 1998. With an appendix by S. Mukai. (Cited on pages 325, 345, 347, and 350.)

[323] S. Kondō. The maximum order of finite groups of automorphisms of K3 surfaces. *Amer. J. Math.*, 121(6):1245–1252, 1999. (Cited on page 345.)

[324] S. Kondō. Maximal subgroups of the Mathieu group M_{23} and symplectic automorphisms of supersingular K3 surfaces. *Int. Math. Res. Not.*, Art. ID 71517, 9, 2006. (Cited on page 345.)

[325] S. Kondō and I. Shimada. The automorphism group of a supersingular K3 surface with Artin invariant 1 in characteristic 3. *Int. Math. Res. Not. IMRN*, (7):1885–1924, 2014. (Cited on page 345.)

[326] S. Kondō and I. Shimada. On a certain duality of Néron–Severi lattices of supersingular K3 surfaces. *Algebr. Geom.*, 1(3):311–333, 2014. (Cited on pages 401, 402, and 403.)

[327] S. Kovács. The cone of curves of a K3 surface. *Math. Ann.*, 300(4):681–691, 1994. (Cited on pages 160, 162, 163, 281, and 316.)

[328] H. Kraft, P. Slodowy, and T. Springer, editors. *Algebraische Transformationsgruppen und Invariantentheorie*, vol. 13 of *DMV Seminar*. Birkhäuser Verlag, Basel, 1989. (Cited on page 94.)

[329] A. Kresch. Hodge-theoretic obstruction to the existence of quaternion algebras. *Bull. Lond. Math. Soc.*, 35(1):109–116, 2003. (Cited on page 412.)

[330] A. Kresch. On the geometry of Deligne–Mumford stacks. In *Algebraic geometry – Seattle 2005. Part 1*, vol. 80 of *Proc. Sympos. Pure Math.*, pages 259–271. Amer. Math. Soc., Providence, RI, 2009. (Cited on page 99.)

[331] M. Kuga and I. Satake. Abelian varieties attached to polarized K3 surfaces. *Math. Ann.*, 169:239–242, 1967. (Cited on page 72.)

[332] S. Kuleshov. A theorem on the existence of exceptional bundles on surfaces of type K3. *Izv. Akad. Nauk SSSR Ser. Mat.*, 53(2):363–378, 1989. (Cited on pages 198 and 212.)

[333] S. Kuleshov. Exceptional bundles on K3 surfaces. In *Helices and vector bundles*, vol. 148 of *Lond. Math. Soc. Lecture Note Ser.*, pages 105–114. Cambridge University Press, 1990. (Cited on page 198.)

[334] S. Kuleshov. Stable bundles on a K3 surface. *Izv. Akad. Nauk SSSR Ser. Mat.*, 54(1): 213–220, 223, 1990. (Cited on page 198.)

[335] V. Kulikov. Degenerations of K3 surfaces and Enriques surfaces. *Izv. Akad. Nauk SSSR Ser. Mat.*, 41(5):1008–1042, 1199, 1977. (Cited on pages 114, 125, and 138.)

[336] V. Kulikov. Surjectivity of the period mapping for K3 surfaces. *Uspehi Mat. Nauk*, 32(4(196)):257–258, 1977. (Cited on page 127.)

[337] V. Kulikov. Surgery of degenerations of surfaces with $\kappa = 0$. *Izv. Akad. Nauk SSSR Ser. Mat.*, 44(5):1115–1119, 1214, 1980. (Cited on page 125.)

[338] V. Kulikov and P. Kurchanov. Complex algebraic varieties: periods of integrals and Hodge structures [MR1060327 (91k:14010)]. In *Algebraic geometry, III*, vol. 36 of *Encyclopaedia Math. Sci.*, pages 1–217, 263–270. Springer, Berlin, 1998. (Cited on page 126.)

[339] A. Kumar. Elliptic fibrations on a generic Jacobian Kummer surface. *J. Algebraic Geom.*, 23(4):599–667, 2014. (Cited on page 253.)

[340] H. Kurke. *Vorlesungen über algebraische Flächen*, vol. 43 of *Teubner-Texte zur Mathematik*. BSB B. G. Teubner Verlagsgesellschaft, Leipzig, 1982. (Cited on page 15.)

[341] M. Kuwata. Elliptic fibrations on quartic K3 surfaces with large Picard numbers. *Pacific J. Math.*, 171(1):231–243, 1995. (Cited on pages 253, 389, and 390.)

[342] M. Kuwata. Elliptic K3 surfaces with given Mordell–Weil rank. *Comment. Math. University St. Paul.*, 49(1):91–100, 2000. (Cited on page 237.)

[343] M. Kuwata. Equal sums of sixth powers and quadratic line complexes. *Rocky Mountain J. Math.*, 37(2):497–517, 2007. (Cited on page 390.)

[344] M. Kuwata and T. Shioda. Elliptic parameters and defining equations for elliptic fibrations on a Kummer surface. In *Algebraic geometry in East Asia–Hanoi 2005*, vol. 50 of *Adv. Stud. Pure Math.*, pages 177–215. Math. Soc. Japan, Tokyo, 2008. (Cited on page 253.)

[345] A. Kuznetsov. Derived categories of cubic fourfolds. In *Cohomological and geometric approaches to rationality problems*, vol. 282 of *Progr. Math.*, pages 219–243. Birkhäuser, Boston 2010. (Cited on page 384.)

[346] A. Lamari. Courants kählériens et surfaces compactes. *Ann. Inst. Fourier (Grenoble)*, 49(1):vii, x, 263–285, 1999. (Cited on pages 135, 137, and 171.)

[347] A. Lamari. Le cône kählérien d'une surface. *J. Math. Pures Appl. (9)*, 78(3):249–263, 1999. (Cited on page 171.)

[348] S. Lang and A. Néron. Rational points of abelian varieties over function fields. *Amer. J. Math.*, 81:95–118, 1959. (Cited on page 5.)

[349] W. Lang and N. Nygaard. A short proof of the Rudakov–Safarevič theorem. *Math. Ann.*, 251(2):171–173, 1980. (Cited on pages 189 and 194.)

[350] A. Langer and S. Saito. Torsion zero-cycles on the self-product of a modular elliptic curve. *Duke Math. J.*, 85(2):315–357, 1996. (Cited on page 270.)

[351] R. Laza. Triangulations of the sphere and degenerations of K3 surfaces. 2008. arXiv:0809.0937. (Cited on page 126.)

[352] R. Laza. The KSBA compactification for the moduli space of degree two K3 pairs. 2012. arXiv:1205.3144. (Cited on page 127.)

[353] M. Lazard. Sur les groupes de Lie formels à un paramètre. *Bull. Soc. Math. France*, 83:251–274, 1955. (Cited on page 429.)

[354] R. Lazarsfeld. Brill–Noether–Petri without degenerations. *J. Differential Geom.*, 23(3):299–307, 1986. (Cited on pages 181, 182, and 184.)

[355] R. Lazarsfeld. A sampling of vector bundle techniques in the study of linear series. In *Lectures on Riemann surfaces (Trieste, 1987)*, pages 500–559. World Sci. Publ., Teaneck, NJ, 1989. (Cited on pages 182 and 183.)

[356] R. Lazarsfeld. Lectures on linear series. In *Complex algebraic geometry (Park City, UT, 1993)*, vol. 3 of *IAS/Park City Math. Ser.*, pages 161–219. Amer. Math. Soc., Providence, RI, 1997. With the assistance of Guillermo Fernández del Busto. (Cited on pages 33, 180, and 183.)

[357] R. Lazarsfeld. *Positivity in algebraic geometry. I, II*, vols. 48, 49 of *Ergebnisse der Mathematik und ihrer Grenzgebiete (3)*. Springer-Verlag, Berlin, 2004. (Cited on pages 24, 33, 34, 149, 158, and 174.)

[358] V. Lazić. Around and beyond the canonical class. In *Birational geometry, rational curves, and arithmetic*, pages 171–203. Springer-Verlag, New York, 2013. (Cited on page 165.)

[359] M. Lehn. Symplectic moduli spaces. In *Intersection theory and moduli*, ICTP Lect. Notes, XIX, pages 139–184 (electronic). Abdus Salam Int. Cent. Theoret. Phys., Trieste, 2004. (Cited on page 216.)

[360] J. Li and C. Liedtke. Rational curves on K3 surfaces. *Invent. Math.*, 188(3):713–727, 2012. (Cited on pages 281, 286, 287, 288, 289, 290, 291, and 292.)

[361] Z. Li and Z. Tian. Picard groups of moduli of K3 surfaces of low degree K3 surfaces. 2013. arXiv:1304.3219. (Cited on pages 90 and 127.)

[362] D. Lieberman. Compactness of the Chow scheme: applications to automorphisms and deformations of Kähler manifolds. In *Fonctions de plusieurs variables complexes, III (Sém. François Norguet, 1975–1977)*, vol. 670 of *Lecture Notes in Math.*, pages 140–186. Springer-Verlag, Berlin, 1978. (Cited on page 337.)

[363] D. Lieberman and D. Mumford. Matsusaka's big theorem. In *Algebraic geometry (Proc. Sympos. Pure Math., Vol. 29, Humboldt State Univ., Arcata, Calif., 1974)*, pages 513–530. Amer. Math. Soc., Providence, RI, 1975. (Cited on page 34.)

[364] M. Lieblich. Groupoids and quotients in algebraic geometry. In *Snowbird lectures in algebraic geometry*, vol. 388 of *Contemp. Math.*, pages 119–136. Amer. Math. Soc., Providence, RI, 2005. (Cited on page 83.)

[365] M. Lieblich. Moduli of twisted sheaves. *Duke Math. J.*, 138(1):23–118, 2007. (Cited on page 383.)

[366] M. Lieblich. Twisted sheaves and the period-index problem. *Compositio Math.*, 144(1): 1–31, 2008. (Cited on page 412.)

[367] M. Lieblich. On the unirationality of supersingular K3 surfaces. 2014. arXiv:1403.3073. (Cited on page 437.)

[368] M. Lieblich. Rational curves in the moduli of supersingular K3 surfaces. 2015. arXiv:1507.08387. (Cited on page 437.)

[369] M. Lieblich and D. Maulik. A note on the cone conjecture for K3 surfaces in positive characteristic. 2011. arXiv:1102.3377v3. (Cited on pages 168, 169, 197, 339, and 340.)

[370] M. Lieblich, D. Maulik, and A. Snowden. Finiteness of K3 surfaces and the Tate conjecture. *Ann. Sci. Éc. Norm. Supér. (4)*, 47(2):285–308, 2014. (Cited on pages 408, 425, and 426.)

[371] M. Lieblich and M. Olsson. Fourier–Mukai partners of K3 surfaces in positive characteristic. 2011. arXiv:1112.5114. (Cited on pages 197, 373, 379, and 380.)

[372] C. Liedtke. Lectures on supersingular K3 surfaces and the crystalline Torelli theorem. 2014. arXiv:1403.2538. (Cited on pages 197, 433, and 437.)

[373] C. Liedtke. Supersingular K3 surfaces are unirational. *Invent. Math.*, 200(3):979–1014, 2015. (Cited on pages 264, 273, 292, and 437.)

[374] C. Liedtke and Y. Matsumoto. Good reduction of K3 surfaces. 2014. arXiv:1411.4797. (Cited on page 127.)

[375] Q. Liu. *Algebraic geometry and arithmetic curves*, vol. 6 of *Oxford Graduate Texts in Mathematics*. Oxford University Press, Oxford, 2002. (Cited on pages 1, 2, 219, 222, and 226.)

[376] Q. Liu, D. Lorenzini, and M. Raynaud. On the Brauer group of a surface. *Invent. Math.*, 159(3):673–676, 2005. (Cited on page 423.)

[377] G. Lombardo. Abelian varieties of Weil type and Kuga–Satake varieties. *Tohoku Math. J. (2)*, 53(3):453–466, 2001. (Cited on page 79.)

[378] E. Looijenga. A Torelli theorem for Kähler–Einstein K3 surfaces. In *Geometry Symposium, Utrecht 1980*, vol. 894 of *Lecture Notes in Math.*, pages 107–112. Springer Verlag, Berlin, 1981. (Cited on pages 115 and 146.)

[379] E. Looijenga. Discrete automorphism groups of convex cones of finite type. *Compositio Math.*, 150(11):1939–1962, 2014. (Cited on page 167.)

[380] E. Looijenga and C. Peters. Torelli theorems for Kähler K3 surfaces. *Compositio Math.*, 42(2):145–186, 1980/1981. (Cited on pages 307, 321, and 324.)

[381] M. Lübke and A. Teleman. *The Kobayashi–Hitchin correspondence*. World Scientific River Edge, NJ, 1995. (Cited on page 192.)

[382] S. Ma. Fourier–Mukai partners of a K3 surface and the cusps of its Kähler moduli. *Internat. J. Math.*, 20(6):727–750, 2009. (Cited on page 384.)

[383] S. Ma. Twisted Fourier–Mukai number of a K3 surface. *Trans. Amer. Math. Soc.*, 362(1):537–552, 2010. (Cited on page 377.)

[384] N. Machida and K. Oguiso. On K3 surfaces admitting finite non-symplectic group actions. *J. Math. Sci. University Tokyo*, 5(2):273–297, 1998. (Cited on pages 336 and 356.)

[385] C. Maclean. Chow groups of surfaces with $h^{2,0} \leq 1$. *C. R. Math. Acad. Sci. Paris*, 338(1):55–58, 2004. (Cited on pages 271, 296, and 297.)

[386] E. Macrì and P. Stellari. Automorphisms and autoequivalences of generic analytic K3 surfaces. *J. Geom. Phys.*, 58(1):133–164, 2008. (Cited on pages 51, 342, 357, and 382.)

[387] K. Madapusi Pera. The Tate conjecture for K3 surfaces in odd characteristic. *Invent. Math.*, 201(2):625–668, 2015. (Cited on pages 89, 99, 129, 211, and 407.)

[388] Y. Manin. Theory of commutative formal groups over fields of finite characteristic. *Uspehi Mat. Nauk*, 18(6(114)):3–90, 1963. (Cited on page 430.)

[389] Y. Manin. The Tate height of points on an Abelian variety, its variants and applications. *Izv. Akad. Nauk SSSR Ser. Mat.*, 28:1363–1390, 1964. (Cited on page 237.)

[390] E. Markman and K. Yoshioka. A proof of the Kawamata–Morrison cone conjecture for holomorphic symplectic varieties of K3$^{[n]}$ or generalized Kummer deformation type. 2014. arXiv:1402.2049. (Cited on page 174.)

[391] G. Mason. Symplectic automorphisms of K3 surfaces (after S. Mukai and V. V. Nikulin). *CWI Newslett.*, 13:3–19, 1986. (Cited on pages 344 and 345.)

[392] É. Mathieu. Mémoire sur l'étude des fonctions de plusieurs quantités, sur la manière de les former et sur les substitutions qui les laissent invariables. *J. Math. Pures et Appl.*, 6:241–323, 1961. (Cited on page 345.)

[393] K. Matsuki. *Introduction to the Mori program.* Universitext. Springer-Verlag, New York, 2002. (Cited on pages 190 and 304.)

[394] K. Matsumoto, T. Sasaki, and M. Yoshida. The monodromy of the period map of a 4 parameter family of K3 surfaces and the hypergeometric function of type $(3, 6)$. *Internat. J. Math.*, 3(1):164, 1992. (Cited on page 73.)

[395] T. Matsusaka. Polarized varieties with a given Hilbert polynomial. *Amer. J. Math.*, 94:1027–1077, 1972. (Cited on page 34.)

[396] T. Matsusaka and D. Mumford. Two fundamental theorems on deformations of polarized varieties. *Amer. J. Math.*, 86:668–684, 1964. (Cited on pages 90 and 380.)

[397] T. Matumoto. On diffeomorphisms of a K3 surface. In *Algebraic and topological theories (Kinosaki, 1984)*, pages 616–621. Kinokuniya, Tokyo, 1986. (Cited on page 142.)

[398] D. Maulik. Supersingular K3 surfaces for large primes. *Duke Math. J.*, 163(13):2357–2425, 2014. With an appendix by A. Snowden. (Cited on pages 89, 99, 127, 407, and 436.)

[399] D. Maulik and R. Pandharipande. Gromov–Witten theory and Noether–Lefschetz theory. In *A celebration of algebraic geometry*, vol. 18 of *Clay Math. Proc.*, pages 469–507. Amer. Math. Soc., Providence, RI, 2013. (Cited on pages x and 127.)

[400] D. Maulik and B. Poonen. Néron–Severi groups under specialization. *Duke Math. J.*, 161(11):2167–2206, 2012. (Cited on page 400.)

[401] A. Mayer. Families of K3 surfaces. *Nagoya Math. J.*, 48:1–17, 1972. (Cited on page 23.)

[402] B. Mazur. Frobenius and the Hodge filtration. *Bull. Amer. Math. Soc.*, 78:653–667, 1972. (Cited on page 431.)

[403] C. McMullen. Dynamics on K3 surfaces: Salem numbers and Siegel disks. *J. Reine Angew. Math.*, 545:201–233, 2002. (Cited on pages x and 357.)

[404] J. Milne. On a conjecture of Artin and Tate. *Ann. of Math. (2)*, 102(3):517–533, 1975. (Cited on page 423.)

[405] J. Milne. *Étale cohomology*, vol. 33 of *Princeton Mathematical Series*. Princeton University Press, Princeton, NJ, 1980. (Cited on pages 13, 78, 383, 410, 411, 412, and 413.)

[406] J. Milne. Zero cycles on algebraic varieties in nonzero characteristic: Rojtman's theorem. *Compositio Math.*, 47(3):271–287, 1982. (Cited on page 256.)

[407] J. Milne. Abelian varieties. In *Arithmetic geometry (Storrs, CT, 1984)*, pages 103–150. Springer-Verlag, New York, 1986. (Cited on page 3.)

[408] J. Milne. Introduction to Shimura varieties. *Course notes*, 2004. (Cited on pages 104 and 105.)

[409] J. Milne. *Elliptic curves.* BookSurge Publishers, Charleston, SC, 2006. (Cited on pages 241 and 245.)

[410] J. Milne. Class field theory. *Course notes 4.02*, 2013. (Cited on page 420.)

[411] J. Milnor and D. Husemoller. *Symmetric bilinear forms*, vol. 73 of *Ergebnisse der Mathematik und ihrer Grenzgebiete (2)*. Springer-Verlag, New York, 1973. (Cited on pages 299 and 305.)

[412] J. Milnor and J. Stasheff. *Characteristic classes*. Princeton University Press, Princeton, NJ, 1974. Annals of Mathematics Studies, No. 76. (Cited on page 12.)

[413] R. Miranda. The moduli of Weierstrass fibrations over $\mathbb{P}^1$. *Math. Ann.*, 255(3):379–394, 1981. (Cited on page 226.)

[414] R. Miranda. *The basic theory of elliptic surfaces*. Dottorato di Ricerca in Matematica. ETS Editrice, Pisa, 1989. (Cited on pages 219, 224, 226, 229, 231, 235, 236, and 238.)

[415] R. Miranda and D. Morrison. The minus one theorem. In *The birational geometry of degenerations (Cambridge, MA, 1981)*, vol. 29 of *Progr. Math.*, pages 173–259. Birkhäuser, Boston, 1983. (Cited on page 126.)

[416] R. Miranda and U. Persson. Configurations of I_n fibers on elliptic K3 surfaces. *Math. Z.*, 201(3):339–361, 1989. (Cited on pages 226 and 253.)

[417] Y. Miyaoka. Deformations of a morphism along a foliation and applications. In *Algebraic geometry, Bowdoin, 1985 (Brunswick, ME, 1985)*, vol. 46 of *Proc. Sympos. Pure Math.*, pages 245–268. Amer. Math. Soc., Providence, RI, 1987. (Cited on page 188.)

[418] Y. Miyaoka and T. Peternell. *Geometry of higher-dimensional algebraic varieties*, vol. 26 of *DMV Seminar*. Birkhäuser, Basel, 1997. (Cited on pages 188, 189, and 190.)

[419] G. Mongardi. Symplectic involutions on deformations of K3$^{[2]}$. *Cent. Eur. J. Math.*, 10(4):1472–1485, 2012. (Cited on page 357.)

[420] B. Moonen. An introduction to Mumford–Tate groups. www.math.ru.nl/personal/bmoonen/Lecturenotes/MTGps.pdf. (Cited on page 53.)

[421] B. Moonen and Y. Zarhin. Hodge classes on abelian varieties of low dimension. *Math. Ann.*, 315(4):711–733, 1999. (Cited on page 71.)

[422] S. Mori. On degrees and genera of curves on smooth quartic surfaces in $\mathbb{P}^3$. *Nagoya Math. J.*, 96:127–132, 1984. (Cited on page 34.)

[423] S. Mori and S. Mukai. The uniruledness of the moduli space of curves of genus 11. In *Algebraic geometry (Tokyo/Kyoto, 1982)*, vol. 1016 of *Lecture Notes in Math.*, pages 334–353. Springer-Verlag, Berlin, 1983. (Cited on page 277.)

[424] D. Morrison. On K3 surfaces with large Picard number. *Invent. Math.*, 75(1):105–121, 1984. (Cited on pages 72, 315, 316, 321, 353, and 354.)

[425] D. Morrison. The Kuga–Satake variety of an abelian surface. *J. Algebra*, 92(2):454–476, 1985. (Cited on pages 70 and 72.)

[426] D. Morrison. The geometry of K3 surfaces. www.math.ucsb.edu/~drm/manuscripts/cortona.pdf, 1988. Cortona Lectures. (Cited on page 33.)

[427] I. Morrison. Stability of Hilbert points of generic K3 surfaces. *Centre de Recerca Matemática Publication*, 401, 1999. (Cited on page 82.)

[428] S. Mukai. Symplectic structure of the moduli space of sheaves on an abelian or K3 surface. *Invent. Math.*, 77(1):101–116, 1984. (Cited on pages 205, 207, and 209.)

[429] S. Mukai. On the moduli space of bundles on K3 surfaces. I. In *Vector bundles on algebraic varieties (Bombay, 1984)*, vol. 11 of *Tata Inst. Fund. Res. Stud. Math.*, pages 341–413. Bombay, 1987. (Cited on pages 185, 206, 208, 211, 212, 317, 373, and 384.)

[430] S. Mukai. Finite groups of automorphisms of K3 surfaces and the Mathieu group. *Invent. Math.*, 94(1):183–221, 1988. (Cited on pages 332, 333, 344, 346, and 350.)

[431] S. Mukai. Biregular classification of Fano 3-folds and Fano manifolds of coindex 3. *Proc. Nat. Acad. Sci. U.S.A.*, 86(9):3000–3002, 1989. (Cited on page 14.)

[432] S. Mukai. Curves and Grassmannians. In *Algebraic geometry and related topics (Inchon, 1992)*, Conf. Proc. Lecture Notes Algebraic Geom., I, pages 19–40. International Press, Cambridge, MA, 1993. (Cited on page 14.)

[433] S. Mukai. New developments in the theory of Fano threefolds: vector bundle method and moduli problems [translation of Sūgaku **47** (1995), no. 2, 125–144; MR1364825 (96m:14059)]. *Sugaku Expositions*, 15(2):125–150, 2002. (Cited on page 14.)

[434] S. Mukai. *An introduction to invariants and moduli*, vol. 81 of *Cambridge Studies in Advanced Mathematics*. Cambridge University Press, 2003. (Cited on page 90.)

[435] S. Mukai. Polarized K3 surfaces of genus thirteen. In *Moduli spaces and arithmetic geometry*, vol. 45 of *Adv. Stud. Pure Math.*, pages 315–326. Math. Soc. Japan, Tokyo, 2006. (Cited on page 99.)

[436] S. Müller-Stach, E. Viehweg, and K. Zuo. Relative proportionality for subvarieties of moduli spaces of K3 and abelian surfaces. *Pure Appl. Math. Q.*, 5(3, Special Issue: In honor of F. Hirzebruch. Part 2):1161–1199, 2009. (Cited on page 99.)

[437] D. Mumford. *Lectures on curves on an algebraic surface*. Annals of Mathematics Studies, No. 59. Princeton University Press, Princeton, NJ, 1966. With a section by G. Bergman. (Cited on pages 4 and 15.)

[438] D. Mumford. Pathologies. III. *Amer. J. Math.*, 89:94–104, 1967. (Cited on page 20.)

[439] D. Mumford. Rational equivalence of 0-cycles on surfaces. *J. Math. Kyoto University*, 9:195–204, 1968. (Cited on pages 256 and 259.)

[440] D. Mumford. Enriques' classification of surfaces in char *p*. I. In *Global Analysis (Papers in Honor of K. Kodaira)*, pages 325–339. University of Tokyo Press, Tokyo, 1969. (Cited on page 221.)

[441] D. Mumford. Varieties defined by quadratic equations. In *Questions on algebraic varieties (C.I.M.E., III Ciclo, Varenna, 1969)*, pages 29–100. Edizioni Cremonese, Rome, 1970. (Cited on page 24.)

[442] D. Mumford. *Algebraic geometry. I*. Classics in Mathematics. Springer-Verlag, Berlin, 1995. Complex projective varieties, Reprint of the 1976 edition. (Cited on page 3.)

[443] D. Mumford. *Abelian varieties*, vol. 5 of *Tata Inst. Fund. Res. Stud. Math.* Published for the Tata Institute of Fundamental Research, Bombay, 2008. With appendices by C. P. Ramanujam and Y. Manin. (Cited on pages 3, 23, 58, and 414.)

[444] D. Mumford, J. Fogarty, and F. Kirwan. *Geometric invariant theory*, vol. 34 of *Ergebnisse der Mathematik und ihrer Grenzgebiete (2)*. Springer-Verlag, Berlin, third edition, 1994. (Cited on pages 85, 89, 90, 94, and 201.)

[445] J. Murre. On the motive of an algebraic surface. *J. Reine Angew. Math.*, 409:190–204, 1990. (Cited on page 265.)

[446] A. Neeman. *Algebraic and analytic geometry*, vol. 345 of *Lond. Math. Soc. Lecture Note Series*. Cambridge University Press, 2007. (Cited on page 9.)

[447] A. Néron. Modèles minimaux des variétés abéliennes sur les corps locaux et globaux. *Inst. Hautes Études Sci. Publ. Math.*, 21:128, 1964. (Cited on page 225.)

[448] H.-V. Niemeier. Definite quadratische Formen der Dimension 24 und Diskriminante 1. *J. Number Theory*, 5:142–178, 1973. (Cited on page 325.)

[449] V. Nikulin. On Kummer surfaces. *Izv. Akad. Nauk SSSR Ser. Mat.*, 39(2):278–293, 471, 1975. (Cited on pages 15, 318, 320, and 321.)

[450] V. Nikulin. Finite groups of automorphisms of Kählerian K3 surfaces. *Trudy Moskov. Mat. Obshch.*, 38:75–137, 1979. (Cited on pages 313, 325, 332, 333, 335, 336, 344, 347, 352, 353, and 387.)

[451] V. Nikulin. Integer symmetric bilinear forms and some of their geometric applications. *Izv. Akad. Nauk SSSR Ser. Mat.*, 43(1):111–177, 238, 1979. (Cited on pages 299, 300, 301, 305, 307, 309, 310, 311, and 313.)

[452] V. Nikulin. Quotient-groups of groups of automorphisms of hyperbolic forms of subgroups generated by 2-reflections. *Dokl. Akad. Nauk SSSR*, 248(6):1307–1309, 1979. (Cited on page 341.)

[453] V. Nikulin. Quotient-groups of groups of automorphisms of hyperbolic forms by subgroups generated by 2-reflections. Algebro-geometric applications. In *Current problems in mathematics, Vol. 18*, pages 3–114. Akad. Nauk SSSR, Vsesoyuz. Inst. Nauchn. i Tekhn. Informatsii, Moscow, 1981. (Cited on pages 341 and 342.)

[454] V. Nikulin. K3 surfaces with a finite group of automorphisms and a Picard group of rank three. *Trudy Mat. Inst. Steklov.*, 165:119–142, 1984. (Cited on pages 173 and 341.)

[455] V. Nikulin. On correspondences between surfaces of K3 type. *Izv. Akad. Nauk SSSR Ser. Mat.*, 51(2):402–411, 448, 1987. (Cited on page 374.)

[456] V. Nikulin. Kählerian K3 surfaces and Niemeier lattices. I. *Izv. Ross. Akad. Nauk Ser. Mat.*, 77(5):109–154, 2013. Also arXiv:1109.2879. (Cited on page 327.)

[457] V. Nikulin. Elliptic fibrations on K3 surfaces. *Proc. Edinb. Math. Soc. (2)*, 57(1):253–267, 2014. (Cited on page 341.)

[458] K. Nishiguchi. Degeneration of K3 surfaces. *J. Math. Kyoto University*, 28(2):267–300, 1988. (Cited on page 125.)

[459] K. Nishiyama. Examples of Jacobian fibrations on some K3 surfaces whose Mordell–Weil lattices have the maximal rank 18. *Comment. Math. University St. Paul.*, 44(2):219–223, 1995. (Cited on page 238.)

[460] K. Nishiyama. The Jacobian fibrations on some K3 surfaces and their Mordell–Weil groups. *Japan. J. Math. (N.S.)*, 22(2):293–347, 1996. (Cited on page 253.)

[461] K. Nishiyama. The minimal height of Jacobian fibrations on K3 surfaces. *Tohoku Math. J. (2)*, 48(4):501–517, 1996. (Cited on page 238.)

[462] N. Nygaard. A *p*-adic proof of the nonexistence of vector fields on K3 surfaces. *Ann. of Math. (2)*, 110(3):515–528, 1979. (Cited on pages 189 and 194.)

[463] N. Nygaard. The Tate conjecture for ordinary K3 surfaces over finite fields. *Invent. Math.*, 74(2):213–237, 1983. (Cited on page 407.)

[464] N. Nygaard and A. Ogus. Tate's conjecture for K3 surfaces of finite height. *Ann. of Math. (2)*, 122(3):461–507, 1985. (Cited on page 407.)

[465] K. O'Grady. On the Picard group of the moduli space for K3 surfaces. *Duke Math. J.*, 53(1):117–124, 1986. (Cited on page 127.)

[466] K. O'Grady. The weight-two Hodge structure of moduli spaces of sheaves on a K3 surface. *J. Algebraic Geom.*, 6(4):599–644, 1997. (Cited on pages 216 and 217.)

[467] K. Oguiso. On Jacobian fibrations on the Kummer surfaces of the product of nonisogenous elliptic curves. *J. Math. Soc. Japan*, 41(4):651–680, 1989. (Cited on page 253.)

[468] K. Oguiso. A note on $\mathbb{Z}/p\mathbb{Z}$-actions on K3 surfaces in odd characteristic *p*. *Math. Ann.*, 286(4):735–752, 1990. (Cited on page 335.)

[469] K. Oguiso. Families of hyperkähler manifolds. 1999. arXiv:math/9911105. (Cited on page 388.)

[470] K. Oguiso. K3 surfaces via almost-primes. *Math. Res. Lett.*, 9(1):47–63, 2002. (Cited on pages 51, 369, and 384.)

[471] K. Oguiso. Local families of K3 surfaces and applications. *J. Algebraic Geom.*, 12(3): 405–433, 2003. (Cited on pages 111, 238, and 356.)

[472] K. Oguiso. A characterization of the Fermat quartic K3 surface by means of finite symmetries. *Compositio Math.*, 141(2):404–424, 2005. (Cited on page 345.)

[473] K. Oguiso. Bimeromorphic automorphism groups of non-projective hyperkähler manifolds – a note inspired by C. T. McMullen. *J. Differential Geom.*, 78(1):163–191, 2008. (Cited on pages 340 and 357.)

[474] K. Oguiso. Free automorphisms of positive entropy on smooth Kähler surfaces. 2012. arXiv:1202.2637. (Cited on page 343.)

[475] K. Oguiso. Some aspects of explicit birational geometry inspired by complex dynamics. 2014. 1404.2982. (Cited on page 357.)

[476] K. Oguiso and T. Shioda. The Mordell–Weil lattice of a rational elliptic surface. *Comment. Math. University St. Paul.*, 40(1):83–99, 1991. (Cited on page 235.)

[477] K. Oguiso and D.-Q. Zhang. On Vorontsov's theorem on K3 surfaces with non-symplectic group actions. *Proc. Amer. Math. Soc.*, 128(6):1571–1580, 2000. (Cited on pages 336 and 337.)

[478] A. Ogus. Supersingular K3 crystals. In: *Journées de Géométrie Algébrique de Rennes (Rennes, 1978), Vol. II. Astérisque*, 64:3–86, 1979. (Cited on pages 91, 197, 338, 391, 402, 403, and 437.)

[479] A. Ogus. A crystalline Torelli theorem for supersingular K3 surfaces. In *Arithmetic and geometry, Vol. II*, vol. 36 of *Progr. Math.*, pages 361–394. Birkhäuser, Boston, 1983. (Cited on pages 152, 153, 168, 338, 403, and 437.)

[480] A. Ogus. Singularities of the height strata in the moduli of K3 surfaces. In *Moduli of abelian varieties (Texel Island, 1999)*, vol. 195 of *Progr. Math.*, pages 325–343. Birkhäuser, Basel, 2001. (Cited on page 435.)

[481] M. Olsson. Semistable degenerations and period spaces for polarized K3 surfaces. *Duke Math. J.*, 125(1):121–203, 2004. (Cited on pages 99 and 127.)

[482] N. Onishi and K. Yoshioka. Singularities on the 2-dimensional moduli spaces of stable sheaves on K3 surfaces. *Internat. J. Math.*, 14(8):837–864, 2003. (Cited on page 217.)

[483] D. Orlov. Equivalences of derived categories and K3 surfaces. *J. Math. Sci. (New York)*, 84(5):1361–1381, 1997. Algebraic geometry, 7. (Cited on page 371.)

[484] J. Ottem. Cox rings of K3 surfaces with Picard number 2. *J. Pure Appl. Algebra*, 217(4):709–715, 2013. (Cited on page 174.)

[485] R. Pandharipande. Maps, sheaves, and K3 surfaces. 2008. arXiv:0808.0253. (Cited on page 297.)

[486] K. Paranjape. Abelian varieties associated to certain K3 surfaces. *Compositio Math.*, 68(1):11–22, 1988. (Cited on page 73.)

[487] K. Paranjape and S. Ramanan. On the canonical ring of a curve. In *Algebraic geometry and commutative algebra, Vol. II*, pages 503–516. Kinokuniya, Tokyo, 1988. (Cited on page 187.)

[488] G. Pareschi. A proof of Lazarsfeld's theorem on curves on K3 surfaces. *J. Algebraic Geom.*, 4(1):195–200, 1995. (Cited on page 184.)

[489] C. Pedrini. The Chow motive of a K3 surface. *Milan J. Math.*, 77:151–170, 2009. (Cited on page 271.)

[490] U. Persson. On degenerations of algebraic surfaces. *Mem. Amer. Math. Soc.*, 11(189):xv+144, 1977. (Cited on page 125.)

[491] U. Persson. Double sextics and singular K3 surfaces. In *Algebraic geometry, Sitges (Barcelona), 1983*, vol. 1124 of *Lecture Notes in Math.*, pages 262–328. Springer Verlag, Berlin, 1985. (Cited on pages 234, 329, and 390.)

[492] U. Persson and H. Pinkham. Degeneration of surfaces with trivial canonical bundle. *Ann. of Math. (2)*, 113(1):45–66, 1981. (Cited on pages 114 and 125.)

[493] I. Pjateckiĭ-Šapiro and I. Šafarevič. Torelli's theorem for algebraic surfaces of type K3. *Izv. Akad. Nauk SSSR Ser. Mat.*, 35:530–572, 1971. (Cited on pages 44, 48, 82, 109, 116, 117, 145, 169, 170, 307, 318, 319, 324, 337, 340, 341, and 343.)

[494] I. Pjateckiĭ-Šapiro and I. Šafarevič. The arithmetic of surfaces of type K3. In *Proc. Internat. Conference on Number Theory (Moscow, 1971)*, vol. 132, pages 44–54, 1973. English translation in: Proc. of the Steklov Institute of Mathematics, No. 132 (1973), pages 45–57. Amer. Math. Soc., Providence, RI, 1975. (Cited on pages 52, 54, 69, 74, 117, 118, 121, and 143.)

[495] B. Poonen. Rational points on varieties. *Course notes*, 2013. (Cited on page 410.)

[496] H. Popp. On moduli of algebraic varieties. I. *Invent. Math.*, 22:1–40, 1973/1974. (Cited on page 83.)

[497] H. Popp. On moduli of algebraic varieties. II. *Compositio Math.*, 28:51–81, 1974. (Cited on page 83.)

[498] H. Popp. On moduli of algebraic varieties. III. Fine moduli spaces. *Compositio Math.*, 31(3):237–258, 1975. (Cited on page 83.)

[499] A. Prendergast-Smith. The cone conjecture for abelian varieties. *J. Math. Sci. University Tokyo*, 19(2):243–261, 2012. (Cited on page 167.)

[500] C. Procesi. *Lie groups*. Universitext. Springer-Verlag, New York, 2007. (Cited on pages 59, 61, and 142.)

[501] S. Rams and M. Schütt. 64 lines on smooth quartic surfaces. *Math. Ann.*, 362(1–2): 679–698, 2015. (Cited on page 297.)

[502] S. Rams and T. Szemberg. Simultaneous generation of jets on K3 surfaces. *Arch. Math. (Basel)*, 83(4):353–359, 2004. (Cited on page 33.)

[503] Z. Ran. Hodge theory and deformations of maps. *Compositio Math.*, 97(3):309–328, 1995. (Cited on page 286.)

[504] M. Rapoport, N. Schappacher, and P. Schneider, editors. *Beilinson's conjectures on special values of L-functions*, vol. 4 of *Perspectives in Mathematics*. Academic Press, Boston, 1988. (Cited on page 261.)

[505] W. Raskind. Torsion algebraic cycles on varieties over local fields. In *Algebraic K-theory: connections with geometry and topology (Lake Louise, AB, 1987)*, vol. 279 of *NATO Adv. Sci. Inst. Ser. C Math. Phys. Sci.*, pages 343–388. Kluwer Academic, Dordrecht, 1989. (Cited on page 270.)

[506] M. Raynaud. Contre-exemple au "vanishing theorem" en caractéristique $p > 0$. In *C. P. Ramanujam – a tribute*, vol. 8 of *Tata Inst. Fund. Res. Stud. Math.*, pages 273–278. Springer-Verlag, Berlin, 1978. (Cited on page 20.)

[507] M. Raynaud. Faisceaux amples et très amples [d'après T. Matsusaka], Séminaire Bourbaki, Exposé 493, 1976/1977, vol. 677 of *Lecture Notes in Math.*, pages 46–58. Springer Verlag, Berlin, 1978. (Cited on page 34.)

[508] M. Raynaud. "p-torsion" du schéma de Picard. In *Journées de Géométrie Algébrique de Rennes (Rennes, 1978), Vol. II*, vol. 64 of *Astérisque*, pages 87–148. Soc. Math. France, Paris, 1979. (Cited on page 397.)

[509] M. Reid. Chapters on algebraic surfaces. In *Complex algebraic geometry (Park City, UT, 1993)*, vol. 3 of *IAS/Park City Math. Ser.*, pages 3–159. Amer. Math. Soc., Providence, RI, 1997. (Cited on pages 15, 18, 21, 28, and 33.)

[510] J. Rizov. Moduli stacks of polarized K3 surfaces in mixed characteristic. *Serdica Math. J.*, 32(2-3):131–178, 2006. (Cited on pages 84, 91, 92, 98, 333, and 338.)

[511] T. Rockafellar. *Convex analysis*. Princeton Mathematical Series, No. 28. Princeton University Press, Princeton, NJ, 1970. (Cited on page 158.)

[512] A. Rojtman. The torsion of the group of 0-cycles modulo rational equivalence. *Ann. of Math. (2)*, 111(3):553–569, 1980. (Cited on page 256.)

[513] A. Rudakov and I. Šafarevič. Inseparable morphisms of algebraic surfaces. *Izv. Akad. Nauk SSSR Ser. Mat.*, 40(6):1269–1307, 1439, 1976. (Cited on pages 189, 193, and 194.)

[514] A. Rudakov and I. Šafarevič. Surfaces of type K3 over fields of finite characteristic. In *Current problems in mathematics, Vol. 18*, pages 115–207. Akad. Nauk SSSR, Vsesoyuz. Inst. Nauchn. i Tekhn. Informatsii, Moscow, 1981. (Cited on pages 193, 391, 401, 402, and 403.)

[515] A. Rudakov, T. Zink, and I. Šafarevič. The effect of height on degenerations of algebraic K3 surfaces. *Izv. Akad. Nauk SSSR Ser. Mat.*, 46(1):117–134, 192, 1982. (Cited on pages 407 and 436.)

[516] I. Šafarevič. On the arithmetic of singular K3 surfaces. In *Algebra and analysis (Kazan, 1994)*, pages 103–108. De Gruyter, Berlin, 1996. (Cited on page 409.)

[517] I. Šafarevič et al. *Algebraic surfaces*. By the members of the seminar of I. R. Šafarevič. Translation edited, with supplementary material, by K. Kodaira and D. C. Spencer. Proceedings of the Steklov Institute of Mathematics, No. 75 (1965). Amer. Math. Soc., Providence, RI, 1965. (Cited on pages 15, 26, 109, and 251.)

[518] B. Saint-Donat. Projective models of K3 surfaces. *Amer. J. Math.*, 96:602–639, 1974. (Cited on pages 23, 24, 26, 27, 30, and 33.)

[519] S. Saito and K. Sato. A finiteness theorem for zero-cycles over *p*-adic fields. *Ann. of Math. (2)*, 172(3):1593–1639, 2010. With an appendix by U. Jannsen. (Cited on page 270.)

[520] D. Salamon. Uniqueness of symplectic structures. *Acta Math. Vietnam.*, 38(1):123–144, 2013. (Cited on page 174.)

[521] P. Salberger. Torsion cycles of codimension 2 and ℓ-adic realizations of motivic cohomology. In *Séminaire de Théorie des Nombres, Paris, 1991–92*, vol. 116 of *Progr. Math.*, pages 247–277. Birkhäuser, Boston, 1993. (Cited on page 270.)

[522] A. Sarti. Group actions, cyclic coverings and families of K3 surfaces. *Canad. Math. Bull.*, 49(4):592–608, 2006. (Cited on page 328.)

[523] A. Sarti. Transcendental lattices of some K3 surfaces. *Math. Nachr.*, 281(7):1031–1046, 2008. (Cited on page 328.)

[524] I. Satake. Clifford algebras and families of abelian varieties. *Nagoya Math. J.*, 27:435–446, 1966. (Cited on page 65.)

[525] I. Satake. *Algebraic structures of symmetric domains*, vol. 4 of *Kanô Memorial Lectures*. Iwanami Shoten, Tokyo, 1980. (Cited on pages 102 and 104.)

[526] F. Scattone. On the compactification of moduli spaces for algebraic K3 surfaces. *Mem. Amer. Math. Soc.*, 70(374):x+86, 1987. (Cited on pages 99 and 127.)

[527] M. Schlessinger. Functors of Artin rings. *Trans. Amer. Math. Soc.*, 130:208–222, 1968. (Cited on pages 194, 195, 416, and 418.)

[528] U. Schlickewei. The Hodge conjecture for self-products of certain K3 surfaces. *J. Algebra*, 324(3):507–529, 2010. (Cited on page 79.)

[529] C. Schnell. Two lectures about Mumford–Tate groups. *Rend. Semin. Mat. University Politec. Torino*, 69(2):199–216, 2011. (Cited on page 53.)

[530] C. Schoen. Zero cycles modulo rational equivalence for some varieties over fields of transcendence degree one. In *Algebraic geometry, Bowdoin, 1985 (Brunswick, ME, 1985)*,

vol. 46 of *Proc. Sympos. Pure Math.*, pages 463–473. Amer. Math. Soc., Providence, RI, 1987. (Cited on pages 266 and 267.)

[531] C. Schoen. Varieties dominated by product varieties. *Internat. J. Math.*, 7(4):541–571, 1996. (Cited on page 80.)

[532] S. Schröer. Kummer surfaces for the self-product of the cuspidal rational curve. *J. Algebraic Geom.*, 16(2):305–346, 2007. (Cited on page 3.)

[533] S. Schröer. On genus change in algebraic curves over imperfect fields. *Proc. Amer. Math. Soc.*, 137(4):1239–1243, 2009. (Cited on pages 29 and 221.)

[534] H.-W. Schuster. Locally free resolutions of coherent sheaves on surfaces. *J. Reine Angew. Math.*, 337:159–165, 1982. (Cited on page 184.)

[535] M. Schütt. The maximal singular fibres of elliptic K3 surfaces. *Arch. Math. (Basel)*, 87(4):309–319, 2006. (Cited on page 234.)

[536] M. Schütt. Elliptic fibrations of some extremal K3 surfaces. *Rocky Mountain J. Math.*, 37(2):609–652, 2007. (Cited on page 234.)

[537] M. Schütt. Fields of definition of singular K3 surfaces. *Commun. Number Theory Phys.*, 1(2):307–321, 2007. (Cited on pages 316, 323, and 398.)

[538] M. Schütt. A note on the supersingular K3 surface of Artin invariant 1. *J. Pure Appl. Algebra*, 216(6):1438–1441, 2012. (Cited on page 403.)

[539] M. Schütt. K3 surfaces with an automorphism of order 11. *Tohoku Math. J. (2)*, 65(4): 515–522, 2013. (Cited on page 335.)

[540] M. Schütt and T. Shioda. Elliptic surfaces. In *Algebraic geometry in East Asia – Seoul 2008*, vol. 60 of *Adv. Stud. Pure Math.*, pages 51–160. Math. Soc. Japan, Tokyo, 2010.

[541] M. Schütt, T. Shioda, and R. van Luijk. Lines on Fermat surfaces. *J. Number Theory*, 130(9):1939–1963, 2010. (Cited on pages 48 and 389.)

[542] B. Segre. The maximum number of lines lying on a quartic surface. *Quart. J. Math., Oxford Ser.*, 14:86–96, 1943. (Cited on page 297.)

[543] P. Seidel. Lectures on four-dimensional Dehn twists. In *Symplectic 4-manifolds and algebraic surfaces*, vol. 1938 of *Lecture Notes in Math.*, pages 231–267. Springer-Verlag, Berlin, 2008. (Cited on page 147.)

[544] E. Sernesi. *Deformations of algebraic schemes*, vol. 334 of *Grundlehren der Mathematischen Wissenschaften*. Springer-Verlag, Berlin, 2006. (Cited on pages 92 and 110.)

[545] J.P. Serre. Applications algébriques de la cohomologie des groupes. II: théorie des algèbres simples. *Séminaire Henri Cartan*, 3:1–11, 1950–1951. (Cited on page 412.)

[546] J.-P. Serre. Géométrie algébrique et géométrie analytique. *Ann. Inst. Fourier, Grenoble*, 6:1–42, 1955–1956. (Cited on page 9.)

[547] J.-P. Serre. *A course in arithmetic*, vol. 7 of *Graduate Texts in Mathematics*. Springer Verlag, New York, 1973. (Cited on pages 12, 164, 220, 293, 299, and 305.)

[548] J.-P. Serre. Représentations *l*-adiques. In *Algebraic number theory (Kyoto Internat. Sympos., Res. Inst. Math. Sci., Univ. Kyoto 1976)*, pages 177–193. Japan Soc. Promotion Sci., Tokyo, 1977. (Cited on page 54.)

[549] J.-P. Serre. *Local fields*, vol. 67 of *Graduate Texts in Mathematics*. Springer-Verlag, New York and Berlin, 1979. (Cited on pages 393, 413, 420, and 422.)

[550] J.-P. Serre and J. Tate. Good reduction of abelian varieties. *Ann. of Math. (2)*, 88:492–517, 1968. (Cited on page 78.)

[551] J. Shah. A complete moduli space for K3 surfaces of degree 2. *Ann. of Math. (2)*, 112(3):485–510, 1980. (Cited on pages 116, 127, and 146.)

[552] J. Shah. Degenerations of K3 surfaces of degree 4. *Trans. Amer. Math. Soc.*, 263(2): 271–308, 1981. (Cited on page 127.)

[553] N. Shepherd-Barron. Extending polarizations on families of K3 surfaces. In *The birational geometry of degenerations (Cambridge, MA, 1981)*, vol. 29 of *Progr. Math.*, pages 135–171. Birkhäuser, Boston, 1983. (Cited on page 127.)

[554] I. Shimada. Transcendental lattices and supersingular reduction lattices of a singular K3 surface. *Trans. Amer. Math. Soc.*, 361(2):909–949, 2009. (Cited on pages 316, 323, and 398.)

[555] I. Shimada. An algorithm to compute automorphism groups of K3 surfaces. 2013. arXiv:1304.7427. (Cited on pages 326, 327, 329, and 343.)

[556] I. Shimada. Holes of the Leech lattice and the projective models of K3 surfaces. 2015. arXiv:1502.02099. (Cited on pages 174 and 329.)

[557] I. Shimada and D.-Q. Zhang. Classification of extremal elliptic K3 surfaces and fundamental groups of open K3 surfaces. *Nagoya Math. J.*, 161:23–54, 2001. (Cited on page 234.)

[558] I. Shimada and D.-Q. Zhang. On Kummer type construction of supersingular K3 surfaces in characteristic 2. *Pacific J. Math.*, 232(2):379–400, 2007. (Cited on page 3.)

[559] T. Shioda. On elliptic modular surfaces. *J. Math. Soc. Japan*, 24:20–59, 1972. (Cited on page 237.)

[560] T. Shioda. An example of unirational surfaces in characteristic *p*. *Math. Ann.*, 211:233–236, 1974. (Cited on pages 273, 394, and 395.)

[561] T. Shioda. Kummer surfaces in characteristic 2. *Proc. Japan Acad.*, 50:718–722, 1974. (Cited on page 3.)

[562] T. Shioda. Some results on unirationality of algebraic surfaces. *Math. Ann.*, 230(2): 153–168, 1977. (Cited on page 437.)

[563] T. Shioda. The period map of Abelian surfaces. *J. Fac. Sci. University Tokyo Sect. IA Math.*, 25(1):47–59, 1978. (Cited on pages 46 and 115.)

[564] T. Shioda. Supersingular K3 surfaces. In *Algebraic geometry (Proc. Summer Meeting, Univ. Copenhagen 1978)*, vol. 732 of *Lecture Notes in Math.*, pages 564–591. Springer Verlag, Berlin, 1979. (Cited on pages 49, 429, and 435.)

[565] T. Shioda. On the Mordell–Weil lattices. *Comment. Math. University St. Paul.*, 39(2): 211–240, 1990. (Cited on pages 232 and 236.)

[566] T. Shioda. Theory of Mordell–Weil lattices. In *Proceedings of the International Congress of Mathematicians, Vols. I, II (Kyoto, 1990)*, pages 473–489. Math. Soc. Japan, Tokyo, 1991. (Cited on page 232.)

[567] T. Shioda. The elliptic K3 surfaces with a maximal singular fibre. *C. R. Math. Acad. Sci. Paris*, 337(7):461–466, 2003. (Cited on page 234.)

[568] T. Shioda and H. Inose. On singular K3 surfaces. In *Complex analysis and algebraic geometry*, pages 119–136. Iwanami Shoten, Tokyo, 1977. (Cited on pages 72, 220, 225, 237, 322, 323, 341, 343, 344, and 398.)

[569] J. Silverman. *The arithmetic of elliptic curves*, vol. 106 of *Graduate Texts in Mathematics*. Springer-Verlag, Dordrecht, second edition, 2009. (Cited on pages 227, 241, 245, and 429.)

[570] Y.-T. Siu. A simple proof of the surjectivity of the period map of K3 surfaces. *Manuscripta Math.*, 35(3):311–321, 1981. (Cited on page 138.)

[571] Y.-T. Siu. Every K3 surface is Kähler. *Invent. Math.*, 73(1):139–150, 1983. (Cited on page 135.)

[572] A. Skorobogatov. The Kuga–Satake variety of a Kummer surface. *Uspekhi Mat. Nauk*, 40(1(241)):219–220, 1985. (Cited on page 72.)

[573] A. Skorobogatov and Y. Zarhin. A finiteness theorem for the Brauer group of abelian varieties and K3 surfaces. *J. Algebraic Geom.*, 17(3):481–502, 2008. (Cited on page 427.)

[574] A. Skorobogatov and Y. Zarhin. A finiteness theorem for the Brauer group of K3 surfaces in odd characteristic. 2014. arXiv:1403.0849. (Cited on page 427.)

[575] P. Stellari. Some remarks about the FM-partners of K3 surfaces with Picard numbers 1 and 2. *Geom. Dedicata*, 108:1–13, 2004. (Cited on pages 218, 369, and 384.)

[576] P. Stellari. Derived categories and Kummer varieties. *Math. Z.*, 256(2):425–441, 2007. (Cited on page 373.)

[577] H. Sterk. Finiteness results for algebraic K3 surfaces. *Math. Z.*, 189(4):507–513, 1985. (Cited on pages 156, 165, 167, 168, 169, and 339.)

[578] H. Sterk. Lattices and K3 surfaces of degree 6. *Linear Algebra Appl.*, 226/228:297–309, 1995. (Cited on page 127.)

[579] J. Stienstra. Cartier–Dieudonné theory for Chow groups. *J. Reine Angew. Math.*, 355:1–66, 1985. (Cited on page 437.)

[580] J. Stienstra and F. Beukers. On the Picard–Fuchs equation and the formal Brauer group of certain elliptic K3 surfaces. *Math. Ann.*, 271(2):269–304, 1985. (Cited on page 229.)

[581] B. Szendrői. Diffeomorphisms and families of Fourier–Mukai transforms in mirror symmetry. In *Applications of algebraic geometry to coding theory, physics and computation (Eilat, 2001)*, vol. 36 of *NATO Sci. Ser. II Math. Phys. Chem.*, pages 317–337. 2001. (Cited on page 145.)

[582] L. Taelman. K3 surfaces over finite fields with given L-function. 2015. arXiv:1507.08547. (Cited on pages 52 and 79.)

[583] S. Tankeev. Algebraic cycles on surfaces and abelian varieties. *Izv. Akad. Nauk SSSR Ser. Mat.*, 45(2):398–434, 463–464, 1981. (Cited on page 408.)

[584] S. Tankeev. Surfaces of type K3 over number fields, and ℓ-adic representations. *Izv. Akad. Nauk SSSR Ser. Mat.*, 52(6):1252–1271, 1328, 1988. (Cited on page 404.)

[585] S. Tankeev. Surfaces of K3 type over number fields and the Mumford–Tate conjecture. *Izv. Akad. Nauk SSSR Ser. Mat.*, 54(4):846–861, 1990. (Cited on page 54.)

[586] S. Tankeev. Surfaces of K3 type over number fields and the Mumford–Tate conjecture. II. *Izv. Ross. Akad. Nauk Ser. Mat.*, 59(3):179–206, 1995. (Cited on page 54.)

[587] A. Tannenbaum. Families of curves with nodes on K3 surfaces. *Math. Ann.*, 260(2):239–253, 1982. (Cited on page 26.)

[588] A. Tannenbaum. A note on linear systems on K3 surfaces. *Proc. Amer. Math. Soc.*, 86(1):6–8, 1982. (Cited on page 33.)

[589] A. Taormina and K. Wendland. The overarching finite symmetry group of Kummer surfaces in the Mathieu group M_{24}. *J. High Energy Phys.*, 8:125, front matter+62, 2013. (Cited on page 327.)

[590] J. Tate. Genus change in inseparable extensions of function fields. *Proc. Amer. Math. Soc.*, 3:400–406, 1952. (Cited on pages 29 and 221.)

[591] J. Tate. Algebraic cohomology classes. In *Summer Institute on Algebraic Geometry Woods Hole 1964*. Amer. Math. Soc., 1964. (Cited on pages 272, 394, 395, 403, and 406.)

[592] J. Tate. Algebraic cycles and poles of zeta functions. In *Arithmetical Algebraic Geometry (Proc. Conf. Purdue Univ., 1963)*, pages 93–110. Harper & Row, New York, 1965. (Cited on pages 272, 394, and 395.)

[593] J. Tate. Algorithm for determining the type of a singular fiber in an elliptic pencil. In *Modular functions of one variable, IV (Proc. Internat. Summer School, Univ. Antwerp, Antwerp, 1972)*, vol. 476 of *Lecture Notes in Math.*, pages 33–52. Springer-Verlag, Berlin, 1975. (Cited on page 225.)

[594] J. Tate. Variation of the canonical height of a point depending on a parameter. *Amer. J. Math.*, 105(1):287–294, 1983. (Cited on page 237.)

[595] J. Tate. Conjectures on algebraic cycles in ℓ-adic cohomology. In *Motives (Seattle, WA, 1991)*, vol. 55 of *Proc. Sympos. Pure Math.*, pages 71–83. Amer. Math. Soc., Providence, RI, 1994. (Cited on pages 403, 404, 406, 422, and 424.)

[596] J. Tate. On the conjectures of Birch and Swinnerton-Dyer and a geometric analog, Séminaire Bourbaki, Exposé 306, 1964/1966. pages 415–440. Soc. Math. France, Paris, 1995. (Cited on pages 250, 404, 406, 421, and 423.)

[597] H. Terakawa. The d-very ampleness on a projective surface in positive characteristic. *Pacific J. Math.*, 187(1):187–199, 1999. (Cited on page 24.)

[598] T. Terasoma. Complete intersections with middle Picard number 1 defined over $\mathbb{Q}$. *Math. Z.*, 189(2):289–296, 1985. (Cited on pages 398 and 399.)

[599] A. Thompson. Degenerations of K3 surfaces of degree two. *Trans. Amer. Math. Soc.*, 366(1):219–243, 2014. (Cited on page 127.)

[600] A. Todorov. The period mapping that is surjective for K3 surfaces representable as a double plane. *Mat. Zametki*, 26(3):465–474, 494, 1979. (Cited on page 116.)

[601] A. Todorov. Applications of the Kähler–Einstein–Calabi–Yau metric to moduli of K3 surfaces. *Invent. Math.*, 61(3):251–265, 1980. (Cited on pages 114, 135, and 138.)

[602] J. Top and F. De Zeeuw. Explicit elliptic K3 surfaces with rank 15. *Rocky Mountain J. Math.*, 39(5):1689–1697, 2009. (Cited on page 237.)

[603] B. Totaro. The cone conjecture for Calabi–Yau pairs in dimension 2. *Duke Math. J.*, 154(2):241–263, 2010. (Cited on pages 165, 167, and 174.)

[604] B. Totaro. Algebraic surfaces and hyperbolic geometry. In *Current developments in algebraic geometry*, vol. 59 of *Math. Sci. Res. Inst. Publ.*, pages 405–426. Cambridge University Press, 2012. (Cited on pages 165, 166, and 357.)

[605] D. Ulmer. Elliptic curves over function fields. In *Arithmetic of L-functions*, vol. 18 of *IAS/Park City Math. Ser.*, pages 211–280. Amer. Math. Soc., Providence, RI, 2011. (Cited on pages 404, 406, and 424.)

[606] D. Ulmer. CRM lectures on curves and Jacobians over function fields. 2012. arXiv:1203.5573. (Cited on page 247.)

[607] G. van der Geer and T. Katsura. On a stratification of the moduli of K3 surfaces. *J. Eur. Math. Soc. (JEMS)*, 2(3):259–290, 2000. (Cited on pages 391, 433, and 437.)

[608] B. van Geemen. Some remarks on Brauer groups of K3 surfaces. *Adv. Math.*, 197(1): 222–247, 2005. (Cited on page 415.)

[609] R. van Luijk. K3 surfaces with Picard number one and infinitely many rational points. *Algebra Number Theory*, 1(1):1–15, 2007. (Cited on pages 400, 401, and 408.)

[610] B. Venkov. On the classification of integral even unimodular 24-dimensional quadratic forms. *Trudy Mat. Inst. Steklov.*, 148:65–76, 273, 1978. Algebra, number theory and their applications. (Cited on page 325.)

[611] M. Verbitsky. Coherent sheaves on general K3 surfaces and tori. *Pure Appl. Math. Q.*, 4(3, part 2):651–714, 2008. (Cited on page 382.)

[612] M. Verbitsky. Mapping class group and a global Torelli theorem for hyperkähler manifolds. *Duke Math. J.*, 162(15):2929–2986, 2013. Appendix A by E. Markman. (Cited on pages 128 and 139.)

[613] J.-L. Verdier. *Des catégories dérivées des catégories abéliennes*, vol. 239 of *Astérisque*. Soc. Math. France, Paris, 1996. With a preface by L. Illusie, edited and with a note by G. Maltsiniotis. (Cited on pages 258 and 358.)

[614] H. Verrill and N. Yui. Thompson series, and the mirror maps of pencils of K3 surfaces. In *The arithmetic and geometry of algebraic cycles (Banff, AB, 1998)*, vol. 24 of *CRM Proc. Lecture Notes*, pages 399–432. Amer. Math. Soc., Providence, RI, 2000. (Cited on page 147.)

[615] E. Viehweg. Weak positivity and the stability of certain Hilbert points. III. *Invent. Math.*, 101(3):521–543, 1990. (Cited on pages 82 and 89.)

[616] E. Viehweg. *Quasi-projective moduli for polarized manifolds*, vol. 30 of *Ergebnisse der Mathematik und ihrer Grenzgebiete (3)*. Springer-Verlag, Berlin, 1995. (Cited on pages 83, 85, and 91.)

[617] È. Vinberg. Discrete linear groups that are generated by reflections. *Izv. Akad. Nauk SSSR Ser. Mat.*, 35:1072–1112, 1971. (Cited on page 152.)

[618] È. Vinberg. The two most algebraic K3 surfaces. *Math. Ann.*, 265(1):1–21, 1983. (Cited on pages 323 and 341.)

[619] È. Vinberg. Classification of 2-reflective hyperbolic lattices of rank 4. *Tr. Mosk. Mat. Obs.*, 68:44–76, 2007. (Cited on page 341.)

[620] C. Voisin. Remarks on zero-cycles of self-products of varieties. In *Moduli of vector bundles (Sanda, 1994; Kyoto, 1994)*, vol. 179 of *Lecture Notes in Pure and Appl. Math.*, pages 265–285. Dekker, New York, 1996. (Cited on page 271.)

[621] C. Voisin. *Théorie de Hodge et géométrie algébrique complexe*, vol. 10 of *Cours Spécialisés*. Soc. Math. France, Paris, 2002. (Cited on pages 58, 102, 108, 111, 122, 143, 171, 255, 256, 259, 260, and 265.)

[622] C. Voisin. A generalization of the Kuga–Satake construction. *Pure Appl. Math. Q.*, 1(3, part 2):415–439, 2005. (Cited on page 79.)

[623] C. Voisin. Géométrie des espaces de modules de courbes et de surfaces K3 [d'après Gritsenko–Hulek–Sankaran, Farkas–Popa, Mukai, Verra, et al.], Séminaire Bourbaki, Exposé 981, 2006/2007. *Astérisque*, 317:467–490, 2008. (Cited on page 99.)

[624] C. Voisin. On the Chow ring of certain algebraic hyper-Kähler manifolds. *Pure Appl. Math. Q.*, 4(3, part 2):613–649, 2008. (Cited on page 271.)

[625] C. Voisin. Symplectic involutions of K3 surfaces act trivially on CH_0. *Doc. Math.*, 17:851–860, 2012. (Cited on pages 271 and 357.)

[626] C. Voisin. Rational equivalence of 0-cycles on K3 surfaces and conjectures of Huybrechts and O'Grady. In *Recent advances in algebraic geometry*, vol. 417 of *London Math. Soc. Lecture Note Ser.*, pages 422–436. Cambridge University Press, Cambridge, 2015. (Cited on pages 270 and 297.)

[627] S. Vorontsov. Automorphisms of even lattices arising in connection with automorphisms of algebraic K3 surfaces. *Vestnik Moskov. University Ser. I Mat. Mekh.*, 2:19–21, 1983. (Cited on page 336.)

[628] C. T. C. Wall. On the orthogonal groups of unimodular quadratic forms. *Math. Ann.*, 147:328–338, 1962. (Cited on page 308.)

[629] C. T. C. Wall. Quadratic forms on finite groups, and related topics. *Topology*, 2:281–298, 1963. (Cited on pages 300 and 307.)

[630] C. T. C. Wall. On the orthogonal groups of unimodular quadratic forms. II. *J. Reine Angew. Math.*, 213:122–136, 1963/1964. (Cited on pages 312 and 314.)

[631] C. T. C. Wall. Quadratic forms on finite groups. II. *Bull. Lond. Math. Soc.*, 4:156–160, 1972. (Cited on page 300.)

[632] J. Wehler. K3 surfaces with Picard number 2. *Arch. Math. (Basel)*, 50(1):73–82, 1988. (Cited on page 343.)

[633] A. Weil. *Variétés abéliennes et courbes algébriques.* Publ. Inst. Math. University Strasbourg, VIII. Actualités Sci. Ind., No. 1064. Hermann & Cie., Paris, 1948. (Cited on page 78.)

[634] A. Weil. *Scientific works. Collected papers. Vol. II (1951–1964).* Springer-Verlag, New York, 1979. (Cited on pages xi, 109, and 144.)

[635] K. Wendland. Consistency of orbifold conformal field theories on K3. *Adv. Theor. Math. Phys.*, 5(3):429–456, 2001. (Cited on page 329.)

[636] U. Whitcher. Symplectic automorphisms and the Picard group of a K3 surface. *Comm. Algebra*, 39(4):1427–1440, 2011. (Cited on page 352.)

[637] A. Wiles. The Birch and Swinnerton-Dyer conjecture. In *The millennium prize problems*, pages 31–41. Clay Math. Inst., Cambridge, MA, 2006. (Cited on pages 250 and 423.)

[638] J. Wolf. *Spaces of constant curvature.* Publish or Perish, Houston, TX, 1984. (Cited on pages 104 and 152.)

[639] G. Xiao. Non-symplectic involutions of a K3 surface. 1995. arXiv:alg-geom/9512007. (Cited on page 336.)

[640] G. Xiao. Galois covers between K3 surfaces. *Ann. Inst. Fourier (Grenoble)*, 46(1):73–88, 1996. (Cited on pages 344, 345, 351, and 352.)

[641] S. Yanagida and K. Yoshioka. Bridgeland's stabilities on abelian surfaces. *Math. Z.*, 276(1–2):571–610, 2014. (Cited on page 218.)

[642] S.-T. Yau and E. Zaslow. BPS states, string duality, and nodal curves on K3. *Nuclear Phys. B*, 471(3):503–512, 1996. (Cited on page 293.)

[643] H. Yoshihara. Structure of complex tori with the automorphisms of maximal degree. *Tsukuba J. Math.*, 4(2):303–311, 1980. (Cited on page 335.)

[644] K. Yoshioka. Some examples of Mukai's reflections on K3 surfaces. *J. Reine Angew. Math.*, 515:97–123, 1999. (Cited on page 217.)

[645] K. Yoshioka. Moduli spaces of stable sheaves on abelian surfaces. *Math. Ann.*, 321(4): 817–884, 2001. (Cited on pages 198, 210, 211, 216, and 217.)

[646] K. Yoshioka. Moduli spaces of twisted sheaves on a projective variety. In *Moduli spaces and arithmetic geometry*, vol. 45 of *Adv. Stud. Pure Math.*, pages 1–30. Math. Soc. Japan, Tokyo, 2006. (Cited on page 383.)

[647] K. Yoshioka. Stability and the Fourier–Mukai transform. II. *Compositio Math.*, 145(1):112–142, 2009. (Cited on pages 211 and 217.)

[648] Y. Zarhin. Hodge groups of K3 surfaces. *J. Reine Angew. Math.*, 341:193–220, 1983. (Cited on pages 52, 53, and 54.)

[649] Y. Zarhin. Transcendental cycles on ordinary K3 surfaces over finite fields. *Duke Math. J.*, 72(1):65–83, 1993. (Cited on page 408.)

[650] Y. Zarhin. The Tate conjecture for powers of ordinary K3 surfaces over finite fields. *J. Algebraic Geom.*, 5(1):151–172, 1996. (Cited on page 408.)

[651] O. Zariski. The theorem of Bertini on the variable singular points of a linear system of varieties. *Trans. Amer. Math. Soc.*, 56:130–140, 1944. (Cited on page 27.)

[652] O. Zariski. *Algebraic surfaces*. Classics in Mathematics. Springer-Verlag, Berlin, 1995. With appendices by S. S. Abhyankar, J. Lipman, and D. Mumford, preface to the appendices by D. Mumford, reprint of the second (1971) edition. (Cited on page 385.)

[653] D.-Q. Zhang. Automorphisms of K3 surfaces. In *Proc. Internat. Conference on Complex Geometery and Related Fields*, vol. 39 of *AMS/IP Stud. Adv. Math.*, pages 379–392. Amer. Math. Soc., Providence, RI, 2007. (Cited on pages 336 and 356.)

[654] M. Zowislok. On moduli spaces of sheaves on K3 or abelian surfaces. *Math. Z.*, 272(3–4):1195–1217, 2012. (Cited on page 210.)

[655] S. Zucker. The Hodge conjecture for cubic fourfolds. *Compositio Math.*, 34(2):199–209, 1977. (Cited on page 38.)

Index

List of Notation

$\underset{\sim}{A}_n, \underset{\sim}{D}_n, \underset{\sim}{E}_n$	ADE lattices, Dynkin diagrams
$\widetilde{A}_n, \widetilde{D}_n, \widetilde{E}_n$	Extended Dynkin diagrams
$A(E)$	Atiyah class
(A_Λ, q_Λ)	Discriminant form of even lattice Λ
$\mathrm{Amp}(X)$	Ample cone
$\mathrm{Aut}(X), \mathrm{Aut}_s(X)$	Group of (symplectic) automorphisms
$\mathrm{Aut}(\mathrm{D}^b(X)), \mathrm{Aut}_s(\mathrm{D}^b(X))$	Group of (symplectic) exact equivalences
$\mathrm{Aut}(\widetilde{H}(X, \mathbb{Z}))$	Group of Hodge isometries
$\mathrm{Br}(X), \mathrm{Br}(X)[n], \mathrm{Br}(X)[\ell^\infty]$	Brauer group of X, n-torsion part, ℓ-primary part
$\widehat{\mathrm{Br}}_X$	Formal Brauer group
$\mathrm{Bs}(L)$	Base locus of line bundle L
$\mathcal{C}_X$	Positive cone
$\mathcal{C}_X^{\mathrm{e}}$	Effective positive cone
$c_X \in \mathrm{CH}^2(X)$	Beauville–Voisin class
$\chi(E, F)$	Euler pairing
$\mathrm{CH}^*(X)$	Chow ring
$\mathrm{Cl}(V), \mathrm{Cl}^\pm(V)$	Clifford algebra
$\mathrm{CSpin}(V)$	Clifford group
$\mathit{Co}_0, \mathit{Co}_1$	Conway groups
$\mathrm{Coh}(X)$	Abelian category of coherent sheaves
$D \subset \mathbb{P}(\Lambda_\mathbb{C})$	Period domain
$\mathrm{D}^b(X) = \mathrm{D}^b(\mathrm{Coh}(X))$	Derived category of coherent sheaves
$\mathrm{D}^b(X, \alpha)$	Derived category of twisted coherent sheaves
$\mathrm{Def}(X)$	Base of universal deformation
$\Delta, \Delta_P, \Delta_+$	Set of (positive) roots
$\Delta(E)$	Discriminant of sheaf E or elliptic curve E
$\mathrm{Diff}(X)$	Diffeomorphism group
$\mathrm{disc}\, \Lambda$	Discriminant of lattice Λ
$\exp(B)$	B-field shift
E_8	E_8-lattice
F, f	Frobenius
f_C	Symplectic automorphism associated with a section
$\mathrm{Fix}(f)$	Fixed point set of automorphism f
$\mathrm{FM}(X)$	Set of isomorphism classes of Fourier–Mukai partners
$\mathbb{G}_a, \mathbb{G}_m$	Additive, multiplicative group

$\mathrm{Gr}^{po}(2,V)$	Grassmannian of positive planes
$h(X)$	Height of X
$\widetilde{H}(X,\mathbb{Z})$	Mukai lattice
$H^{1,1}(X,\mathbb{Z})$	$= H^{1,1}(X) \cap H^2(X,\mathbb{Z})$
$H^*(X,\mathbb{Q})_p$	Primitive cohomology
$H^*_{\mathrm{cr}}(X/W)$	Crystalline cohomology
$\mathrm{Hdg}(V)$	Hodge group of Hodge structure V
Hilb, Hilb^P, $\mathrm{Hilb}^n(X)$	Hilbert schemes
$\mathrm{I}_n, \mathrm{II}, \mathrm{III}, \mathrm{IV}, \mathrm{I}_n^*, \mathrm{II}^*, \mathrm{III}^*, \mathrm{IV}^*$	Singularity type of fibres of elliptic fibration
$\mathrm{I}_{n_+,n_-}, \mathrm{II}_{n_+,n_-}$	Odd/even unimodular lattice of signature (n_+,n_-)
$\mathrm{JH}(E)$	Graded object of Jordan–Hölder filtration
$\mathrm{J}(X), \mathrm{J}^d(X)$	Jacobian fibration (of degree d) of elliptic $X \longrightarrow \mathbb{P}^1$
K	Kummer lattice
$K(X) = K(\mathrm{Coh}(X))$	Grothendieck group
$K(X)$	Function field of X
$\mathcal{K}_X$	Kähler cone
$K(T)$	Endomorphism ring of Hodge structure T
$\mathrm{KS}(V)$	Kuga–Satake variety
k_s	Separable closure of k
L_G	Orthogonal complement of invariant part
Λ	A lattice, often the K3 lattice
Λ_d	$= \ell^\perp$ for polarized K3 surface (X,ℓ) with $(\ell)^2 = 2d$
$\Lambda(n)$	Twist of lattice Λ
$\ell(\Lambda)$	Number of generators of discriminant A_Λ of lattice Λ
$\mu(E)$	Slope
$\mathrm{Mon}(X)$	Subgroup of $\mathrm{O}(H^2(X,\mathbb{Z}))$ generated by monodromies
$\mathrm{MT}(V)$	Mumford–Tate group of Hodge structure V
$\mathrm{MW}(X)$	Mordell–Weil group of elliptic fibration $X \longrightarrow \mathbb{P}^1$
$\mathcal{M}_d$	Moduli functor, stack of polarized K3 surfaces
M_{23}, M_{24}	Mathieu groups
M_d	Moduli space of polarized K3 surfaces
M_d^{lev}	Moduli space of polarized K3 surfaces with level structure
$M(v), M(v)^s$	Moduli space of (semi)stable sheaves with Mukai vector v
N, N_d	Moduli spaces of marked (polarized) K3 surfaces (X,φ), (X,L,φ)
$N_d(\rho) \subset N_d$	… with $\rho(X) \geq \rho$
N_0	Leech lattice, Niemeier lattice without roots
$N_{p,\sigma}$	Rudakov–Šafarevič lattice
$N(X)$	Numerical Grothendieck group
$\mathrm{Nef}(X)$	Nef cone
$\mathrm{Nef}^e(X)$	Effective nef cone
$\mathrm{NL}(X/S)$	Noether–Lefschetz locus
$\mathrm{NS}(X)$	Néron–Severi lattice
$\mathrm{Num}(X)$	Divisor group modulo numerical equivalence
$\mathrm{NE}(X)$	Mori cone
$\mathrm{O}(\Lambda)$	Orthogonal group of lattice Λ
$\mathrm{O}^+(\Lambda) \subset \mathrm{O}(\Lambda)$	Subgroup preserving orientation of positive directions
$\widetilde{\mathrm{O}}(\Lambda_d)$	Orthogonal group fixing a class $e + df$
$\mathcal{P}: S \longrightarrow \mathbb{P}(\Lambda_\mathbb{C})$	Period map
$P(E,m), p(E,m)$	(Reduced) Hilbert polynomial

$\mathrm{Pic}(X)$	Picard group
$\Phi_{\mathcal{P}}$	Fourier–Mukai transform with kernel $\mathcal{P}$
$\Phi_{\mathcal{P}}^K, \Phi_{\mathcal{P}}^N, \Phi_{\mathcal{P}}^H$	Action of Fourier–Mukai transform on $K(X), N(X), H^*(X)$
$\varphi(n)$	Euler function
Quot^P	Quot-scheme
$\rho(g, r, d)$	Brill–Noether number
$\rho(X)$	Picard number
$R(X) \subset \mathrm{CH}^*(X)$	Beauville–Voisin subring
$s_\delta, s_{[C]}$	Reflection in $\delta^\perp, [C]^\perp$
$\mathrm{sign}\,\Lambda$	Signature of lattice Λ
$\mathrm{Spin}(V)$	Spin group
$\mathrm{sp}: \mathrm{Pic}(X_\eta) \longrightarrow \mathrm{Pic}(X_t)$	Specialization morphism
$\mathrm{III}(E), \mathrm{III}(X)$	Tate–Šafarevič group of elliptic curve, elliptic K3 surface
$\sigma(X)$	Artin invariant of supersingular K3 surface
T_E	Spherical twist
$T(X)$	Transcendental lattice
$T(F)$	Torsion of sheaf F
$T_W, T(\alpha)$	Twistor line to positive three-space, Kähler class
$T_\ell \mathrm{Br}(X)$	Tate module of Brauer group
$\mathcal{T}_X$	Tangent bundle (of K3 surface)
U	Hyperbolic plane
$V(1)$	Tate twist of Hodge structure V
$v(E), v^{\mathrm{CH}}(E)$	Mukai vector
W	Weyl group
$(X, H), (X, L)$	Polarized K3 surface
$\mathbb{Z}(n)$	Tate twist or twist of trivial rank one lattice
$Z(X, t)$	Zeta function
$(.)$	Intersection pairing
$\langle \, , \, \rangle$	Mukai pairing
$(\alpha)^2$	$= (\alpha.\alpha)$

Printed in the United States
By Bookmasters